THE RIVERS HANDBOOK

The Rivers Handbook

HYDROLOGICAL AND ECOLOGICAL PRINCIPLES

EDITED BY

PETER CALOW

DSc, PhD, CBiol, FIBiol
Department of Animal and Plant Sciences
The University of Sheffield

AND

GEOFFREY E. PETTS

BSc, PhD
Department of Geography
Loughborough University of Technology

IN TWO VOLUMES
VOLUME ONE

OXFORD
BLACKWELL SCIENTIFIC PUBLICATIONS
LONDON EDINBURGH BOSTON
MELBOURNE PARIS BERLIN VIENNA

© 1992 by
Blackwell Scientific Publications
Editorial Offices:
Osney Mead, Oxford OX2 0EL
25 John Street, London WC1N 2BL
23 Ainslie Place, Edinburgh EH3 6AJ
238 Main Street, Cambridge
    Massachusetts 02142, USA
54 University Street, Carlton
    Victoria 3053, Australia

Other Editorial Offices:
Librairie Arnette SA
2, rue Casimir-Delavigne
75006 Paris
France

Blackwell Wissenschafts-Verlag
Meinekestrasse 4
D-1000 Berlin 15
Germany

Blackwell MZV
Feldgasse 13
A-1238 Wien
Austria

First published in 1992

Set by Setrite Typesetters, Hong Kong
Printed in Great Britain at
The Alden Press, Oxford,
and bound by Hartnolls Ltd,
Bodmin, Cornwall

DISTRIBUTORS

Marston Book Services Ltd
PO Box 87
Oxford OX2 0DT
(*Orders*: Tel: 0865 791155
         Fax: 0865 791927
         Telex: 837515)

USA
Blackwell Scientific Publications, Inc.
238 Main Street
Cambridge, MA 02142
(*Orders*: Tel: 800 759−6102
           617 876−7000)

Canada
Oxford University Press
70 Wynford Drive
Don Mills
Ontario M3C 1J9
(*Orders*: Tel: 416 441−2941)

Australia
Blackwell Scientific Publications
(Australia) Pty Ltd
54 University Street
Carlton, Victoria 3053
(*Orders*: Tel: 03 347−0300)

A catalogue record for this book is
available from the British Library

ISBN 0−632−02832−7

Library of Congress
Cataloging in Publication Data

The Rivers handbook:
hydrological and ecological principles/
    edited by Peter Calow, Geoffrey E. Petts.
        p.   cm.
    Includes bibliographical references
    and index.
    1. Stream ecology.   2. Hydrology
    I. Calow, Peter.   II. Petts, Geoffrey E.
QH541.5.S7R59   1992
574.5′26323 − dc20

Contents

List of Contributors

C.AMOROS *Laboratoire d' Ecologie des Eaux Douces, Université Claude Bernard, Lyon 1, F 69622 Villeurbanne Cedex, France*

P.B.BAYLEY *Illinois Natural History Survey, Champaign, IL 61820, USA*

J.P.BRAVARD *Laboratoire de Geographie Rhodanienne, Université Jean Moulin, Lyon 3, F 69239 Lyon, Cedex, France*

T.P.BURT *School of Geography, University of Oxford, UK*

P.CALOW *Department of Animal and Plant Sciences, University of Sheffield, PO Box 601, Sheffield S10 2UQ, UK*

P.A.CARLING *Institute of Freshwater Ecology, Windermere Laboratory, Cumbria LA22 0LP, UK*

M.CHURCH *Department of Geography, University of British Columbia, Vancouver, British Columbia V6T 1Z2, Canada*

K.W.CUMMINS *Pymatuning Laboratory of Ecology and Department of Biological Sciences, University of Pittsburgh, PA 15260, USA*

A.M.FOX *Center for Aquatic Plants, University of Florida, FL 32606, USA*

A.GUSTARD *Institute of Hydrology, Wallingford, Oxon OX10 8BB, UK*

P.P.HARPER *Département de Sciences Biologiques, Université de Montréal, CP 6128, Montréal, Québec H3C 3J7, Canada*

A.G.HILDREW *School of Biological Sciences, Queen Mary & Westfield College, Mile End Road, University of London, London E1 4NS, UK*

J.LEWIN *Institute of Earth Studies, University College of Wales, Aberystwyth SY23 3DB, UK*

H.W.LI *US Fish and Wildlife Service, Oregon State University, OR 97331, USA*

E.P.McELRAVY *Department of Entomological Sciences, University of California, Berkeley, CA 94720, USA*

L.MALTBY *Department of Animal and Plant Sciences, University of Sheffield, PO Box 601, Sheffield S10 2UQ, UK*

J.D.NEWBOLD *Stroud Water Research Center, Academy of Natural Sciences of Philadelphia, 512 Spencer Road, Avondale, PA 19311, USA*

R.H.NORRIS *Water Research Centre, Canberra University of Canberra, PO Box 1, Belconnen ACT, Australia 2616*

V.H.RESH *Department of Entomological Sciences, University of California, Berkeley, CA 94720, USA*

J.L.REYGROBELLET *Laboratoire d' Ecologie des Eaux Douces, Université Claude Bernard, Lyon 1, F 69622 Villeurbanne Cedex, France*

C.S.REYNOLDS *Institute of Freshwater Ecology, Windermere Laboratory, Cumbria LA22 0LP, UK*

A.L.ROUX *Laboratoire d' Ecologie des Eaux Douces, Université Claude Bernard, Lyon 1, F 69622 Villeurbanne Cedex, France*

E.VÁSQUEZ *Fundación La Salle de Ciencias Nat., Estación Hidrobiológica de Guayana, Apdo. 51, San Félix, Edo. Bolivar, Venezuela*

K.F.WALKER *River Murray Laboratory, Department of Zoology, University of Adelaide, S. Australia, Australia 5000*

D.E.WALLING *Department of Geography, University of Exeter, Exeter EX4 4RJ, UK*

A.K.WARD *Department of Biology, University of Alabama, Tuscaloosa, AL 35487–0344, USA*

J.V.WARD *Department of Biology, Colorado State University, Fort Collins, CO 80523, USA*

B.W.WEBB *Department of Geography, University of Exeter, Exeter EX4 4RJ, UK*

R.G.WETZEL *Department of Biology, University of Alabama, Tuscaloosa, AL 35487–0344, USA*

W.WILBERT *Fundación La Salle de Ciencias Nat., Instituto Caribe de Antropología y Sociología, Apdo. 1930, Caracas 1010-A, Venezuela*

Preface

To be effective, river management requires the improved application of environmental sciences — hydrology, geomorphology and biology. A major stumbling block in the advancement of the case for conservation has been the failure of scientists to communicate their knowledge to the practitioners, managers and decision-makers. This is one of the functions of *The Rivers Handbook*. The two-volume work was conceived to provide a synthesis of scientific knowledge on rivers and to provide the basis for achieving scientifically-sound and environmentally-sensitive river management.

Volume 1 provides an up-to-date scientific background. It emphasizes the limitations of our knowledge which reflect the complexities of the interactions within running-water ecosystems and the logistical difficulties of sampling rivers adequately either in space or in time. As noted by Norris *et al* (Chapter 13), 'sampling in rivers differs from many other environments because the unidirectional flow creates problems that are otherwise uncommon, such as difficulties in finding controls and replicating sample units'. In such situations, scientific judgement is important for data interpretation to support statistical hypothesis testing. Thus, whilst the chapters in Volume 1 seek systematically to describe the physical, chemical and biological features of rivers they also provide critical appraisals of the approaches for their investigation and the models available for their analysis.

Volume 1 has four major parts. Part 1 focuses on the hydrological, physicochemical, and geomorphological characteristics of rivers. It begins by considering the hydrological, physical and chemical processes operative within the production and transfer zones and moves on to provide a detailed analysis of channel hydraulics which is crucial to an understanding of the distribution and abundance of biota. The first Part ends with a consideration and classification of channel morphology and floodplains.

Part 2 moves on to consider the biota and is organized to reflect the trophic hierarchy. It begins with the microbial decomposers, progresses through the primary producers and invertebrate consumers, and ends with the fishes. Recognizing that information on the ecology of river biota relies on adequate sampling, this Part ends with a treatment of the design of sampling programmes and the effective deployment of statistical procedures within them.

Part 3 develops from the structural organization of river communities of microbes and multicellular plants and animals to consider the functioning of river ecosystems. Following analyses of functional links (food webs), of the functional biology of the primary sources of energy and matter, and of the ways that energy and matter flow within and through communities, the Part ends with a review of the general patterns of river ecosystem structure and function.

The final Part comprises five case studies. In seeking to apply the general principles examined in the previous Parts, mainly by reference to small temperate streams, the case studies focus on large rivers in climatic regions ranging from subarctic to tropical. These chapters clearly demonstrate that there is still much to be done in understanding the ecology of these large systems!

Volume 2 develops the principles and philosophy presented in Volume 1 into the management sphere, organizing the approach around *problems, diagnosis* and *treatment*. Together, the volumes will provide a reference work on the application of ecologically-sound principles, approaches, methods and tools, using experience in river management gained throughout Europe and North America over the past three decades.

Geoffrey E. Petts
Peter Calow

Abbreviations and Symbols

These are arranged in alphabetical order with Roman first, followed by Greek. One letter can be used for several different terms; we have not attempted to rationalize these since it would have been counter to common usage and potentially would have caused more confusion than it avoided. Because of the widely different ways in which terms with common letters are used, there is little danger of misunderstanding.

A	Area
A	Food absorbed
AE	Absorption efficiency
ANOVA	Analysis of variance
ATP	Adenosine triphosphate
a	Constant
asl	Above sea level
B	Biomass
b	Exponent
BFI	Base flow index
BOD	Biochemical oxygen demand
C	Concentration
C	Connectance
c	Integration constant
CL	Confidence limit
CMC	Carboxymethyl cellulose
CPI	Cohort production interval
CPOM	Coarse particulate organic matter
CV	Coefficient of variation
D	Grain size
D	Coefficient of longitudinal dispersion
D	Standard error of mean (also SE)
DF	Degrees of freedom (also v)
DIC	Dissolved inorganic carbon
DIN	Dissolved inorganic nitrogen
DIP	Dissolved inorganic phosphorus

DNA	Deoxyribonucleic acid
DOC	Dissolved organic carbon
DOM	Dissolved organic matter
DON	Dissolved organic nitrogen
DOP	Dissolved organic phosphorus
d	Water depth (also h)
E	Evaporation/evapotranspiration
E	Excretion
EPC	Equilibrium phosphorus concentration
ETS	Electron transport system
F	Critical value from statistical tables used in ANOVA and ANCOVA
F	Food defecated (egested)
F	Nutrient flux
FR	Fitness ratio
Fr	Froude number
FPOM	Fine particulate organic matter
f	Flow resistance; frictional factor
G	Groundwater discharge
G	Growth rate
g	Acceleration due to gravity
HPLC	High-performance liquid chromatography
h	Water depth (see also d)
I	Light intensity
I	Food ingested
I_b	Bedload transport rate
IBI	Index of biological integrity
IFIM	In-stream flow incremental methodology
IGR	Instantaneous growth rate
IOF	Infiltration-excess overland flow
J	Joule
k	Constant
k_L	Proportion of downstream flux of nutrients taken up by biota

Symbol	Definition
k_s	Equivalent roughness
n	Population size/density
N	Net of sources and sinks of nutrients
N	Number of replicates
n	Manning's roughness
P	Precipitation
P	Production
P	Wetted perimeter of channel
P	Desired probability
POM	Particulate organic matter
Q	Stream discharge
Q_{10}	Temperature coefficient
R	Respiration/respiratory heat loss
R	Hydraulic radius
R	Nutrient regeneration flux
R_d	Deposition rate
RCC	River continuum concept
Re	Reynolds number
RGR	Relative growth rate
RIV	Relative importance value
RNA	Ribonucleic acid
r	Population rate of increase
r_p	Height of an object from the bed
S	Number of species (richness)
S	Energy scope
S	Spiralling length
S_B	Average distance moved by a nutrient atom
S_C	Turnover length
SE	Standard error of mean (also D)
ΔS	Change in storage
S^2	Variance
SRP	Soluble reactive phosphorus
SSF	Subsurface stormflow
SOF	Saturation-excess overland flow
T	Water temperature
T	Seal thickness
t	Student's statistical test
t	Time
TDP	Total dissolved phosphorus
TDS	Total dissolved solids
TR	Turnover ratio

Symbol	Definition
U	Velocity
U	Biotic utilization of nutrients
U	Energy loss in excretion and secretion
U_z	Velocity at height z
U_∞	Free-stream velocity
u_*	Shear velocity
u	Downstream velocity component
V	Kinematic viscosity
v	Transverse velocity component
$\langle v \rangle$	Mean velocity of channel
W	Shapiro Wilk test
w	Settling velocity
w	Vertical velocity component
X	Quantity of nutrient
x	Exponent
$\overline{Y}$	Arithmetic mean
z	Height above bed
z_0	Roughness length
α	Level of significance
β	Power of statistical test $(1-P)$
β	Exponent
δ	Theoretical thickness of laminar layer
δ	True standard deviation
ϵ	Coefficient of monochromatic light extinction
Φ	Form resistance factor
η	Correction factor
θ	Shields parameter
κ	von Karman's constant
τ	Stress on bed; shear stress
λ	Porosities
Λ	Total load of solute
ν	Kinematic viscosity
ν	Degrees of freedom (also DF)
ρ, ρ_s	Density of water; sediment
χ	Exponent
ψ	Exponent
ω	Exponent

Part 1: Hydrological and Physicochemical Characteristics

Possibly more than any other ecological system, river ecosystems are moulded by physical forces; they flow and their flow rates can vary dramatically even over short spaces and periods of time. The *Handbook* therefore begins here with a detailed consideration of physical and chemical properties of rivers and their immediate surroundings. However, each chapter seeks not only to describe these features, but to explain how they can be measured and to develop models that provide causal links between relevant components. Thus, logically, the first two chapters begin with where the water comes from; the first deals with headwaters and the second with lower reaches. Both link hydrological regimes with catchment characteristics. The quality of water that collects in the channels is discussed first from a physical point of view in Chapter 3 and then from a related chemical point of view in Chapter 4. Neither of these chapters attempts exhaustive treatments of variables since these are available in textbooks; instead they focus on those physical properties that are considered particularly important and on the behaviour of both dissolved and adsorbed chemicals in terms of the processes and pathways they are involved in and their variation through space and time. The physical description is continued in Chapter 5 with a detailed analysis of in-stream hydraulics—crucial to an understanding of the distribution and abundance of biota   and this leads to a consideration and classification of channel morphometry in Chapter 6. A recurrent theme throughout the *Handbook* is the intimate relationship between channel and surrounding catchment and this is not only in terms of the supply of water and materials to the channel but, as the final chapter of this Part (Chapter 7) discusses, in terms of the way activity within the channel influences the form of the surrounding area.

1: The Hydrology of
Headwater Catchments

T.P.BURT

1.1 BACKGROUND

The need for process studies

Until recently, hydrologists took the view that the headwaters of a drainage basin were nothing more than source areas for runoff. Since their concern was largely with forecasting water resources or floods at downstream locations, they felt able to ignore the physical characteristics of the headwaters and the exact processes responsible for generating the runoff. This philosophy was reflected in the lumped models traditionally employed to forecast flood runoff and is perhaps best exemplified by the unit hydrograph model of Sherman (1932), which is still much used today. At about the same time that Sherman introduced his model, Horton (1933) proposed his classical theory of hillslope hydrology in which he assumed that the sole source of flood runoff was excess water which was unable to infiltrate the soil. Water that could infiltrate would eventually recharge groundwater and so provide a source for baseflow. Horton's theory of runoff production provided a physical basis for Sherman's model. Together, their ideas dominated hydrology for several decades.

It became apparent, however, that Horton's model was inappropriate in many areas because the infiltration capacity of many soils was too high to produce infiltration-excess overland flow. In particular, researchers such as Hursh and Hewlett at the Coweeta Hydrologic Laboratory in North Carolina, demonstrated that subsurface stormflow was dominant in forested basins with permeable soils and that overland flow, if it did occur at all, was limited in extent and generated by quite different processes than those described by Horton (Swank & Crossley 1988).

Since the 1960s, many field studies have been conducted in headwater catchments, with the purpose of describing the full range of runoff processes that may occur, and identifying and explaining their location in time and space. Such studies have made possible the development of distributed runoff models (Beven 1985); these have the capability of forecasting the spatial pattern of hydrological conditions within the catchment as well as simple outflows and storage volumes. Field studies have also identified clear links between hydrological pathways and streamwater quality; these are described in detail in Chapters 4 and 5.

How large is a headwater catchment?

Any simple definition of a 'headwater catchment' is likely to be inadequate. The approach taken here is to assume that the pattern of outflow discharge from headwater catchments is strongly related to runoff production at the hillslope scale. In 'large' river basins, flow patterns will be dominantly controlled by channel and floodplain storage and by the routing of runoff through the channel network. In addition, mixing of runoff from widely differing source areas (e.g. in terms of precipitation inputs or land cover) will obscure the process–response relations that can be identified in smaller basins. Most field research on runoff production has been conducted in small catchments, often less than 10 km^2 in area; detailed study of small runoff plots on hillslopes has provided the foundation for such work. There is surprisingly little work in physical hydrology that has attempted to relate runoff production in small catchments to the flood response of larger basins. Most of the examples discussed here refer to basins of less than 100 km^2 in area, but this must be seen as an arbitrary division.

Headwater catchments can be found in any geographical location; they are not necessarily small upland valleys despite their frequent portrayal as such. Equally, the headwaters of a large river could be found in flat agricultural land or even within an urban area. Whatever its characteristics, the importance of the headwater region is that it determines both the quantity and quality of water received downstream and as such may impose important constraints on resource development and hazard control on the main river.

Terminology

In studying headwater catchments, it is important to distinguish between unconcentrated flows on hillslopes and concentrated flows in channels. Atkinson (1978) identified three methods of measuring moisture movement on hillslopes: methods involving interception of the flow, in which all or part of the flow is intercepted into a measuring device in order to determine its discharge; methods involving the addition of tracers in order to determine the velocity of flow; and indirect methods whereby measurements of moisture content and hydraulic potential over the slope profile or experimental plot are used to calculate moisture flux in the soil matrix. These methods are fully reviewed in Atkinson (1978) and Goudie (1990).

A variety of terms have been used to describe the movement of water in channels: streamflow, runoff and discharge being the most common. The rate of flow (i.e. discharge) is the product of a cross-sectional area of flowing water and its velocity; it is usually expressed in terms of volume per unit time (e.g. cubic metres per second, or 'cumecs'). Often, the discharge is divided by the area of the catchment so that the runoff can be expressed as depth per unit time (e.g. mm per hour); this allows runoff rates to be compared easily with other variables such as rainfall intensity or infiltration rate which are also expressed

in depth units. Gauged flows are usually calculated by converting the record of stage, or water level, using a stage–discharge relation, often referred to as the rating curve. Stage is measured and recorded through time by instruments placed in a stilling well; traditionally floats have been used but pressure transducers are becoming increasingly common. Stage is recorded continuously by pen and chart or digitally by a datalogger. The stage–discharge relation is obtained either by installing a gauging structure, usually a weir or flume, with known hydraulic characteristics, or by measuring the stream velocity and cross-sectional area of flow at a site where the river channel is stable. Some methods of gauging streamflow are particularly well suited for use on small streams in headwater catchments (e.g. dilution gauging, thin-plate v-notch weirs). Methods for stream gauging are fully reviewed in Gregory and Walling (1973), Rodda *et al* (1976), Herschy (1978) and British Standards Institution BS 3680 (1964).

A graph of discharge plotted against time is called a hydrograph. It is clear from the above discussion of runoff generation and from Fig. 1.1

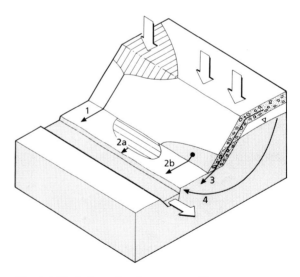

Fig. 1.1 Hydrological pathways. 1, Infiltration-excess overland flow (IOF). 2, Saturation-excess overland flow (SOF); 2a, direct runoff from saturated soil; 2b, return flow. 3, Subsurface stormflow (SSF). 4, Groundwater flow.

that many different pathways exist by which hill-slope runoff can reach a stream. Some precipitation (or snowmelt) takes a rapid route to the stream channel and is often described as quick-flow or stormflow; quickflow is usually associated with high discharge. Subsurface flow moves at much lower velocities, often by longer flow paths, and, although it may contribute to stormflow as noted above, its main effect is to maintain streamflow during dry periods through the sustained release of water stored within soil and bedrock; this is termed low flow or baseflow. As a result, a hydrograph typically consists of episodes of high discharge separated by longer periods of low, gradually declining, flow. For some purposes it is necessary to separate arbitrarily stormflow and baseflow; given that stormflow can be produced by subsurface as well as by surface runoff, this line of separation has no physical basis, but may be helpful nevertheless (e.g. see the discussion relating to Fig. 1.8). The value selected by Hewlett and Hibbert (1967) of 0.05 cubic feet per second per square mile per hour (0.000546 cumecs per square kilometre per hour, or 0.0472 mm per day; Ward & Robinson 1990) has often been adopted, although this may not be appropriate for all basins. Figure 1.2 shows the terms normally used to describe a storm hydrograph and its (arbitrary) separation from the baseflow component.

When studying quickflow hydrographs in small basins, it is usually necessary to plot instantaneous discharge values; on larger rivers hourly mean or daily mean flows may suffice. However, even when studying small basins, it is often more convenient to analyse certain aspects of streamflow (e.g. seasonal variations in flow) using discharge totals or mean discharge rather than instantaneous values.

1.2 STORM RUNOFF MECHANISMS

The nature of the soil and bedrock determine the pathways by which hillslope runoff will reach a stream channel. The paths taken by water (see Fig. 1.1) determine many of the characteristics of the landscape, the uses to which land can be put and the strategies required for wise land-use management (Dunne 1978). Much work has been done in the temperate and warmer latitudes where rainfall is the primary hydrological input; hill-slope hydrology during the snowmelt season has received less attention and the hillslopes of frigid regions are rarely considered (Church & Woo 1990).

The dominance of runoff theories based on the occurrence of infiltration-excess overland flow (IOF) (Horton 1933, 1945) meant that research into subsurface flow mechanisms was neglected. However, despite modifications such as the partial area concept proposed by Betson (1964), it

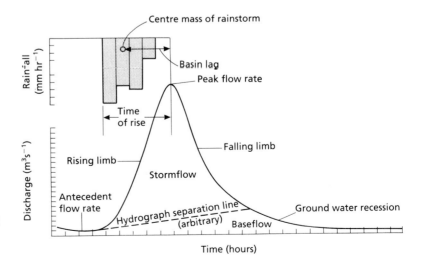

Fig. 1.2 Terms commonly used to describe a storm hydrograph (after Hewlett 1982).

became apparent that Horton's model was inappropriate in many locations. Where permeable soils overlie impermeable bedrock, subsurface stormflow (SSF) within the soil can account for most of the flood runoff leaving the catchment. When the soil profile becomes completely saturated, saturation-excess overland flow (SOF) may also occur (Dunne 1978). Both SOF and SSF may occur at rainfall intensities well below those required to generate IOF. SOF and SSF will be produced from source areas that are limited, although variable, in size and different in location to the source areas for IOF. The notion of localized sources of storm runoff which may vary in area both seasonally and during precipitation events provides the basis for the variable source area model first outlined by Hewlett (1961). Since then, this concept has come to dominate hillslope hydrology; subsurface flow is regarded as *the* major mechanism controlling the generation of storm runoff, both because of its influence on the generation of 'return flow' (a component of SOF; Dunne & Black 1970) and as an important process in its own right (Anderson & Burt 1978; Burt 1986). Figure 1.3 classifies storm runoff mechanisms on the basis of the two main models; these mechanisms are reviewed briefly below.

Infiltration-excess overland flow (IOF)

Central to the theory of hillslope hydrology put forward by Horton (1933) was his view concerning the role of infiltration processes at the soil surface. Horton considered that infiltration divides rainfall into two parts, which thereafter pursue different courses. One part goes via overland flow to the stream channel as surface runoff; the other goes initially into the soil and thence through the groundwater flow again to the stream or else is returned to the air by evaporative processes.

Horton established that infiltration capacity, the maximum rate at which a soil can absorb falling rain (or meltwater), decreases asymptotically over time as saturation of the surface soil causes a reduction in hydraulic gradients near the surface. Changes in the surface of the soil (e.g. swelling of clay particles, in-washing of fine particles into pores, compaction by rainbeat) may also reduce infiltration capacity through the course of a storm. Figure 1.4 shows the way in which rainfall intensity and infiltration capacity may interact during a storm to produce overland flow. At the beginning of the storm, infiltration capacity exceeds rainfall intensity and there is no surface ponding. However, when later the rainfall

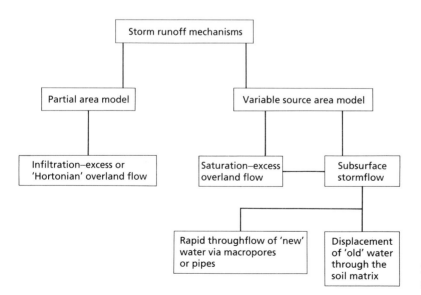

Fig. 1.3 Storm runoff mechanisms.

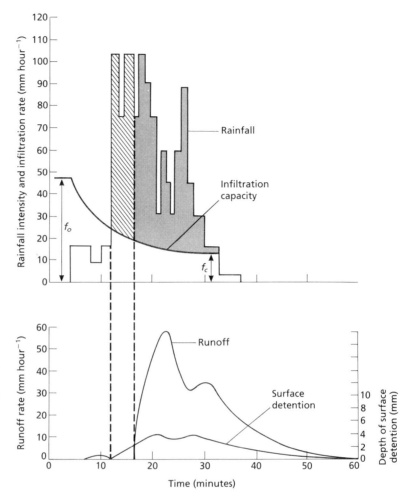

Fig. 1.4 Relationship of infiltration-excess overland flow (IOF) and surface detention to rainfall intensity and infiltration capacity for a storm of varying intensity. The lightly shaded portion represents depression storage which must be full before runoff, darkly shaded, is generated (after Horton 1940; Dunne 1978).

intensity exceeds infiltration capacity, the excess begins to fill up surface depressions. When these are full, the excess rainfall overflows downslope and surface runoff begins.

Infiltration capacity is not necessarily constant even within a small catchment, and Betson (1964) used the idea that IOF might be produced from only part of the basin area to improve his predictions of the volumes of storm runoff. In identifying relatively impermeable surfaces within a basin, Betson's partial area model remains the best guide to the location of source areas for IOF. Recent evidence suggests that IOF is not necessarily so rare as proponents of SSF have argued in the past (Heathwaite *et al* 1990). Where IOF is the

dominant producer of storm runoff, overland flow may be generated across large areas of hillside.

Subsurface stormflow (SSF)

As Fig. 1.3 shows, SSF may be generated by two mechanisms: by non-Darcian flow through large voids such as macropores or pipes, and by Darcian flow through the micropores of the soil matrix. Hillslope hydrologists have, until recently, emphasized micropore flow, although there has been a continued interest in pipeflow (Gilman & Newson 1980; Jones 1981) and in the influence of underdrainage on the flood response of small catchments (Robinson & Beven 1983). Field evidence relating to macropores (e.g. Whipkey 1965) was somewhat ignored until studies such as those of Mosley (1979), Beven and Germann (1982) and Kneale (1986) described the hydrological effects of rapid infiltration down macropores. Kneale and White (1984) studied infiltration into 9-cm cores of dry cracked clay-loam soil. Bypassing flow occurred down the cracks once the rainfall intensity exceeded the infiltration capacity of the soil peds (2.2 mm per hour). For the Oxford (UK) region where the soils were collected, their results imply that 10–20% of summer rainfall would bypass the root zone. Coles and Trudgill (1985) and Germann (1986) have identified important thresholds governing macropore flow. In addition to the infiltration threshold of the peds already described, they show that antecedent soil moisture is also an important control: if the soil is too dry, no macropore flow will happen whatever the rainfall rate; if the soil is at field capacity, then all rainfall must bypass the soil peds. Such results have clear implications for the production of storm runoff; Robinson and Beven (1983) showed that flow through cracks in a clay soil produced higher peak flows compared with an uncracked soil, in summer when the soil was dry. One unresolved question remains the connectivity of macropores in the downslope direction.

Lateral subsurface flow through the soil matrix will occur in any soil in which the hydraulic conductivity declines with depth (Zaslavsky & Sinai 1981). Where a layer of low permeability is found at depth in the soil profile, the flow direction in the more permeable soil above is parallel to the slope (Burt 1986). On the other hand, if both soil and bedrock are permeable at depth, percolation remains vertical and no lateral flow occurs within the soil layers; under these circumstances, infiltrating water serves only to recharge groundwater storage and to provide baseflow.

In itself, the occurrence of subsurface flow is not enough to produce storm runoff and it used to be assumed that micropore flow was too slow to provide this. However, rapid subsurface flow can occur in several ways: if the hydraulic conductivity of the soil is high, infiltration can lead to rapid recharge of the saturated zone at the base of the soil profile; macropore flow can have the same effect. Soils close to saturation (the 'capillary fringe'; Abdul & Gillham 1984) may require only a small amount of infiltration to produce a significant rise in the water-table. In all cases, if the saturated hydraulic conductivity of the soil is high, then large amounts of SSF will be produced as a result of a rapid rise in the water-table. In addition to this immediate effect, SSF may also occur in the form of a delayed hydrograph peaking several days after the rainfall input. Anderson and Burt (1978) and Burt and Butcher (1985a, 1985b) showed that convergent flow of soil moisture into hillslope hollows causes extensive development of soil saturation at such locations (the 'saturated wedge'). Figure 1.5 shows changes in soil water potential during the period of a double-peaked storm hydrograph. The mapped hillslope measured 70 × 50 m; soil moisture conditions are shown at a depth of 600 mm for a hillslope hollow and its adjacent spurs. The maps show that rainfall causes a widespread but shallow zone of saturation to develop at the base of the slope. The generation of the delayed peak is associated with convergence of soil water into the lower part of the hollow from further upslope and from the adjacent slopes. This causes the deepest saturation during this period, although the saturated wedge is now spatially less extensive than during the rainfall. SSF contributes to the first peak in stream discharge; the second peak is entirely SSF. Such runoff is strongly seasonal, being largely confined to winter when soil moisture deficits have been recharged. That SSF can dominate the stormflow response of some catchments is a point discussed further below.

Saturation-excess overland flow (SOF)

It is impossible to divorce the generation of SSF from the production of SOF. The variable source area model (Hewlett 1961; Troendle 1985) is based on the assumption that water moves downslope through the soil. The source areas for SSF and SOF are therefore essentially identical (Burt & Arkell 1986). Three locations within a catchment may be identified where maximum soil moisture levels will be reached (Kirkby & Chorley 1967; Burt 1986): at the foot of any slope, particularly those that are concave in profile; in areas of thin soil where soil moisture storage is reduced; and,

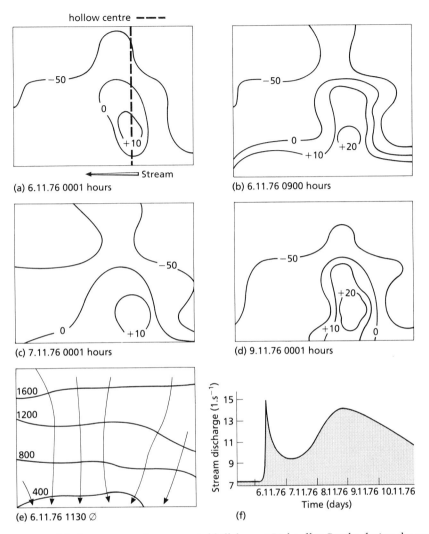

Fig. 1.5 Soil-water potential changes on an instrumented hillslope at Bicknoller Combe during the period of a double-peaked hydrograph: maps (a)–(d) show soil-water potential changes at a depth of 600 mm; map (e) shows the flow net; map (f) shows the stream hydrograph (after Burt 1978).

most importantly, in hillslope hollows where, as described above, convergence of flow lines favours the accumulation of soil water. The extent of the saturated area depends upon soil wetness and so source areas vary seasonally and during storms (Dunne 1978). Several authors have attempted to predict the distribution of soil moisture in relation to topography; in all cases upslope drainage area is the crucial control (Kirkby 1978; O'Loughlin 1981, 1986; Burt & Butcher 1985a; Thorne *et al*

1987). Where surface saturation occurs to any great extent (e.g. where wide valley bottoms exist), SOF will dominate the stormflow response with higher peak discharges and lower lag times than are characteristic for SSF (Dunne 1978). If soils are permeable, the source areas for SOF are likely to remain a relatively small percentage of basin area and the ratio of storm runoff to rainfall may be low (commonly below 5%). Where soils are impermeable (e.g. peat), surface saturation will be extensive, with much more rainfall translated into runoff (Burt & Gardiner 1984). SOF will be a mixture of return flow ('old' soil water) and direct runoff (water unable to infiltrate into saturated soil) so that its solute and sediment load may differ widely from that produced by IOF (which is essentially direct runoff alone).

Snowmelt and storm runoff

Snowmelt runoff occurs when energy is added to a snowpack that is at 0°C. The principal source of energy for snowmelt is direct solar radiation, but long-wave radiation and fluxes of sensible and latent heat may also be important on occasions. A strong diurnal rhythm dominates snowmelt. The form of the diurnal melt flood is controlled by the pattern of melt at the surface, by the passage of water through the snowpack, and by the saturated flow at the base of the snow to an open channel (Church 1988). In the UK, snowmelt occurs relatively infrequently and, additionally, its effect is often overlooked because the flood is often associated only with rainfall at lower elevations; even so, melt-related events have generated the most extreme floods recorded on many basins. Over much of Canada, snow cover persists sufficiently long to alter the pattern of runoff from that which would otherwise be determined by the distribution of precipitation. Here, engineering designs and water management procedures must be modified to take account of snow accumulation and snowmelt (Church 1988).

Runoff from a snowpack is the last occurrence in a series of events beginning with snowfall itself. The density of new snow varies from 0.05 to 0.20 depending on air temperature during its precipitation. The density of the snowpack gradually increases due to settling and compaction, melting and recrystallization, and rainfall, and may be 0.30–0.50 by springtime. An important element in snowpack evolution is 'ripening' by which the temperature of the snowpack rises to reach 0°C and its liquid content is maximized. A deep layer of snow may be well below freezing initially, but when milder weather sets in some melting will take place at the surface. This meltwater percolates into the pack where much of it may refreeze, liberating latent heat of fusion in so

doing; at this stage little of the melt may form runoff. In addition, heat is added to the snowpack from the overlying air. The pack warms continually and eventually reaches 0°C. With continued melting, the liquid content of the snowpack rises to its field capacity and is now said to be 'ripe'; further melt will generate significant runoff (Dunne & Leopold 1978). Most snowpacks contain strata of varying texture and permeability; during the ripening stage, these serve to diffuse the meltwave and reduce peak flow somewhat. In shallow snow, or when the snowpack is ripe, runoff delay due to passage through the snow becomes less important, and stream hydrographs become more peaked (Church 1988). Church considers potential meltwater production rates (in the absence of rainfall) and shows that the size of a meltwater flood is strictly limited by the energy available. Near the solstices, up to 40 mm per day may melt if all available radiant energy is utilized; in late winter between 10 and 20 mm per day is more likely. Such rates may be important in a large basin such as the Fraser River which is so large that only snowmelt can generate runoff over most of it at any one time; rainfall events and other synoptic weather variations serve only to modify details of the seasonal melt hydrograph. On the other hand, in small basins, although melt may continue over several months, individual peak flows are usually rain-on-snow events (Church 1988). In addition, thermally induced snowmelt may produce a distinctive diurnal rhythm of runoff in small drainage areas in direct response to energy inputs for melt; Church (1988) provides examples.

Once the snowpack is ripe, rapid snowmelt is most frequently brought about by a sudden influx of warm, moist, unstable air, and rain often accompanies the melt. With an air mass of high humidity, much of the energy for melt may be provided by the release of latent energy when there is condensation on to the snow surface. None the less, even though the heat energy supplied by warm, moist air and the rainfall itself can be important in a large, warm rainstorm, the volume of water directly contributed by rainfall will often far outweigh the amount of melting (Dunne & Leopold 1978). Critical situations can arise when rainfall combines with snowmelt, especially if the soil below is frozen, producing a flood which may be extreme even in a small basin; on larger rivers it may even be the major design criterion (NERC 1975, pp. 502−4). This is particularly the case in temperate, coastal mountains where autumn and early winter storms are the most vigorous of the year and frequently bring heavy rain on to shallow snow (Church 1988). For example, the UK Flood Studies Report (NERC 1975) showed that 14−20% of all floods on the South Tyne were snowmelt related over the period 1959−69; for 53 of 329 gauging stations, the record peak discharge was associated with snowmelt.

Once runoff from the snowpack begins, the pathways by which meltwater may reach a stream are identical to those already discussed. Where the topsoil has remained frozen during the winter, infiltration capacity may be effectively zero and large quantities of IOF will be generated (Dunne & Black 1971; Dunne *et al* 1976). However, in many cases the ground below a snowpack is unfrozen or covered only in porous 'needle ice', in which case infiltration capacity is unaffected and meltwater is able to enter the soil (Stephenson & Freeze 1974).

Peak runoff from glacierized basins is displaced into middle or late summer when seasonal snow is much reduced, glacial drainage paths are well integrated and, most importantly, glacier ice with low albedo is exposed to melt (Church 1988). Episodic runoff events are associated either with meteorological events identical to those described for snowmelt, or with the catastrophic release of water from within, or dammed up by the glacier.

Controls of storm runoff

Climate, soil type and bedrock lithology control which type of runoff mechanism will dominate in a particular catchment, with vegetation cover and topography as an important secondary control at the hillslope scale (Dunne 1978; Whipkey & Kirkby 1978). Kirkby (1978) recognized the crucial role of hydraulic conductivity in relation to rainfall intensity and identified two runoff domains: where rainfall intensities commonly exceed infiltration capacity, IOF will be the dominant storm runoff mechanism; where the infiltration capacity of the soil is higher than the rainfall rate, SSF and SOF will be important, the balance between the two being determined by the permeability of soil and bedrock, and by catchment topography (Dunne 1978; Anderson & Burt 1990). Church and Woo (1990) have noted that spatial variation in rainfall intensity is much less than the range of soil permeabilities, so that soil may be a more important control of runoff generation than climate. IOF is often associated with arid environments, encouraged by a combination of intense rainfall and bare ground (Horton 1945; Langbein & Schumm 1958). Conversely, in humid temperate environments, the combination of low rainfall intensity, complete vegetation cover and high infiltration capacity favours SSF and SOF. This is, however, an oversimplification; runoff may be absent in arid areas if rainfall intensity is low or if soils are permeable (Yair & Lavee 1985). In humid areas, many soil types have a naturally low infiltration capacity (e.g. peat; Burt & Gardiner 1984), and in other cases the soil surface may have become less permeable due to compaction by poor management (Heathwaite *et al* 1990), so that IOF may be more common than is often thought. Thus, although it has been represented as a zonal phenomenon, the propensity for IOF to take place may be mainly a function of soil properties (Church & Woo 1990).

Figure 1.6 summarizes these points. The occurrence of IOF depends on the relative magnitude of rainfall intensity and the hydraulic conductivity of the upper soil layer. The occurrence of SSF depends on the balance between rainfall intensity and the hydraulic conductivity of the lower layer. SOF is most likely to occur in soils of medium to low hydraulic conductivity where drainage is slow and storage limited, in thin soils, or in deep permeable soils where flow convergence leads to soil saturation. Vegetation cover may significantly influence production of IOF, soils in forests tending to have much higher infiltration capacities than those of farmland. Land use and techniques of land management may strongly influence the hydrological response of a catchment; some effects of a change in land use are described at the end of this section.

The information given in Fig. 1.6 defines the domains for stormflow, but gives no indication of

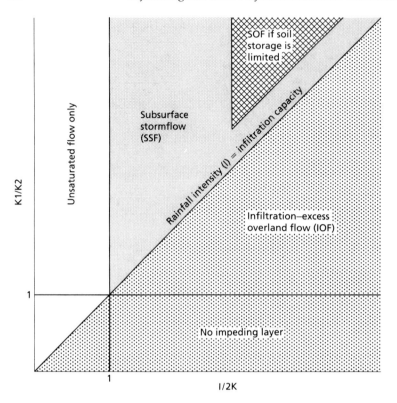

Fig. 1.6 Regimes of flow for a two-layer soil in relation to the hydraulic conductivity of the upper (K1) and lower (K2) layers and rainfall intensity (I) (after Whipkey & Kirkby 1978; Burt 1986).

the magnitude and timing of storm runoff production; Fig. 1.7 provides this information. In general, the largest and quickest responses are generated by surface runoff. Where source areas for surface runoff are very limited in extent or absent altogether, stormflow is generated by SSF alone. Since SSF can effectively provide two peaks in stream discharge (see Fig. 1.5), the lag times to peak on Fig. 1.7 show a wide range for a given basin area.

1.3 BASEFLOW GENERATION

The two main sources of baseflow are ice and snow melt, and drainage of soil and ground water. Stream response to melt depends on meteorological inputs to the snowpack (energy and matter), the state of the snowpack, the delivery of melt to the ground surface beneath it, and the routing of meltwater over or through the soil to the stream. In small basins, flood peaks involving snowmelt are most likely to be generated by rain-on-snow events. Since continued melt in mountain en-

vironments is essentially a thermal process, the resulting pattern of streamflow may bear little relation to the temporal pattern of precipitation input, although it may be diurnally quite regular. Given the low rates of melt that are likely to occur (up to 40 mm per day; Church 1988) in comparison with the infiltration capacity of many soils, it is likely that snowmelt will augment baseflow since it is likely to recharge soil and ground water, rather than generate surface or near-surface runoff.

Water that percolates to groundwater moves at much lower velocities by longer paths and reaches the stream slowly over long periods of time, sustaining streamflow during rainless periods (see Fig. 1.1; Dunne & Leopold 1978). Any layer of rock or unconsolidated material that can yield significant quantities of water is known as an aquifer; a stratum through which water cannot move except at negligible rates is known as an aquiclude or aquitard. Although its porosity determines how much water a rock can store, its specific yield, the amount of water released by

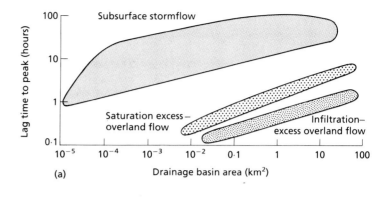

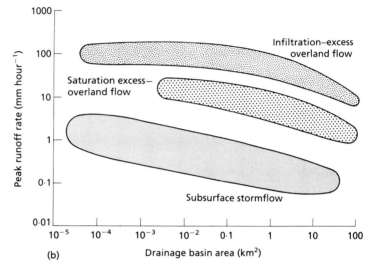

Fig. 1.7 Responses of catchments to hillslope flow processes: (a) lag times; and (b) peak runoff rates (after Dunne 1978; Kirkby 1985; Anderson & Burt 1990).

drainage under gravity, is of more relevance in supplying water to local rivers. This, together with its permeability and thickness, determines the importance of any aquifer. As described below, the nature of the porosity may also be influential; a porous aquifer may have a very different pattern of drainage compared with a well-fractured or pervious rock. In headwater catchments, the main mode of recharge is infiltration of water from the ground surface; influent seepage from river beds and lakes is likely to assume greater importance downstream. The main process of groundwater discharge is by spring flow and effluent seepage into rivers and lakes.

Depending on the nature of the aquifer, peak baseflow discharge may lag significantly behind precipitation inputs. Impermeable rocks are as-

sociated with strongly peaked hydrographs because there is little subsurface storage and rapid surface or near-surface runoff. Basins composed of highly permeable formations such as limestones or basalts tend to have flatter, broader, and more delayed responses (Ford & Williams 1989) There is a continuum of rainfall-runoff responses, with cavernous (karst) limestones providing the most rapid, peaked groundwater hydrographs, while deep, porous aquifers, such as the chalk limestone of southern England, yield the most subdued hydrographs. Except for cavernous limestones, it is unusual for aquifers to provide flood hydrographs. If there are no source areas for overland flow, floods may be generated by groundwater. Otherwise high discharges are generated by surface and near-surface processes, and groundwater provides the prolonged recession flow associated with the falling limb of the hydrograph (see Fig. 1.2). The flood hydrographs of vadose (above the water-table) cave systems tend to be peaked and similar to surface streams, but should the cave streams flow into a flooded phreatic (below the water-table) zone before their waters emerge at a spring, then the influence of the inflow hydrograph on the composite outflow hydrograph of a spring is similar to that of a tributary flowing into a lake — the outflow is a muted, delayed reflection of the inflow (Ford & Williams 1989). Karstic limestones, being massive, low-porosity limestones with mainly conduit permeability, have low baseflow because little water is held in storage. Deep, porous aquifers such as chalk limestone have low flows, only slightly lower than the mean flow because the response time for discharging water in storage is comparable with the time between wet and dry seasons. In such aquifers, there will be a long lag before a response occurs, a slow rate of rise to peak groundwater discharge, and a low rate of recession. Flow regimes which reflect the influence of groundwater discharge are exemplified below in section 1.4.

Hewlett and Hibbert (1963) showed that, in the absence of an aquifer, drainage from a soil profile can maintain streamflow over long periods. Drainage of an isolated block of soil during the first few days represents outflow from a thinning and flattening saturated wedge. Thereafter, discharge is produced by drainage from the unsaturated zone which is transmitted by the thin saturated zone at the outlet (R.D. Moore, personal communication). Anderson and Burt (1977) argued that in some situations unsaturated flow remains insignificant so that the outflow is provided primarily by drainage of the saturated wedge. As noted in section 1.4 below, in catchments that predominantly produce subsurface flow through the soil layers, the runoff regime is more seasonally variable than that of an aquifer.

Nevertheless, many catchments lack significant storage of soil or ground water. Even where rainfall is relatively uniform throughout the year, large rates of evaporation in summer mean that many headwater catchments have a characteristic drought, even in humid climates (see also comments on river regimes and flow duration curves

in section 1.4 below). Conditions are most critical in late summer, or late in the dry season in the tropics and subtropics. Agricultural underdrainage and reclamation of wetlands have exacerbated this problem in highly developed areas such as the UK. As a result, many headwater streams are seasonally ephemeral. This is well known even on reliable aquifers such as the chalk: prolonged summer discharge lowers the water-table to such an extent that the source of the river migrates several kilometres downstream. Autumn recharge eventually raises the water-table and the intermittent ('bourne') stream reappears, although perhaps not until several months after the main period of rainfall. Where the abstraction of groundwater occurs, such natural effects may be aggravated. A numerical model of the Lambourn, a typical bourne stream on the chalk in southern England, showed that if abstracted water was piped directly out of the basin, a pronounced downstream migration of the perennial head would occur. However, if the abstracted water was piped into the river at the perennial streamhead, although less efficient in supply terms, this preserved the location of the perennial streamhead, a highly desirable result on ecological and amenity grounds (Oakes & Pontin 1976). Land-use changes in headwater catchments may have similar effects. Afforestation will significantly decrease baseflow during the summer to such an extent that headwater streams may dry up. This effect is further elaborated in the last paragraphs of section 1.4 below.

1.4 PATTERNS OF STREAMFLOW IN HEADWATER CATCHMENTS

In the following paragraphs, a selection of commonly used methods of analysing streamflow are presented; in general, as the timescale of analysis increases, the use of aggregate discharge figures becomes more convenient. Some of the data used below were compiled by the Surface Water Archive (IH 1989), an extremely useful source of hydrological information for UK catchments, both large and small.

Daily mean flow

This is calculated by averaging the instantaneous flows occurring throughout the water-day (in the UK from 09.00 to 09.00 hours Greenwich Mean Time) obtained from the continuous record. For comparatively small basins that respond rapidly to rainfall or melt events, hydrographs of daily mean flow provide a useful summary of flow variations during a given year. In particular, quickflow and baseflow contributions are easily separated by eye on such plots. As discussed below, the relative proportions of quickflow and baseflow are an important diagnostic feature of headwater catchments.

Figure 1.8 shows the annual hydrograph for 1978 for two lowland basins (Foston Beck and Blackfoss Beck) that are about 30 km apart in east Yorkshire, England; total precipitation for the year was about 700 mm in both cases. In addition, flow figures are shown for a third basin (Snaizeholme Beck), a small upland basin in the same region where total rainfall was 1600 mm over the same period. For the two lowland basins, the large contrast in their annual hydrographs must relate to differences in catchment characteristics rather than to differences in climate. Blackfoss Beck has very subdued topography; an extensive network of ditches and tile drains provides the necessary drainage of the clay soil; the clay bedrock below is totally impermeable. Infiltration capacities are reasonably high, because of the presence of macropores in the clay, but may fall to very low levels if compacted by agriculture. Foston Beck lies on chalk limestone: high infiltration capacities and permeable bedrock mean that little storm runoff is generated despite the hillier terrain; groundwater provides almost all of the runoff. The flashy response of Blackfoss Beck is typical of a basin which produces little baseflow but much quickflow; the high peak discharge and symmetry of the quickflow hydrographs suggest that much surface or near-surface runoff is generated during storm events. Once the quickflow has finished, there is almost no baseflow to maintain dry weather flow. By contrast, the annual hydrograph for Foston Beck is typical of a catchment dominated by baseflow production with very small quickflow hydrographs and a very protrac-ted rise and fall of groundwater discharge during the year. Thus, streamflow at Blackfoss Beck is intimately correlated with rainfall inputs whilst at Foston Beck the pattern of streamflow bears little relation to the pattern of rainfall, peak baseflow occurring some 2 months after the main period of rainfall. The hydrograph for Snaizeholme Beck is also typical of a catchment which produces much surface runoff and little baseflow; soils here are peaty and permanently saturated for most of the year. The higher number of quickflow peaks reflects the higher number of precipitation events in the Pennine uplands, compared with the much drier Vale of York which lies to the east in its rain shadow.

The storm hydrograph

As might be expected from section 1.2, the shape and dimensions of a storm hydrograph in a headwater catchment reflect closely the runoff mechanisms operating. Figure 1.2 shows the elements of the storm hydrograph. Surface runoff tends to produce the highest peak runoff rate and the shortest lag time to peak (see Fig. 1.7); in cases where little surface runoff is produced, SSF will dominate the flood hydrograph. In dry periods the recession limb of the storm hydrograph may be very steep because of the lack of subsurface flow. However, as successive rainstorms recharge the soil moisture and groundwater, the recession becomes more gradual and baseflow increases. In winter when recharge is complete, SSF may be important enough to generate delayed peaks in storm discharge (see Fig. 1.5) and groundwater discharge can maintain baseflow at high levels for several months (Fig. 1.8).

Figure 1.9 shows a variety of hydrographs for small, headwater catchments. The flashy response of Shiny Brook, another Pennine stream, is typical for a small basin that produces little but surface runoff; the symmetry of the hydrograph shows the almost complete lack of subsurface flow, a contribution that normally causes the falling limb to be more gentle than the rising limb. Peat-covered catchments produce very little baseflow indeed, a point well illustrated in this example (*cf.* Snaizeholme Beck in Fig. 1.8). The runoff response of the Shiny Brook catchment

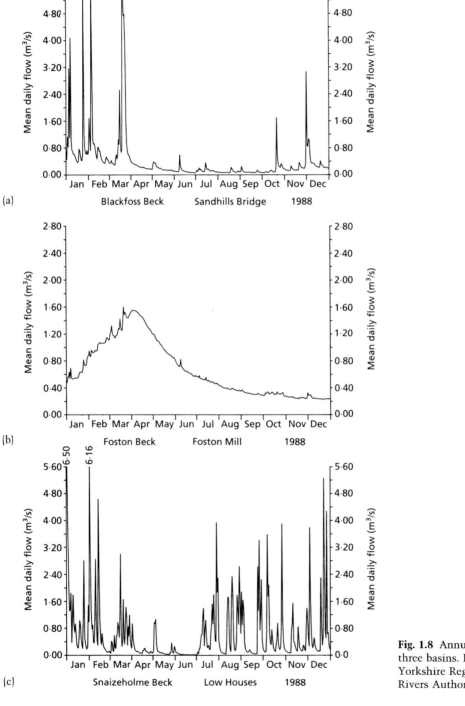

Fig. 1.8 Annual hydrographs for three basins. Data supplied by Yorkshire Region, National Rivers Authority, UK.

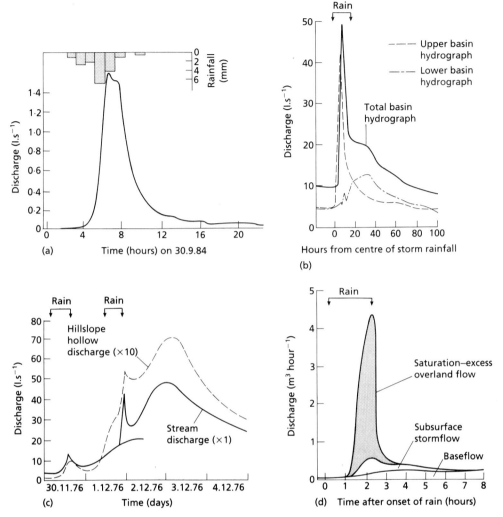

Fig. 1.9 Storm hydrographs for four catchments: (a) Shiny Brook; (b) East Twin Brook; (c) Bicknoller Combe; and (d) Sleepers River. See text for details and source references.

has been described in some detail by Burt and Gardiner (1984) and by Burt *et al* (1990). The hydrograph for the East Twin Brook (Weyman 1973) shows that this small basin may be divided into two distinct sections: in the upper basin, which is infilled by peat, the response is very like Shiny Brook in that much surface runoff is generated; in the lower basin, where permeable loamy soils overlie impermeable sandstone, the subsurface runoff is much delayed compared with the surface flow. Together, the total basin hydrograph reflects these two contributions, the gentle re-

cession denoting the contribution of subsurface drainage. At Bicknoller Combe (Anderson & Burt 1978), little surface runoff is produced and the runoff response is dominated by the delayed peak in stream discharge (see also Burt & Butcher 1985a, 1985b and Fig. 1.5). The subsurface discharge from the hillslope hollow mirrors that of the stream and shows that even the 'first' peak in stream discharge for each event must contain subsurface as well as surface runoff. In the Sleepers River basin (Dunne & Black 1970), SSF is restricted by complete saturation of the soil profile, so that SOF dominates the runoff response; once again, where surface runoff is most important, a sharp peak in discharge tends to result.

Clearly, these four examples can only begin to demonstrate the range of hydrographs which are possible in headwater catchments. In all four cases, interpretation was made possible by knowledge of the hillslope hydrology. Where only the stream hydrograph is available, little can be said with confidence about the *exact* runoff mechanisms operating since many different processes might generate a given response.

Floods

As noted in the introduction to this section, forecasting the flood hydrograph, either in its entirety or specific aspects such as the peak discharge, is a long-established practice in hydrology. Hydrological forecasting has been fully reviewed by Anderson and Burt (1985). Standard methods, such as the unit hydrograph or extreme frequency analysis, are described in most hydrology texts, including Shaw (1988). For the UK, the Flood Studies Report (NERC 1975) has provided a comprehensive guide to flood prediction in ungauged basins throughout the country, although the Report did not use much information from very small basins (<10 km^2). Often, the unit hydrograph or even a regionalized rational method, stratified by cover type (Hewlett *et al* 1977), is the only practical possibility in ungauged small basins. The Flood Studies Report provides detailed information on selected major floods in the UK (see also Newson 1975; Rodda *et al* 1976). Although the causes of floods are largely climatic for headwater catchments, the conditions which tend to intensify floods are primarily related to characteristics of the catchment such as soil type, topography, degree of land drainage, and so on. Only in large catchments does the nature of the channel network exert a significant influence on the shape of the flood hydrograph (Ward 1978).

Runoff regime

The regime of a river may be defined as the seasonal variation in its runoff response and is usually portrayed by a curve based on monthly mean flow. Seasonal variations in the natural runoff of a drainage basin depend primarily on the relations between climate, vegetation, soils and rock structure, of which only the last can be strictly independent of climate (Beckinsale 1969). Beckinsale noted that there are such large areas of the world within which the annual pattern of runoff for small and moderately sized basins closely reflect the regional climatic rhythm that areal differentiation of hydrological regions is easily achieved by adapting Koppen's climatic divisions (Fig. 1.10). In Beckinsale's classification, Koppen's terms retain their climatic meaning:

A = tropical rainy climates; all months with mean over 18°C
B = dry climates with an excess of potential evaporation over precipitation
C = warm, temperate rainy climates
D = seasonally cold, snowy climate; mean temperature of the coldest month being below -3°C
Beckinsale applied the rainfall symbols of Koppen to provide the second capital letter in the code:
F = appreciable runoff all year
W = marked winter low flow
S = marked summer low flow
He added a further class to take into account regimes that occur in the snow and ice environments of high mountains outside the polar ice caps, codes EN and EG (HN and HG in Beckinsale's scheme) denoting nival and glacial regimes respectively. He also added a third category of small letters (not shown in Fig. 1.10) to make allowance for temperature regimes which have some relevance to hydrological regimes, such as the occurrence of high evaporation losses in summer.

Ward (1968) analysed river regimes for 37 British basins and showed that 29 were clearly characterized by winter maxima and summer minima. A reasonably uniform precipitation input combines with modest summer evaporation to reduce flows significantly, although not drastically, in that season. Thus, Beckinsale's code CF applies. However, even in a small country like the UK, variations in river regime do occur, some being climatic in origin and others relating to the physical characteristics of the basin itself. Figure 1.11 shows river regimes for several British basins; flow for each month is expressed as a dimension-

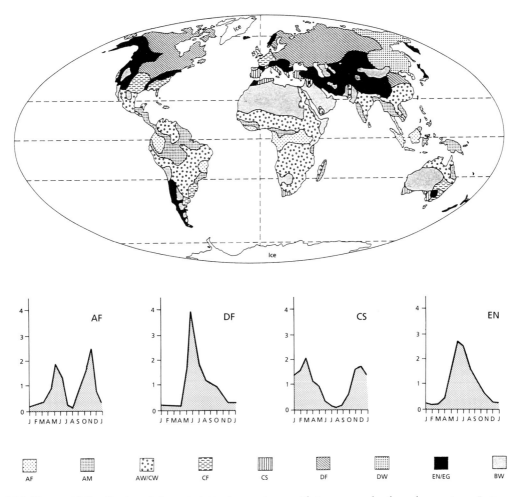

Fig. 1.10 The world distribution of characteristic river regimes with type examples from four regions. Letters are specified in the text. (Adapted and simplified from Beckinsale 1969.)

less index – the ratio of the monthly mean flow to the mean monthly flow. Regimes for Cheriton and Foston Beck are typical of catchments where groundwater dominates: monthly flows vary little from the mean monthly flow, suggesting both minimal quickflow inputs and sustained groundwater flow in summer; peak discharge is relatively late (January or February) indicating the delay involved during recharge. Regimes for Snaizeholme Beck and Blackfoss Beck are more extreme: lack of baseflow means that flows are very low in summer; winter flows are high and peak earlier because quickflow dominates the runoff response. The Slapton Wood regime is also

surprisingly extreme: though a catchment dominated by subsurface stormflow, it is evident that flows are quite low in summer after the soil has drained. In winter when much subsurface stormflow is generated, flows are very much higher than the mean flow. The regime for the River Falloch in Scotland is typical of a mountainous stream with little groundwater storage: summer flows are relatively low whilst winter flows are high and peak early in November in response to high rainfall at that time; a second peak in March (which seems to be typical of many upland streams in Britain) perhaps indicates the influence of snowmelt. Apart from the groundwater-fed

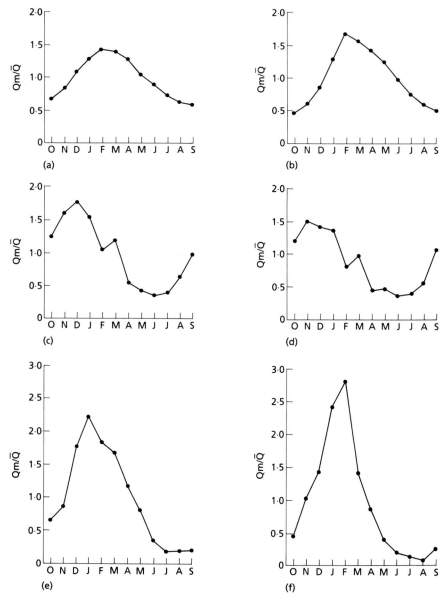

Fig. 1.11 Runoff regimes for selected headwater catchments in Britain. Numbers refer to the UK Surface Water Archive classification of gauging stations. Qm/Q̄, monthly mean flow as a ratio of annual mean flow. (a) Cheriton (42008); (b) Foston Beck (26003); (c) Snaizeholme Beck (27047); (d) Falloch (85003); (e) Blackfoss Beck (27044); and (f) Slapton Wood.

streams, all regimes show the characteristic regime drought referred to in section 1.3 above: despite its humid climate, many British streams have seasonally recurrent low flows. Such low flows could provide the limiting conditions of the aquatic habitat, and so be of particular interest for managers of stream biota.

Water balance

Increasingly, headwater catchments are being used for water supply, so that, even here, aggregation of flow figures into discharge totals is important to show the total resource available. The water balance over a selected time period can be evaluated as follows:

$$P - Q - G - \Delta S - E = 0$$

where P is precipitation, Q is stream discharge, G is groundwater discharge, E is evaporation and ΔS is change in storage. In many cases it is assumed that the catchment is watertight and that no inflow or outflow of groundwater occurs; however, this may not be the case where there are aquifers. On an annual basis it is also often assumed that no change in storage takes place from year to year, although this may well not be the case. If a water balance is required for a shorter period, then changes in storage must be measured. Ferguson and Znamensky (1981) discussed the errors associated with water balance computations and showed that where measurements are inaccurate, the balance period needs to be longer. Probably most uncertainty has surrounded the estimation of basin-wide totals of precipitation and evaporation: these problems are reviewed fully by Rodda *et al* (1976) and Shaw (1988).

Figure 1.12 shows the monthly water balance for the Slapton Wood stream from October 1974 to September 1977, which includes a period of major drought. Potential evaporation was calculated using the Penman formula and these data were then used to calculate actual evaporation using Penman's 'root constant' concept. The full procedure is given in Shaw (1988, Chapter 11). It should be noted that such figures are now routinely available in the UK via the Meteorological Office's MORECS scheme. Particularly notable on Fig. 1.13 is the rainfall deficit in 1975 and 1976, especially during the winter months. Accordingly, very low runoff resulted in the winter of 1975–76. Soil moisture deficits were consistently high during the summers of 1975 and 1976, and soils were barely recharged during the intervening winter. In both summers soil moisture deficits were sufficiently high to preclude evaporation at the potential rate. By contrast, the period from September 1976 onwards was one of the wettest on record, and soil moisture and streamflow recovered quickly; high flow in the winter was associated with a major period of nitrate leaching (Burt *et al* 1988). Assuming an arbitrary initial storage of 500 mm at the end of September 1974, catchment storage had fallen to 368 mm by October 1975 and to 277 mm at the end of August 1976; nevertheless, by the end of February 1977 storage had fully recovered to 577 mm.

Flow frequency and duration

The contrast between baseflow- and quickflow-dominated basins may also be seen clearly using flow duration curves. These are prepared by grouping daily mean flows into selected discharge classes, starting with the lowest values. The cumulative frequency, expressed as a percentage of the total, is then the basis for the flow duration curve, which gives the percentage of time during which any selected discharge may be equalled or exceeded (Shaw 1988). Such cumulative frequency curves are most conveniently plotted on probability paper (on which normal distributions plot as a straight line). In Fig. 1.13 flow duration curves are plotted using a dimensionless flow axis (daily mean flow divided by mean daily flow) to allow comparison between basins. The shape of the flow duration curve gives a good indication of the catchment's runoff response to precipitation: a steeply sloping curve indicates very variable flow, usually from catchments with much quickflow and little baseflow; flow duration curves with a flat slope result from the dampening effects of high infiltration and groundwater storage. Thus, in Fig. 1.13, the Coln, Windrush and Foston Beck all indicate groundwater catchments with sustained, reliable baseflow (minimum flow is still a relatively high percentage of mean flow) and low flood flows (maximum flow is at most only 3–4 times the mean flow). By contrast, Snaizeholme Beck and Shiny Brook, both peat-covered catchments in the Pennine uplands, have very steep flow duration curves; this indicates minimal baseflow in summer, since there is virtually no groundwater and little drainage from the peat (Burt *et al* 1990), and very high flood

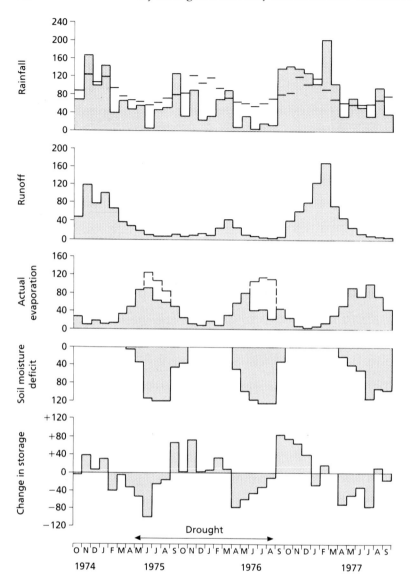

Fig. 1.12 Monthly water balance for the period from October 1974 to September 1977 for the Slapton Wood stream. Dashes on the rainfall histogram indicate mean monthly totals; dashed lines on the evaporation histogram indicate potential (Penman) evaporation. All units are in millimetres.

runoff, given the widespread production of surface flow in peat-covered catchments (Burt & Gardiner 1984). The Slapton Wood stream is intermediate to these two groups: relatively low flow in summer but high winter runoff due both to surface quickflow contributions from a variety of source areas (zones of permanent surface saturation, roads, tracks and fields of low infiltration capacity; Heathwaite *et al* 1990) and to delayed peaks in subsurface stormflow (Burt & Butcher 1985a, 1985b).

Low flow

Only recently has much attention been focused on low flows, with the result that standard definitions and methods of analysis are lacking. Unlike floods, low flows tend to be prolonged so that analyses based on mean flow over intervals ranging from a week to a month tend to be more useful than those based on daily mean flow; also, flow on a single day is too readily affected by abstractions. UK river agencies have often

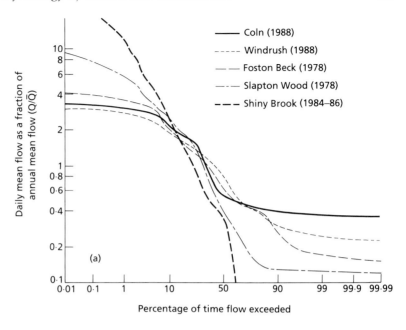

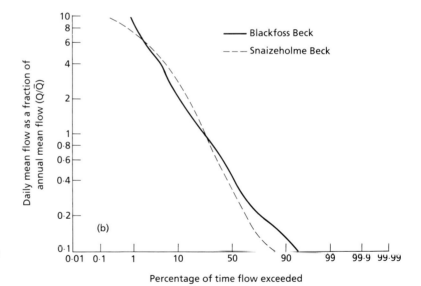

Fig. 1.13 Flow duration curves plotted as a dimensionless index for selected headwater catchments in Britain.

adopted the mean annual minimum 7-day flow as their index of dry weather flow; the Low Flow Studies Report (IH 1980) used the mean annual minimum 10-day flow. The Institute of Hydrology (IH 1989) noted that the daily mean flow exceeded for 95% of the time remains a useful low flow parameter for the assessment of river water quality consent conditions. Like floods, low flows may be analysed in terms of flow frequency or flow duration; for the latter, a flow duration curve of low slope indicates a groundwater basin with reliable low flows (*cf.* Fig. 1.13). Given its regularity, graphs of recession flow may be very useful for low-flow prediction purposes, although their use as indicators of the flow processes operating during recession is probably very limited (Anderson & Burt 1980). Recent droughts in the UK have been described by Doornkamp *et al* (1980) and by Marsh and Lees (1985).

Long-term variations in flow

Long-term variations in flow are caused by climatic variation and by human impact. Human influence on streamflow may be classified into two groups: *direct* impacts such as construction and operation of reservoirs, direct abstractions for domestic supply or irrigation, and diversions of streamflow. Urbanization changes permeable land surfaces to impermeable and probably has one of the most dramatic effects of any land-use change on hydrology, particularly on the volume and speed of storm runoff. Even in rural catchments, roads and villages may constitute a significant area of impermeable surface. In headwater catchments *indirect* impacts related to the condition of the land surface are probably more important, however. They are crucial because they can significantly change the conditions governing runoff formation in the basin. They include agricultural practices, deforestation, urbanization and drainage of wetlands. A quantitative evaluation of human impact on streamflow is complicated because numerous causal factors operate simultaneously. Moreover, deliberate human actions may overlap in the short or intermediate term with natural variations in flow, the amplitude of which may considerably exceed the magnitude of the cultural changes.

Given the intimate connection between runoff processes and streamflow in headwater catchments, it has proved possible to conduct hydrological experiments which allow the evaluation of the role and importance of each anthropogenic factor individually. Classically, such experiments have involved a 'paired catchment' approach, following the pioneering example of Bates and Henry (1928) at Wagon Wheel Gap, Colorado, USA. In this, two nearby or adjacent catchments are selected on the basis of their similarity in size, cover type, aspect and suitability for streamflow gauging. Runoff from the two basins is compared during a period of 'calibration' when both basins

remain unchanged. Then, one catchment is 'treated' while leaving the other unchanged as a control. The purpose of such experiments is to establish absolute differences in water yield between different basins, rather than to establish the absolute water balance for each basin (Hewlett 1982). Notable paired catchment studies in forest hydrology have been conducted at the Coweeta Hydrologic Laboratory in North Carolina, USA (Swank & Crossley 1988) and at the Institute of Hydrology's catchments at Plynlimon, mid-Wales (Kirby *et al* 1991). Plot studies, such as that of Law (1956), may provide useful evidence relating to process rates and the direction of change following a treatment, but basin scale experiments are preferable in that, depending on the basin size and the character of the manipulation, they reduce 'oasis' effects and also avoid the experimental difficulties associated with attempting to compute a full water balance. In some cases, specific studies are established on individual catchments in a process of standardization (or calibration of a catchment on itself; Hursh, quoted in Swank & Crossley 1988). However, this requires climate to remain constant during the treatment period if the effects of a treatment are to be evaluated with confidence. Techniques such as double-mass analysis (Dunne & Leopold 1978) may also be useful to indicate changes in flow where only the record from a single basin is available.

Space does not allow a full account of the effects of all possible human influences on streamflow; extensive reviews are given in Ward and Robinson (1990), for example. By way of brief illustration, some examples from forest hydrology will be described. Most paired catchment experiments have used water-balance methods to study changes in water yield at annual or monthly timescales (see review by Swank *et al* 1988). Changes in the size and shape of storm hydrographs have been examined by Swank *et al* (1982), and by Hewlett and Helvey (1970) who found that the main change following forest clearance was an increase in quickflow volume; peak discharge and time to peak were not significantly different. Hewlett and Helvey argued that forest clearance would produce moister soils; this would encourage the generation of increased volumes of sub-

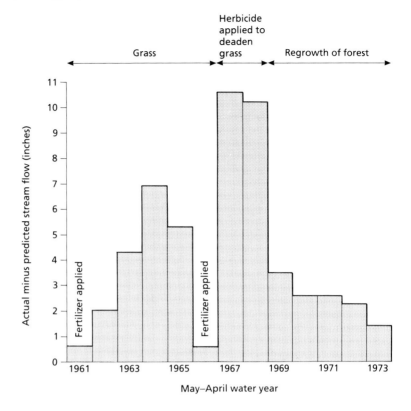

Fig. 1.14 Changes in annual water yield on watershed 6 at Coweeta, North Carolina, USA, in response to changes in land surface cover (see text for details).

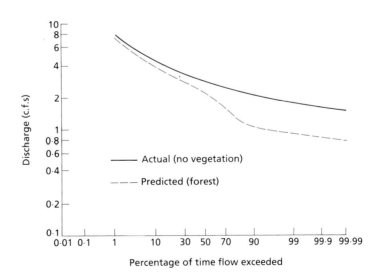

Fig. 1.15 Flow duration curves for mature forest and a surface of deadened grass; watershed 6, Coweeta, North Carolina, USA. c.f.s., cubic feet per second.

surface stormflow on the cleared catchment so that the mean increment to the hydrograph due to the treatment occurred on the falling limb of the storm hydrograph. Changes in flow duration have been examined (Burt & Swank 1992) for an experiment first reported by Hibbert (1969). Changes in annual water yield over the study period are shown in Fig. 1.14. Following forest clearance and its replacement by grass, water yields increase, the actual amount depending on the vigour of grass growth; as grass production declines, water yields rise. When herbicide is applied to deaden the grass, the water yield increases significantly. Thereafter, forest regrowth is allowed to progress naturally and water yields gradually decline towards the expected level. Figure 1.15 shows the predicted and actual flow duration curves for the 1967 water year. This shows that the increase in water yield at that time was associated particularly with increases in baseflow but that the discharge for a given exceedance probability was higher throughout the range of flows. Such results show that forest clearance provides an increase in water yield and that, although flood peaks may increase, the main effect is that of more baseflow. On this evidence, claims by some environmentalists that forest clearance causes a *decrease* in water yield would seem to be false.

ACKNOWLEDGEMENTS

I am grateful to Professor Mike Church for a typically thoughtful and thorough review of an earlier draft of this chapter. Figure 1.8 was kindly supplied by Dr Tony Edwards of the National Rivers Authority, Yorkshire Region, UK.

REFERENCES

Abdul AS, Gillham RW. (1984) Laboratory studies of the effects of the capillary fringe on streamflow generation. *Water Resources Research* **20**: 691–8. [1.2]

Anderson MG, Burt TP. (1977) A laboratory model to investigate the soil moisture conditions on a draining slope. *Journal of Hydrology* **33**: 383–90. [1.3]

Anderson MG, Burt TP. (1978) The role of topography in controlling throughflow generation. *Earth Surface Processes* **3**: 331–44. [1.2]

Anderson MG, Burt TP. (1980) Interpretation of re-

cession curves. *Journal of Hydrology* **46**: 89–101. [1.4]

Anderson MG, Burt TP. (1985) *Hydrological Forecasting*. John Wiley & Sons, Chichester. [1.4]

Anderson MG, Burt TP. (1990) Subsurface runoff. In: Anderson MG, Burt TP (eds) *Process Studies in Hillslope Hydrology*, pp 365–400. John Wiley & Sons, Chichester. [1.2]

Atkinson TC. (1978) Techniques for measuring subsurface flow on hillslopes. In: Kirkby MJ (ed.) *Hillslope Hydrology*, pp 73–120. John Wiley & Sons, Chichester. [1.1]

Bates CG, Henry AJ. (1928) Forest and streamflow experiments at Wagon Wheel Gap, Colorado. *Monthly Weather Review* Supplement 30. [1.4]

Beckinsale RP. (1969) River regimes. In: Chorley RJ (ed.) *Water, Earth and Man*, pp 176–92. Methuen, London. [1.4]

Betson RP. (1964) What is watershed runoff? *Journal of Geophysical Research* **69**: 1541–52. [1.2]

Beven KJ. (1985) Distributed models. In: Anderson MG, Burt TP (eds) *Hydrological Forecasting*, pp 405–36. John Wiley & Sons, Chichester. [1.1]

Beven KJ, Germann PF. (1982) Macropores and water flow in soils. *Water Resources Research* **18** (5): 1311–25. [1.4]

British Standards Institution (1964) Methods of measurement of liquid flow in open channels. *BS 3680*. [1.1]

Burt TP. (1978) *Runoff processes in a small upland catchment with special reference to the role of hillslope hollows*. Unpublished PhD thesis, University of Bristol, UK. [1.2]

Burt TP. (1986) Runoff processes and solutional denudation rates on humid temperate hillslopes. In: Trudgill ST (ed.) *Solute Processes*, pp 193–249. John Wiley & Sons, Chichester. [1.2]

Burt TP, Arkell BP. (1986) Variable source areas of stream discharge and their relationship to point and non-point sources of nitrate pollution. *International Association of Hydrological Sciences Publication* **157**: 155–64. [1.2]

Burt TP, Butcher DP. (1985a) On the generation of delayed peaks in stream discharge. *Journal of Hydrology* **78**: 361–78. [1.2, 1.4]

Burt TP, Butcher DP. (1985b) Topographic controls of soil moisture distribution. *Journal of Soil Science* **36**: 469–76. [1.2, 1.4]

Burt TP, Gardiner AT. (1984) Runoff and sediment production in a small peat-covered catchment: some preliminary results. In: Burt TP, Walling DE (eds) *Catchment Experiments in Fluvial Geomorphology*, pp 133–52. GeoBooks, Norwich. [1.2, 1.4]

Burt TP, Swank WT. (1992) Flow frequency responses to hardwood-to-grass conversion and subsequent succession. *Hydrological Processes* **6**: (in press). [1.4]

Burt TP, Arkell BP, Trudgill ST, Walling DE. (1988) Stream nitrate in a small catchment in south west England over a period of 15 years (1970–1985). *Hydrological Processes* **2**: 267–84. [1.4]

Burt TP, Heathwaite AL, Labadz JC. (1990) Runoff production in peat-covered catchments. In: Anderson MG, Burt TP (eds) *Process Studies in Hillslope Hydrology*, pp 463–500. John Wiley & Sons, Chichester. [1.4]

Calder IR, Newson MD. (1979) Land-use and upland water resources in Britain—a strategic look. *Water Resources Bulletin* **15**: 1628–39. [1.4]

Church MA. (1988) Floods in cold climates. In: Baker VR, Kochel RC, Patton PC (eds) *Flood Geomorphology*, pp 205–29. Wiley Interscience, Chichester. [1.2, 1.3]

Church MA, Woo MK. (1990) Geography of surface runoff: some lessons for research. In: Anderson MG, Burt TP (eds) *Process Studies in Hillslope Hydrology*, pp 299–325. John Wiley & Sons, Chichester. [1.2]

Coles N, Trudgill ST. (1985) The movement of nitrate fertilizer from the soil surface to drainage waters by preferential flow in weakly structured soils, Slapton, south Devon. *Agriculture, Ecosystems and Environment* **13**: 241–59. [1.2]

Doornkamp JC, Gregory KJ, Burn AS. (1980) *Atlas of Drought in Britain 1975–76*. Institute of British Geographers, London. [1.4]

Dunne T. (1978) Field studies of hillslope flow processes. In: Kirkby MJ (ed.) *Hillslope Hydrology*, pp 227–93. John Wiley & Sons, Chichester. [1.2]

Dunne T, Black RD. (1970) An experimental investigation of runoff production in permeable soils. *Water Resources Research* **6**: 478–90. [1.2, 1.4]

Dunne T, Black RD. (1971) Runoff processes during snowmelt. *Water Resources Research* **10**: 119–23. [1.2]

Dunne T, Leopold LB. (1978) *Water in Environmental Planning*. Freeman, San Francisco. [1.2, 1.3, 1.4]

Dunne T, Price AG, Colbeck SC. (1976) The generation of runoff from subarctic snowpacks. *Water Resources Research* **12**: 677–85. [1.2]

Ferguson HL, Znamensky VA. (1981) Methods of computation of the water balance in large lakes and reservoirs. *UNESCO Studies and Reports in Hydrology 31*. UNESCO, Paris. [1.4]

Ford DC, Williams PW. (1989) *Karst Geomorphology and Hydrology*. Unwin Hyman, London. [1.3]

Germann PF. (1986) Rapid drainage response to precipitation. *Hydrological Processes* **1**: 3–14. [1.2]

Gilman K, Newson MD. (1980) *Soil pipes and pipeflow: a hydrological study in upland Wales*. British Geomorphological Research Group Research Monograph No. 1. GeoBooks, Norwich. [1.2]

Gregory KJ, Walling DE. (1973) *Drainage Basin Form and Process*. Edward Arnold, London. [1.1]

Goudie AS (ed.) (1990) *Geomorphological Techniques* 2nd edn. Unwin Hyman, London. [1.1]

Heathwaite AL, Burt TP, Trudgill ST. (1990) Land-use controls on sediment production in a lowland catchment, south-west England. In: Boardman J, Foster IDL, Dearing JA (eds) *Soil Erosion on Agricultural Land*, pp 69–86. John Wiley & Sons, Chichester. [1.2, 1.4]

Herschy RW (ed.) (1978) *Hydrometry, Principles and Practice*. Wiley Interscience, Chichester. [1.1]

Hewlett JD. (1961) Watershed management. In: *Report for 1961 Southeastern Forest Experiment Station*. pp. 62–66 US Forest Service, Asheville, North Carolina. [1.2]

Hewlett JD. (1982) *Principles of Forest Hydrology*. University of Georgia Press, Athens, USA. [1.4]

Hewlett JD, Helvey JD. (1970) Effects of forest clear-felling on the storm hydrograph. *Water Resources Research* **6**: 768–82. [1.4]

Hewlett JD, Hibbert AR. (1963) Moisture and energy conditions within a sloping soil mass during drainage. *Journal of Geophysical Research* **68**: 1080–7. [1.3]

Hewlett JD, Hibbert AR. (1967) Factors affecting the response of small watersheds to precipitation in humid areas. In: Sopper WE, Lull HW (eds) *Proceedings of the International Symposium of Forest Hydrology*, pp 275–90. Pergamon, Oxford. [1.1]

Hewlett JD, Cunningham GB, Troendle CA. (1977) Predicting stormflow and peakflow from small basins in humid areas by the R-index method. *Water Resources Bulletin* **13**: 231–54. [1.4]

Hibbert AR. (1969) Water yield changes after converting a forested catchment to grass. *Water Resources Research* **5**: 634–40. [1.4]

Horton RE. (1933) The role of infiltration in the hydrological cycle. *Transactions of the American Geophysical Union* **14**: 446–60. [1.1, 1.2]

Horton RE. (1940) An approach towards a physical interpretation of infiltration capacity. *Proceedings of the Soil Science Society of America* **4**: 399–417. [1.2]

Horton RE. (1945) Erosional development of streams and their drainage basins; hydrophysical approach to quantitative morphology. *Bulletin of the Geological Society of America* **56**: 275–370. [1.2]

Institute of Hydrology (1980) *Low Flow Studies Report*. Institute of Hydrology, Wallingford, UK. [1.4]

Institute of Hydrology (1989) *Hydrological Data UK, 1988*. Surface Water Archive, Institute of Hydrology, Wallingford, UK. [1.4]

Jones JAA. (1981) *The nature of soil piping: a review of research*. British Geomorphological Research Group Research Monograph No. 3. GeoBooks, Norwich. [1.2]

Kirby C, Newson MD, Gilman K. (1991) Plynlimon research: the first two decades. *Institute of Hydrology Report* **109**: pp 188. Wallingford, UK. [1.4]

Kirkby MJ. (1978) Implications for sediment transport.

In: Kirkby MJ (ed.) *Hillslope Hydrology*, pp 325−63. John Wiley & Sons, Chichester. [1.2]

Kirkby MJ. (1985) Hillslope hydrology. In: Anderson MG, Burt TP (eds) *Hydrological Forecasting*, pp 37−75. [1.2]

Kirkby MJ, Chorley RJ. (1967) Throughflow, overland flow and erosion. *Bulletin of the International Association of the Science of Hydrology* **12**: 5−21. [1.2]

Kneale WR. (1986) The hydrology of a sloping, structured clay soil at Wytham, near Oxford, England. *Journal of Hydrology* **85**: 1−14. [1.2]

Kneale WR, White RE. (1984) The movement of water through cores of a dry (cracked) clay-loam grassland topsoil. *Journal of Hydrology* **67**: 361−5. [1.2]

Langbein WB, Schumm SA. (1958) Yield of sediment in relation to mean annual precipitation. *Transactions of the American Geophysical Union* **39**: 1076−84. [1.2]

Law F. (1956) The effect of afforestation upon the yield of water catchment areas. *Journal of the British Waterworks Association* **38**: 489−94. [1.4]

Marsh T, Lees M. (1985) *Hydrological Data UK, The 1984 Drought*. Institute of Hydrology/British Geological Survey, Wallingford, UK. [1.4]

Mosley MP. (1979) Streamflow generation in a forested watershed, New Zealand. *Water Resources Research* **15**: 795−806. [1.2]

NERC (1975) *Flood Studies Report*. Natural Environment Research Council, London. [1.2, 1.4]

Newson MD. (1975) *Floods and Flood Hazard in the United Kingdom*. Oxford University Press, Oxford. [1.4]

Oakes DB, Pontin JMA. (1976) Mathematical modelling of a chalk aquifer. *Water Research Centre Technical Report* **24**: 37 pp. Water Research Centre, Medmenham, UK. [1.3]

O'Loughlin EM. (1981) Saturated regions in catchments and their relations to soil and topographic properties. *Journal of Hydrology* **53**: 229−46. [1.2]

O'Loughlin EM. (1986) Prediction of surface saturation zones in natural catchments by topographic analysis. *Water Resources Research* **22**: 794−804. [1.2]

Robinson M, Beven KJ. (1983) The effect of mole drainage on the hydrological response of a swelling clay soil. *Journal of Hydrology* **64**: 205−23. [1.2]

Rodda JC, Downing RA, Law FM. (1976) Systematic Hydrology. Butterworth, London. [1.1, 1.4]

Shaw EM. (1988) *Hydrology in Practice* 2nd edn. Van Nostrand Reinhold, London. [1.4]

Sherman LK. (1932) Streamflow from rainfall by the unit hydrograph method. *Engineering News Record* **108**: 501−5. [1.1]

Stephenson GR, Freeze RA. (1974) Mathematical simulation of subsurface flow contributions to snowmelt runoff, Reynolds Creek, Idaho. *Water Resources Research* **10**: 284−94. [1.2]

Swank WT, Crossley DA. (1988) *Forest Hydrology and Ecology at Coweeta*. Ecological Studies 66. Springer-Verlag, New York. [1.1, 1.4]

Swank WT, Douglass JE, Cunningham GB. (1982) Changes in water yield and storm hydrographs following commercial clearcutting on a southern Appalachian catchment. In: *Proceedings of the Symposium on Hydrological Research Basins*, pp 583−94. Sonderh. Landeshydrologie, Bern. [1.4]

Swank WT, Swift LW, Douglass JE. (1988) Streamflow changes associated with forest cutting, species conversions, and natural disturbances. In: Swank WT, Crossley DA (eds) *Forest Hydrology and Ecology at Coweeta*, pp 297−312. Springer-Verlag, New York. [1.4]

Thorne CR, Zevenbergen LW, Burt TP, Butcher DP. (1987) Terrain analysis for quantitative description of zero order basins. In: *Proceedings of the International Symposium on Erosion and Sedimentation*, **165**: 121−130. Corvallis, Oregon. IAHS Publication. [1.2]

Troendle CA. (1985) Variable source area models. In: Anderson MG, Burt TP (eds) *Hydrological Forecasting*, pp 347−404. John Wiley & Sons, Chichester. [1.2]

Ward RC. (1968) Some runoff characteristics of British rivers. *Journal of Hydrology* **6**: 358−72. [1.4]

Ward RC. (1978) *Floods: A Geographical Perspective*. Macmillan, London. [1.4]

Ward RC, Robinson M. (1990) *Principles of Hydrology* 3rd edn. McGraw-Hill, London. [1.1, 1.4]

Weyman DR. (1973) Measurements of the downslope flow of water in a soil. *Journal of Hydrology* **20**: 267−84. [1.3]

Whipkey RZ. (1965) Subsurface stormflow from forested watersheds. *Bulletin of the International Association of Scientific Hydrology* **10**: 74−85. [1.2]

Whipkey RZ, Kirkby MJ. (1978) Flow within the soil. In: Kirkby MJ (ed.) *Hillslope Hydrology*, pp 121−44. John Wiley & Sons, Chichester. [1.2]

Yair A, Lavee H. (1985) Runoff generation in arid and semi-arid zones. In: Anderson MG, Burt TP (eds) *Hydrological Forecasting*, pp 183−220. John Wiley & Sons, Chichester. [1.2]

Zaslavsky D, Sinai G. (1981) Surface hydrology: 5 parts. *Journal of the Hydraulics Division, Proceedings of the American Society of Civil Engineers* **107**: 1−93. [1.2]

2: Analysis of River Regimes

A. GUSTARD

2.1 INTRODUCTION

The previous chapter described the hydrology of headwater catchments where the river regime can be closely related to catchment processes. Research at this scale has generally focused on small (less than 100 km^2), topographically well-defined, impermeable upland valleys. However, in terms of source areas for larger river systems, flat agricultural areas and catchments with a significant groundwater component are of equal importance for runoff generation. With increasing scale and heterogeneity of catchment properties the link between process and hydrological response becomes obscured. Furthermore, processes such as channel, bank and floodplain storage, and the interaction between the river and the alluvial aquifer and the greater diversity of climate and hydrogeology, increase their significance in larger catchments. For example, a flood wave moving down a large river system will result in an increase in storage within each reach. With permeable alluvial deposits, river water will move from the river into bank storage and if the flood is of sufficient magnitude water will move across and be stored on the floodplain adjacent to the channel. These processes will influence the river regime by attenuating the flood wave, reducing both the magnitude of the flood peak and the velocity of the wave.

The regime of large rivers will be controlled by the diverse nature of upstream subcatchments which will have a range of climate, topography, geology, soils and land use. The Rhine, with a catchment area of 185 300 km^2, has a range of precipitation from 3000 mm per annum in the alpine headwaters to 600 mm where it enters the North Sea (Friedrich & Müller 1984). In the Alpine region 50% of the precipitation falls as snow compared with less than 10% in the lower Rhine. This pattern results in an Alpine flow regime in the Rhine upstream of the confluence with the Main, with snowmelt producing a higher mean flow in the summer than the winter months. Further downstream below the confluence with the Mosel the mean flow in the winter is higher than in the summer as a result of the increased catchment area with a maritime climate with less snowfall. There is a similar diversity in land use in the Rhine catchment including forestry, viticulture, arable and pastoral agriculture, and urbanization. Although each land use will produce a distinct hydrological impact at the catchment scale of less than 100 km^2, in the large river system of the Rhine individual controls of particular land use (or topography, soils and hydrogeology) will be obscured. Furthermore the river regime will be influenced by a number of direct artificial influences, including abstractions and discharges, groundwater pumping, reservoir impoundments and flood protection schemes.

One approach to modelling these systems has been to develop relationships between the hydrological regime and the main catchment characteristics that control regime variability. The hydrological regime is defined by a number of flow statistics derived from daily flow data. These include the mean flow, the annual variability and seasonal pattern of runoff, the cumulative frequency distribution of daily flows and the distribution of annual minimum series of given durations. Flood frequency distributions are also derived from daily data for large catchments, but from hourly or shorter time-interval data for catchments with rapid response times. Catchment characteristics are single number indices of a particular physical property of the catchment. Thus, scale may be measured by topographic or groundwater catchment area or by the length of the main stream. Precipitation may be indexed by mean annual precipitation, the proportion of snowmelt or the depth of rainfall for a specified duration and return period. Catchment characteristics have been derived for a number of other variables including drainage density, channel slope, land use and soil type. Relationships have been developed between hydrological regimes and the controlling catchment characteristics using multiple regression. For example, the mean annual flood has been related to catchment characteristics which can be derived from readily available maps (NERC 1975). Such methods can then be used to estimate design floods at sites which do not have recorded flow data.

A knowledge of different river regimes and the techniques used to analyse them provides an important basis for understanding other environmental aspects of larger river systems. For example, flood frequency analysis has been found useful in understanding the spatial pattern of pollution in floodplain soils on the Meuse river (Rang & Schouten 1989). The time and location of geomorphological change on the Rhone river (Vivian 1989) have also been assisted by analysing the seasonal changes in river regimes from 1920. Thus, the impact of hydropower developments and abstractions for irrigation and water supply are related to the natural variability of river flows based on an analysis of a number of continuous flow series.

Where river flow data are available at the site of interest, they can be used to analyse aspects of the hydrological regime directly. However, there is frequently a need for estimating flow characteristics at ungauged sites. Use must then be made of physically based, or conceptual, models or of regional relationships between flow indices and catchment characteristics. The following sections describe a number of different ways of defining hydrological regimes, concluding with a summary of approaches used to estimate river regimes when recorded flow data are not available.

2.2 CLASSIFICATION OF HYDROLOGICAL REGIMES

Global classifications

A number of different procedures have been developed for the classification of hydrological regimes. At the global scale these have been based primarily on climate, using monthly rainfall and temperature, or based on mean values of the water or energy budget. Climatic classifications are, however, difficult to apply to larger river basins which cross climatic divides. This problem has been addressed by Parde (1955), who developed a classification based on the seasonal variation of river flow, and by Beckinsale (1969), who modified Koppen's climatic classification for hydrological regime definition.

A recent hydrological classification (Falkenmark & Chapman 1989) is based on a threefold subdivision of potential evaporation into cold, temperate and warm regions, and subdivisions into dry and humid according to the ratio of precipitation to potential evaporation (Unesco 1979). Although the importance of a number of hydrological processes in modifying climatic inputs was recognized, the need to provide a simple global classification resulted in using only two hydrological categories. These were 'areas with catchment response', with an organized natural drainage network typical of sloping land, and 'flatlands', defined as areas with a less organized natural drainage network.

Within the context of comparative hydrology, this classification defined temperate regions with a mean annual potential evaporation between 500 and 1000 mm, dry regions having a ratio of precipitation to potential evaporation of less than 0.75 and humid regions a ratio greater than 0.75. Table 2.1 gives examples of temperate regions typical of each category. The concept of hydrological regions has been used for regional analysis of both flood and low flow frequency with the objective of developing design methods for estimation at the ungauged site. The traditional approach was to combine flow statistics, for example the annual maximum flood from a number of gauging stations within a geographically defined area (NERC 1975). However, this

Table 2.1 Examples of temperate regions (Falkenmark & Chapman 1989)

Humid temperate sloping land	Dry temperate sloping land	Humid temperate flatlands	Dry temperate flatlands
Western Europe South-west South America Pacific north-west coast of North America Japan New Zealand Tasmania	Steppes of USSR Mongolia North China Patagonia in South America Coastal strip of Australia between Adelaide and Melbourne	East Europe Asia and North America centred approximately on latitude 50°	Buenos Aires province of Argentina, Caspian and Hungarian Plain

has resulted in some regions being defined that are not significantly different from one another (Wiltshire 1986) with very heterogeneous flood frequency characteristics being included within a single region. The search for a more objective definition of a homogeneous hydrological region has led to geographical regions being replaced by grouping basins that are physically similar to each other (Acreman & Wiltshire 1989), such as small, steep, wet catchments or large, flat, dry catchments. A problem with this (and the traditional) approach is that unambiguous assignment of the ungauged basins to one or other region creates discontinuities at the boundaries. This may be overcome by allowing fractional membership to more than one region. However, Acreman and Wiltshire (1989) have argued that once fractional membership is included in the procedure, regions become redundant. Each site can comprise its own region with estimation at the ungauged site being achieved by a weighted average of the observed flood frequency curves from a number of sites. This is a major departure from the traditional approach to defining regions, but with the ease of access to large hydrological and thematic databases it represents a very probable approach for design estimation in the future.

Hydrological regions, however they are defined, provide a basis for identifying areas where there is a similarity in hydrological response. Although this does not necessarily infer that there is a similarity in hydrological process, regions can be used to guide the extent to which hydrological models developed and calibrated in one location

can be used for hydrological prediction in another. At the global scale, regions can be used to ensure that hydrological results and methodologies are not transferred into areas where they are inappropriate. For relationships between hydrological regimes and catchment properties to be developed, statistical definitions of river regimes are required. These are normally derived from long-term observations of the daily mean flow hydrograph or peak flow statistics, and include a number of different measures which describe the mean, seasonal variability and extremes of flood and low flows, and other properties of the river hydrograph such as recession (defining the rate of decrease of discharge) and base flow (the component of river discharge with a slow response to precipitation which maintains river flow in dry periods) characteristics. Hydrological regions enable flow characteristics of different rivers to be compared and contrasted; they assist in evaluating historical changes in river regimes and provide a hydrological basis for water quality and ecological research.

Statistical definition of river regimes

This section summarizes different methods for defining river regimes, ranging from indices of the average annual regime to 'inter-annual' and 'intra-annual' variability. The analysis of the streamflow hydrograph to estimate extremes of both flood and drought are considered, together with methods for determining hydrograph recession and base flow characteristics.

Mean flow

The mean flow is the most fundamental variable for comparing the regime of different rivers, as well as for evaluating available water resources, for estimating changes in historic flow sequences and for determining the impact of human activity. By expressing the mean flow as an average depth over the topographic catchment area, comparisons can be made between catchments with different areas and between precipitation and runoff (assuming that there are no losses to groundwater and that topographic and groundwater catchment areas are the same).

Variations in mean flow are controlled primarily by variations in annual precipitation and annual evaporation, although differences in land use and groundwater flow will impose local differences on any regional trend. In humid temperate regions, annual precipitation increases with altitude and proximity to the sea, being, for example, in excess of 2000 mm in upland areas of western and central Europe and below 600 mm in continental humid temperate areas. Variation in annual potential evaporation is less marked, ranging between 500 and 700 mm with regional variations controlled by latitude, altitude and wind speed. Actual evaporation over most of humid temperate Europe is between 500 and 600 mm (Unesco 1978) with variations resulting from differences in potential evaporation, land use and soil type.

Mean runoff in humid temperate regions typically ranges from over 2000 mm in mountainous areas (e.g. western Britain) to approximately 100 mm (e.g. in north central France, eastern England and eastern Germany) where precipitation and evaporative losses are nearly in balance. However, these regional values obscure local variability. For example, the Val de Bonce catchment (topographic catchment area 203 km^2) gauged at Montboissier, south of Paris, France, has a mean annual runoff based on its topographic catchment area of only 8 mm. This observed value is much lower than the expected runoff of approximately 100 mm and results from a considerably reduced mean river discharge. This is primarily caused by a very high groundwater component of flow which flows through the chalk aquifer and thus 'bypasses' the gauging station. River regimes are considerably influenced by the hydrogeological characteristics of the underlying aquifer. This can result in the location of the topographic catchment boundary being different from the groundwater boundary which may give rise to differences in topographic and groundwater catchment areas, and between the observed and expected mean runoff. Local variability in runoff also arises from different losses by evaporation and transpiration as a result of land-use differences. For example, reductions in mean runoff equivalent to 290 mm for a fully forested coniferous catchment compared with a grassland catchment have been identified from studies of catchments of approximately 10 km^2 in central Wales (Kirkby & Newson 1991). As one moves downstream to larger catchments, the influence of a particular hydrogeological unit or land-use change is diminished as a result of the river system draining a more diversified region with a greater variability of hydrogeology and land use.

The annual variability of river runoff is a simple index of river regimes and is also an important variable for water resource assessment. The FREND (Flow Regimes from Experimental and Network Data) project analysed the coefficient of variation (CV) of annual runoff from over 500 catchments (Fig. 2.1) from 13 European countries (Gustard *et al* 1989) with catchment areas generally less than 500 km^2. The mean value was 0.28 and the spatial distribution over western and northern Europe indicated consistently low values of less than 0.20 over maritime western areas of the British Isles and south-western Norway and higher values between 0.40 and 0.60 in parts of eastern England, eastern France and central Germany. These differences were found to relate to the value of mean runoff with high annual variability in drier catchments, a result also found by McMahon (1979a). However, within more limited geographical regions, other variables may be significant. For example, Kovacs (1989), studying catchments in Hungary, found that the CV increased as the catchment area decreased.

The variability of annual runoff in this European region is compared with that of a sample of 126 'Unesco' rivers from different continents in Fig. 2.2. It can be seen that a wide range of

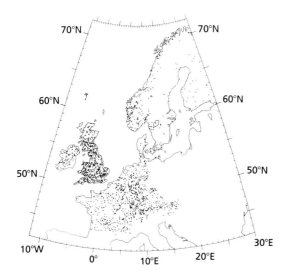

Fig. 2.1 Location of FREND gauging stations in Europe.

Thames, which for the period 1883–1986 was 0.29, and that of the River Darling in Australia for the period 1881–1959 which was 1.46. Figure 2.3 illustrates this greater range in the variability in the Australian flow series and the characteristic way in which there are sustained periods of low annual runoff interspersed with short periods of high runoff. This highlights the difficulties in estimating simple flow statistics such as the mean runoff even from long records in arid or semiarid regimes.

Seasonal variability in runoff

In discussing the definition of distinct seasonal flow regimes, Arnell (1989) highlights the difficulties of classifying a continuous process into discrete regions and presents examples of monthly runoff histograms from eight European rivers (Fig. 2.4, Table 2.2). Most of the examples are characterized by maximum flows in the winter and this is enhanced in western Europe where winter precipitation is highest. In continental areas where most precipitation falls in summer when evapotranspiration is highest, there is less difference between summer and winter flows. These patterns are modified in mountainous and continental areas where snowmelt is an important seasonal influence, as illustrated in the histo-

annual variability is exhibited by European rivers. Of course, the very stable rivers of the world typical of the humid tropics with CVs of less than 0.1 and the very variable rivers with CVs in excess of 0.7 that are typical of arid and semiarid environments are not represented in humid temperate Europe. Ward and Robinson (1990) make similar contrasts between that of the River

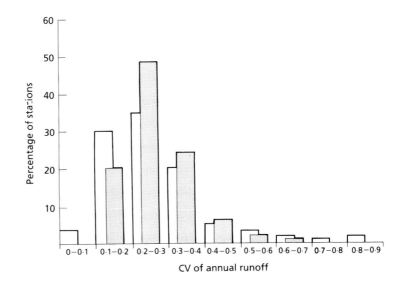

Fig. 2.2 Variability of annual runoff for 126 rivers of the world and European rivers (after Gustard *et al* 1989) ☐ 126 Unesco rivers (McMahon & Mein 1986); ☐ 577 European rivers.

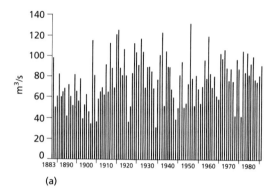
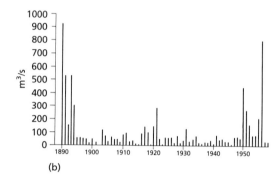

(a)   (b)

Fig. 2.3 Comparison of annual runoff between River Thames, UK and Darling River, Australia (after Ward & Robinson 1990). (a) Thames at Kingston (Teddington) 1883–1986; and (b) Darling at Menindee 1890–1958.

grams for the Lutschine, Grosser Regen and Versanan catchments.

The importance of hydrological processes in modifying hydrological regimes is clearly illustrated by comparing the histograms for the Kym and the Pang. Both catchments have similar maritime humid climates but they exhibit significantly different regimes. The Pang catchment which is underlain by permeable chalk has greater storage capacity enabling high flows to be maintained throughout the summer. This contrasts with the Kym catchment which is underlain by impermeable clay, and thus with limited storage the summer flows cannot be sustained. Larger catchments with areas in excess of 800 km^2 will be influenced by a number of different subcatchments, generally with a wide range of hydrogeology. The response of the larger catchments is thus the sum of a number of different regimes, each with their own seasonal variability in runoff.

Ward (1968) analysed 37 British flow records and demonstrated that 29 of them fell clearly into a simple summer minimum–winter maximum regime. Furthermore, regional patterns in the

Table 2.2 Location and area of example catchments (Arnell 1989)

River and gauging station	Area (km^2)	Location
Kym at Meagre Farm	137.5	East England
Pang at Pangbourne	170.9	South-east England
Severn at Plynlimon	8.8	Central Wales
Ammer at Oberammergau	114.0	Bavaria, Germany
Grosser Regen at Zweisel	177.0	Lower Danube, Germany
Reiche Ebrach at Herrnsdorf	169.0	Central Germany
Schussen at Magenhaus	246.0	Bavaria, Germany
Wurm at Randerath	305.0	North-west Germany
Zusam at Pfaffenhofen	505.0	Bavaria, Germany
Gardon de Mialet at Roucan	239.0	South France
Layon at St. George	250.0	West France
Maumont at La Chanourdie	162.0	South-west France
Orgeval at Le Theil	104.0	North central France
Versanan at Halaback	4.7	Southern Sweden
Lutschine at Gsteig	379.0	Central Switzerland
Yeongsan at Naju	2060.0	South-west Republic of Korea

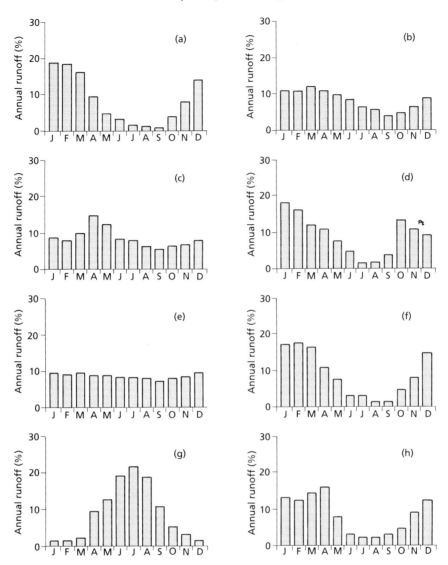

Fig. 2.4 Monthly runoff histograms, showing average monthly runoff as a percentage of average annual runoff for eight of the catchments listed in Table 2.2 (after Arnell 1989). (a) Kym at Meagre Farm; (b) Pang at Pangbourne; (c) Grosser Regen at Zweisel; (d) Gardon de Mialet at Roucan; (e) Wurm at Randerath; (f) Orgeval at le Thiel; (g) Lutschine at Gsteig; and (h) Versanan at Halaback.

month of maximum and minimum runoff were related to both variations in climatic factors and the presence of aquifers in the catchment. A study of more than 500 flow records in the UK (Institute of Hydrology 1980a) stratified the flow records into three groups defined by the 95th percentile discharge expressed as a percentage of the mean flow (ADF). An analysis was thus carried out on all 'impermeable catchments' (Q95 less than 15%), 'average catchments' (Q95 between 15% ADF and 30% ADF) and 'permeable catchments' (Q95 greater than 30% ADF). Having reduced the influence of catchment hydrogeology it was possible to use standard contouring routines to evaluate the spatial pattern of monthly runoff. Figure 2.5 illustrates the distribution of the mean runoff in January and August for impermeable catchments. The figures illustrate the clear regional variability due to climatic controls which is obscured when catchments with contrasting hydrogeologies are included in the analysis. The complete set of maps provides a useful procedure for estimating the seasonal distribution of monthly flows at ungauged sites.

Flow duration curve

The cumulative frequency distribution of daily mean flows (Fig. 2.6) shows the percentage of time during which specified discharges are equalled or exceeded during the period of record. The relationship is normally referred to as the flow duration curve, and although it does not convey information about the sequencing properties of flows it is one of the most informative methods of displaying the complete range of river discharges from low to flood flows. A normal probability scale is commonly used for the frequency axis and a logarithmic scale for the discharge axis. If the logarithms of the daily discharges are distributed normally then the plotted points will lie on a straight line, so aiding esti-

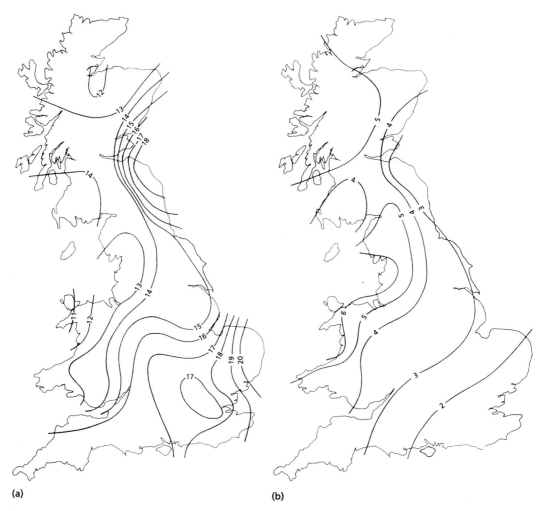

(a)   (b)

Fig. 2.5 Spatial variability of the percentage mean annual runoff in January (a) and August (b) (from Institute of Hydrology 1980a).

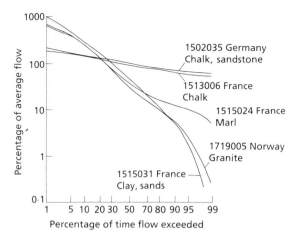

Fig. 2.6 One-day flow duration curves for catchments with contrasting geology (after Gustard *et al* 1989).

note from Fig. 2.6 the similarity of the flow duration curves from the same lithology but from very different locations in Europe. Such clear contrasts are normally evident only in small catchments of less than 500 km² with homogeneous hydrogeology. This suggests that within an area that is broadly 'humid temperate', local catchment controls in catchments of less than 500 km² exert a greater influence on short-term runoff variability than regional climatic conditions. With increasing catchment area, the contributions of a number of diverse catchments lead to an 'averaging' of catchment response. As a result there is greater similarity of flow duration curves from large basins in a similar climatic area, particularly those basins in excess of 10 000 km².

mation from the curve of discharges for different frequencies. Note that the discharge axis in Fig. 2.6 is standardized by the average flow. This facilitates comparison between catchments, because it reduces differences in the location of the curve caused by differences in the mean annual runoff.

Figure 2.6 illustrates the strong control of catchment geology in determining the distribution of daily flows. Rivers supported by aquifers have well-sustained low flows, whilst flood discharges are reduced as a result of the storage provided by permeable soils and geology. In contrast the gradients of curves from impermeable catchments are much steeper, reflecting the higher flood flows and lower low flows. It is also interesting to

Flow duration curves can be derived from individual months or groups of months (Institute of Hydrology 1980b) to provide a more detailed analysis of the seasonal distribution of river flows. These can be presented in the form of a flow duration surface where the discharge (or water level) is plotted on the vertical axis, the month on the horizontal axis and the flow duration curve as a parameter. The flow duration surface (Kovacs 1989) is a useful method for characterizing the seasonal variability of the full range of flows. Figure 2.7 illustrates the water level regime from two gauging stations 100 km apart on the two main rivers with very different regimes in the Carpathian basin. The extensive alpine areas above the permanent snow-line are the main source of water for the Danube at Budapest, and this gives rise to a prolonged snowmelt period

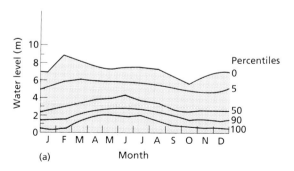

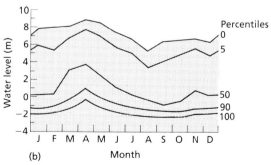

Fig. 2.7 Level duration surfaces characterizing different regimes (after Kovacs 1989). (a) River Danube at Budapest (1921–50); and (b) River Tisza at Szolnok (1921–50).

and a similar probability of flooding between March and August. The low flows are similarly influenced by this alpine region, with low flows being lower in January and February than in the drier months of September and October. In contrast, the headwaters of the Tisza river are much lower and are below the permanent snow-line, resulting in the high floods due to spring snow-melt being confined to April, which is also the time of the highest low flows.

Flow frequency curve

While the flow duration curve displays information about the proportion of time during which a flow is exceeded, the flow frequency curve shows the proportion of years when a flow is exceeded, or equivalently the average interval in years that the river falls below a given threshold discharge. Figure 2.8 illustrates the plot derived from an analysis of 10-day annual minimum discharges from six flow records in western Europe. The analysis can be derived from minima of other durations, for example 1, 30, 60, 90 and 180 days. The annual minima are standardized by the mean flow to enable frequency curves derived from both large and small catchments and catchments with high and low mean precipitation to be more easily compared. The figure is based on using a Weibull extreme value distribution which has been used in a number of low-flow investigations in the USA (Matalas 1963; Joseph 1970), in Malawi (Drayton *et al* 1980) and in the UK (Institute of Hydrology 1980b). Figure 2.8 illustrates the strong control that the geology of the catchment imposes on the slope and the position of the flow frequency curve. There are also close similarities between the curves from limestone lithologies from catchments in the UK and those from the karst area of Yugoslavia.

The similarity of river regimes defined by their flow frequency curve was investigated in the FREND project (Gustard *et al* 1989) by pooling flow frequency curves from 643 European rivers. Annual minima series were first grouped according to the value of their mean annual 10-day minima, MAM(10), which was expressed as a percentage of the mean flow. For example, all relatively impermeable catchments with a

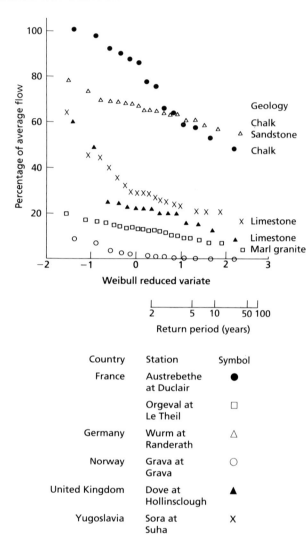

Country	Station	Symbol
France	Austrebethe at Duclair	●
	Orgeval at Le Theil	□
Germany	Wurm at Randerath	△
Norway	Grava at Grava	○
United Kingdom	Dove at Hollinsclough	▲
Yugoslavia	Sora at Suha	X

Fig. 2.8 Annual minima series for rivers with contrasting geology (after IAHS 1990).

MAM(10) between 10% and 20% of the mean flow were analysed and then all permeable catchments with MAM(10) between 30% and 40% of the mean flow. The analysis was carried out by dividing these groups of stations into smaller subsets from different geographical regions of Europe and deriving pooled flow frequency curves. The pooled curves were derived by plotting the mean value of the *x* and *y* co-ordinates for class intervals of the Weibull reduced variate, e.g.

0.0−0.05, 0.05−1.0 etc. Figure 2.9 illustrates the results of this analysis which demonstrates the similarity of flow frequency curves in different geographical regions. For example, the ratio of the 50- to the 2-year return period 10-day annual minima is very similar in the Seine basin to that of Norway for rivers with the same value of MAM(10). The full analysis showed that although there was great spatial variability in low flow statistics such as MAM(10) there was similarity in the relationships between frequent and extreme drought events across Europe.

Flood frequency curve

A similar procedure for estimating the frequency of low flows can be applied to annual maximum flood discharges to derive flood frequency curves. A number of different distributions including log-normal, Pearson, log-Pearson type 3, Wakeby and extreme value distributions have been fitted to flood data (Cunnane 1988). In assessing which was the most appropriate distribution to use, Cunnane concluded that the choice of distribution is dependent upon the fitting technique, the goodness of fit test, and the assumptions made about the real world distribution.

Arnell (1989) used a general extreme-value distribution fitted by the method of probability-weighted moments to compare the flood regimes of eight European catchments. Figure 2.10 shows the relationship between annual maximum flood (standardized by the mean annual flood) and return period. Flood frequency curves typical of lowland regions with low annual rainfall are illustrated by the three steepest curves. These result from data series with a large number of years with low annual maxima but with some extreme events caused by summer thunderstorms. Lowland catchments with more maritime climates have flatter frequency curves as shown by the Ammer and Zusman data. The Zusman

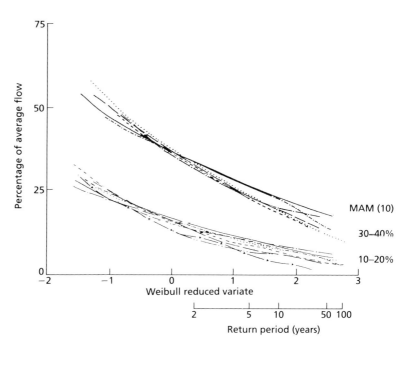

Fig. 2.9 Pooled 10-day mean annual minimum series expressed as a percentage of the average flow. Stations grouped by value of 10-day mean annual minima and hydrometric areas. (after Gustard & Gross 1989).

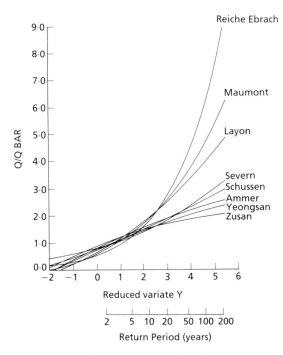

Fig. 2.10 Flood frequency curves for eight catchments listed in Table 2.2 (after Arnell 1989). Discharge standardized by mean annual flood (QBAR).

Table 2.3 Flow regimes in north-west Europe (from Gustard *et al* 1989)

Regime type	Description
Glacier	Dominant high flows in July and August, mostly caused by snowmelt in combination with precipitation
Mountain	Dominant spring flood in April–July, low flows in winter
Inland	Dominant spring flood in March–June, low flows in winter, but higher flows in the autumn
Transition	High runoff in spring and autumn, low-flow spells in both summer and winter
Maritime	High runoff in autumn and winter, with lowest flows in summer

(505 km^2) and Yeongsen (2060 km^2) are the catchments with the flattest frequency curves and the largest catchment areas, a trend found by Farquharson *et al* (1987) for catchments in excess of 100 000 km^2 in area. Steeper curves are generally found in more upland catchments (Severn), although flatter curves may be observed in upland catchments with lakes or more significant snowmelt (Schussen).

An analysis of more than 1500 flow records investigated the seasonal aspects of flood regimes in north-west Europe which were classified into one of five groups based on the number of distinct flood seasons and the relative magnitude of the spring and autumn floods (Gustard *et al* 1989). A summary of this classification, which is based on daily flow data, is presented in Table 2.3. The analysis of daily mean flow data is valuable for both flood and low flows on large river basins in excess of 500 km^2. However, details of catchment processes are often not apparent when average daily discharge is analysed, and Fig. 2.11 illus-

trates the rapid fluctuations in discharge which can occur even on large catchments. It illustrates the diurnal variation in runoff draining an area of 195 km^2 on the Aletschgletscher basin, of which 67% is glaciated. The hydrograph is from a typical melt period in July 1975 during a period of high temperatures and high radiation.

A detailed analysis of annual maximum data (Roald 1989) enabled flood frequency distributions to be subdivided into ten groups. These were based on a cluster analysis of station values of the specific mean annual flood and the CV of the annual maximum flood. The cluster analysis partitions the data into homogeneous groups,

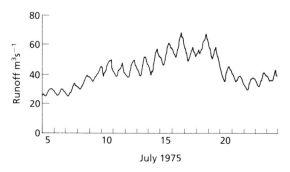

Fig. 2.11 Diurnal variations of runoff from the Aletschgletscher River (after Lang 1989).

for each of which a pooled frequency curve was derived. Although there was some spatial coherence in the regional distribution of the ten groups, it was not possible accurately to predict group membership from basin characteristics. However, the method was considered to have advantages over geographically defined regions which are not necessarily homogeneous in terms of flood frequency distributions.

Figure 2.12 shows a range of pooled flood frequency distributions based on a world flood study using data from 1121 gauging stations in 70 different countries (Farquharson *et al* 1987). They illustrate the lower variability of flood discharges from large river basins in the same region. For example, in Iran the average frequency curve was flatter for catchments with an area greater than 75 000 km² compared with smaller catchments.

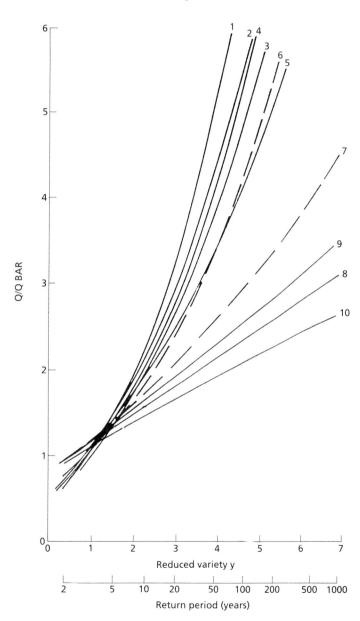

Fig. 2.12 Regional flood frequency distributions—Asia and Australasia; standardized by mean annual flood (General Extreme Value (GEV) fitted by Probability Weighted Moments (PWM)) (from Farquharson *et al* 1987). 1, Saudi Arabia and Yemen; 2, Jordan; 3, Iran (all); 4, Iran (area <7500 km²); 5, Iran (area >7500 km²); 6, Sri Lanka; 7, India, Kerala State; 8, India and Bangladesh (all); 9, India and Bangladesh (area <100 000 km²); 10, India and Bangladesh (area >100 000 km²).

This trend was also found in data from India and Bangladesh where catchments greater in area than $100\,000$ km^2 had the flattest frequency curve. The steep flood growth curves typical of arid and semiarid environments in Saudi Arabia, Yemen, Jordan and Iran are also illustrated.

Other measures of flow regime

There are a number of additional methods of describing the flow regime including, for example, the volume of flood response calculated using a separation between quick response flow and base flow (Hewlett & Hibbert 1967). A number of techniques have also been developed for estimating the base flow discharge using daily flow data. A method developed by Kille (1970) is based on the analysis of monthly minimum flows and has been applied in Germany and Poland. More recently, Demuth (1989) has automated this procedure and applied it to 102 small research basins in western and northern Europe.

The Low Flow Studies Report (Institute of Hydrology 1980b) recommends using the base flow index (BFI) derived from mean daily flow data for classifying the low-flow response of a catchment and for indexing the storage at the catchment scale. The index is the ratio of base flow (calculated from a hydrograph separation procedure) to total flow. The procedure for calculating the index is as follows:

1 Divide the mean daily flow data into non-overlapping blocks of 5 days and calculate the minima for each of these blocks, and let them be called $Q_1, Q_2, Q_3, \ldots, Q_n$.

2 Consider in turn (Q_1, Q_2, Q_3), (Q_2, Q_3, Q_4), $\ldots$, (Q_{n-1}, Q_n, Q_{n+1}), etc. In each case, if $0.9 \times$ central value is less than outer values, then the central value is a turning point for the baseflow line. Continue this procedure until all the data have been analysed to provide a derived set of baseflow ordinates $QB_1, QB_2, QB_3, \ldots QB_i$ which will have different time periods between them.

3 By linear interpolation between each QB_i value, estimate each daily value of $QB_1 \ldots QB_n$.

4 If $QB_i > Q_i$ then set $QB_i = Q_i$.

5 Calculate V_A, the volume beneath the recorded mean daily flows Q_n.

6 Calculate V_B, the volume beneath the baseflow line between the first and last baseflow turning points $Q_1 \ldots Q_n$.

7 The BFI is then V_B/V_A.

Relationships were derived relating BFI to low-flow statistics, and recommendations were made concerning the estimation of BFI at the ungauged site from catchment geology. The index has been applied in Canada (Pilon & Condie 1986), Zimbabwe and Malawi (Wright 1989), New Zealand (National Water and Soil Conservation Authority 1984) and Norway (Tallaksen 1986). These studies have indicated that some modification of the index and care in interpretation are required when applying the separation procedure in different river regimes. However, the use of a standard procedure has facilitated the comparison of catchment response from a number of different countries. The BFI has also been used for calibrating the hydrological response of soils in the UK (Boorman & Hollis 1990) and in mainland Europe (Gustard & Gross 1989).

Recession properties calculated either from short time-interval data for flood events or daily data for low-flow investigations are also an important property of the flow hydrograph. Toebes and Strang (1964) and Hall (1968) reviewed a number of different procedures for estimating recession characteristics. A number of analytical procedures have been developed for the rapid calculation of recession parameters based on daily flow data. These include studies by Demuth (1989) who related recession properties to basin characteristics in Europe, Tallaksen (1989) who investigated the temporal variability in recession properties in Norway, and Schwarze et al (1989) who used a computer-aided analysis of flow recessions in a water-balance investigation of 30 catchments in Germany.

2.3 ESTIMATING REGIMES AT UNGAUGED SITES

The previous sections have described a number of methods for defining particular aspects of river regimes based on the analysis of recorded flow data. There are, however, occasions when this approach is not appropriate. First, when there are no recorded flow data at or in the vicinity of the site of interest. Second, when estimates of the change in flow regime are required following a change in catchment land use, for example urbanization. Third, when a change in flow regime will occur as a result of water resource development, for example groundwater abstraction or reservoir impoundment.

A wide range of hydrological models has been developed for evaluating these problems. The most recent modelling development has been in the area of physically based distributed models. The component processes of the hydrological system, for example the unsaturated zone and the saturated zone, are described by partial differential equations representing the conservation of mass and momentum operating on a grid cell basis. Beven and O'Connell (1982) have reviewed the general concepts of physically based models, examples of which include IHDM (Beven *et al* 1987) and SHE (Abbott *et al* 1986a, 1986b). Although this family of models has demanding data requirements it provides a physically based simulation of catchment behaviour and is therefore able to model a number of different changes occurring simultaneously on a catchment.

There has been a wider use of simpler conceptual models. Such models may be lumped or distributed depending on whether or not the spatial distribution of hydrological variables within the catchment is considered. Figure 2.13 illustrates a typical structure of a conceptual model which has been used extensively in land-use change modelling (Blackie & Eeles 1985). Both physically based and conceptual models can be used to simulate a time series from which statistics of the flow regime can be calculated.

A number of regional studies have been carried out with the objective of estimating flow statistics at ungauged sites. These studies relate flow variables to catchment characteristics, for example mean annual precipitation, catchment area, slope, land use and soil type, using multiple regression analysis techniques. These statistical models are generally calibrated on all the available flow data within a region, and are thus based on a large number of station-years of data, which minimizes the errors in estimating hydrological extremes. However, because the models do not incorporate catchment processes they are not

appropriate for addressing problems such as land-use change or the impact of water resource development on downstream flow regimes. Examples of regional low flow studies include Simmers (1975) for New Zealand, Musiake *et al* (1975) for Japan, Knisel (1963), Mitchell (1957), Hines (1975) and Riggs (1973, 1990) in the USA, and McMahon (1969) for Australia. In Europe, studies have been carried out by Martin and Cunnane (1976) in Ireland, by Wittenberg (1989) in Germany, by Moltzau (1990) in Norway, and by the Institute of Hydrology (1980a, 1980b) in the UK. Similarly, a number of flood studies have been completed, including those in the UK (NERC 1975), New Zealand (Beable & McKerchar 1982) and North America (Chong & Moore 1983).

2.4 CONCLUSIONS

Recent advances in the analysis of river regimes have been in three main areas. The first of these has been the further development of methods for the definition of river regimes. For example, Nathan and McMahon (1990) have developed automated techniques for the estimation of recession and base flow which have been applied to 186 catchments in south-eastern Australia. Second, the definition of hydrological regions has developed from the concept of geographical regions defined by administrative or topographic boundaries to the 'region of influence approach' (Burn 1990). This concept considers that each site on a river defines its own unique region, and estimation of flow variables at an ungauged site should be based on a weighted average of observed flow variables at a number of other sites (Acreman & Wiltshire 1989). Third, the development of international databases has led to a classification of regimes over much wider areas. For example, the FREND project has produced a time series and thematic database for much of western and northern Europe (Gustard *et al* 1989). The analysis of these data has enabled regionalization techniques to be developed across national boundaries using consistent methods of analysis. Similarly, a world flood study (Farquharson *et al* 1987) has led to a greater understanding of the main global controls on flood frequency. Haines and colleagues (1988) have carried out a global classifi-

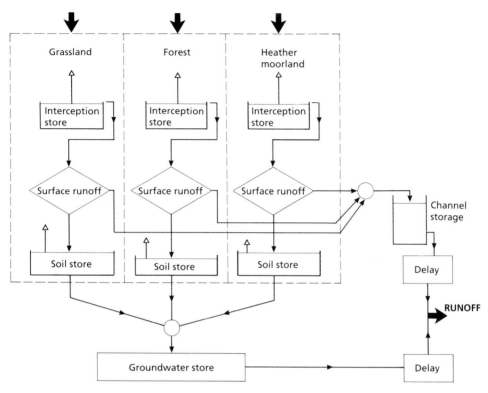

Fig. 2.13 The Institute of Hydrology integrated land-use model.→Water movement; ↑ evaporation/transpiration; −○−areal summation; ↓ precipitation.

cation of regimes, using 32 000 station-years of monthly streamflow data from 969 stream-gauging stations. This study defined 15 seasonal flow regimes and presented the first world map of regime type based on hydrograph analysis alone. The availability of an increasing number of these national and international databases will enable continued advances to be made both in the definition of river regimes and in analysing their spatial distribution. This will be complemented by the application of geographical information systems which will enhance the speed, number and accuracy of catchment characteristic calculation, eventually leading to improvements in estimating flow regimes at the ungauged site.

GLOSSARY

Base flow index the ratio of base flow to total flow derived from a hydrograph separation of daily flows.

Coefficient of variation of annual runoff a measure of the year-to-year variability of runoff calculated by dividing the standard deviation of annual runoff by the mean.

Flood frequency curve shows the relationship between the average interval between years (return period) in which the river exceeds a given flood discharge. The curve is normally derived from hourly or shorter time-interval data but can be derived from daily flow data for large catchments. Comparison between curves from different catchments is assisted by standardizing the discharge ordinate by the mean of the annual maximum flood series.

Flow duration curve the cumulative frequency distribution of daily (or monthly) mean flows. The curve shows the relationship between discharge and the percentage of time for which a given discharge is exceeded. The curve can be derived for the period of record, individual years, monthly or groups of months. Comparisons between curves from different catchments are assisted by dividing the discharge ordinates by the mean flow or by the catchment area.

Low flow frequency curve the curve shows the average

interval between years (return period) in which the river falls below a given discharge. It can be derived from daily or monthly data and from one or D-day consecutive flows, for example the 10-day or 90-day annual minimum. Comparisons between curves from different catchments are assisted by standardizing the discharge ordinates by the mean flow or catchment area.

Q95 the 95th percentile discharge derived from the flow duration curve is the discharge exceeded for 95% of the days of the period analysed. The 95th percentile is equivalent to the discharge which is exceeded 18 days a year on average. Other percentiles can be derived from the flow duration curve such as Q5, a high discharge, or Q50, the median discharge.

Recession the rate of decrease in river flow normally derived from hourly (or shorter time-interval) data for flood events or daily data for low-flow recessions.

REFERENCES

Acreman MC, Wiltshire SE. (1989) The regions are dead. Long live the regions. Methods of identifying and dispensing with regions for flood frequency analysis. In: *FRIENDS in Hydrology* (Proceedings of the Bolkesjo Symposium, Norway). IAHS Publication No. 187, pp 175–88. Wallingford, UK. [2.2, 2.4]

Abbott MB, Bathurst JC, Cunge JA, O'Connell PE, Rasmussen J. (1986a) An introduction to the European Hydrological System — Système Hydrologique Européen 'SHE' (1): History and philosophy of a physically based, distributed modelling system. *Journal of Hydrology* 87: 45–59. [2.3]

Abbott MB, Bathurst JC, Cunge JA, O'Connell PE, Rasmussen J. (1986b) An introduction to the European Hydrological System — Système Hydrologique Européen 'SHE' (2): Structure of a physically based, distributed modelling system. *Journal of Hydrology* 87: 61–77. [2.3]

Arnell NW. (1989) Humid temperate sloping land. In: Falkenmark M, Chapman T (eds) *Comparative Hydrology*, pp 163–207. Unesco, Paris. [2.2]

Beable ME, McKerchar AI. (1982) Regional flood estimation in New Zealand. *Water and Soil Technical Publication* 20, National Water and Soil Conservation Organisation. Wellington, New Zealand. [2.3]

Beckinsale RP. (1969) River Regimes. In: Chorley RJ (ed.) *Water, Earth and Man*, pp 455–71. Methuen, London. [2.2]

Beven KJ, O'Connell PE. (1982) *On the role of physically based distributed modelling in hydrology*. IH Report No. 81. Institute of Hydrology, Wallingford, UK. [2.3]

Beven K, Calver A, Morris EM. (1987) *The Institute of Hydrology distributed model*. IH Report No. 98. Institute of Hydrology, Wallingford, UK. [2.3]

Blackie JR, Eeles CWO. (1985) Lumped catchment models. In: Anderson MG, Burt TP (eds) *Hydrological Forecasting*, pp 311–45. John Wiley and Sons, Chichester. [2.3]

Boorman DB, Hollis JM. (1990) *Hydrology of soil types: a hydrologically-based classification of the soils of England and Wales*. Ministry of Agriculture, Fisheries and Food. Conference of River and Coastal Engineers, Loughborough University, July 1990. Wallingford, UK. [2.2]

Burn DH. (1990) Evaluation of regional flood frequency analysis with a region of influence approach. *Water Resources Research* 26: 2257–65. [2.4]

Chong SK, Moore SM. (1983) Flood frequency analysis for small watersheds in southern Illinois. *Water Resources Bulletin* 19 (2): 277–82. [2.3]

Cunnane C. (1988) Methods and merits of regional flood frequency analysis. *Journal of Hydrology* 100: 269–90. [2.2]

Demuth S. (1989) The application of the West German IHP Recommendations for the analysis of data from small research basins. In: *FRIENDS in Hydrology* (Proceedings of the Bolkesjo Symposium, Norway). IAHS Publication No. 187, pp 46–60. Wallingford, UK. [2.2]

Drayton RS, Kidd CHR, Mandeville AN, Miller JB. (1980) *A regional analysis of river floods and low flows in Malawi*. IH Report No. 72. Institute of Hydrology, Wallingford, UK. [2.2]

Falkenmark M, Chapman T. (1989) *Comparative Hydrology*. Unesco, Paris. [2.1]

Farquharson FAK, Green CS, Meigh JR, Sutcliffe JV. (1987) Comparison of flood frequency curves for many different regions of the world. In: Singh VP (ed.) *Regional Flood Frequency Analysis*, pp 223–56. Reidel, Dordrecht, Holland. [2.2, 2.4]

Friedrich G, Müller D. (1984) Rhine. In: Whitton BA (ed.) *Ecology of European Rivers*, pp 263–315. Blackwell Scientific Publications, Oxford. [2.1]

Gustard A, Gross R. (1989) Low flow regimes of northern and western Europe. In: *FRIENDS in Hydrology* (Proceedings of the Bolkesjo Symposium, Norway). IAHS Publication No 187, pp 205–12. Wallingford, UK. [2.2]

Gustard A, Roald LA, Demuth S, Lumadjeng HS, Gross R. (1989) *Flow Regimes from Experimental and Network Data* (FREND). Institute of Hydrology, Wallingford, UK. [2.2, 2.4]

Haines AT, Finlayson BL, McMahon TA. (1988) A global classification of river regimes. *Applied Geography* 8: 255–72. [2.4]

Hall FR. (1968) Base flow recessions — a review. *Water Resources Research* 4 (5): 973–98. [2.2]

Hewlett JD, Hibbert AR. (1967) Factors affecting the response of small watersheds to precipitation in humid areas. In: Sopper WE, Lull HW (eds) *Forest*

Hydrology, pp 275–90. Pergamon, Oxford. [2.2]

Hines MS. (1975) Flow-duration and low-flow frequency determinations of selected Arkansas streams. *United States Geological Survey, Water Resources Circular No. 12.* Government Printing Office, Washington, DC. [2.3]

Institute of Hydrology (1980a) *Seasonal flow duration curve estimation manual.* Low Flow Studies Report No. 2.4, Institute of Hydrology, Wallingford, UK. [2.2, 2.3]

Institute of Hydrology (1980b) *Research report.* Low Flow Studies Report No. 1, Institute of Hydrology, Wallingford, UK. [2.2, 2.3]

IAHS (1990) *Front cover regionalisation in hydrology* (Proceedings of the Ljubljana Symposium, Yugoslavia). IAHS Publication No. 191. Wallingford, UK. [2.2]

Joseph ES. (1970) Probability distribution of annual droughts. *Journal of the Irrigation Division ASCE* **96** (IR): 461–74. [2.2]

Kille K. (1970) Das Verfahren MoNMQ, ein Beitrag zur Berechnung der mittleren Langjahrigen Grundwasser–neubildung mit Hilfe der monatlichen Niedrigwasser–abflusse. *Zeitschrift der deutschen geologischen Gesellschaft*, Sonderheft Hydrologie und Hydrochemie, pp 89–95. [2.2]

Kirkby C, Newson MD. (1991) *Plynlimon research: the first two decades.* IH Report No. 109. Institute of Hydrology, Wallingford, UK. [2.2]

Knisel WG. (1963) Baseflow recession analyses for comparison of drainage basins and geology. *Journal of Geophysical Research* **68** (12): 3649–53. [2.3]

Kovacs K. (1989) Measurement and estimation of hydrological processes. In: Falkenmark M, Chapman T (eds) *Comparative Hydrology.* 75–104 Unesco, Paris. [2.2]

Lang H. (1989) Sloping land with snow and ice. In: Falkenmark M, Chapman T (eds) *Comparative Hydrology.* Unesco, Paris. pp 146–62. [2.2]

McMahon TA. (1969) Water resources research: aspects of a regional study in the Hunter Valley, New South Wales. *Journal of Hydrology* **7**: 14–38. [2.3]

McMahon TA. (1979a) Hydrological characteristics of arid zones. In: *The Hydrology of Areas of Low Precipitation* (Proceedings of the Canberra Symposium, Australia). IAHS Publication No. 128, pp 105–23. Wallingford, UK. [2.2]

McMahon TA, Mein RG. (1986) *River and Reservoir Yield.* Water Resources Publication, Colorado. [2.2]

Martin JV, Cunnane C. (1976) *Analysis and prediction of low-flow and drought volumes for selected Irish rivers.* Institution of Engineers of Ireland. Dublin. [2.3]

Matalas NC. (1963) *Probability distribution of low flows.* United States Geological Survey Professional Paper 434–A. Washington, DC. [2.2]

Mitchell WD. (1957) *Flow duration of Illinois streams.* Department of Public Works and Buildings, State of Illinois. [2.3]

Moltzau B. (1990) *Low flow analysis: a regional approach for low flow calculation in Norway.* Department of Geography Report No. 23, University of Oslo. [2.3]

Musiake K, Inokuti S, Takahasi Y. (1975) Dependence of low flow characteristics on basin geology in mountainous areas of Japan. In: *The Hydrological Characteristics of River Basin* (Proceedings of the Tokyo Symposium, Japan). IAHS Publication No. 117, pp 147–56. Wallingford, UK. [2.3]

Nathan RJ, McMahon TA. (1990) Evaluation of automated techniques for base flow and recession analysis. *Water Resources Research* **26** (7):1465–73. [2.4]

National Water and Soil Conservation Authority (1984) *An index for base flows.* Streamland 24, Water and Soil Directorate, Ministry of Works and Development, Wellington, New Zealand. [2.2]

NERC (1975) *Flood Studies Report.* Natural Environment Research Council, London. [2.1, 2.2, 2.3]

Parde, M. (1955) *Fleuves et Rivières* 3rd edn. Armand Colin, Paris. [2.2]

Pilon PJ, Condie R. (1986) Median drought flows at ungauged sites in Southern Ontario. *Canadian Hydrology Symposium* (CHS86), National Research Council of Canada, Regina. [2.2]

Rang MC, Schouten CJ. (1989) Evidence for historical heavy metal pollution in floodplain soils: the Meuse. In: Petts GE (ed.) *Historical Change of Large Alluvial Rivers*, pp 127–42. John Wiley and Sons, Chichester. [2.1]

Riggs HC. (1973) Regional analyses of streamflow techniques. *Techniques of Water Research Investigations*, Book 4, Chapter B3. USGS Washington, DC. [2.3]

Riggs HC. (1990) Estimating flow characteristics at ungauged sites. In: *Regionalization in Hydrology* (Proceedings of the Ljubljana Symposium, Yugoslavia). IAHS Publication No. 191, pp 159–70. Wallingford, UK. [2.3]

Roald LA. (1989) Application of regional flood frequency analysis to basins in North West Europe. In: *FRIENDS in Hydrology* (Proceedings of the Bolkesjo Symposium, Norway). IAHS Publication No. 187, pp 163–74. Wallingford, UK. [2.2]

Schwarze R, Grünewald U, Becker A, Frülich W. (1989) Computer-aided analysis of flow recessions and coupled basin water balance investigations. In: *FRIENDS in Hydrology* (Proceedings of the Bolkesjo Symposium, Norway). IAHS Publication No. 157, pp 75–84. Wallingford, UK. [2.2]

Simmers I. (1975) The use of regional hydrology concepts for spatial translation of stream data. In: *The Hydrological Characteristics of River Basins* (Proceedings of the Tokyo Symposium, Japan). IAHS Publication

No. 117, pp 109–18. Wallingford, UK. [2.3]

Tallaksen L. (1986) *An evaluation of the base flow index (BFI)*. Department of Geography, University of Oslo. [2.2]

Tallaksen L. (1989) Analysis of time variability in recession. In: *FRIENDS in Hydrology* (Proceedings of the Bolkesjo Symposium, Norway). IAHS Publication No. 187, pp 85–96. Wallingford, UK. [2.2]

Toebes C, Strang DD. (1964) On recession curves. 1 – Recession equations. *Journal of Hydrology, New Zealand* **3** (2): 2–15. [2.2]

Unesco (1978) World water balance and water resources of the earth. *Studies and Reports in Hydrology 25*. Unesco, Paris. [2.2]

Unesco (1979) Map of the world distribution of arid regions. *MAB Technical Note 7*. Unesco, Paris. [2.2]

Vivian H. (1989) Hydrological changes of the Rhône river. In: Petts GE (ed.) *Historical Changes of Large Alluvial Rivers*, pp 57–78. John Wiley and Sons, Chichester. [2.1]

Ward RC. (1968) Some runoff characteristics of British rivers. *Journal of Hydrology* **vi** (4): 358–372. [2.2]

Ward RC, Robinson M. (1990) *Principles of Hydrology*. McGraw Hill, London. [2.2]

Wiltshire SE. (1986) Regional flood frequency analysis II: Multivariate classification of drainage basins in Britain. *Hydrological Sciences Journal* **31**: 334–46. [2.2]

Wittenberg H. (1989) Regional analysis of flow duration curves, case studies on catchments in north-west Germany. In: *FRIENDS in Hydrology* (Proceedings of the Bolkesjo Symposium, Norway). IAHS Publication No. 187, pp 213–20. Wallingford, UK. [2.3]

Wright EP. (1989) The basement aquifer research project 1984–1989: final report to the Overseas Development Administration. *British Geological Survey Technical Report WD/89/15*. Wallingford, UK. [2.2]

3: Water Quality
I. Physical Characteristics

D.E.WALLING AND B.W.WEBB

A review of existing manuals and texts covering the field of water quality indicates that the *physical* characteristics of water flowing in rivers and streams have been variously defined to embrace a wide range of parameters extending from suspended sediment concentration, turbidity and the presence of foam, through colour, taste, odour and dissolved oxygen content, to electrical conductivity, temperature and radioactivity. Some blurring of the distinction between physical and chemical properties clearly exists. For example, electrical conductivity primarily reflects the total dissolved solids content, which is a chemical parameter. Likewise, the content of dissolved gases could be viewed as a chemical characteristic. In this section attention will be limited to four parameters that are physical in nature and that have important implications for river ecology and for water use. These are suspended sediment concentration, colour, temperature and dissolved oxygen content. Furthermore, although all of these aspects of physical water quality may be strongly influenced by river pollution, the present chapter focuses on uncontaminated watercourses. The influence of pollution on physical water quality will be considered in more detail in Volume 2 of this *Handbook*.

3.1 SUSPENDED SEDIMENT

The presence of suspended sediment or solids in river water is an important physical characteristic. Such sediment can have both a direct effect on aquatic life through damage to organisms and their habitat (e.g. Ritchie 1972; Muncy *et al* 1979) and an indirect effect through its influence on turbidity and light penetration. Decreased light penetration reduces primary production and hinders the growth of benthic macrophytes (e.g. US Environmental Protection Agency 1979). Increased sediment concentrations will also necessitate increased treatment of domestic and industrial water supplies (e.g. Castorina 1980).

Definitions, sources and budgets

The suspended sediment load of a river represents the fine-grained material transported in suspension, with its weight supported by the upward component of fluid turbulence. It is conventionally separated from material in solution by filtration through a 0.45-μm filter and may therefore include some colloidal material. Particles are commonly less than 0.2 mm in diameter and in most rivers the suspended load will be dominated by clay- and silt-sized particles (i.e <0.062 mm in diameter). A distinction is often made between the *wash load*, which comprises the finer material washed in from the catchment slopes and which commonly remains in suspension as it is transported downstream, and the *suspended bed material* which represents the coarser, sand-sized particles, mobilized from within the channel. The latter particles frequently move in and out of suspension and interact with the bedload component of sediment transport, according to the local hydraulic conditions within the channel.

The suspended sediment load is dominated by material eroded from a variety of sources within the upstream basin, and the precise nature and location of these will exert an important influence on both the character and the behaviour of the load. Potential sources include channel and bank erosion, gully erosion, erosion of tracks and unmetalled roads, sheet and rill erosion on the catchment slopes and the removal of sediment delivered to the channel by mass movement. In agricultural areas, soil loss from cultivated fields often accounts for a major proportion of the suspended sediment load. Other sources of sediment of ecological significance include mining, quarrying, construction work and effluent discharges.

When attempting to link the suspended sediment load of a drainage basin to the erosion processes and sources operating within it, it is important to recognize that only a proportion (probably a small one) of the sediment mobilized by erosion will find its way to the basin outlet. Much will be deposited within the system. The *sediment delivery ratio*, which expresses the sediment yield at the basin outlet as a proportion of the gross erosion and sediment mobilization within the basin, provides a simple basis for representing this link, although the many uncertainties involved in estimating and in interpreting this parameter should be recognized (*cf.* Walling 1983). Existing work indicates that sediment delivery ratios decline as drainage basin size increases, in response to the increased opportunities for deposition, and values are frequently as low as 10% in medium-sized basins. This trend is shown clearly by the relationship presented in Fig. 3.1 (a), although the range of delivery ratio values evident for a given basin size emphasizes the importance of other controls.

The sediment delivery ratio provides only a simple lumped, black-box representation of the linkages interposed between the erosion processes operating within a drainage basin and the downstream suspended sediment yield. An alternative approach to elucidating and quantifying these linkages is the establishment of a *sediment budget* (*cf.* Fig. 3.1(b)). With this, an attempt is made to quantify the various sediment sources and sinks, in order to provide an improved understanding of the linkages involved. The sediment budget approach advocated by Dietrich and Dunne (1978) and developed by Lehre (1982) and Swanson *et al* (1982) is important in focusing attention on the processes governing the suspended sediment yield at the basin outlet, but it should be seen as an essentially conceptual approach, since it is difficult to assemble precise

information on the rates and fluxes involved for anything but a relatively small drainage basin. Recent advances in the use of caesium-137 for investigating rates and patterns of erosion and deposition within a drainage basin (*cf.* Walling & Bradley 1988; Walling 1990) and the use of specific sediment properties to fingerprint major sources (*cf.* Wall and Wilding 1976; Peart & Walling 1986) would, however, appear to offer considerable potential in this field.

Most of the sediment sinks or stores depicted in the examples of sediment budgets presented in Fig. 3.1(b) can be viewed as essentially permanent in the context of the short-term operation of the sediment conveyance system. Remobilization may, however, occur in the longer term as slope, valley and floodplain deposits are reworked (*cf.* Trimble 1976, 1981). The potential for short-term channel storage to attenuate the transmission of sediment through the channel system must nevertheless be considered (*cf.* Duijsings 1985, 1986), since this could introduce a distinct phase difference between sediment supply and sediment output, with sediment supplied to the channel during one season appearing at the basin outlet several months later.

The nature of suspended sediment

Traditionally, investigations of the suspended sediment loads of streams have focused on collection of gross data concerning both concentrations and loads. Increasing awareness of the wider environmental significance of the fine sediment transported by streams has, however, emphasized the need for information on the physical and chemical properties of the sediment. In this context it is important to note the distinction which is sometimes made between *suspended solids* and *suspended sediment*, the former referring to the total material in suspension and the latter to the mineral sediment or inorganic fraction. The term suspended sediment is, however, widely used to refer to both the inorganic and the organic components, and this interpretation has been adopted here. The properties of suspended sediment reflect clearly those of the source material from which it was derived, but any attempt to link the two must take account of the effects of

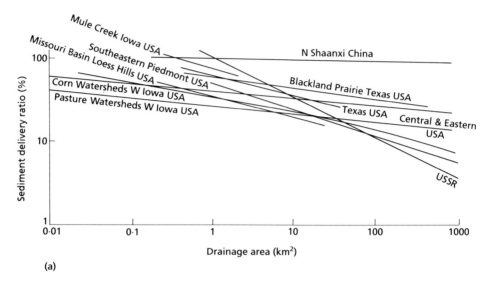

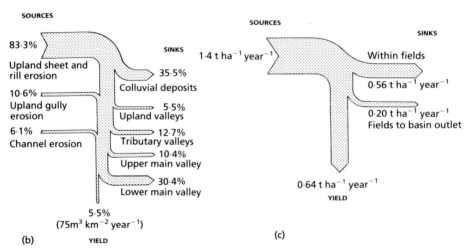

Fig. 3.1 (a) Relationships between sediment delivery ratio and drainage basin area proposed for various areas of the world (based on Walling 1983). Examples of sediment budgets established for (b) Coon Creek, Wisconsin, USA (360 km²) by Trimble (1981) and (c) for Jackmoor Brook, Devon, UK (9.8 km²) by Walling (1990).

selective erosion and deposition mechanisms operating within the sediment delivery system. Such mechanisms frequently increase the content of both fines and organic matter in the suspended sediment, relative to the source material.

Information on the organic matter content of suspended sediment is relatively sparse, but, as might be expected, existing data exhibit substantial spatial variation in response to the sources

involved and the nature of the source material itself. For example, Burton *et al* (1977) cited organic matter percentages of 30.6, 17.2 and 14.8, respectively, for sediment from forested-agricultural, suburban and urban drainage basins in northern Florida; Walling and Kane (1982) reported mean organic matter contents of between 8% and 14% for suspended sediment from four rivers in Devon, UK; whilst Skvortsov (1959) documented much lower values of between 1.5% and 3.1% for the Rion River draining to the Black Sea in the USSR.

Particle size analyses of suspended sediment are traditionally undertaken on the mineral fraction after chemical treatment to remove organic material. Available data demonstrate consider-able variation in the particle size characteristics of sediment from different rivers and streams in response to variations in source material and other physiographic controls (*cf.* Walling & Moorehead 1989). The potential range of particle size distributions associated with suspended sediment is illustrated in Fig. 3.2(a) which presents data for a

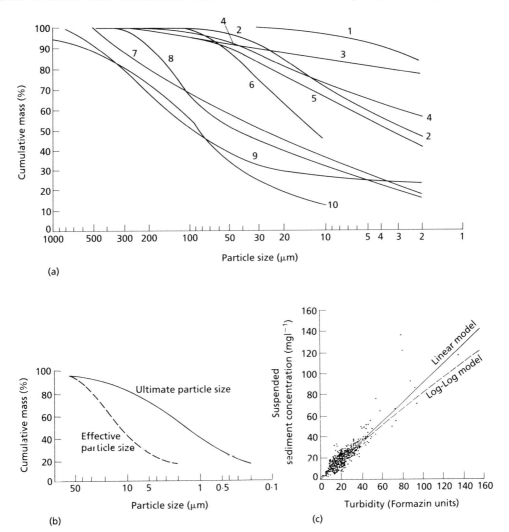

Fig. 3.2 (a) Examples of characteristic grain-size distributions of suspended sediment for various world rivers. (b) Comparison of typical ultimate and effective particle-size distributions of suspended sediment from the River Exe, Devon, UK. (c) Relationship between turbidity and suspended sediment concentration established for Geebing Creek, New South Wales, Australia. (Based (a) on Walling & Moorehead 1989; and (c) on Gippel 1989). 1 Barwon River, New South Wales, Australia; 2 Citarum River at Nanjung, Indonesia; 3 Sanaga River at Nachtigal, Cameroon; 4 River Nile at Cairo, Egypt (1925–6); 5 Amazon River at Obidos, Brazil; 6 Sao Francisco at S. Romao, Brazil; 7 Chulitna River near Talkeetna, Alaska; 8 Colorado River at Lees Ferry Arizona, USA; 9 Limpopo River, Zimbabwe; and 10 Huangfu River, China.

number of world rivers. Thus, for example, the situation in the Barwon River in Australia, where the majority of the suspended sediment is clay sized and <0.002 mm in diameter, may be contrasted with the Middle Yellow River in China where up to 60% of the suspended load may be composed of sand-sized particles (>0.062 mm) and clay-sized material represents less than 10% of the total load. Particle size in turn exerts an important influence on the basic mineralogy and geochemistry of fine sediment. The <0.002-mm fraction will, for example, be composed primarily of secondary silicate minerals, whereas quartz will dominate in the larger fractions. The specific surface area of sediment, which is a major control on its surface chemistry, also increases markedly with decreasing particle size, such that typical values for clay ($20-800$ m^2 g^{-1}) are several orders of magnitude greater than those for silt and sand.

Traditional laboratory determinations of particle size, involving chemical dispersion of the mineral fraction, may be unrepresentative of the *in situ* particle-size characteristics of suspended sediment transported by a river. This is likely to include a substantial proportion of aggregates comprising smaller particles (*cf.* Droppo & Ongley 1989). The processes responsible for producing such aggregates are poorly understood but are likely to involve a variety of mechanisms including electrochemical attraction and bonding associated with organic substances and bacteria. Figure 3.2(b) provides an example from a river in Devon, UK of the potential contrast between the *ultimate* particle-size distribution provided by traditional laboratory analysis and the *effective* particle-size distribution of *in situ* sediment. In this case the median particle sizes of the two distributions differ by almost an order of magnitude. Measurement of the effective size distribution inevitably involves significant problems and uncertainties, and in this example the measurements were made in the field using a sedimentation tube, immediately after collecting the sample.

Although the measured turbidity of river water commonly exhibits a close positive relationship with suspended sediment concentration (e.g. Fig. 3.2(c)), the precise form of the relationship will vary both spatially and temporally in

response to variations in sediment composition and water colour. Particle size, shape and composition can all be expected to influence light attenuation and turbidity, and any attempt to use turbidity measurements as a surrogate for direct determinations of suspended sediment concentration should take careful account of such factors (*cf.* Gippel 1989).

Spatial variation of suspended sediment transport

In many applications it is the concentrations of suspended sediment occurring in a river (i.e. mg l^{-1}) rather than the specific sediment yield (i.e. t km^{-2} year^{-1}) that is of prime interest. However, the two are closely linked, since rivers with high suspended sediment yields are likely to exhibit high sediment concentrations. The average concentration in a river is by definition the mean annual sediment yield divided by the mean annual runoff volume. In Britain, for example, where suspended sediment yields are typically in the range $50-100$ t km^{-2} year^{-1} (*cf.* Walling 1990), maximum concentrations rarely exceed 5000 mg l^{-1} and average concentrations are of the order of $50-100$ mg l^{-1}, whereas in the USA where sediment yields can exceed 1000 t km^{-2} year^{-1}, average concentrations are frequently in excess of 2000 mg l^{-1}. Most existing work on the spatial variability of suspended sediment transport has, however, been undertaken on annual suspended sediment yields. At the global scale, the sediment yields of small river basins (i.e. about 100 km^2) are known to range from less than 2 t km^{-2} year^{-1} to in excess of 10 000 t km^{-2} year^{-1} (Walling 1987), and the generalized global pattern described by Walling and Webb (1983) reflects control by a number of factors, including climate, rock type, relief, tectonic activity and land use. Similar factors control the pattern of average suspended sediment concentrations in US rivers mapped by Meade and Parker (1985) and presented in Fig. 3.3(a). Here the effect of climate is clearly evident, with average sediment concentrations showing a general increase across the country, in response to increasing aridity. At the more local scale, factors such as geology, soil type and land use assume increasing importance in controlling spatial patterns, and Fig. 3.3(b), based on the

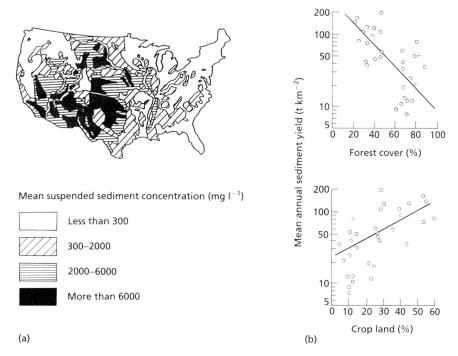

Fig. 3.3 (a) Spatial variation of mean suspended sediment concentrations within the conterminous USA (after Meade & Parker 1985). (b) Relationships between mean annual suspended sediment yield and land use established for catchments within the Potomac River basin by Wark and Keller (1963).

classic work of Wark and Keller (1963) in the basin of the Potomac River in eastern USA, high-lights the role of land use in influencing soil loss and sediment yield.

Temporal variation in suspended sediment transport

General characteristics

Summary statistics such as annual suspended sediment yields and average concentrations can fail to convey a sense of the extreme temporal variability of suspended sediment transport in most rivers. For much of the time rivers may transport very little suspended sediment and the water will be essentially clear. A large proportion of the suspended sediment transport occurs during storm events when rainfall and storm runoff mobilize sediment from the upstream watershed and channel network. Figure 3.4(a), based on a 7.75-year programme of detailed monitoring

undertaken by the authors in the 262 km^2 drain-age basin of the River Creedy in Devon, UK, characterizes this variability by means of concen-tration–duration and load–duration curves. These indicate that most of the total sediment load is transported during only 5% of the time when sediment concentrations exceed about 100 mg l^{-1}. Equally, if suspended sediment con-centrations below 20 mg l^{-1} are considered to be minimal, the concentration–duration curve indi-cates that significant concentrations occur only for 10% of the time. Most rivers can be expected to exhibit similar behaviour.

Maximum suspended sediment concentrations and loads are commonly transported during flood events and geomorphologists have frequently attempted to assess the relative importance of high-magnitude, low-frequency flood events and smaller more frequent events to the long-term sediment load. Lack of long-term records and the difficulties of measuring sediment transport during rare high-magnitude events severely

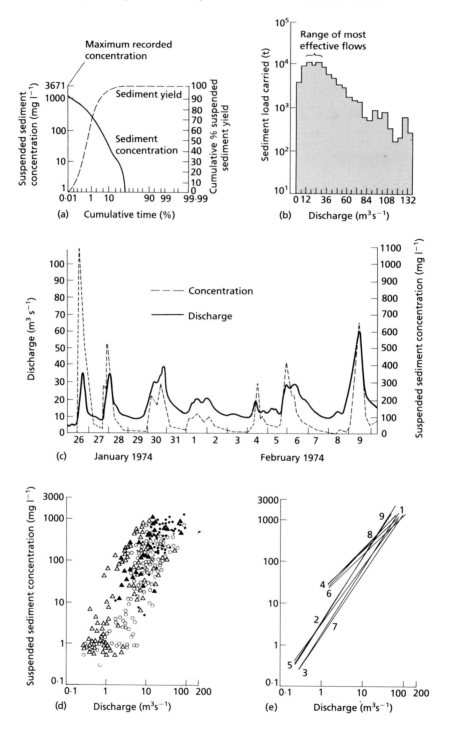

(a)   Cumulative time (%)

(b)   Discharge (m³s⁻¹)

(c)   January 1974   February 1974

(d)   Discharge (m³s⁻¹)

(e)   Discharge (m³s⁻¹)

hampers such analysis. The authors have, how-ever, analysed the 7.75-year record of suspended sediment transport for the River Creedy in Devon, UK, referred to above, in an attempt to evaluate the relative importance of events of varying frequency (*cf.* Webb & Walling 1982). Figure 3.4(b) indicates the total load carried during this 7.75-year period by 23 equal discharge classes spanning the total flow range. Three flow classes between 12 m^3 s^{-1} and 30 m^{3}s^{-1} each have loads in excess of 10000 tonnes and represent the most effective discharge range for the transport of suspended solids. It appears that the more extreme flows experienced by this basin are less effective in removing suspended sediment. This conclusion accords with that advanced by Wolman and Miller (1960) on the basis of a study of the suspended sediment loads of several rivers in the USA, but the true significance of extreme catastrophic events, for which records are rarely available, still remains somewhat uncertain. For example, Meade and Parker (1985) referred to the impact of an extreme storm which struck north-western California in December 1964. In 3 days the Eel River, which drains an 8063-km^2 drainage basin, transported more sediment than it had carried in the previous 7 years, and in 10 days it transported a total sediment load equivalent to that for 10 average years.

Short-term variations

In considering the major factors governing short-term temporal variations in suspended sediment concentrations, most workers have emphasized the dominant role of water discharge or flow magnitude. High suspended sediment concentrations are in most cases associated with periods of high discharge, as illustrated by Fig. 3.4(c), but closer inspection of this example indicates that suspended sediment concentration is not a simple

function of water discharge and that other factors such as the time elapsed since a previous storm event are also important. Simple positive straight-line logarithmic relationships between suspended sediment concentration (C) and water discharge (Q) have nevertheless been established for many rivers (*cf.* Fig. 3.4(d)). These relationships can be described by the equation $C = aQ^b$, with the exponent b typically falling in the range 1–2. The existence of a relationship between C and Q should not, however, be interpreted as a simple transport function, whereby increases in water discharge are associated with increased shear velocities and turbulence and therefore increased capacity to erode and transport sediment. In most cases the amount of suspended sediment carried by a river is several orders of magnitude less than its maximum transport capacity and the dominant control on suspended sediment concentration is the *supply* of material to the river. The existence of a relationship between concentration and discharge is rather a reflection of the fact that sediment supply increases during periods of storm rainfall and storm runoff and these periods are generally characterized by high discharges. Suspended sediment loads must therefore be treated as *non-capacity loads* and the logarithmic plot of concentration versus discharge will typically exhibit very considerable scatter (*cf.* Fig. 3.4(d)). Concentrations associated with a given level of discharge will frequently range over several orders of magnitude in response to variations in sediment supply.

There have been many attempts to account for the scatter apparent in sediment concentration/discharge relationships such as that portrayed in Fig. 3.4(d). Seasonal differentiation of the plot is often apparent and a distinction can also frequently be made between samples collected during periods of rising and falling stage (Fig. 3.4(e)). Hysteresis has also been widely ident-

Fig. 3.4 (*Opposite*) Characteristics of suspended sediment transport in the River Creedy, Devon, UK, 1972–80. (a) Cumulative frequency curves of suspended sediment concentration and suspended sediment yield. (b) The effectiveness of different flow classes for transporting suspended sediment. (c) A typical record of the variation of suspended sediment concentration during a sequence of storm runoff events. (d) A scatterplot for the relationship between suspended sediment concentration and discharge subdivided according to season and stage condition. (Winter: ● rising stage; ○ falling stage. Summer: ▲ rising stage; △ falling stage.) (e) Straight-line logarithmic relationships fitted to various subsets of the data plotted in (d); (1 All data; 2 winter; 3 summer; 4 rising stage; 5 falling stage; 6 winter rising; 7 winter falling; 8 summer rising; and 9 summer falling.)

ified in relationships between sediment concentration and discharge established for individual events, and may reflect both phase differences between the water discharge and sediment concentration response or a tendency for concentrations to be higher on the rising stage than on the falling stage or *vice versa*. Williams (1989) has attempted to classify the form of such hysteretic relationships into five types which are shown schematically in Fig. 3.5(a).

Longer-term trends

Erosion processes and sediment yields are particularly sensitive to changes in land use, and long-term sediment records will frequently exhibit marked fluctuations in response to land clearance and intensification of land use. Table 3.1 provides several examples of the magnitude of the impact of land-use change on sediment yields from small drainage basins. These examples relate to

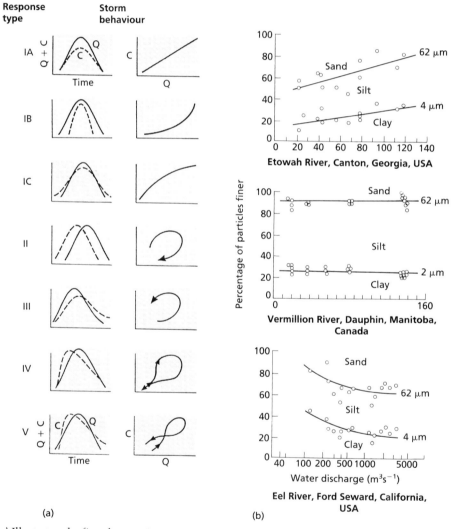

Fig. 3.5 (a) Illustrates the five characteristic response types associated with suspended sediment concentration/discharge relationships proposed by Williams (1989). (b) Presents contrasting examples of the response of the particle-size composition of suspended sediment to changing discharge (after Walling & Moorehead 1989).

Table 3.1 Results from experimental basin studies of the impact of land-use change on sediment yield

Region	Land-use change	Increase in sediment yield	Reference
Westland, New Zealand	Clearfelling	×8	O'Loughlin *et al* (1980)
Oregon, USA	Clearfelling	×39	Fredriksen (1970)
Texas, USA	Forest clearance and cultivation	×310	Chang *et al* (1982)
Maryland, USA	Building construction	×126−375	Wolman & Schick (1967)

increases, but other land-use changes, particularly the implementation of conservation measures, can lead to reduced sediment yields (*cf.* Lal 1982).

Temporal variation in sediment properties

Suspended sediment properties can also be expected to vary temporally in response to variations in sediment sources and the operation of erosion and sediment delivery processes. Such variations in properties are likely to be particularly marked in streams experiencing floods generated by both spring melt and summer rainstorms, since different sediment sources and erosion processes are likely to be associated with these events. Considering more short-term variations in response to fluctuations in discharge, the organic matter content of suspended sediment commonly exhibits an inverse relationship with discharge (e.g. Walling & Kane 1982). The behaviour of particle-size composition is, however, more complex and Fig. 3.5(b)) illustrates situations where particle size increases, decreases and remains essentially constant with increasing discharge. The tendency for suspended sediment to become coarser as flow increases may be accounted for in terms of increasing transport capacity and shear stress within the channel, whereas the reverse case is generally explained in terms of increased supply of fine sediment eroded from the slopes of the watershed and reflecting the interaction of runoff dynamics and sediment sources.

3.2 NATURAL COLOUR

Measurements of water colour conventionally distinguish between 'true colour' which represents that due to dissolved matter, and 'apparent colour' which also includes the effects of suspended solids. Measurements of true colour are made on water samples filtered through a 0.45-μm membrane filter and are the most widely cited (*cf.* Department of the Environment 1981). Colour can be seen as an important physical property of water primarily because of its implications for water supply and the need in some areas to reduce it to acceptable levels by water treatment. Recent increases in the colour of water in reservoirs in the Pennines, UK have, for example, resulted in increases in treatment costs by up to 10 times (*cf.* Naden & McDonald 1987). Colour also has a more indirect significance because of the complexing and adsorptive behaviour of dissolved organic colour with heavy metals (*cf.* Reuter & Perdue 1977; Malcolm 1985), which frequently results in a positive relationship between colour and the concentration of toxic metals such as mercury.

Causes

Colour in natural water usually results from the leaching of organic materials and is primarily a result of dissolved and colloidal humic substances, principally humic and fulvic acids. Highly coloured water occurs in many different environments where decaying vegetation is plentiful, as in swamps and bogs, and is frequently associated with acid conditions. In the UK, for example, it is primarily associated with upland moorland areas and there is evidence that colour increases in areas where the peat cover is being actively eroded or disturbed by burning or deep ploughing and ditching for afforestation.

Temporal variation

Little information currently exists on the detailed temporal behaviour of colour in streams, but recent concern for the increasing discoloration of water obtained from reservoirs fed by runoff from the peat moorlands of Yorkshire has focused attention on the need for detailed investigations of this parameter (*cf.* Edwards *et al* 1987). Available records indicate a pattern of annual variation, with highest levels of colour occurring during the autumn 'flush', when the peat is wetted, and accumulated soluble organic material is leached into the streams. The supply of readily soluble coloured organics may be increased after periods of extreme drought which promote aeration of the peat and its aerobic decomposition. The onset of enhanced colour levels usually occurs in October, although in wet summers it may begin as early as July, and it generally lasts for about 3 months. The decline in colour during the winter may reflect both the exhaustion of available supply or the influence of temperature or snowfall in limiting microbial activity. An analysis of monthly colour data for several catchments in Nidderdale reported by Naden and McDonald (1987) also demonstrated a positive relationship with monthly rainfall (Fig. 3.6(a)), indicating that maximum levels of colour are associated with periods of increased leaching and runoff. The same authors also reported a significant relationship with an antecedent moisture index representing the moisture deficit 3 months earlier (Fig. 3.6(b)). In this case samples collected following dry periods were seen to exhibit substantially higher levels of colour than those collected after wet periods.

In the longer term, evidence collected from the peat moorlands of Yorkshire also indicates that colour levels may demonstrate significant upward shifts through time. Figure 3.6(c), based on the work of McDonald *et al* (1989) presents monthly mean values of colour for water from Thornton Moor reservoir during the period 1968–88. The influence of drought conditions is clearly seen, since the severe droughts of 1976 and 1984 are both followed by significant upward shifts in colour. This is probably a result of the lowering of the water-table and the desiccation of the peat, which promotes aerobic decomposition and increases the availability of readily leached organic substances responsible for the discoloration once the moisture content has been replenished. Elsewhere, land-use changes, particularly moorland drainage, have resulted in more highly coloured runoff and there have also been suggestions that changes in precipitation chemistry, particularly increased acidity, have also been responsible for increased coloration of moorland streams.

3.3 WATER TEMPERATURE

Temperature represents one of the most important physical characteristics of river water. It affects other physical properties of rivers, such as dissolved oxygen and suspended solids content, and it influences the chemical and biochemical reactions which take place in lotic systems. The evolution, distribution and ecology of aquatic organisms is also fundamentally affected by river temperature (Rose 1967), and an enormous volume of research has been undertaken to investigate the thermobiology of fishes and invertebrates (e.g. Brett 1960; Fry 1964; Langford 1972; Ward & Stanford 1982; Crisp 1988a).

Temperature is a significant consideration in the domestic utilization of water since it has a bearing on the toxicity of contaminants, the efficacy of water treatment and the presence of tastes and odours (Everts 1963). Furthermore, temperature characteristics can determine the quality of water supplies for industrial and agricultural purposes, and are of particular economic importance in the provision of cooling water for the electricity generation industry (Langford 1983) and other industrial processes, and in the formation of ice in navigable waterways (Smith 1972). Human activities in utilizing water can also radically alter the thermal behaviour of a water-

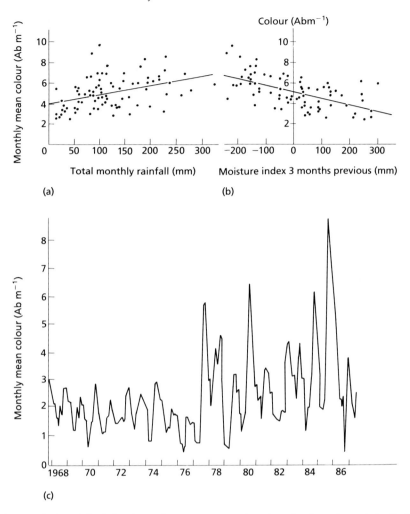

Fig. 3.6 Some characteristics of the behaviour of stream-water colour in upland catchments in Yorkshire, UK. The relationships between monthly mean colour and monthly rainfall (a) and catchment moisture status (b) are based on Naden and McDonald (1987), and the long-term trend in the colour of water from Thornton Moor Reservoir (c) is after McDonald *et al* (1989).

course, and direct impacts due to discharge of heated effluents (e.g. Parker & Krenkel 1969) and indirect modification associated with impoundment (e.g. Petts 1984) and land-use changes (e.g. Holtby 1988) have been extensively documented.

Temperature is a very easily and widely measured parameter of water quality, and further details of the instruments available, field applications and procedures, and problems of data collection and processing can be found in standard reference works (e.g. Stevens *et al* 1975).

Concepts

Energy budget

The temperature of a river is fundamentally determined by the transfers of heat energy which affect the water-course (Pluhowski 1970). Inputs of heat energy to a particular river reach arise as short-wave solar radiation and long-wave atmospheric and forest radiation, from condensation and precipitation, and through advection of heat from groundwater, from upstream and from tributary inflows (Fig. 3.7(a)). Heat energy is lost from the water-body through reflection of solar, atmospheric and forest radiation, by back radiation from the water surface itself, in the process of evaporation, and as the heat content of streamflow leaving a reach. Some terms in the energy budget, including convection and conduction between the water-body and its environment, can represent gains or losses depending on the thermal status of the channel margin, the water and the overlying air. The net heat exchange affecting a river, together with the volume of water, will determine the direction and extent of temperature change, although detailed studies (e.g. Brown 1969; Comer *et al* 1976; Troxler & Thackston 1977) indicate that the energy budget components can be highly variable in time, for example over a daily period, and in space, for example between forested and open watersheds (Fig. 3.7(b)). In

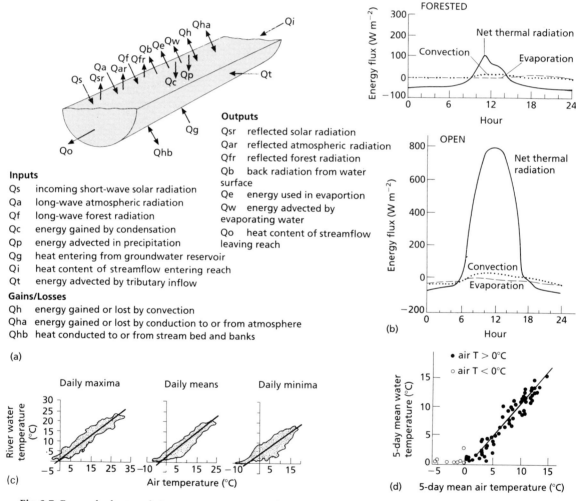

Inputs

Qs   incoming short-wave solar radiation
Qa   long-wave atmospheric radiation
Qf   long-wave forest radiation
Qc   energy gained by condensation
Qp   energy advected in precipitation
Qg   heat entering from groundwater reservoir
Qi   heat content of streamflow entering reach
Qt   energy advected by tributary inflow

Gains/Losses

Qh   energy gained or lost by convection
Qha energy gained or lost by conduction to or from atmosphere
Qhb heat conducted to or from stream bed and banks

Outputs

Qsr reflected solar radiation
Qar reflected atmospheric radiation
Qfr reflected forest radiation
Qb   back radiation from water surface
Qe   energy used in evaportion
Qw   energy advected by evaporating water
Qo   heat content of streamflow leaving reach

Fig. 3.7 Energy budgets and air–water temperature relationships. (a) Principal components of the energy budget of a water-course. (b) Temporal variation in heat budget components of Deer Creek, USA (forested) and Berry Creek, USA (open catchment) (after Brown 1969). (c) Water–air temperature relationships in the River Clyst catchment, Devon, UK and (d) in a Pennine stream, Mattergill Sike, UK (after Crisp & Howson 1982).

general, however, evaporative, convective and conductive fluxes are found to be less important than the solar, atmospheric and back radiation terms in the energy budget (Brown 1969).

Air/water temperature relationship

Fluctuations in meteorological conditions together with the relatively high heat capacity of water, which damps the response of river temperature to changing energy gains and losses, ensure that the water surface is constantly striving to attain an equilibrium temperature. The latter is defined as the temperature at which there is no net energy exchange with the atmosphere (Edinger *et al* 1968; Dingman 1972). Air temperature is commonly taken as an approximation of the equilibrium temperature and is frequently used as a surrogate for the complexities of the energy budget components in predicting river water temperatures (Smith 1975; Walker & Lawson 1977). Strong linear relationships can often be established between air and river water temperatures (Smith 1981), although analysis of 2192 observations of daily maxima, means and minima in the River Clyst in Devon, UK (Webb 1987) has indicated that correlations are weakest for daily minimum values (Fig. 3.7(c)). Furthermore, departure from a linear relationship between air and water temperatures can arise in streams subject to winter freezing. This phenomenon has been demonstrated from an analysis of 5-day means of air and water temperatures during 1977–8 for Mattergill Sike in the northern Pennines, UK (Fig. 3.7(d)), and has been ascribed to latent heat effects associated with ice formation (Crisp & Howson 1982). Several studies have demonstrated the similarities of air and water temperature behaviour (e.g. Johnson 1971; Calandro 1973), although the latter often fluctuates over a much narrower range and tends to lag in its response behind the former (Smith 1968; Fowles 1975).

Temporal behaviour

The nature of water temperature behaviour varies significantly according to the temporal perspec-

tive under consideration, and river temperatures tend to exhibit more dramatic fluctuations over short rather than long time periods.

Longer-term trends

Although water temperature data have been collected in some Austrian and Russian rivers for more than 100 years, the availability of long-term and high-quality data series, on which to base an analysis of trends in river water temperatures, is generally limited. An increase in the temperature of the Mississippi River over a period of 40 years, however, has been inferred by a comparison of annual records collected in 1923 (Collins 1925) with those taken in 1962 (Blakey 1966). Mean annual water temperature was 1.6°C higher in the latter year despite the fact that mean air temperature was 2.2°C lower. The increase was thought to reflect the growing impact of human activity over the time period, although it was also recognized that these differences might be partly ascribed to the climatic vagaries of the individual years involved and to contrasts in measurement techniques (Blakey 1966). Smaller rivers, draining rural environments, may not evidence significant longer-term changes in temperature characteristics. Continuous monitoring at three stations within the Exe Basin, Devon, UK over a decade (1974–83) revealed the coefficient of variation for annual mean water temperatures to be in the range of 3–5% but did not indicate any significant upward or downward trend over the period (Webb & Walling 1985a).

Collection of records from a substantial run of years is necessary to define properly extreme temperature conditions and to establish adequately duration characteristics and other temperature statistics of ecological importance, such as accumulated degree hours above threshold temperatures (e.g. Macan 1958; Crisp & Le Cren 1970). An example of a temperature duration curve is shown in Fig. 3.8(a) for the River Exe at Thorverton, Devon, UK, which drains a rural catchment of 601 km^2. It is evident from this curve, which is based on 10 years of continuous data, that low water temperatures (<1°C) occur very infrequently (<0.5% of the time) in this

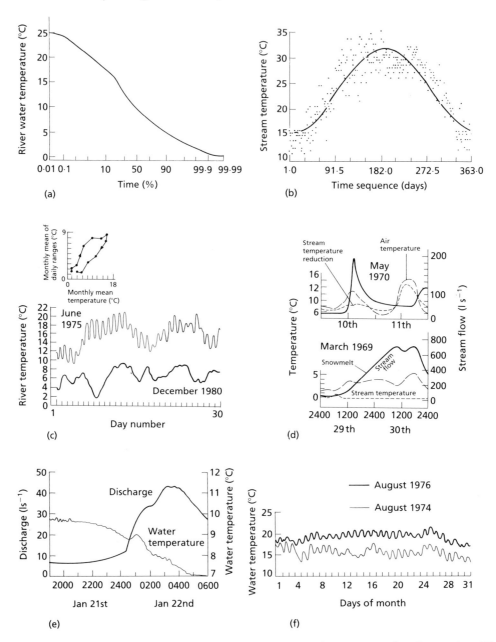

Fig. 3.8 Temporal fluctuations in water temperature. (a) A temperature duration curve for a Devon river; River Exe, Thorverton, UK (1974–83). (b) A multiyear harmonic to define annual temperature regime for Soan River, Dhok Pathan, Pakistan (after Steele 1982). (c) Daily and diurnal fluctuations in temperature of a Devon river; River Barle, Brushford, UK (inset depicts annual hysteresis in monthly average daily range in water temperature of a Pennine stream (Great Egglesthorpe Beck, UK, 1984), after Crisp 1988b). (d) Effects of rain- and snowmelt-generated floods on water temperature in Upper Weardale, UK (after Smith & Lavis 1975). (e) Water temperature fluctuations in response to changing sources of storm runoff in a Georgia headwater stream; Lower Gage, Panola Mountain, USA (after Shanley & Peters 1988). (f) The impact of the 1976 drought on August water temperatures in the River Exe, Thorverton, UK (after Walling & Carter 1980).

temperate system, although a considerably greater occurrence of freezing conditions would be expected in smaller streams at higher altitudes and latitudes.

Annual regime

Most rivers exhibit a marked seasonal variation in water temperature, although the amplitude of the annual regime tends to be less in watercourses which receive a large groundwater flow component (Hopkins 1971; Crisp *et al* 1982). A simple harmonic curve of the form:

$$T(x) = A \left(\sin(bx + C) \right) + M \qquad (1)$$

where:

$T(x)$ = water temperature on day x of year (°C)
  A = amplitude of the harmonic (°C)
  b = 0.0172 radians per day ($2\pi/365$ days)
  x = number of days since beginning of year
  C = phase angle of harmonic (radians)
  M = mean of the harmonic (°C)

has often been used to characterize the seasonal cycle of river temperature variation (e.g. Collings 1969; Johnson 1971; Mosley 1982). This approach has been employed to define the seasonal regime in individual years (Ward 1963) and to derive a general annual cycle from observations made intermittently over a series of years (Steele *et al* 1974). An example of a multiyear harmonic function fitted to data collected between 1964 and 1972 for the Soan River at Dhok Pathan, Pakistan (Steele 1982) is given in Fig. 3.8(b). In some cases where the annual temperature regime is interrupted by a period of freezing (MacKichan 1967) or where the seasonal cycle is asymmetrical in nature (Moore 1967), it may be more appropriate to fit a higher order harmonic in order to characterize the annual march in temperature.

Daily and diurnal fluctuations

Superimposed upon the annual temperature regime of rivers are variations which take place from day to day and within a 24-hour period. In winter months, water temperature often fluctuates in response to changes in weather conditions, such as the passage of fronts, whereas in the summer period a clear diurnal cycle is generally

evident. This contrast in behaviour is apparent in continuous records of water temperature collected from the River Barle, Devon, UK (Webb & Walling 1985b) during June 1975 and December 1980 (Fig. 3.8(c)). A sine curve can again be employed to characterize the diurnal cycle of water temperature when it follows a regular and symmetrical pattern (Smith 1981). The magnitude of the diel range may exhibit hysteresis in the pattern of its variation throughout the year. Studies in the Great Egglesthorpe Beck and other streams of northern England (Crisp 1988b), for example, have shown that the monthly mean of daily range is higher relative to monthly mean temperature during periods of rising water temperatures in spring and early summer than during periods of falling water temperatures in late summer and autumn (Fig. 3.8(c)).

Storm and other events

Water temperatures also respond to changing flow conditions in the river channel. Investigations in a small Pennine stream in northern England (Smith & Lavis 1975) have shown that temperatures can be significantly reduced in summer months by rainfall which increases river discharge and in turn increases the thermal capacity of the watercourse. In the winter period, more prolonged depression of temperature is produced by hydrographs which are generated by snowmelt (Fig. 3.8(d)). Detailed monitoring of storm events at Panola Mountain, Georgia, USA (Shanley & Peters 1988) also indicates that stream temperatures may exhibit complex patterns during the passage of a hydrograph (Fig. 3.8(e)) which reflect the balance of surface and subsurface contributions to flow and the magnitude of difference in temperature between these two sources.

Extreme meteorological conditions also affect water temperature. The occurrence of prolonged drought with attendant high air temperatures and low volumes of flow can encourage unusually high water-temperature maxima and increase the range of diurnal fluctuation, especially in smaller rock-floored channels. In larger rivers, however, diel variation in water temperature may be reduced under drought conditions as a result of extended residence times and slow downstream transfer of channel flow (Walling & Carter 1980). This effect is evident when records from the River Exe at Thorverton, Devon, UK for August of the 1976 drought year are compared with those for August 1974 which was not affected by a severe drought (Fig. 3.8(f)).

Anomalously low water temperatures can also be generated in small streams through the chilling effects of heavy snowfall and wind-driven ice floes (Pluhowski 1972).

Spatial variation and controls

The thermal behaviour of rivers also varies considerably in space as well as in time. A variety of controlling factors determine the extent of variation and these controls, in turn, differ according to the spatial scale under consideration.

Macro-scale

Significant contrasts in the temperature characteristics of rivers are evident when differences over large geographical areas, encompassing global, continental, national and regional variations, are considered. Ward (1985) has suggested that water-courses in the southern hemisphere might exhibit more extreme temperatures, a lower incidence of freezing conditions and more unpredictable thermal behaviour than those in the northern hemisphere in response to the greater incidence of intermittent streams, to the occurrence of higher silt loads and more variable flows, and to the different distribution of land masses south of the Equator. Within the southern hemisphere (Fig. 3.9(a)), equatorial rivers are generally characterized by a restricted annual range in temperature, tropical rivers by very high maximum temperatures and intermediate annual ranges, and temperate rivers by large annual variability and in a few cases the occurrence of freezing (Ward 1985).

Macro-scale variations in water temperature behaviour are strongly controlled by latitude, altitude and continentality. These factors exert their influence by governing climatic conditions and especially air temperature characteristics. The effects of latitude on average annual water temperature are well seen at a continental scale

in the case of the conterminous USA (Blakey 1966), at a national scale for New Zealand (Mosley 1982), and at a regional scale for the eastern coast of the USA (Steele 1983), where the mean and amplitude of the annual water temperature harmonic has a strong correlation with latitudinal position (Fig. 3.9(b)). An example of a significant relationship between water temperature and altitude on a countrywide scale is shown for Pakistani rivers (Steele 1982) in Fig. 3.9(c), although in some regions a clearcut relationship between these variables is obscured by other factors. A thermal profile of Bolivian rivers (Wasson *et al* 1989), for example, shows that annual average and maximum temperatures vary considerably at given altitudes between the western and eastern slopes of the Cordillera Oriental (Fig. 3.9(d)).

Meso-scale

Temperature behaviour also varies between streams in a local area and within the tributaries and mainstream of individual river systems. Mean annual temperature can be related to altitude for a single catchment (Walker & Lawson 1977), although data from the Exe Basin, Devon, UK indicate that significant positive and negative residuals from this relationship can occur (Fig. 3.9(e)). The former are related to mainstream sites in large open valleys, whereas the latter are caused by vegetational shading and the influence of groundwater (Webb & Walling 1986). The moderating effects on thermal regime of subsurface flows are particularly apparent in catchments underlain by chalk lithologies (Crisp *et al* 1982). The temperature behaviour of a particular river catchment is often a function of several controls rather than a single factor. For example, the diurnal range of river temperature in the Nagara River, Japan has been related to drainage area, stream slope, rice-field area, mean altitude and relief, basin shape and catchment orientation (Miyazawa *et al* 1982).

Within an individual river system, a contrast can often be drawn between the thermal behaviour of the mainstream river and that of tributary streams (Boon & Shires 1976; Smith 1979). The latter tend to exhibit greater annual and diurnal

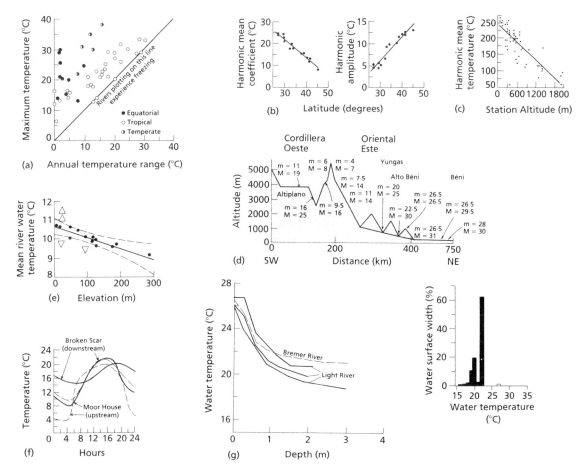

Fig. 3.9 Spatial variability in water temperature. (a) The relationship between maximum temperature and annual temperature range for water-courses in the southern hemisphere (after Ward 1985). (b) The relationship of harmonic mean temperature and amplitude of the temperature harmonic to latitude for streams of the East Coast, USA (after Steele 1983). (c) The relationship of harmonic mean temperature and altitude in Pakistani rivers (after Steele 1982). (d) Thermal profile of Bolivian rivers (after Wasson *et al* 1989); m = annual average temperature (°C), M = maximum temperature (°C). (e) The relationship of mean annual water temperature and elevation in the Exe Basin, UK. (f) Contrasts in diurnal temperature fluctuations at upstream and downstream locations in the River Tees, UK (after Smith 1979) (mean hourly variation 7–13 June 1969. —— River, – – – air). (g) Variation of temperature and depth in Southern Australian water-courses (after Morrissy 1971) (inset depicts lateral variability in water temperature in the Ashley River, New Zealand 16.12.82) (after Mosley 1983).

variation because their smaller volume of flow and lower thermal capacity make them more responsive to heat exchange processes. An increase in flow volume from river source to mouth also promotes systematic downstream trends in water temperature characteristics (e.g. Schmitz 1954; Smith 1972). In particular, diurnal fluctuations in water temperatures become less pronounced and lag increasingly behind air temperature fluctuations in the downstream direction, as results from the River Tees in northern England (Smith 1979) clearly demonstrate (Fig. 3.9(f)).

Micro-scale

Temperature characteristics may also vary along individual river reaches in response to factors such as bank-side vegetation (Weatherley & Ormerod 1990), groundwater seepage (Smith & Lavis 1975), channel depth (Sioli 1964), shape (Macan 1958) and orientation (Brown *et al* 1971), substrate conditions (Geijskes 1942) and silt content of the water (Reid & Wood 1976).

Within individual river reaches, there may be significant vertical and lateral variations in water temperature. Studies of pools in several rivers of South Australia (Fig. 3.9(g)) reveal an average temperature decrease with increasing depth of 1.77°C m^{-1} (Morrissy 1971), whereas an investigation of temperature in the distributaries of the braided Ashley River, New Zealand has revealed considerable lateral contrasts (Fig. 3.9(g)) between the main channel and smaller side channels kept cooler by seepage of groundwater (Mosley 1983).

3.4 DISSOLVED OXYGEN

Controls

Several gases, including nitrogen and carbon dioxide, may be dissolved in river water but of these oxygen is arguably the most significant because of its vital importance to aquatic organisms. The dissolved oxygen content of a watercourse is subject to physical, chemical and biological controls, and at any one time the oxygen concentration reflects a balance between the various sources and sinks for this gas in a stream system.

The ultimate source of oxygen in water is the atmosphere, although the solubility of oxygen in river water is a function of temperature and pressure conditions (Hem 1970). At a temperature of 0°C and normal atmospheric pressure at sea level (760 mmHg), the solubility of oxygen is 14.6 mg l^{-1}, assuming equilibrium is achieved between the water-body and overlying air, but this value falls to 7.63 mg l^{-1} at a temperature of 30°C (Fig 3.10(a)). Reduction in the atmospheric partial pressure of oxygen with increasing altitude above sea level will also reduce the equilibrium concentration of O_2, as the relationships between the equilibrium solubility of oxygen and water temperature for sea level and 1000 m above sea level clearly demonstrate (Fig. 3.10(a)).

The attainment of oxygen equilibrium or saturation conditions in a river is often strongly mediated by biological processes within aquatic ecosystems. Autotrophic plants, including phyto-

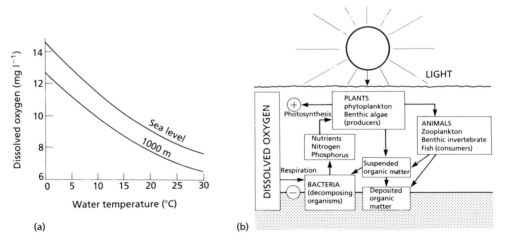

Fig. 3.10 Factors controlling dissolved oxygen concentrations in rivers. (a) The relationship between the equilibrium solubility of oxygen and water temperature (atmospheric pressure at sea level = 760 mm). (b) Interactions affecting dissolved oxygen in aquatic ecosystems (after Gras *et al* 1983).

plankton and fixed plants, produce oxygen in a water-body by photosynthesis. Photosynthetic activity by macrophytes and algae may cause an oxygen supersaturation during daylight hours. Butcher and colleagues (1930) indicated that supersaturation is more likely to be generated by algae, which produce oxygen that goes straight into solution, than by the higher plants, which produce bubbles of the gas, although daily oxygen contributions of up to $2.27 \ g^{-2} \ day^{-1}$ have been recorded for plants on weed beds in small British rivers (Owens & Edwards 1961). In contrast to photosynthesis, oxygen is consumed in a river by plant and bacterial respiration. It has been found, for example, that the respiration of a thick carpet of the benthic filamentous green alga *Cladophora glomerata* on the bed of the River Tees, UK had the effect of reducing the oxygen saturation to nearly 50% at night (Butcher *et al* 1937). Bacterial decomposition of organic matter, such as decaying water plants and dead leaves blown into the channel, and microbially accelerated oxidation of NH_3 to NH_4^+ also depletes dissolved oxygen concentrations. The influence of plants and

bacteria on dissolved oxygen levels in rivers is complexly interrelated with grazing and carnivorous aquatic animals (heterotrophic consumers), and with suspended and deposited organic matter and dissolved nutrients in the river system. The main interactions in aquatic ecosystems (Fig. 3.10(b)) have been summarized by Gras *et al* (1983).

Variations

Dissolved oxygen concentrations in rivers may also vary markedly in time and over space. The annual regime of dissolved oxygen is inversely related to the annual cycle of water temperature. Concentrations are lowest in the summer months because of the decreased equilibrium solubility of oxygen, as results from the Red Cedar River, Michigan, USA (Ball & Bahr 1975) have clearly demonstrated (Fig. 3.11(a)). The impact of photosynthesis and respiration may cause a marked diurnal cycle in stream oxygen levels. This phenomenon is especially pronounced in rivers where high nutrient concentrations have stimu-

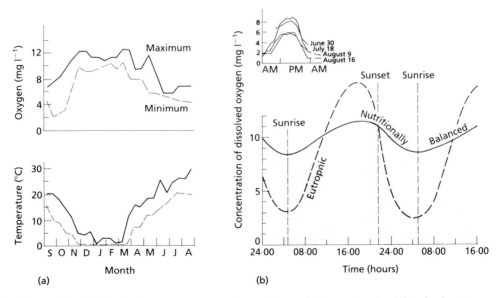

Fig. 3.11 Fluctuations in dissolved oxygen concentrations of rivers; (a) Annual cycle of dissolved oxygen concentration in a Michigan river; Red Cedar River, USA (after Ball & Bahr 1975). (b) Diurnal cycle of dissolved oxygen concentration in a nutritionally balanced and a eutrophic stream (after Gower 1980) (inset depicts diel variation of dissolved oxygen concentration in the Red Cedar River) (after Ball & Bahr 1975).

lated excessive plant growth (Simonsen & Harremoës 1978; Gower 1980) and photosynthetic activity causes high dissolved oxygen levels during the late afternoon, which is followed by oxygen depletion at night due to continued respiratory activity (Fig. 3.11(b)).

Results from the Red Cedar River (Ball & Bahr 1975) show that the magnitude of the diurnal variation in oxygen levels may vary from day to day according to flow rate and temperature (Fig 3.11(b)). The diel cycle of oxygen levels may also differ between headwater and downstream environments (Macan 1974). Near the source, oxygen is lowest by day when the water is warmest and solubility is reduced, whereas further downstream the effects of decomposition of greater quantities of organic matter leads to minimum saturation levels at night.

Spatial variation in dissolved oxygen content between and within river systems will reflect the influence of physiographic and especially hydraulic factors. In particular, oxygen exchange between air and water is reduced at lower water velocities, and oxygen balance is affected by water depth and residence time (Gras *et al* 1983). Lower concentrations of dissolved oxygen could be anticipated in headwater streams, because of the influence of altitude on equilibrium concentration (see Fig. 3.10(a)), but this effect is often outweighed by the greater turbulence of upland streams compared with lowland rivers. At a micro-scale level, variations in dissolved oxygen concentration across the sediment–water interface and in relation to stones on the stream bed may be of particular importance to the habitats of many benthic invertebrates. Milne and Calow (1990), for example, reported a mean oxygen concentration of 5.4 mg l^{-1} under stones in a river during May 1987 when temperature ranged from 9.0 to 9.5°C. Oxygen concentrations in the water column were >11.6 mg l^{-1}, and it was found that the under-stone microhabitat of glossiphoniid leeches may become severely hypoxic during the breeding season.

The re-aeration properties of rivers are also strongly controlled by hydraulic conditions, since recovery from any oxygen deficit occurs more rapidly in shallow and turbulent reaches, where there is maximum contact between the water and the overlying air, than in deep and sluggish channel sections (Walling 1980). Natural and artificial changes in channel gradient, such as waterfalls and weirs, lead to an increase in dissolved oxygen concentrations (Gameson 1957). In many major lowland rivers, the spatial pattern of dissolved oxygen levels is also strongly affected by domestic and industrial point-source pollution.

REFERENCES

Ball RC, Bahr TG. (1975) Intensive survey: Red Cedar River, Michigan. In: Whitton BA (ed.) *River Ecology*, pp 431–60. Blackwell Scientific Publications, Oxford. [3.4]

Blakey JF. (1966) Temperature of surface waters in the conterminous United States. *United States Geological Survey Hydrological Investigations Atlas* HA-235, Washington, DC. [3.3]

Boon PJ, Shires SW. (1976) Temperature studies on a river system in north-east England. *Freshwater Biology* **6**: 23–32. [3.3]

Brett JR. (1960) Thermal requirements of fish—three decades of study, 1940–1970. In: *Biological Problems in Water Pollution*, Robert A. Taft Sanitary Engineering Centre Technical Report W60–3, pp 110–17. Cincinnati, Ohio. [3.3]

Brown GW. (1969) Predicting temperatures of small streams. *Water Resources Research* **5**: 68–75. [3.3]

Brown GW, Swank GW, Rothacher J. (1971) Water temperature in the Steamboat Drainage. *United States Department of Agriculture Forest Service Research Paper* PNW-119, pp 1–17. [3.3]

Burton TM, Turner RR, Harriss RC. (1977) Suspended and dissolved solids export from three North Florida watersheds in contrasting land use. In: Correll DL (ed.) *Watershed Research in Eastern North America*, pp 471–85. Chesapeake Bay Centre for Environmental Studies. [3.1]

Butcher RW, Pentelow FTK, Woodley JWA. (1930) Variations in composition of river waters. *International Reviews in Hydrobiology* **24**: 47–80. [3.4]

Butcher RW, Longwell J, Pentelow FTK. (1937) Survey of the River Tees. III—The non-tidal reaches—chemical and biological. *Technical Papers in Water Pollution Research*, London **6**: 189 pp. [3.4]

Calandro AJ. (1973) Analysis of stream-temperatures in Louisiana. *Louisiana Department of Public Works, Technical Report* 6. [3.3]

Castorina AR. (1980) Reservoir improvements as a key to source quality control. *Journal of the American Water Works Association* **72**: 28. [3.1]

Chang M, Roth FA, Hunt EV. (1982) Sediment production under various forest-site conditions. In:

Recent Developments in the Explanation and Prediction of Erosion and Sediment Yield. International Association of Hydrological Sciences Publication No. 137, pp 13–22. Wallingford, UK. [3.1]

Collings MR. (1969) Temperature analysis of a stream. *United States Geological Survey Professional Paper* 650-B, pp B174–B179. [3.3]

Collins WD. (1925) Temperature of water available for industrial use in the United States. *United States Geological Survey Water-Supply Paper* 520-F, pp 97–104. [3.3]

Comer LE, Greeney WJ, Dimhirn I. (1976) Stream temperature modeling. In: *Proceedings of the Symposium and Speciality Conference on Instream Flow Needs*, Rodeway Inn-Boise, Idaho, 3–6 May 1976, Volume II, pp 527–39. American Fisheries Society, Bethesda, Maryland. [3.3]

Crisp DT. (1988a) Prediction, from temperature, of eyeing, hatching and 'swim-up' times for salmonid embryos. *Freshwater Biology* 19: 41–8. [3.3]

Crisp DT. (1988b) Water temperature data from streams and rivers in north east England. *Freshwater Biological Association Occasional Publication* 26, pp 1–60. [3.3]

Crisp DT, Howson G. (1982) Effect of air temperature upon mean water temperature in streams in the north Pennines and English Lake District. *Freshwater Biology* 12: 359–67. [3.3]

Crisp DT, Le Cren ED. (1970) The temperatures of three different small streams in North West England. *Hydrobiologia* 35: 305–23. [3.3]

Crisp DT, Matthews AM, Westlake DF. (1982) The temperatures of nine flowing waters in southern England. *Hydrobiologia* 89: 193–204. [3.3]

Department of the Environment (1981) *Examination of water and associated materials: colour and turbidity of waters.* HMSO, London. [3.2]

Dietrich WE, Dunne T. (1978) Sediment budget for a small catchment in mountainous terrain. *Zeitschrift fur Geomorphologie* 29: 191–206. [3.1]

Dingman SL. (1972) Equilibrium temperatures of water surfaces as related to air temperature and solar radiation. *Water Resources Research* 8: 42–9. [3.3]

Droppo IG, Ongley ED. (1989) Flocculation of suspended solids in southern Ontario rivers. In: *Sediment and the Environment*, International Association of Hydrological Sciences Publication No. 184, pp 95–103. [3.1]

Duijsings JJHM. (1985) *Streambank contribution to the sediment budget of a forest stream.* PhD thesis, University of Amsterdam. [3.1]

Duijsings JJHM. (1986) Seasonal variations in the sediment delivery ratio of a forested drainage basin in Luxembourg. In: *Drainage Basin Sediment Delivery*, International Association of Hydrological Sciences Publication No. 159, pp 153–64. [3.1]

Edinger JE, Duttweiler DW, Geyer JC. (1968) The response of water temperatures to meteorological conditions. *Water Resources Research* 4: 1137–43. [3.3]

Edwards A, Martin D, Mitchell G. (eds) (1987) *Colour in upland waters. Proceedings of a workshop held at Yorkshire Water, Leeds, September 1987.* Yorkshire Water, Leeds; Water Research Centre, Medmenham. [3.2]

Everts CM. (1963) Temperature as a water quality parameter. In: *Water Temperature — Influences, Effects, and Control* (Proceedings of the 12th Pacific North West Symposium on Water Pollution Research, November 1963), pp 2–5. Corvallis, Oregon. [3.3]

Fowles CR. (1975) Temperature records from a small Canterbury stream. *Mauri Ora* 3: 89–94. [3.3]

Fredriksen RL. (1970) Erosion and sedimentation following road construction and timber harvest on unstable soils in three small western Oregon watersheds. *US Forest Service Research Paper No. PNW 104.* [3.1]

Fry FEJ. (1964) Animals in aquatic environments: fishes. In: Dill DB, Adolph EF, Wilber CG (eds) *Adaptation to the Environment*, Handbook of Physiology Section 4, pp 715–28. American Physiology Society, Washington, DC. [3.3]

Gameson ALH. (1957) Weirs and the aeration of rivers. *Journal of the Institution of Water Engineers* 11: 477–90. [3.4]

Geijskes DC. (1942) Observations on temperature in a tropical river. *Ecology* 23: 106–10. [3.3]

Gippel CJ. (1989) *The Use of Turbidity Instruments to Measure Stream Water Suspended Sediment Concentration.* Monograph Series No. 4, Department of Geography and Oceanography, University College, Australian Defence Force Academy, Canberra. [3.1]

Gower AM. (1980) Ecological effects of changes in water quality. In: Gower AM (ed.) *Water Quality in Catchment Ecosystems*, pp 145–71. John Wiley & Sons, Chichester. [3.4]

Gras R, Albignat JP, Gosse Ph. (1983) The effects of hydraulic projects and their management on water quality. In: *Dissolved Loads of Rivers and Surface Quantity/Quality Relationships* (Proceedings of the Hamburg Symposium, August 1983), pp 313–32. IAHS Publication No. 141. Wallingford, UK. [3.4]

Hem JD. (1970) Study and interpretation of the chemical characteristics of natural water. Second edition. *United States Geological Survey Water-Supply Paper 1473.* [3.4]

Holtby LB. (1988) Effects of logging on stream temperatures in Carnation Creek, British Columbia, and associated impacts on the Coho salmon (*Oncorhynchus kisutch*). *Canadian Journal of Fisheries and Aquatic Sciences* 45: 502–15. [3.3]

Hopkins CL. (1971) The annual temperature regime of a small stream in New Zealand. *Hydrobiologia* 37:

397–408. [3.3]

Johnson FA. (1971) Stream temperatures in an alpine area. *Journal of Hydrology* **14**: 322–36. [3.3]

Lal R. (1982) Effects of slope length and terracing on runoff and erosion on a tropical soil. In: *Recent Developments in the Explanation and Prediction of Erosion and Sediment Yield.* International Association of Hydrological Sciences Publication No. 137, pp 23–31. [3.1]

Langford TE. (1972) A comparative assessment of thermal effects in some British and North American rivers. In: Oglesby RT, Carlson CA, McCann JA (eds) *River Ecology and Man*, pp 319–51. Academic Press, New York. [3.3]

Langford TE. (1983) *Electricity Generation and the Ecology of Natural Waters.* Liverpool University Press, Liverpool. [3.3]

Lehre AK. (1982) Sediment budget in a small coast range drainage basin in North-Central California. In: *Sediment Budgets and Routing in Forested Drainage Basins*, US Forest Service General Technical Report PNW-141. [3.1]

Macan TT. (1958) The temperature of a small stony stream. *Hydrobiologia* **12**: 89–106. [3.3]

Macan TT. (1974) *Freshwater Ecology* 2nd edn. Longman, London. [3.4]

McDonald AT, Edwards AMC, Naden PS, Martin D, Mitchell G. (1989) Discoloured runoff in the Yorkshire Pennines. In: *Proceedings of the Second National Hydrology Symposium*, pp 1.59–1.64, British Hydrological Society. [3.2]

MacKichan KA. (1967) Diurnal temperature fluctuations of three Nebraska streams. *United States Geological Survey Professional Paper 575-B*, B233–B234. [3.3]

Malcolm RLC. (1985) Geochemistry of stream fulvic and humic substances. In: Aiken GR, McKnight DM, Wershaw RL, Maccarthy P (eds) *Humic Substances in Soil and Water*, pp 181–209. John Wiley & Sons, New York. [3.2]

Meade RH, Parker RS. (1985) Sediment in rivers of the United States. In: *National Water Summary 1984*, US Geological Survey Water-Supply Paper 2275, pp 49–60. [3.1]

Milne IS, Calow P. (1990) Costs and benefits of brooding in glossiphoniid leeches with special reference to hypoxia as a selection pressure. *Journal of Animal Ecology* **59**: 41–56. [3.4]

Miyazawa T, Yamashita S, Kitagawa M, Sekine K. (1982) An evaluation and mapping of topographic factors affecting river water temperature in the upstream area of the Nagara River, Japan. *Beiträge zur Hydrologie Sonderheft* **3**: 83–95. [3.3]

Moore AM. (1967) Correlation and analysis of water-temperature data for Oregon streams. *United States Geological Survey Water-Supply Paper 1819 K.* [3.3]

Morrissy NM. (1971) Temperature relationships in small bodies of freshwater with special reference to trout streams in South Australia. *Bulletin of the Australian Society of Limnology* **4**: 8–20. [3.3]

Mosley MP. (1982) *New Zealand River Temperature Regimes.* Water and Soil Miscellaneous Publication No. 36, Water and Soil Division, Ministry of Works and Development for the National Water and Soil Conservation Organisation, Christchurch, New Zealand. [3.3]

Mosley MP. (1983) Variability of water temperatures in the braided Ashley and Rakaia rivers. *New Zealand Journal of Marine and Freshwater Research* **17**: 331–42. [3.3]

Muncy RJ, Atchison GJ, Bulkley RV, Menzel BW, Perry LG, Summerfelt RC. (1979) *Effects of Suspended Solids and Sediment on the Reproduction and Early Life of Freshwater Fishes: a Review.* US Environmental Protection Agency, Corvallis, Oregon. [3.1]

Naden PS, McDonald AT. (1987) Statistical modelling of water colour in the uplands: the Upper Nidd catchment 1979–1987. *Environmental Pollution* **60**: 141–63. [3.2]

O'Loughlin CL, Rowe LK, Pearce AJ. (1980) Sediment yield and water quality responses to clear felling of evergreen mixed forests in western New Zealand. In: *The Influence of Man on the Hydrological Regime with Special Reference to Representative and Experimental Basins.* International Association of Hydrological Sciences Publication No. 130, pp 285–92. [3.1]

Owens M, Edwards RW. (1961) The effects of plants on river conditions. II. Further studies and estimates of net productivity of macrophytes in a chalk stream. *Journal of Ecology* **49**: 119–26. [3.4]

Parker FL, Krenkel PA. (eds) (1969) *Engineering Aspects of Thermal Pollution.* Vanderbilt University Press, Portland, Oregon. [3.3]

Peart MR, Walling DE. (1986) Fingerprinting sediment source: the example of a drainage basin in Devon, UK. In: *Drainage Basin Sediment Delivery*, International Association of Hydrological Sciences Publication No. 159, pp 41–55. [3.1]

Petts GE. (1984) *Impounded Rivers: Perspectives for Ecological Management.* John Wiley & Sons, Chichester. [3.3]

Pluhowski EJ. (1970) Urbanization and its effect on the temperature of the streams on Long Island, New York. *United States Geological Survey Professional Paper 627-D.* [3.3]

Pluhowski EJ. (1972) Unusual temperature variations in two small streams in northern Virginia. In: *Geological Survey Research, 1972, United States Geological Survey Professional Paper 800-B*, pp B255–B258. [3.3]

Reid GK, Wood RD. (1976) *Ecology of Inland Waters and Estuaries.* Van Nostrand, New York. [3.3]

Reuter JH, Perdue EM. (1977) Importance of heavy

metal−organic matter interaction in natural waters. *Geochimica Cosmochimica Acta* **41**: 325−34. [3.2]

Ritchie JC. (1972) Sediment, fish and fish habitat. *Journal of Soil and Water Conservation* **27**: 125. [3.1]

Rose AH. (ed.) (1967) *Thermobiology*. Academic Press, London. [3.3]

Schmitz W. (1954) Grundlagen der Untersuchung der Temperaturverhältnisse in der Fliessgewässern. *Berliner Limnologie der Flusstation Freundenthal* **6**: 29−50. [3.3]

Shanley JB, Peters NE. (1988) Preliminary observations of streamflow generation during storms in a forested Piedmont watershed using temperature as a tracer. *Journal of Contaminant Hydrology* **3**: 349−65. [3.3]

Simonsen JF, Harremoës P. (1978) Oxygen and pH fluctuations in rivers. *Water Research* **12**: 477−89. [3.4]

Sioli H. (1964) General features of the limnology of Amazónia. *Verh. int. Ver. Limnol.* **15**: 1053−8. [3.3]

Skvortsov AF. (1959) River suspension and soils. *Soviet Soil Science* **4**: 409−16. [3.1]

Smith K. (1968) Some thermal characteristics of two rivers in the Pennine area of northern England. *Journal of Hydrology* **6**: 405−16. [3.3]

Smith K. (1972) River water temperatures: an environmental review. *Scottish Geographical Magazine* **88**: 211−20. [3.3]

Smith K. (1975) Water temperature variations within a major river system. *Nordic Hydrology* **6**: 155−69. [3.3]

Smith K. (1979) Temperature characteristics of British rivers and the effects of thermal pollution. In: Hollis GE (ed.) *Man's Impact on the Hydrological Cycle in the United Kingdom*, pp 229−42. GeoBooks, Norwich. [3.3]

Smith K. (1981) The prediction of river water temperatures. *Hydrological Sciences Bulletin* **26**: 19−32. [3.3]

Smith K, Lavis ME.(1975) Environmental influences on the temperature of a small upland stream. *Oikos* **26**: 228−36.

Steele TD. (1982) A characterization of stream temperatures in Pakistan using harmonic analysis. *Hydrological Sciences Journal* **4**: 451−67. [3.3]

Steele TD. (1983) A regional analysis of water quality in major streams of the United States. In: *Dissolved Loads of Rivers and Surface Water Quantity/Quality Relationships* (Proceedings of the Hamburg Symposium, August 1983), pp 131−43. International Association of Hydrological Sciences Publication No. 141. Wallingford, UK. [3.3]

Steele TD, Gilroy EJ, Hawkinson RO. (1974) An assessment of areal and temporal variations in streamflow quality using selected data from the National Stream Quality Accounting Network. *United States Geological Survey Open-File Report 74−217*. [3.3]

Stevens HH Jr, Ficke JF, Smoot GF. (1975) Water temperature − influential factors, field measurement, and data presentation. *United States Geological Survey Techniques of Water-Resources Investigations*, Book 1, Chapter D1. [3.3]

Swanson FJ, Janda RJ, Dunne T, Swanson DN. (eds) (1982) *Sediment Budgets and Routing in Forested Drainage Basins*. US Forest Service General Technical Report PNW-141. [3.1]

Trimble SW. (1976) Sedimentation in Coon Creek Valley, Wisconsin. In: *Proceedings of the Third Federal Interagency Sedimentation Conference*, pp 5-100−5-112. US Water Resources Council, Washington, DC. [3.1]

Trimble SW. (1981) Changes in sediment storage in Coon Creek Basin, Driftless Area, Wisconsin, 1853−1975. *Science* **214**: 181−3. [3.1]

Troxler RW, Thackston EL. (1977) Predicting the rate of warming of rivers below hydro-electric installations. *Journal of the Water Pollution Control Federation* August 1977, 1902−12. [3.3]

US Environmental Protection Agency (1979) *Impacts of Sediment and Nutrients on Biota in Surface Waters of the United States*. US Environmental Protection Agency, Athens, Georgia. [3.1]

Walker JH, Lawson JD. (1977) Natural stream temperature variation in a catchment. *Water Research* **11**: 373−7. [3.3]

Wall CJ, Wilding LP. (1976) Mineralogy and related parameters of fluvial suspended sediments in Northwestern Ohio. *Journal of Environmental Quality* **5**: 168−73. [3.1]

Walling DE. (1980) Water in the catchment ecosystem. In: Gower AM (ed.) *Water Quality in Catchment Ecosystems*, pp 1−47. John Wiley & Sons, Chichester. [3.4]

Walling DE. (1983) The sediment delivery problem. *Journal of Hydrology* **65**: 209−37. [3.1]

Walling DE. (1987) Rainfall, runoff and erosion of the land: a global view. In: Gregory KJ (ed.) *Energetics of the Physical Environment*, pp 89−117. John Wiley & Sons, Chichester. [3.1]

Walling DE. (1990) Linking the field to the river: sediment delivery from agricultural land. In: Boardman J, Foster IDL, Dearing JA (eds) *Soil Erosion on Agricultural Land*, pp 129−52. John Wiley & Sons, Chichester. [3.1]

Walling DE, Bradley SB. (1988) The use of caesium-137 measurements to investigate sediment delivery from cultivated areas in Devon, UK. In: *Sediment Budgets*, International Association of Hydrological Sciences Publication No. 174, pp 325−35. [3.1]

Walling DE, Carter R. (1980) River water temperatures. In: Doornkamp JC, Gregory KJ (eds) *Atlas of Drought in Britain 1975−76*, p 49. Institute of British Geographers, London. [3.3]

Walling DE, Kane P. (1982) Temporal variation of suspended sediment properties. In: *Recent Developments in the Explanation and Production of Erosion and Sediment Yield*, International Association of Hydrological Sciences Publication No. 137, pp 409–19. [3.1]

Walling DE, Moorehead PW. (1989) The particle size characteristics of fluvial suspended sediment: an overview. *Hydrobiologia* **176/177**: 125–49. [3.1]

Walling DE, Webb BW. (1983) Patterns of sediment yield. In: Gregory KJ (ed.) *Background to Palaeohydrology*, pp 69–100. John Wiley & Sons, Chichester. [3.1]

Ward JC. (1963) Annual variation of stream water temperature. *American Society of Civil Engineers Journal of the Sanitary Engineering Division* **89** (SA6): 1–16. [3.3]

Ward JV. (1985) Thermal characteristics of running waters. *Hydrobiologia* **125**: 31–46. [3.3]

Ward JV, Stanford JA. (1982) Thermal responses in the evolutionary ecology of aquatic insects. *Annual Reviews in Entomology* **27**: 97–117. [3.3]

Wark JW, Keller FJ. (1963) *Preliminary Study of Sediment Sources and Transport in the Potomac River Basin.* Interstate Commission on the Potomac River Basin. Edgewater, Maryland, USA. [3.1]

Wasson JG, Guyot JL, Dejoux C, Roche MA. (1989) *Regimen termico de los rios de Bolivia.* ORSTOM IHH-UMSA PHICAB IIQ-UMSA Senamhi Hidrobiologia-UMSA. La Paz, Bolivia. [3.3]

Weatherley NS, Ormerod SJ. (1990) Forests and the temperature of upland streams in Wales: a modelling exploration of the biological effects. *Freshwater Biology* **24**: 109–22. [3.3]

Webb BW. (1987) The relationship between air and water temperatures for a Devon river. *Reports and Transactions of the Devonshire Association for the Advancement of Science, Literature and Art* **119**: 197–222. [3.3]

Webb BW, Walling DE. (1982) The magnitude and frequency characteristics of fluvial transport in a Devon drainage basin and some geomorphological implications. *Catena* **9**: 9–23. [3.1]

Webb BW, Walling DE. (1985a) Temporal variation of river water temperatures in a Devon river system. *Hydrological Sciences Journal* **30**: 449–64. [3.3]

Webb BW, Walling DE. (1985b) Temperature characteristics of Devon rivers. *Proceedings of the Ussher Society* **6**: 237–45. [3.3]

Webb BW, Walling DE. (1986) Spatial variation of water temperature characteristics and behaviour in a Devon river system. *Freshwater Biology* **16**: 585–608. [3.3]

Williams GP. (1989) Sediment concentration versus water discharge during single hydrologic events in rivers. *Journal of Hydrology* **111**: 89–106. [3.1]

Wolman MG, Miller JC. (1960) Magnitude and frequency of forces in geomorphic processes. *Journal of Geology* **68**: 54–74. [3.1]

Wolman MG, Schick AP. (1967) Effects of construction on fluvial sediment, urban and suburban areas of Maryland. *Water Resources Research* **3**: 451–64. [3.1]

4: Water Quality
II. Chemical Characteristics

B.W.WEBB AND D.E.WALLING

A large number of different properties and parameters are available to describe the chemical characteristics of rivers. These range from general descriptors, such as measures of salinity and acidity, to composition in terms of major cation and anion content, and to the concentration of organic and inorganic micropollutants. The present chapter focuses on the chemical characteristics of rivers free from major point-source contamination by human activities, so that discussion of chemical parameters, which are most strongly influenced by river pollution and include trace metal concentrations, organic constituents and pesticide levels, is deferred to Volume 2. Attention will be given to the behaviour of chemical properties in uncontaminated rivers in terms of both the concentrations and transport (flux) of materials. The emphasis is on a comparison *between* river systems from global to local scales. Chapter 18, on the other hand, deals with the dynamics of chemicals *within* river systems.

A fundamental distinction is drawn in the present chapter between chemical constituents that are present in dissolved form and those that are transported in sediment-associated form. Discussion of dissolved chemical species will include those elements that are vital to the health of plants and animals, are strongly involved in cycling processes between the inorganic and organic compartments of rivers and their drainage basins, and are often referred to by aquatic ecologists as nutrients. The definition of what constitutes a nutrient species is somewhat dependent on the perspective from which the river is being studied. For example, some geochemists (e.g. Meybeck 1982) view the major nutrients in rivers as comprising various species of nitrogen and phosphorus

together with organic carbon, whereas investigators of forested and other stream ecosystems (e.g. Sanders 1972; Likens *et al* 1977) may define dissolved nutrient substances more widely. In the present chapter, the term solutes is employed to refer to dissolved chemical species, including nutrients.

The techniques available for determining the concentration and flux of chemical constituents in rivers have been comprehensively described elsewhere (e.g. Golterman & Clymo 1969; American Public Health Association 1971; Walling 1984). Considerable efforts have also been made to develop and refine laboratory procedures for reliably detecting simple and more complex chemical constituents present in low concentrations (e.g. Pereira *et al* 1987; Taylor 1987) and to standardize and harmonize sampling methods and analytical techniques between different laboratories and agencies involved in river quality investigations (e.g. Skougstad *et al* 1979; Simpson 1980).

4.1 SOLUTE BEHAVIOUR

Sources, processes and pathways

The transfer of chemical elements through the hydrological cycle involves a complex interaction of chemical, biological, and hydrological systems and processes (Neal & Hornung 1990). The nature of solute behaviour in river systems ultimately reflects the various sources and stores of dissolved material that are present in the drainage area and the different processes that mobilize and modify chemical constituents found in draining waters. Although in-stream transformations of solutes through biological, chemical and physical processes can exert a considerable influence on water chemistry, solute behaviour is determined largely by the interaction of hydrological and biogeochemical processes at a basin-wide rather than a channel scale (Walling 1980). This interaction is schematically represented in Fig. 4.1(a), and of particular influence on stream chemistry are the different pathways, volumes and flux rates of water which may be involved in the transformation of precipitation to river runoff.

Atmospheric inputs

Rivers may receive a substantial part of their solute load from the atmosphere. This can be in solid, liquid or gaseous form, and is derived from a variety of different sources. A distinction is often made between dry and wet fallout of material from the atmosphere (Walling 1980; Cryer 1986). The former involves deposition of relatively large particles (>20 μm in diameter) under the influence of gravity, whereas the latter comprises the chemical constituents of wet precipitation, which occur through the solution of particles that have acted as condensation nuclei for raindrops (rain-out) or through solution of atmospheric particles below cloud level which have been impacted by falling raindrops (washout). An important source of atmospheric aerosols and gases is marine-derived material, which is transferred from the sea surface via spray and gas exchange. This is the predominant source of Na^+ and Cl^-, is a major contributor to Mg^{2+} and to K^+, and is a significant source of sulphur in the atmos-

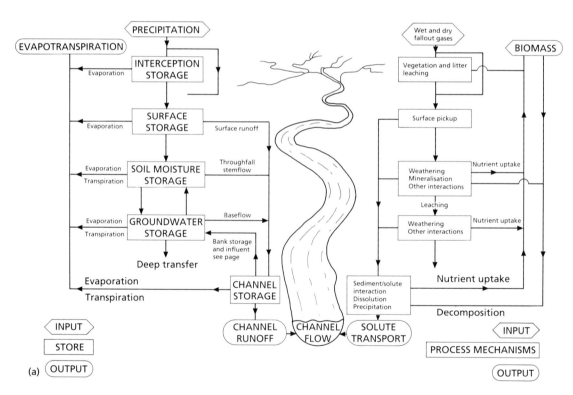

Fig. 4.1 Sources and controls of stream and river solutes. (a) Simplified representation of the interaction of hydrological and biogeochemical processes operating in a drainage basin. (b) Evolution of the ionic content of atmospheric precipitation in France, south of the Cherbourg–Basel line (from Meybeck 1986). (c) Streamwater quality for a Virginian stream, Mill Run, USA, superimposed on the $K_2O-Al_2O_3-SiO_2$ solubility diagram (from Afifi & Bricker 1983). (d) Annual calcium budget for an aggrading (northern hardwood) forested ecosystem at Hubbard Brook; standing crop values (boxed) and calcium fluxes expressed as kg ha^{-1} and kg ha^{-1} year^{-1}, respectively; values in parentheses represent annual accretion rates (after Likens *et al* 1977).

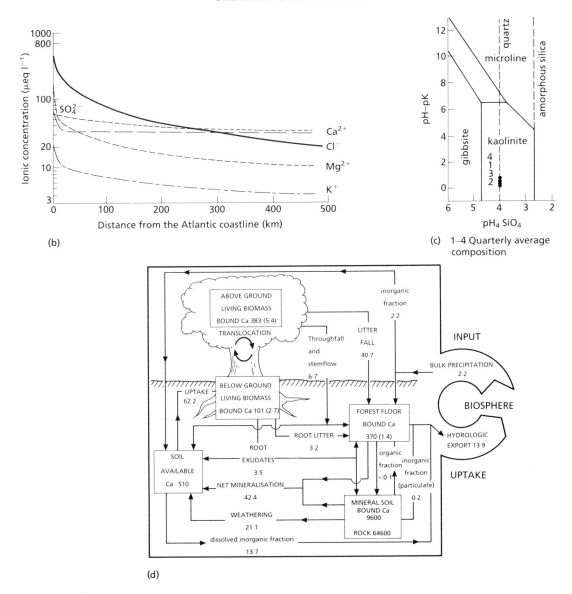

(b)

(c) 1–4 Quarterly average composition

(d)

Fig. 4.1 (*contd*)

phere. Material derived from the terrestrial environment in the form of blown soil and dust, and through biological emissions from living vegetation and the burning of organic matter, is also an important source of aerosols and gases in the atmosphere, especially in the case of Ca^{2+}, NH_4^+, NO_3^-, NO_2^-, HCO_3^- and SO_4^{2-}. Emissions from anthropogenic sources and from volcanoes also

may significantly contribute to atmospheric fallout of material.

There is a general tendency for the atmospheric input of terrestrially derived material to increase and for the fallout of cyclic salts (marine-derived material) to decrease with increasing distance from the coastal zone (e.g. Eriksson 1960; Stallard & Edmond 1981), as data from a number of stations in France (Meybeck 1986) have clearly demonstrated (Fig. 4.1(b)). Complex physico-chemical processes, which are involved in the transport and deposition of atmospheric materials and include fractionation and relative enrichment mechanisms (e.g. Bloch *et al* 1966; Belot *et al* 1982), can significantly modify the composition of atmospheric inputs received at a given location. Spatial patterns of solute input to rivers (e.g. Junge & Werby 1958; Stevenson 1968; Munger & Eisenreich 1983) may be further diversified by the particular storm directions and air-mass trajectories that affect a river basin. Atmospheric inputs may also vary over time in response to seasonal changes in meteorological conditions and to the type and duration of precipitation events (Walling 1980). Interception of wet and dry atmospheric fallout by a vegetation canopy can, through solution and cation exchange processes (e.g. Best & Monk 1975; White 1981; Foster & Grieve 1984; Reynolds & Pomeroy 1988; Ferrier *et al* 1990; Rodà *et al* 1990), further modify the content and composition of solute input to river systems.

The significance of atmospheric inputs to stream and river solute content varies depending on the ionic constituent involved and the environmental setting of the river basin. At a global scale, Meybeck (1983) calculated that natural wet and dry fallout contributes only 4% and 6% of river transport to the oceans for Si and Ca^{2+} respectively but 53% and 72% in the case of Na^+ and Cl^- respectively. Although information on worldwide variations in the nutrient content of unpolluted precipitation is relatively limited, it has been estimated that in the case of some nutrient elements the atmospheric input far exceeds river output. Fallout of dissolved inorganic and organic nitrogen, for example, is 170% of total dissolved nitrogen transport in world rivers (Meybeck 1983). The impact of atmospheric inputs at more local scales will depend on the importance of other solute sources together with hydrological conditions, so that the atmospheric influence will be greatest in drainage basins underlain by rock types resistant to chemical breakdown and where precipitation is routed quickly to the river channel.

The contribution of atmospherically derived solutes to rivers under natural conditions may be obscured by the occurrence of atmospheric pollution. Considerable attention has been given in recent years to increasing industrial emissions of SO_2 and NO_x which have resulted from combustion of fossil fuels, smelting of non-ferrous metals, fertilizer and sulphuric acid manufacture, and other industrial processes, and have given rise to problems of acid precipitation and the acidification of freshwater environments (e.g. Seip & Tollan 1985; Buijsman *et al* 1987; Young *et al* 1988; Meybeck *et al* 1989). Anthropogenic emissions of oxidized sulphur, for example, are estimated to have risen for the globe as a whole from $75-80 \times 10^6$ t year^{-1} in the late 1970s to about 90×10^6 t year^{-1} in 1985 (Möller 1984; Varheyli 1985).

A distinction can be drawn between those constituents of atmospheric fallout, both natural and anthropogenic, that affect river solutes in a passive manner and those that have an impact in an active way (Cryer 1986). In the former case, soluble material from the atmosphere is simply added in an unchanged form to the water-course, although its concentration may be increased by the effects of evapotranspiration losses. In the latter case, the components of atmospheric input may react chemically with the vegetation, soil and rock materials of a catchment area and mobilize new solute species in weathering and other processes. The role of complex biogeochemical ion exchanges and weathering reactions between precipitation water and permeable material in catchments has been highlighted, for example, in recent studies of stream acidification (Edwards *et al* 1990; Neal *et al* 1990b; Rosenqvist 1990).

Chemical reactions

In many cases, a large proportion of the solute load carried by streams and rivers is derived from chemical reactions between water and rock and soil minerals. At the global scale, it is estimated that 60% of the major dissolved constituents carried by rivers to the oceans is derived from chemical weathering (Walling & Webb 1986a). Hem (1970) identified four groups of chemical reactions which are important in establishing and maintaining the composition of natural water:

1 Reversible solution and deposition reactions,

including ion-exchange processes, in which water is not chemically altered. Cation exchange and, to a lesser extent, anion exchange, which involve reversible transfer of ionic constituents between soil particles and draining water, represent important mechanisms whereby the chemistry of precipitation is modified in its translation to river runoff.

2 Reversible solution and deposition reactions in which water molecules break down into H^+ and OH^- ions. These include simple hydrolysis reactions, such as the solution of carbonates.

3 Reversible solution and deposition reactions and ion reactions which involve changes in oxidation state, such as the reduction of ferric to ferrous iron.

4 Less readily reversible reactions which include complex processes of hydration and hydrolysis such as those involved in the breakdown of aluminosilicate rock-forming minerals. Absorption of ions by clay minerals and sesqui-oxides/hydroxides and the complexing of metal cations by organic matter ligands are processes that may also be placed in this category.

The release of solutes in weathering processes is fundamentally governed by the thermodynamics and kinetics of the reactions involved as well as by the residence time of water in the soil or rock body (Walling 1980). The latter factor influences whether weathering reactions can proceed to an equilibrium stage. Given sufficient data on soil and rock mineralogy, precipitation and streamflow volume, and water chemistry, it is possible to quantify the contribution of individual weathering reactions to river solutes (e.g. Garrels & Christ 1965; Waylen 1979). A study of Mill Run (Afifi & Bricker 1983), a small forested drainage basin in Virginia, USA, demonstrated, for example, that stream-water chemistry is controlled by the dissolution of orthoclase, plagioclase, chlorite, amphibole and pyrite minerals, by the precipitation of microcrystalline gibbsite and FeOOH, and by the formation of an SiO_2-rich residue. Mineral equilibria diagrams (Fig. 4.1(c)) suggest that cation concentrations in the stream water of this catchment are controlled by the kinetics of dissolution–precipitation reactions, rather than by thermodynamic equilibria with primary rock-forming minerals.

Biological processes

Water chemistry may also be strongly affected by biotic processes associated with the soil and vegetation cover of a drainage basin ecosystem. Plant respiration and bacterial degradation, for example, may influence CO_2 concentrations in soil pores, and terrestrial insect populations may regulate solute transfer processes through an enhancement of litter decomposition (Swank 1986). A variety of microbial transformations may strongly influence processes of mobilization and immobilization of nutrients and other elements in catchment ecosystems. The dynamics of nitrogen in the soil, for example, provides a very good example of the influence that bacterial mineralization has on the balance of nitrogen species in soil water which, in turn, supplies the river channel. Organic nitrogen contained in organic matter can be converted to salts of NH_4^+ through decomposition by heterotrophic bacteria. NH_4^+ ions, in turn, may be nitrified by autotrophic bacteria; this involves oxidation initially to NO_2^- by *Nitrosomonas* and finally to highly soluble NO_3^- by *Nitrobacter*. Other biologically mediated transformations may also affect the cycling of nitrogen in catchments under forest or grassland (Roberts *et al* 1983; Royal Society 1983), and these include the take-up and immobilization by microflora and fauna of NH_4^+ ions released in the process of mineralization; reduction of nitrate into nitrogen oxides and dinitrogen by a wide variety of microorganisms in the process of denitrification (Swank 1986), and nitrogen fixation which involves reduction of dinitrogen in the atmosphere to NH_3 through the action of certain prokaryotic microorganisms.

Solute uptake and incorporation in biomass above and below ground is a particularly important mechanism in forested drainage basins (Stevens *et al* 1988). A study in a small granitic basin in Colorado, USA (Lewis & Grant 1979) showed that most chemical elements (Ca, Na, K, N, P, S) are stored within the soils and vegetation of the catchment, and an investigation of chemical mass balances in a small stream in central Java, Indonesia (Bruijnzeel 1983) indicated that a vigorously growing plantation forest of *Agathis dammara* acts as a long-term sink for chemical elements, with annual uptake ranging from 71.9 to 1.1 kg ha^{-1} year^{-1} in the case of SiO_2 and Mn, respectively. In the Hubbard Brook Experimental Forest, the importance of biomass accumulation in the vegetation and forest floor has been shown to be greatest for nitrogen and least for sodium, with storage accounting for 80% and <3% respectively of the inputs from weathering and the atmosphere (Likens *et al* 1977). In forested drainage basins underlain by resistant lithologies, storage of solutes and processes of internal cycling within the ecosystem, involving litter fall, canopy leaching, root mortality and resorption mechanisms, may be more important in regulating stream chemistry than inputs from the atmosphere or from weathering. In such circumstances, construction of detailed chemical budgets (Fig. 4.1(d)) of the kind formulated from long-term studies of watersheds in the Hubbard Brook Experimental Forest (Likens *et al* 1977) and the Coweeta Hydrologic Laboratory (Swank & Douglass 1977) may be required to interpret fully the biological, physical and chemical factors controlling stream solute behaviour.

Biological processes operating within the river channel itself may also modify water chemistry (Swank 1986). These include uptake of nutrients and silica by aquatic organisms, nitrogen fixation, nitrification of ammonia and denitrification by bacteria living in stream sediments, and the processing of detrital material by macroinvertebrates, which may add dissolved ions, such as K$^+$ and Ca^{2+}, to the solute load of the stream (Webster & Patten 1979). Recent studies of the Afon Hafren in mid-Wales (Fiebig *et al* 1990) have also highlighted the role that the stream bed biota have in immobilizing dissolved organic carbon which enters the water-course in soil waters.

Spatial variation and controls

The solute content and composition of rivers varies greatly in space depending on the particular sources, processes and pathways that are dominant in a given drainage basin. In general, Meybeck and Helmer (1989) have suggested that six environmental factors, comprising the occurrence of highly soluble or easily weatherable minerals, distance to the marine environment,

aridity, terrestrial primary productivity, ambient temperature and rates of tectonic uplift, are the most important environmental controls on river chemistry. In detail, the extent of variation and the controlling factors responsible will vary depending on the scale of the area under consideration.

World and continental variations

The load-weighted average total dissolved solids content of world river water has been calculated as 120 mg l^{-1} from a sample of 496 rivers in which pollution is of minor significance (Walling & Webb 1986a). World averages of individual ionic constituents in river water have also been computed (Meybeck & Helmer 1989), both as a global discharge-weighted natural content and as a global most-common natural content (Table 4.1). The latter is defined as the median of a cumulative distribution curve of concentrations assembled from information available for major world rivers and their tributaries. The dissolved content of rivers is dominated by HCO$_3^-$, SO$_4^{2-}$, Ca^{2+} and SiO$_2$, and 97.3% of global runoff has been classified as being of the calcium bicarbonate type (Meybeck 1981).

Despite the dominance of a few chemical constituents in world rivers, enormous spatial variability can be encountered in the chemistry of river water. Total dissolved solids concentrations, for example, are less than 10 mg l^{-1} in some Amazonian tributaries (Meybeck 1976) but reach 60 000 mg l^{-1} in the Leben River, Tunisia (Colombani 1983). Furthermore, more than 20 water types, defined by the relative importance of the major cation and anion constituents, have been recognized across the globe (Meybeck *et al* 1989). Worldwide variations in water chemistry have been accounted for by Gibbs (1970) in terms of dominance by rock weathering, by atmospheric precipitation and by evaporation–crystallization processes, and climatic and geological factors are often cited as the most important environmental controls on stream solutes at a global scale. The influence of climate acting through water availability is seen in an inverse relationship between discharge-weighted total dissolved solids concentration and mean annual runoff (Fig. 4.2(b)) which

Table 4.1 Average chemical composition of unpolluted world rivers and variations in chemistry of pristine streams according to rock type (from Meybeck & Helmer 1989)

Parameter (units)	Global averages		Common rock types*						Rarer rock types†			
	DWNC	MCNC	1	2	3	4	5	6	A	B	C	D
Tz$^+$ (μeq l^{-1})	200	800	166	207	435	223	770	3247	2700	40700	312000	4130
pH	–	–	6.6	6.6	7.2	6.8	–	7.9	–	–	8.0	7.4
SiO$_2$ (μmol l^{-1})	170	180	150	130	200	150	150	100	332	116	20	1660
Ca^{2+} (μeq l^{-1})	670	400	39	60	154	88	404	2560	150	4350	30350	245
Mg^{2+} (μeq l^{-1})	275	200	31	57	161	63	240	640	2490	10000	5640	40
Na$^+$ (μeq l^{-1})	225	160	88	80	105	51	105	34	52	26100	276000	3830
K$^+$ (μeq l^{-1})	33	27	8	10	14	21	20	13	5	260	197	184
Cl$^-$ (μeq l^{-1})	162	110	0	0	0	0	20	0	93	420	266000	1653
SO$_4^{2-}$ (μeq l^{-1})	172	100	31	56	10	95	143	85	472	29000	27700	290
HCO$_3^-$ (μeq l^{-1})	850	500	128	136	425	125	580	3195	2020	10700	3000	2230
DOC (mg l^{-1})	5.75	4.2										
N-NH$_4^+$ (mg l^{-1})	–	0.015										
N-NO$_3^-$ (mg l^{-1})	–	0.10										
N organic (mg l^{-1})	–	0.26										
P-PO$_4^{3-}$ (mg l^{-1})	–	0.010										

Tz$^+$, cation sum; DOC, dissolved organic carbon; DWNC, global discharge-weighted natural concentration; * MCNC, most common natural concentration; * corrected for oceanic cyclic salts: 1 granite, 2 gneiss, 3 volcanics, 4 sandstone, 5 shale, 6 carbonates; † uncorrected for oceanic salts: A batholith, B coal shale, C salt rock, D hydrothermal.

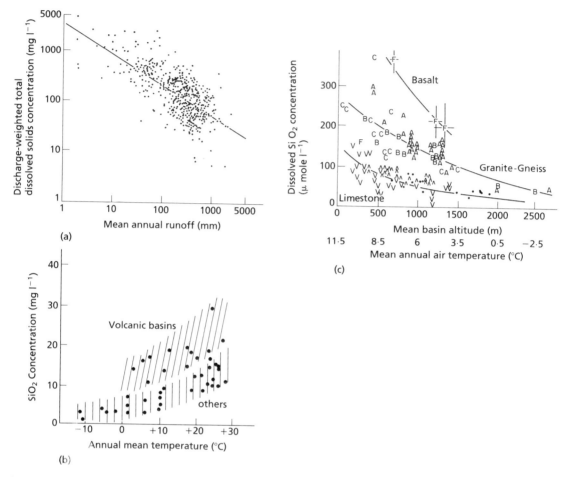

Fig. 4.2 Spatial variations in river solute concentrations. (a) The relationship between mean total dissolved solids concentration and mean annual runoff for a sample of 496 world rivers. (b) The relationship between SiO_2 content of river water and mean annual temperature (after Meybeck 1981). (c) Variation of dissolved silica concentration as a function of mean drainage basin altitude and corresponding mean annual air temperature in France (from Meybeck 1986). A granite; B gneiss; C micaschist; F basalt; v limestone (chalk excluded); ∧ Jura limestone; ● limestone watersheds.

indicates a general dilution effect as runoff volumes increase. Mean annual temperature is also an important control on the SiO_2 content of rivers (Fig. 4.2(b)), and the impact of climate on chemical composition of river water at a global level is demonstrated by the distinctiveness of the ionic balance in rivers of the arid zone compared with other morphoclimatic regions (Walling & Webb 1986b). A general climatic control is also evident in the worldwide variation of dissolved organic as well as dissolved inorganic constituents of river

runoff. Median dissolved organic carbon concentrations, for example, have been found to be 10 mg l^{-1} in taiga rivers, 6 mg l^{-1} in humid tropical environments and 3 mg l^{-1} in temperate and semiarid areas (Meybeck 1982), and this probably reflects variations in the organic content of soils in different climatic zones.

Contrasts in rock type may often moderate the effects of climate on river solutes as indicated by the contrast in SiO_2 concentration between volcanic and other rock types over a range of annual mean temperatures (Fig. 4.2(b)). The variability of dissolved major elements in pristine streams draining common and rarer rock types is presented in Table 4.1, although worldwide contrasts in solute content associated with lithological differences are less marked than those associated with contrasting climates (Walling & Webb 1986a).

Spatial variability of stream chemistry has also been investigated across large geographical areas at a subglobal scale. For example, nutrient and total alkalinity levels have been mapped for the surface waters of the conterminous USA and general correlations established with land-use type (Omernik 1977; Omernik & Powers 1983); Hu and colleagues (1982) investigated the major ion composition of some large Chinese rivers and found their chemistry to be generally dominated by the weathering of carbonates and evaporites rather than aluminosilicates, and Stallard and Edmond (1983) demonstrated that substrate lithology and the nature of the erosional regime exert the strongest control on the solute content of rivers within the Amazon Basin.

National, regional and local variations

A narrowing of the focus to consider spatial variability of river solutes at countrywide and regional scales often reveals the occurrence of more complex patterns of variation and the influence of additional controls compared with global and continental scales. Marked geographical patterns of total and individual solute concentrations have been mapped for mainland Britain, where specific electrical conductance has been found to vary between 35 and 1200 µS cm^{-1}, nitrate-nitrogen to vary between 0.1 and 15.0 mg l^{-1} and chloride to vary between 5 and 200 mg l^{-1} in small and unpolluted tributaries (Walling & Webb 1981). The pattern of spatial variation in the case of conductance levels was found to be strongly related to geological controls, in the case of nitrate-nitrogen levels to the intensity of agricultural activity, and in the case of chloride concentrations to the distance from the coast and the balance between precipitation and runoff (Walling & Webb 1981). In France, the effect of increasing basin altitude and, in turn, decreasing mean annual air temperature on the dissolved SiO_2

content of streams is apparent (Fig. 4.2(c)), and this trend is independent of catchment geology (Meybeck 1986).

The likely occurrence of acid waters, which has been mapped for Wales (United Kingdom Acid Waters Review Group 1988), provides a good example of how soils, geology and land use interact to control water chemistry at a regional scale. In this area, a transition from acidic soils overlying rocks with little or no buffering capacity to acidic soils on rocks with low, moderate and infinite buffering capacities, and to non-acidic soils over any lithology, sees a transition from the occurrence of acid waters at all flow levels to the absence of acidic conditions at any discharge value.

Stream chemistry may also vary considerably between the tributaries of a single river system. Mapping of total and individual solute concentrations in the Exe Basin, Devon, UK under stable low-flow conditions, for example, has revealed marked and distinctive geographical patterns (Webb 1983; Webb & Walling 1983). At this scale of investigation, the influence of topographic factors, such as basin size, slope and altitude, may operate in addition to geological and land-use controls (Miller 1961; Foggin & Forcier 1977).

The influence of flow conditions

Solute concentrations in river channels are strongly influenced by flow conditions and may respond dramatically to temporal changes in river discharge levels. River flow exerts a strong control on water chemistry by determining the volume of water available to dilute dissolved concentrations (Hem 1970), although the precise relationship between solute behaviour and streamflow discharge will also reflect the origins and routes by which runoff is produced within the drainage basin system (Burt 1986).

Rating relationships

The response of total or individual solute levels to changing river flow is often represented by the construction of a rating relationship. Most commonly, this consists of a simple power function:

$$C = aQ^b \qquad (1)$$

where C is solute concentration, Q is discharge, and a and b are the constant and exponent of the function. Most rivers exhibit decreasing total dissolved solids (TDS) concentrations with increasing flow (Fig. 4.3(a)) and therefore are characterized by a negative value for the b exponent of the rating relationship (Gregory & Walling 1973). For example, more than 97% of relationships between TDS concentration and discharge reported from a global sample of 370 rivers (Walling & Webb 1983) evidenced a dilution effect, and this widespread phenomenon may be explained in general terms by reference to the sources and routes by which water and solutes are supplied to the river channel under different flow conditions. At low flows, solute concentrations are high because of evapotranspirational losses from the channel and because runoff is supplied from the lower soil profile and groundwater reservoir where water has a long residence time and solute release is promoted. During higher discharges, runoff is generally translated much more rapidly to the river channel, has less opportunity for solute pickup, and therefore has a lower dissolved solids content.

The slope of solute-rating relationships can vary considerably between catchments and between individual chemical constituents. Several factors, including the nature of groundwater circulation (Talsma & Hallam 1982; Andrews 1983), the availability and location of readily soluble material in vegetation, soil and rock (Edwards 1973; Carling 1983), and soil buffering processes (Johnson *et al* 1969) may influence the slope of the rating relationship. Positive exponent values, which indicate the occurrence of a 'concentration' rather than a 'dilution' effect at higher discharges, are occasionally found for TDS concentrations, especially in rivers where solute concentrations are low and atmospheric fallout is the dominant source (Cryer 1976; Foster 1979a) or in drainage basins where unusual geological or hydrological conditions cause surface runoff to be more highly mineralized than subsurface flows (Iorns *et al* 1965; Cryer 1980). A positive relationship between concentration and flow is more frequently encountered for some individual ionic species, especially those such as NO_3^-, PO_4^{3-} and

K^+ which are actively involved in the nutrient cycles of forested and moorland ecosystems and are mobilized in surface runoff through the action of vegetation leaching (Bond 1979; Waylen 1979). The extension of the drainage network to areas not drained under dry conditions, where a store of readily soluble material produced by the breakdown of surface material by biological and microbiological processes can be tapped, may also contribute to increasing concentrations of nutrients at higher discharges. Concentrations of dissolved organic matter and dissolved organic carbon commonly exhibit a positive correlation with discharge level (Baker *et al* 1974; Grieve 1984), which is explained by the greater dissolved organic content of flows from the upper and more organic horizons of the soil that supply runoff at times of high discharge (Grieve 1990). Investigation of the variation in rating curves on a micro-scale in subcatchments of the Hubbard Brook Experimental Forest, New Hampshire, USA (Lawrence *et al* 1988; Lawrence & Driscoll 1990) has shown that the form of dissolved concentration–discharge relationships for parameters such as H^+, Al and dissolved organic carbon may vary markedly over short distances from the source of small forested streams, and this behaviour is explained by vertical variations in the subsurface flow paths involved and areal variations in soil solution chemistry and flow response.

Although the relationship between solute concentration and river discharge is reasonably well defined for many rivers, scattering of data points around the rating line does occur. Scatter may be reduced by subdividing data values on the basis of season (Foster 1978; Oborne *et al* 1980; Al-Jabbari *et al* 1983), stage conditions (Oxley 1974; Loughran & Malone 1976) and flow components (Walling 1974). In some catchments, however, the simple power function relationship is an oversimplified model of the general response of solute concentrations to changing flow conditions, and more complex segmented and polynomial functions (e.g. Steele 1968; Finlayson 1977; Foster 1980), and looped and trapezoidal rating plots (O'Connor 1976; Carbonnel & Meybeck 1975; Collins 1979) have to be adopted.

Storm-period responses

Complexity and variability of solute response to changing flow conditions is often apparent when fluctuations in water chemistry are examined during storm events (e.g. Miller & Drever 1977; Cornish 1982; Muraoka & Hirata 1988; Jenkins 1989). Intra- and intercatchment contrasts in storm-period response may be evident for particular dissolved constituents (e.g. Foster 1979b, Webb & Walling 1983; Reynolds *et al* 1989), and strong differences are often apparent between the chemographs of individual ions during flood events. The latter type of variability reflects the different origins and storage locations of different chemical species in a drainage basin and the extent to which they are accessed by runoff from different sources. Reid and colleagues (1981), for example, have shown in the Glendye catchment of north-east Scotland that dissolved species derived from chemical weathering (e.g. Ca^{2+}, HCO_3^-) are strongly diluted during storm events by waters originating in the upper layers of the soil, but that runoff from these surface horizons increases the concentration of Fe and related species in the river channel. The different flow pathways, their chemistry, and the residence time of water along them have been invoked to explain the episodic pattern of pH depression, which is associated with storm events in many acidic headwater streams. Monitoring of fluctuations of the major chemical constituents in the Svartberget catchment of northern Sweden (Fig. 4.3(b)), together with measurements of hillslope hydrology and chemistry, have suggested that the 'acid episodes' in the stream arise from the passage of runoff through organic-rich forest mor and streambank vegetation (Bishop *et al* 1990). Solute response may also vary between storm events for an individual chemical constituent at a particular river site. Such variation is often typical of streams dominated by atmospheric inputs of soluble material (Fig. 4.3(c)), where interstorm differences in precipitation chemistry strongly influence stream solute concentration (Dupraz *et al* 1982). Marked seasonal variation in storm-period response of nitrate-nitrogen concentrations have been recorded in the River Dart, Devon, UK, where

marked dilution occurs in winter months but concentration effects are typical of the summer period. This contrast in behaviour reflects the fact that in winter months storm runoff is low but baseflows are relatively high in nitrate-nitrogen, whereas in the summer period, the nitrate-nitrogen content of baseflow is low but that of storm runoff, which originates as through-flow in the soil, is relatively high (Webb & Walling 1985).

Complexity in solute behaviour during storm events is often associated with the occurrence of hysteretic effects, whereby chemical concentrations differ markedly at the same value of discharge on the rising and falling limbs of the hydrograph, and a looped trend in the relationship between concentration and flow results for an individual storm event (e.g. Hendrickson & Krieger 1964; Webb *et al* 1987; Meybeck *et al* 1989). Hysteresis reflects contrasts in solute and discharge responses during storm events with respect to both their timing and form (Walling & Webb 1986a), and this may be promoted by first and subsequent flushes of soluble material accumulated in a prestorm period (Walling & Foster 1975; Klein 1981; Muscutt *et al* 1990), especially after periods of prolonged drought (Slack 1977; Walling & Foster 1978), by the exhaustion of solute stores within a drainage basin during a sequence of flood hydrographs (Walling 1978), and by variations throughout a storm in the contribution of flows from vertically and areally differentiated catchment sources, which differ in their chemical characteristics (Burt 1979; Anderson & Burt 1982; Johnson & East 1982; Burt & Arkell 1986; Lawrence & Driscoll 1990). Additional complexity may arise through the modification of soil-water chemistry in its passage through the riparian zone to the stream channel (Stanford & Ward 1988; Fiebig *et al* 1990). Increasingly, stream-water chemistry is being viewed as the result of a variable mixture of soil-water end-members (Lawrence *et al* 1988; Christophersen *et al* 1990a), and Hooper and colleagues (1990), for example, have explained fluctuations in stream chemistry at the Panola Mountain catchment, Georgia, USA as a varying mixture of groundwater, water draining from the

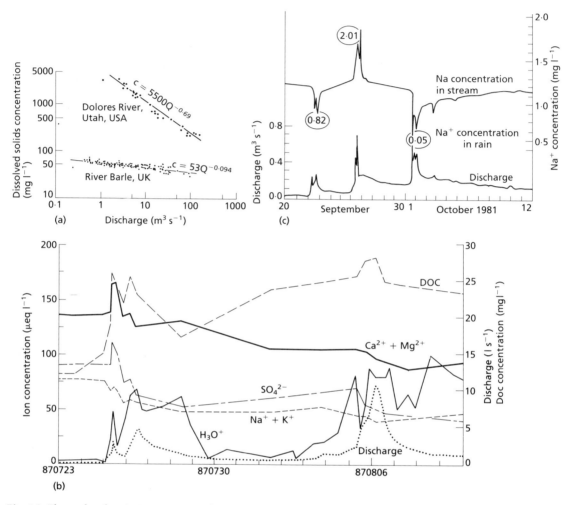

Fig. 4.3 Flow-related variations in river solute concentrations. (a) Solute concentration in discharge rating relationships (data for Dolores River; from Iorns *et al* 1965). (b) Storm-period changes in water chemistry in the Swedish catchment, Svartberget (after Bishop *et al* 1990). (c) Variation in storm-period solute response of Na⁺ according to precipitation chemistry in a French drainage basin; Valet des Cloutasses (after Dupraz *et al* 1982).

hillslope and flows from the organic horizon.

Storm-period solute responses may show additional complexity when large, rather than small, drainage basins are considered. Studies in the Exe Basin, Devon, UK (Walling & Webb 1980; Webb & Walling 1982a), have demonstrated that fluctuations in water chemistry at downstream sites during storm events are strongly affected by the spatial origins of storm runoff in the upstream catchment area and by the routing of flows through the channel network which results in a

kinematic differential between floodwave and floodwater velocities (Glover & Johnson 1974).

Time trends

Solute levels also exhibit variations through time which are not related to the occurrence of storm events. Marked diurnal oscillations have been recorded for total and individual solute concentrations (Walling 1975) and have been accounted for by changes in discharge, which are caused by increasing evaporation and a lowering of the water-table during the day and in turn promote a daily cycle of accumulation and redissolution of soluble material (Hem 1948). An investigation of diurnal variations of electrical conductivity in the Kassjoan basin in Central Sweden suggests that the effects of evapotranspiration may be supplemented by the influence of earth tides, which are assumed to affect fissures within the bedrock and in turn change the composition of groundwater contributing to river runoff (Calles 1982).

A clear annual cycle of solute behaviour is apparent for many rivers (e.g. Sutcliffe & Carrick 1973; Foster & Walling 1978; Houston & Brooker 1981; Neal *et al* 1990a) and can be readily identified when chemical concentrations of river samples collected throughout the year are plotted up against discharge in the form of 'elliptical doughnut' and 'Q-c-t' (Fig. 4.4(a)) diagrams (Gunnerson 1967; Davis & Keller 1983). In some catchments, the annual march of stream solute levels is a simple inverse reflection of the discharge regime, but in other rivers an annual hysteresis in concentrations will be present, reflecting such factors as autumn flushing of soluble material accumulated over summer months, spring release of meltwater from a winter snow-pack, and seasonal variations in the chemistry of incoming precipitation (Feller & Kimmins 1979; Williams *et al* 1983). The annual cycle of biological activity may also strongly influence water chemistry through the impact of autumnal vegetation dieback and leaf fall (Slack & Feltz 1968), the seasonal uptake by plants and animals (Edwards 1974; Casey & Ladle 1976; Casey *et al* 1981) and the effects of microbial populations on the reactions that build up and decompose organic matter in catchment soils (e.g. Blackie & Newson 1986; Stevens *et al* 1989). Dissolved organic matter concentrations may exhibit seasonal changes which are much more pronounced than those occurring during storm events. Results from the Loch Fleet catchment in south-west Scotland (Grieve 1990), for example, indicate storm-period variations in dissolved organic carbon of about 2 mg l^{-1} but a seasonal fluctuation with an amplitude of 8–9 mg l^{-1}. Dissolved organic carbon and dissolved organic matter often reach maximum levels in the summer and autumn

period, which can be explained by a concentration effect due to summer drying (Moore 1987) but also more strongly by increased decomposition of soil organic materials at higher temperatures during the summer period (Grieve 1990). The nature of the annual cycle of water chemistry may also vary in character from year to year. Stream concentrations of nitrate-nitrogen may show large contrasts between the autumn periods of different years, depending on the amount of accumulation in the catchment over the summer months and on the sequence by which soils are wetted up and nitrogen is mobilized to water-courses with the onset of the winter season (Webb & Walling 1985). Data on dissolved reactive phosphate concentrations, collected on a weekly basis over a 16-year period in the River Frome, Dorset, UK (Casey & Clarke 1986), although indicating a spring depression in levels also reveal large year-to-year variations in the annual regime of this nutrient (Fig. 4.4(b)).

Significant changes in solute concentrations may also be evident over longer time periods where records of sufficient quality and length are available to establish a trend. Smith and colleagues (1987) examined water quality trends in major US rivers over the period 1974–81, and found in the case of alkalinity (Fig. 4.4(c)), for example, that significant decreases were more common than significant increases. In general, it was observed that the reductions in alkalinity did not result from the introduction of strong acids associated with acid deposition from the atmosphere or with acid mine drainage, but rather reflected other processes occurring in the soils and vegetation of drainage basin, including the reduction of soil carbon dioxide and uptake of base cations during tree growth. A steady decline in calcium (1.1–1.4 µeq l^{-1} year^{-1}) and magnesium (0.6–0.8 µeq l^{-1} year^{-1}) have been recorded for the period 1972–87 in the Birkenes catchment of southern Norway and ascribed to soil acidification in the drainage area (Christophersen *et al* 1990b). Significant increases in nitrate-nitrogen concentrations have been monitored in the surface waters of many western European countries over the past 30 years (Fig. 4.4(d)) and an intensification of agricultural practice, involving land-use changes and the in-

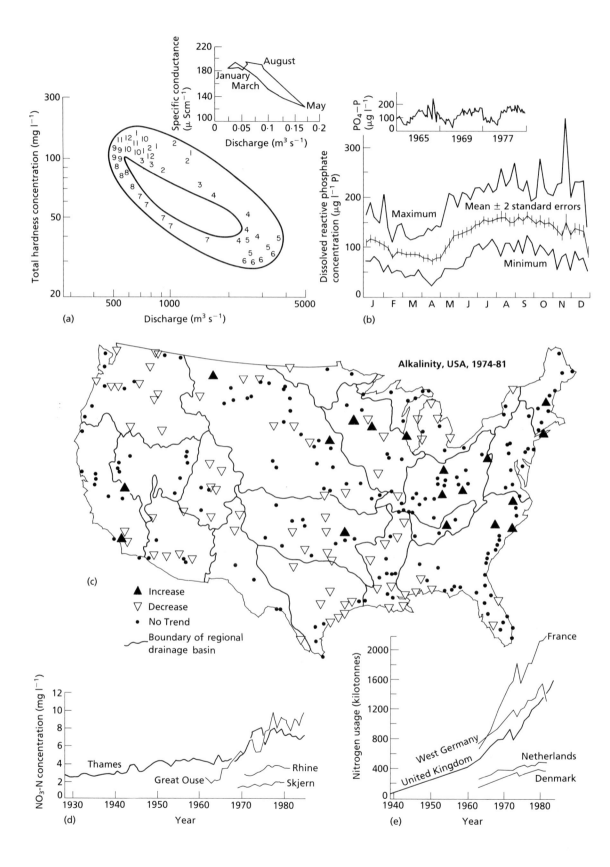

(a)

Specific conductance (μ Scm⁻¹)

January
March
August
May
Discharge (m³ s⁻¹)

Total hardness concentration (mg l⁻¹)

Discharge (m³ s⁻¹)

(b)

PO₄–P (μg l⁻¹)

1965   1969   1977

Dissolved reactive phosphate concentration (μg l⁻¹ P)

Maximum   Mean ± 2 standard errors

Minimum

J F M A M J J A S O N D

(c)

Alkalinity, USA, 1974-81

▲ Increase
▽ Decrease
• No Trend
— Boundary of regional drainage basin

(d)

NO₃-N concentration (mg l⁻¹)

Thames
Great Ouse
Rhine
Skjern

1930 1940 1950 1960 1970 1980
Year

(e)

Nitrogen usage (kilotonnes)

France
West Germany
United Kingdom
Netherlands
Denmark

1940 1950 1960 1970 1980
Year

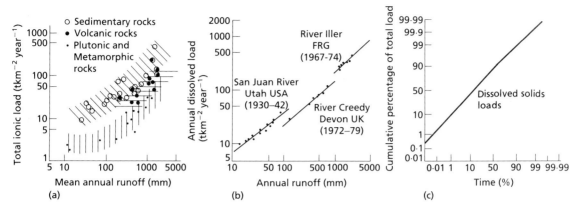

Fig. 4.5 Variations in river solute loads. (a) World rivers: the relationship between total ionic load and mean annual runoff for different rock types (from Meybeck 1981). (b) Selected catchments: the relationship between dissolved load and annual runoff for three rivers in different climatic zones. (c) River Creedy, UK: a dissolved load duration curve for a Devon river.

creased use of nitrogenous fertilizers (Fig. 4.4(e)), is identified as a major contributor to this trend (Dykzeul 1982; Hagebro *et al* 1983; Roberts & Marsh 1987; Burt *et al* 1988; Meybeck *et al* 1989). Analysis of changes in nitrate-nitrogen concentrations at 383 rivers across the USA over a shorter period (1974–81) also indicated widespread increases, but suggested that increased emissions of NO$_x$ gases may have played a large role in elevating stream nitrate-nitrogen levels in the midwestern and mid-Atlantic regions (Smith *et al* 1987).

Solute fluxes

Considerable quantities of dissolved material can be transported over time by a river system. On a global scale, it is estimated that total dissolved load transport to the oceans is 3.7×10^9 t year^{-1} (Walling & Webb 1987), although for individual rivers solute loads may range from <1.0 t km^{-2}

year^{-1} in catchments underlain by rocks resistant to chemical attack to values of 6000 t km^{-2} year^{-1} in basins underlain by halite deposits (Meybeck 1984). The mean annual solute load tends to increase with increasing mean annual runoff (Walling & Webb 1983), although there is considerable scatter in this relationship (Fig. 4.5(a)) which can be largely attributed to variation in basin geology (Meybeck 1981). Spatial patterns of solute transport become more complex and are explicable in terms of additional environmental factors, such as land use and topographic characteristics, when variations over more restricted geographical areas are considered (e.g. Walling & Webb 1981).

Solute flux also fluctuates over time as well as in space. A close relationship is often evident between annual dissolved load and annual runoff as results from three contrasted rivers indicate (Fig. 4.5(b)). Transport of solutes in river systems is generally less biased towards episodic and high

Fig. 4.4 (*Opposite*) Temporal trends in river solute concentrations. (a) Annual hysteresis in total hardness concentrations over period from October 1961 to September 1962 in the Snake River, Wawawai, USA (after Gunnerson 1967) (numbers 1–12 indicate calendar month during which sample was taken). (Inset depicts a 'Q-c-t' diagram which shows the annual cycle of specific conductance in Basin 3, Alptal, Switzerland; after Davis & Keller 1983.) (b) Variation in the annual regime of dissolved reactive phosphate concentration in a Dorset river; River Frome, UK, 1965–80 (after Casey & Clarke 1986). (c) Trends in flow-adjusted concentrations of alkalinity (1974–81) at NASQAN stations in the USA (after Smith *et al* 1987). (d) Long-term trends in nitrate-nitrogen concentrations of selected European rivers (after Dykzeul 1982; Hagebro *et al* 1983; Roberts & Marsh 1987). (e) Fertilizer usage in western Europe (after Roberts & Marsh 1987).

flows than is the case for particulate material. A long-term record from the River Creedy, Devon, UK (Fig. 4.5(c)) has indicated that 50% and 90% of solute flux occurred in 12% and 56%, respectively, of the 7.75-year study period (Webb & Walling 1982b). Storm events may influence the transport of dissolved organic material more strongly than that of dissolved inorganic constituents because flows from surface or near-surface horizons of the soil are most effective in mobilizing organic material in solution. Results from a small catchment in the Ochil Hills in eastern Scotland suggested that half of the annual load of dissolved organic matter is carried by flows occurring less than 6% of the time (Grieve 1984).

4.2 SEDIMENT-ASSOCIATED TRANSPORT

The preceding discussion of water chemistry and material transport by rivers relates to substances carried in solution. This is conventionally defined as material passing through a 0.45-μm filter and may in reality include colloidal material as well as true solutes. It is, however, important to emphasize that the suspended sediment carried by a river, which was discussed in section 3.1 and which is traditionally studied in isolation as a physical water-quality characteristic, also has important implications for water chemistry and material transport. Suspended sediment may exhibit a complex chemical composition both as a mineral or organic substrate and in terms of the substances 'attached' to or 'incorporated' into the sediment particles. Interactions between material in solution and the sediment may exert an important control on water chemistry. Thus, for example, sediment-associated substances may be released into solution and, conversely, substances in solution may be sorbed by sediment. In some circumstances, increases and decreases in concentration associated with the dissolved phase could therefore be buffered by uptake or release associated with the particulate phase. Any assessment of material transport by rivers must also consider both the sediment-associated and the dissolved components, because the former may dominate the transport of many substances. Förstner (1977) has, for example,

estimated that 98% of the Fe and Al transport by rivers in the USA and Germany occurs in association with the particulate load.

Sediment geochemistry and bonding mechanisms

The natural suspended sediment load of a river will comprise both inorganic and organic material, and a distinction may be made between the allochthonous and autochthonous components. The former component is commonly dominant and represents material washed into the river from the surrounding drainage basin, whereas the latter component comprises material formed in the water-body itself. The inorganic material will be largely allochthonous in origin and comprises clay minerals, mineral and rock fragments and precipitates (e.g. SiO_2 and $CaCO_3$). The organic matter consists of micro-organisms (phytoplankton, zooplankton and bacteria), the remains of macrophytes and other large-sized organisms, detritus derived from decaying material, and organic matter associated with eroded soil. A large proportion of the organic material may therefore be autochthonous in origin. The composition of individual particles may be complex, involving both organic and inorganic coatings of the particle substrate.

Bonding of substances to sediment occurs as a result of two major groups of processes, namely physical and chemical sorption (*cf.* Golterman *et al* 1983). The former primarily involves electrostatic attraction and ion exchange, whereas the latter includes exchange interactions with the OH groups of clay minerals and metal hydroxides and with the COOH and OH groups of organic substances, chelation, interactions with Fe/Mn oxyhydrates and coprecipitation.

Sediment–water quality interactions

With some sorption processes operating in natural waters, adsorption equilibria may be attained between suspended particulates and fine-bed material and certain dissolved components. These equilibria are commonly regulated by temperature, pH and the nature of the clay minerals and they can control the trace element concentration in the water. Thus, for example, Golterman *et al* (1983) indicated that if large quantities of zinc were to be released into a river to raise concentrations to in excess of 1 mg l^{-1}, these would return to the normal level of $10-100$ µg l^{-1}, within a downstream distance of 10 km. Similar equilibria have also been widely reported for phosphorus, such that a reduction in dissolved phosphate concentrations can trigger desorption and vice versa (e.g. Stumm & Leckie 1971). This behaviour can be quantified experimentally using adsorption isotherms (*cf.* Olsen & Watanabe 1957). In the case of phosphorus, Froelich (1988) suggested that the process of sorption and desorption by sediment involves two components that operate at different speeds. With sorption, the first 'fast' reaction occurs rapidly over timescales of minutes to hours and represents adsorption on to an exchangeable ion surface, whilst the second 'slow' reaction operates on timescales of days to months or perhaps years, and represents a slow diffusion into the particle. Because of this 'slow' second step, reversal of the process and therefore desorption may occur less readily.

As indicated above, sorption of substances on to sediment frequently operates as a reversible process. In some circumstances, however, sorbed or incorporated substances are more firmly fixed. Nevertheless, biological activity and changes in the river environment may cause release from the particulate to the dissolved phase and in this way the particulate phase may again influence the dissolved phase (*cf.* Table 4.2).

When considering both sediment bonding and interaction between the dissolved and particulate phase, the nature and composition of the sediment particles will exert an important control on their precise significance. In general, the 'chemical activity' of sediment increases with decreasing grain size due to the increased specific surface area (m^2g^{-1}) of fine clay particles (*cf.* Fig. 4.6(a)). Such fine particles will, for example, therefore exhibit greatly increased levels of cation exchange capacity (*cf.* Fig. 4.6(b)). Analyses of the copper content of several size fractions of suspended sediment from the River Amazon reported by Gibbs (1977) and presented in Fig. 4.6(c) demonstrate the preferential association of this element with the finest fractions of the sediment trans-

ported by this river. Similar results for the total concentration of a variety of major and trace elements in the bed sediments of the Niagara River in Canada have been reported by Mudroch and Duncan (1986), although in this case the cutoff for the finest fraction was much larger (13 µm). Variations in particle composition and the presence or absence of surface coatings may also be important. Thus in the case of phosphorus sorption, Froelich (1988) suggested that the initial 'fast' step is largely dependent on the surface area and charge balance of the sediment particles, whereas the second 'slow' diffusion step is very dependent on the composition of the solid phase. Sediments containing iron and aluminium oxides, or with surfaces coated with these substances, display a much higher 'slow' sorption capacity, probably due to the reaction of phosphate with the oxides. Thus 'pure' clays (e.g. kaolinite) have only a limited ability to sorb phosphorus beyond the initial step, while gibbsite or clays containing natural oxide coatings have a much higher capacity for phosphorus sorption. As noted in Section 3.1, the particle-size distribution and other properties of suspended sediment can be expected to vary temporally, in response to season and flow magnitude (*cf.* Fig. 4.6(d)), and these may in turn have important implications for sediment–water interactions.

Sediment-associated loads

When investigating catchment nutrient budgets or other aspects of material export from drainage basins, it is important to recognize the need to consider sediment-associated loads as well as the transport of material in solution. In some cases the former may represent a major proportion of the total load. For example, Duffy *et al* (1978) reported that the sediment phase accounts for $64-76\%$ of the total phosphorus yield from five small forested watersheds in north-central Mississippi, USA, and Schuman and colleagues (1976) indicated that sediment losses from small contour-cropped watersheds at Treynor, Iowa, USA account for 85% of the total discharge of phosphorus. Meybeck (1984) has also emphasized the important role of suspended sediment in geochemical budgets and in the overall flux of el-

Table 4.2 Fate of particulate phases when submitted to various environmental changes

Environmental changes	Electrostatically adsorbed	Specifically adsorbed	Bound to particulate organic matter	Bound to carbonates	Occluded in Fe, Mg oxides and hydroxide	Bound to sulphides	Silicates and other residuals
Bacterial degradation	0	0	−	0	0	+ anaerobic − aerobic	0
Establishment of oxidizing conditions	0	0	0	0	+	−	0
Establishment of reducing conditions	0	0	0	0	−	+	0
Small pH variations	+/−	+/−	0	+/−	0	+/−	0, −(1)
Transfer to brackish or saline water	−	0	0	0	0	0	0
Intestinal tract of organisms after ingestion	−	−	−	−	−	−	0

0, no marked change; −, release from particulate matter to dissolved phase; +, gain from dissolved phase to particulate phase; +/−, release or gain; (1), in the case of diatoms. Based on Unesco (1978).

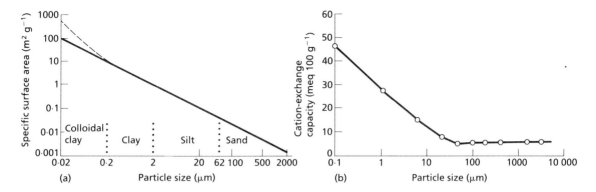

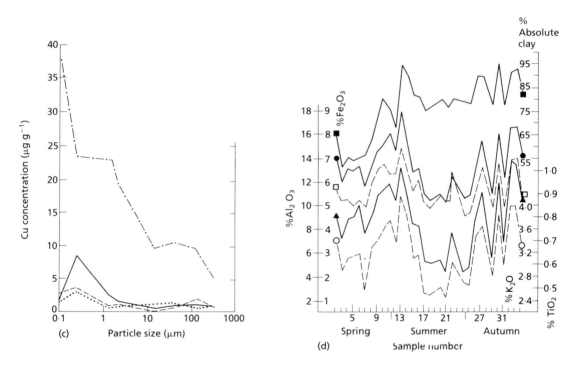

Fig. 4.6 The significance of the particle-size composition of suspended sediment in controlling (a) specific surface area. (b) Cation exchange capacity, Mattole River, California, USA. (c) Trace metal content in the River Amazon. —·—· crystal; —— coat; ---- organic; ······ exchange. (d) Seasonal variation of the chemistry and clay content of suspended sediment transported by Wilton Creek, Ontario, Canada. ■ Clay; ○ Al_2O_3; ▲ TiO_2; ● Fe_2O_3; □ K_2O. (After (b) Malcolm & Kennedy 1970; (c) Gibbs 1977; and (d) Ongley *et al* 1981).

ements from the land surface of the globe to the oceans. He estimated the typical elemental composition of suspended sediment transported by world rivers and used this to calculate the likely contribution of sediment-associated transport to the total land−ocean flux of individual elements. Some of his results presented in Table 4.3 provide information on the typical composition of suspended sediment and indicate, for example, that suspended sediment probably accounts for more than 90% of the transport of Al, Fe, Ti, Mn, Si and P, and a major proportion of the flux of several other elements. Suspended sediment also accounts for a major proportion of the transport of organic substances, such as organic nitrogen and organic phosphorus. Such estimates can clearly be only approximate at the present time, since detailed data on sediment composition are still lacking for many rivers of the world, but their general order of magnitude is likely to be correct.

Because of the episodic nature of high sus-pended sediment loads, sediment-associated transport also tends to be episodic and dominated by storm events. For this reason it is important that sampling programmes are designed carefully, to document the main periods of transport. Programmes of infrequent regular sampling are unlikely to provide meaningful load data (*cf.* Walling & Webb 1985). The need to collect substantial quantities of sediment for chemical analysis may also necessitate the use of specialized equipment such as a continuous flow centrifuge, which permits the recovery of sediment from bulk water samples (e.g. Horowitz *et al* 1989).

When evaluating sediment-associated loads of specific substances, it is frequently useful to assemble information on the types of sediment bonding involved. This could, for example, be particularly important in studies of sediment−water quality interactions, when it is necesary to predict the potential interchange between the particulate and dissolved phases. Equally, when

Table 4.3 Role of suspended sediment in the global transport of major elements from the land to the oceans by rivers (after Meybeck 1984).

Element	Sediment-associated transport (%) Total transport	Typical sediment content $(mg\ g^{-1})$
Al	99.87	90.0
Fe	99.8	51.5
Ti	99.6	5.8
Mn	98	1.0
Si	96	281
P	95	1.15
K	86.5	21.0
F	80	0.97
B	60	0.07
N	57	1.2
Mg	56	11.0
Ca	43	24.5
C	34	20.0
Sr	32	0.15
Na	28	7.1
S	13	2.15
Cl	1	0.5
Inorganic C	28	10.0
Organic C	45	10.0
Organic N	68	1.2
Organic P	94	0.45
Inorganic P	97	0.70

investigating the fate of sediment deposited in a lake or other depositional sinks it could be important to determine whether a substance may be released from the sediment or whether it is permanently bound and therefore unlikely to participate in future biogeochemical cycling. Bioavailability assessments assume particular importance in the case of toxic contaminants, but such pollutants lie beyond the scope of this discussion.

To take the example of sediment-associated phosphorus, fractionation procedures exist to separate the apatite phosphorus (A-P), the non-apatite inorganic phosphorus (NAI-P) and the organic phosphorus (O-P) (*cf.* Allan 1979). The A-P may be viewed as mineral phosphorus associated with rock particles and this is essentially inert and unavailable for biological uptake or dissolved-particulate phase interactions. The O-P includes all phosphorus in organic forms and may have both autochthonous and allochthonous origins. This is not immediately available for biological uptake, but is a pool of potentially available phosphorus. Bacterial degradation of algae will, for example, release orthophosphorus. NAI-P consists of orthophosphorus adsorbed by or occluded within sediment particles and is the fraction which is readily bioavailable. Other workers have developed procedures for further fractionating the NAI-P, for example into loosely sorbed phosphorus and Fe- and Al-bound phosphorus (*cf.* Pettersson 1986). Similar fractionation procedures exist for determining the speciation of sediment-associated metals (e.g. Tessier *et al* 1979) and Table 4.4 demonstrates their potential utility. Some of these procedures have been used by

Gibbs (1977) to fractionate the copper content of suspended sediment transported by the River Amazon, as shown in Fig. 4.6(c). Where interest focuses primarily on the potential bioavailability of substances such as phosphorus, rather than their precise chemical speciation, bioassay techniques may also be employed (*cf.* Lee *et al* 1980; de Pinto *et al* 1981). In this case the crystalline fraction may be viewed as essentially inert, whereas the other fractions are more readily available, although to different degrees.

Temporal variations in sediment sources and erosional processes can be expected to cause variations in both the total content of specific sediment-associated substances and the relative importance of the various forms in which they are held within the sediment. To use again the example of phosphorus, Fig. 4.7 demonstrates seasonal variation in both the total-P content of suspended sediment and the contribution of O-P, A-P and NAI-P to the total-P content, as reported by Ongley (1978) for Wilton Creek, Ontario, Canada. A-P does not demonstrate a marked seasonal trend but reaches its highest concentrations in spring when the erosion associated with spring melt contributes large quantities of detrital material. The O-P content of suspended sediment evidences a marked seasonal cycle and rises rapidly in the spring and maintains high values throughout the summer in response to increased algal productivity. NAI-P similarly increases through the summer, probably reflecting the increased residence time of sediment within the channel during periods of low flow.

Table 4.4 Sequential extraction procedures used for the fractionation of sediment-associated trace metals

Extraction method	Extracted phase
H_2O	Easily soluble fraction
$BaCl_2/TEA$ pH 8.1	Exchangeable cations
NaOH	Humates, fulvates
Acidic cation exchange	Carbonate fraction
$NH_2OH/HCl/HNO_3$	Mn-oxides, amorphous Fe-oxides
H_2O_2/NH_4OAc	Organic residues and sulphides
$NH_2OH/HCl/$acetic acid	Hydrous Fe-oxides
$HF/HClO_4$ digestion	Inorganic residues

After Allan (1979).

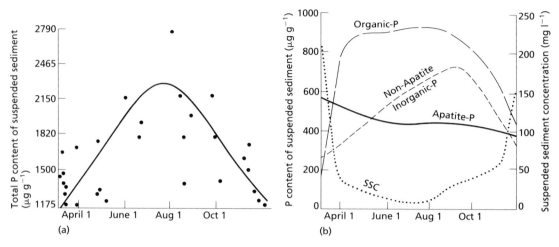

Fig. 4.7 Seasonal variation of (a) the total-P content of suspended sediment, and (b) sediment-associated P forms in Wilton Creek, Ontario, Canada (based on data from Ongley 1978, reported in Allan 1979).

REFERENCES

Afifi AA, Bricker OP. (1983) Weathering reactions, water chemistry and denudation rates in drainage basins of different bedrock types: 1 — Sandstone and shale. In: *Dissolved Loads of Rivers and Surface Water Quantity/Quality Relationships* (Proceedings of the Hamburg Symposium, August 1983), pp 193–203. International Association of Hydrological Sciences Publication No. 141. Wallingford, UK. [4.1]

Al-Jabbari MH, Al-Ansari NA, McManus J. (1983) Variation in solute concentration within the River Almond and its effect on the estimated dissolved load. In: *Dissolved Loads of Rivers and Surface Water Quantity/Quality Relationships* (Proceedings of the Hamburg Symposium, August 1983), pp 21–9. International Association of Hydrological Sciences Publication No. 141. Wallingford, UK. [4.1]

Allan RJ. (1979) Sediment-related fluvial transmission of contaminants: some advances by 1979. *Inland Waters Directorate, Environment Canada, Scientific Series No. 107*. Ottawa, Canada. [4.2]

American Public Health Association (1971) *Standard methods for the examination of water and waste water* 13th edn. APHA, Washington, DC. [4, 4.1]

Anderson MG, Burt TP. (1982) The contribution of throughflow to storm runoff: an evaluation of a chemical mixing model. *Earth Surface Processes and Landforms* 7: 565–74. [4.1]

Andrews ED. (1983) Denudation of the Piceance Creek Basin, Colorado. In: *Dissolved Loads of Rivers and Surface Water Quantity/Quality Relationships* (Proceedings of the Hamburg Symposium, August 1983), pp 205–15. International Association of Hydro-

logical Sciences Publication No. 141. Wallingford, UK. [4.1]

Baker CD, Bartlett PD, Farr IS, Williams GI. (1974) Improved methods for the measurement of dissolved and particulate organic carbon in fresh water and their application to chalk streams. *Freshwater Biology* 4: 467–81. [4.1]

Belot Y, Caput C, Gauthier C. (1982) Transfer of americium from sea water to atmosphere by bubble bursting. *Atmosphere and Environment* 16: 1463–6. [4.1]

Best GR, Monk CD. (1975) Cation flux in hardwood and white pine watersheds. In: Howell FG, Gentry JB, Smith MH (eds) *Mineral Cycling in South Eastern Ecosystems*, pp 847–61. United States Energy Research and Development Administration, Washington, DC. [4.1]

Bishop KH, Grip H, O'Neill A. (1990) The origins of acid runoff in a hillslope during storm events. In: Neal C, Hornung M (eds) *Special Issue: Transfer of Elements through the Hydrological Cycle. Journal of Hydrology* 116: 35–61. [4.1]

Blackie JR, Newson MD. (1986) The effects of forestry on the quantity and quality of runoff in upland Britain. In: Solbe JF de LG (ed.) *Effects of Land Use on Fresh Waters*, pp 398–412. Ellis Horwood Series in Water and Wastewater Technology, Water Research Centre Publication, Chichester. [4.1]

Bloch MRD, Kaplan V, Kertes V, Schnerb J. (1966) Ion separation in bursting air bubbles: an explanation for the irregular ion ratios in atmospheric precipitations. *Nature* 209: 802–3. [4.1]

Bond HW. (1979) Nutrient concentration patterns in a stream draining a montane ecosystem in Utah.

Ecology **60**: 1184–96. [4.1]

Bruijnzeel LA. (1983) The chemical mass balance of a small basin in a wet monsoonal environment and the effect of fast-growing plantation forest. In: *Dissolved Loads of Rivers and Surface Water Quantity/Quality Relationships* (Proceedings of the Hamburg Symposium, August 1983), pp 229–39. International Association of Hydrological Sciences Publication No. 141. Wallingford, UK. [4.1]

Buijsman E, Mass HFM, Asman WAH. (1987) Anthropogenic NH₃ emissions in Europe. *Atmosphere and Environment* **21**: 1009–22. [4.1]

Burt TP. (1979) The relationship between throughflow generation and the solute concentration of soil and stream water. *Earth Surface Processes* **4**: 257–66. [4.1]

Burt TP (1986) Runoff processes and solutional denudation rates on humid temperate hillslopes. In: Trudgill ST (ed.) *Solute Processes*, pp 193–249. John Wiley and Sons, Chichester, UK. [4.1]

Burt TP, Arkell BP. (1986) Variable source areas of stream discharge and their relationship to point and non-point sources of nitrate pollution. In: *Monitoring to Detect Changes in Water Quality Series* (Proceedings of the Budapest Symposium, July 1986), pp 155–64. International Association of Hydrological Sciences Publication No. 157. Wallingford, UK. [4.1]

Burt TP, Arkell BP, Trudgill ST, Walling DE. (1988) Stream nitrate levels in a small catchment in southwest England over a period of 15 years (1970–1985). *Hydrological Processes* **2**: 267–84. [4.1]

Calles UM. (1982) Diurnal variations in electrical conductivity of water in a small stream. *Nordic Hydrology* **13**: 157–64. [4.1]

Carbonnel JP, Meybeck M. (1975) Quality variations of the Mekong River at Phnom Penh, Cambodia, and chemical transport in the Mekong Basin. *Journal of Hydrology* **27**: 249–65. [4.1]

Carling PA. (1983) Particulate dynamics, dissolved and total load, in two small basins, northern Pennines, UK. *Hydrological Sciences Journal* **28**: 355–75. [4.1]

Casey H, Clarke RT. (1986) The seasonal variation of dissolved reactive phosphate concentrations in the River Frome, Dorset, England. In: *Monitoring to Detect Changes in Water Quality Series* (Proceedings of the Budapest Symposium, July 1986), pp 257–65. International Association of Hydrological Sciences Publication No. 157. Wallingford, UK. [4.1]

Casey H, Ladle M. (1976) Chemistry and biology of the South Winterbourne, Dorset, England. *Freshwater Biology* **6**: 1–12. [4.1]

Casey H, Clarke RT, Marker AFH. (1981) The seasonal variation in silicon concentration in chalk-streams in relation to diatom growth. *Freshwater Biology* **11**: 335–44. [4.1]

Christophersen N, Neal C, Hooper RP, Vogt RD, Andersen S. (1990a) Modelling streamwater chemistry as a mixture of soilwater end-members—a step towards second-generation acidification models. In: Neal C, Hornung M (eds) *Special Issue: Transfer of Elements through the Hydrological Cycle. Journal of Hydrology* **116**: 307–20. [4.1]

Christophersen N, Robson A, Neal C, Whitehead PG, Vigerust B, Henriksen A. (1990b) Evidence for long-term deterioration of streamwater chemistry and soil acidification at the Birkenes catchment, southern Norway. In: Neal C, Hornung M (eds) *Special Issue: Transfer of Elements through the Hydrological Cycle. Journal of Hydrology* **116**: 63–76. [4.1]

Collins DN. (1979) Hydrochemistry of meltwaters draining from an alpine glacier. *Arctic and Alpine Research* **11**: 307–24. [4.1]

Colombani J. (1983) Evolution de la concentration en matières dissolutes en Afrique. Deux exemples opposés: les fleuves du Togo et la Medjerdah en Tunisie. In: *Dissolved Loads of Rivers and Surface Water Quantity/Quality Relationships* (Proceedings of the Hamburg Symposium, August 1983), pp 51–69. International Association of Hydrological Sciences Publication No. 141. [4.1]

Cornish PM. (1982) The variations of dissolved ion concentration with discharge in some New South Wales streams. In: O'Loughlin EM, Bren LJ (eds) *The First National Symposium on Forest Hydrology*, Melbourne, 11–13 May 1982, pp 67–71.

Cryer R. (1976) The significance and variation of atmospheric nutrient inputs in a small catchment system. *Journal of Hydrology* **29**: 121–37. [4.1]

Cryer R. (1980) The chemical quality of some pipeflow waters in upland mid-Wales and its implications. *Cambria* **6**: 28–46. [4.1]

Cryer R. (1986) Atmospheric solute inputs. In: Trudgill ST (ed.) *Solute Processes*, pp 15–84. John Wiley and Sons, Chichester, UK. [4.1]

Davis JS, Keller HM. (1983) Dissolved loads in streams and rivers—discharge and seasonally related variations. In: *Dissolved Loads of Rivers and Surface Water Quantity/Quality Relationships* (Proceedings of the Hamburg Symposium, August 1983), pp 79–89. International Association of Hydrological Sciences Publication No. 141. Wallingford, UK. [4.1]

De Pinto JV, Young TC, Martin SC. (1981) Algal-available phosphorus in suspended sediments from lower Great Lakes tributaries. *Journal of Great Lakes Research* **7**: 311–25. [4.2]

Duffy PD, Schreiber JD, McClurkin DC, McDowell LL. (1978) Aqueous and sediment phase phosphorus yield from five southern pine watersheds. *Journal of Environmental Quality* **7**: 45–50. [4.2]

Dupraz C, Lelong F, Trop JP, Dumazet B. (1982) Comparative study of the effects of vegetation on the hydrological and hydrochemical flows in three minor catchments of Mount Lozère (France) — methodological aspects and first results. In: *Hydrological Research Basins and their Use in Water Resources Planning*, pp 671–82. Landeshydrologies, Berne. [4.1]

Dykzeul A. (1982) *The water quality of the River Rhine in the Netherlands over the period 1970–81*. Government Institute for Waste Water Treatment, Report No. 82–061, pp 43–6. [4.1]

Edwards AMC. (1973) The variation of dissolved constituents with discharge in some Norfolk rivers. *Journal of Hydrology* 18: 219–42. [4.1]

Edwards AMC. (1974) Silicon depletions in some Norfolk rivers. *Freshwater Biology* 4: 267–74. [4.1]

Edwards RW, Gee AS, Stoner JH. (1990) *Acid Waters in Wales*. Kluwer Academic Publishers, Dordrecht, Netherlands. [4.1]

Eriksson E. (1960) The yearly circulation of chloride and sulfur in nature; meteorological, geochemical and pedological implications, part II. *Tellus* 12: 63–109. [4.1]

Feller MC, Kimmins JP. (1979) Chemical characteristics of small streams near Haney in Southwestern British Columbia. *Water Resources Research* 15: 247–58. [4.1]

Ferrier RC, Walker TAB, Harriman R, Miller JD, Anderson HA. (1990) Hydrological and hydrochemical fluxes through vegetation and soil in the Allt a'Mharcaidh, Western Cairngorms, Scotland: their effect on streamwater quality. In: Neal C, Hornung M (eds) *Special Issue: Transfer of Elements through the Hydrological Cycle*. *Journal of Hydrology* 116: 251–66. [4.1]

Fiebig DM, Lock MA, Neal C. (1990) Soil water in the riparian zone as a source of carbon for a headwater stream. In: Neal C, Hornung M (eds) *Special Issue: Transfer of Elements through the Hydrological Cycle*. *Journal of Hydrology* 116: 217–37. [4.1]

Finlayson B. (1977) *Runoff contributing areas and erosion*. Research Papers, School of Geography, University of Oxford, No. 18. [4.1]

Foggin GT, Forcier LK. (1977) *Using topographic characteristics to predict total solute concentrations in streams draining small forested watersheds in western Montana*. University of Montana Joint Water Resources Research Center Report No. 89. [4.1]

Förstner U. (1977) Trace metals. In: Shear H, Watson AEP (eds) *The Fluvial Transport of Sediment-Associated Nutrients and Contaminants*, pp 219–33. International Joint Commission on the Great Lakes, Windsor, Ontario. [4.2]

Foster IDL. (1978) Seasonal solute behaviour of stormflow in a small agricultural catchment. *Catena* 5: 151–63. [4.1]

Foster IDL. (1979a) Chemistry of bulk precipitation throughfall, soil water and stream water in a small catchment in Devon, England. *Catena* 6: 145–55. [4.1]

Foster IDL. (1979b) Intra-catchment variability in solute response, an East Devon example. *Earth Surface Processes* 4: 381–94. [4.1]

Foster IDL. (1980) Chemical yields in runoff, and denudation in a small arable catchment, East Devon, England. *Journal of Hydrology* 47: 349–68. [4.1]

Foster IDL, Grieve IC. (1984) Some implications of small catchment solute studies for geomorphological research. In: Burt TP, Walling DE (eds) *Catchment Experiments in Fluvial Geomorphology*, pp 359–98. GeoBooks, Norwich. [4.1]

Foster IDL, Walling DE. (1978) The effects of the 1976 drought and autumn rainfall on stream solute levels. *Earth Surface Processes* 3: 393–406. [4.1]

Froelich PN. (1988) Kinetic control of dissolved phosphate in natural rivers and estuaries: a primer on the phosphate buffer mechanism. *Limnology and Oceanography* 33: 649–68. [4.2]

Garrels RM, Christ CL. (1965) *Solutions, Minerals and Equilibria*. Harper and Row, New York. [4.1]

Gibbs R. (1970) Mechanisms controlling world water chemistry. *Science* 170: 1088–90. [4.1]

Gibbs RJ. (1977) Transport phases of transition metals in the Amazon and Yukon Rivers. *Geological Society of America Bulletin* 88: 829–43. [4.2]

Glover BJ, Johnson P. (1974) Variations in the natural chemical concentrations of river water during flood flows and the lag effect. *Journal of Hydrology* 22: 303–16. [1.4]

Golterman HL, Clymo RS (eds) (1969) Methods for chemical analysis of fresh waters. *International Biological Programme Handbook 8*. [4]

Golterman HL, Sly PG, Thomas RL. (1983) *Study of the Relationship Between Water Quality and Sediment Transport*. Unesco Technical Papers in Hydrology No. 26, Unesco, Paris. [4.2]

Gregory KJ, Walling DE. (1973) *Drainage Basin Form and Process: A Geomorphological Approach*. Edward Arnold, London. [4.1]

Grieve IC. (1984) Concentrations and annual loading of dissolved organic matter in a small moorland stream. *Freshwater Biology* 14: 533–7. [4.1]

Grieve IC. (1990) Seasonal, hydrological, and land management factors controlling dissolved organic carbon concentrations in the Loch Fleet catchments, southwest Scotland. *Hydrological Processes* 4: 231–9. [4.1]

Gunnerson CG. (1967) Streamflow and quality in the Columbia River Basin. *Proceedings American Society of Civil Engineers, Journal of the Sanitary Engineering Division* 93 (SA6): 1–16. [4.1]

Hagebro C, Bang S, Somer E. (1983) Nitrate load/discharge relationships and nitrate load trends in Danish rivers. In: *Dissolved Loads of Rivers and Surface Water Quantity/Quality Relationships* (Proceedings of the Hamburg Symposium, August 1983), pp 377–86. International Association of Hydrological Sciences

Publication No. 141. Wallingford, UK. [4.1]

Hem JD. (1948) Fluctuations in concentrations of dissolved solids in some southwestern streams. *Transactions of the American Geophysical Union* **29**: 80–4. [4.1]

Hem JD. (1970) Study and interpretation of the chemical characteristics of natural water. *United States Geological Survey Water-Supply Paper 1473*. [4.1]

Hendrickson GE, Krieger RA. (1964) Geochemistry of natural waters of the Blue Grass Region, Kentucky. *United States Geological Survey Water-Supply Paper 1700*. [4.1]

Hooper RP, Christophersen N, Peters NE. (1990) Modelling streamwater chemistry as a mixture of soilwater end-members — an application to the Panola Mountain catchment, Georgia, USA. In: Neal C, Hornung M (eds) *Special Issue: Transfer of Elements through the Hydrological Cycle. Journal of Hydrology* **116**: 321–43. [4.1]

Horowitz AJ, Elrick KA, Hooper RC. (1989) A comparison of instrumental dewatering methods for the separation on concentration of suspended sediment for subsequent trace element analysis. *Hydrological Processes* **3**: 163–84. [4.2]

Houston JA, Brooker MP. (1981) A comparison of nutrient sources and behaviour in two lowland subcatchments of the River Wye. *Water Research* **15**: 49–57. [4.1]

Hu M, Stallard RF, Edmond JM. (1982) Major ion chemistry of some large Chinese rivers. *Nature* **298**: 550–3. [4.1]

Iorns WV, Hembree CH, Oakland GL. (1965) Water resources of the Upper Colorado basin. *United States Geological Survey Professional Paper 441*. [4.1]

Jenkins A. (1989) Storm period hydrochemical response in an unforested Scottish catchment. *Hydrological Sciences Journal* **34**: 393–404. [4.1]

Johnson FA, East JW. (1982) Cyclical relationships between river discharge and chemical concentration during flood events. *Journal of Hydrology* **57**: 93–106. [4.1]

Johnson NM, Likens GE, Bormann FH, Fisher DW, Pierce RS. (1969) A working model for the variation in stream water chemistry at the Hubbard Brook Experimental Forest, New Hampshire. *Water Resources Research* **5**: 1353–63. [4.1]

Junge CE, Werby RT. (1958) The concentration of chloride, sodium, potassium, calcium and sulfate in rain water over the United States. *Journal of Meteorology* **15**: 417–25. [4.1]

Klein M. (1981) Dissolved material transport — the flushing effect in surface and subsurface flow. *Earth Surface Processes and Landforms* **6**: 173–8. [4.1]

Lawrence GB, Driscoll CT. (1990) Longitudinal patterns of concentration — discharge relationships in stream water draining the Hubbard Brook Experimental Forest, New Hampshire. In: Neal C, Hornung M (eds) *Special Issue: Transfer of Elements through the Hydrological Cycle. Journal of Hydrology* **116**: 147–65. [4.1]

Lawrence GB, Driscoll CT, Fuller RD. (1988) Hydrologic control of aluminium chemistry in an acidic headwater stream. *Water Resources Research* **24**: 659–69. [4.1]

Lee GF, Jones RA, Rast W. (1980) Availability of phosphorus to phytoplankton and its implications for phosphorus management strategies. In: Loehr RC, Martin CS, Rast W (eds) *Phosphorus Management Strategies for Lakes*. Ann Arbor Science Publishers. [4.2]

Lewis WM, Grant MC. (1979) Changes in the output of ions from a watershed as a result of the acidification of precipitation. *Ecology* **60**: 1093–7. [4.1]

Likens GE, Bormann FH, Pierce RS, Eaton JS, Johnson NM. (1977) *Biogeochemistry of a Forested Ecosystem*. Springer-Verlag, New York. [4.1]

Loughran RJ, Malone KJ. (1976) *Variations in some stream solutes in a Hunter Valley catchment*. Research Papers in Geography, University of Newcastle, New South Wales, No. 8. [4.1]

Malcolm RL, Kennedy VC. (1970) Variation of cation exchange capacity and rate with particle size in stream sediment. *Journal of the Water Pollution Control Federation* **2**: 153–60. [4.2]

Meybeck M. (1976) Total dissolved transport by world major rivers. *Hydrological Sciences Bulletin* **21**: 265–84. [4.1]

Meybeck M. (1981) Pathways of major elements from land to ocean through rivers. In: *River Inputs to Ocean Systems*, pp 18–30. UNEP/Unesco Report. [4.1]

Meybeck M. (1982) Carbon, nitrogen and phosphorus transport by world rivers. *American Journal of Science* **282**: 401–50. [4, 4.1]

Meybeck M. (1983) Atmospheric inputs and river transport of dissolved substances. In: *Dissolved Loads of Rivers and Surface Water Quantity/Quality Relationships* (Proceedings of the Hamburg Symposium, August 1983), pp 173–92. International Association of Hydrological Sciences Publication No. 141. Wallingford, UK. [4.1]

Meybeck M. (1984) *Les fleuves et le cycle geochimique des éléments*. Thèse de Doctorat d'Etat Université Pierre et Marie Curie, Paris. [4.1, 4.2]

Meybeck M. (1986) Composition chimique des ruisseaux non pollués de France. *Sci. Géol. Bull.* **39**: 3–77. [4.1]

Meybeck M, Helmer R. (1989) The quality of rivers: from pristine stage to global pollution. *Palaeogeography, Palaeoclimatology, Palaeoecology (Global and Planetary Change Section)* **75**: 283–309. [4.1]

Meybeck M, Chapman D, Helmer R. (eds) (1989) *Global Freshwater Quality: A First Assessment*. Blackwell Reference, Oxford. [4.1]

Miller JP. (1961) Solutes in small streams draining single rock types, Sangre de Cristo Range, New Mexico. *United States Geological Survey Water-Supply Paper 1535-F*. [4.1]

Miller WR, Drever JI. (1977) Water chemistry of a stream following a storm, Absaroka Mountains, Wyoming. *Geological Society of America Bulletin* **88**: 286–90. [4.1]

Möller D. (1984) Estimation of the global man-made sulphur emission. *Atmosphere and Environment* **18**: 19–27. [4.1]

Moore TR. (1987) Patterns of dissolved organic matter in sub-Arctic peatlands. *Earth Surface Processes and Landforms* **12**: 387–97. [4.1]

Mudroch A, Duncan GA. (1986) Distribution of metals in different size fractions of sediment from the Niagara River. *Journal of Great Lakes Research* **12**: 117–26. [4.2]

Munger JW, Eisenreich SJ. (1983) Continental-scale variations in precipitation chemistry. *Environmental Science and Technology* **17**: 32A–42A. [4.1]

Muraoka K, Hirata T. (1988) Streamwater chemistry during rainfall events in a forested basin. *Journal of Hydrology* **102**: 235–53. [4.1]

Muscutt AD, Wheater HS, Reynolds B. (1990) Stormflow hydrochemistry of a small Welsh upland catchment. In: Neal C, Hornung M (eds) *Special Issue: Transfer of Elements through the Hydrological Cycle. Journal of Hydrology* **116**: 239–49. [4.1]

Neal C, Hornung, M. (1990) Special Issue: Transfer of Elements through the Hydrological Cycle. *Journal of Hydrology* **116**. [4.1]

Neal C, Smith CJ, Walls J, Billingham P, Hill S, Neal M. (1990a) Hydrogeochemical variations in Hafren forest stream waters, mid-Wales. In: Neal C, Hornung M (eds) *Special Issue: Transfer of Elements through the Hydrological Cycle. Journal of Hydrology* **116**: 185–200. [4.1]

Neal C, Mulder J, Christophersen N *et al* (1990b) Limitations to the understanding of ion-exchange and solubility controls for acidic Welsh, Scottish and Norwegian sites. In: Neal C, Hornung M (eds) *Special Issue: Transfer of Elements through the Hydrological Cycle. Journal of Hydrology* **116**: 11–23. [4.1]

Oborne AC, Brooker MP, Edwards RW. (1980) The chemistry of the River Wye. *Journal of Hydrology* **52**: 59–70. [4.1]

O'Connor DJ. (1976) The concentration of dissolved solids and river flow. *Water Resources Research* **12**: 279–94. [4.1]

Olsen SR, Watanabe FS. (1957) A method to determine a phosphorus adsorption isotherm of soils as measured by the Langmuir adsorption isotherm. *Soil Science Society of America Proceedings* **21**: 144–9. [4.2]

Omernik JM. (1977) *Nonpoint source — stream nutrient level relationships: a nationwide study*. EPA-600/ 3–77–105. Corvallis Environmental Research Laboratory, United States Environmental Protection Agency, Corvallis, Oregon. [4.1]

Omernik JM, Powers CF. (1983) Total alkalinity of surface waters — a national map. *Annals of the Association of American Geographers* **73**: 133–6. [4.1]

Ongley ED. (1978) *Sediment-related phosphorus, trace metals and anion flux in two rural southeastern Ontario catchments*. Inland Waters Directorate Research Subvention Program Progress Report (unpublished). [4.2]

Ongley ED, Bynoe MC, Percival JB. (1981) Physical and geochemical characteristics of suspended solids, Wilton Creek, Ontario. *Canadian Journal of Earth Science* **18**: 1365–79. [4.2]

Oxley NC. (1974) Suspended sediment delivery rates and the solute concentration of stream discharge in two Welsh catchments. In: Gregory KJ, Walling DE (eds) *Fluvial Processes in Instrumented Watersheds*, pp 141–54. Institute of British Geographers Special Publication No. 6. [4.1]

Pereira WE, Rostad CE, Updegraff DM, Bennett JL. (1987) Anaerobic microbial transformations of azaarenes in groundwater at hazardous-waste sites. In: Averett RC, McKnight DM (eds) *Chemical Quality of Water and the Hydrologic Cycle*, pp 111–23. Lewis Publishers, Chelsea, Michigan. [4]

Pettersson K. (1986) The fractional composition of phosphorus in lake sediments of different characteristics. In: Sly PG (ed.) *Sediments and Water Interactions*, pp 149–155. Springer-Verlag, New York. [4.2]

Reid JM, MacLeod DA, Cresser MS. (1981) Factors affecting the chemistry of precipitation and river water in an upland catchment. *Journal of Hydrology* **50**: 129–45. [4.1]

Reynolds B, Pomeroy AB. (1988) Hydrogeochemistry of chlorine in an upland catchment in mid-Wales. *Journal of Hydrology* **99**: 19–32. [4.1]

Reynolds B, Hornung M, Hughes S. (1989) Chemistry of streams draining grassland and forest catchments at Plynlimon, mid-Wales. *Hydrological Sciences Journal* **34**: 667–86. [4.1]

Roberts G, Marsh T. (1987) The effects of agricultural practices on the nitrate concentrations in the surface water domestic supply sources of Western Europe. In: *Water for the Future: Hydrology in Perspective* (Proceedings of the Rome Symposium, April 1987), pp 365–80. International Association of Hydrological Sciences Publication No. 164. Wallingford, UK. [4.1]

Roberts G, Hudson JA, Blackie JR. (1983) Nutrient cycling in the Wye and Severn at Plynlimon. *Institute of Hydrology Report* No. 86. Wallingford, UK. [4.1]

Rodà F, Avila A, Bonilla D. (1990) Precipitation, throughfall, soil solution and streamwater chemistry in a holm-oak (*Quercus ilex*) forest. In: Neal C,

Hornung M (eds) *Special Issue: Transfer of Elements through the Hydrological Cycle. Journal of Hydrology* **116**: 167–83. [4.1]

Rosenqvist ITh. (1990) From rain to lake: water pathways and chemical changes. In: Neal C, Hornung M (eds) *Special Issue: Transfer of Elements through the Hydrological Cycle. Journal of Hydrology* **116**: 3–10. [4.1]

Royal Society (1983) *The Nitrogen Cycle of the United Kingdom.* Report of a Royal Society Study Group, London. [4.1]

Sanders WM III. (1972) Nutrients. In: Oglesby RT, Carlson CA, McCann JA (eds) *River Ecology and Man*, pp 389–415. Academic Press. New York. [4]

Schuman GE, Piest RF, Spomer RG. (1976) Physical and chemical characteristics of sediment originating from Missouri valley loess. In: *Proceedings of the Third Federal Interagency Sedimentation Conference,* 3-28–3-40. US Water Resources Council, Washington, DC. [4.2]

Seip HM, Tollan A. (1985) Acid Deposition. In: Rodda JC (ed.) *Facets of Hydrology Volume II*, pp 69–98. John Wiley & Sons, Chichester, UK. [4.1]

Simpson EA. (1980) The harmonization of the monitoring of the quality of rivers in the United Kingdom. *Hydrological Sciences Bulletin* **25**: 13–23. [4]

Skougstad MW, Fishman MJ, Friedman LC, Erdmann OE, Duncan SS. (1979) Methods for the determination of inorganic substances in water and fluvial sediments. *United States Geological Survey Techniques of Water-Resources Investigations 5*, Chapter A1. [4]

Slack JG. (1977) River water quality in Essex during and after the 1976 drought. *Effluent and Water Treatment Journal* **17**: 575–8. [4.1]

Slack KV, Feltz HR. (1968) Tree leaf control on lowflow water quality in a small Virginia stream. *Environmental Science and Technology* **2**: 126–31. [4.1]

Smith RA, Alexander RB, Wolman MG. (1987) Analysis and interpretation of water-quality trends in major US rivers, 1974–81. *United States Geological Survey Water-Supply Paper 2307.* [4.1]

Stallard RF, Edmond JM. (1981) Geochemistry of the Amazon 1. Precipitation chemistry and the marine contribution to the dissolved load at the time of peak discharge. *Journal of Geophysical Research* **86** (C10): 9844–58. [4.1]

Stallard RF, Edmond JM. (1983) Geochemistry of the Amazon 2. The influence of geology and weathering environment on the dissolved load. *Journal of Geophysical Research* **88** (C14): 9671–88. [4.1]

Stanford JA, Ward JV. (1988) The hyporheic habitat of river ecosystems. *Nature* **355**: 64–5. [4.1]

Steele TD. (1968) Digital-computer applications in chemical quality studies of surface water in a small watershed. *Publications de l'Association Internationale d'Hydrologie Scientifique* **80**: 203–14. [4.1]

Stevens PA, Adamson JK, Anderson MA, Hornung M. (1988) Effects of clearfelling on surface water quality and site nutrient status. In: Usher MB, Thomson DBA (eds) *Ecological Changes in the Uplands*, pp 289–93. Special Publication M. British Ecological Society, Blackwell, Oxford. [4.1]

Stevens PA, Hornung M, Hughes S. (1989) Solute concentrations in a mature sitka spruce plantation in Beddgelert forest, North Wales. *For. Ecol. Management* **27**: 1–20. [4.1]

Stevenson CM. (1968) An analysis of the chemical composition of rain-water and air over the British Isles and Eire for the years 1959–1964. *Royal Meteorological Society Quarterly Journal* **94**: 56–70. [4.1]

Stumm W, Leckie JO. (1971) Phosphate exchange with sediment: its role in the productivity of surface waters. *Advances in Water Pollution Research* **2** (3): 26/1–26/16. [4.2]

Sutcliffe DW, Carrick TR. (1973) Studies on mountain streams in the English Lake District II. Aspects of water chemistry in the River Duddon. *Freshwater Biology* **3**: 543–60. [4.1]

Swank WT. (1986) Biological control of solute losses from forest ecosystems. In: Trudgill ST (ed.) *Solute Processes*, pp 85–139. John Wiley & Sons, Chichester, UK. [4.1]

Swank WT, Douglass JE. (1977) Nutrient budgets for undisturbed and manipulated hardwood forest ecosystems in the mountains of North Carolina. In: Correll DL (ed.) *Watershed Research in Eastern North America: A Workshop to Compare Results*, pp 343–63. Smithsonian Institution, Edgewater, Maryland. [4.1]

Talsma T, Hallam PM. (1982) Stream water quality of forest catchments in the Cotter Valley, ACT. In: O'Loughlin EM, Bren LJ (eds) *The First National Symposium on Forest Hydrology*, Melbourne, 11–13 May 1982, pp 50–9. [4.1]

Taylor HE. (1987) Analytical methodology for the measurement of the chemical composition of snow cores from the Cascade/Sierra Nevada mountain ranges. In: Averett RC, McKnight DM (eds) *Chemical Quality of Water and the Hydrologic Cycle*, pp 55–69. Lewis Publishers, Chelsea, Michigan. [4]

Tessier A, Campbell P, Bisson M. (1979) Sequential extraction procedure for the speciation of particulate trace metals. *Analytical Chemistry* **51**: 844–51. [4.2]

Unesco (1978) *Monitoring of particulate matter quality in rivers and lakes. Recommendations of a workshop on the assessment of particulate matter contamination in rivers and lakes.* Unesco (GEMS/WATER; Med-IX) BUD/Report 1. [4.2]

United Kingdom Acid Waters Review Group (1988) *Acidity in United Kingdom Fresh Waters.* Second report of the UK Acid Waters Review Group. Her

Majesty's Stationery Office, London. [4.1]

Varheyli G. (1985) Continental and global sulphur budgets 1: anthropogenic SO_2 emissions. *Atmosphere and Environment* **19**: 1029–40. [4.1]

Walling D.E. (1974) Suspended sediment and solute yields from a small catchment prior to urbanization. In: Gregory KJ, Walling DE (eds) *Fluvial Processes in Instrumented Watersheds*, pp 169–92. Institute of British Geographers Special Publication No. 6. [4.1]

Walling DE. (1975) Solute variations in small catchment streams: some comments. *Transactions of the Institute of British Geographers* **64**: 141–7. [4.1]

Walling DE. (1978) Suspended sediment and solute response characteristics of the River Exe, Devon, England. In: Davison-Arnott R, Nickling W (eds) *Research in Fluvial Geomorphology*, pp 169–97. GeoBooks, Norwich. [4.1]

Walling DE. (1980) Water in the catchment ecosystem. In: Gower AM (ed.) *Water Quality in Catchment Ecosystems*, pp 2–47. John Wiley and Sons, Chichester, UK. [4.1]

Walling DE (1984) Dissolved loads and their measurement. In: Hadley RF, Walling DE (eds) *Erosion and Sediment Yield: Some Methods of Measurement and Modelling*, pp 111–77. GeoBooks, Norwich. [4]

Walling DE, Foster IDL. (1975) Variations in the natural chemical concentration of river water during flood flows, and the lag effect: some further comments. *Journal of Hydrology* **26**: 237–44. [4.1]

Walling DE, Foster IDL. (1978) The 1976 drought and nitrate levels in the River Exe Basin. *Journal of the Institution of Water Engineers and Scientists* **32**: 341–52. [4.1]

Walling DE, Webb BW. (1980) The spatial dimension in the interpretation of stream solute behaviour. *Journal of Hydrology* **47**: 129–49. [4.1]

Walling DE, Webb BW. (1981) Water quality. In: Lewin J (ed.) *British Rivers*, pp 126–69. George Allen & Unwin, London. [4.1]

Walling DE, Webb BW. (1983) The dissolved loads of rivers: a global overview. In: *Dissolved Loads of Rivers and Surface Water Quantity/Quality Relationships* (Proceedings of the Hamburg Symposium, August 1983), pp 3–20. International Association of Hydrological Sciences Publication No. 141. Wallingford, UK. [4.1]

Walling DE, Webb BW. (1985) Estimating the discharge of contaminants to coastal waters by rivers. *Marine Pollution Bulletin* **16**: 488–92. [4.2]

Walling DE, Webb BW. (1986a) Solutes in river systems. In: Trudgill ST (ed.) *Solute Processes*, pp 251–327. John Wiley & Sons, Chichester, UK. [4.1]

Walling DE, Webb BW. (1986b) Solute transport by rivers in arid environments: an overview. *Journal of Water Resources* **5**: 800–22. [4.1]

Walling DE, Webb BW. (1987) Material transport by the world's rivers: evolving perspectives. In: *Water for the Future: Hydrology in Perspective* (Proceedings of the Rome Symposium, April 1987), pp 313–29. International Association of Hydrological Sciences Publication No. 164. Wallingford, UK. [4.1]

Waylen MJ. (1979) Chemical weathering in a drainage basin underlain by Old Red Sandstone. *Earth Surface Processes* **4**: 167–78. [4.1]

Webb BW. (1983) Factors influencing spatial variation of background solute levels in a Devon river system. *Reports and Transactions of the Devonshire Association for the Advancement of Science Literature and Arts* **115**: 51–69. [4.1]

Webb BW, Walling DE. (1982a) Catchment scale and the interpretation of water quality behaviour. In: *Hydrological Research Basins and their use in Water Resources Planning*, pp 759–70. Landeshydrologie, Berne. [4.1]

Webb BW, Walling DE. (1982b) The magnitude and frequency characteristics of fluvial transport in a Devon drainage basin and some geomorphological implications. *Catena* **9**: 9–23. [4.1]

Webb BW, Walling DE. (1983) Stream solute behaviour in the River Exe basin, Devon, UK. In: *Dissolved Loads of Rivers and Surface Water Quantity/Quality Relationships* (Proceedings of the Hamburg Symposium, August 1983), pp 153–69. International Association of Hydrological Sciences Publication No. 141. Wallingford, UK. [4.1]

Webb BW, Walling DE. (1985) Nitrate behaviour in streamflow from a grassland catchment in Devon, UK. *Water Research* **19**: 1005–16. [4.1]

Webb BW, Davis JS, Keller HM. (1987) Hysteresis in stream solute behaviour. In: Gardiner V (ed.) *International Geomorphology 1986 Part I*, pp 767–82. John Wiley & Sons, Chichester, UK. [4.1]

Webster JR, Patten BC. (1979) Effects of watershed perturbation on stream potassium and calcium dynamics. *Ecological Monographs* **49**: 51–72. [4.1]

White EM. (1981) Nutrient contents of precipitation and canopy throughfall under corn, soybeans and oats. *Water Resources Bulletin* **17**: 708–12. [4.1]

Williams AG, Ternan JL, Kent M. (1983) Stream solute sources and variations in a temperate granite drainage basin. In: *Dissolved Loads of Rivers and Surface Water Quantity/Quality Relationships* (Proceedings of the Hamburg Symposium, August 1983), pp 299–310. International Association of Hydrological Sciences Publication No. 141. Wallingford, UK. [4.1]

Young JR, Ellis EC, Hidy GM. (1988) Deposition of airborne acidifiers in the western environment. *Journal of Environmental Quality* **17**: 1–26. [4.1]

5: In-stream Hydraulics and Sediment Transport

P. A. CARLING

5.1 INTRODUCTION

Rivers are linear features and so it might be supposed that the flow hydraulics would be uni-directional and relatively simple to describe. To some extent this is true. Except for the largest of the world rivers, the effects of wind-waves and geostrophic circulation can be ignored, as can periodically reversing flow which dominates tidal reaches. However, to complicate matters, rivers are characterized by relatively rapid changes in flow rate and consequently can be regarded as stationary in behaviour only over short time spans. The relative shape and roughness of the channel may change as water levels fluctuate, and the twisting and turning of meandering channels means only short reaches can be considered as linear conduits.

Despite complexities, the relatively two-dimensional nature of many rivers has meant that engineers have been able to describe the flow structure present at any one moment by two-dimensional or even one-dimensional flow models with sufficient accuracy for most practical applications. However, all rivers are characterized by turbulent flow, and large rivers in particular have a complex three-dimensional structure that can be well described only by sophisticated mathematical modelling, scaled by laboratory data obtained using delicate apparatus. The latter are usually not suitable or are difficult and expensive to use in real rivers.

It is clear, then, that there are a number of levels of increasing complexity at which the natural river system can be described. It is very important to decide at which level of complexity observations should be made to obtain answers to problems and to consider whether genuine understanding can be obtained at any one level. For example, at the most basic level a good correlation may be obtained between the behaviour of an organism and mean current speed when in fact it is the level of turbulence intensity that is really controlling behaviour. However, moving to a higher level of explanation may not be productive if too many assumptions are required to produce a 'working' model of hydraulic structure. For many applications, sophisticated explanation is neither necessary nor cost effective. As Peters and Goldberg (1989) observed, average data may well describe the environment, and standards exist that assume temporal stability of simple phenomena (Gore 1978; Newbury 1984). Peters and Goldberg (1989) further noted that the field scientist frequently 'has to rely on rough, robust apparatus, often lacking sensitivity' and consequently it is important to consider what can realistically be achieved when designing any field monitoring or experimental programme. The tools available must be capable of providing at least a degree of insight into the problem of interest.

With these limitations in mind, a considerable degree of understanding can be obtained by those who appreciate the complexities of natural flow structure even if they are unable to describe fully many phenomena mathematically. To this end, this chapter aims to provide a practical approach to dealing with the intricacy of hydraulics and sediment transport within the context of recent research related to real rivers, and to emphasize those methods that are likely to be fruitful in practical applications. It is not possible within the space available to describe all procedures fully but details can be found in the references cited.

5.2 THE NEAR-BED BOUNDARY LAYER

Fluvial currents are driven by gravitational gradients, either imposed by the nature of the terrain or as modified by fluvial erosion and deposition and the quantity of water delivered to the channel. In turn, the structure of the flow is mediated by the friction induced by the channel boundary. In deep rivers the region where the frictional effects are felt, the *boundary layer*, may occupy only a small proportion of the total depth whilst in shallow rivers it may extend to the water surface. Laminar (non-turbulent) flow never occurs throughout the full depth in natural rivers, so that it is turbulence which transfers frictional forces throughout the fluid and redistributes suspended particles. Further, turbulence intensity mediates the momentary level of shear stress exerted on the boundary which may result in the movement of bed and bank sediments so modifying the shape and capacity of the river channel. A consideration of the velocity structure is therefore of prime importance.

Turbulent flow may be divided into smooth, transitional and rough hydrodynamic regimes. Such a consideration is required to select appropriate equations to describe the velocity structure. For flat sand-beds the division may be given by considering the roughness Reynolds number, a non-dimensional ratio defined by the shear velocity $(u_*,$ defined below), the grain size (D) and the kinematic viscosity (v):

Smooth turbulent: $u_* D/v < 3.5$ (1a)

Transitional: $3.5 < u_* D/v < 68$ (1b)

Rough turbulent: $u_* D/v > 68$ (1c)

The ratio expresses the balance of inertial and viscous forces. Where bed roughness is due to gravel or ripples, for example, then D (the grain size) needs replacing by some other characteristic roughness length $(k_s;$ Table 5.1). The Reynolds number, with the length defined by flow depth, will be referred to again in section 5.3.

The time-averaged velocity (denoted as U, ignoring the usual over-bar) usually increases from zero at the bed to the free-stream velocity (U_∞) at the edge of the boundary layer where the water is sufficiently deep to exceed the boundary layer thickness (Fig. 5.1(a)). However, in shallow flow the boundary layer may extend close to the surface (Fig. 5.1(b)). The theoretical structure of the flow can be divided into sublayers. The layer closest to the bed, the *bed layer*, is usually thin, often only a few millimetres thick. In slow flow over smooth beds (such as clay) it may be termed the *laminar sublayer*. In natural flows, however, the laminar nature may be disrupted. Local distortions in the velocity profile and turbulence levels then exist within the bed layer. The theoretical thickness of the laminar layer (δ) in *ideal* flow can be estimated using the relationship:

$$\delta = 11.5 v/u_*$$ (2)

The value of the constant can vary between 8 and 20 (Chriss & Caldwell 1983) but in rough turbulent flow over a gravel bed, where the calculated thickness is only millimetres, the layer is

Table 5.1 Assessment of probable hydrodynamic regime

Bed type	k_s (cm)	u_* (cm s^{-1})	$u_* k_s/v$	Regime
Smooth mud	0.006	1.2	0.5	Smooth
Smooth sand	0.03	2.2	5	Transitional
Dunes	15	2.8	3000	Rough
Flat gravel bed	1.5	3.6	400	Rough
Rocks	>30	4.6	$>10^4$	Rough

k_s is the equivalent roughness, u_* is the shear velocity, and v is the kinematic viscosity (0.013 cm^2 s^{-1}, 10°C, freshwater). Reproduced with modification from Soulsby (1983).

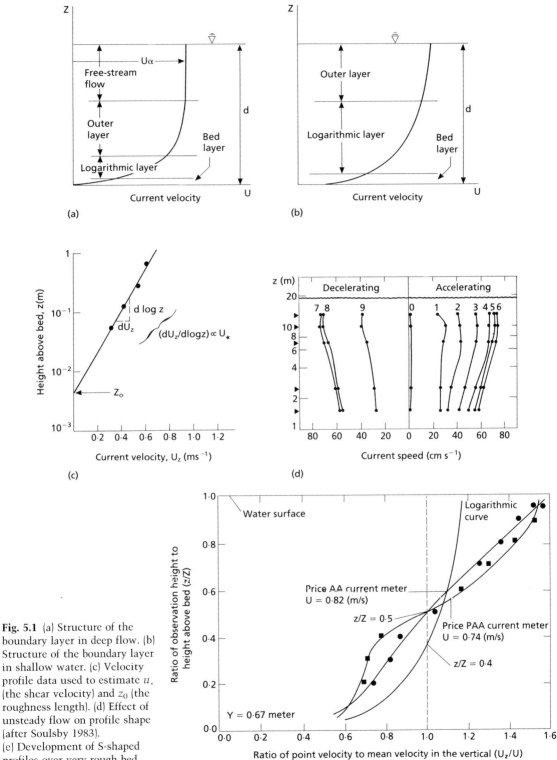

Fig. 5.1 (a) Structure of the boundary layer in deep flow. (b) Structure of the boundary layer in shallow water. (c) Velocity profile data used to estimate u_* (the shear velocity) and z_0 (the roughness length). (d) Effect of unsteady flow on profile shape (after Soulsby 1983). (e) Development of S-shaped profiles over very rough bed (after Jarrett 1990).

disrupted and often absent. In general, if the bed roughness value is greater than the calculated thickness of δ, then the latter is absent.

Many ecological texts argue that invertebrates are adapted morphologically to live within a laminar layer (see references in Statzner & Holm 1982), but Carling (1991) has argued that in many streams invertebrates that venture out of the interstitial environment are subject to low current speed but *high shear stress and turbulence levels*, a point made by Décamps and co-workers (1972, 1975) but largely ignored in contemporary literature.

Above the bed layer is the *logarithmic layer* (Figs 5.1(a) & 5.1(b)), the basic form of which is neither affected by the local roughness of the bed nor by the free-stream flow structure. As its name implies, the structure can be described by logarithmic functions. This layer is very important, as measurements taken within it allow estimation of the shear stress acting on the bed. Usually it extends over 10–15% of the depth and may extend to the surface. To estimate the boundary shear stress, it is important not to obtain current readings in the outer layer (Figs 5.1(a) & 5.1(b)), where the flow may be non-logarithmic. The usual relationship for the log-layer in the rough turbulent regime is:

$$U_z = (u_*/\kappa) \ln (z/z_0) \qquad (3)$$

where von Kármán's constant (κ) equals 0.40, and the roughness length (z_0) scales with the roughness of the river bed (Table 5.1). An estimate of z_0 can be obtained assuming $z_0 = D/30$ for sand or $D/15$ for well-sorted gravel, whilst the roughness of periodic bed-forms depends on height and spacing (e.g. Wooding *et al* 1973). Estimates of u_* and z_0 can be obtained from a regression analysis of current speed (U_z) against the height (z) above the bed (Fig. 5.1(c)). The shear stress (τ_0) is related to u_* as $\tau_0 = \rho u_*^2$. However, profiles should be plotted to ensure that the data conform to the logarithmic model; otherwise incorrect values of u_* and z_0 may be obtained.

The structure of the outer layer may be influenced by the free-stream velocity and consequently may not be described by any universal relationship. It is worth noting that the common practice of assuming that the depth-averaged vel-

ocity in the profile exists at 0.6 of the depth assumes that the logarithmic profile extends to the surface (Walker 1988).

Although Fig. 5.1(c) is a true representation of an idealized velocity profile, the logarithmic profile may be distorted by such effects as acceleration or deacceleration (Fig. 5.1(d)), extreme bed roughness (dunes or large rocks; Jarrett 1990), or bank drag which may suppress the filament of maximum velocity to below the water surface (Fig. 5.2). In the case of large-scale roughness an S-shaped profile may be present with a logarithmic section close to the bed and a further log-profile at some distance from the bed (Fig. 5.1(e)). If this reflects the influence of two scales of roughness (Dyer 1971), respectively that induced by the size of the bed material and that owing to the size of larger projections such as dunes, then each section of the log-profile may be treated separately to estimate the frictional effect of both roughness scales (Paris 1989). However, such an approach is open to criticism and where the height of the bed roughness is of the same order as the water depth no accepted theory exists to describe the vertical current structure.

5.3 BULK FLOW IN STRAIGHT CHANNELS

Theoretical consideration of bulk flow divides the flow properties into three categories: uniform, gradually varied and rapidly varied flow. Uniform flow is constant in depth, so the pressure distribution remains constant. Consequently the streamlines, water surface and bed profile are all parallel. Gradually varied flow is typified by gradual changes in cross-section and hence depth, so that streamlines may diverge or converge. The water surface and bed may not be parallel everywhere and the pressure distribution varies. Rapidly varied flow is typified by such phenomena as hydraulic drops and is common locally in steep, rough-bedded mountain streams (Whittaker 1987).

In many rivers, section properties change only slowly so that locally the flow can be treated as uniform. In addition, flow may be considered as steady (no change in discharge or velocity with time) or unsteady. Examples are the constant compensation flow immediately downstream of a reservoir which is dependent on a steady release of water from a reservoir, and the varying discharge owing to the passage of a storm-generated flood wave through the channel. If flow does not change too rapidly with respect to the time necessary to sample it, then it is often acceptable to treat it as if steady conditions pertain.

There are a number of fundamental non-dimensional ratios in basic hydraulics. Two important ones describe the state of flow. These are the Reynolds number (*Re*) and the Froude number (*Fr*). A form of the *Re* has already been defined in equation 1. Often the characteristic length is expressed as *d* (the depth) or as *R* (the hydraulic radius, i.e. sectional area/section perimeter) which is appropriate when width to mean depth ratios are less than 13. Using this definition of length, in natural rivers laminar flow exists when $Re < 500$ and fully turbulent flow exists when $Re > 2500$.

Fr expresses the relationship between the inertial force and the swiftness of a gravity wave:

$$Fr = U/\sqrt{gd} \tag{4}$$

where *g* is the acceleration due to gravity. If $Fr < 1.0$ the flow is regarded as subcritical (tranquil) and where $Fr > 1.0$ the flow is supercritical (rapid). In the latter case surface waves appear and break repeatedly in an upstream direction. Supercritical flow in most rivers only occurs locally. Even in mountain torrents the reach-averaged Froude numbers rarely exceed 0.6 (Jarrett 1984).

If uniform flow exists, the driving force (the component of the weight of the water acting downslope) is balanced by the resisting force (the bed resistance). If the resisting force varies as the square of velocity:

$$\tau_0 = kU^2 \tag{5}$$

then the balance of forces may be written as:

$$U^2 = (\rho g/k)RS \tag{6}$$

(where τ_0 is the shear stress, *U* is the depth-averaged velocity, ρ is the density of water, *g* is the acceleration due to gravity, *R* is the hydraulic radius, *S* is the energy slope and the square root

of the term in parentheses is often termed the Chezy coefficient) or as:

$$\tau_0 = \rho g R S = \gamma R S \tag{7}$$

where γ is the specific weight of water. This indicates that the bed shear stress is balanced by the product of the unit weight of water, the hydraulic radius and the bed slope.

The friction associated with the flow resistance can be expressed by the Darcy–Weisbach friction factor:

$$f = 8(\sqrt{gRS}/U)^2 = 8(u_*/U)^2 \tag{8}$$

which is dimensionless. In equation 8, u_* is the shear velocity. The Chezy C coefficient can be regarded as a form of resistance coefficient and alternative formulations may be used such as the Manning number. These may be related one to another (e.g. Richards 1982).

Equations 7 and 8 express the total shear resistance imposed by the boundary over a given reach and give no information as to how this is distributed locally around the perimeter. The local shear resistance associated with individual velocity profiles may differ across the bed and needs to be integrated to yield an average term. Variation in roughness across the bed and between bed and banks can induce significant variation across the section such that the velocity structure of straight channels is distorted.

This observation highlights the fact that natural flow is three dimensional. In natural channels with width to depth ratios greater than about 13, the effects of the banks are not felt in the central two-thirds of the flow. However, as channels narrow and deepen the importance of bank resistance cannot be ignored. Roughening the walls of a hydraulic flume alters the flow structure, depressing the maximum filament of velocity to below the surface. The same effect is seen in natural channels (Fig. 5.2) and strong upwelling and secondary flow cells develop along the banks (Bhomik 1982). Once the river goes over-bank the effect may intensify with momentum transfer distorting the flow structure in the vicinity of the banks. In wide channels, the flow may be characterized by more than one flow cell (Fig. 5.2), each exhibiting a degree of secondary 'corkscrew' spiralling flow (Fig. 5.3).

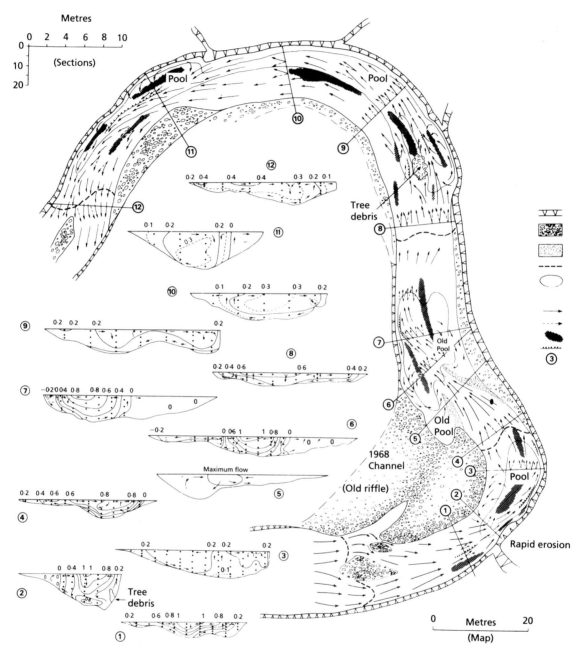

Fig. 5.2 Channel morphology, secondary flows and cross-section isovels through a meander loop. Velocities in m s^{-1} (after Hooke & Harvey 1983).

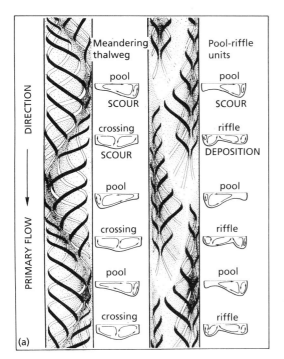

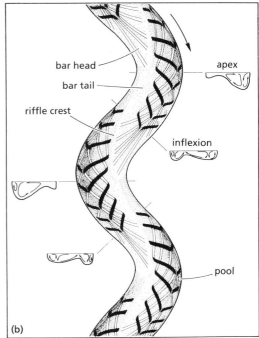

Fig. 5.3 Models of flow structure in (a) straight and (b) meandering channels (after Thompson 1986).

The resistance to flow is generated by the nature and scale of the boundary material (Table 5.2). The relationship is not simple, however, and large-scale distortion of the channel alignment, such as bank protrusions or large bars on the bed, will add resistance to flow over that associated with a flat bed of sand or gravel. Where the resulting flow distortion is gross then energy dissipation will be intense, resulting in additional flow resistance.

It is usual to recognize four kinds of flow resistance: the *skin* (or grain) *resistance* (f') associated with the individual sand or gravel particles making up the bed; the *form resistance* (f'') associated with the presence of ripples or dunes; the *internal distortion resistance* associated with banks and changes in channel alignment; the *spill resistance*, a localized phenomenon associated with rapid flow, for example between and over boulders in a mountain stream. In this latter case Froude numbers are high (>0.5) and the roughness coefficient is a function of the Froude number. For low Froude numbers, the resistance

factors sum to give the total roughness, i.e.:

$$f = f' + f'' + \dots$$

It is possible to determine the relative contributions of skin and form roughness by considering pressure distributions (Shen *et al* 1990), but in the field it is usually accomplished by analysing compound velocity profiles or by estimating the skin roughness (Prestegaard 1983) using a relative roughness relationship such as proposed by Limerinos (1970):

$$1/\sqrt{f'} = 1.16 + 2 \log (R/D_{84}) \tag{9}$$

where D_{84} is the 84th percentile at the coarse end of a size distribution of the bed material. In principle, in straight, wide channels the difference between this estimate of f' and a field measure of the total resistance is an estimate of the form of drag contribution. The skin friction factor is of fundamental importance when estimating the threshold shear stress required to initiate sediment transport.

In straight uniform channels there is important

Table 5.2 Typical values of the roughness length (z_0) (see text for more details)

Bed type	z_0 (cm)
Mud	0.02
Mud/sand	0.07
Silty sand	0.005
Flat-bed sand	0.04
Rippled sand	0.60
Small dunes	0.12
Large gravelly dunes	<0.50
Sand/gravel	0.03
Mud/sand/gravel	0.03
Fine gravel	0.30
Medium gravel	0.65
Coarse gravels	1.64
Slumped banks	2.42
Moss bed	<5.00
Prone ranunculus	<20.00

Data are from a variety of sources including Soulsby (1983), Bridge and Jarvis (1977), Dyer (1970, 1972) and unpublished data of author. Values are means. Standard errors of estimates are typically up to 1.0, indicating wide variation induced, for example, in mixtures where fines can infill surface void space smoothing rougher textures.

variation in the velocity and flow resistance across the section (Fig. 5.3(a)). If the flow is competent to move sediment (scour and fill), then a degree of spatial sorting may occur, changing the local roughness characteristic. This in turn can be exaggerated by aquatic vegetation growing in the more suitable habitats, and by the presence of discontinuities in bank alignment or the presence of tributary junctions. These areas may have well-developed secondary flow cells with reverse flow flowing up-stream along the bank (see Fig. 5.2). Where the river is wide enough, remote sensing can readily demonstrate this variability (Plate 5.1, opposite). The instigation of scour and fill, altering the topography, will inevitably result in a gradual variation in flow from one section to another.

A good example of gradually varied flow is provided by pool and riffle sequences. Straight channels are frequently characterized by deeper sections called *pools* and shallow sections termed *riffles* (Fig. 5.4(a)). There is some debate about how these are initiated but, once formed, they are hydrodynamically stable and regularly spaced along the river (typically every three to ten widths; Ferguson 1981). The surface sediment in riffles is often coarser than that in pools, which may have a blanket of fines on the bottom (Hirsch & Abrahams 1981). During low stage, riffles have faster currents over them than the pools owing to the reduced cross-sectional area (Carling 1991). In some streams, riffles are wider than pools so that the flow diverges. A central shallow bar may develop with deep channels impinging against and eroding the banks (Hooke 1986). This divergence and convergence may induce secondary spiralling flow and complex shear stress distribution patterns on the bed. As the discharge increases the relative difference in pool and riffle cross-sectional area reduces so that the mean velocities and mean shear stress in each section begin to converge (Fig. 5.4(b)). Convergence usually occurs close to bankfull, but local variations in the shear stress in pool or riffle may persist.

5.4 BULK FLOW IN CURVED CHANNELS

In a straight channel, variation in the strength of secondary circulation and the cross-channel velocity component results in a meandering zone of bedload transport (Hey & Thorne 1975) which can lead to the deposition of alternate channel-side bars (see Fig. 5.3(a)) and the instigation of meandering (Leopold 1982). Flow through a bend is three dimensional (Okoye 1989), and although a useful description of flow structure can be obtained using impeller current meters (Bridge & Jarvis 1977), for a full description the secondary flow cells need to be defined by using electromagnetic flowmeters. Typically, the flow is dominated by a large cell directing fast-flowing surface water towards the outer bank and slower near-bed water towards the inner bank (point bar) (Figs 5.2 & 5.3(b)). A small inner cell of opposite polarity may occur close to the outer bank (Thorne & Hey 1979; Thorne & Rais 1984). Such complex flow results in spatial variation in the shear stress on the bed. Peak values obtain where the isovels are compressed near the bed owing to high velocities or downwelling, whilst low shear stresses

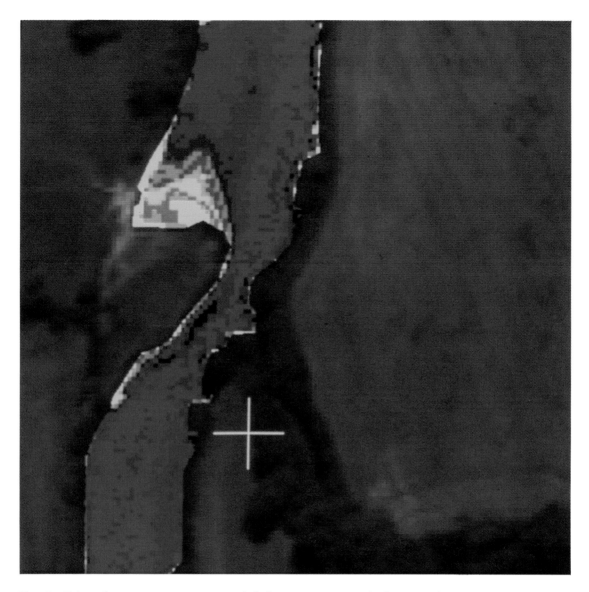

Plate 5.1 False colour remote sensing image of algal concentrations in 'dead zone' in the River Severn, UK, during low summer flow. Dark blue represents low concentrations in very turbulent flow; red represents intermediate concentrations; and the light colours represent high concentrations in almost stagnant water. River is up to 60 m wide. Flow from bottom to top.

[*facing page 108*]

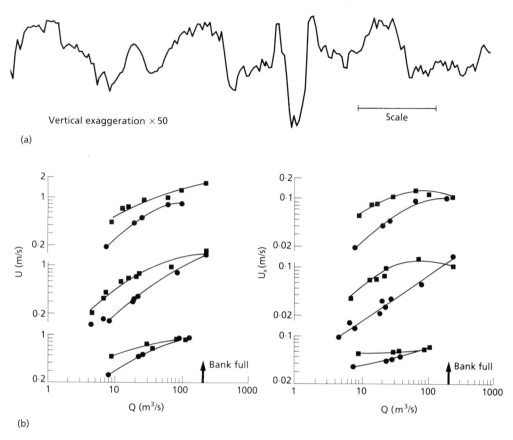

Fig. 5.4 (a) Longitudinal bed profile of the River Severn near Welshpool, UK, showing distinct pool and riffle structure. (b) Convergence of mean current velocity and shear velocity in riffles (squares) and pools (dots) as discharge increases to bankfull.

obtain where there is upwelling (Hey 1984). Peak values associated with the main flow line are typically 1.5–2.5 times the sectionally averaged value, but local downwelling may result in factors of four applying.

The locus of the maximum velocity depends on the discharge (Fig. 5.5), but at the entrance to the bend the swiftest flow is usually located close to the inner bank, whilst near the apex of the bend the flow has moved to close the outer bank. Here the flow lines are highly distorted. Near the surface, upwelling and flow away from the bank results in low applied shear on the bank, but near the bed the two cells flow towards the bank and the stress on the bank-foot may be high, resulting in basal scour and bank recession. The track of

the bed-load zone tends to mimic the variation in the locus of the maximum velocity filament (Dietrich *et al* 1984) and, as the outward flow away from the inside of the meander results in decreased shear stress close to the inside of the bend, deposition occurs there (Richards 1982; Dietrich & Smith 1984; Parker & Andrews 1985). The flow in nature can be shown to compare well with recent theoretical models (Bridge 1984; Dietrich & Smith 1984; Lee & Hsieh 1989), but improvement in the models will require a proper understanding of the turbulent structure associated with highly variable topography.

Turbulence in rivers has been little studied (McQuivey 1973; Heslop & Allen 1989) and much more is known from laboratory studies and from

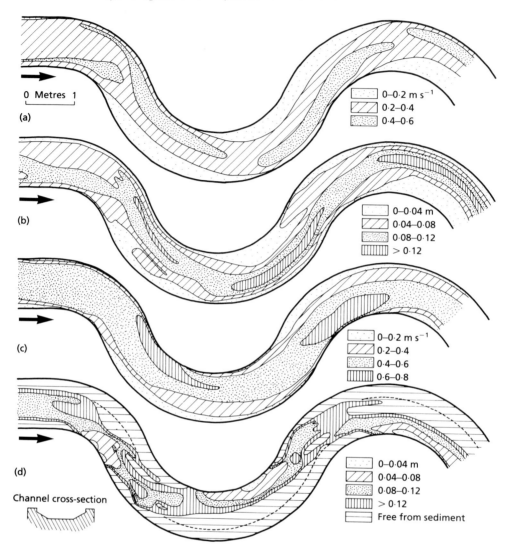

Fig. 5.5 Patterns of surface velocity and flow depth observed by Martvall and Nilsson (1972) in a laboratory meandering sand-bedded channel. (a) Surface velocity:discharge = 0.014 m³ s⁻¹; (b) flow depth:discharge = 0.014 m³ s⁻¹; (c) surface velocity:discharge = 0.05 m³ s⁻¹; (d) flow depth:discharge = 0.05 m³ s⁻¹ (after Allen 1984).

the shallow marine environment. If x is the downstream direction, y the transverse direction and z the vertical direction, then the quantities of interest are the time-averaged variances of the velocities in these respective directions; u^2, v^2 and w^2. The Reynolds shear stresses, averaged over a few minutes, are also frequently used to summarize turbulence information; these are written as $-\rho uw$, $-\rho vw$, $-\rho uv$. The term $-\rho vw$,

for example, represents the rate of turbulent transfer across the $x-y$ plane in the y (cross-channel) direction. Intermittent fluctuations over longer timescales give rise to unsteady flow events in the near-bed region, known as 'bursting' (Kline *et al* 1967). Accelerated flow events acting towards the bed are termed 'sweeps' and rapid fluid movements away from the bed are called 'ejections'. In the laboratory such behaviour has a profound influence on the entrainment of non-cohesive sediments (Fig. 5.6).

Although it is not clear whether the phenomena observed in the field are directly analogous to the laboratory experience, unsteady bedload transport in the sea is driven by 'bursting-type' cycles (Thorne *et al* 1989), whilst suspension of sand may be periodic and related to 'ejection' events (Soulsby *et al* 1987). Unsteady transport of gravel bedload is now known to occur widely in rivers and in some cases may be related to the inherent (turbulent) structure of river flow. Consequently, further effort will be directed to understand the

intermittence in turbulence structure (Williams *et al* 1989).

Secondary currents and turbulence induce mixing across the river, but it must not be implied that complete mixing always results. Many dispersion models often underestimate the residence time of a tracer (Day & Wood 1976; Bencala & Walters 1983), in part because of inadequate consideration of the effect of slow-flowing water close to the banks. However, more extensive patches of slow-flowing or 'still' water can occur (see Fig. 5.2) which mix only slowly across a shear zone with the main advective flow (Chatwin &

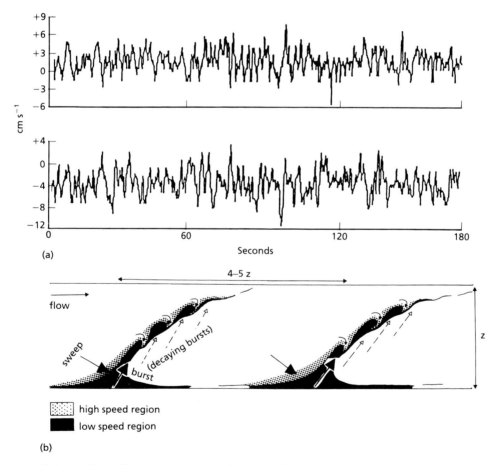

Fig. 5.6 (a) Turbulent velocity fluctuations in a straight reach of the River Severn, UK, 0.5 m above the bed. Upper trace shows vertical velocities and the lower trace shows the cross-stream component (from Heslop & Allen 1989). (b) Schematic diagram depicting high speed turbulent ejection of fluid and suspended solids away from the sediment bed (after Leeder 1983).

Allen 1985). The effect of these 'dead zones' is to delay the passage through a reach of solutes and suspended solids (Young & Wallis 1987) and so they are important in the context of chemical dispersion processes. They also may act as refuges for organisms during high flows and as temporary sinks for fine sediments during low flows. Recently, *in situ* fluorimetry and airborne remote sensing have been used to detect the spatial variation in flow structure (Reynolds *et al* 1991; see Chapter 9).

Similar complex flow patterns can be expected at channel confluences (Best 1987; Roy & Bergeron 1990) and in braided stream environments with multiple channels (Ashmore & Parker 1983), where flow separation and slack-water areas in cut-off channel reaches are common.

5.5 INITIAL MOTION OF LOOSE BED SEDIMENT

The stability of a particle depends primarily upon its immersed weight and the drag and lift forces imposed by the flow (Fig. 5.7(a)). The actual value of the forces depends on fluctuations in the current speed as well as the variation in particle

shape, density and degree of exposure upon the surface (see James 1990 for a review). Consequently, equilibrium can be expressed as a force balance (e.g. White 1940; Komar & Li 1988) between the shear stress promoting entrainment and the particle size, density and gravity resisting entrainment, expressed as a non-dimensional ratio:

$$\theta = \tau_0 / (\rho_s - \rho) g D \qquad (10)$$

where θ is Shields parameter and ρ_s is the density of sediment; all other terms are defined above. Variation in θ with the particle Reynolds number (Fig. 5.7(b)) produces a curve known as the Shields function. There are few data to validate the curve for fine silts at low Reynolds numbers (<2) where cohesive forces (Mantz 1978; 1980) and negative lift forces tend to augment the resistance force. At high Reynolds numbers (>200), viscous forces become unimportant and the ratio tends to a constant value that depends on such factors as particle shape, degree of protrusion from the bed, and the overall degree of particle sorting and bed roughness. Traditionally this value has been 0.06 for well-sorted sediment and 0.047 for poorly sorted material (Miller *et al* 1977). The actual

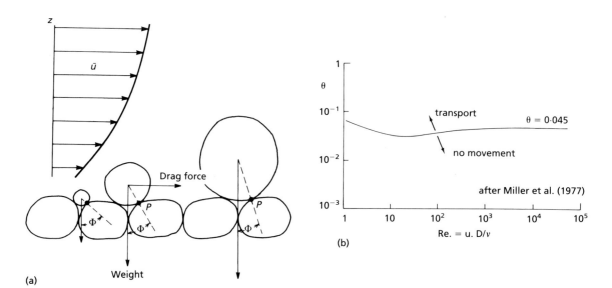

Fig. 5.7 (a) Particles in pivoting position for entrainment, illustrating that the larger the grain the greater its exposure to the flow and the smaller the pivoting angle (ϕ) (after Komar 1989). (b) Shields entrainment function (equation 10) plotted as a function of the particle Reynolds number.

value depends in part on how initial motion is defined, so that a region of uncertainty exists, but the function reaches a minimum ($\theta \sim 0.03$) in the smooth turbulent regime, which could be representative of fine sand in natural rivers.

Considerable interest has centred recently on the correct function for mixtures of coarse gravel of irregular shape in natural rivers (Komar 1987, 1989). Flattened pebbles may become imbricated (i.e. overlapped like shingling on a roof), increasing resistance to entrainment, but usually variation in particle exposure (Fig. 5.7(a)) and shape, for example, result in differing sizes of rounded pebbles becoming entrained for a given value of θ. Consequently for some mixtures many susceptible pebbles may be regarded as equally mobile, whilst others may be entrained only as the force on the bed increases (Ashworth & Ferguson 1989). The importance of equal mobility or selective entrainment depends upon the nature of particle sorting, packing and bed roughness (Komar & Carling 1991; Carling *et al* 1992). For example, gravel particles can roll freely over a flat sand-bed and so it should not be assumed axiomatically that small particles will be entrained before larger ones.

The critical shear stress needed to move a well-mixed sediment is usually related to the $D_{50\%}$ of the grading curve, but many coarse gravel beds become size segregated in the vertical so that a coarse layer, devoid of fines, overlies finer gravels or sand beneath the surface. This may be up to three times as coarse as the subsurface, and is usually called the *armour layer*, because a high shear stress is needed to destroy it and so (during moderate discharges) it protects the finer sediments from entrainment. To add an element of confusion, in the USA especially, this layer is often referred to as *pavement* whilst the term armour is reserved for surfaces that cannot readily be eroded (Parker *et al* 1982). Where segregated surfaces are present, initial motion of the armour can be calculated using either surface samples (Wolman 1954) or the $d_{84\%}$ of a bulk bed–sediment sample. Despite the interest in defining the threshold for initial motion, little attention has been given to the cessation of transport. Reid and Frostick (1984) and Lisle (1989) noted that gravel bedload transport continued until the shear

stress on the bed was only a fifth of that required to initiate motion.

The cohesive materials most likely to be found in rivers are muds and clays. These may be entrained by fluid shear or by corrasion, that is erosion owing to the scouring action of gravels or blocks of cohesive sediment swept over the surface (Allen 1984). As with non-cohesive materials, once a threshold has been exceeded, erosion increases as the applied stress increases. Factors that affect the yield strength of the mud include particle size, mineralogy, chemistry, compaction history and water temperature, so that satisfactory universal relationships for initial entrainment have not been devised. In a similar vein, hydraulic patterns and sediment transport have been little studied in bedrock channels. Erodibility depends on the nature of the imposed bedload, as well as weathering and solution rates (Howard 1987).

5.6 BED-FORM PHASE EXISTENCE FIELDS

A unidirectional flow of water over loose sediment may deform the bed to create a variety of bedforms (Fig. 5.8(a)). The most extensively studied material is fine sand, but bedforms may also be generated in silt and gravels. Starting with a flat planar bed of fine sand and low competent shear stress, some sediment transport may take place over the flat surface (low-stage plane bed). *Ripples* develop in fine sand (up to 0.6 mm in size) once a critical shear stress has been exceeded. Ripples are sharp crested, roughly triangular in profile, with gentle upstream (stoss) slopes and steeper downstream (lee) slopes close to the angle of repose of granules in water. Ripple crestlines are more or less transverse to the flow direction but may be sinuous and bifurcate. Migration downstream is by erosion on the stoss slope and deposition over the lee slope. Heights are less than 0.04 m and wavelengths are below 0.6 m. The literature concerning ripples is vast and complex, but a useful summary has been provided by Allen (1984).

At higher velocities ripples are replaced by superficially similar but much larger forms called *dunes*. These may reach metres in height with

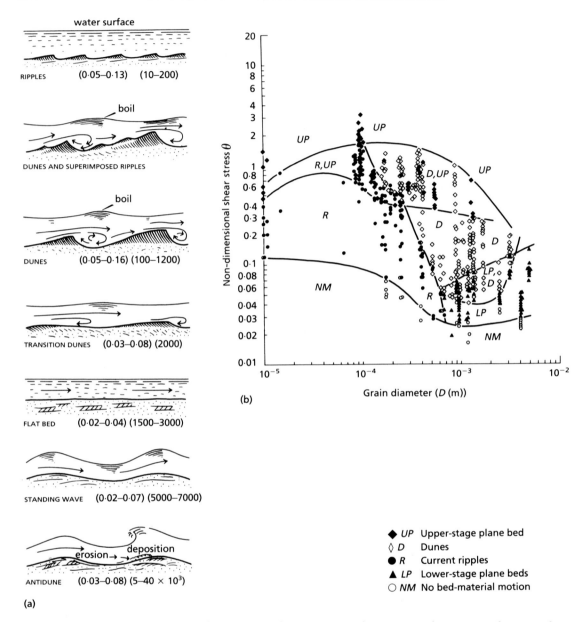

Fig. 5.8 (a) The sequence of sandy bed-forms generated as current speed increases. Values in parentheses are the friction factor (f) and the bedload transport rate (p.p.m.) (after Richards 1982). (b) Bed-form existence field constructed using laboratory data for predominantly sandy sediment (after Allen 1984).

wavelengths of tens of metres. Unlike ripples, the scale of which is related to grain size, dunes scale with depth of flow. Consequently, they may be related to large-scale turbulent eddies with dimensions comparable to the flow depth or boundary-

layer thickness. The transition from ripples to dunes is usually abrupt, although conditions can exist where ripples form on the stoss-side of dunes. With increasing velocity, the dunes be-come more rounded and lower (so-called *hump-*

back dunes; Saunderson & Lockett 1983) until the bed-forms are washed out and an upper-stage flat bed forms, characterized by intense bedload transport. A tremendous variety of names has been applied to various dune-like features (Ashley 1990): some of the confusion has arisen because of the variety and scales of non-equilibrium geometries that can occur in natural accelerating and deaccelerating flows (Allen 1983).

As the velocity increases further ($Fr > 0.8$), standing waves form at the water surface, and the bed is remolded into a train of sediment waves (*anti-dunes*) in phase with the water waves. As *Fr* increases, so the amplitude of the waves increases. These bed-forms may migrate upstream with the upstream slope steepening until the geometry is unstable and the features are washed out catastrophically only to begin to form again (Hand 1974; Allen 1984; Yagishita & Taira 1989).

A great variety of bed-form stability or existence fields have been proposed and some of these are appropriate for particular purposes (reviewed by Allen 1984). One approach is to consider the grain size of the bed material and the mean flow velocity (e.g. Southard & Boguchwal 1973), but the most successful diagrams, in terms of adequately delineating the existence fields, consider the shear velocity, non-dimensional shear stress (equation 10) or the streampower ($\tau_0 U_z$) and particle size (Fig. 5.8(b)). These diagrams are constructed largely from laboratory data collected under idealized conditions.

Little is known of the stability fields of large bed-forms in rivers (with scales equivalent to river width), especially where sediments are coarser than 1 mm (gravels). At the smallest scale, a few individual gravel particles may be grouped into *clusters* (Brayshaw 1983, 1984) which resist entrainment, whilst at larger scales (equivalent to the dune scale) a variety of low-amplitude *bars* develops. The larger features are frequently affected by the proximity of the channel margins and subunits may be 'welded on' as flows fluctuate, resulting in a variety of sediment associations (Brierley 1989). Consequently an alarming proliferation of nomenclature has resulted to describe features that may not be hydrodynamically distinct (Smith 1978). Nomenclature often refers to the position of the bars in the channel, for example

alternating bar, *medial bar*, *transverse bar*, *point bar* (Richards 1982). It is not certain how these features relate to flow dynamics but they are long lived relative to the period of formative discharges. Crowley (1983), for example, considered that such features formed in sandy sediment are not hydrodynamically equivalent to the regime bed-forms considered in Fig. 5.8(b), which generally respond rapidly to hydrodynamic regime. Ashley (1990) noted that as these bars are channel-scale features they should scale with bankfull (formative) discharge, but this observation is not particularly illuminating. Examples of gravel bars to compare with Crowley's sand-bed features are provided by Bluck (1976).

Bluck (1976) considered that the surface facies arrangement of bars can be divided into bar-head (upstream) and bar-tail (downstream) deposits. The bar-head gravels form during high-flow stages, whilst as flow drops the focus for deposition shifts to the bar tail, and this has its own suite of deposits. Sediment is delivered to the bar head by either the migration of low-amplitude gravel dunes (Carling 1990), which leave distinct cross-bedded deposits over the pre-existing sediment surface, or by the progression of mobile diffuse gravel sheets (Kuhnle & Southard 1988; Whiting *et al* 1988), which leave indistinct stratigraphic structures (Wells & Dohrenwend 1985).

Bed-forms also occur in cohesive materials such as muds (Allen 1984). These include ripples and corrasion marks that can develop into distinctive flutes and scallops. Allen (1971) noted a distinct relationship between the wavelength of mud ripples and the hydraulic radius, but importantly found no relationship with flow velocity.

5.7 SUSPENSION DYNAMICS

The study of suspended sediments is vital for understanding siltation processes, but also the control of sediment-associated contamination (Golterman *et al* 1983). Suspended grains are held above the bed by turbulent fluctuations in the fluid (Brush *et al* 1962). Although turbulent structure cannot be regarded as isotropic, as a general rule the variances of the turbulent fluctuations of velocity in the *x*, *y* and *z* planes are of similar magnitude and close to the bed are roughly equivalent to u_*. Consequently, a general criterion for suspension is that the shear velocity should exceed the velocity (w) at which the grain would settle in still clear water:

$$u_* > w \tag{11}$$

or in terms of the shear stress:

$$\tau_0 > \rho w^2 \tag{12}$$

Middleton (1976) gave support to this general function from the results of hydraulic measurements in natural rivers. Sometimes more conservative estimates have been applied to high concentration turbulent flows (Richardson & Zaki 1954), in which case u_* is multiplied by a factor of up to 1.6.

An alternative rough approximation was suggested by Graf (1971), who argued that as the upward velocity component and the Reynolds number are indicative of the turbulence level, then a function relating these should act as a guideline for suspension. When calibrated using empirical data, this reduced to:

$$u_z = 0.17(Ud)^{0.46} \tag{13}$$

If the settling velocity is known (see Vanoni 1977) then this can be compared with u_z. Clearly, turbulence will still bring some suspended grains into close contact with the bed, and if there are open pore spaces, as in a gravel bed, then suspended particles may be entrapped even if quiescent deposition is not possible (Carling 1984).

The actual prediction of sediment fluxes requires measured suspended sediment profiles (for techniques see Jansen *et al* 1979; Mangelsdorf *et al* 1990) or a prediction of the vertical distribution, together with velocity profile data. Inadequate sampling designs can result in variable results (Horowitz *et al* 1990). Coarse sediments tend to be found close to the bed with only the finest sediment in the outer layer. For any given grain size, the distribution throughout the total depth (d) can be approximated if a reference concentration (C_a) is known at a given height (a) above the bed:

$$\frac{C_z}{C_a} = \left[\frac{d-z}{z}\frac{a}{d-a}\right]^{\beta} \tag{14}$$

The exponent $\beta \sim w/0.4u_*$ for fine sediment but

takes larger values for coarse sediments. For low β values the distribution is nearly uniform, whereas for large values little sediment is found near the surface (Graf 1971). A comprehensive treatment of suspension mechanics is given by Vanoni (1977) whilst simpler introductions are provided by Henderson (1966) and Graf (1971).

Once the stress on the bed (τ_0) falls below a critical value (τ_c) suspended sediments can begin to settle out. Not all the sediment will fall out at once because of its distribution by size in the vertical and because there is still a turbulent current flowing. The rate of deposition (R_d) for any size fraction can be estimated where the near-bed concentration (C_a) is known as follows:

$$R_d = C_a w[1 - (\tau_0/\tau_c)] \tag{15}$$

The above overview is extremely simplified but provides a practical means to make reasonable assessments of the likelihood of resuspension or deposition.

5.8 TRANSPORT OF BED MATERIALS

Once a particle resting on the bed is entrained it may go into suspension. However, if it is too heavy for suspension it may travel along the bed surface. Particles may roll or slide along the bed (the *traction load*) or hop as they rebound on impact with the bed. In the latter case, ballistic trajectories occur and the material is said to move by *saltation*. Many particles moving by either of the mechanisms are collectively referred to as *bedload* and it is of considerable importance in many investigations to be able to estimate or measure this load.

A great number of bedload transport equations have been proposed over the years. Many of the more sophisticated are calibrated only with flume data and are difficult to apply in the field. Where equations have been calibrated using field data, reasonable results can be expected when the equations are applied to rivers very similar to those for which they were devised, but performance can be very poor in other environments. A major problem is that the equations assume that the flow will transport as much material as it is *competent* to do so, so that a *capacity* load is assumed. However, because of the armouring effect in many rivers and vagaries in supply of sediment to rivers from the land, sediment loads are often lower than predicted. Of the simpler formulae for gravel-bed rivers that of Meyer-Peter and Mueller (1948) gives reliable results:

$$\eta \, \rho \, g \, R \, S - 0.047 \, (\rho - \rho_s) \, g \, D_m = 0.25 \, \rho^{1/3} \, I_b^{2/3} \tag{16}$$

$$\eta = (n'/n)^{1.5} \tag{17}$$

$$n' = 0.038 \, D_{84}^{1/6} \tag{18}$$

$$n = \frac{R^{2/3} \, S^{1/2}}{U} \tag{19}$$

$$I_b = \left[\frac{\text{LHS}}{0.25\rho^{1/3}} \right]^{3/2} \tag{20}$$

where I_b is the bedload transport rate, n is Manning's roughness; all other terms have been defined above.

Equation 16 has three terms. The first on the left represents total shear stress acting on the bed grains; the second, shear at initial motion; and the right-hand side, the transport rate. The correction factor, η (equation 17), is the ratio of the grain roughness coefficient to the total roughness coefficient, raised to a power of 1.5. For gravel-bed rivers where bed-forms are absent it will be close to unity. The grain roughness is calculated from a Strickler-type function (equation 18) whilst the total roughness is calculated from stream characteristics (equation 19). The second term is the critical shear stress in terms of Shields entrainment function (equation 10). Thus for total shear stress less than the value of the second term, the left-hand side (LHS) of the equation is negative indicating zero transport, whilst for a greater total stress it is positive and gives increasing transport rates with increasing flow. The weight per unit width and time, I_b, can be calculated using equation 20, and dividing the result by g yields the mass per unit width and time.

Alternative formulae are available, such as that of Ackers and White (1973). The Ackers–White approach was developed from analysis of about 1000 sets of flume data, based on dimensionless expressions for sediment transport and derived from the streampower $(\tau_0 U_z)$ concept of Bagnold

(1966). More recently Parker and colleagues (1982) introduced a method for gravel-bed rivers with well-developed armour layers. Analysis showed that once the armour was broken up, all grain sizes were equally mobile. This latter approach has not been tested widely, but when applied to several gravel-bed rivers in north-west England gave results consistently greater (by a factor of four) than were obtained using the Meyer–Peter and the Ackers–White formulations. The latter two approaches yielded almost identical results of a magnitude commensurate with transport rates which have been sampled directly in similar British rivers. Although somewhat more complex than the Meyer–Peter approach, the Ackers–White model can also be applied to sand-bedded rivers, and deserves attention.

Many transport equations have been proposed for the calculation of bedload transport in sand-bedded channels. A major difference between these expressions and those for gravel-bed rivers is in the expression of the shear stress available for transporting material. In gravel rivers the total roughness depends largely on grain roughness, but in sand-bed streams the total roughness is influenced by form resistance, limiting the shear available for transport. For sand-bed rivers the formula of Yang (1973), based on the concept of unit stream power, has been widely compared with data-sets from both field and laboratory (Stevens & Yang 1989) and shown to be one of the most reliable available. Other formulae suitable for sand-bed rivers are those of Toffaleti (1968) and Ackers and White (1973).

An alternative to estimating the transport rate is to measure it directly. This can be done by trapping sediment in pits (Reid *et al* 1980; Bathurst 1987; Carling 1989), or by electro-magnetic detection (Reid *et al* 1984; Custer *et al* 1987). Another alternative is to lower a sampler directly on to the bed from a boat or gantry and to trap material for a given time period. The problems here include representative sampling of a spatially and time-variable load across the complete bed-width, together with the risk of the sampler not sampling efficiently, either by disrupting the flow field and altering the local transport rate or indeed by digging into the sediment surface or not sitting evenly on the bed. Today, the most popular sampler in the USA is the Helley–Smith pressure difference sampler (Helley & Smith 1971) with a uniform entrance section (7 × 7 cm or 17 × 15 cm). Extensively calibrated (see Hubbell 1987) for sand and fine gravel, it is of limited value for very coarse gravels. For the latter, other samplers, notably the full-size VUV (Novak 1957; Gibbs & Neill 1973), with a large entrance section (20 cm high × 50 cm wide) are required (Carling 1989). The particular sampling requirements needed to map bedload transport vectors are given by Dietrich and Smith (1984).

The use of a sampling or estimation procedure assumes that a unique relationship exists between the characteristics of the flow at the time of sampling and the mass of bedload in transit past that point, although this is rarely the case (Church 1985). It has already been noted that supply limitations may reduce the amount actually transported. In addition it is now accepted that temporal variations in transport rates occur (Fig. 5.9(a)) at a variety of timescales even when the flow is uniform and steady (Gomez *et al* 1989). Instantaneous fluctuations (seconds) in transport rates may be related to the inherent stochastic nature of the entrainment and transport process (e.g. bursting processes), whilst periodicities scaled in minutes and hours might be related to the controlling influence of the bed armour (Gomez 1983) or to clusters of particles forming or breaking up (Naden & Brayshaw 1987) or to the periodic passage of bed-forms. Reid and colleagues (1985), for example, noted that bedload transport was more intense on the falling limb of a hydrograph because, after the discharge peak, the bed surface armour had been disrupted. Longer-term periodicity or variation can result from intra-event (Reid *et al* 1985) or seasonal exhaustion of supply (Carling & Hurley 1987), or longer-term controls on sediment availability (e.g. Meade 1985; Roberts & Church 1986).

Given the unsteady nature of the sediment transport process, it is not surprising that the nature of the particle sorting and depositional processes over bed-forms is not well understood. Although considerable attention has been given to the flow structure (Nelson & Smith 1989), particle sorting and the formation of sedimentary structures over sandy ripples and dunes (e.g. Bridge & Best 1988), the same cannot be said for coarser sediments. Carling (1990) described a sorting process over gravel dunes, whilst Iseya and Ikeda (1987) and Dietrich *et al* (1989) demonstrated that the longitudinal sorting processes can result in rhythmic fluctuations in the bedload and characteristic depositional facies.

Better studied is the sorting of particles through meander bends in association with the formation of point bars (Dietrich 1987). At the bend entrance, where shoaling is significant, the largest particles in motion are near the inside of the bank and the finest are near the outer bank (Fig. 5.9(b)). However, because of the interaction of the near-bed flow and bar slope, the direction that particles move relative to the cross-stream plane through the bend depends on the magnitudes of the lift and drag forces relative to frictional resistance of the bed. The net result is that large particles generally exhibit a tendency to roll outwards obliquely across the bar surface towards deeper water whilst finer particles move into shallow waters on the upper bar surface. There are two views on how equilibrium is reached. The traditional one is that equilibrium is achieved when the cross-stream component of the particles' weight and lift force are balanced by the inward component of the drag force owing to secondary circulation. In this model particles travel along lines defined by equal depth. The alternative view is that equilibrium is achieved when the outward zone of maximum shear stress is balanced by convergent sediment transport owing to a net outward flux of bedload. The two models are summarized by Richards (1982) and Dietrich (1987), respectively. In either case the processes result in the characteristic fining inwards sequence of deposits observed on point bars (Fig. 5.9(c); Allen 1971).

Increasingly, the field investigator and the modeller are turning their attention to the transport dynamics of mixtures of particle sizes, including distinctly bimodal mixes of sand and gravel (Ferguson *et al* 1989). Close attention to the dynamics of the complete particle size distribution will be required to elucidate complex sorting patterns and entrainment and deposition processes (Shih & Komar 1990). A process that is

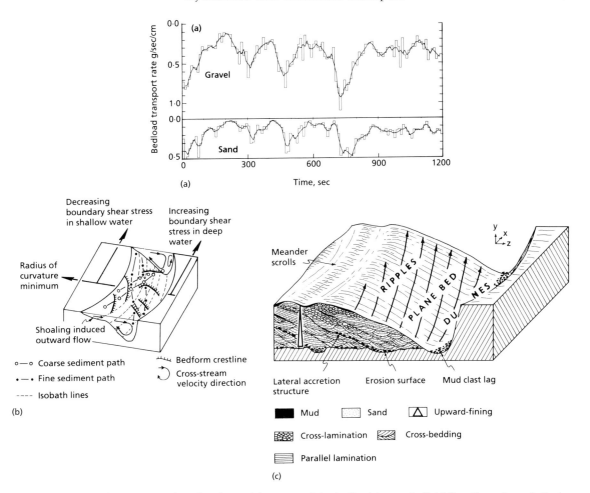

Fig. 5.9 (a) Unsteady transport of sand and gravel fractions of the bedload in steady fluid flow (from Iseya & Ikeda 1987). (b) Sketch showing the primary transport pathways of fine and coarse bedload through a channel bend (from Dietrich 1987). (c) Cross-section through sandy point-bar deposits, showing general direction of fine particle transport, the locus of bed-forms and the characteristic sedimentary deposits (after Allen 1984).

still not well understood is the one by which fine sediments are deposited and infiltrate down into the void space in gravel beds. This is not only important in understanding downstream sorting and the construction of sedimentary bodies, but also has practical significance for fish spawning habitat (Everest *et al* 1987). The effect on the latter has become prominent because human activities in many environments have increased the runoff of fine sediments into rivers which previously transported mainly coarse bed-material (Scrivener & Brownlee 1989), or river regulation has precluded high flows which previously

flushed fines away so that silt accumulates. Salmonid fish in particular deposit their eggs in gravels by excavating and then backfilling a depression in the bed (Gustafson-Marjanen & Moring 1984; Crisp & Carling 1989). A flow of oxygenated water through the hummocky structure occurs (Thibodeaux & Boyle 1987; Jobson & Carey 1989) and this is needed to ensure successful development to hatching. The young fish (alevins) need to be able to swim up through the interstitial spaces to emerge. Consequently, heavy siltation (>20% 1 mm sand in bed) will reduce the recruitment of young fish to the population (Lisle 1989). The depth to which the fines will settle depends on their size relative to the size of the bed material (Einstein 1968; Carling 1984; Diplas & Parker 1985) and to its vertical grading (Frostick *et al* 1984). This is because the particle size of the bed material mediates the size of the intervening pores. However, scour and fill will alter the active bed surface level so that sand lenses may occur at depth (Lisle 1989). Some particles trickling down will inevitably be larger than the gaps between pebbles. Consequently, lodgement occurs, hindering the passage of other particles, and as a result a sand-seal may develop at any depth. The finest materials may infiltrate much deeper into the bed, although fine clays (<3 μm) can become attached to and coat bed particles, a process influenced by water chemistry (Matlack *et al* 1989). The inner point-bar environment is particularly typified by clay infiltration deposits (Fig. 5.9(c)) because not only does the bedload transport process move the finest sediments into this region, but fine suspended sediments are delivered to this area only on high flows when concentrations are high. Fluctuating water levels also induce intra-gravel flow which aids the infiltration process, and as the point-bar is rarely subject to extensive reworking, deposits are persistent.

Empirical criteria have been developed to describe the likelihood of fine particles filtering down into gravel beds. These methods are based on the relative sizes of the two populations of sediment and have been reviewed by Carey (1991). Should lodgement occur, so that a sand-seal develops, the thickness of the seal can be estimated using a relationship proposed by Lisle (1989).

$$T = 0.1A[\rho_s\lambda_g(1-\lambda_m)]^{-1} \qquad (21)$$

where T is the seal thickness, A is the area considered, ρ_s is the sediment density, and λ_g and λ_m are gravel and sand porosities, respectively.

In addition, flow rates through fish spawning beds can be measured using standpipes and dilution techniques (Carling & Boole 1986) whilst the nature of the silted sediments can be studied using a freeze-coring technique, which retrieves undisturbed cores of bed gravel complete with the sandy infill (Crisp & Carling 1989). The method is also useful for investigating the depo-

sitional history of polluted sediments (Petts *et al* 1989). However, in regulated streams which are not subject to intensive periodic scour by high flows, fines are rarely winnowed from below the surface layer (one or two grain-diameters thick). This is because laboratory (Carling 1984) and field experience (Kondolf *et al* 1987) have demonstrated that the bed sediments need to be disturbed to remove the fines (Reiser *et al* 1985). In many cases this can be achieved only by mechanical cleansing (see collected papers in Cassidy 1981).

REFERENCES

Ackers P, White WR. (1973) Sediment transport: new approach and analysis. *Journal of the Hydraulics Division of American Society of Civil Engineers* **99**: 2041–60. [5.8]

Allen, JRL. (1971) *Physical Processes of Sedimentation.* Allen and Unwin, London. [5.6, 5.8]

Allen JRL. (1983) River bedforms: progress and problems. In: Collinson JD, Lewin J (eds) *Modern and Ancient Fluvial Systems.* Special Publication of the International Association of Sedimentologists, **6**: 19–33. [5.6]

Allen JRL. (1984) Sedimentary structures: their character and physical basis. *Developments in Sedimentology 30.* Elsevier, Amsterdam. p. 1256 [5.5, 5.6, 5.8]

Ashley GM. (1990) Classification of large-scale subaqueous bedforms: a new look at an old problem. *Journal of Sedimentary Petrology* **60**: 160–72. [5.6]

Ashmore P, Parker G. (1983) Confluence scour in coarse braided streams. *Water Resources Research* **19**: 392–402. [5.4]

Ashworth PJ, Ferguson RI. (1989) Size-selective entrainment of bed load in gravel bed streams. *Water Resources Research* **25**: 627–34. [5.5]

Bagnold RA. (1966) An approach to the sediment transport problem from general physics. *Professional Paper, US Geological Survey 422I.* Washington, DC. [5.8]

Bathurst JC. (1987) Measuring and modelling bedload transport in channels with coarse bed materials. In: Richards KS (ed.) *River Channels: Environment and Process,* pp 272–94. Blackwell, Oxford. [5.8]

Bencala KE, Walters RA. (1983) Simulation of solute transport in a mountain pool-and-riffle stream: a transient storage model. *Water Resources Research* **9**: 718–24. [5.4]

Best JL. (1987) Flow dynamics at river channel confluences: implications for sediment transport and bed morphology. In: Ethridge FG, Flores RM, Harvey MD (eds). *Recent Developments in Fluvial Sedimentology*, pp 27–35. Society of Economic Paleontologists and Mineralogists Publication No. 39. Tulsa, Oklahoma. [5.4]

Bhomik NG. (1982) Shear stress distribution and secondary currents in straight open channels. In: Hey RD, Bathurst JC, Thorne CR (eds) *Gravel-bed Rivers*, pp 31–61. John Wiley & Sons, Chichester. [5.3]

Bluck BJ. (1976) Sedimentation in some Scottish rivers of low sinuosity. *Transactions of the Royal Society of Edinburgh* **69**: 425–56. [5.6]

Brayshaw AC. (1983) The hydrodynamics of particle clusters and sediment entrainment in coarse alluvial channels. *Sedimentology* **30**: 137–43. [5.6]

Brayshaw AC. (1984) Characteristics and origin of cluster bedforms in coarse-grained alluvial channels. *Canadian Society of Petroleum Geologists Memoir* **10**: 77–85. [5.6]

Bridge JS. (1984) Flow and sedimentary processes in river bends: comparisons of field observations and theory. In: Elliot CM (ed). *River Meandering*, pp 857–72. *American Society of Civil Engineers*.

Bridge JS, Best JL. (1988) Flow, sediment transport and bedform dynamics over the transition from dunes to upper-stage plane beds: implications for the formation of planar laminae. *Sedimentology* **35**: 753–63. [5.8]

Bridge JS, Jarvis J. (1977) Velocity profiles and bed shear stress over various bed configurations in a river bend. *Earth Surface Processes* **2**: 281–94. [5.3, 5.4]

Brierley G. (1989) River planform facies models: the sedimentology of braided, wandering and meandering reaches of the Squamish River, British Columbia. *Sedimentary Geology* **6**: 17–35. [5.6]

Brush LM, Ho H-W, Singamsetti R. (1962) A study of sediment in suspension. In: *Symposium of the International Association of Hydrological Sciences. Commission on Land Erosion*. IAHS Publication No. 59, pp 293–310. [5.7]

Carey WP. (1991) Evaluating filtration processes in alluvial channels. *Proceedings of the Fifth Federal Interagency Sedimentation Conference*. (in press). [5.8]

Carling PA. (1984) Deposition of fine and coarse sand in an open-work gravel bed. *Canadian Journal of Fisheries and Aquatic Sciences* **41**: 263–70. [5.7, 5.8]

Carling PA. (1989) Bedload transport in two gravel-bedded streams. *Earth Surface Processes and Landforms* **14**: 27–39. [5.8]

Carling PA. (1990) Particle over-passing on depth-limited gravel bars. *Sedimentology* **37**: 345–55. [5.8]

Carling PA. (1991) An appraisal of the velocity reversal hypothesis for stable pool-riffle sequences in the River Severn, England. *Earth Surface Processes and Landforms* **16**: 19–31. [5.2, 5.3, 5.6]

Carling PA. (In Press) The nature of the fluid boundary layer and the selection of parameters for benthic ecology. *Freshwater Biology*.

Carling PA, Boole P. (1986) An improved conduction-metric standpipe technique for measuring interstitial seepage velocities. *Hydrobiologia* **135**: 3–8. [5.8]

Carling PA, Hurley MA. (1987) A time-varying stochastic model of the frequency and magnitude of bedload transport events in two small trout streams. In: Thorne CR, Bathurst JC, Hey RD (eds) *Sediment Transport In Gravel-bed Rivers*, pp 897–920. Wiley, Chichester. [5.8]

Carling PA, Kelsey A, Glaister MS. (1992) Effect of bed roughness, particle shape and orientation on initial motion criteria. In: Billi P *et al* (eds) *Dynamics of Gravel-bed Rivers*, pp 23–37. John Wiley & Sons, Chichester. [5.5]

Cassidy JJ (ed.) (1981) *Salmon-spawning gravel: a renewable resource in the Pacific Northwest?* Conference held in Seattle, Washington 6–7 October 1980. Washington Water Research Center, Pullman, Washington. [5.8]

Chatwin PD, Allen CM. (1985) Mathematical models of dispersion in rivers and estuaries. *Annual Review of Fluid Mechanics* **17**: 119–49. [5.4]

Chriss TM, Caldwell DR. (1983) Universal similarity and the thickness of the viscous sublayer at the ocean floor. *Journal of Geophysical Research* **89** (part C4): 6403–14. [5.2]

Church M. (1985) Bedload in gravel-bed rivers: observed phenomena and implications for computation. *Proceedings of the Annual Meeting of the Canadian Society for Civil Engineering, Saskatoon, 1985.* **2**: 17–37. [5.8]

Crisp DT, Carling PA. (1989) Observations on siting, dimensions and structure of salmonid redds. *Journal of Fish Biology* **34**: 119–34. [5.8]

Crowley KD. (1983) Large-scale bed configurations (macroforms), Platte River Basin, Colorado and Nebraska: primary structures and formative processes. *Geological Society of America Bulletin* **94**: 117–33. [5.6]

Custer SG, Bugosh N, Ergenzinger PE, Anderson BC. (1987) Electromagnetic detection of pebble transport in streams: a method for measurement of sediment-transport waves. In: Ethridge F, Flores R. (eds) *Recent Developments in Fluvial Sedimentology*, pp 21–6. Society of Economic Paleontologists and Mineralogists Publication No. 39. Tulsa, Oklahoma. [5.8]

Day TJ, Wood LR. (1976) Similarity of the mean motion of fluid particles dispersing in a natural channel. *Water Resources Research* **12**: 655–66. [5.4]

Décamps H, Capblancq J, Hirigoyen JP. (1972) Etude des conditions découlement près du substrat en canal expérimental. *Verhandlungen Internationale Vereinigung für Theoretische und Angewandt Limnologie* **18**: 718–25. [5.2]

Décamps H, Larrony G, Trivellato D. (1975) Approche

hydrodynamique de la microdistribution d'invertebrates benthiques en eau courante. *Annales de Limnologie* **11**: 79–100. [5.2]

Dietrich WE. (1987) Mechanics of flow and sediment transport in river bends. In: Richards KS (ed.) *River Channels: Environment and Process*, pp 179–227. Blackwell, Oxford. [5.8]

Dietrich WE, Smith JD. 1984. Bedload transport in a river meander. *Water Resources Research* **20**: 1355–80. [5.4, 5.8]

Dietrich WE, Smith JD, Dunne T. (1984) Boundary shear stress, sediment transport and bed morphology in a sand-bedded river meander during high and low flow. In: Elliot CM (ed.) *River Meandering*, pp 632–9. ASCE, New York. [5.4]

Dietrich WE, Kirchner JW, Ikeda H, Iseya F. (1989) Sediment supply and the development of the coarse surface layer in gravel bedded rivers. *Nature* 340, 215–17. [5.8]

Diplas P, Parker G. (1985) Pollution of gravel spawning grounds due to fine sediment. *St. Anthony Falls Hydraulic Laboratory Project Report 240*. University of Minnesota, Minneapolis. [5.8]

Dyer KR. (1970) Current velocity profiles in a tidal channel. *Geophysical Journal of the Royal Astronomical Society* **22**: 153–61. [5.3]

Dyer KR. (1971) Current velocities in a tidal channel. *Geophysical Journal of the Royal Astronomical Society* **22**: 153–61. [5.2]

Dyer KR. (1972) Bed shear stresses and the sedimentation of sandy gravels. *Marine Geology* **13**: M31–M36. [5.3]

Einstein HA. (1968) Deposition of suspended particles in a gravel bed. *Journal of the Hydraulics Division of American Society of Civil Engineers* **94**: 1197–205. [5.8]

Everest FH, Beschta RL, Scrivener JC, Koshi KV, Sedell JR, Cederholm CJ. (1987) Fine sediment and salmonid production: a paradox. In: Salo EO, Cundy TW (eds) *Streamside Management: Forestry and Fisheries Interactions*, pp 98–142. Symposium Proceedings, University of Washington, February, 1986. Institute of Forest Resources, University of Washington. [5.8]

Ferguson RI. (1981) Channel forms and channel changes. In: Lewin J (ed.) *British Rivers*, pp 90–125. Allen and Unwin, London. [5.3]

Ferguson RI, Prestegaard KL, Ashworth PJ (1989) Influence of sand on hydraulics and gravel transport in a braided gravel bed river. *Water Resources Research* **25**: 635–43. [5.8]

Frostick LE, Lucas PM, Reid I. (1984) The infiltration of fine matrices into coarse-grained alluvial sediments and its implications for stratigraphical interpretation. *Journal of the Geological Society of London* **141**: 955–65. [5.8]

Gibbs CJ, Neill CR. (1973) Laboratory testing of model

VUV bedload sampler. Research Council of Alberta Report REH./72/2. Alberta, Canada. [5.8]

Golterman HL, Sly PG, Thomas RL. (1983) Study of the relationship between water quality and sediment transport. *Technical Paper in Hydrology No. 26*. Unesco, Paris. [5.7]

Gomez B. (1983) Temporal variations in bedload transport rates: the effect of progressive armouring. *Earth Surface Processes and Landforms* **8**: 41–54. [5.8]

Gomez B, Naff RL, Hubbell DW. (1989) Temporal variations in bedload transport rates associated with the migration of bedforms. *Earth Surface Processes and Landforms* **14**: 135–56. [5.8]

Gore JA. (1978) A technique for predicting in-stream flow requirements of benthic macroinvertebrates. *Freshwater Biology* **8**: 141–51. [5.1]

Graf WH. (1971) *Hydraulics of Sediment Transport*. McGraw-Hill, New York. [5.7]

Gustafson-Marjanen KI, Moring JR. (1984) Construction of artificial redds for evaluating survival of Atlantic salmon eggs and alevins. *North American Journal of Fish Management* **4**: 455–6. [5.8]

Hand BM. (1974) Supercritical flow in density currents. *Journal of Sedimentary Petrology* **44**: 637–48. [5.6]

Helley EJ, Smith W. (1971) Development and calibration of a pressure-difference bedload sampler. *US Geological Survey Open-File Report*. Washington, DC. [5.8]

Henderson FM. (1966) *Open Channel Flow*. MacMillan, New York. [5.7]

Heslop SE, Allen CM. (1989) Turbulence and dispersion in larger UK rivers. In: *Proceedings of the International Association of Hydraulic Research Congress*, Ottawa, Canada, pp D75–D82. [5.4]

Hey RD. (1984) Plan geometry of river meanders. In: Elliot CM (ed.) *River Meandering*, pp 30–43. ASCE, New York. [5.4]

Hey RD, Thorne CR. (1975) Secondary flows in river channels. *Area* **7**: 191–6. [5.4]

Hirsch PJ, Abrahams AD. (1981) The properties of bed sediments in pools and riffles. *Journal of Sedimentary Petrology* **51**: 757–76. [5.3]

Hooke JM. (1986) The significance of mid-channel bars in an active meandering river. *Sedimentology* **33**: 839–50. [5.3]

Hooke JM, Harvey AM. (1983) Meander changes in relation to bend morphology and secondary flows. In: Collinson JD, Lewin J (eds) *Modern and Ancient Fluvial Systems*, pp 121–32. Special Publication No. 6, International Association of Sedimentologists. Blackwell, Oxford. [5.2]

Horowitz AJ, Rinella FA, Lamothe P *et al* (1990) Variations in suspended sediment and associated trace element concentrations in selected riverine cross-sections. *Environmental Science and Technology* **24**: 1313–20. [5.7]

Howard AD. (1987) Modelling fluvial systems: rock-,

gravel- and sand-bed channels. In: Richards KS (ed.) *River channels: Environment and Process*, pp 69–94. Blackwell, Oxford. [5.5]

Hubbell DW. (1987) Bedload sampling and analysis. In: Thorne CR, Bathurst JC, Hey RD (eds) *Sediment Transport in Gravel-bed Rivers*, pp 89–120. Wiley, Chichester. [5.8]

Iseya F, Ikeda H. (1987) Pulsations in bedload transport rates induced by a longitudinal sediment sorting: a flume study using sand and gravel mixtures. *Geografiska Annaler* **69A**: 15–27. [5.8]

James CS. (1990) Prediction of entrainment conditions for nonuniform, noncohesive sediments. *Journal of Hydraulic Research* **28**: 25–41. [5.5]

Jansen PP, van Bendegom L, van den Berg J, de Vries M, Zanen A. (1979) *Principles of River Engineering: The Non-tidal Alluvial River*. Pitman, London. [5.7]

Jarrett RD. (1984) Hydraulics of high gradient streams. *Journal of Hydraulic Engineering* **110**: 1517–18. [5.3]

Jarrett RD. (1990) Hydrological and hydraulic research in mountain streams. *Water Resources Research* **26**: 419–29. [5.2]

Jobson HE, Carey WP. (1989) Interaction of fine sediment with alluvial streambeds. *Water Resources Research* **25**: 135–40. [5.8]

Kline SJ, Reynolds WC, Schraub FA, Rundstadler PW. (1967) The structure of the turbulent boundary layer. *Journal of Fluid Mechanics* **30**: 741–73. [5.4]

Komar PD. (1987) Selective gravel entrainment and the empirical evaluation of flow competence. *Sedimentology* **34**: 1165–76. [5.5]

Komar PD. (1989) Flow-competence evaluations of the hydraulic parameters of floods: an assessment of the technique. In: Beven K, Carling P (eds) *Floods: Hydrological, Sedimentological and Geomorphological Implications*, pp 107–34. Wiley, Chichester. [5.5]

Komar PD, Carling PA. (1991) Grain sorting in gravel-bed streams and the choice of particle sizes for flow-competence evaluations. *Sedimentology* **38**: 489–502. [5.5]

Komar PD, Li Z. (1988) Applications of grain pivoting and sliding analysis to selective entrainment of gravel and to flow competence evaluations. *Sedimentology* **35**: 681–95. [5.5]

Kondolt GM, Cada GF, Sale MJ. (1987) Assessing flushing-flow requirements for brown trout spawning gravels in steep streams. *Water Resources Bulletin* **23**: 927–35. [5.8]

Kuhnle RA, Southard JB. (1988) Bedload transport fluctuations in a gravel bed laboratory channel. *Water Resources Research* **24**: 247–60. [5.6]

Lee H-J, Hsieh K-C. (1989) Flow characteristics in an alluvial river bend. In: *Proceedings of Technical Session B. Fluvial Hydraulics*, pp B25–B31. IAHR, Ottawa, Canada, 21–25 August 1989. [5.4]

Leeder MR. (1983) On the interactions between turbu-lent flow, sediment transport and bedform mechanics in channelized flows. In: Collinson JD, Lewin J (eds) *Modern and Ancient Fluvial Systems*, pp 121–132. Special Publication No. 6, International Association of Sedimentologists. Blackwell, Oxford. [5.4]

Leopold LD. (1982) Water surface topography in river channels and the implications for meander develop-ment: In: Hey RD, Bathurst JC, Thorne CR (eds) *Gravel-bed Rivers*, pp 359–88. John Wiley & Sons, Chichester. [5.4]

Limerinos JT. (1970) Determination of the Manning coefficient from measured bed roughness in natural channels. *US Geological Survey Water Supply Paper 1898-B*. Washington, DC. [5.3]

Lisle TE. (1989) Sediment transport and resulting depo-sition in spawning gravels, North Coastal California. *Water Resources Research* **25**: 1303–19. [5.5, 5.8]

McQuivey RS. (1973) Summary of turbulence data from rivers, conveyance channels and laboratory flumes. *US Geological Survey Professional Paper 802B*. Washington, DC. [5.4]

Mangelsdorf J, Scheurmann K, Weiss F-H. (1990) *River Morphology*. Springer-Verlag, Berlin. [5.7]

Mantz PA. (1978) Bedforms produced by fine, cohesion-less, granular and flaky sediments under subcritical water flows. *Sedimentology* **25**: 83–103. [5.5]

Mantz PA. (1980) Laboratory flume experiments on the transport of cohesionless silica silts by water streams. *Proceedings of the Institution of Civil Engineers* **69**: 977–94. [5.5]

Martvall S, Nilsson G. (1972) *Experimental Studies of Meandering*. University of Uppsala, UNGI Report No. 20. [5.4]

Matlack KS, Houseknecht DW, Applin KP. (1989) Emplacement of clay into sand by infiltration. *Journal of Sedimentary Petrology* **59**: 77–87. [5.8]

Meade R. (1985) Wave-like movement of bedload sedi-ment, East Fork River, Wyoming. *Environmental Geology and Water Science* **7**: 215–25. [5.8]

Meyer-Peter E, Mueller R. (1948) Formulas for bed-load transport. In: *Proceedings of the International Association of Hydraulic Research*, Third Annual Conference, Stockholm, pp 39–64. IAHR, Delft, Ottawa, Canada. [5.8]

Middleton GV. (1976) Hydraulic interpretation of sand size distributions. *Journal of Geology* **84**: 405–26. [5.7]

Miller MC, McCave IN, Komar PD. (1977) Threshold of sediment motion under unidirectional currents. *Sedimentology* **24**: 507–27. [5.5]

Naden PS, Brayshaw AC. (1987) Small and medium-scale bedforms in gravel-bed river. In: Richards KS (ed.) *River Channels: Environment and Process*, pp 249–71. Blackwell, Oxford. [5.8]

Nelson JM, Smith JD. (1989) Mechanics of flow over ripples and dunes. *Journal of Geophysical Research*

94: 8146–62. [5.8]

Newbury RW. (1984) Hydrologic determinants of aquatic insect habitats. In: Resh VH, Rosenberg DM (eds) *The Ecology of Aquatic Insects*, pp 323–57. Praeger, New York. [5.1]

Novak P. (1957) Bedload meters—development of a new type and determination of their efficiency with the aid of scale models. In: *Transactions of the International Association of Hydraulic Research*, Vol. 1, pp A9-1–A-11. Seventh General Meeting, Lisbon. [5.8]

Okoye KG. (1989) Mean flow structure in model alluvial channel bends. In: *Proceedings of Technical Session B. Fluvial Hydraulics*, pp B-81–B-90. IAHR, Ottawa, Canada, 21–25 August 1989. Delft, Ottama, Canada. [5.4]

Paris E. (1989) Velocity distribution over macroscale roughness: preliminary results. In: *Fourth International Symposium on River Sedimentation*, pp 625–32. Water Resources and Electric Power Press, Beijing. [5.2]

Parker G, Andrews ED. (1985) Sorting of bedload sediment by flow in meander bends. *Water Resources Research* **21**: 1361–73. [5.4]

Parker G, Klingeman PC, McLean DG. (1982) Bedload and size distribution in paved gravel-bed streams. *Journal of the Hydraulics Division, American Society of Civil Engineers* **108**: 544–71. [5.5, 5.8]

Peters JJ, Goldberg A. (1989) Flow data in large alluvial channels. In: Malesimović C, Radojković M (eds) *Computational Modelling and Experimental Methods in Hydraulics*, pp 75–85. Elsevier, London. [5.1]

Petts GE, Thoms MC, Brittan K, Atkin B. (1989) A freeze-coring technique applied to pollution by fine sediment in gravel-bed rivers. *The Science of the Total Environment* **84**: 259–72. [5.8]

Prestegaard KL. (1983) Bar resistance in gravel bed streams at bankfull stage. *Water Resources Research* **19**: 472–76. [5.3]

Reid I, Frostick LE. (1984) Particle interaction and its effect on the thresholds of initial and final bedload motion in coarse alluvial channels. In: Koster EH, Steel RJ (eds) *Sedimentology of Gravels and Conglomerates*, pp 61–8. Canadian Society of Petroleum Geologists Memoir 10. [5.5]

Reid I, Layman JT, Frostick LE. (1980) The continuous measurement of bedload discharge. *Journal of Hydraulics Research* **18**: 243–9. [5.8]

Reid I, Brayshaw AC, Frostick LE. (1984) An electromagnetic device for automatic detection of bedload motion and its field applications. *Sedimentology* **31**: 269–76. [5.8]

Reid I, Frostick LE, Layman JT. (1985) The incidence and nature of bedload transport during flood flows in coarse-grained alluvial channels. *Earth Surface Processes and Landforms* **10**: 33–44. [5.8]

Reiser DW, Ramey MP, Lambert TR. (1985) Review of flushing flow requirements in regulated streams. In: *Report to Pacific Gas and Electric Company No. Z19–5–120–84*. San Roman, California. [5.8]

Reynolds CS, Carling PA, Beven KJ. (1991) Flow in river channels: new insights into hydraulic retention. *Archiv fuer Hydrobiologie* **121**: 171–179. [5.4]

Richards KS. (1982) *Rivers: Form and Process in Alluvial Channels*. Methuen, London. [5.3, 5.4, 5.6, 5.8]

Richardson JF, Zaki WN. (1954) Sedimentation and fluidization—Part 1. *Transactions of the Institute of Chemical Engineers* **32**: 35–53. [5.7]

Roberts RG, Church M. (1986) The sediment budget of severely disturbed watersheds, Queen Charlotte Ranges, British Columbia. *Canadian Journal of Forest Research* **16**: 1092–96. [5.8]

Roy AG, Bergeron N. (1990) Flow and particle patterns at a natural river confluence with coarse bed material. *Geomorphology* **3**: 99–112. [5.4]

Saunderson HC, Lockett FP. (1983) Flume experiments on bedforms and structures at the dune–plane bed transition. In: Collinson JD, Lewin J (eds) *Modern and Ancient Fluvial Systems*, pp 49–58. Special Publication No. 6, International Association of Sedimentologists. Blackwell Scientific Publications, Oxford. [5.6]

Scrivener JC, Brownlee MJ. (1989) Effects of forest harvesting on spawning gravel and incubation survival of chum (*Oncorhynchus beta*) and coho salmon (*D. Kisutch*) in Carnation Creek, British Columbia. *Canadian Journal of Fisheries and Aquatic Sciences* **46**: 681–96. [5.8]

Shen HW, Fehlman HM, Mendoza C. (1990) Bedform resistance in open channel flows. *Journal of Hydraulic Engineering* **116**: 799–815. [5.3]

Shih S-M, Komar PD. (1990) Differential bedload transport rates in a gravel-bed stream: a grain-size distribution approach. *Earth Surface Processes and Landforms* **15**: 539–52. [5.8]

Smith ND. (1978) Some comments on terminology for bars in shallow rivers. In: Miall AD (ed.) *Fluvial Sedimentology*, pp 85–8. Canadian Society of Petroleum Geologists Memoir 5. [5.6]

Soulsby (RL). (1983) The bottom boundary layer of shelf seas. In: Johns B (ed.) *Physical Oceanography of Coastal and Shelf Seas*, pp 189–266. Elsevier, Amsterdam. [5.2, 5.3]

Soulsby RL, Atkins R, Salkield AP. (1987) Observations of the turbulent structure of a suspension of sand in a tidal current. In: *Euromech 215, Mechanics of Sediment Transport in Fluvial and Marine Environments*, pp 88–91. Balkema, Rotterdam. [5.4]

Southard JB, Boguchwal LA. (1973) Flume experiments on the transition from ripples to lower flat bed with increasing size. *Journal of Sedimentary Petrology* **43**: 1114−21. [5.4]

Statzner B, Holm TF. (1982) Morphological adaptation of benthic invertebrates to stream flow — an old question studied by means of a new technique (Laser A Doppler Anemometry). *Oecologia* **53**: 290−2. [5.2]

Stevens HH, Yang CT. (1989) Summary and use of selected fluvial sediment-discharge formulae. *US Geological Survey Water Resources Investigations Report 89−4026*. Washington, DC. [5.8]

Thibodeaux LJ, Boyle JD. (1987) Bedform-generated convective transport in bottom sediment. *Nature* **325**: 341−3. [5.8]

Thompson A. (1986) Secondary flows and the pool−riffle unit: a case study of the processes of meander development. *Earth Surface Processes and Landforms* **11**: 631−41. [5.3]

Thorne CR, Hey RD. (1979) Direct measurements of secondary currents at a river inflexion point. *Nature* **280**: 226−8. [5.4]

Thorne CR, Rais S. (1984) Secondary current measurements in a meandering river. In: Elliot CM (ed.) *River Meandering*, pp 675−86. ASCE, New York. [5.4]

Thorne PD, Williams JJ, Heathershaw AD. (1989) *In situ* acoustic measurements of marine gravel threshold and transport. *Sedimentology* **36**: 61−74. [5.4]

Toffaleti FB. (1968) *A procedure for computation of the total river sand discharge and detailed distribution bed to surface*. Committee on Channel Stabilization, US Army Corp of Engineers Waterways Experiment Station Technical Report 5. Vicksburg. [5.8]

Vanoni VA. (1977) *Sedimentation Engineering*. ASCE, New York. [5.7]

Walker JF. (1988) General two-point method for determining velocity in open channel. *Journal of Hydraulic Engineering* **114**: 801−5. [5.2]

Wells SG, Dohrenwend JC. (1985) Relief sheetflood bedforms on Late Quaternary alluvial-pan surfaces in the southwestern United States. *Geology* **13**: 512−13. [5.6]

White CM. (1940) The equilibrium of grains on the bed of a stream. *Proceedings of the Royal Society of London* **A174**: 332−8. [5.5]

Whiting PJ, Dietrich WE, Leopold LB, Drake TG, Shreve RL. (1988) Bedload sheets in heterogeneous sediments. *Geology* **16**: 105−8. [5.6]

Whittaker JG. (1987) Sediment transport in step-pool streams. In: Thorne CR, Bathurst JC, Hey RD (eds) *Sediment Transport in Gravel-bed Rivers*, pp 545−79. John Wiley & Sons, Chichester. [5.3]

Williams JJ, Thorne PD, Heathershaw AD. (1989) Measurements of turbulence in the benthic boundary layer over a gravel bed. *Sedimentology* **36**: 959−71. [5.4]

Wolman MG. (1954) A method of sampling coarse river-bed material. *Transactions of the American Geophysical Union* **35**: 951−6. [5.5]

Wooding RA, Bradley EF, Marshall JK. (1973) Drag due to arrays of roughness elements of varying geometry. *Boundary-Layer Meteorology* **5**: 285−308. [5.2]

Yagishita K, Taira A. (1989) Grain fabric of a laboratory antidune. *Sedimentology* **36**: 1001−5. [5.6]

Yang CT. (1973) Incipient motion and sediment transport. *Journal of the Hydraulics Division of American Society of Civil Engineers* **99**: 1679−1704. [5.8]

Young PC, Wallis SG. (1987) The aggregated dead zone model for dispersion. In: *BHRA: Proceedings of a conference on water-quality modelling in the inland natural environment*, pp 421−33. BHRA, Cranfield. [5.4]

6: Channel Morphology and Typology

M. CHURCH

6.1 INTRODUCTION

Several factors govern the physical processes in rivers and hence their morphology. The primary ones are: the volume and time distribution of water supplied from upstream; the volume, timing and character of sediment delivered to the channel; the nature of the materials through which the river flows; the local geological history of the riverine landscape. Secondary factors that can be important determinants of channel morphology include local climate (particularly the occurrence of a freezing winter or an extended dry season), the nature of riparian vegetation, and land use in the drainage basin. Direct modification of the channel by humans is a further important factor in many streams in settled regions.

The size of the channel is determined by the water flow through it, particularly by flood peak flows that effect erosion and channel-shaping sediment transport. Many investigators have advocated floods that recur once in about 1.5−2.5 years (i.e. approximately the mean annual flood) as the bankfull, 'channel-forming flow', but there appears to be no universally consistent correlation between flow frequency and bankfull, nor between flood frequency and effectiveness in creating morphological change. In fact, large rivers flowing in relatively fine, easily mobilized sediments may be shaped predominantly by frequently recurring flows, whereas headwater boulder or cobble-gravel channels may be subjected to major disturbance only during much more extreme events. Certainly, regionally consistent combinations of these circumstances may occur.

The duration of inundation — and possibly the season when inundation occurs — may be significant in determining the character of channel edge and riparian habitats. Figure 6.1 presents a simple classification of channel sections that reflects qualitatively flow depth and duration. Correlations have been demonstrated between riparian plant communities and elevation above some reference water level; hence between plant species and duration of inundation (e.g. Teversham & Slaymaker 1976). A significant channel boundary is the 'lower limit of continuous terrestrial vegetation' — the limit of the 'active channel' in Fig. 6.1 — which is more or less well defined on most stream banks. In some jurisdictions, this forms the limit of the river channel for legal purposes.

River morphology reflects the concentration and calibre of sediment moving down the channel. When sediment delivered to the channel is predominantly fine-grained it is largely carried in suspension, and much is deposited in slack water overbank during floods. This builds relatively high, cohesive banks. The result is a relatively narrow, single-thread channel that habitually meanders. Coarse sediment is transported on or near the bed. Consequently, it is deposited in bars which fill the channel and deflect the river in a less regular pattern of lateral activity. According to the supply of sediment, such channels may meander irregularly, wander, or become braided. They are characteristically wide and shallow, with non-cohesive lower banks formed in the coarse material. The division between 'fine' and 'coarse' sediment depends in some measure on the energy of the stream, but for purposes of this discussion it can be placed within the range 0.3−1.0 mm, that is, in the medium to coarse sand range.

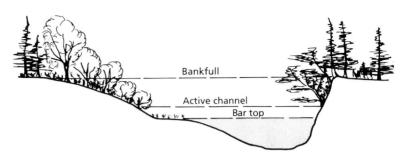

Fig. 6.1 Sketch of channel sections according to morphology (after Osterkamp & Hedman 1982). Within the 'active channel', vegetation is restricted to herbs and to species able to survive extended periods of inundation.

Channels that flow through sediments which they have previously deposited are termed *alluvial channels*. They certainly are competent to modify their form, since they previously moved the sediment that makes up their bed and banks. Here, we find consistent relations between flow and the width, depth and velocity. Such relations are termed *hydraulic geometry*. The actual form of the channel is, of course, influenced by the strength of the material that makes up the bed and banks, so distinctive hydraulic geometries occur for different materials. In fact, the strength of granular sediments has only a moderate range. In comparison, the range of flows may vary over many orders of magnitude as one moves down a river system, and over several orders of magnitude at one place through the seasons. The depth of a stable channel is limited by material strength — hence its ability to withstand the shearing force of the flow — so as rivers become larger they chiefly become wider. The scale of alluvial river channels can be summarized conveniently in the relation between channel width and characteristic flow (Fig. 6.2).

Channel size increases systematically through a river system as the increasing drainage area contributes larger flows to the trunk channel. The morphological scale of the channel changes accordingly (Fig. 6.3). Rivers flowing through non-alluvial material (bedrock or non-alluvial sediments) may depart from the scale relations implied in Figs 6.2 and 6.3.

Because of the significant correlation between channel scale and position in a drainage system, classification of river channels on the basis of their position in the drainage system is of some interest. For this purpose rivers are defined as sequences of *links*, an individual link comprising the channel between two successive tributary confluences (*internal link*), or between the channel source and the first tributary confluence (*external link*). The most commonly cited link classification system defines external links as 'order 1' streams, and defines a link of next higher order whenever two links of *equal* order join. Because of the irregular structure of drainage networks, order is not a reliable indicator of channel size. A more consistent index is link *magnitude* (Shreve 1967), each link being defined as having magnitude equal to the sum of all external links draining to it. The magnitude of a drainage basin is a useful surrogate of its discharge potential. (However, drainage networks are usually defined from maps, which often do not depict the smallest headward tributaries, so that consistent definition often remains a problem; *cf.* Dunne & Leopold 1978). In small drainage basins, there is also a regionally reliable correlation between drainage area and discharge characteristics.

Geological history and physiographic setting impose constraints on river morphology and behaviour. Most rivers flow in valleys that exert some lateral and vertical control over the channel. Rivers may be confined or entrenched in narrow valleys. Conversely, on broad plains — especially river-constructed deltas or alluvial fans — there are few constraints on lateral movement of the channel. The gradient is theoretically self-adjusted in a purely alluvial channel, but very few rivers conform with this ideal. In most cases, rivers flow in topographic defiles determined by a long geological history, and valley gradient is more or less imposed. River adjustment in the medium term is restricted to the possibility to reduce gradient locally by adoption of a meandered habit.

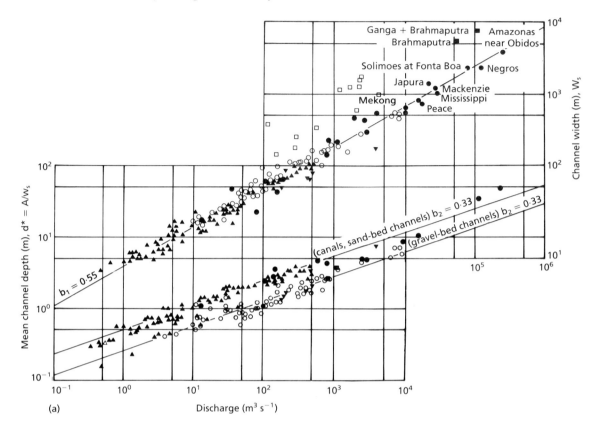

(a)

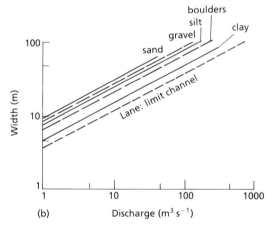

(b)

Fig. 6.2 Scale relations for channel width and depth versus flow. (a) Channels that are dynamic models of each other scale as $Q^{0.4}$ in both width and depth, indicating that similarity is preserved. However, alluvial channels are distorted toward relatively greater width as flow increases; that is, the ratio of width to depth increases as channels become larger. ● sand-bed rivers (Leopold & Wolman 1957; Neill 1973); ○ gravel-bed rivers (Bray 1973); ▼ lake outlet channels (gravel) (Kellerhals 1967); ■ □ braided channels (sand, gravel); ▲ flumes (sand). (b) Variation in channel width due to material properties. The 'limit channel' is the narrowest mechanically stable channel for a given flow.

This constraint often affects even major 'alluvial' rivers in those parts of the world that were subject to Pleistocene glaciation, and is generally true of headwaters.

Human activity can affect river channels through direct structural interference, as when artificial channels are constructed or weirs placed to control water levels, or it may result from effects on runoff, sediment calibre and sediment yield produced by the land use or change thereof

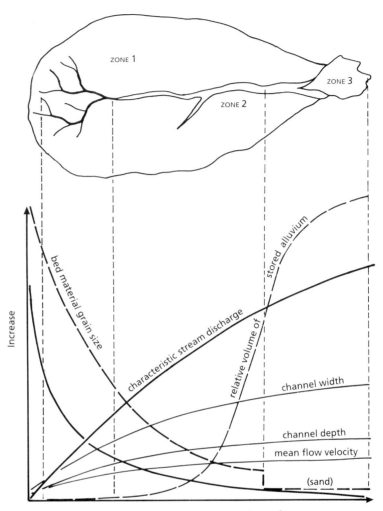

Fig. 6.3 Schematic representation of the variation in channel properties through a drainage basin (based on a concept of Schumm 1977).

in the contributing drainage basin (Kellerhals & Church 1989). Human activities manifest a characteristic intensity on the landscape; it is a reasonable generalization that the impacts of land use are most severely visited upon smaller, head-water channels. None the less, the cumulative impact of widespread human activity can be observed in even the greatest rivers (*cf.* Fremling & Claflin 1984, on the Upper Mississippi River), and individual engineering projects directly control some of the world's largest channels.

The controls described above change along a river. Some changes may be subtle, as sediment load is modified in a long, slowly aggrading channel. Others are abrupt, as at major tributary junctions. Before a river can be classified successfully it must be divided into homogeneous units, so-called *reaches*, within which the controlling factors do not change appreciably. This normally can be done *a priori* on air photographs and maps, at least for larger channels. However, certain reach limits, associated perhaps with changes in geological history or with the history of human activity, may become obvious only after substantial study.

The granular materials that make up the river

bed and banks customarily vary downstream according to the ease with which they can be transported by the river. Accordingly, in headwaters we find cobble and boulder gravels (Fig. 6.3), the larger fractions of which cannot be moved very far, whereas the far downstream reaches of large rivers are formed in sand and silt. In headwaters, then, channel scale converges toward the size of individual boundary elements, whereas downstream the channel becomes very much larger than the characteristic sediment grains. This is the major determinant of the behaviour and morphology of river channels, and provides a framework for classification. In such a scheme, the most fundamental division is between 'small channels', ones in which channel scale is comparable with the scale of individual sediment grains, which therefore are individually significant elements of the channel boundary, and larger channels in which the boundary is made up of aggregate structures of grains.

Channels in bedrock do not easily lend themselves to definition in this scheme. Generally, small channels would be ones defined by individual bedding planes or joint partings.

6.2 SMALL CHANNELS

In coarse material, the scale of small channels is of the order of one to ten particles only. Small channels flowing through cobble (>64 mm) to boulder (>256 mm) sized material are significant in a management context. Channel depth characteristically is of the order of 0.1–1 large grain, so the relative roughness (the ratio, D/d, between grain diameter and flow depth) is usually greater than 1.0. Individual clasts constitute significant form elements of the channel (Fig. 6.4). (Note that definition on the basis of relative size of channel and bounding sediment removes any absolute scale; a sand channel on a beach over which one could step with impunity would not, in this view, be a 'small channel'. In fine materials, small channels are restricted to hillside rills, which are not considered here.) The smallest headwater channels in forested environments are often difficult to identify because the streams flow under vegetation mats and roots. Here and in moorlands, large 'pipes' often feed the channel from concentrated subsurface flow.

The channel forms a sequence of pools dammed

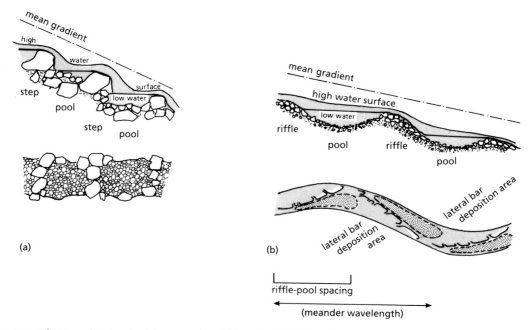

Fig. 6.4 Definition sketches for (a) step-pool and (b) pool–riffle channels.

behind individual rocks, or a line of rocks, and falls over the same features. In channels with surface width $w_s \geq 3D$, lines or cells of boulders form, characteristically held by a keystone (Fig. 6.5(a)). In forests, major pieces of wood may be incorporated into the steps. Banks are composed of bedrock, boulders, turf, or heavy root development. Such channels are restricted to headwater positions in drainage systems, so they are characteristically steep. Gradients fall in the range 2° to >20° (4° to >35%). Consequently, steps are frequent and the pools are short; pool length often is of the order of 3–4 w_s only. Gradients are commonly bedrock or debris controlled, and

streambed material may be only one to a few grains deep. Not infrequently, bedrock-controlled cascades or falls occur.

Such channels do not often re-form. Exceptional floods, characteristically with recurrence intervals measured in decades, may break the steps. Experimental observations (Whittaker & Jaeggi 1982) have demonstrated that step-pool features form in extreme flows ($D_{max}/d \approx 1.0$). Initial bed deformation in the flume tends to produce wave-like (antidune) features, but the positions of the largest stones, which move only sporadically, anchor the pattern and promote the development by local scour and deposition of step-pool features

Fig. 6.5 Morphological features in small and intermediate channels. (a) Details of a step; the keystone is under the water fountain. (b) Riffle–pool morphology; the pool upstream of the riffle is filled with cobbles and would be classified as a 'glide' for habitat purposes. (c) Step-pool channel forming a boulder cascade at low flow. (d) Channel-spanning log jam (vertical aerial view); the jam is highly permeable and would be classified as a mesojam in the typology of Fig. 6.7 despite the relatively large volume of debris present.

with characteristic roughness spacing (height to length ratio) of the order of 0.1–0.15. These are much steeper than ordinary anti-dunes and represent a configuration that presents maximum flow resistance. It is obvious from field observation that the locking together of the largest boulders is an important mechanism to stabilize them and control the step-pool formation.

If the height of the steps is determined by the diameter of the largest stones, then roughness spacing of 0.1–0.15 implies that a gradient of 6–9° can be contributed by the dominant grains alone. Steeper gradients can be achieved by aggregate structures. A large portion of the total drop in step-pool channels is controlled by the placement of the largest stones.

In channels with gradients >10–12°, extreme flows may mobilize the entire mass of sediment in the channel and create a debris flow (review in Costa 1984). Afterwards normal flows re-establish the step-pool sequence by moving the larger stones into stable, interlocked positions. Boulders are moved mainly by being undermined so that they roll forward under the influence of gravity.

The frequency of channel re-formation is determined not only by flood history but also by sediment supply, channels with abundant debris influx being subject to frequent reorganization. Many headwater channels are confined in steep gulleys or narrow valleys (so they are directly coupled with hillslope sources of debris). Vertical aggradation and degradation form the major style of instability here. Channels on alluvial cones characteristically become choked with debris and then avulse (Kellerhals & Church 1990); the entire channel shifts abruptly to a new, usually steeper line of descent.

Many small channels, as defined here, have no direct fishery value because they are too steep to be colonized (and often lie beyond impassable barriers). In the lower end of the range, pool-dwelling salmonids occur. Such channels may, however, be important for production of invertebrates and for recruitment of organic material that may drop into the water, both of which may subsequently be transported downstream. Conditions in these headwater channels also influence water quality downstream. Finally, the frequently steep, moist banks of such streams often provide sites for survival of otherwise endangered plants.

6.3 INTERMEDIATE CHANNELS

Intermediate channels have $w_s \gg D$, but they still may be influenced by blockage across a substantial portion or all of their cross-section by unusual sediment accumulations or, more usually, by fallen trees or branches. This criterion places an absolute but approximate upper limit on the width, which in most forest regions is in the range of 20–30 m. In open country, the limit may be observed in much smaller channels, depending on the character of the sediments and sediment transport. Relative roughness usually falls in the range $1.0 > D/d > 0.1$, so that flows are wake dominated or transitional toward deep shear flows. Individual large boulders may still represent significant elements on the boundary of intermediate channels, but they are subordinate to structures formed by many grains, including accumulations into bars (Church & Jones 1982), to elements formed from large parts or root boles of trees (Keller & Swanson 1979), and accumulations of large organic material termed 'jams'. In the absence of significant log jams, the pool–riffle–bar unit (see Fig. 6.4) forms the major physical element in such channels, and it has a characteristic but not invariant spacing of about five to seven channel widths. Significant variation of this basic unit has important implications for habitat quality.

Riffles take several forms (the following terminology is from Grant *et al* 1990). Riffles *sensu stricto* are zones of relatively shallow, rapid flow in comparison with pools. $D/d < 1.0$, in general, and flow remains subcritical or near critical. Local channel gradient characteristically remains less than 1° (2%). Riffles are the dominant rapid in larger intermediate channels with average gradient less than about 0.5° (1%) and occur where the channel is dominated by a sequence of alternating bars with intervening crossovers on the riffles (Fig. 6.5(b)).

Rapids exhibit irregular boulder lines and cells, have $D/d \approx 1.0$, and flow near critical. Some boulders are emergent at low flows. Gradient is typically in the range of 1–2° (2–4%) and flow occurs as a series of poorly defined filaments or discrete *chutes* through the bed structures.

Cascades are steep reaches in which flow occurs over a sequence of emergent boulders, often organized into steps (Fig. 6.5(c)). Flow occurs as supercritical jets between the boulders or falls over them. Gradient is greater than 2°. Cascades may be defined by boulders or by bedrock. When cascades dominate the channel, it is usual for the reach to be classified as a 'small channel' (but see below).

Glides (also 'runs') are extended riffles, often replacing a pool that has been filled by sediment (Fig. 6.5(b)). Pool—riffle or pool—rapid transition zones have a similar character. Glides are not usually distinguished in strictly morphological classifications, but they represent significant habitat units (Bisson *et al* 1981).

Pools can be classified in two major groups (Bisson *et al* 1981; Sullivan 1986). *Backwater pools* occur where major obstructions occur in the channel. These include boulders, bars and large organic debris. Flow diverges into these pools, with declining velocity and energy gradient. Consequently, sediment tends to be deposited in these locations and so they are commonly relatively shallow, with finer than average substrate. Within this class are *separation zones* on the lee side of boulders, logs, or emergent debris accumulations (backwater pools of Bisson; eddy pools of Sullivan). Separation zones also occur at channel confluences and at abrupt bankline angles. *Dammed pools* occur upstream of boulder lines in rapids and cascades and, on a larger scale, upstream of channel-spanning debris jams or gravel accumulations. *Scour pools* (drawdown pools of Sullivan) occur where flow converges over or around a constriction. These are the deepest places in the channel, but velocities are relatively high during high flows even though they may become backwater units at low flow. A number of variants have been described, including *plunge pools* on the downstream side of steps, *lateral scour pools* where flow is deflected by obstructions such as root boles, logs and channel bars, *vertical scour pools* under hanging logs or roots, and *trench pools* where flow is confined for some distance against erosion resistant substrate, such as bedrock.

Lateral scour pools adjacent to sediment accumulations are the dominant feature in pool—riffle sequences. Most of the other features defined above are distinctive mainly by virtue of local arrangements of boulders, logs or sediment accumulations. However, they constitute significant individual habitat units. For example, certain of the scour pool types provide hiding places under overhanging banks or hanging logs. Figure 6.6 summarizes riffle and pool features in a hierarchical diagram which emphasizes that the major channel units are composed of groupings of the smaller features.

Finally, it is worth emphasizing that the hydraulic character of individual units, and their morphological appearance, changes dramatically as flow varies. Although they invariably are observed and classified at low flow, they are of course formed at flows competent to move sediment and debris.

The third element of the pool—riffle—bar unit is the sediment accumulation zone termed a *bar*. These occur in areas of flow divergence where the sediment transporting competence of the stream declines, typically upon riffles, or downstream from scour pools (Church & Jones 1982). Bars strictly defined as accumulations of sediment grains require $D/\langle d \rangle < 0.5$, where $\langle d \rangle$ indicates mean depth, in order that the grains may pile up. Typically $D/\langle d \rangle < 0.1$. Since material must be moved by traction on to the bar, we require, following Shield's criterion for competence (see Chapter 5) $\tau_c \rho_s' g D_{90} \approx 0.03$, where $\tau_c = \rho g d S$ is the critical tractive force to move sediment of size D, ρ_s' is submerged sediment density, ρ is the density of water, g is the acceleration of gravity, S is the energy gradient, and D is a relatively large grain size (here D_{90}). If $\rho_s'/\rho = 1.65$, we have $S \leq 0.05 D_{90}/d$. For $D_{90}/d \approx 1.0$, $S \leq 3°$ (5%). In fact, bars become well developed at gradients of less than 2° in association with rapids and riffles. Riffles often constitute the front faces of bars. Bars are often anchored at fixed locations by the patterns of flow divergence and convergence determined by the overall channel form (Lisle 1986).

Like other channel elements, bars exhibit a variety of morphologies (see following section on large channels). In intermediate channels, a

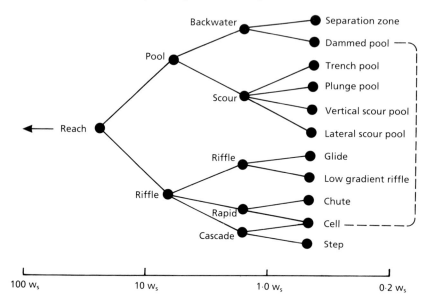

Fig. 6.6 A hierarchical representation of channel morphological units (partly after Sullivan 1986).

variety of relatively ill-defined bar features occurs in association with debris accumulations.

Channel morphology is modified along intermediate channels by normal fluvial sediment transport, most of the sediment being moved from bar to bar. However, in reaches containing large amounts of organic debris, steps may be common over individual, transversely oriented logs, or through debris accumulations up to several metres in height (Fig. 6.5(d)). Log steps and jams in forest streams typically account for between 5% and 15% of the total drop (Marston 1982; Hogan 1986) (some studies have reported much higher proportions — to more than 50% — but there remain some questions about equivalence of measurements). The formation and decay of log jams regulates the transfer of bed material downstream. The upstream side of a jam typically consists of a sediment-filled backwater reach. Downstream may be an extended, sediment-starved riffle or rapid. D.L. Hogan (personal communication) has defined a characteristic process/morphology history for log jams (Fig. 6.7). Since jams develop at random along the channel, they often obscure or serve to relocate the normal sequence of bars. Jams also encourage the occurrence of channel avulsions where the channel is not confined, so promoting relocation of the stream channel and the development of blind side-channels and backwaters.

Channels that flow along the base of cliffs or steep hillslopes often become filled with large boulders which the stream is not capable of moving at any stage. Such a channel may technically be a 'small channel', but where the overall gradient is less than about 2° and flow remains continuous between the boulders, the channel is more appropriately classified as intermediate. Similarly, deep, narrow channels are common in alluvial floodplains and deltas. Whilst their scale may qualify them as intermediate channels, their overall morphology and hydraulic conditions identify them as scaled-down large channels, with which they are more appropriately grouped. The introduction of 'regime groups' of channels in the next section of this chapter will provide a hydraulic basis for these assignments.

Intermediate channels constitute the optimum spawning and rearing habitat for a range of migratory and resident fishes, particularly salmonids. Riffles with clean gravels provide spawning sites, whilst pools and side-channels contribute rearing habitat. Microhabitat units within the channel are particularly important in rearing inasmuch as they provide hiding and resting places. Intermediate channels in forest environments may still be dominated by overhanging riparian vegetation which is important as shelter, as a source of drop-in food, and as an agent of streambank stability.

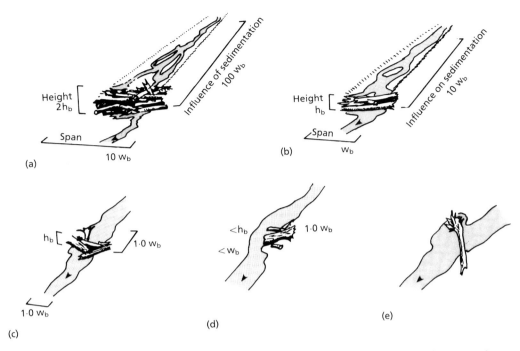

Fig. 6.7 Schematic diagram of log jam morphologies and their effect on channel morphology (courtesy of D.L. Hogan). (a) Megajam; (b) macrojam; (c) mesojam; (d) microjam; (e) individual log pieces. h_b, W_b, are bankfull depth, width of the channel (for scale).

6.4 LARGE CHANNELS

Large rivers are ones in which purely fluvial processes and geological constraints determine the morphology. Riparian effects do not dominate the channel, although laterally unstable channels in forests may recruit large volumes of wood debris which accumulate locally to influence bar and channel developments. (The transition to large channels occurs in many environments somewhere near $w \approx 20-30$ m and bankfull discharge, $Q_b \approx 20-50$ m^{3}s^{-1}, but much smaller channels can exhibit the characteristics of 'large' channels.) $D/d < 0.1$ in most cases, and flow is a deep shear flow with a well-defined velocity profile. Bed structure is dominated by pool–riffle sequences amongst major bar-forms, or by meandered bend–pool and crossover sequences. Large rivers are familiar from textbook descriptions. However, most textbooks present a very poor, one-dimensional classification of them.

Because the calibre and volume of sediment supply have an important influence on channel morphology (section 6.1), channel pattern and the style of within-channel sediment storage are two useful bases for morphological classification (Fig. 6.8). Since the style of lateral instability may also be important in determining the character of riverine habitat, this provides a third useful dimension for classification. A classification that uses these dimensions has been developed by Kellerhals *et al* (1976; see also Kellerhals & Church 1989). It is presented in outline in Figs 6.9–6.11.

The hydraulic relations underlying the systematic variation of river channel morphology have been only partly quantified (see Ackers 1988). The most important control of channel morphology clearly is the water flow regime. Scale relations for river channels (as illustrated in Fig. 6.2) are empirically summarized in the equations of hydraulic geometry:

$$P = a_p Q_n^{b_1}$$

$$R = a_r Q_n^{b_2}$$

in which P is the wetted perimeter of the channel, $R = A/P$ is the hydraulic radius, A is channel

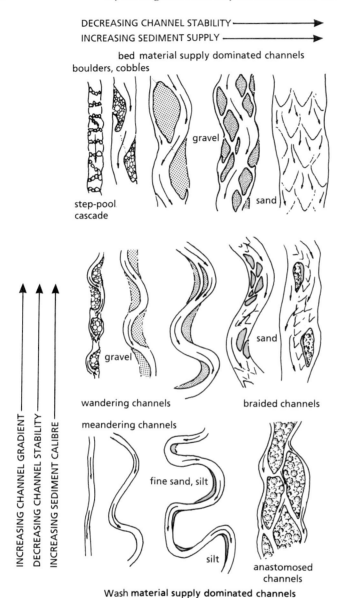

Fig. 6.8 Conceptual pattern of morphological types of large channels (based on concepts of Mollard 1973 and Schumm 1985). (a) Bed material supply-dominated channels; (b) wash material supply-dominated channels.

cross-section, and Q_n is the flow for some fixed return period (such as mean annual flow, mean annual flood, etc.). Because of measurement constraints, P is often replaced by w_s, the surface width of the river, and R then becomes $d\cdot = A/w_s$, the 'hydraulic mean depth'.

The coefficients a_p and a_r (or a_w and a_d) are known to depend upon the sedimentary materials that make up the bed and lower banks of the channel, since they control the resistance of the bed and banks to erosion, hence determine the channel shape (*cf.* Fig. 6.2, inset). Therefore, channels flowing in the same materials conform to the same hydraulic geometry and form a *regime*

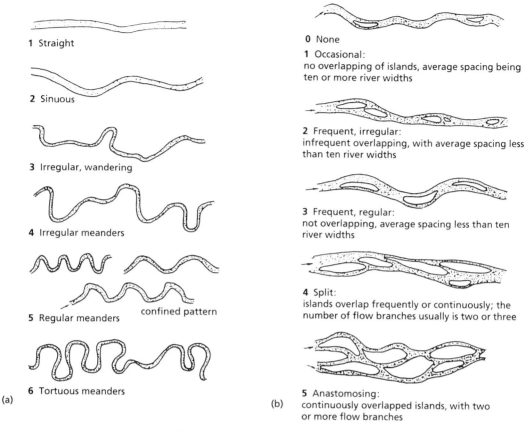

Fig. 6.9 Morphological classification of large channels: plan-form pattern (Kellerhals & Church 1989). (a) Channel pattern; (b)channel islands.

group. The exponent $b_1 \approx 0.55$, but b_2 appears to take either the value 0.40 or the value 0.33. The former indicates a proportional increase in forces on the bed as the channels become larger (the condition of 'Froude similarity'), whereas the latter implies that the limit strength of the bed constrains the increase in channel depth. In these conditions bed-forms become important as additional energy dissipators.

These equations would describe the change in channel form downstream in uniform sediments as drainage area—hence discharge—becomes greater. The reference flow is one competent to move sediment, hence to form the channel. Such equations can be developed for all classes of channels discussed in this chapter, but have been investigated systematically only on large chan-

nels. Many small and intermediate channels follow the scale $b_2 = 0.4$, whereas most large channels follow $b_2 = 0.33$.

Flow continuity associates a third equation with these two, a 'kinematic' equation for the flow:

$$\langle v \rangle = (1/a_w a_d) Q_n^{(1-b_1-b_2)}$$

in which $\langle v \rangle$ is the mean velocity for the channel. This equation reveals the necessity for channel size, shape, plan-form pattern, and bed-forms to become mutually adjusted so that the total resistance to flow permits just the velocity $\langle v \rangle$ to be maintained in the stable channel.

The actual relation between sediment calibre and channel pattern depends upon the discharge and the gradient of the channel. Together, they

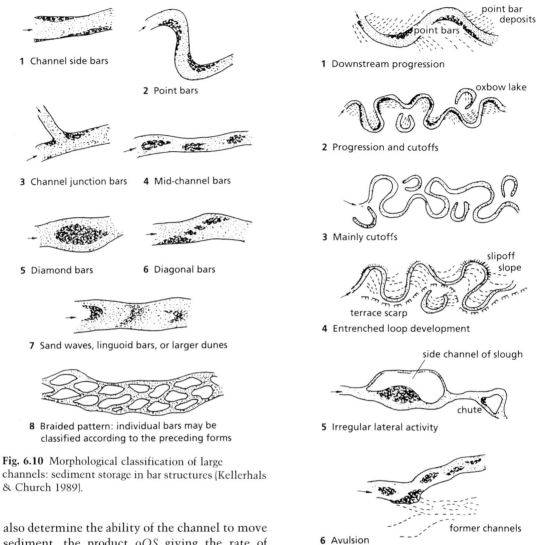

Fig. 6.10 Morphological classification of large channels: sediment storage in bar structures (Kellerhals & Church 1989).

Fig. 6.11 Classification of large channels: lateral activity (Kellerhals & Church 1989).

also determine the ability of the channel to move sediment, the product ρQS giving the rate of energy expenditure along the channel. The limit gradient for a river is the valley gradient, S_v. However, the river might take up a lower gradient, S, by doing erosional and sediment transporting work to become sinuous. The sinuosity is the quotient S_v/S = river length/valley length. The sinuosity introduces an additional mechanism for energy dissipation since the water must be accelerated radially through the bends, so that eventually a statistically stable plan-form will be reached for a given formative flow and sediment supply. At this stage, the channel will continue to move sediment and shift laterally, but will not

become more sinuous, on average. If sediment supply becomes relatively large, the river may not transport all of it, even when flowing on the limit gradient, S_v. Then, the river will raise its bed (aggrade) by net deposition of sediment; in the very long term it will thereby increase its gradient. In such a state, the usual pattern of lateral instability is a sudden shift of course, termed an *avulsion*, just as in smaller channels that are episodically blocked by debris. If the imbalance of sediment load is sufficiently great, the river will perform frequent avulsions within a broad channel zone, thereby taking up a braided habit through the material. An empirical summary of conditions for various channel patterns is given in Fig. 6.12. Where major rivers are confined or entrenched, they often are constrained to flow in a single channel with gradient S_v. Such a channel is almost never alluvial.

Where large rivers flow in large valleys, or without constraint on plains or deltas, they characteristically are associated with a 'floodplain'. In hydraulic literature the term floodplain denotes a surface adjacent to a river that is episodically flooded, without implication about its formation. In ecological and geomorphological literature, a floodplain almost always denotes a genetic floodplain—a surface constructed by the present river as the result of sediment deposition during lateral shifting and overbank flooding. Such floodplains sometimes contain lakes or side-channels, seasonally or permanently connected with the main channel, which themselves constitute important aquatic habitats.

Braided and split channels present special habitat conditions. As flows change, the main channel—as in single-thread channels—experiences deeper and faster flows, but additional side-channels tend to become active, so that the low-flow habitat tends to be replicated toward the channel margins. The centre of a main channel often is a relatively hostile environment, where high sediment transport maintains a sterile substrate and high velocities extract large energy tolls from organisms to maintain themselves. The channel margin usually is much more favourable, both hydraulically and for food resources. Side-channels may, then, provide the best habitat. In braided systems, secondary channels change rapidly, but the persistent secondary channels of anastomosed systems may provide the most stable habitat units of all. In circumstances where the coarser part of the sediment load, travelling on or near the bed, is excluded at the side-channel entrance, they may also present conditions that are qualitatively distinct from those in the main channel.

6.5 PHYSICAL BASIS OF AQUATIC HABITAT

The equations of hydraulic geometry given above were developed to describe downstream changes in channel form at some flow of constant return period. However, they can be adapted to provide a description of hydraulic changes in a single cross-section as flow varies. In this context, they are

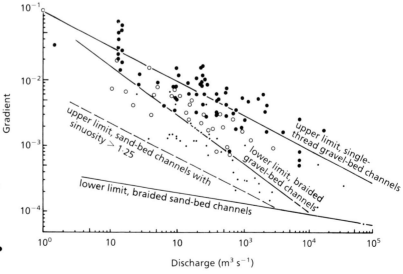

Fig. 6.12 Channel pattern gradient-discharge limits based on data compiled by the writer. Limits for sand-bed channels based on work by Bray (1973) and Neill (1973). • Braided gravel-bed channel; ○ wandering gravel-bed channel; • braided sand-bed channel.

purely descriptive; coefficients and exponents are not constrained in the manner discussed above, and a single equation may not necessarily describe the entire range of flows. However, they are very helpful in habitat description, since P (or w_s) approximates available substrate area per unit length of channel, $PR = w_s d. = A$ defines water volume per unit length of channel, and $\langle v \rangle = Q/A$ provides a measure of water velocity.

Figure 6.13(a) shows the at-a-station hydraulic

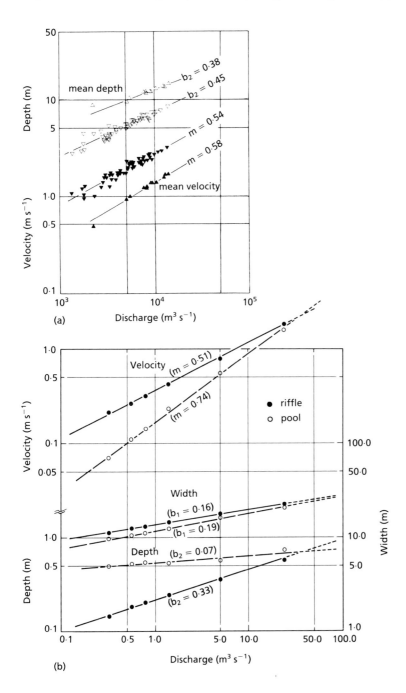

Fig. 6.13 At-a-station hydraulic geometry. (a) Fraser River, British Columbia at Agassiz (gravel bed (∇ ▼)) and at Mission (sand bed ($\triangle$ ▲)); $m = 1 - b_1 - b_2$. (b) Hangover Creek, Queen Charlotte Islands, British Columbia, a boulder−gravel stream of intermediate size, averaged over major morphological units (Hogan & Church 1989)

geometry for two stations on Fraser River, British Columbia. One section is in a steep, gravel-bed reach with wandering channel morphology, whereas the other is in a straight, sand-bed reach with much lower gradient. In both cases, channel width is about the same. The steep reach has much higher velocity and smaller depth at all flows. The at-a-station concept can be extended to consider the average hydraulic geometry based on a number of cross-sections within a reach. Figure 6.13(b) shows the average hydraulic geometry for two major morphological units in an intermediate boulder–gravel channel. Some distinctions between the units are evident; of particular interest, however, is the loss of distinctiveness at high flows.

The summary hydraulic geometry obscures some aspects of channel shape that are significant

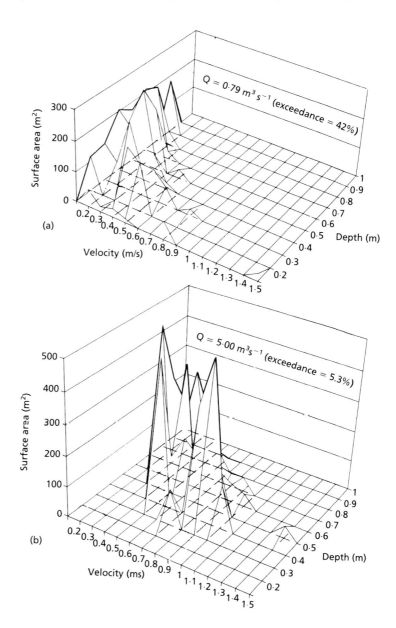

Fig. 6.14 Disaggregated hydraulic geometry; mean depth and mean velocity for Hangover Creek, Queen Charlotte Islands (Hogan & Church 1989).

in habitat appraisal. In a trapezoidal cross-section with steep banks, such as the gauge section at Mission on Fraser River, most of the bed area has a very similar depth at any given stage. In comparison, a triangular cross-section, which the Agassiz section approximates, maintains a varied distribution of depths at all stages, which usually is desirable for habitat maintenance. Natural channels tend often to take up triangular shapes, whereas engineered channels are nearly always trapezoidal.

Complete habitat assessment usually requires a more refined assessment of depth and velocity distributions. Fish prefer particular combinations of velocity and depth in the channel. Accordingly, the best description of physical habitat would be a completely disaggregated bivariate distribution of velocities and depths everywhere in a reach. Figure 6.14 shows the distribution of surface area by mean depth–mean velocity classes along a 260-m reach in a cobble–gravel bed stream. This information is recoverable from a topographic map of the reach provided that water surface elevations are available for each flow and the flow is measured in one section (a gauge section). This is relatively easy to arrange, but it still entails a degree of undesirable averaging within sections. The alternatives require a very onerous measurement programme to determine velocities at many points in the reach at each flow (Collings *et al* 1972; Sullivan 1986), or some means to estimate velocities. In large channels with regular velocity profiles, simulation procedures may allow estimation (Stalnaker *et al* 1989), but in intermediate to small channels with high relative roughness this is not feasible. The question of whether characteristic distributions about the mean of velocity can be assumed in such channels remains a question at the research frontier.

ACKNOWLEDGEMENTS

D.L. Hogan has generously shared many results on rivers dominated by organic debris, a much underappreciated determinant of morphology in smaller channels. He has also provided helpful comments on the text. The regional bias of this text is determined by the writer's experience, but

that constrains us all and readers should always be aware of it.

REFERENCES

Ackers P. (1988) Alluvial channel hydraulics. *Journal of Hydrology* **100**: 177–204. [6.4]

Bisson PA, Nielsen JL, Palmason RA, Grove LE. (1981) A system of naming habitat types in small streams with examples of habitat utilization by salmonids during low streamflow. In: Armantrout NB (ed.) *Acquisition and Utilization of Aquatic Habitat Inventory Information*, pp 62–73. American Fisheries Society, Western Division. [6.3]

Bray DI. (1973) Regime relations for Alberta gravel-bed rivers. *Proceedings of the 7th Canadian Hydrology Symposium*, pp 440–52. [6.1, 6.4]

Church M, Jones DP. (1982) Channel bars in gravel-bed rivers In: Hey RD, Bathurst JC, Thorne CR (eds) *Gravel-bed Rivers*, pp 291–336. John Wiley and Sons, Chichester. [6.3]

Collings MR, Smith RW, Higgins GT. (1972) The hydrology of four streams in western Washington as related to several salmon species. *United States Geological Survey Water Supply Paper 1968*. [6.5]

Costa JE. (1984) Physical geomorphology of debris flows. In: Costa JE, Fleisher PJ (eds) *Developments and Applications of Geomorphology*, pp 268–317. Springer-Verlag, Berlin. [6.2]

Dunne T, Leopold LB. (1978) *Water in Environmental Planning*. WH Freeman, San Francisco. [6.1]

Fremling CR, Claflin TO. (1984) Ecological history of the Upper Mississippi River. In: Wiener JG, Anderson RV, McConville DR (eds) *Contaminants in the Upper Mississippi River*, pp 5–24. Butterworth, Stoneham, Massachusetts. [6.1]

Grant GE, Swanson FJ, Wolman MG. (1990) Pattern and origin of stepped-bed morphology in high-gradient streams, Western Cascades, Oregon. *Geological Society of America Bulletin* **102**: 340–52. [6.3]

Hogan DL. (1986) Channel morphology of unlogged, logged and torrented streams in the Queen Charlotte Islands. *British Columbia Ministry of Forests Land Management Report 49*. British Columbia Ministry of Forests, Victoria. [6.3]

Hogan DL, Church M. (1989) Hydraulic geometry in small, coastal streams: progress toward quantification of salmonid habitat. *Canadian Journal of Fisheries and Aquatic Sciences* **46**: 844–52. [6.5]

Keller EA, Swanson FJ. (1979) Effects of large organic material on channel form and fluvial processes. *Earth Surface Processes* **4**: 361–80. [6.3]

Kellerhals R. (1967) Stable channels with gravel-paved

beds. *American Society of Civil Engineers. Proceedings, Journal of the Waterways and Harbours Division* 93 (WW1), pp 63–84. [6.1]

Kellerhals R, Church M. (1989) The morphology of large rivers: characterization and management. In: Dodge DP (ed.) *Proceedings of the International Large River Symposium. Canadian Special Publication of Fisheries and Aquatic Sciences* **106**: 31–48. [6.1, 6.4]

Kellerhals R, Church M. (1990) Hazard management on fans, with examples from British Columbia. In: Rachocki A, Church M (eds) *Alluvial Fans: A Field Approach*, pp 335–54. John Wiley & Sons, Chichester, UK. [6.2]

Kellerhals R, Church M, Bray DI. (1976) Classification and analysis of river processes. *American Society of Civil Engineers Proceedings, Journal of the Hydraulics Division* **102**: 813–29. [6.4]

Leopold LB, Wolman MG. (1957) River channel patterns: braided, meandering and straight. *United States Geological Survey Professional Paper* 282–8.

Lisle TE. (1986) Stabilization of a gravel channel by large streamside obstructions and bedrock bends, Jacoby Creek, northwestern California. *Geological Society of America Bulletin* **97**: 999–1011. [6.3]

Marston RN. (1982) The geomorphic significance of log steps in forest streams. *Association of American Geographers Annals* **72**: 99–108. [6.3]

Mollard JD. (1973) Air photo interpretation of fluvial features. *Proceedings of the 7th Canadian Hydrology Symposium* 341–80. [6.4]

Neill CR. (1973) Hydraulic geometry of sand rivers in Alberta. *Proceedings of the 7th Canadian Hydrology Symposium* 453–61. [6.1, 6.4]

Osterkamp WR, Hedman ER. (1982) Perennial-stream-flow characteristics related to channel geometry and sediment in Missouri River basin. *United States Geological Survey Professional Paper 1242*. [6.1]

Schumm SA. (1977) *The Fluvial System*. John Wiley & Sons, New York. [6.1]

Schumm SA. (1985) Patterns of alluvial rivers. *Annual Reviews of Earth and Planetary Science* **13**: 5–27. [6.4]

Shreve RL. (1967) Infinite topologically random channel networks. *Journal of Geology* **75**: 178–86. [6.1]

Stalnaker CB, Milhous RT, Bovee KD. (1989) Hydrology and hydraulics applied to fishery management in large rivers. In: Dodge DP (ed.) *Proceedings of the International Large River Symposium. Canadian Special Publication of Fisheries and Aquatic Sciences* **106**: 13–30. [6.5]

Sullivan K. (1986) *Hydraulics and fish habitat in relation to channel morphology*. PhD dissertation, The Johns Hopkins University, Baltimore. [6.3, 6.5]

Teversham JM, Slaymaker O. (1976) Vegetation composition in relation to flood frequency in Lillooet River valley, British Columbia. *Catena* **3**: 191–201. [6.1]

Whittaker JG, Jaeggi MNR. (1982) Origin of step-pool systems in mountain streams. *American Society of Civil Engineers Proceedings, Journal of the Hydraulics Division* **108**: 758–73. [6.2]

7: Floodplain Construction and Erosion

J.LEWIN

There are three major reasons why floodplain development and characteristics need to be understood for river management purposes. At high flows, floodplains become part of the surface flow system, and even at low water, groundwater returns can be a highly significant baseflow component of river discharge. A knowledge both of the surface forms that affect the passage of flooding waters, and of how these forms can change, is thus important as is a knowledge of subsurface floodplain materials that affect groundwater storage, flow and quality. A second reason involves the dynamic nature of rivers and floodplains. In mid-latitudes where there are mobile river channels, a high proportion of river sediments may be derived from the erosion of floodplain materials so that their physical and chemical characteristics are determinants of channel material flows. The same applies to other environments, such as semi-arid systems where vertical cut and fill sequences over a timescale of decades to centuries makes floodplains alternately into sources or sinks for transported sediments. Finally, because floodplains are primarily constructed through past river sedimentation, they can preserve an extended record of the changing suite of formative factors involving climatic change and variable human influence which has affected valley floor environments over some thousands of years.

7.1 FLOODPLAIN CONSTRUCTION

Fluvial deposition results from the major processes of *sedimentation* of finer material which has been suspended in flowing or ponded water, or where material travelling along at or near the river bed (which is usually coarser) comes to rest.

The latter commonly occurs in *accretionary bed forms* or bars. Lag deposits, coarser material left behind or let down on to river beds, may also form significant accumulations in some environments.

Processes of accretion and sedimentation may take place within river channels, or beyond them out on floodplains at periods of high river flows. It is usual to associate suspension deposits with overbank flows and accretionary deposits with within-channel flows, but in practice the depositional environments in which these processes occur can be very varied. Flood gravels can be deposited overbank, and fine sedimentation can take place quite commonly in within-channel slack-water zones. The latter may of course be reactivated again, and there may be seasonal recycling of fines within river channels according to river stage, with low flow build-up and flood flow removal.

River deposition processes of these kinds, which may create more permanent floodplain construction, reflect variability in three groups of factors. The first concerns sediment supply, both in terms of volume and sediment size. Floodplain materials reflect the bedrock and superficial deposits of their catchments, the size and quantity of such materials released into rivers by weathering and erosion (including accelerated soil erosion) on hillslopes, and the upstream erosional activity of river channels themselves. In-stream hydraulic sorting and abrasion processes also reduce the size of materials available downstream.

It is extremely difficult to generalize about the rates at which these processes operate. Figure 7.1 shows some sediment budget studies in which the sources of sediments are identified together

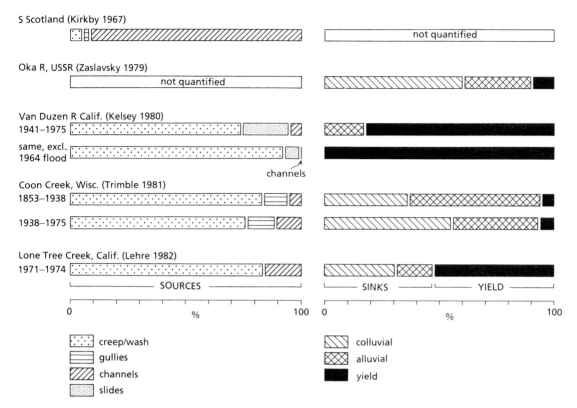

Fig. 7.1 Selected sediment budgets for contrasted catchments. Bar lengths represent the proportions of eroded sediment being derived from the source types indicated, and the percentages which are deposited on slopes (colluvial) or on floodplains (alluvial), or which are transported out of the studied catchments.

with the proportion of such materials which remain in alluvial 'storage'. This may differ according to the time period assessed, with individual flood events having particular and distinctive effects (Van Duzen River, California, USA), and human-induced changes producing effects which differ over time (Coon Creek, Wisconsin, USA). Some catchment materials, such as wind-deposited loess, may give catchments very high sediment throughput and deposition rates. The major example is the Huang He (Yellow River) in China with its tributaries draining thick loess deposits particularly in Shaanxi Province. Loess deposits have also been important sources for floodplain deposition in the southeastern USA and southern Britain. For practical purposes it is therefore necessary to make an assessment of

catchment sediment sources, actual and potential, in order to explain or anticipate floodplain sedimentation response.

A second factor concerns the availability of subaqueous voids which must be available for thick accumulations to occur. On a local scale, this can be provided by repeated overbank flooding such that deposits may accumulate in laminae, layer on layer, as long as overbank flows can be maintained across the thickening deposit. For accretionary deposits to persist, lateral channel mobility is usually required so that bank erosion may create channel enlargement which is balanced by infilling. Both of these processes may be self-limiting, but there can also in the longer run be external controls on sedimentation, such as tectonic subsidence. Surveys of floodplain geometry, and of channel mobility patterns and rates, are required to set floodplain sedimentation patterns in context.

The third factor concerns hydraulics. These have been discussed in Chapters 5 and 6, and it is important here mainly to appreciate that available river energy may set limits on the sizes of material that can be transported (and hence subsequently deposited). Energy availability may vary on a catchment basis (from high to low in the downstream direction), or locally within sinuous channel patterns, and out on the floodplain itself, with distance away from the channel. Both local and regional hydraulic factors give variety to the size suite of accretionary and suspension deposits that are eventually incorporated into floodplains.

On the assumptions that the sizes of deposited sediments are competence- or transport-limited, and that sediment size or size distributions can be uniquely related to measures of streamflow, it has proved possible to reconstruct former conditions via a range of palaeohydrological equations (see Williams 1983; Church 1987; Maizels 1987). A more generalized approach to floodplain material description has been the characterization of facies (units of uniform sediment type) using a simple coding system developed by A. Miall. A version of this is given in Table 7.1. Sediments are broadly grouped as gravel-sized, sandy or fines, with units being further categorized by sedimentary structure. Using bank sections or borehole data it is also possible to record facies thicknesses, vertical sequences, and possibly lateral extent. Again, without recourse to interpretation, it is also desirable to record measures of floodplain form, for example width, gradient and local relief, possibly using profiles or three-dimensional mapping (see, for example, Lewin & Manton 1975; Chisholm & Collin 1980) on a scale that is much more detailed than usually available from published maps. From these it commonly emerges that floodplains have significant if small-scale relief. For example Lewin and Manton (1975) showed that relief of up to 2.5 m was to be found on the floodplains of Welsh rivers, with the present river banks the highest part of the 'plain'. Figure 7.2 shows a portion of the floodplain of the Afon Tywi at Llandeilo, Dyfed, Wales (SN 6423 and SN 6524). The topographic 'lows' lead to early entry of floodwaters out on to the floodplain, with depressions being waterlogged for long periods. Such floodplain features are created in a variety of sedimentary environments.

7.2 SEDIMENTARY ENVIRONMENTS

At a scale approximating that of river channel width, it is possible to identify six fluvial sedimentary environments in which sedimentation or accretion may take place (Table 7.2). Over time or as rivers shift their courses, these environments may relocate with the development of enlarging sediment units which come together to make up the floodplain itself. The six environments may each develop characteristic or indeed contrasting facies and geometries, and it is helpful to appreciate what these may be.

On the beds of active channels, *lag deposits* may line the basal erosion surface produced by migrating streams. Because stream depth varies, this surface can be undulating. For example, as meanders develop and tighten their radius of curvature, there may be a corresponding deepening of the channel so that basal erosion surfaces are lowered. Furthermore, confinement of channels against unerodible valley sides or artificial structures also leads to channel deepening, so that coarser lags may be found at increased depths in such locations.

Channel deposits assume a variety of geometries, depending initially on active bar geometry. These may be classified with regard to formation position in the channel (mid-channel, lateral or point bars attached to one bank, or diagonal, passing from one side of the channel to the other); according to relative dimension (longitudinal, where the downstream dimension is several times the cross-stream dimension, or transverse where they are more nearly equivalent); or according to some descriptor of shape (lingoid, crescentic, lobate, rhomboid, scroll etc.). In time, sediment bars may build up as a succession of accreting sheets, or may prograde laterally or downchannel with avalanche fronts producing steeply dipping beds. Accretionary surfaces may thus be at various inclinations, possibly quite steep, as on the inside of curving channels where migration of the channel on one side may

Table 7.1 A lithofacies/structure coding for fluvial sediments (from Miall 1978)

Facies code	Lithofacies	Sedimentary structures	Interpretation
Gms	Massive, matrix-supported gravel	None	Debris flow deposits
Gm	Massive or crudely bedded gravel	Horizontal bedding, imbrication	Longitudinal bars, lag deposits
Gt	Gravel, stratified	Trough crossbeds	Minor channel fills
Gp	Gravel, stratified	Planar crossbeds	Linguoid bars or growths from older bar remnants
St	Sand, medium to very coarse, may be pebbly	Solitary or grouped trough crossbeds	Dunes (lower flow regime)
Sp	Sand, medium to very coarse, may be pebbly	Solitary or grouped planar crossbeds	Linguoid, transverse bars, sandwaves (lower flow regime)
Sr	Sand, very fine to coarse	Ripple marks	Ripples (lower flow regime)
Sh	Sand, very fine to very coarse, may be pebbly	Horizontal lamination, parting or streaming lineation	Planar bed flow (lower and upper flow regime)
Sl	Sand, fine	Low angle (<10°) crossbeds	Scour fills, crevasse splays, anti-dunes
Se	Erosional scours with intraclasts	Crude crossbedding	Scour fills
Ss	Sand, fine to coarse, may be pebbly	Broad, shallow scours including cross-stratification	Scour fills
Sse, She, Spe	Sand	Analogous to Ss, Sh, Sp	Eolian deposits
Fl	Sand, silt, mud	Fine lamination, very small ripples	Overbank or waning flood deposits
Fsc	Silt, mud	Laminated to massive	Backswamp deposits
Fcf	Mud	Massive, with freshwater molluscs	Backswamp pond deposits
Fm	Mud, silt	Massive, desiccation cracks	Overbank or drape deposits
Fr	Silt, mud	Rootlets	Seatearth
C	Peat, carbonaceous mud	Plants, mud films	Swamp deposits
P	Carbonate	Pedogenic features	Soil

be accompanied by lateral accretion on the other, to form so-called 'epsilon' cross-stratification. As bars develop over time, their sediments may be incorporated into floodplains as much more extensive contiguous bodies whose limits are set by switches in channel location and erosion. Accretionary sediments, perhaps crudely strat-ified sandy gravels or alternatively cross-bedded sands, may thus come to dominate the floodplain materials of some mid-latitude migrating-stream floodplains, where active sediment sources and available stream powers are appropriate.

Many larger and some smaller rivers may possess larger sedimentary bed-forms than the

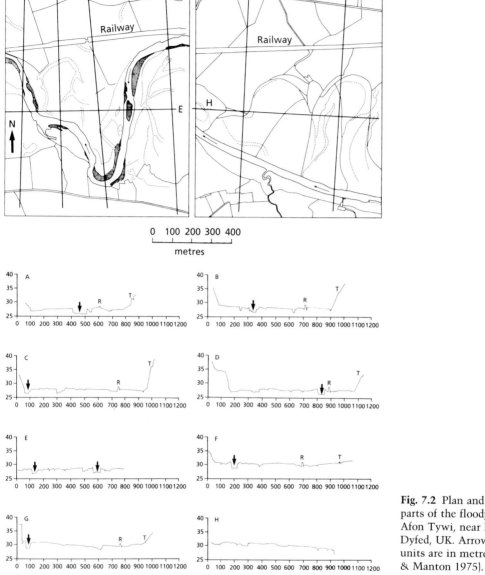

Fig. 7.2 Plan and profiles for parts of the floodplain of the Afon Tywi, near Llandeilo, Dyfed, UK. Arrow marks river; units are in metres. (after Lewin & Manton 1975).

'unit bar' features of about channel-width dimensions so far discussed. For example, Crowley (1983) reported 200–400 m long sandy bed-forms on the Platte river, whilst large sandwaves occur on the lower Brahmaputra (Coleman 1969; Bristow 1987).

Channel fills occupy abandoned channels; their cross-section and plan follows the dimensions of the former channel, but their lengths may vary from short arcuate reaches to very extensive linear traces where much longer channel stretches have been deserted. For example, former courses of the Huang He in China, some hundreds of kilometres in length, have been abandoned and infilled,

Table 7.2 Fluvial sedimentary environments

Environment	Major facies
Active channels, lag deposits	G
Active channel bars	Gm, Gt, Gp, St, Sp, Sr
Abandoned channel fills	Sr, Sh, Fl, Fsc
Channel margin (levee, crevasse, crevasse splay)	Sh, Sr, Sl
Floodplains and backswamps	Fl, Fsc, Fm, C, P
Non-fluvial materials (aeolian, glacial, colluvial)	Sse, She, Spe, Gms

whilst branches of former multiple channel systems may be found in many mid-latitude environments where now only a single channel is to be found (a number of examples are illustrated in Petts 1989). Fill materials can be very varied: rapidly accreting sand plugs adjacent to the new channel, organic fills where the supply of mineral flood sediments is less dominant, or thick laminated flood clays. Such clay plugs may later restrict lateral channel movement across the floodplain, as indeed may fine-grained deposits more generally (Ikeda 1989).

In the nature of things, abandoned channels are ephemeral features which are liable to be infilled over a period of years. Temporarily they may provide important wetland habitats, but under natural conditions such habitats would eventually be infilled, to be replaced by others elsewhere.

Characteristic *channel marginal* environments include levees and crevasse chutes and splays. The former are channel-side ridges, usually with sediments of sand size or coarser, fining away from the river with decreasing elevation, and produced by accretion/sedimentation in the critical zone where overbank flows first pass out on to the floodplain. Hydraulics in this zone have recently been modelled, with travelling voltices in the shearing zone along channel banks being involved in sedimentation processes (see Knight 1989). Levees may be dissected and their deposits (together with direct channel-derived additional material) spread out across the floodplain in crevasse splay features.

In this zone, sedimentation/accretion features

may also be crucially related to accretionary bar features, infilling hollows (*swales*) and other topographic depressions. Bar-top deposition thus blurs the surface form of bars with finer infills.

Floodplain/backswamp environments are dominated by sedimentation, generally with materials that are increasingly finer with distance from the channel. Such environments may be temporarily or permanently waterlogged with ponded waters. Four factors crucially affect the nature of sedimentation: overbank suspension loads, the geometry of the floodplain, the sequence of channel developments (see Farrell 1987), and the nature of vegetation growth. It is thus possible to have at one end of the scale relatively planar floodplains rapidly building up with silty materials (Schmudde 1963); at the other, sandy levees or rivers on raised ribbon-like alluvial ridges may be backed by swamps and lakes in which organic or lacustrine sedimentation dominates (Blake & Ollier 1971). Under natural conditions these basins may be very persistent, although in settled landscapes they may be a prime target for land drainage works.

Finally *colluvial* or other non-fluvial sediments may form an appreciable component of floodplain deposits, especially at the margins under conditions of active slope erosion. This is particularly true of steep, narrow headwater valleys (Lattman 1960), and was true of Pleistocene cold-climate environments which, as we shall see, contributed a great deal to the valley fills of many lowland mid-latitude areas. In dry conditions, wind-blown material, either derived from arid-zone deflation, glacial materials or from river deposits, may also be important components of floodplains.

Floodplain sedimentation as a whole reflects very strongly the ways in which these six sediment types are combined within individual floodplain segments; this in turn relates to stream sediment loads and the pattern of channel change and migration. For alluvial rivers (the 'large channels' of chapter 6, section 6.4), sedimentation style is commonly associated with channel pattern prototypes (braided, meandering or anastomosing) which have themselves been further differentiated largely on grounds of sediment size. For example, Miall (1977) suggested six *braided-stream* facies assemblage types, with dominant sand or gravel components, whilst Jackson (1978) similarly identified five classes for *meandering* streams dominated respectively by muddy, sandy or gravelly sediments and mud/sand or sand/gravel combinations. A couplet of finer sedimentation deposits overlying accretionary deposits is very common for meandering streams. To these may be added an anastomosing prototype (Smith & Smith 1980), a low-gradient, aggrading, multiple-channel and sinuous system dominated by ribbon channel sediments and organic wetland flood basins. Miall (1985) similarly provided 'illustrations' of 12 models of fluvial sedimentation style (which he stressed was not a comprehensive set).

Identification of these numerous prototypes is an advance on the pioneering but simplistic distinction between braided and meandering models of earlier workers, but there may still be other models to consider. For example, Graf (1988) has questioned the applicability of the term 'floodplain' to dryland rivers where the whole valley floor between terraces is ephemerally occupied by a braided channel, with parts of this becoming infilled during periods when a transition to meandering occurs. A meandering style may thus be set within largely braiding sediments. Rust and Nanson (1986) have reported coexisting mud-braids (activated during exceptional floods) and sand-bed anastomosing channels, the latter operating at moderate discharges. Others have also reasonably argued for a continuum of channel patterns (e.g. Ferguson 1987) responding to such factors as bank strength and sediment competence.

From a floodplain point of view, it is also necessary to appreciate that channel deposits are indeed only one of the six sedimentary environments discussed above. Their relative significance is extremely important in determining the nature of floodplains: many are dominated by overbank and backswamp processes beyond a relatively narrow ribbon of lateral channel activity (e.g. Blake & Ollier 1971), although others may largely reflect accretion processes (Wolman & Leopold 1957).

Channel pattern and migration style are of interest in that they determine how floodplains may develop contrasted patterns of lateral and vertical age sequencing, and the 'preservation times' that alluvial units may have before they are re-eroded. For example, meandering streams may migrate in different styles producing facies of contrasted grain sizes as they evolve. Where quality factors are important, as where pollutants may have been incorporated into floodplains during particular time periods, or into particular sediment size ranges, it is important to identify where on floodplains such 'risk' zones are concentrated. Equally, particular floodplain sites may have particular 'lifetimes' which vary according to the rate and pattern of channel migration.

Figure 7.3 shows a reach of the River Teme, UK in 1947 and 1973. The earlier photography, obtained after the major 1947 floods, shows complex evolving loops and cutoffs and the large quantity of overbank flood sediments (white areas). By 1973, further loop development and abandonment occurred, with channel fills and overbank sedimentation being of considerable relative importance. Use of air photographs in this way, and also of historical maps and dating of sediments, may allow age zoning of floodplains to be undertaken for particular sites of interest.

Figure 7.4 shows accretion patterns in the last century for three Welsh river reaches. Although in meandering streams, point bar sedimentation is usually considered to dominate, this is not in practice always the case. For example, cutoff fills occupy 14% of the recent sediments on the Teme, with point bar sediments on both Teme and Dee amounting to only about half the total of recent accretion sediments.

7.3 RATES OF ACTIVITY

The complex store of sediments which make up floodplains may be augmented or eroded at varying rates and with varying patterns. We can identify these rates on three scales. First are the erosion/deposition rates for particular sedimentary environments. Then come input–output budgets for 'architectural assemblages' as they are combined in particular alluvial valley reaches. Finally, there are the responses of alluvial systems within whole drainage basins.

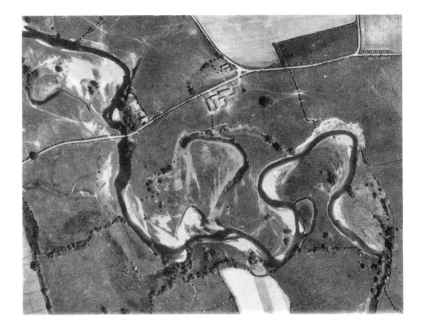

(a)

(b)

Fig. 7.3 The River Teme near Ludlow, Hereford and Worcester, UK, in (a) 1947 and (b) 1973.

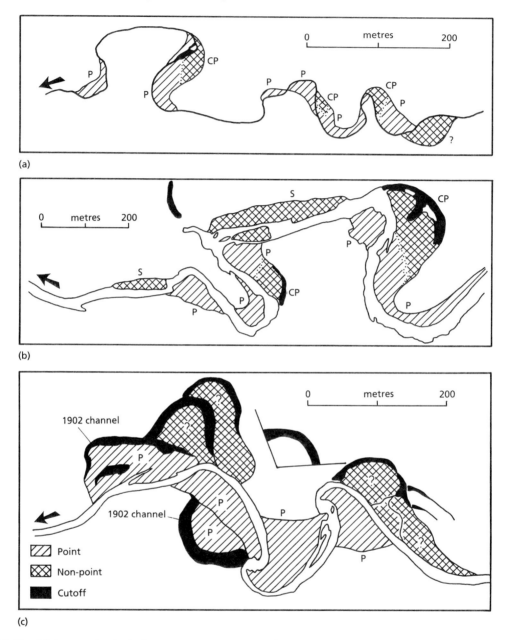

Fig. 7.4 Floodplain accretion in the last century on rivers in Wales. (a) The Afon Twymyn, Dyfed (SN 885998); (b) the Dee, Gwynedd (SH 980369); (c) the Teme, Powys (SO 384730) (from Lewin 1983).

Erosion and deposition rates

Rates of overbank sedimentation that have been recorded were usefully reviewed by Bridge and Leeder (1979); there have been several more recent studies for particular events at specific sites using a range of direct measurement or dating techniques (e.g. Lambert & Walling 1987). A diffusion model developed by Pizzuto (1987) appears to predict quite well both rates of sedimentation and the decrease in grain size which occurs away from channels in the post-settlement alluvial overbank deposits in Pennsylvania, USA. The model predicts a more rapid decrease in sand deposition than is apparent in the deposits, suggesting non-diffusion processes may be additionally important for sand dispersion.

Long-term rates given by Bridge and Leeder (1979) range from 0.01 m year^{-1} to 0.0002 m year^{-1}; Brown (1987) suggested a rate of 0.0014 for the main floodplain of the Lower Severn, UK. Cutoff and within-channel sedimentation fills can be more rapid than this, as can deposition from single extreme events. Stear (1985) recorded 2–3.5 m of laminated fine-grained sand and silt deposited overbank by an extreme flood in a semi-arid environment. The deposits of a spring 1964 flood on the Ohio River, USA ranged from 0.46 m on a levee to less than 3 mm away from the river (Alexander & Prior 1971). Rates of sedimentation are very clearly both site and event specific.

Lateral accretion rates can in some circumstances be determined by meander migration rate assessment (Wolman & Leopold 1957), either through direct measurement using ground survey, or from historical maps (see also Lewin *et al* 1977; Hooke 1980). Migration rates (and thus rates of floodplain accretion) have been linked to stream power, outer bank height and a resistance coefficient in a regression model developed by Hickin and Nanson (1984). Such rates may be variously expressed, as m year^{-1} or as a percentage of channel width per year, for example, with annual rates in the order of metres per year being common in higher lateral accretion rate environments.

Cutoff and avulsion rates are significant for the termination of meandering sequences. Cutoffs may be of several types: in particular neck cutoffs produced by the breaching of narrow gaps between adjacent channel bends, and chute cutoffs which involve more extensive scouring of new channels across floodplains and point bars. High sinuosity and low floodplain slope appear to favour the former, steep slopes and erodible non-cohesive and poorly vegetated surfaces the latter. In a study of some 1000 km of valley floor in Wales and the Borderland, Lewis and Lewin (1983) found some 145 recent cutoffs of which 55% were simple

chute, and 16% simple neck types. Others involved cutoffs related to bar development and multiloop forms.

Avulsions are important in the development of larger floodplains, for example in the Holocene development of the lower Mississippi Valley, USA, in the evolution of anastomosing systems (Smith *et al* 1989), and also in other types of multichannel system (Ferguson & Werritty 1983). However, such avulsions are extremely difficult to predict. Avulsion may be accompanied by a transformation in channel patterns (Brizga & Finlayson 1990).

Input–output budgets

It is possible to model valley floors and floodplains as complexes of sediment storage units of different ages. For example, a study of recent floodplain history of a reach of the Little Missouri involved the age zoning of the valley floor by tree age (Everitt 1968), with a negative exponential relationship between area and age of deposition. Reservoir theory may be applied to such data (see Dietrich *et al* 1982), and computation of mean residence time, age distribution of sediment leaving storage, and sediment transport rates may be undertaken. These may vary for different valley floor components (Kelsey *et al* 1986; Nakamura *et al* 1987). Breaks in cumulative age–area distribution curves may be interpreted as resulting from significant events reordering parts of the valley floor. 'Inset' valley floors of different age, some on different topographic levels, may also be distinguished, as shown in Fig. 7.5.

Drainage basin alluvial responses

At the catchment scale, downstream sediment transfers and upstream recession of erosion phases come into effect. Upstream incision may lead to downvalley aggradation, whilst there may be downvalley movement of 'slugs' of sediment. These kinds of 'complex response' activities (Schumm 1977) may operate at a variety of timescales. Rains and Welch (1988) have discussed the existence of non-synchronous Holocene alluviation which could be explained in this way. On a shorter timescale, alluviation and incision in

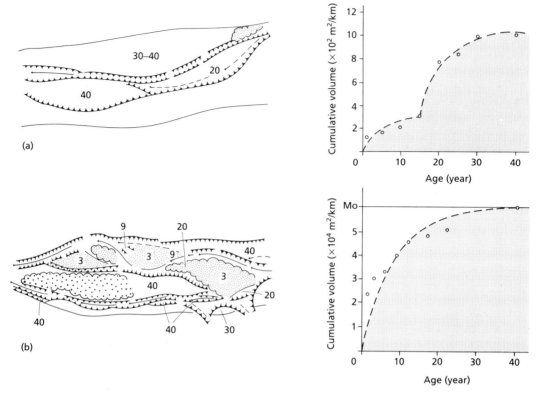

Fig. 7.5 Valley floor area/age relationships for two valley types on the River Furano, Hokkaido, Japan (after Nakamura *et al* 1987).

different parts of catchment systems have been documented with respect to recent historical alluviation (e.g. Trimble 1981; Knox 1987).

7.4 LATE QUATERNARY HISTORIES

Whilst floodplain sediments accumulate, environments may change. Over the past several thousand years, there have been six important external factors that have modified floodplain processes to the extent that floodplain sediments themselves record a considerable variety of process system changes. These are:

1 Major climatic oscillations between glacial and interglacial conditions. The most recent maximum advance of ice sheets in mid-latitudes some 18 000 years ago directly over-ran many valleys and provided large volumes of sediment in the form of glacigenic materials. Peripheral to ice sheets, periglacial conditions also led to considerable slope sediment production, infilling smaller valleys and aggrading larger ones. Low sea levels steepened and incised rivers close to the sea.

2 Subsequent climatic recovery re-established forest vegetation with a developing soil cover. Rising sea levels drowned the lower parts of valleys. These filled, in part with fluvial and marine sediments at a rate determined by local postglacial sediment inputs. Some drowned valleys remain as estuaries; others have infilled and are now floodplains. Removal of the ice-sheet load also led to isostatic (tectonic) movements, and the complex interaction of sea-level rise, tectonics and sediment supply has produced complex alluvial histories. On the whole, postglacial conditions have led to decreasing sediment supply rates under natural conditions (see Church & Rider 1972).

3 Within the Holocene (the last 10 000 years) there have been second-order climatic fluctuations. For example, alternating flood- and drought-dominated river regimes have been identified in coastal rivers of New South Wales, Australia (Erskine & Warner 1988), whilst dry and humid periods can also be identified in Europe (Probst 1989). The so-called Litte Ice Age, *c*. 1600–1850, led to recorded increases in floods, slope erosion and glacier advance, and these achieved transformations in channel pattern (Bravard 1989). El Nino events have produced phases of aggradation in South America (Wells 1990). Finally, humid–arid transitions in North America have produced marked changes in alluviation patterns (see Knox 1983; Schumm & Brackenridge 1987).

4 In tectonically unstable orogenic environments, rivers have been affected by uplift and subsidence. This may provide both deepening alluvial basins and steepening or incising rivers where they are accommodating to uplift (see Schumm 1986).

5 Forest clearance, agriculture and urbanization, over the past several thousand years in some places but only decades or less in others, have severely modified sediment systems. 'Post-settlement' alluvium is a common feature of settled landscapes, resulting particularly from accelerated soil erosion. This has been widely recorded in Europe in the longer term (so-called Haugh loams) and North America in the shorter (see review by Butzer 1981).

6 To such inadvertent environmental effects of human activities, must be added deliberate efforts to improve land drainage and to control river runoff and erosion. This includes field underdrainage, ditching for upland afforestation, wetland drainage, cutoff infilling and other landfill exercises, river channelization, and river impounding for water supply, power generation, flood control, navigation and irrigation. In few settled environments, therefore, are rivers now free to continue the processes of sedimentation which generated the floodplains through which they run.

The nature and timing of these environmental changes and constraints has varied very considerably, and even at well-studied sites it may be an open question as to whether a set of deciphered changes in sedimentation style is a response to 'natural' or 'human' causes in the catchment. In some studies climatic causes have been pinpointed (e.g. Brackenridge 1980). By contrast, for the Upper Thames, UK, Robinson and Lambrick (1984) were convinced that human activities were an 'entirely adequate' explanation for Holocene changes. They suggested a lower unit of Late Devensian (last glaciation) gravels that had not subsequently been reworked, with a relatively dry valley floor existing until the late Neolithic and Bronze Age. This was followed by a period with increased evidence of flooding (by around 2000 BP), and then the deposition of alluvial clays sealing in prehistoric remains. Later alluviation phases followed in Saxon and medieval times to produce the floodplain we see today.

In the Lower Severn also, Brown (1987) has suggested stable earlier Holocene channels, possibly anastomosing in pattern. Simplification of this pattern by abandonment with some avulsion and cutoff activity then followed, together with deposition of a thick layer of overbank sediments of anthropogenic origin. In what are now the Norfolk Broads, UK, Holocene organic sedimentation dominated in backswamp environments, and the peat deposits were cut and then the depression left drowned to create the lakes we see today (Lambert *et al* 1960).

In upland Britain, with higher power streams and steeper gradients, Holocene rivers have been incised into glacigenic and periglacial materials, and they have continued to rework them. This has created vertical flights of valley-floor terraces at some locations (e.g. Macklin & Lewin 1986), the most recent of which may be associated with 19th century aggradation. Activity in such gravel-bed rivers may be concentrated also in limited 'sedimentation zones' separated by more stable reaches, with a spatially rather intricate response to environmental change (Macklin & Lewin 1989). Deforestation can also produce sediment waves moving downvalley (Roberts & Church 1986). Extreme events can also have a greater effect in steepland headwater catchments (see for example Carling 1986; Harvey 1986). Some environments may indeed be dominated by such events, as in parts of south-east Australia where

Nanson (1986) has documented a catastrophic stripping of large parts of floodplains during extreme floods, followed by rebuilding and sedimentation over hundreds or even thousands of years. Useful instances and reviews of flood effects are provided by Beven and Carling (1989).

In the light of all these environmental changes, and the interpretations briefly reviewed above, three conclusions are possible. The first is that present channel activity may give only a poor guide as to the ways in which adjacent floodplains have developed. This is especially true of long-settled and managed environments, ones where small environmental perturbations can trigger large alluvial effects, and where the record of alluviation is an extended one. Second, rivers and their floodplains are not, contrary to some assumptions, environmentally equilibrated. Channels may be incising, aggrading, sedimenting and accreting in response to longer-term patterns of environmental change to which the record of floodplain sediments provides the key. Third,

analysis of individual valley-floor reaches is required because the history that any location has undergone is as yet not 'retrodictable', nor is its future predictable, on the basis of present knowledge. Figure 7.6 illustrates, for example, the vertical sequence that a Welsh river is believed to have undergone since Late Devensian times. This sequence is site specific; it draws attention to the potential for vertical instability as well as the lateral channel mobility that floodplains have been created by (see for example Fig. 7.3). It illustrates both the alluvial 'recovery' from partial infilling of the valley with glacial outwash, but also the significant effects of human-induced effects. Figure 7.7 shows part of the San River valley near Jaroslaw in Poland. Here considerable changes in sedimentation style can again be seen: large-scale meandering in the Late Glacial (*c.* 10 000 BP), smaller meanders in the earlier Holocene, and then braiding most recently. The present river channel is incised into these deposits.

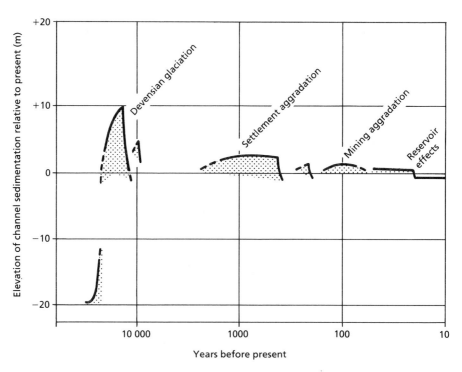

Fig. 7.6 Vertical tendencies in alluviation of the Afon Rheidol, Dyfed, UK, based on data from Macklin and Lewin (1986). Note the logarithmic timescale.

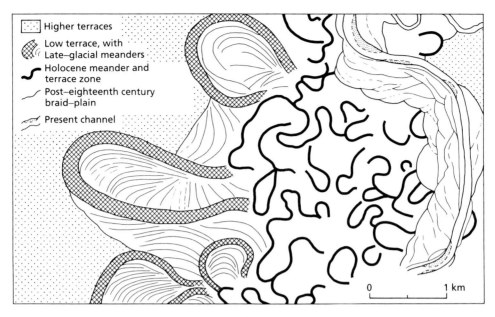

Higher terraces
Low terrace, with Late–glacial meanders
Holocene meander and terrace zone
Post–eighteenth century braid–plain
Present channel

0   1 km

Fig. 7.7 A part of the San River valley near Jaroslaw, Poland (after Szumanski 1983).

7.5 FLOODPLAIN QUALITY

Floodplains can be important accumulation zones for pollutants that are dispersed by fluvial activity from point or areal sources. This can lead to soil pollution *in situ*, or to further river contamination as floodplain sediments are mechanically re-eroded and incorporated into active river systems. Metal mining pollution has been a particular focus of concern and study, but other substances derived from agricultural treatments, urban effluent and industrial and mining activities may also be incorporated into floodplain materials. Under 'natural' conditions floodplains benefit from nutrient enrichment, but pollution effects may become significant in certain environments downvalley from source areas.

Metal dispersal patterns in alluvial sediments have been widely studied in the UK (Lewin & Macklin 1987), in Belgium and the Netherlands (Leenaers *et al* 1988), in Poland (Klimek & Zawilinska 1985), in Australia (East *et al* 1988), and North America (Miller & Wells 1986; Knox 1987; Marron 1989; Graf 1990). A prime aim has been to establish downvalley distance/concentration relationships together with the sediment size ranges or sedimentary environments which may be particularly enriched. Knowledge of the historical development of floodplains in relation to the timing of pollution inputs is also helpful. In addition to the passive dispersal of metals, the volume of wastes may also transform floodplains, causing aggradation and then possibly incision once mining activity and waste input ceases (Lewin *et al* 1983). Particularly dramatic were the effects of hydraulic mining during the 19th century California gold rush (James 1989).

Figure 7.8 shows a schematic cross-section of the South Tyne, UK, with levels of sedimentation and alluvial metal content illustrated. This valley aggraded in the 19th century when metal mining was active in the catchment; the river has subsequently become incised, with a decrease in the metal content of deposited sediment (Macklin & Lewin 1989). Downstream from urban, industrial or mining activities, floodplains are liable to suffer quality problems of this kind. Fine suspension deposits may be particularly affected, to the extent that remobilization by river erosion or indeed the use of particular soils for activities such as market gardening may be undesirable.

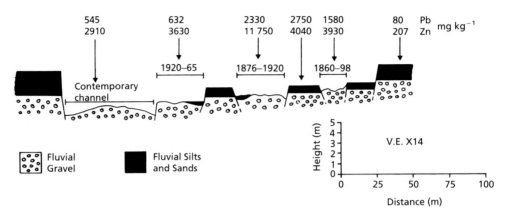

Fig. 7.8 A sketch section of the South Tyne valley floor at Broomhouse, Northumberland, UK (NY 6962), showing alluvial sediments and metal concentrations (after Macklin & Lewin 1989).

7.6 CONCLUSIONS

Floodplains are in large measure produced by physical processes of river deposition in a range of sedimentary subenvironments whose varying dominance produces a considerable global and regional variety to floodplain forms. Studies in recent years have also stressed that floodplains have a history, recording environmental changes of recent centuries and millenia, and not least the effects of human actions on geomorphological activity and sediment quality. There are no simple global models for floodplain formation: instead it is important that individual site investigations should be undertaken and the variety of forms, sediments and processes be recorded. Present and potential future interactions between rivers and floodplains may then be properly understood.

REFERENCES

Alexander CS, Prior JC. (1971) Holocene sedimentation rates in overbank deposits in the Black Bottom of the Lower Ohio River, Southern Illinois. *American Journal of Science* **270**: 361–72. [7.3]

Beven K, Carling P. (1989) *Floods: Hydrological, Sedimentological and Geomorphological implications.* John Wiley & Sons, Chichester. [7.4]

Blake DH, Ollier CD. (1971) Alluvial plains of the Fly River, Papua. *Zeitschrift für Geomorphologie* **12** (Suppl.): 1–17. [7.2]

Brackenridge GR. (1980) Widespread episodes of stream erosion during the Holocene and their climatic cause.

Nature **283**: 655–6. [7.4]

Bravard JP. (1989) La metamorphose des rivières des Alpes Françaises à la fin du moyenage et à l'époque moderne. *Bulletin de la Société Géographique du Liège* **25**: 145–57. [7.4]

Bridge JS, Leeder MR. (1979) A simulation model of alluvial stratigraphy. *Sedimentology* **26**: 617–44. [7.3]

Bristow CS. (1987) Brahmaputra River: channel migration and deposition. In: Ethridge FG, Flores RM, Harvey MD (eds) *Recent Developments in Fluvial Sedimentology*, pp 63–74. Society of Economic Paleontologists and Mineralogists, Tulsa, Oklahoma. [7.2]

Brizga SO, Finlayson BL. (1990) Channel avulsion and river metamorphosis: the case of the Thomson River, Victoria, Australia. *Earth Surface Processes and Landforms* **15**: 391–404. [7.3]

Brown AG. (1987) Holocene floodplain sedimentation and channel response of the lower River Severn, United Kingdom. *Zeitschrift für Geomorphologie* **31**: 293–310. [7.3, 7.4]

Butzer KW (1981) Holocene alluvial sequences: problems of dating and correlation. In: Cullingford RA, Davidson DA, Lewin J (eds) *Time-scales in Geomorphology*, pp 131–42. John Wiley & Sons, Chichester. [7.4]

Carling PA (1986) The Noon Hill flash floods: July 1983. Hydrological and geomorphological aspects of a major formative event in an upland catchment. *Transactions of the Institute of British Geographers* **NS 11**: 105–18. [7.4]

Chisholm NWT, Collin RL. (1980) Topographic definition for washland mapping. *Photogrammetric Record* **10**: 215–27. [7.1]

Church M. (1987) Palaeohydrological reconstructions from a Holocene valley fill. In: Miall AD (ed.) *Fluvial Sedimentology*, pp 743–72. Canadian Society of Petroleum Geologists, Calgary. [7.1]

Church M, Rider JM. (1972) Paraglacial sedimentation; a consideration of fluvial processes conditioned by glaciation. *Geological Society of America Bulletin* **83**: 3059–72. [7.4]

Coleman JM. (1969) Brahmaputra River: channel processes and sedimentation. *Sedimentology and Geology* **3**: 129–239. [7.2]

Crowley KD. (1983) Large-scale bed configurations (macroforms), Platte River Basin, Colorado and Nebraska: primary structures and formative processes. *Geological Society of America Bulletin* **94**: 117–33. Washington, DC. [7.2]

Dietrich WE, Dunne T, Humphrey NF, Reid LM. (1982) Construction of sediment budgets for drainage basins. In: Swanson FJ, Jands RJ, Dunne T, Swanson DN (eds) *Sediment Budgets and Routing in Forested Drainage Basins*, pp 5–23. US Department of Agriculture, Forest Service General Technical Report PNW-141. [7.3]

East TJ, Cull RF, Murray AS, Duggan K. (1988) Fluvial dispersion of radioactive mill tailings in the seasonally wet tropics, northern Australia. In: Warner RF (ed.) *Fluvial Geomorphology of Australia*, pp 303–22. Academic Press, Sydney. [7.5]

Erskine WD, Warner RF. (1988) Geomorphic effects of alternating flood- and drought-dominated regimes on NSW coastal rivers. In: Warner RF (ed.) *Fluvial Geomorphology of Australia*, pp 223–44. Academic Press, Sydney. [7.4]

Everitt BL. (1968) Use of the cottonwood in an investigation of the recent history of a floodplain. *American Journal of Science* **266**: 417–39. [7.3]

Farrell KM. (1987) Sedimentology and facies architecture of overbank deposits of the Mississippi River, False River Region Louisiana. In: Ethridge FG, Flores RM, Harvey MD (eds) *Recent Developments in Fluvial Sedimentology*, pp 111–20. Society of Economic Paleontologists and Mineralogists, Tulsa, Oklahoma. [7.2]

Ferguson RI. (1987) Hydraulic and sedimentary controls of channel pattern. In: Richards KS (ed.) *River Channels: Environment and Process*, pp 129–58. Blackwell, Oxford. [7.2]

Ferguson RI, Werritty A. (1983) Bar development and channel changes in the gravelly River Feshie, Scotland. In: Collinson JD, Lewin J (eds) *Modern and Ancient Fluvial Systems*, pp 181–93. Blackwell, Oxford. [7.3]

Graf WL. (1988) *Fluvial Processes in Dryland Rivers*. Springer-Verlag, New York. [7.2]

Graf WL. (1990) Fluvial dynamics of thorium-230 in the Church Rock Event, Puerco River, New Mexico. *Annals of the Association of American Geographers* **80**: 327–42. [7.5]

Harvey A. (1986) Geomorphic effects of a 100-year storm in the Howgill Fells, northwest England. *Zeitschrift für Geomorphologie* **30**: 71–91. [7.4]

Hickin EJ, Nanson GC. (1984) Lateral migration rates of river bends. *Journal of Hydraulic Engineering* **110**: 1557–67. [7.3]

Hooke JM. (1980) Magnitude and distribution of rates of river bank erosion. *Earth Surface Processes* **5**: 143–57. [7.3]

Ikeda H. (1989) Sedimentary controls on channel migration and origin of point bars in sand-bedded meandering rivers. In: Ikeda S, Parker G (eds) *River Meandering*, pp 51–68. American Geophysical Union, Water Resources Monograph No. 12. Washington, DC. [7.2]

Jackson RG. (1978) Preliminary evaluation of lithofacies models for meandering alluvial streams. In: Miall AD (ed.) *Fluvial Sedimentology*, pp 543–76. Canadian Society of Petroleum Geologists, Calgary. [7.2]

James LA. (1989) Sustained storage and transport of hydraulic gold mining sediment in the Bear River, California. *Annals of the Association of American Geographers* **79**: 570–92. [7.5]

Kelsey HM. (1980) A sediment budget and an analysis of geomorphic processes in the Van Duzen River basin, north coastal California, 1941–1975. *Geological Society of America Bulletin* Part II, **91**: 1119–216. [7.1]

Kelsey HM, Lamberson R, Madej MA. (1986) Modeling the transport of stored sediment in a gravel bed river, northwestern California. In: *Basin Sediment Delivery*, pp 367–91. IASH Publication No. 159. Wallingford, UK. [7.3]

Kirkby MJ. (1967) Measurement and theory of soil creep. *Journal of Geology* **75**: 359–78.

Klimek K, Zawilinska I. (1985) Trace elements in alluvium of the Upper Vistula as indicators of palaeohydrology. *Earth Surface Processes and Landforms* **10**: 273–80. [7.5]

Knight DM. (1989) Hydraulics of flood channels. In: Beven K, Carling P (eds) *Floods: Hydrological, Sedimentological and Geomorphological Implications*, pp 83–105. John Wiley & Sons, Chichester. [7.2]

Knox JC. (1983) Responses of river systems to Holocene climates. In: Wright HE Jr (ed.) *Late Quaternary Environments of the United States*, pp 26–41. University of Minnesota Press, Minneapolis. [7.4]

Knox JC. (1987) Historical valley floor sedimentation in the Upper Mississippi valley. *Annals of the Association of American Geographers* **77**: 224–44. [7.3, 7.5]

Lambert CP, Walling DE. (1987) Floodplain sedimentation: a preliminary investigation of contemporary deposition rates within the lower reaches of the River Culm, Devon, UK. *Geografiska Annaler* **69A**: 393–404. [7.3]

Lambert JM, Jennings JN, Smith CT, Green C, Hutchinson JN. (1960) *The Making of the Broads*. R.G.S. Research Series 3. [7.4]

Lattman LH. (1960) Cross section of a floodplain in a moist region of modest relief. *Journal of Sedimentary Petrology* **30**: 275–82. [7.2]

Leenaers H, Schouten CJ, Rang MC. (1988) Variability of the metal content of flood deposits. *Environmental Geology and Water Science* **11**: 95–106. [7.5]

Lehre AK. (1982) Sediment budget of a small Coast Range drainage basin in north-central California. In: Swanson FJ, Janda RJ, Dunne T, Swanston DN (eds) *Sediment Budgets and Routing in Forested Drainage Basins*, pp 67–77. US Department of Agriculture, Forest Service General Technical Report PNW-141. Washington, DC. [7.1]

Lewin J. (1983) Changes of channel patterns and floodplains. In: Gregory KJ (ed.) *Background to Palaeohydrology*, pp 303–19. John Wiley & Sons, Chichester. [7.2]

Lewin J, Macklin MG. (1987) Metal mining and floodplain sedimentation in Britain. In: Gardiner V (ed.) *International Geomorphology 1986, Part I*, pp 1009–27. John Wiley & Sons, Chichester. [7.5]

Lewin J, Manton MMM. (1975) Welsh floodplain studies: the nature of floodplain geometry. *Journal of Hydrology* **25**: 37–50. [7.1]

Lewin J, Hughes D, Blacknell C. (1977) Incidence of river erosion. *Area* **9**: 177–80. [7.3]

Lewin J, Bradley SB, Macklin MG. (1983) Historical valley alluviation in mid-Wales. *Geological Journal* **18**: 331–50. [7.5]

Lewis GW, Lewin J. (1983) Alluvial cutoffs in Wales and the Borderland. In: Collinson JD, Lewin J (eds) *Modern and Ancient Fluvial Systems*, pp 145–54. Blackwell, Oxford. [7.3]

Macklin MG, Lewin J. (1986) Terraced fills of Pleistocene and Holocene age in the Rheidol Valley, Wales. *Journal of Quaternary Science* **1**: 21–34. [7.4]

Macklin MG, Lewin J. (1989) Sediment transfer and transformation of an alluvial valley floor: the River South Tyne, Northumbria, UK. *Earth Surface Processes and Landforms* **14**: 233–46. [7.4, 7.5]

Maizels J. (1987) Large-scale flood deposits associated with the formation of coarse-grained braided terrace sequences. In: Ethridge FG, Flores RM, Harvey MD (eds) *Recent Developments in Fluvial Sedimentology*, pp 135–48. Society of Economic Paleontologists and Mineralogists, Tulsa, Oklahoma. [7.1]

Marron DC. (1989) Physical and chemical characteristics of a metal contaminated overbank deposit, west-central South Dakota, USA. *Earth Surface Processes and Landforms* **14**: 419–32. [7.5]

Miall AD. (1977) A review of the braided river depositional environment. *Earth Science Reviews* **13**: 1–62. [7.2]

Miall AD. (1978) Lithofacies types and vertical profile models in braided river deposits: a summary. In: Miall AD (ed.) *Fluvial Sedimentology*, pp 597–604.

Canadian Society of Petroleum Geologists, Calgary. [7.1]

Miall AD. (1985) Architectural-element analysis: a new method of facies analysis applied to fluvial deposits. *Earth Science Reviews* **22**: 261–308. [7.2]

Miller JR, Wells SG. (1986) Types and processes of short term sediment and uranium tailings storage in arroyos: an example from the Rio Puerco of the West, New Mexico. In: *Basin Sediment Delivery*, pp 335–53. IASH Publication No. 159. Wallingford, UK. [7.5]

Nakamura F, Araya T, Higashi S. (1987) Influence of river channel morphology and sediment production on residence time and transport distance. In: *Erosion and Sedimentation in the Pacific Rim*, pp 355–64. IASH Publication No. 165. Wallingford, UK. [7.3]

Nanson GC. (1986) Episodes of vertical accretion and catastrophic stripping: a model of disequilibrium flood-plain development. *Geological Society of America Bulletin* **97**: 1467–75. [7.4]

Petts GE. (ed.) (1989) *Historical Change of Large Alluvial Rivers: Western Europe*. John Wiley & Sons, Chichester. [7.2]

Pizzuto JE. (1987) Sediment diffusion during overbank flows. *Sedimentology* **34**: 301–17. [7.3]

Probst J-L. (1989) Hydroclimatic fluctuations of some European rivers since 1800. In: Petts GE (ed.) *Historical Changes of Large Alluvial Rivers—Western Europe*, pp 41–55. John Wiley & Sons, Chichester. [7.4]

Rains B, Welch J. (1988) Out-of-phase Holocene terraces in part of the North Saskatchewan River basin, Alberta. *Canadian Journal of Earth Sciences* **25**: 454–64. [7.3]

Roberts RG, Church M. (1986) The sediment budget in severely disturbed watersheds, Queen Charlotte Ranges, British Columbia. *Canadian Journal of Forest Research* **16**: 1092–106. [7.4]

Robinson MA, Lambrick GH. (1984) Holocene alluviation and hydrology in the upper Thames basin. *Nature* **308**: 809–14. [7.4]

Rust BR, Nanson GC. (1986) Contemporary and paleochannel patterns and the Late Quaternary stratigraphy of Cooper Creek, Southwest Queensland, Australia. *Earth Surface Processes and Landforms* **11**: 581–90. [7.2]

Schmudde JH. (1963) Some aspects of the landforms of the lower Missouri river floodplain. *Annals of the Association of American Geographers* **53**: 60–73. [7.2]

Schumm SA. (1977) *The Fluvial System*. John Wiley & Sons, New York. [7.3]

Schumm SA. (1986) Alluvial river response to active tectonics. In: Wallace R (ed.) *Active Tectonics*, pp 80–94. Geophysical Research Forum, National Academic Press, Washington. [7.4]

Schumm SA, Brackenridge GR. (1987) River responses.

In: Ruddiman WF, Wright HE (eds) *North America and Adjacent Oceans during the Last Deglaciation*, pp 221–40. Geological Society of America, Boulder, Colorado. [7.4]

Smith DG, Smith ND. (1980) Sedimentation in anastomosed river systems: examples from alluvial valleys near Banff, Alberta. *Journal of Sedimentary Petrology* **50**: 157–64. [7.2]

Smith ND, Cross TA, Dufficy JP, Clough SR. (1989) Anatomy of an avulsion. *Sedimentology* **36**: 1–23. [7.3]

Stear WM. (1985) Comparison of the bedform distribution and dynamics of modern and ancient sandy ephemeral flood deposits in the south-western Karoo region, South Africa. *Sedimentary Geology* **45**: 209–30. [7.3]

Szumanski A. (1983) Paleochannels of large meanders in the river valleys of the Polish lowlands. *Quaternary Studies in Poland* **4**: 207–16. [7.4]

Trimble SW. (1981) Changes in sediment storage in the Coon Creek basin, Driftless area, Wisconsin 1953–1975. *Science* **214**: 181–3. [7.3]

Wells LE. (1990) Holocene history of the El Niño phenomenon as recorded in flood sediments of northern coastal Peru. *Geology* **18**: 1134–7. [7.4]

Williams GP. (1983) Paleohydrological methods and some examples from Swedish fluvial environments, 1 Cobble and boulder deposits. *Geografiska Annaler* **65A**: 227–43. [7.1]

Wolman MG, Leopold LB. (1957) River floodplains: some observations on their formation. *US Geological Survey Professional Paper 282-C, 83–109.* [7.2, 7.3]

Part 2: The Biota

Part 1 emphasized the physicochemical properties of rivers; this part moves to a consideration of the biota. It is organized in a trophic hierarchy. The first three chapters are concerned with organisms that are primary sources of materials and energy in flowing water systems: first (Chapter 8) the crucial, but still poorly studied, microbial decomposers; second the primary producers, beginning with the algae (Chapter 9); and then the macrophytes (Chapter 10). Invertebrate consumers are considered in Chapter 11, not only in terms of which taxa are involved but also in terms of how they are distributed amongst functional feeding groups — and how they are distributed spatially. Chapter 12 considers the fishes; it addresses issues of distribution and abundance, life-history adaptation, and management.

Recognizing that information on the ecology and general biology of the river biota relies on adequate sampling, this Part ends with a treatment of sampling programmes (Chapter 13). However, there are no detailed descriptions of techniques since these will either be addressed in appropriate chapters or will be found in specialist textbooks and monographs. Instead the emphasis is on the design of sound programmes and the effective deployment of statistical procedures within them.

8: Heterotrophic Microbes

L.MALTBY

8.1 INTRODUCTION

This chapter is concerned with heterotrophic micro-organisms (i.e. bacteria and fungi) involved in the breakdown of detritus in rivers and streams. It will consider the types of techniques used to study these organisms (section 8.2), their ecology and distribution (sections 8.3 and 8.4) and their interactions with detritus (section 8.5). Finally, the effect of water quality on heterotrophic bacteria and fungi will be considered (section 8.6). The reason for considering techniques first is that, to investigate microbial ecology, it is necessary to obtain reliable estimates of the populations present. Although this is a modest objective, it is fraught with problems (Herbert 1990) which greatly limit our understanding of the role and importance of these organisms in freshwater systems.

8.2 TECHNIQUES

Many methods have been used to study the fungi and bacteria associated with detritus in fresh waters (Ingold 1975; Jones 1977, 1979; Poindexter & Leadbetter 1986; Austin 1988; Grigorova & Norris 1990). The aim of this section is not to provide a detailed description of each of these techniques, but rather to provide an introduction to the most commonly used ones (Table 8.1), their advantages and limitations. Methods used to enumerate microbial populations can be classified as either direct or indirect; direct methods being further divided into those that record total or viable organisms. As there is no universal growth medium for all micro-organisms, the main disadvantage of viable counts is that they under-

estimate microbial populations, often by large amounts (Staley & Konopka 1985). Indirect methods involve measurement of chemicals that act as indicators of biomass. Hence, a major potential limitation of these is that they require that the indicator chemical has a known, constant, ratio to cell biomass.

Fungi

Direct microscopic observations

The advantage of using direct observation to assess fungal assemblages is that it relies on the production of reproductive structures over a short incubation time and therefore only records active biomass (Bärlocher & Kendrick 1974). Generally, detrital material is collected from the field and incubated in sterile water for a few days before being microscopically examined and the spores present identified. The disadvantage of this technique is that not all active biomass will sporulate during the incubation period, or will produce spores which are easy to identify. Direct microscopic observations will favour conidial fungi that produce large characteristic spores (i.e. hyphomycetes).

Direct observations can provide information on the diversity, distribution and abundance of the mycoflora associated with detritus. Shearer and Lane (1983) described a technique for assessing the relative importance value (RIV) of fungal species based on recording the occurrence of species on discs cut from several leaves. The RIV of a species ranges from zero (absent) to two (only species present) (Shearer & Webster 1985a) and provides a measure of the abundance and distri-

Table 8.1 Methods used for studying heterotrophic microbes associated with detritus

Unit of study	Method
Fungi	Direct observations
	Baiting
	Dilution plating
	Particle plating
	Foam/water samples
	Hyphal biomass
	Chitin assay
	Ergosterol analysis
Bacteria	Direct counts
	Viable counts
	^{3}H-thymidine, ^{3}H-adenine, ^{3}H-leucine incorporation
Microbial community	ATP concentration
	Oxygen consumption
	^{14}C utilization
	ETS activity
	Microbial fatty acids

bution of a species both between and within leaves; this is important given the patchy distribution of aquatic fungi on leaves. The disadvantage of the technique is that it is time consuming and tedious. Shearer and Lane (1983) concluded that the optimum number of discs that should be examined per leaf was ten. However, the optimum number of leaves per site will be dependent on the site itself; more diverse communities require more samples.

Baiting

Two types of baits have been used: 'natural' baits (e.g. leaves, wood) for use in the field, and seeds for baiting material in the laboratory (Park 1972; Duddridge & Wainwright 1980). The most common application of baiting techniques is to collect organic material from the field and place it into containers containing sterile water and a number of baits. After 2–3 weeks' incubation, the baits are removed and either examined microscopically or plated on agar and the resulting colonies identified. This technique favours species which have motile spores including members of the genera *Pythium*, *Achlya* and *Saprolegnia* (i.e. Mastigomycotina). Park (1972) assessed the effectiveness of a range of different seed baits. He concluded

that small baits, such as *Brassica* half-cotyledons, yielded a higher variety of fungi than larger baits such as hemp seeds.

Although wood baits have been used to sample fungi in the field (Shearer 1972; Willoughby & Archer 1973; Lamore & Goos 1978; Sanders & Anderson 1979; Shearer & Von Bodman 1983), most studies have been concerned with the mycoflora of decaying leaves and have therefore used leaf baits (e.g. Bärlocher & Kendrick 1974; Suberkropp & Klug 1976; Bärlocher 1980; Chamier & Dixon 1982a; Suberkropp 1984; Shearer & Webster 1985a, 1985b, 1985c). Leaves are incubated in the river, either as leaf packs or in mesh bags, for a set period of time (>2 weeks) after which the fungal species colonizing them are identified, either by direct observation or after particle plating on agar. Factors to consider when using natural baits are bait size and deployment method (i.e. bags or packs). Leaf packs attached to an anchor (e.g. Cummins *et al* 1980) are more representative of natural leaf accumulations, although they are vulnerable to consumption by macroinvertebrates and physical abrasion. To reduce these losses, leaf baits are often placed in mesh bags; but here the size of mesh needs to be considered. Whereas fine mesh (<1 mm) may be desired to exclude macroinvertebrates, which may consume the material and associated microbes, small mesh size will limit the flow of water though the bag, reducing the exchange of nutrients and oxygen and thereby influencing colonization patterns and decomposition rates (Bird & Kaushik 1981; Webster & Benfield 1986).

The effect of bait size on the number of species isolated from it has been assessed for both wood (Sanders & Anderson 1979) and leaves (Bärlocher & Schweizer 1983). In both cases there was a positive correlation between bait size and number of aquatic hyphomycetes isolated. Leaf baits used by Bärlocher and Schweizer (1983) were small (0.16–4.00 cm²). Usually, however, whole leaves are used as baits and therefore size differences are probably less important in influencing diversity estimates.

The possible limitations of using leaf baits for ecological studies include: (i) they tend to consist of a single leaf type and are often deployed at the same time; hence they may underestimate the diversity of the fungal community; (ii) grouping leaves together, either in packs or within bags, tends to facilitate within-group colonization, again influencing diversity and distribution patterns. Despite these reservations, Shearer and Webster (1985c) found a high degree of similarity (67–87%) between the assemblages of fungi colonizing leaf baits (packs) and naturally occurring leaves. Such results suggest that leaf baits may provide a useful tool for determining the composition of leaf mycoflora.

Plating methods

There are several types of plating methods used. For dilution plating, detrital material is homogenized in sterile water and serial dilutions of the homogenate are spread over agar plates which are then incubated (Park 1972). In particle plating, small pieces of detrital material are placed directly on to agar plates. A modified particle plating technique has recently been proposed by Kirby *et al* (1990). As with dilution plating, the detritus is initially homogenized. The resulting particles are then sieved through 600 µm and 500 µm mesh sieves. Particles retained by the smaller sieve are resuspended and spread over agar plates. The advantage of this method is that the particles are small and standardized. Particle plating techniques provide information on the fungal species present on a particle but not their relative abundances. As particle size increases, both the number of species sampled per particle and the similarity between particles will increase, until eventually all species will be present on all particles. Therefore, with large particles it may not be possible to calculate the relative abundances of species in a sample nor the diversity of the asssemblage. If, on the other hand, particles were so small that they were colonized by a single species, sampling a large number of particles would provide an accurate and sensitive measurement of abundance and therefore diversity. Kirby *et al* (1990) found that particles in the size range 500–600 µm were colonized by an average of 1.12 species and the maximum number of species colonizing a given particle was two.

A limitation of plating techniques is that not all the fungal species that develop on the agar plates will have been active on the detritus. For example, Kaushik and Hynes (1968) plated leaf discs, which had previously been incubated in river water, directly on to agar plates that were then incubated at room temperature. All the fungi that grew on the plates were hyphomycetes typical of terrestrial, rather than aquatic, habitats. Inactive propagules may be removed from detritus by surface sterilization with mercuric chloride or serial washing (Kirby 1987).

Foam/water samples

The branched and sigmoid conidia of aquatic hyphomycetes (Fig. 8.1) are easily trapped in foam commonly observed in fast-flowing rivers and streams. The fungal spores present in foam samples have been used to provide information on the distribution and diversity of river mycoflora (e.g. Iqbal & Webster 1973, 1977; Sanders & Anderson 1979; Bärlocher & Rosset 1981; Wood-Eggenschwiler & Bärlocher 1983; Aimer & Segedin 1985a; Shearer & Webster 1985a, 1985b, 1985c). However, the presence of spores in river foam is not proof that the species is aquatic (i.e. can grow and sporulate under water), nor are all species equally represented in foam samples. Foam samples are biased in favour of highly

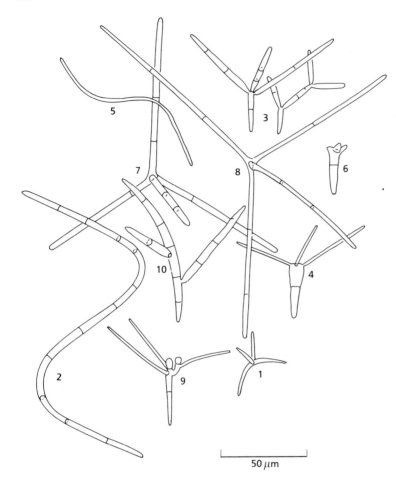

Fig. 8.1 Conidia of some common aquatic hyphomycetes. **1**: *Alatospora acuminata*; **2**: *Anguillospora longissima*; **3**: *Articulospora tetracladia*; **4**: *Clavariopsis aquatica*; **5**: *Flagellospora curvula*; **6**: *Heliscus lugdunensis*; **7**: *Lemonniera aquatica*; **8**: *Tetracladium elegans*; **9**: *Tetrachaetum marchalianum*; **10**: *Tricladium splendans* (After Ingold 1975).

branched spores, which are more easily removed from suspension by air bubbles than less branched spores (Iqbal & Webster 1973).

An alternative method of sampling water-borne spores is filtration. Water samples are filtered through millipore filters, the volume filtered being dependent on spore density (Iqbal & Webster 1973; Shearer & Lane 1983; Aimer & Segedin 1985b). In rivers of high spore density a 250-ml water sample may be satisfactory; however, in those of low spore density it may be necessary to filter up to 5 litres of water (Webster & Descals 1981). To prevent germination, spores should be fixed and stained in the field (e.g. by Lugol's iodine, lactophenol–fuchsin or cotton blue or trypan blue in either lactic acid or lactophenol (Iqbal & Webster 1973; Suberkropp & Klug 1976; Akridge & Koehn 1987; Chamier 1987; Findlay & Arsuffi 1989)). The water samples are filtered and the millipore filters rendered transparent by heating at 60°C for 1 h.

There are a number of problems with the filtration technique (Shearer & Lane 1983); they include:

1 Blocking of filters when sampling waters high in suspended solids.

2 Variability between samples (Shearer & Webster 1985c, but see Iqbal & Webster 1973).

3 Difficulty in identifying spores due to either obstruction by debris or deformation during filtering. Many species have similar conidial morphologies.

4 Relationship between the number of conidia produced by a species and its actual occurrence and activity is generally unknown (Suberkropp 1991).

5 Selectivity, i.e. spores of different sizes and branching patterns may be removed differentially.
6 Number of conidia produced by many species is dependent on flow rate; therefore, in slow-flowing systems, species that produce few fungi may go undetected.
7 Origin of spores. The propagule pool receives contributions from many sources, e.g. decaying leaves, wood or macrophytes and spores may have been transported considerable distances. The presence of spores in a site is not evidence of active growth; in fact, several studies have noted a discrepancy between the species list obtained from filtering of water samples and that obtained by direct examination of organic matter (Iqbal & Webster 1977; Chamier & Dixon 1982a; Bärlocher & Schweizer 1983; Shearer & Lane 1983; Shearer & Webster 1985c).

Hyphal biomass

The problem of measuring fungal biomass in detritus is that it is embedded within an opaque solid structure. Several approaches have been developed to overcome this; homogenizing the detritus and either: (i) embedding it within a thin film of agar (Jones & Mollison 1948; Frankland *et al.* 1978; Bengtsson 1982); or (ii) staining it with a fluorescent dye and filtering on to black filters (Paul & Johnson 1977; van Frankenhuyzen & Geen 1986; Findlay & Arsuffi 1989); or (iii) chemically clearing the detritus and measuring the hyphae within intact substrate (Hering & Nicholson 1964; Bärlocher & Kendrick 1974).

Newell and Hicks (1982) compared the effectiveness of these three approaches for assessing fungal biovolume in dead leaves. They concluded that agar-films provided the best method for direct estimation of quantities of mycelium. However, this method does have limitations, including difficulty in identifying fungal fragments in litter homogenate; inability of homogenization to release all fungal hyphae; and problems in recognizing fungal reproductive structures making fungal identification difficult. The problem of separating fungal mycelium from plant tissue may be overcome by staining fungal material. A commonly used stain is aniline blue which causes mycelia to fluoresce when excited by ultraviolet light. However, the degree to which hyphae fluoresce varies both between and within species (Newell & Hicks 1982; Findlay & Arsuffi 1989), thereby introducing variability in estimates of hyphal abundance.

Hyphal dimensions, and hence biovolume, have usually been measured either directly (Newell & Hicks 1982) or from photographs (Findlay & Arsuffi 1989), although it is possible to use image-analysis techniques (Morgan *et al* 1991). Biovolume estimates may be converted to biomass using appropriate conversion factors (Newell & Statzell-Tallman 1982).

Indirect methods

There are several indirect methods for estimating fungal biomass in plant material (Whipps *et al* 1982) including measuring chitin (Ride & Drysdale 1972; Cochran & Vercellotti 1978) or ergosterol (Seitz *et al* 1979; Lee *et al* 1980).

Chitin assay Hydrolysis of chitin, a major constituent of most fungal cell walls (Hudson 1986), produces hexosamines such as glucosamine that can be assayed colorimetrically (Ride & Drysdale 1972; Jones 1979) or by gas chromatography (Hicks & Newell 1983). As the amount of glucosamine per unit weight of mycelium varies between species and between different ages of mycelium, this can reliably be used only to assess the fungal biomass of material colonized by fungi of known age and composition. Another disadvantage is that glucosamine polymers are also present in some bacterial cell walls (Gottschalk 1986; West *et al* 1987).

Ergosterol analysis Ergosterol (ergosta-5,7,22-trien-3β-ol) is the prominent membrane-bound sterol in most fungi but is absent from vascular plant material and bacteria. It has been used to estimate fungal biomass in terrestrial (Matcham *et al* 1985; West *et al* 1987), marine (Lee *et al* 1980; Newell & Fallon 1991) and, more recently, freshwater systems (Gessner & Schwoerbel 1991; Sinsabaugh *et al* 1991). It can be extracted from detrital material by saponification with methanol and quantified by high-performance liquid chromatography with an ultraviolet detector

(Grant & West 1986; Newell *et al* 1988). Because ergosterol has a high rate of mineralization upon the death of fungi, it is indicative of living biomass. This is in contrast to chitin, which is more resistant to breakdown.

Selectivity of techniques

All the techniques described above are selective and, therefore, several must be used to produce a comprehensive list of fungi present in a sample. For example, direct microscopic examination of detritus is selective for those groups with large distinctive spores (i.e. aquatic hyphomycetes), whereas particle plating may be selective towards different genera depending on the incubation conditions used. Incubating material on nutrient-rich agar at room temperature will favour terrestrial hyphomycetes, whereas incubating it on nutrient-poor agar at lower temperatures will favour aquatic hyphomycetes (Bärlocher & Kendrick 1974; Suberkropp & Klug 1976; Chergui & Pattee 1988). A list of the media used to culture aquatic fungi is presented in Table 8.2. To reduce bacterial contamination, lactic acid or antibiotics (e.g. penicillin, streptomycin) are frequently added to culture media (Kaushik & Hynes 1968; Webster & Descals 1981; Chamier *et al* 1984).

Conidial frequencies in water samples have been extensively used to study aquatic hyphomycete communities (e.g. Iqbal & Webster 1973, 1977; Bärlocher & Rosset 1981; Shearer &

Table 8.2 Media used for culturing aquatic fungi

Medium	Reference
Cellulose	Park (1972, 1974b)
Cooke agar	Kirby *et al* (1990)
Cornmeal agar	Duddridge & Wainwright (1980)
Czapek agar	Kaushik & Hynes (1968)
Goos' medium	Webster *et al* (1976)
Glucose-peptone/Rose Bengal/aureomycin agar	Park (1974b)
Glucose−yeast extract−starch medium	Kirby (1987)
Inorganic salts agar	Suberkropp & Klug (1976) Suberkropp (1984)
Malt extract agar	Abel & Bärlocher (1984) Bärlocher & Kendrick (1974) Butler & Suberkropp (1986) Chamier *et al* (1984) Kirby (1987) Kirby *et al* (1990) Rosset & Bärlocher (1985) Thornton (1963)
Malt extract broth	Bärlocher & Kendrick (1973)
Potato dextrose agar	Godfrey (1983)
Starch−casein−nitrate medium	Suberkropp & Klug (1976)
Water agar	Bärlocher & Kendrick (1974)
Water agar + leaf material	Bärlocher & Kendrick (1974) Chamier *et al* (1984)
Yeast extract−glucose broth	Arsuffi & Suberkropp (1984)

Webster 1985a, 1985b). However, except for the most common species, there is little agreement between the abundance of species in the propagule pool and their abundance on leaf litter (Shearer & Lane 1983; Shearer & Webster 1985c).

Although there is good agreement between the types of aquatic hyphomycetes colonizing leaf baits and naturally occurring leaves (Shearer & Lane 1983), baiting in the laboratory with seeds tends to be selective towards aquatic Oomycetes (e.g. *Pythium*, *Achlya* and *Saprolegnia*) which have motile zoospores (Park 1972).

Bacteria

Direct microscopic counts

Bacteria present on detrital material may be observed, *in situ*, using scanning electron microscopy (e.g. Suberkropp & Klug 1974; Armstrong & Bärlocher 1989; Groom & Hildrew 1989). Alternatively, they can be dislodged from detritus by sonification (Chamier 1987; Bott & Kaplan 1990) or homogenization (Suberkropp & Klug 1976; van Frankenhuyzsen & Geen 1986; Henebry & Gorden 1989). However, the efficiency of removal of attached bacteria from particles by sonification ranges from 20% to 50% (Benner *et al* 1988), and homogenization results in samples of high detritus content which can cause problems when using epifluorescence microscopy. The dislodged bacterial cells can be stained with a fluorescent dye, usually acridine orange, and filtered on to black 0.2-μm pore size filters before being counted using epifluorescent microscopy. If acridine orange preparations are excited with blue light at 470 nm, debris fluoresces red, orange or yellow whereas bacteria fluoresce green.

Epifluorescence direct counting can also be used for estimating the size and biomass of bacteria (Fry 1990). In order to estimate biomass, bacteria are measured and biovolume:carbon conversion factors applied. The measurement of bacterial cells is not without problems (Fry 1988) and conversion factors can vary by a factor of two (Benner *et al* 1988). Image analysis techniques can be used both to count and measure bacteria in epifluorescence images (Bjørnsen 1986).

Difficulties may be encountered when using acridine orange for direct counts especially if the sample contains a large amount of detritus since this may pick up the stain and autofluoresce. Porter and Feig (1980) described the use of a DNA-specific fluorescing stain, 4′,6-diamidino-2-phenylindole (DAPI), which can be used for epifluorescent counts. When exposed to ultraviolet light (365 nm) the DNA−DAPI complex fluoresces bright blue whereas unbound DAPI and DAPI bound to non-DNA material may fluoresce a weak yellow. This technique has been used successfully for studying stream bacteria (Kaplan & Bott 1983, 1985; Bott & Kaplan 1990). However, Edwards (1987) found the technique to be unsuitable for studying bacteria in a blackwater river. DAPI is specific for A−T-rich DNA and may not therefore be suitable for all types of bacterial assemblages.

Viable counts

Not all the bacterial cells recorded by direct microscopic counts will be viable. Viable counting procedures involve providing bacteria with a nutrient source and recording the number of active cells (i.e. those that have grown and/or divided) after incubation. Viable counts can be obtained either by plating samples on to agar plates and recording the number of colonies produced (i.e. colony counts) or by 'most probable number' (MPN) procedures (Jones 1979; Herbert 1990). To obtain MPN counts, samples are serially diluted and aliquots of each dilution are inoculated into, or on to, a growth medium. Most MPN methods use liquid media which become turbid due to growth of bacterial cells.

The results obtained from viable count methods are dependent upon a number of factors, including dilution procedures, incubation conditions and duration, and the type of medium used. Jones (1979) suggests the use of casein−peptone−starch (CPS) medium for freshwater bacteria, although other media including peptone−yeast extract−glucose (PYG) agar (Suberkropp & Klug 1976) and nutrient agar (Baker & Farr 1977; Chamier *et al* 1984) have been used to study bacteria associated with detritus. The type of medium used will influence both the diversity and abundance of bacteria isolated from environmental samples.

For example, Baker and Farr (1977) consistently isolated one to three times more bacteria from stream water using CPS medium than nutrient agar. Other problems with viable plate counts include: (i) bacteria remain aggregated and therefore individual colonies are derived from many rather than single cells; (ii) antagonistic interaction occurs between proximate bacterial colonies; (iii) they encourage the growth of cells that are metabolically inactive *in situ*.

Staley and Konopka (1985) concluded that with viable plate counts, the maximum recovery of heterotrophic bacteria from a variety of oligotrophic and mesotrophic habitats was 1% of the total direct count. This is in sharp contrast with results from studies using tritiated amino acids and thymidine which have concluded that at least 85% of the total direct count are metabolically active (Tabor & Neihof 1984).

Because of these limitations, Fry (1982) stated that 'viable counts should only be used with care in natural habitats in limited circumstances'. Boulton and Boon (1991) go further and state that the continued use of viable counts for the determination of bacterial abundance is 'indefensible'. Despite this, viable counts are still used (e.g. Paul *et al* 1983; Hornor & Hilt 1985; Colwell *et al* 1989).

Other methods of distinguishing between metabolically active and inactive bacteria include autoradiography (O'Carroll 1988), use of inhibitors such as 2-(p-iodophenyl)-3-(p-nitrophenyl)-5-phenyl tetrazolium chloride (INT) Tabor & Neihof 1982), and the use of vital fluorogenic stains such as fluorescein diacetate (FDA) (Schnürer & Rosswall 1982). However, none of these methods is ideal for studying bacteria associated with detritus (Boulton & Boon 1991). For example, not all bacteria fluoresce with FDA (Fry 1990). It is especially poor for Gram-negative bacteria, which tend to be the predominant group in unpolluted rivers and streams (Chrzanowski *et al* 1984).

^{3}H-thymidine incorporation

Several radioisotopic methods have been developed to assess bacterial biomass, including ^{3}H-adenine incorporation into DNA and RNA, 5-^{3}H-uridine incorporation into RNA, ^{3}H-thymi-dine incorporation into DNA and ^{3}H-leucine incorporation into protein (Karl 1980; Moriarty 1986; Riemann & Bell 1990). The most commonly used isotope is ^{3}H-thymidine.

The incorporation of ^{3}H-thymidine into DNA has been used to assess bacteria in the seston of both marine (Fuhrman & Azam 1980, 1982) and freshwater (Riemann *et al* 1982; Bell *et al* 1983) systems as well as marine sediments (Moriarty & Pollard 1981). Recently its usefulness as a technique for quantifying bacteria associated with benthic detritus in river systems has been evaluated (Findlay *et al* 1984a). The technique has been criticized on the grounds that compounds other than DNA are labelled. For example, ^{3}H-thymidine may be catabolized and the label incorporated into protein (Staley & Konopka 1985); it may also be incorporated into RNA (O'Carroll 1988). However, these problems can be overcome by using pulse-label techniques (Moriarty 1986; 1990). Using short incubation times (e.g. 10 min at 25°C, 20 min at 15°C) reduces the chance that thymidine will be degraded and the label incorporated into other macromolecules. Another approach is to isolate and purify the DNA before assaying radioactivity. Several methods have been used to extract DNA including: acid−base hydrolysis, phenol−chloroform extraction and enzymatic fractionation. However, the effectiveness of these methods depends on the type of material being analysed and this must therefore be considered when selecting an extraction procedure (Torréton & Bouvy 1991).

^{3}H-thymidine incorporation has been used to evaluate bacterial communities on riverine detritus in both the laboratory (e.g. Findlay *et al* 1984b, 1986; Findlay & Arsuffi 1989) and the field (e.g. Edwards & Meyer 1986; Palumbo *et al* 1987; Peters *et al* 1987). Palumbo and colleagues (1987) used this technique to assess microbial communities on leaf litter in streams and found it to be a more sensitive method than either oxygen consumption or ATP concentration.

Microbial community

ATP content

ATP is present in all living material. Its use as a means of estimating total microbial biomass was initially proposed by Holm-Hansen and Booth in 1966, but since then it has been widely used in both laboratory (Federle *et al* 1982; Suberkropp *et al* 1983; Peters *et al* 1987; Findlay & Arsuffi 1989) and field (Forbes & Magnuson 1980; Rounick & Winterbourn 1983; Taylor & Roff 1984; Palumbo *et al* 1987) based studies. Although several methods can be used to assay ATP (Karl 1980), the most commonly used techniques are those based on the luciferin−luciferase reaction. This involves a number of assumptions including: (i) ATP is not adsorbed on to detrital material, and (ii) there is a constant ratio of ATP to total cell carbon for all microbial taxa. The first assumption is possibly false and the second assumption is certainly false. ATP is not just associated with living biomass but occurs as free dissolved ATP (D-ATP) in both marine and freshwater environments (Karl 1980). The extent to which D-ATP is adsorbed on to detrital material, and the implications of this for the use of ATP as an indicator of microbial biomass, have not been fully investigated.

Frequently a carbon:ATP ratio of 250:1 is used (e.g. Kaplan & Bott 1983; Findlay & Arsuffi 1989). This was an average value derived by Holm-Hansen and Booth (1966) from a wide range of microbial cells. However, carbon:ATP ratios vary both within and between species and may range from less than 100:1 to more than 2000:1 (Jones 1979; Karl 1980). Therefore the application of the 250:1 ratio to unidentified field samples at best can provide only a crude indication of microbial biomass. ATP-biomass estimates will be most accurate when the natural microbial assemblage approaches a monoculture of a known species for which a specific carbon:ATP ratio can be determined.

Oxygen consumption

Microbial respiration has frequently been used as an index of community biomass and activity and is the basis of the 'biochemical oxygen demand' (BOD) test. It is usually measured by quantifying the oxygen uptake of a sample either by using chemical techniques (e.g. Groom & Hildrew 1989), an oxygen electrode (e.g. Findlay & Arsuffi 1989; Sinsabaugh & Linkins 1990) or a Gilson Differential Respirometer (e.g. Palumbo *et al*

1987; Elwood *et al* 1988; Garden & Davies 1988; Petersen *et al* 1989).

Microbial communities contain both heterotrophic and autotrophic organisms. Although oxygen is utilized by all these organisms during aerobic respiration, it is also generated by autotrophs during photosynthesis. Therefore, to obtain a measure of respiration alone, samples must be incubated in the dark.

^{14}C utilization

When organic compounds are mineralized by heterotrophic micro-organisms, the carbon they contain is either assimilated into microbial biomass or respired and converted to carbon dioxide. The activity of the heterotrophic microbial community can therefore be estimated by measuring the rate at which organic carbon, supplied as ^{14}C-glucose or ^{14}C-acetate, is either converted to carbon dioxide (Sorokin & Kadota 1972) or assimilated (Wright & Hobbie 1966). Hall *et al* (1990) reviewed studies using ^{14}C to estimate microbial activity in fresh waters. Recent studies using this technique to assess heterotrophic microbial activity associated with detritus include Carpenter *et al* (1983), Fairchild *et al* (1983) and Peters *et al* (1987).

Electron transport system activity

Electron transport system (ETS) activity is a characteristic of respiring cells and can be measured by quantifying the reduction of 2-(*p*-iodophenyl)-3-(*p*-nitrophenyl)-5-phenyl tetrazolium chloride (INT) to its formazan derivative, a water-insoluble precipitate which can be observed microscopically or quantified spectrophotomically (Tabor & Neihof 1982; Trevors 1984). For example, Elwood *et al* (1988) used ETS activity to assess the microbial community on leaf litter. Leaf discs were incubated in a 0.25% solution of INT for 30 min, after which the formazan derivative was extracted in 90% acetone and its absorbance read at 520 nm on a spectrophotometer. Whereas Trevors (1984) found a strong correlation between ETS activity and oxygen consumption for freshwater sediments ($r > 0.74$), no such correlation was detected by Elwood *et al* (1988). This led them to conclude that the technique was an unreliable measure of microbial activity on decomposing leaves. It is possible that not all species of bacteria can reduce the tetrazolium salt or that it inhibits or suppresses respiratory activity in some species. The technique has also been used to study micro-organisms associated with biofilms (Freeman *et al* 1990; Ladd & Costerston 1990) and can be combined with epifluorescence microscopy as a means of assessing the abundance of metabolically active micro-organisms (Tabor & Neihof 1982).

Microbial fatty acids

The abundance and composition of fatty acids extracted from detritus has been suggested as a means of quantifying and characterizing benthic microbial communities (Bobbie & White 1980). Some fatty acids, such as polyenoic acids, are restricted to fungi, whereas others, for example *cis*-vaccenic acid, are restricted to bacteria. A list of fatty acids that can be used as biomarkers is given in Fry (1988). Van Frankenhuyzen and Geen (1986) used fatty acids to investigate the effect of pH on the microbial community of leaf litter in streams. They found the technique to be less satisfactory than direct microscopic counts.

8.3 FUNGI

The fungi found in fresh waters include members of the Mastigomycotina (e.g. *Pythium*, *Saprolegnia*), Deuteromycotina (e.g. *Lemonniera*, *Helicoon*), Ascomycotina (e.g. *Penicillium*, *Fusarium*) and Zygomycotina (e.g. *Mucor*). Although no fungi have been found to produce basidiocarps in freshwater, conidial basidiomycetes do occur (Hudson 1986). In a study of fungi associated with leaves in streams, Bärlocher and Kendrick (1974) identified 55 different species, of which 48 were deuteromycetes (29 species were terrestrial hyphomycetes, 14 species were aquatic hyphomycetes and 5 species were coelomycetes), the remaining being members of the Zygomycotina and Mastigomycotina. A similar number of species were recorded by Lamore and Goos (1978) from woody substrates. Of the 59 species they identified, 40 were deuteromycetes, 15 were ascomycetes and four were zygomycetes.

Aquatic hyphomycetes

The term 'aquatic hyphomycetes' has traditionally been used to describe conidial aquatic fungi abundant on organic material in streams and rivers. As some members of this group are neither exclusively aquatic (Park 1974a; Singh & Musa 1977; Bandoni 1981) nor hyphomycetes, some being anamorphs of ascomycetes and basidiomycetes, Webster and Descals (1981) have proposed that they be referred to as Ingoldian fungi, after C.T. Ingold who pioneered their study in the 1940s. However, the use of the term 'aquatic hyphomycetes' is widespread and is retained here.

Since the initial studies by Ingold (Ingold 1942), more than 150 species of aquatic hyphomycetes have been described. They tend to be concentrated on coarse, as opposed to fine, particulate organic matter and are particularly abundant on submerged decaying leaves (Bärlocher & Kendrick 1981). As a group they are widely distributed, and many species (e.g. *Alatospora acuminata*, *Anguillospora longissima*) have a global distribution (Nilsson 1964). Recent studies into the ecology and distribution of aquatic hyphomycetes have been conducted in Europe (Bärlocher & Rosset 1981; Chamier & Dixon 1982a; Wood-Eggenschwiler & Bärlocher 1983; Czeczuga & Próba 1987; Chergui & Pattee 1988), New Zealand (Aimer & Segedin 1985a, 1985b), Australia (Thomas *et al* 1989), India (Sridhar & Kaveriappa 1982, 1989; Gupta & Mahrotra 1989; Mer & Sati 1989), North America (Suberkropp 1984; Akridge & Koehn 1987), Hawaii (Ranzoni 1979), Puerto Rico (Padgett 1976), North Africa (Chergui 1990) and the subarctic (Müller-Haeckel & Marvanová 1979).

The conidia of aquatic hyphomycetes are large and are predominantly tetraradiate (e.g. *Alatospora*, *Tetrachaetum*, *Lemonniera*) or sigmoid (e.g. *Anguillospora*, *Flagellospora*, *Lunulospora*) in shape (Ingold 1975; Fig. 8.1). The tetraradiate shape is thought to be an adaptation to a flowing water environment, increasing the effectiveness with which the conidia become trapped on underwater surfaces (Webster 1981). It may also enhance their dispersal in surface films of water such as might occur between leaves in terrestrial habitats (Bandoni 1974, 1975).

Seasonal fluctuations in conidial numbers have been recorded in many parts of the world; peak conidial numbers occurring shortly after peak leaf fall (Iqbal & Webster 1973, 1977; Müller-Haekel & Marvanová 1979; Bärlocher & Rosset 1981; Chamier & Dixon 1982a; Aimer & Segedin 1985a, 1985b; Shearer & Webster 1985b; Thomas *et al* 1989). The conidial numbers of most species mirror the availability of substrate, but a few show distinctive patterns of seasonal abundance not explained in terms of substrate availability. In such cases, temperature is invoked as the causal factor.

Although negative relationships between temperature and species richness have been observed in the field (Mer & Sati 1989), evidence from laboratory studies on the effect of temperature on growth and sporulation of individual species is far from convincing.

The effect of temperature on the seasonal occurrence of aquatic hyphomycetes at several sites in a North American stream was investigated by Suberkropp (1984). Two groups of fungi were identified: those present only in summer (e.g. *Lunulospora curvula*, *Flagellospora penicillioides*, *Trisicelophorus monosporus* and *Clavatospora tentacula*) and a winter group (e.g. *Alatospora acuminata*, *Flagellospora curvula* and *Lemonniera aquatica*) which declined in importance during the summer. During the summer, water temperatures in the study sites ranged between 9° and 24°C, whereas in winter the temperature range was between 0° and 15°C. Experiments investigating the effect of temperature on growth concluded that members of the summer group had an optimum for growth of 25–30°C and minimum for growth between 5° and 10°C, whereas those of the winter group had an optimum for growth of 20°C and continued to grow at 1°C. Although these data may provide an explanation for the absence of the summer assemblage of aquatic hyphomycetes during the colder winter months, they do not explain the decline of the winter assemblage. Also, they do not explain the decline of the summer assemblage at sites where the temperature did not fall below 9–10°C.

Typical 'summer' and 'winter' species also occur in British streams; e.g. conidia of *Lunulospora* *curvula* were detected in the River Creedy only from August to November in contrast to *Tricladium chaetocladium* (incorrectly identified as *T. gracile*, Webster *et al* 1976) which was detected only from December to April (Iqbal & Webster 1973, 1977). The possibility that these patterns were due to the two species having different temperature optima was investigated by Webster *et al* (1976). When grown in single-species culture in the laboratory, the optima for growth and sporulation of *L. curvula* was found to be higher (25°C) than that for *T. chaetocladium* (20°C). Interestingly, when these two species were grown together in mixed culture, the optimum temperature for sporulation decreased to 5°C for *T. chaetocladium* and 10°C for *L. curvula*, even though in single-species culture *T. chaetocladium* failed to sporulate at 5°C. It would therefore appear that the temperature responses of these species can be modified by interspecific interactions, although the mechanism by which this occurs is unclear. The possible importance of interspecific interactions in determining temporal occurrence is suggested by the observation that if leaves are already colonized by summer species, the ability of winter species to colonize them is reduced (Suberkropp 1984).

In general, the majority of species found in temperate regions have an optimum temperature for growth in the laboratory of between 15° and 25°C, and continue to grow when the temperature drops below 5°C, (Thornton 1963; Suberkropp & Klug 1981). Those species that are classified as 'tropical species' (Nilsson 1964), or are common in the summer in temperate regions, have higher optimum temperatures and do not grow below 5°C (e.g. *Clavatospora tentacula*, *Flagellospora penicilliodies*, *Lunulospora curvula*).

Spatial variation in aquatic hyphomycete abundances has been observed both between (e.g. Iqbal & Webster 1973, 1977; Wood-Eggenschwiler & Bärlocher 1983) and within (e.g. Shearer & Webster 1985a) rivers. Whereas major differences in riparian vegetation (e.g. change from moorland to woodland) are correlated with changes in aquatic hyphomycete diversity (Webster 1981; Shearer & Webster 1985a), despite the apparent substrate specificity of some species (e.g. Bengtsson 1982, 1983), changes in riparian

176   *The Biota*

vegetation within a habitat type (e.g. deciduous woodland) are not (Wood-Eggenschwiler & Bärlocher 1983; Aimer & Segedin 1985; Thomas *et al* 1989).

Aero-aquatic hyphomycetes

Aero-aquatic hyphomycetes are distinguished by the fact that they generally sporulate only when exposed to moist air and produce conidia which trap air (Fisher 1977; Fig. 8.2). They are commonly found in slow-flowing or static water and many appear adapted to exist in hypoxic conditions (Webster 1981). However, as aero-aquatics grow well (Fisher & Webster 1979) and reproduce under well-aerated conditions, their distribution may be limited more by their poor competitive ability than by physiological constraints (Fisher & Webster 1981).

Aero-aquatics are found on both leaves (deciduous and coniferous) and wood (Lamore & Goos 1978; Fisher 1979; Jones 1981), and common species include *Helicodendron fractum* and *Helicoon sessile*.

Enzymatic capabilities

As aquatic hyphomycetes can utilize dissolved organic matter (e.g. Thornton 1963; Bengtsson 1982), to what extent are they capable of degrading the detrital material to which they are attached? Although they colonize both leaf and wood substrates (Willoughby & Archer 1973; Sanders & Anderson 1979), most studies have concentrated on the mycoflora of decaying leaves (e.g. Suberkropp & Klug 1980, 1981; Jones 1981; Chamier & Dixon 1982b; Fisher *et al* 1983; Suberkropp *et al* 1983; Chamier *et al* 1984; Butler & Suberkropp 1986; Abdullah & Taj-Aldeen 1989).

In general, plant cell walls consist of approximately 35% (w/w) peptic polysaccharides, 24% (w/w) hemicellulose and 23% (w/w) cellulose (Darvill *et al* 1980). Cell-wall-degrading enzymes of aquatic hyphomycetes were reviewed recently by Chamier (1985) and a summary of the enzymatic capabilities of some aquatic hyphomycetes is presented in Table 8.3. As a group, these fungi can utilize a wide range of substrates and the

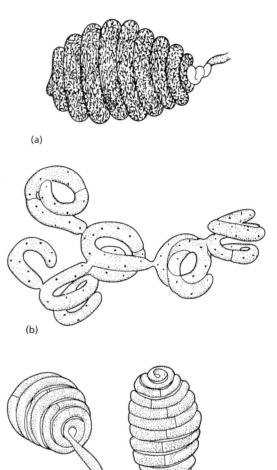

Fig. 8.2 Spores of aero-aquatic hyphomycetes. (a) *Helicoon richonis*; (b) *Helicodendron triglitziense*; (c) *Helicodendron conglomeratum* (after Webster 1980).

majority of species can degrade at least one of the major components of plant cell walls. In addition to utilizing plant cells as a carbon source, fungi may also be saprophytic on other fungi. All of the 14 species of aquatic hyphomycetes tested by Chamier (1985) could utilize lamarin and a number could break down chitin. Both lamarin and chitin are constituents of fungal cell walls. Interestingly, those species with weak cellulase

Table 8.3 Enzymatic capabilities of aquatic hyphomycetes

Species	Substrate							Reference
	Cellulose	Xylan	Starch	Protein	Tannic acid	Pectin	Lignin	
Alatospora acuminata	✓	✓	–	x	–	✓	–	beh
Anguillospora crassa	✓	–	✓	✓	✓	–	–	ag
Anguillospora furtiva	–	–	–	–	✓	–	✓	f
Anguillospora longissima	–	–	✓	x	–	–	x	ag
Anguillospora pseudolongissima	✓	✓	–	✓	–	✓	–	e
Articulospora tetracladia	✓	✓	–	–	–	✓	x	cdg
Articulospora inflata	–	–	–	x	–	✓	–	e
Clavariopsis aquatica	✓	–	–	✓	–	✓	x	beh
Clavatospora tentacula	✓	–	–	–	–	✓	–	h
Dactyleila aquatica	x	–	x	–	x	✓	✓	a
Dendrospora tenella	–	–	–	–	–	–	x	f
Dimorphospora foliicola	–	–	–	–	–	–	✓	f
Filosporella annelidica	✓	–	–	–	–	–	–	b
Flagellospora curvula	✓	–	–	–	–	✓	–	beh
Flagellospora penicilloides	✓	–	–	–	–	✓	–	h
Gyoerffyella rotula	–	–	–	–	–	–	✓	f
Heliscus lugdunensis	–	✓	✓	✓	–	–	x	abefgh
Lemonniera aquatica	✓	–	✓	✓	–	–	x	acdefgh
Lemonniera cornuta	–	–	–	✓	–	–	x	f
Lemonniera terrestris	✓	–	✓	✓	–	–	–	e
Lunulospora curvula	✓	–	–	–	–	–	x	afh
Margaritispora aquatica	–	–	–	–	–	–	x	f
Mycocentrospora angulata	✓	–	–	–	–	✓	–	cd
Scorpiosporium gracile	–	–	–	–	–	✓	x	f
Tetrachaetum elegans	✓	–	–	x	–	–	–	cde
Tetracladium marchalianum	✓	✓	–	–	✓	✓	–	dch
Tetracladium setigerum	x	–	✓	–	✓	✓	✓	acf
Tricladium giganteum	–	–	✓	–	✓	✓	✓	f
Tricladium splendans	✓	–	–	–	–	✓	–	cdg
Varicosporium elodeae	✓	✓	–	✓	x	–	–	acdg

(✓, detected; x, not detected; –, not assessed). a, Abdullah & Taj-Aldeen (1989); b, Butler & Suberkropp (1986); c, Chamier & Dixon (1982b); d, Chamier *et al* (1984); e, Suberkropp *et al* 1983); f, Fisher *et al* (1983); g, Jones (1981); h, Suberkropp & Klug (1981).

activity had greater laminase and chitinase activity than those species with high cellulase activity, suggesting that they may utilize both fungal and plant material as an energy source.

There is some experimental evidence in support of the utilization of wood by hyphomycetes. Fisher *et al* (1983) concluded that 15 of the 23 aero-aquatics, and six of the 16 aquatic hyphomycetes tested could degrade lignin, and Jones (1981) demonstrated that four hyphomycetes, *Anguillospora crassa*, *Anguillospora longissima*, *Lunulospora curvula* and *Tricladium splendans*, caused significant weight loss of wood under laboratory conditions.

Fungi associated with leaf litter

When leaves from deciduous trees enter rivers and streams in the autumn they are already colonized by a number of fungi, including *Alternaria*, *Aureobasidium*, *Cladosporium*, *Epicoccum* and *Humicola* (Bärlocher & Kendrick 1975). Once submerged, the mycoflora of the leaf changes to one dominated by aquatic hyphomycetes (Bärlocher & Kendrick 1975; Suberkropp & Klug 1976; Chamier & Dixon 1982a). Some members of the terrestrial mycoflora of leaves have the ability to degrade leaf material when it is submerged. For example, *Cladosporium* and *Epicoccum* have strong enzymatic activity towards cellulose, xylan and pectin (Chamier *et al* 1984) and can degrade both alder leaves and filter paper in water (Godfrey 1983). The decrease in abundance of the resident mycoflora once the leaf material becomes submerged is thought to be due to their inability to grow at low water temperatures (Bärlocher & Kendrick 1974).

Colonization of a leaf by aquatic hyphomycetes is initially random, with hyphae growing mainly on the leaf surface. After a few weeks, the leaves become covered with a mosaic of fungal colonies (Fig. 8.3; Shearer & Lane 1983; Chamier *et al* 1984). Although habitat selection has been reported for some species of aquatic hyphomycetes (Bengtsson 1983), resource specificity is generally low and the same species tend to be present on different leaf types (Bärlocher & Kendrick 1974; Suberkropp & Klug 1980; Shearer & Lane 1983; Butler & Suberkropp 1986). There is also little

evidence of a succession of aquatic hyphomycetes colonizing leaf litter. Instead, the community appears to be dominated by four to eight species that arrive early and persist (Bärlocher 1982). Genera commonly isolated from leaf litter include *Alatospora*, *Anguillospora*, *Articulospora*, *Cladosporium*, *Clavatospora*, *Epicoccum*, *Flagellospora*, *Fusarium*, *Lemonniera*, *Tetracladium*, *Tricladium* and *Varicosporium*.

Most studies on the decomposition of vascular plant material in flowing waters have concentrated on aquatic hyphomycetes colonizing leaf litter from deciduous trees; however, other species of fungi (e.g. *Pythium*; Park & McKee 1978; Park 1980) and substrate types (e.g. conifer needles; Bärlocher & Michaelides 1978; Bärlocher & Oertli 1978; Bärlocher & Kendrick 1981) may be important in the economy of stream systems (see Chapter 15).

Fungi associated with wood

A significant proportion of allochthonous material enters streams and rivers as twigs, branches and logs. In a study of wood-inhabiting fungi in a North American stream, Lamore and Goos (1978) identified 59 species of fungi, of which 15 were ascomycetes, 36 were hyphomycetes and four were zygomycetes. Ascomycetes such as *Ceratastomella*, *Massarina* and *Nectria lugdunensis* were also common colonizers of wood in a British stream, although again the dominant group was the aquatic hyphomycetes which accounted for 23 of the 37 species identified (Willoughby & Archer 1973). Submerged wood decomposes more slowly than leaves (Shearer & von Bodman 1983; Chergui & Patee 1991) and, in contrast to leaf material, there is some evidence of species succession (Shearer & Webster 1991).

Common colonizers of wood in flowing waters include *Anguillospora longissima*, *Anguillospora crassa*, *Aureobasidium pullulans*, *Camposporium pellucidium*, *Clavariopsis aquatica*, *Dactylella aquatica*, *Fusarium*, *Geotrichum candidum*, *Helicoon sessile*, *Heliscus lugdunensis*, *Massarina* and *Tricladium splendens* (Willoughby & Archer 1973; Lamore & Goos 1978; Jones 1981; Shearer & Webster 1991).

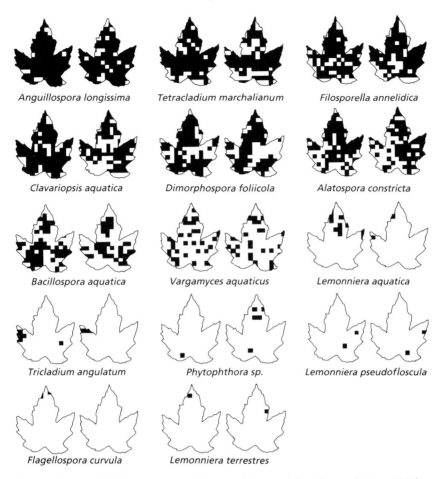

Fig. 8.3 Distribution of aquatic hyphomycetes on silver maple leaves (after Shearer & Lane 1983).

8.4 BACTERIA

Little is known about the species composition and variability of natural bacterial communities, principally because conventional identification requires pure cultures and less than 1% of natural bacteria are cultivatable (Lee & Fuhrman 1990). Jones (1987) stated that 'since we can only isolate, readily, approximately 0.25% of the bacteria seen in freshwater and sediments, there is little hope of assessing the diversity of a microbial community at present'.

Freshwater bacteria may occur either as free-living or attached cells suspended in the water column or associated with sediments. The bacterial flora of detritus consists of rods, cocci, filamentous and mycelial forms as well as myxobacteria (Fenchel & Jørgensen 1977; Armstrong & Bärlocher 1989). In unpolluted streams, Gram-negative rods predominate, although stalked bacteria such as *Hyphomicrobium*, *Caulobacter* and *Gallionella* are often present. For example, Baker and Farr (1977), studying the bacterial flora of two chalk streams in southern England, found that 87% of the bacteria isolated from water samples were Gram-negative rods dominated by *Pseudomonas*, *Flavobacterium*, *Acinetobacter*, *Moraxella*, *Xanthominas* and *Aeromonas*. In nutrient-rich waters, the numbers of *Flavobacterium* and *Achromobacter* species decrease and the bacterial flora becomes dominated by members of the Pseudomonadaceae, Bacillaceae and Enterobacteriaceae (Rheinmeimer 1985).

Suberkropp and Klug (1976) conducted a detailed study of the bacteria associated with leaves during their processing in a North American woodland stream. Bacterial colonization of leaves increased rapidly during the first 2 weeks of exposure to river water and then continued to increase exponentially during the 6-month study period (Fig. 8.4). Although a wide variety of bacterial colonies was observed during the study, no general trend or correlation of colony type with stage of decomposition was detected. The most common genera isolated from leaf material were *Flavobacterium*, *Flexibacter*, *Pseudomonas*, *Acinetobacter*, *Achromobacter*, *Chromobacterium*, *Serriata*, *Bacillus*, *Alcaligenes*, *Cytophaga*, *Sporocytophaga* and *Arthrobacter*.

Whereas the number of bacterial cells in the seston of rivers may reach $10^9 - 10^{11}$ cells l^{-1} (e.g. Edwards 1987), not all of these cells will be active. Studies on the Ogeechee River in North America have distinguished two groups of bacteria: a small group of actively growing cells attached to particles and a much larger group of slowly growing or inactive free-living cells (Edwards & Meyer 1986). A similar observation was made by Kirchman (1983) for a pond system where less than 10% of the total bacterial population was attached to particles at any one time. There is a strong correlation between discharge and bacterial density in the Ogeechee River, suggesting that many of the free-living bacteria are inactive soil bacteria washed in from adjacent land (Edwards 1987; Carlough & Meyer 1989), although no positive correlation between bacterial numbers and discharge was detected by Goulder (1980) for the River Hull (UK). Such inwashing of bacteria may explain the differences in seasonal patterns observed for different systems. In the Ogeechee River the number of suspended bacteria increased in autumn and remained high during winter before decreasing in spring and summer (Carlough & Meyer 1989). In contrast, in the River Hull, numbers of suspended bacteria were highest in summer and lowest in winter (Goulder 1980). Both of these patterns differ from that recorded for bacteria associated with leaf litter by Suber-

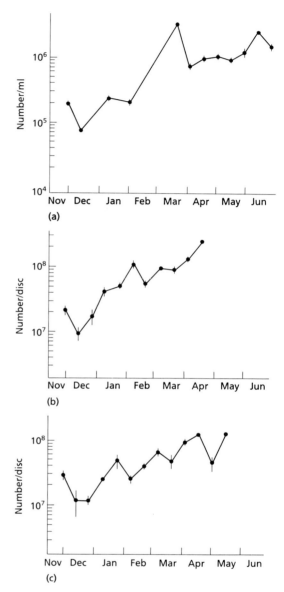

Fig. 8.4 Bacterial numbers, determined by direct microscopic counts, in the water (a) and associated with oak (b) and hickory leaves (c) (after Suberkropp & Klug 1976).

kropp and Klug (1976). Here, bacterial numbers increased exponentially from the time when litter entered the system (November) until the point when the litter had decomposed, 5–7 months later (i.e. April–June). In streams where the major source of suspended bacteria is surfaces and sediments within the stream, bacterial abundance is positively correlated with distance from the source (Goulder 1984; Rimes & Goulder 1986).

Bacteria are an important component of biofilms coating stone surfaces. The number of bacteria in biofilms is often greater than suspended bacterial populations (Geesey *et al* 1978; Lock *et al* 1984) and may reach densities of 10^7–10^8 cells cm^{-2} (Croker & Meyer 1987; Bärlocher & Murdoch 1989). These bacterial communities are diverse, consisting of rods, cocci and spiral-shaped cells (Armstrong & Bärlocher 1989).

The question that must be addressed is whether bacteria associated with particulate detritus are actively involved in its decomposition or are simply using it as a substrate for attachment. In order to decompose vascular plant material, microbes must have the ability to break down cell wall materials such as cellulose, cellulobiose and lignin. Although freshwater bacteria can utilize a range of substrates, including proteins (e.g. *Pseudomonas*), urea (e.g. *Proteus vulgaris*), uric acid (e.g. *Vibrio*) and starch (e.g. *Bacillus, Clostridium*), cellulose is utilized mainly by myxobacteria (e.g. *Cytophaga, Sporocytophaga*)

(Suberkropp & Klug 1976; Rheinmeimer 1985).

Several studies have shown that bacteria in freshwaters can utilize dissolved organic matter (DOM) (Geesey *et al* 1978; Haack & McFeters 1982; Rounick & Winterbourn 1983; Winterbourn *et al* 1986). DOM may originate from many sources (Fig. 8.5; Miller 1987), and is an important component of the energy budget of many stream systems (see Chapter 15). DOM may be leached from detritus or be released by the enzymatic activity of micro-organisms associated with it. Non-detrital DOM may be released by algae and macrophytes.

Heterotrophic bacteria are the major utilizers of DOM in freshwater systems (see section 8.5 and Chapter 15). Their biomass is positively correlated with dissolved organic carbon concentration and responds rapidly to changes in its availability (Kaplan & Bott 1983; Henebry & Gorden 1989).

8.5 BREAKDOWN OF ORGANIC MATTER

The decomposition of particulate organic material in fresh waters has been reviewed by several

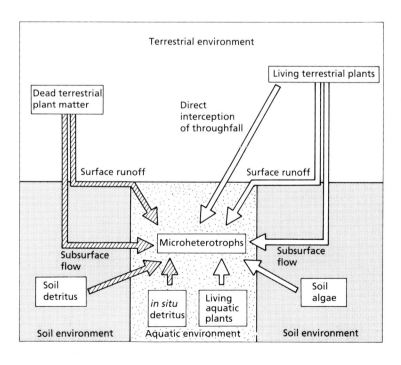

Fig. 8.5 Sources of detrital (hatched arrows) and non-detrital (open arrows) dissolved organic matter (DOM) (after Miller 1987).

authors, including Willoughby (1974), Zdanowski (1977), Anderson and Sedell (1979), Webster and Benfield (1986), and is the subject of Chapter 15.

In lotic environments, cellulose is broken down mainly by myxobacteria and higher fungi. Because of the ability of their hyphae to penetrate the interior of leaves, Bärlocher and Kendrick (1976) concluded that aquatic hyphomycetes were the most important group involved in the degradation of leaf litter. More recent studies of the enzymatic capabilities of these fungi support this view (section 8.3). Amongst bacteria, members of the genera *Cytophaga* and *Sporocytophaga* are the predominant cellulose decomposers (Rheinheimer 1985). Members of the genus *Cytophora* also have the ability to 'tunnel' through plant cell walls (Porter *et al* 1989). Observational studies and laboratory experiments using bacteriocides suggest that fungi are more important than bacteria in the decomposition of leaf litter in flowing water systems (e.g. Triska 1970; Kaushik & Hynes 1971; Suberkropp & Klug 1976). However, bacteria increase in abundance during the latter stages of leaf decomposition and are the major group of micro-organisms on fine particulate organic matter (Mason 1976; Suberkropp & Klug 1976; Chamier *et al* 1984; Sinsabaugh & Linkins 1990). They may also play an important role in the decomposition of conifer needles and more recalcitrant leaves (Iversen 1973; Davis & Winterbourn 1977).

Bacteria are the major group utilizing DOM. Benthic bacteria associated with biofilms appear to be more important than suspended bacteria in the processing of DOM (Lock & Hynes 1976; Dahm 1981; McDowell 1985). Of the suspended bacteria, attached bacteria account for the majority of the heterotrophic activity (Edwards & Meyer 1986). This is in contrast to open water systems (i.e. lakes, oceans) where free-living bacteria are mainly responsible for DOM uptake (Azam *et al* 1990).

8.6 EFFECT OF POLLUTANTS ON BACTERIA AND FUNGI

The distribution and activity of heterotrophic micro-organisms is influenced by a range of pollutants, including cadmium (Giesy 1978), copper (Leland & Carter 1985), zinc (Hornor & Hilt 1985), acid mine drainage (Carpenter *et al* 1983; Maltby & Booth 1991), pesticides (Milner & Goulder 1986; Chandrashekar & Kaveriappa 1989; Cuffney *et al* 1990), organic enrichment (Suberkropp *et al* 1988) and acidity (Traaen 1980; Chamier 1987; Palumbo *et al* 1987a; Thompson & Bärlocher 1989).

Hardness, pH

Bärlocher and Rosset (1981), in a study of the mycoflora of rivers in West Germany and Switzerland, recorded a greater diversity of aquatic hyphomycetes in softwater than in hardwater streams. These differences, which were related to pH (and associated variables) rather than to riparian vegetation, were also apparent in a subsequent, more extensive, study (Wood-Eggenschwiler & Bärlocher 1983). In an attempt to explain the observed distribution patterns, Rosset and Bärlocher (1985) studied the effects of pH, calcium and bicarbonate ions on the growth of aquatic hyphomycetes isolated from hardwater and softwater streams. When tested over the range pH 3–9, all species of fungi investigated grew best at pH values of between 4 and 5. However, the ecological relevance of these results is put into question by the observation that several of the fungi tested exhibited little or no growth at pH values corresponding to those of their native stream. Although species-specific effects on growth were observed for bicarbonate ions, these were not related to the distribution of the species in the field. Calcium stimulated growth of all species tested.

Several studies have noted an accumulation of organic material in acidified waters (Friberg *et al* 1980; Traaen 1980), although there are exceptions (Schindler 1980). Such accumulations are assumed to be due to reductions in microbial and/or faunal processing rates. Although reductions in the microbial activity of leaf litter in acid streams has been recorded (Chamier 1987; Palumbo *et al* 1987b), the results from field and laboratory studies on the effect of pH on aquatic fungi are variable. Whereas some studies have found a negative effect of low pH on fungal abundance (e.g. Hall *et al* 1980; McKinley & Vestal 1982) others have recorded a positive effect (e.g. van Frankenhuyzen *et al* 1985). Thompson and Bärlocher (1989) found that, whereas weight loss of leaf discs was positively correlated with pH and temperature in the field, in a laboratory stream there was a negative correlation between weight loss and pH over the same pH range. One possible reason for the apparent contradiction between laboratory and field results is that under natural conditions, low pH is frequently accompanied by high concentrations of aluminium. Aluminium is concentrated by leaf litter and may be toxic to the associated microbes (Chamier *et al* 1989).

Changes in pH also alter the structure and functioning of bacterial communities. The metabolic activity of both suspended bacteria and that associated with leaf litter is lower in acidic (pH 4.5–5.7) than in circumneutral (pH 6.4–7.8) streams (Rimes & Goulder 1986; Mulholland *et al* 1987). Low pH stress can also result in morphological changes of bacterial cells (Rao *et al* 1984).

The effect of changes in pH on the structure of bacterial communities is illustrated by the study performed by Guthrie *et al* (1978). The structure of a bacterial community was determined before and after the addition of fly ash. The addition of fly ash reduced the pH from 6.5 to 4.6 and resulted in a change in both the number of bacterial species present and their relative abundances (Table 8.4.).

Table 8.4 Dominance of bacterial groups before and after acidification (from Guthrie *et al* 1978)

pH 6.5	pH 4.6
Bacillus	*Pseudomonas*
Sarcina	*Flavobacterium*
Achromobacter	*Chromobacterium*
Flavobacterium	*Bacillus*
Pseudomonas	*Brevibacterium*

Metals

Alterations in fungal assemblages and reduced decomposition rates have been observed below

mine effluent discharges (Carpenter *et al* 1983; Maltby & Booth 1991). For example, Carpenter *et al* (1983) found a significant reduction in the decomposition rate of leaves in a site receiving acid-mine drainage. Although bacterial biomass was not reduced at the polluted site, heterotrophic activity was reduced, owing to a reduction in either bacterial or fungal activity. The effect of coal-mine discharges on aquatic fungi has recently been investigated in British rivers (Maltby & Booth 1991; Bermingham S, unpublished data). Results of these studies suggest that the group of fungi most susceptible to mine discharges are the aquatic hyphomycetes, whose diversity reduces by almost 50% (Fig. 8.6).

Reductions in diversity may occur as a result of either the lethal (death) or sublethal (reductions in growth, reproduction) effects of pollutants. Heavy metals have been shown to inhibit both the growth and sporulation of aquatic fungi (Duddridge & Wainwright 1980; Abel & Bärlocher 1984; Bermingham S, unpublished data). Abel and Bärlocher (1984) compared the sensitivity to cadmium of aquatic hyphomycetes from different sites. Sporulation was ten times more sensitive than growth and, although toxicity was greater for softwater hyphomycete communities, this was due to differences in water chemistry rather than genetic differences between the fungi. Of the five species studied in detail, *Heliscus lugdunensis* and *Tetracladium marchalianum* were the least sensitive, *Alatospora acuminata* was the most sensitive, and *Clavariopsis aquatica* and *Flagellospora curvula* were intermediate.

Metal-induced changes in microbial communities in favour of tolerant species and strains has been observed for heterotropic bacteria (Hornor & Hilt 1985). Bacteria from zinc-polluted sites were more tolerant of zinc stress than those from clean sites (Fig. 8.7) and selection for tolerant bacterial communities was achieved after 20 days' exposure to zinc in artificial stream systems (Colwell *et al* 1989).

Organic enrichment

Suberkropp *et al* (1988) studied the aquatic hyphomycete community above and below the

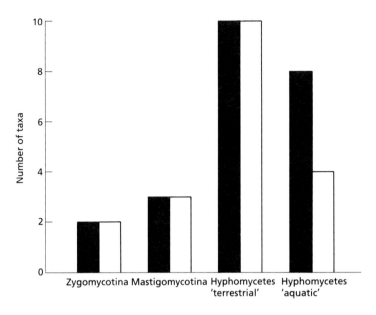

Fig. 8.6 Frequency of fungal taxa on leaf litter collected above and below an effluent discharge from a disused coal mine (■ upstream; □ downstream) (data from Maltby & Booth 1991).

effluent outfall of a sewage treatment works. The effluent resulted in an elevation in biochemical oxygen demand, nitrate, nitrite, ammonia and phosphate; however, it only caused a slight depression in oxygen concentration. The total number and species composition of aquatic hyphomycetes was similar above and below the

effluent input. In contrast, Czeczuga and Próba (1987), studying the River Narew and its tributaries (Poland), found a strong negative correlation between fungal species richness and nitrate concentration. The nitrate values measured during their study ranged from 0.05 to 2.5 mg l^{-1}.

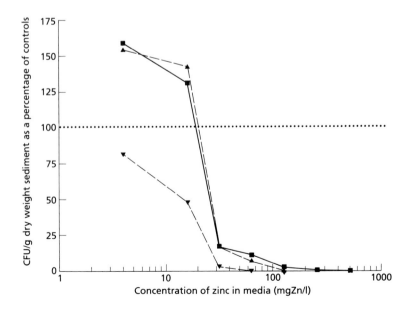

Fig. 8.7 Effect of zinc on the heterotrophic bacterial communities from three sites on the Cedar Run, Virginia, which were exposed to varying degrees of zinc contamination site (CFU, colony forming units) (data from Hornor & Hilt 1985). ■———■ High contamination site (3124.6 μg Zn/g); ▲ – – – ▲ medium contamination site (291.4 μg Zn/g); ▼ ---- ▼ low contamination site (109.0 μg Zn/g).

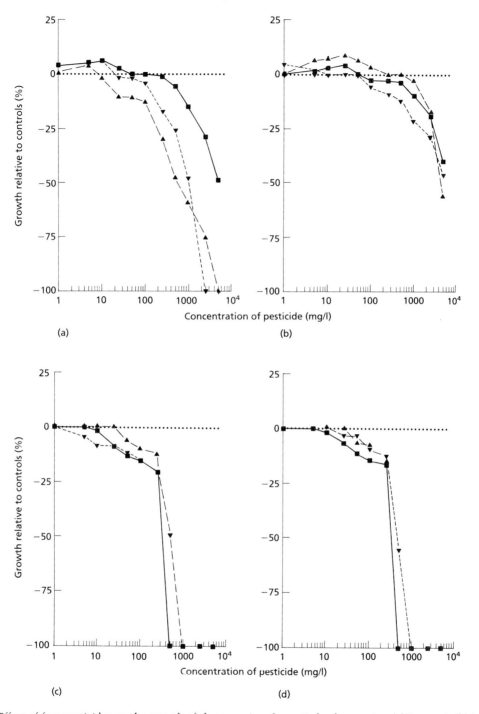

Fig. 8.8 Effect of four pesticides on the growth of three species of aquatic hyphomycetes. (a) Paraquat; (b) 2,4-DB; (c) Mancozeb; (d) Captafol. ■——■ *Flagellospora penicillioides*; ▲ – – – ▲ *Lunulospora curvula* and ▼----▼ *Phalangispora constricta* (data from Chandrashekar & Kaveriappa 1989).

Pesticides

Chandrashekar and Kaveriappa (1989) investigated the effect of four pesticides on the growth of three species of aquatic hyphomycetes: *Flagellospora penicillioides*, *Lunulospora curvula* and *Phalangispora constricta*. The two herbicides tested, paraquat and 2,4-dichlorophenoxybutyric acid (2,4-DB), stimulated growth whereas the two fungicides tested, mancozeb and captafol, either had no effect or inhibited growth (Fig. 8.8). Although species differed in their response to these chemicals, their relative sensitivities were not consistent across all toxicants. For example, whereas *P. constricta* was the most tolerant to the two fungicides, it was the most susceptible to the two herbicides. Other pesticides which have been studied include DDT (Dalton *et al* 1970), diquat (Fronda & Kendrick 1985) and methoxychlor (Cuffney *et al* 1990).

8.7 SUMMARY

Studies on heterotrophic micro-organisms associated with detritus have concentrated upon fungi, particularly aquatic hyphomycetes. Part of the reason for this is the generally held belief that aquatic hyphomycetes play a major role in the decomposition of coarse particulate organic matter, in particular leaves from deciduous trees, and are therefore important in the functioning of allochthonous-based systems. Despite this concentration of effort, relatively little information is available on the ecology and ecophysiology of specific species. Even less is known about other members of the leaf microflora or of the micro-organisms associated with other forms of detritus. There is an obvious need for research into these areas, especially in view of the fact that recent research has highlighted the possible importance of dissolved organic matter in river systems and its utilization by bacteria.

To understand fully the ecology of heterotrophic microbes, it is necessary to be able to sample and identify them. Although techniques are available for studying microbial communities, most of them have limitations. For example, it is possible to quantify total fungal biomass and to identify the species present in a particular fungal assemblage. However, techniques are not available to quantify the biomass of specific individual species within that assemblage. Similarly, because of taxonomic problems, most ecological studies of bacteria use the bacterial, or microbial, community as the unit of study. Techniques under development which may address some of these problems include gene probes (Steffan *et al* 1989), immunofluorescence (Currin *et al* 1990) and DNA hybridization (Lee & Fuhrman 1990).

Until techniques are developed to overcome these problems, the question of the role and importance of specific heterotrophic micro-organisms will remain unanswered. Such information is necessary if we are to predict the impact that changes in water quality and land use will have on the structure and functioning of river systems.

REFERENCES

Abdullah SK, Taj-Aldeen SJ. (1989) Extracellular enzymatic activity of aquatic and aero-aquatic conidial fungi. *Hydrobiologia* **174**: 217–23. [8.3]

Abel TH, Bärlocher F. (1984) Effects of cadmium on aquatic hyphomycetes. *Applied and Environmental Microbiology* **48**: 245–51. [8.2, 8.6]

Aimer RD, Segedin BP. (1985a) Some aquatic hyphomycetes from New Zealand streams. *New Zealand Journal of Botany* **23**: 273–99. [8.2, 8.3]

Aimer RD, Segedin BP. (1985b) Fluctuation in spore numbers of aquatic hyphomycetes in a New Zealand stream. *Botanical Journal of the Linnean Society* **91**: 61–6. [8.2, 8.3]

Akridge RE, Koehn RD. (1987) Amphibious hyphomycetes from San Marcos River in Texas. *Mycologia* **79**: 228–33. [8.2, 8.3]

Anderson NH, Sedell JR. (1979) Detritus processing by macroinvertebrates in stream ecosystems. *Annual Review of Entomology* **24**: 351–77. [8.5]

Armstrong SM, Bärlocher F. (1989) Adsorption and release of amino acids from epilithic biofilms in streams. *Freshwater Biology* **22**: 153–9. [8.2, 8.4]

Arsuffi TL, Suberkropp K. (1984) Leaf processing capabilities of aquatic hyphomycetes: interspecific differences and influence on shredder feeding preferences. *Oikos* **42**: 144–54. [8.2]

Austin B. (1988) *Methods in Aquatic Bacteriology*. John Wiley & Sons, Chichester. [8.2]

Azam F, Cho BC, Smith DC, Simon M. (1990) Bacterial cycling of matter in the pelagic zone of aquatic ecosystems. In: Tilzer MM, Serruya C (eds) *Large Lakes, Ecological Structure and Function*, pp 477–88.

Springer-Verlag, Berlin. [8.5]

Baker JH, Farr IS. (1977) Origins, characterization & dynamics of suspended bacteria in two chalk streams. *Archiv für Hydrobiologie* **80**: 308–26. [8.2, 8.4]

Bandoni RJ. (1974) Mycological observations on the aqueous films covering decaying leaves and other litter. *Transactions of the Mycological Society of Japan* **15**: 309–15. [8.3]

Bandoni RJ. (1975) Significance of the tetraradiate form in dispersal of terrestrial fungi. *Reports of Tottori Mycology Institute* **12**: 105–13. [8.3]

Bandoni RJ. (1981) Aquatic hyphomycetes from terrestrial litter. In: Wicklow DT, Carroll GC (eds) *The Fungal Community. Its Organization and Role in the Ecosystem*, pp 693–708. Marcel Dekker, New York. [8.3]

Bärlocher F. (1980) Leaf-eating invertebrates as competitors of aquatic hyphomycetes. *Oecologia* **47**: 303–6. [8.2]

Bärlocher F. (1982) On the ecology of Ingoldian fungi. *BioScience* **32**: 581–6. [8.3]

Bärlocher F, Kendrick B. (1973) Fungi and food preferences of *Gammarus pseudolimnaeus*. *Archiv für Hydrobiologie* **72**: 501–16. [8.2]

Bärlocher F, Kendrick B. (1974) Dynamics of the fungal populations on leaves in a stream. *Journal of Ecology* **62**: 761–91. [8.2, 8.3]

Bärlocher F, Kendrick B. (1975) Assimilation efficiency of *Gammarus pseudolimnaeus* (Amphipoda) feeding on fungal mycelium or autumn-shed leaves. *Oikos* **26**: 55–9. [8.3]

Bärlocher F, Kendrick B. (1976) Hyphomycetes as intermediaries of energy flow in streams. In: Jones EBG (ed.) *Recent Advances in Aquatic Mycology*, pp 435–46. Elek Science, London. [8.5]

Bärlocher F, Kendrick B. (1981) Role of aquatic hyphomycetes in the trophic structure of streams. In: Wicklow DT, Carroll GC (eds) *The Fungal Community. Its Organization and Role in the Ecosystem*, pp 743–60. Marcel Dekker, New York. [8.3]

Bärlocher F, Michaelides J. (1978) Colonization and conditioning of *Pinus resinosa* needles by aquatic hyphomycetes. *Archiv für Hydrobiologie* **81**: 462–74. [8.3]

Bärlocher F, Murdoch JH. (1989) Hyporheic biofilms—a potential food source for interstitial animals. *Hydrobiologia* **184**: 61–7. [8.4]

Bärlocher F, Oertli JJ. (1978) Colonization of conifer needles by aquatic hyphomycetes. *Canadian Journal of Botany* **56**: 57–62. [8.3]

Bärlocher F, Rosset J. (1981) Aquatic hyphomycete spora of two Black Forest and two Swiss Jura streams. *Transactions of the British Mycological Society* **76**: 479–83. [8.2, 8.3, 8.6]

Bärlocher F, Schweizer M. (1983) Effects of leaf size and decay rate on colonization by aquatic hyphomycetes.

Oikos **41**: 205–10. [8.2]

Bell RT, Ahlgren GM, Ahlgren I. (1983) Estimating bacterioplankton production by measuring [³H] thymidine incorporation in a eutrophic Swedish lake. *Applied and Environmental Microbiology* **45**: 1709–21. [8.2]

Bengtsson G. (1982) Patterns of amino acid utilization by aquatic hyphomycetes. *Oecologia* **55**: 355–63. [8.2, 8.3]

Bengtsson G. (1983) Habitat selection in two species of aquatic hyphomycetes. *Microbial Ecology* **9**: 15–26. [8.3]

Benner R, Lay J, K'nees E, Hodson RE. (1988) Carbon conversion efficiency for bacterial growth on lignocellulose: implications for detritus-based food webs. *Limnology and Oceanography* **33**: 1514–26. [8.2]

Bird GA, Kaushik NK (1981) Coarse particulate organic matter in streams. In: Lock MA, Williams DD (eds) *Perspectives in Running Water Ecology*, pp 41–68. Plenum Press, New York. [8.2]

Bjørnsen PK. (1986) Automatic determination of bacterioplankton biomass by image analysis. *Applied and Environmental Microbiology* **51**: 1199–204. [8.2]

Bobbie RJ, White DC. (1980) Characterization of benthic microbial community structure by high resolution gas chromatography of fatty acid methyl esters. *Applied and Environmental Microbiology* **39**: 1212–22. [8.2]

Bott TL, Kaplan LA. (1990) Potential for protozoan grazing of bacteria in streambed sediments. *Journal of the North American Benthological Society* **9**: 336–45. [8.2]

Boulton AJ, Boon PI. (1991) A review of methodology used to measure leaf litter decomposition in lotic environments: time to turn over an old leaf? *Australian Journal of Marine and Freshwater Research* **42**: 1–43. [8.2]

Butler SK, Suberkropp K. (1986) Aquatic hyphomycetes on oak leaves: comparison of growth, degradation and palatability. *Mycologia* **78**: 922–8. [8.2, 8.3]

Carlough LA, Meyer JL. (1989) Protozoans in two southeastern blackwater rivers and their importance to trophic transfer. *Limnology and Oceanography* **34**: 163–77. [8.4]

Carpenter J, Odum WE, Mills A. (1983) Leaf litter decomposition in a reservoir affected by acid mine drainage. *Oikos* **41**: 165–72. [8.2, 8.6]

Chamier A-C. (1985) Cell-wall-degrading enzymes of aquatic hyphomycetes: a review. *Botanical Journal of the Linnean Society* **91**: 67–81. [8.3]

Chamier A-C. (1987) Effect of pH on microbial degradation of leaf litter in seven streams of the English Lake District. *Oecologia* **71**: 491–500. [8.2, 8.6]

Chamier A-C, Dixon PA. (1982a) Pectinases in leaf degradation by aquatic hyphomycetes I: the field study. The colonization pattern of aquatic hyphomycetes on leaf packs in a Surrey stream. *Oecologia* **52**: 109–15. [8.2, 8.3]

Chamier A-C, Dixon PA. (1982b) Pectinases in leaf degradation by aquatic hyphomycetes: the enzymes and leaf maceration. *Journal of General Microbiology* **128**: 2469–83. [8.3]

Chamier A-C, Dixon PA, Archer SA. (1984) The spatial distribution of fungi on decomposing alder leaves in a freshwater stream. *Oecologia* **64**: 92–103. [8.2, 8.3, 8.5]

Chamier A-C, Sutcliffe DW, Lishman JP. (1989) Changes in Na, K, Ca, Mg and Al content of submersed leaf litter, related to ingestion by the amphipod *Gammarus pulex* (L.). *Freshwater Biology* **21**: 181–9. [8.6]

Chandrashekar KR, Kaveriappa KM. (1989) Effects of pesticides on the growth of aquatic hyphomycetes. *Toxicology Letters* **48**: 311–15. [8.6]

Chergui H. (1990) The dynamics of aquatic hyphomycetes in an eastern Moroccan stream. *Archiv für Hydrobiologie* **118**: 341–52. [8.3]

Chergui H, Pattee E. (1988) The dynamics of Hyphomycetes on decaying leaves in the network of the River Rhone (France). *Archiv für Hydrobiologie* **114**: 3–20. [8.2, 8.3]

Chergui H, Pattee E. (1991) The breakdown of wood in the side arm of a large river: preliminary observations. *Verhandlungen Internationale Vereinigung für Theoretische und Angewandt Limnologie* **24**: 1785–1788. [8.3]

Chrzanowski TH, Crotty RD, Hubbard JG, Welch P. (1984) Applicability of the fluorescein diacetate method of detecting active bacteria in freshwater. *Microbial Ecology* **10**: 179–85. [8.2]

Cochran TW, Vercellotti JR. (1978) Hexosamine biosynthesis and accumulation by fungi in liquid and solid media. *Carbohydrate Research* **61**: 529–43. [8.2]

Colwell FS, Hornor SG, Cherry DS (1989) Evidence of structural and functional adaptations in epilithon exposed to zinc. *Hydrobiologia* **171**: 79–90. [8.2, 8.6]

Croker MT, Meyer JL. (1987) Interstitial dissolved organic carbon in sediments of a southern Appalachian headwater stream. *Journal of the North American Benthological Society* **6**: 159–67. [8.4]

Cuffney TF, Wallace B, Lugthart GJ. (1990) Experimental evidence quantifying the role of benthic invertebrates in organic matter dynamics of headwater streams. *Freshwater Biology* **23**: 281–99. [8.6]

Cummins KW, Spengler GL, Ward GM, Speaker RM, Ovink RM, Mahan DC *et al* (1980) Processing of confined and naturally entrained leaf litter in a woodland stream ecosystem. *Limnology and Oceanography* **25**: 952–7. [8.2]

Currin CA, Paerl HW, Suba GK, Alberte RS. (1990) Immunofluorescence detection and characterization of N_2-fixing microorganisms from aquatic environments. *Limnology and Oceanography* **35**: 59–71. [8.7]

Czeczuga B, Próba D. (1987) Studies of aquatic fungi. VII Mycoflora of the upper part of the River Narew and its tributaries in a differentiated environment. *Nova Hedwigia* **44**: 151–61. [8.3, 8.6]

Dahm PR. (1981) Pathways and mechanisms for removal of dissolved organic carbon from leaf leachate in streams. *Canadian Journal of Fisheries and Aquatic Sciences* **38**: 68–76. [8.5]

Dalton SA, Hodkinson M, Smith KA. (1970) Interactions between DDT and river fungi. *Applied Microbiology* **20**: 662–6. [8.6]

Darvill A, McNeil M, Albersheim P, Delmer DP (1980) The primary cell walls of flowering plants. In: Tolbert NE (ed.) *The Plant Cell*, pp 92–157. Academic Press, New York. [8.3]

Davis SF, Winterbourn MJ. (1977) Breakdown and colonization of *Nothofagus* leaves in a New Zealand stream. *Oikos* **28**: 250–5. [8.5]

Duddridge JE, Wainwright M. (1980) Heavy metal accumulation by aquatic fungi and reduction in viability of *Gammarus pulex* fed Cd^{2+} contaminated mycelium. *Water Research* **14**: 1605–11. [8.2, 8.6]

Edwards RT. (1987) Sestonic bacteria as a food source for filtering invertebrates in two southeastern blackwater rivers. *Limnology and Oceanography* **32**: 221–34. [8.2, 8.4]

Edwards RT, Meyer JL. (1986) Production and turnover of planktonic bacteria in two southeastern blackwater rivers. *Applied and Environmental Microbiology* **52**: 1317–23. [8.2, 8.4, 8.5]

Elwood JW, Mulholland PJ, Newbold JD. (1988) Microbial activity and phosphorus uptake on decomposing leaf detritus in a heterotrophic stream. *Verhandlungen Internationale Vereinigung für Theoretische und Angewandt Limnologie* **23**: 1198–208. [8.2]

Fairchild JF, Boyle TP, Robinson-Wilson E, Jones JR. (1983) Microbial action in detritus leaf processing and the effects of chemical perturbation. In: Fontaine TD, Bartell SM (eds) *Dynamics of Lotic Ecosystems*, pp 437–56. Ann Arbor Science, Michigan. [8.2]

Federle TW, McKinley VL, Vestal JR. (1982) Physical determinants of microbial colonization and decomposition of plant litter in an arctic lake. *Microbial Ecology* **8**: 127–38. [8.2]

Fenchel TM, Jørgensen BB. (1977) Detrital food chains of aquatic ecosystems: the role of bacteria. *Advances in Microbial Ecology* **1**: 1–58. [8.4]

Findlay SEG, Arsuffi TL. (1989) Microbial growth and detritus transformations during decomposition of leaf litter in a stream. *Freshwater Biology* **21**: 261–9. [8.2]

Findlay SEG, Meyer JL, Edwards RT. (1984a) Measuring bacterial production via rate of incorporation of [³H] thymidine into DNA. *Journal of Microbiological*

Methods **2**: 57–72. [8.2]

Findlay S, Meyer JL, Smith PJ. (1984b) Significance of bacterial biomass in the nutrition of a freshwater isopod (*Lirceus* sp.). *Oecologia* **63**: 38–42. [8.2]

Findlay S, Meyer JL, Smith PJ. (1986) Incorporation of microbial biomass by *Peltoperla* sp. (Plecoptera) and *Tipula* sp. (Diptera). *Journal of the North American Benthological Society* **5**: 306–10. [8.2]

Fisher PJ. (1977) New methods for detecting and studying saprophytic behaviour of aero-aquatic hyphomycetes from stagnant water. *Transactions of the British Mycological Society* **68**: 407–11. [8.3]

Fisher PJ. (1979) Colonization of freshly abscissed and decaying leaves by aero-aquatic hyphomycetes. *Transactions of the British Mycological Society* **73**: 99–102. [8.3]

Fisher PJ, Webster J. (1979) Effect of oxygen and carbon dioxide on growth of four aero-aquatic hyphomycetes. *Transactions of the British Mycological Society* **72**: 57–61. [8.3]

Fisher PJ, Webster J. (1981) Ecological studies on aero-aquatic hyphomycetes. In: Wicklow DT, Carroll GC (eds) *The Fungal Community. Its Organization and Role in the Ecosystem*, pp 709–30. Marcel Dekker, New York. [8.3]

Fisher PJ, Davey RA, Webster J. (1983) Degradation of lignin by aquatic and aero-aquatic hyphomycetes. *Transactions of the British Mycological Society* **80**: 166–8. [8.3]

Forbes AM, Magnuson JJ. (1980) Decomposition and microbial colonization of leaves in a stream modified by coal ash effluent. *Hydrobiologia* **76**: 263–7. [8.2]

Frankland JC, Lindley DK, Swift MJ. (1978) A comparison of two methods for the estimation of biomass in leaf litter. *Soil Biology and Biochemistry* **10**: 323–33. [8.2]

Freeman C, Lock MA, Marxsen J, Jones SE. (1990) Inhibitory effects of high molecular weight dissolved organic matter upon metabolic processes in biofilms from contrasting rivers and streams. *Freshwater Biology* **24**: 159–66. [8.2]

Friberg F, Otto C, Svensson BS. (1980) Effects of acidification on the dynamics of allochthonous leaf material and benthic invertebrate communities in running waters. In: Drablos D, Tollan A. (eds) *Proceedings of the International Conference on the Ecological Impact of Acid Precipitation, Norway 1980*, pp 304–5. SNSF project. [8.6]

Fronda A, Kendrick B. (1985) Uptake of diquat by four aero-aquatic hyphomycetes. *Environmental Pollution* **37**: 229–44. [8.6]

Fry JC. (1982) The analysis of microbial interactions and communities *in situ*. In: Bull AT, Slater JH (eds) *Microbial Interactions and Communities*, Vol. 1, pp 103–52. Academic Press, London. [8.2]

Fry JC (1988) Determination of Biomass. In: Austin B (ed.) *Methods in Aquatic Bacteriology*, pp 27–72. John Wiley & Sons, New York. [8.2]

Fry JC. (1990) Direct methods and biomass estimation. *Methods in Microbiology* **22**: 41–85. [8.2]

Fuhrman JA, Azam F. (1980) Bacterioplankton secondary production estimates for coastal waters of British Columbia, Antarctica and California. *Applied and Environmental Microbiology* **39**: 1085–95. [8.2]

Fuhrman JA, Azam F. (1982) Thymidine incorporation as a measure of heterotrophic bacterioplankton production in marine surface waters: evaluation and field results. *Marine Biology* **66**: 109–20. [8.2]

Garden A, Davies RW. (1988) Decay rates of autumn and spring leaf litter in a stream and effects on growth of a detritivore. *Freshwater Biology* **19**: 297–303. [8.2]

Geesey GG, Mutch R, Costerton JW, Green RB. (1978) Sessile bacteria: an important component of the microbial population in small mountain streams. *Limnology and Oceanography* **23**: 1214–23. [8.4]

Gessner MO, Schwoerbel J. (1991) Fungal biomass associated with decaying leaf litter in a stream. *Oecologia* **87**: 602–603. [8.2]

Giesy JP. (1978) Cadmium inhibition of leaf decomposition in an aquatic microcosm. *Chemosphere* **6**: 467–75. [8.6]

Godfrey BES. (1983) Growth of two terrestrial microfungi on submerged alder leaves. *Transactions of the British Mycological Society* **81**: 418–21. [8.2, 8.3]

Gottschalk G. (1986) *Bacterial Metabolism*, 2nd edn. Springer-Verlag, New York. [8.2]

Goulder R. (1980) Seasonal variation in heterotrophic activity and population density of planktonic bacteria in a clean river. *Journal of Ecology* **68**: 349–63. [8.4]

Goulder R. (1984) Downstream increase in the abundance and heterotrophic activity of suspended bacteria in an intermittent calcareous headstream. *Freshwater Biology* **14**: 611–19. [8.4]

Grant WD, West AW. (1986) Measurement of ergosterol, diaminopimelic acid and glucosamine in soil: evaluation as indicators of microbial biomass. *Journal of Microbial Methods* **6**: 47–53. [8.2]

Grigorova R, Norris JR. (1990) *Methods in Microbiology. Vol 22, Techniques in Microbial Ecology*. Academic Press, London. [8.2]

Groom AP, Hildrew AG. (1989) Food quality for detritivores in streams of contrasting pH. *Journal of Animal Ecology* **58**: 863–81. [8.2]

Gupta AK, Mahrotra RS. (1989) Seasonal periodicity of aquatic fungi in tanks at Kurukshetra, India. *Hydrobiologia* **173**: 219–29. [8.3]

Guthrie RK, Cherry DS, Singleton FL. (1978) Responses of heterotrophic bacterial populations to pH changes in coal ash effluent. *Water Resources Bulletin* **14**: 803–8. [8.6]

Haack TK, McFeters GA. (1982) Nutritional relationships among microorganisms in an epilithic biofilm community. *Microbial Ecology* **8**: 115–26. [8.4]

Hall GH, Jones JG, Pickup RW, Simon BM. (1990) Methods to study the bacterial ecology of freshwater environments. *Methods in Microbiology* **22**: 182–209. [8.2]

Hall RJ, Likens GE, Fiance SB, Hendrey GR. (1980) Experimental acidification of a stream in the Hubbard Brook Experimental Forest, New Hampshire. *Ecology* **61**: 976–89. [8.6]

Henebry MS, Gorden RW. (1989) Summer bacterial populations in Mississippi River Pool 19: implications for secondary production. *Hydrobiologia* **182**: 15–23. [8.2, 8.4]

Herbert RA. (1990) Methods for enumerating micro-organisms and determining biomass in natural environments. *Methods in Microbiology* **22**: 1–39. [8.1, 8.2].

Hering TF, Nicholson PB. (1964) A clearing technique for the examination of fungi in plant tissue. *Nature* **201**: 942–3. [8.2]

Hicks RE, Newell SY. (1983) An improved gas chromatographic method for measuring glucosamine and muramic acid concentrations. *Analytical Biochemistry* **128**: 438–45. [8.2]

Holm-Hansen O, Booth CR. (1966) The measurement of adenosine triphosphate in the ocean and its ecological significance. *Limnology and Oceanography* **11**: 510–19. [8.2]

Hornor SG, Hilt BA. (1985) Distribution of zinc-tolerant bacteria in stream sediments. *Hydrobiologia* **128**: 155–60. [8.2, 8.6]

Hudson HJ. (1986) *Fungal Biology*. Edward Arnold, London. [8.2, 8.3]

Ingold CT. (1942) Aquatic Hyphomycetes of decaying alder leaves. *Transactions of the British Mycological Society* **25**: 339–417. [8.3]

Ingold CT. (1975) *Guide to Aquatic Hyphomycetes*. Freshwater Biological Association (Publication No. 30), Cumbria. [8.2, 8.3]

Iqbal SH, Webster J. (1973) Aquatic hyphomycete spora of the River Exe and its tributaries. *Transactions of the British Mycological Society* **61**: 331–46. [8.2, 8.3]

Iqbal SH, Webster J. (1977) Aquatic hyphomycete spora of some Dartmoor streams. *Transactions of the British Mycological Society* **69**: 233–41. [8.2, 8.3]

Iversen TM. (1973) Decomposition of autumn-shed beech leaves in a springbrook and its significance for the fauna. *Archiv für Hydrobiologie* **72**: 305–12. [8.5]

Jones EBG. (1981) Observations on the ecology of lignicolous aquatic hyphomycetes. In: Wicklow DT, Carroll GC (eds) *The Fungal Community. Its Organization and Role in the Ecosystem*, pp 731–42. Marcel Dekker, New York. [8.3]

Jones JG. (1977) The study of aquatic microbial communities. In: Skinner FA, Shewan JM (eds) *Aquatic Microbiology*, pp 1–30. Academic Press, London. [8.2]

Jones JG. (1979) *A Guide to Methods for Estimating Microbial Numbers and Biomass in Fresh Water*. Freshwater Biological Association (Publication No. 39), Cumbria. [8.2]

Jones JG. (1987) Diversity of freshwater microbiology. In: Fletcher M, Gray TRG, Jones JG (eds) *Ecology of Microbial Communities*, pp 236–59. Cambridge University Press, Cambridge. [8.4]

Jones PCT, Mollison JE. (1948) A technique for the quantitative estimation of soil microorganisms. *Journal of General Microbiology* **2**: 54–69. [8.2]

Kaplan LA, Bott TL. (1983) Microbial heterotrophic utilization of dissolved organic matter in a Piedmont stream. *Freshwater Biology* **13**: 363–77. [8.2, 8.4]

Kaplan LA, Bott TL. (1985) Acclimation of stream-bed heterotrophic microflora: metabolic responses to dissolved organic matter. *Freshwater Biology* **15**: 479–92. [8.2]

Karl DM. (1980) Cellular nucleotide measurements and applications in microbial ecology. *Microbiological Reviews* **44**: 739–96. [8.2]

Kaushik NK, Hynes HBN. (1968) Experimental study on the role of autumn shed leaves in aquatic environments. *Journal of Ecology* **56**: 229–43. [8.2]

Kaushik NK, Hynes HBN. (1971) The fate of dead leaves that fall into streams. *Archiv für Hydrobiologie* **68**: 465–515. [8.5]

Kirby JJH. (1987) A comparison of serial washing and surface sterilization. *Transactions of the British Mycological Society* **88**: 559–62. [8.2]

Kirby JJH, Webster J, Baker JH. (1990) A particle plating method for analysis of fungal community composition and structure. *Mycological Research* **94**: 621–6. [8.2]

Kirchman D. (1983) The production of bacteria attached to particles suspended in a freshwater pond. *Limnology and Oceanography* **28**: 858–72. [8.4]

Ladd TI, Costerton JW. (1990) Methods for studying biofilm bacteria. *Methods in Microbiology* **22**: 285–307. [8.2]

Lamore BJ, Goos RD. (1978) Wood-inhabiting fungi of a freshwater stream in Rhode Island. *Mycologia* **70**: 1025–34. [8.2, 8.3]

Lee C, Howarth RW, Howes BL. (1980) Sterols in decomposing *Spartina alterniflora* and the use of ergosterol in estimating the contribution of fungi to detrital nitrogen. *Limnology and Oceanography* **25**: 290–303. [8.2]

Lee S, Fuhrman JA. (1990) DNA hybridization to compare species compositions of natural bacterioplankton assemblages. *Applied and Environmental Microbiology* **56**: 739–46. [8.4, 8.7]

Leland HV, Carter JL. (1985) Effects of copper on production of periphyton, nitrogen fixation and processing of leaf litter in a Sierra Nevada, California stream. *Freshwater Biology* **15**: 155–73. [8.6]

Lock MA, Hynes HBN. (1976) The fate of dissolved organic carbon derived from autumn-shed maple leaves (*Acer saccharum*) in a temperate hard water stream. *Limnology and Oceanography* **21**: 436–43. [8.5]

Lock MA, Wallace RR, Costerton JW, Ventullo RM, Charlton SE. (1984) River epilithon: towards a structural-functional model. *Oikos* **42**: 10–22. [8.4]

McDowell WH. (1985) Kinetics and mechanisms of dissolved organic carbon retention in a headwater stream. *Biogeochemistry* **2**: 329–53. [8.5]

McKinley VL, Vestal JR. (1982) Effects of acid on plant litter decomposition in an arctic lake. *Applied and Environmental Microbiology* **43**: 1188–95. [8.6]

Maltby L, Booth R. (1991) The effect of coal-mine effluent on fungal assemblages and leaf breakdown. *Water Research* **25**: 247–50. [8.6]

Mason CF. (1976) Relative importance of fungi and bacteria in the decomposition of phragmites leaves. *Hydrobiologia* **51**: 65–9. [8.5]

Matcham SE, Jordan BR, Wood DA. (1985) Estimation of fungal biomass in a solid substrate by three independent methods. *Applied Microbiology and Biotechnology* **21**: 108–12. [8.2]

Mer GS, Sati SC. (1989) Seasonal fluctuations in species composition of aquatic hyphomycetous flora in a temperate freshwater stream of central Himalaya, India. *Internationale Revue de Gesamten Hydrobiologie* **74**: 433–7. [8.3]

Miller JC. (1987) Evidence for the use of non-detrital dissolved organic matter by microheterotrophs on plant detritus in a woodland stream. *Freshwater Biology* **18**: 483–94. [8.4]

Milner CR, Goulder R. (1986) The abundance, heterotrophic activity and taxonomy of bacteria in a stream subject to pollution by chlorophenols, nitrophenols and phenoxyalkanoic acids. *Water Research* **20**: 85–90. [8.6]

Morgan P, Cooper CJ, Battersby NS, Lee SA, Lewis ST, Machin TM, Graham SC, Watkinson RJ. (1991) Automated image analysis method to determine fungal biomass in soils and on solid matrices. *Soil Biology and Biochemistry* **23**: 609–16. [8.2]

Moriarty DJW. (1986) Measurement of bacterial growth rates in aquatic systems from rates of nucleic acid synthesis. *Advances in Microbial Ecology* **9**: 245–92. [8.2]

Moriarty DJW. (1990) Techniques for estimating bacterial growth rates and production of biomass in aquatic ecosystems. *Methods in Microbiology* **22**: 212–34. [8.2]

Moriarty DJW, Pollard PC. (1981) DNA synthesis as a measure of bacterial productivity in seagrass sediments. *Marine Ecology Progress Series* **5**: 151–6. [8.2]

Mulholland PJ, Palumbo AV, Elwood JW, Rosemond AD. (1987) Effects of acidification on leaf decomposition in streams. *Journal of the North American Benthological Society* **6**: 147–58. [8.6]

Müller-Haeckel A, Marvanová L. (1979) Periodicity of aquatic hyphomycetes in the subarctic. *Transactions of the British Mycological Society* **73**: 109–16. [8.3]

Newell SY, Arsuffi TL, Fallon RD. (1988) Fundamental procedures for determining ergesterol content of decaying material by liquid chromatography. *Applied and Environmental Microbiology* **54**: 1876–1879. [8.2]

Newell SY, Fallon RD. (1991) Toward a method for measuring instantaneous fungal growth rates in field samples. *Ecology* **72**: 1547–1559. [8.2]

Newell SY, Hicks RE. (1982) Direct-count estimates of fungal and bacterial biovolume in dead leaves of smooth cordgrass (*Spartina alterniflora* Loisel). *Estuaries* **5**: 246–60. [8.2]

Newell SY, Statzell-Tallman A. (1982) Factors for conversion of fungal biovolume values to biomass, carbon and nitrogen: variation with mycelial ages, growth conditions and stains of fungi from a salt marsh. *Oikos* **39**: 261–8. [8.2]

Nilsson S. (1964) Freshwater hyphomycetes. Taxonomy, morphology and ecology. *Symbolae Botanicae Upsalienses* **XVIII**(2): 1–130. [8.3]

O'Carroll K. (1988) Assessment of bacterial activity. In: Austin B (ed.) *Methods in Aquatic Bacteriology*, pp 347–66. John Wiley & Sons, Chichester. [8.2]

Padgett DE. (1976) Leaf decomposition by fungi in a tropical rainforest stream. *Biotropica* **8**: 166–78. [8.3]

Palumbo AV, Bogle MA, Turner RR, Elwood JW, Mulholland PJ. (1987a) Bacterial communities in acidic and circumneutral streams. *Applied and Environmental Microbiology* **53**: 337–44. [8.6]

Palumbo AV, Mulholland PJ, Elwood JW. (1987b) Microbial communities on leaf material protected from macroinvertebrate grazing in acidic and circumneutral streams. *Canadian Journal of Fisheries and Aquatic Sciences* **44**: 1064–70. [8.2, 8.6]

Park D. (1972) Methods of detecting fungi in organic detritus in water. *Transactions of the British Mycological Society* **58**: 281–90. [8.2]

Park D. (1974a) Aquatic hyphomycetes in non-aquatic habitats. *Transactions of the British Mycological Society* **63**: 183–7. [8.3]

Park D. (1974b) Accumulation of fungi by cellulose exposed in a river. *Transactions of the British Mycological Society* **63**: 437–47. [8.2]

Park D. (1980) A two-year study of numbers of cellulolytic *Pythium* in river water. *Transactions of the British Mycological Society* **74**: 253–8. [8.3]

Park D, McKee W. (1978) Cellulolytic *Pythium* as a component of the river mycoflora. *Transactions of the British Mycological Society* **71**: 251–9. [8.3]

Paul EA, Johnson RL. (1977) Microscopic counting and

adenosine 5'-triphosphate measurement in determining microbial growth in soils. *Applied and Environmental Microbiology* **34**: 263–9. [8.2]

Paul RW, Benfield EF, Cairns J. (1983) Dynamics of leaf
processing in a medium-sized river In: Fontaine TD,
Bartell SM (eds) *Dynamics of Lotic Ecosystems*,
pp 403–23. Ann Arbor Science, Michigan. [8.2]

Peters GT, Webster JR, Benfield EF. (1987) Microbial
activity associated with seston in headwater streams:
effects of nitrogen, phosphorus and temperature.
Freshwater Biology **18**: 405–13. [8.2]

Petersen RC, Cummins KW, Ward GM. (1989) Microbial
and animal processing of detritus in a woodland
stream. *Ecological Monographs* **59**: 21–39. [8.2]

Poindexter JS, Leadbetter ER. (1986) *Bacteria in Nature.
Vol. 2. Methods and Special Applications in Bacterial
Ecology*. Plenum Press, New York. [8.2]

Porter KG, Feig YS. (1980) The use of DAPI for identifying and counting aquatic microflora. *Limnology
and Oceanography* **25**: 943–8. [8.2]

Porter D, Newell SY, Lingle WL. (1989) Tunneling bacteria in decaying leaves of a seagrass. *Aquatic Botany*
35: 395–401. [8.5]

Ranzoni FW. (1979) Aquatic hyphomycetes from
Hawaii. *Mycologia* **71**: 786–95. [8.3]

Rao SS, Paolini D, Leppard GG. (1984) Effects of low-pH
stress on the morphology and activity of bacteria
from lakes receiving acid precipitation. *Hydrobiologia*
114: 115–21. [8.6]

Rheinheimer G. (1985) *Aquatic Microbiology*. John
Wiley & Sons, Chichester. [8.4, 8.5]

Ride JP, Drysdale RB. (1972) A rapid method for the
chemical estimation of filamentous fungi in plant
tissue. *Physiological Plant Pathology* **2**: 7–15. [8.2]

Riemann B, Bell RT. (1990) Advances in estimating
bacterial biomass and growth in aquatic systems.
Archiv für Hydrobiologie **118**: 385–402. [8.2]

Riemann B, Fuhrman J, Azam F. (1982) Bacterial secondary production in freshwater measured by (^{3}H)-
thymidine incorporation method. *Microbial Ecology*
8: 101–14. [8.2]

Rimes CA, Goulder R. (1986) Suspended bacteria in
calcareous and acid headstreams: abundance, heterotrophic activity and downstream change. *Freshwater
Biology* **16**: 663–51. [8.4, 8.6]

Rosset J, Bärlocher F. (1985) Aquatic hyphomycetes:
influences of pH, Ca^{2+} and HCO_3^- on growth *in vitro*.
Transactions of the British Mycological Society **84**:
137–45. [8.2, 8.6]

Rounick JS, Winterbourn MJ. (1983) The formation,
structure and utilization of stone surface organic
layers in two New Zealand streams. *Freshwater Biology* **13**: 57–72. [8.2, 8.4]

Sanders PF, Anderson JM. (1979) Colonization of wood
blocks by aquatic hyphomycetes. *Transactions of the
British Mycological Society* **73**: 103–7. [8.2, 8.3]

Schindler DW. (1980) Experimental acidification of a
whole lake: a test of the oligotrophication hypothesis.
In: Drabløs D, Tolan A (eds) *Proceedings International Conference on the Ecological Impact of Acid
Precipitation, Norway 1980*, pp 340–4. SNSF Project.
[8.6]

Schnürer J, Rosswall T. (1982) Fluorescein diacetate
hydrolysis as a measure of total microbial activity in
soil and litter. *Applied and Environmental Microbiology* **43**: 1256–61. [8.2]

Seitz LM, Saver DB, Burroughs R, Mohr HE, Hubbard
JD. (1979) Ergosterol as a measure of fungal growth.
Phytopathology **69**: 1202–3. [8.2]

Shearer CA. (1972) Fungi of the Chesapeake Bay and its
tributaries. III The distribution of wood-inhabiting
Ascomycetes and fungi imperfecti of the Patuxent
River. *American Journal of Botany* **59**: 961–9. [8.2]

Shearer CA, Lane LC. (1983) Comparison of three techniques for the study of aquatic hyphomycete communities. *Mycologia* **75**: 498–508. [8.2, 8.3]

Shearer CA, Webster J. (1985a) Aquatic hyphomycete
communities in the River Teign. I. Longitudinal distribution patterns. *Transactions of the British Mycological Society* **84**: 489–501. [8.2, 8.3]

Shearer CA, Webster J. (1985b) Aquatic hyphomycete
communities in the River Teign. II. Temporal distribution patterns. *Transactions of the British Mycological Society* **84**: 503–7. [8.2, 8.3]

Shearer CA, Webster J. (1985c) Aquatic hyphomycete
communities in the River Teign. III. Comparison of
sampling techniques. *Transactions of the British
Mycological Society* **84**: 509–18. [8.2]

Shearer CA, von Bodman SB. (1983) Patterns of occurrence of Ascomycetes associated with decomposing
twigs in a mid western stream. *Mycologia* **75**: 518–
30. [8.2]

Shearer CA, Webster J. (1991) Aquatic hyphomycete
communities in the river Teign. IV Twig colonization.
Mycological Research **95**: 413–420. [8.2]

Singh N, Musa TM. (1977) Terrestrial occurrence and
the effect of temperature on growth, sporulation and
spore production of some tropical aquatic hyphomycetes. *Transactions of the British Mycological
Society* **68**: 103–6. [8.3]

Sinsabaugh RL, Golladay SW, Linkins AE. (1991)
Comparison of epilithic and epixylic biofilm development in a boreal river. *Freshwater Biology* **25**: 179–
187. [8.2]

Sinsabaugh RL, Linkins AE. (1990) Enzymic and chemical analysis of particulate organic matter from a boreal
river. *Freshwater Biology* **23**: 301–9. [8.2, 8.5]

Sorokin YI, Kadota H. (1972) *Techniques for the Assess-
ment of Microbial Production and Decomposition of
Fresh Waters*. IBP Handbook No. 23, Blackwell,
Oxford. [8.2]

Sridhar KR, Kaveriappa KM. (1982) Aquatic fungi of the

western Ghat forests in Karnataka. *Indian Phytopathology* **35**: 293–6. [8.3]

Sridhar KR, Kaveriappa KM. (1989) Colonization of leaves by water-borne hyphomycetes in a tropical stream. *Mycological Research* **92**: 392–6. [8.3]

Staley JT, Konopka A. (1985) Measurement of *in situ* activities of nonphotosynthetic microorganisms in aquatic and terrestrial habitats. *Annual Review of Microbiology* **39**: 321–46. [8.2]

Steffan RJ, Breen A, Atlas RM, Sayler GS. (1989) Application of gene probe methods for monitoring specific microbial populations in freshwater ecosystems. *Canadian Journal of Microbiology* **35**: 681–5. [8.7]

Suberkropp K. (1984) Effect of temperature on seasonal occurrence of aquatic hyphomycetes. *Transactions of the British Mycological Society* **82**: 53–62. [8.2, 8.3]

Suberkropp K. (1991) Relationship between growth and sporulation of aquatic hyphomycetes on decomposing leaf litter. *Mycological Research* **95**: 843–850. [8.2]

Suberkropp KF, Klug MJ. (1974) Decomposition of deciduous leaf litter in a woodland stream. I. A scanning electron microscopic study. *Microbial Ecology* **1**: 96–103. [8.2]

Suberkropp K, Klug MJ. (1976) Fungi and bacteria associated with leaves during processing in a woodland stream. *Ecology* **57**: 707–19. [8.2, 8.3, 8.4, 8.5]

Suberkropp K, Klug MJ. (1980) The maceration of deciduous leaf litter by aquatic hyphomycetes. *Canadian Journal of Botany* **58**: 1025–31. [8.3]

Suberkropp K, Klug MJ. (1981) Degradation of leaf litter by aquatic hyphomycetes. In: Wicklow DT, Carroll GC (eds) *The Fungal Community. Its Organization and Role in the Ecosystem*, pp 761–76. Marcel Dekker, New York. [8.3]

Suberkropp K, Arsuffi TL, Anderson JP. (1983) Comparison of degradative ability, enzymatic activity and palatability of aquatic hyphomycetes grown on leaf litter. *Applied and Environmental Microbiology* **46**: 237–44. [8.2, 8.3]

Suberkropp K, Michelis A, Lorch HJ, Ottow JCG. (1988) Effect of sewage treatment plant effluents on the distribution of aquatic hyphomycetes in the River Erms, Schwäbische Alb, FGR. *Aquatic Biology* **32**: 141–53. [8.6]

Tabor PS, Neihof RA. (1982) Improved method for determination of respiring individual microorganisms in natural waters. *Applied and Environmental Microbiology* **43**: 1249–55. [8.2]

Tabor PS, Neihof RA. (1984) Direct determination of activities for microorganisms of Chesapeake Bay populations. *Applied and Environmental Microbiology* **48**: 1012–19. [8.2]

Taylor BR, Roff JC. (1984) Use of ATP and carbon: nitrogen ratio as indicators of food quality of stream detritus. *Freshwater Biology* **15**: 195–201. [8.2]

Thomas K, Chilvers GA, Norris RH. (1989) Seasonal occurrence of conidia of aquatic hyphomycetes (fungi) in Lees Creek, Australian Capital Territory. *Australian Journal of Marine and Freshwater Research* **40**, 11–23. [8.3]

Thompson PL, Bärlocher F. (1989) Effect of pH on leaf breakdown in streams and in the laboratory. *Journal of the North American Benthological Society* **8**: 203–10. [8.6]

Thornton DR. (1963) The physiology and nutrition of some aquatic hyphomycetes. *Journal of General Microbiology* **33**: 23–31. [8.2, 8.3]

Torréton JP, Bouvy M. (1991) Estimating DNA synthesis from [³H] thymidine incorporation: discrepancies among macromolecular extraction procedures. *Limnology and Oceanography* **36**: 299–306. [8.2]

Traaen TS. (1980) Effects of acidity on decomposition of organic matter in aquatic environments. In: Drabløs D, Tollan A (eds) *Proceedings International Conference on the Ecological Impact of Acid Precipitation, Norway 1980*, pp 340–1. SNSF Project. [8.6]

Trevors JT. (1984) The measurement of electron transport system (ETS) activity in freshwater sediments. *Water Research* **18**: 581–4. [8.2]

Triska FJ. (1970) *Seasonal distribution of aquatic hyphomycetes in relation to the disappearance of leaf litter from a woodland stream.* Unpublished PhD thesis, University of Pittsburg. [8.5]

van Frankenhuyzen K, Geen GH. (1986) Microbe-mediated effects of low pH on availability of detrital energy to a shredder, *Clistoronia magnifica* (Trichoptera, Limnephilidae). *Canadian Journal of Zoology* **64**: 421–6. [8.2]

van Frankenhuyzen K, Geen GH, Koivisto C. (1985) Direct and indirect effects of low pH on the transformation of detrital energy by the shredding caddisfly, *Clistoronia magnifica* (Banks) (Limnephilidae). *Canadian Journal of Zoology* **63**: 2298–304. [8.6]

Webster J. (1980) *Introduction to Fungi*, 2nd edn. Cambridge University Press, Cambridge. [8.3]

Webster J. (1981) Biology and ecology of aquatic hyphomycetes. In: Wicklow DT, Carroll GC (eds) *The Fungal Community. Its Organization and Role in the Ecosystem*, pp 681–91. Marcel Dekker, New York [8.3]

Webster JR, Benfield EF. (1986) Vascular plant breakdown in freshwater ecosystems. *Annual Review of Ecology and Systematics* **17**: 567–94. [8.2, 8.5]

Webster J, Descals E. (1981) Morphology, distribution and ecology of conidial fungi in freshwater habitats. In: Cole GT, Kendrick B (eds) *Biology of Conidial Fungi*, Vol 1, pp 295–355. Academic Press, New York. [8.2, 8.3]

Webster J, Moran ST, Davey RA. (1976) Growth and sporulation of *Tricladium chaetocladium* and *Lunulospora curvula* in relation to temperature. *Transactions of the British Mycological Society* **67**:

491–5. [8.2, 8.3]

West AW, Grant WD, Sparling GP. (1987) Use of ergosterol, diaminopimelic acid and glucosamine contents of soils to monitor changes in microbial populations. *Soil Biology and Biochemistry* **19**: 607–12. [8.2]

Whipps JM, Haselwandter K, McGee EEM, Lewis DH. (1982) Use of biochemical markers to determine growth, development and biomass of fungi in infected tissues, with particular reference to antagonistic and mutualistic biotrophs. *Transactions of the British Mycological Society* **79**: 385–400. [8.2]

Willoughby LG. (1974) Decomposition of litter in fresh water. In: Dickinson CH, Pugh GJF (eds) *Biology of Plant Litter Decomposition*, Vol 2, pp 659–81. Academic Press, London. [8.5]

Willoughby LG, Archer JF. (1973) The fungal spora of a freshwater stream and its colonization pattern on wood. *Freshwater Biology* **3**: 219–39. [8.2, 8.3]

Winterbourn MJ, Rounick JS, Hildrew AG. (1986) Patterns of carbon resource utilization by benthic invertebrates in two British river systems: a stable isotope study. *Archiv für Hydrobiologie* **3**: 349–61. [8.4]

Wood-Eggenschwiler S, Bärlocher F. (1983) Aquatic hyphomycetes in sixteen streams in France, Germany and Switzerland. *Transactions of the British Mycological Society* **81**: 371–9. [8.2, 8.3, 8.6]

Wright RT, Hobbie JE. (1966) Use of glucose and acetate by bacteria and algae in aquatic ecosystems. *Ecology* **43**: 447–64. [8.2]

Zdanowski MK. (1977) Microbial degradation of cellulose under natural conditions. A review. *Polskie Archiwum Hydrobiologii* **24**: 215–25. [8.5]

9: Algae

C.S.REYNOLDS

9.1 INTRODUCTION

Although they often comprise a conspicuous component of fluvial plant life and are sometimes the dominant primary producers, algae scarcely have an easy time in rivers. Most of the problems relate to the supposedly persistent and unidirectional passage of water (Fig. 9.1). It is not simply a matter of susceptibility to irreversible removal and downstream transport, for variability in the flow may lead alternatively to movement or suspension of substratum material, with a concomitant increase in turbidity and deprivation of light, to siltation and burial by disentrained sediment and, perhaps, to desiccation. Even when the physical environment is apparently less exacting, such biotic factors as comparative rates of arrival, growth or selective removal (as food) by animals have profound effects upon the composition, abundance and temporal change in the algal assemblages present and, potentially, to the functional organization of the fluvial ecosystem.

The purpose of this chapter is to explore these selective mechanisms in greater detail and to make some general deductions about their impact upon the structure and maintenance of the algal assemblages of flowing waters.

Notwithstanding the recent rapid growth of ideas about ecosystem function in running waters (Boudou & Ribeyre 1989) or the long history of diligent phycological investigations of the algal species constituting the fluvial communities (e.g. Round 1981), the factors that regulate their spatial and temporal distributions remain incompletely understood. Neither is there yet a generally recognized paradigm to accommodate past findings or one that will identify priorities for future study.

There are good reasons why this should be the case (see Blum 1960; Whitton 1975); not least of these are the complexities of the interactions and the logistical difficulties of sampling rivers adequately, either in space or in time. So, while it is timely to propose some sort of synthesis, it is necessary to emphasize that the opinions expressed are not unequivocally established and some may well prove to be controversial.

The chapter does not set out to provide a comprehensive list of algal species in rivers, neither is it an intention necessarily to review the key studies in detail. Several major works (Blum 1956; Hynes 1970; Whitton 1975; Round 1981) already fulfil these roles and should be consulted in preference. Here, the approach is to identify the interactions and limitations governing algal production in the principal habitats of the river, only incidentally introducing the important algal genera and species, and then to emphasize those features impinging upon the contribution of algae to fluvial ecosystems.

9.2 HABITATS

In much the same way as it is possible to distinguish among the fragments of fluvial environments — open channel flow, the containing substratum, weedy backwaters etc. — and the inherent variability attaching to each, so there are clearly delimited microhabitats in rivers, each supporting its own distinctive association of biota. A recurrent theme of this chapter is that, because rivers are able to maintain a spatially — and temporally — diverse array of microhabitats, they collectively offer an almost infinitely 'patchy' environment. It is therefore important to

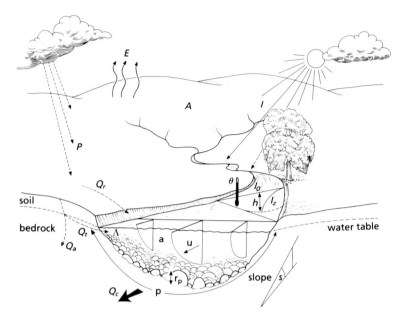

Fig. 9.1 Environmental characters of rivers and their catchments influencing the ecology of fluvial algae. The terms are identified and their quantitative ranges are interrelated in the text.

recognize the various categories of attached, benthic algae—from the larger 'macrophytic' forms to the variety of microscopic 'algal turfs' and films growing on a variety of available surfaces—as well as those planktonic species suspended in the water.

'Macrophytic algae'

To survive in a given location of flowing water, benthic algae are assumed to require, obviously, the means of physical attachment to the solid bounds of channel but, less obviously perhaps, a resistance to physical damage by the flow itself. River algae seem to have adopted two quite different approaches to these problems. The 'macrophytic algae', so styled because their sizes are sometimes quite comparable to those of aquatic mosses, liverworts and small flowering plants, are the most conspicuous when present in any quantity. These are, morphologically, relatively complex, either essentially tubular (e.g. *Lemanea*, *Enteromorpha*) or coenocytic threads (e.g. *Cladophora*), dense tufts of uniseriate filaments (e.g. *Oedogonium*, *Ulothrix*, *Stigeoclonium*) or conspicuously branched structures reinforced with biomineral deposits of carbonate (e.g. *Chara*, *Nitella*).

Generally, they become firmly attached to stones and rocks as well as other available solid surfaces but they are sufficiently flexible to allow water to flow through and around their distal ends. In this way, filaments of *Cladophora*, for example, can grow to lengths sometimes exceeding 4 m. Most macrophytic species are presumed to have upper limits of tolerance to velocity and turbulence. Although these are not necessarily quantified, there are differing views about whether these species are selectively favoured by their tolerance or even their reliance upon fast flows (Whitton 1975) or whether the ability to survive in, and rapidly recolonize from, the adjacent slow-moving boundary layers is crucial.

Epiliths and endoliths

The second approach to damage-resistant, self-maintenance is to exploit continuously the microphytic boundary-layer habit. Whereas the crustose or felt-like turfs of (usually) unicellular algae attain a thickness of a few tens of micrometres, the 'bed-layer' (see Chapter 3) adjacent to the solid surface on which they grow may sometimes extend several millimetres beyond (Fig. 9.2.). Within this microzone, the flow is retarded

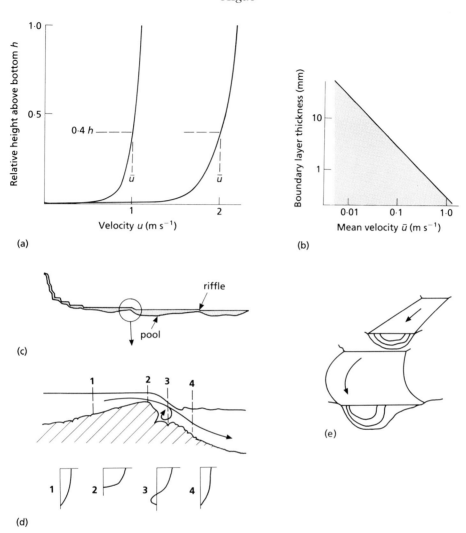

Fig. 9.2 Velocity profiles in rivers. (a) Distribution of velocity (u) against relative water depth of flows over smooth surfaces averaging $\bar{u} = 1$ or 2 m s^{-1}; note that u at 0.4 h $\simeq \bar{u}$. (b) The thickness of the viscous bed layer as a function of mean velocity (after Smith 1975). (c) Long profile of a river, to show changes in gradient and depth. (d) A portion is enlarged to represent a pool–riffle–pool sequence, through which the velocity profiles (1–4) are distorted and, in places, reversed. (e) Sagittal sections of a river channel, with velocity contours to show eccentric displacement towards the outside bend of the channel. Based on various sources.

and becomes laminar so that, even at main channel velocities approaching 2 m s^{-1}, it is possible for these microphytic algae to maintain their positions. Nevertheless, the epilithic growths on hard surfaces forming the 'wetted perimeter' of the channel have to be firmly attached, either by sticky, gelatinous secretions, as in the diatom *Cocconeis* and the cyanophyte *Chamaesiphon*, or by means of a stalked 'hold-fast', exemplified by the pennate diatoms, *Achnanthes*, *Cymbella* and *Gomphonema*. In the encrusting Rhodophytes (e.g. *Hildenbrandia*, *Lithoderma*), a flattened thallus is closely applied to the stone surface, while low-growing cushions of filamentous cyanophytes, such as *Phormidium* and *Rivularia*, exploit and, if sufficiently compact, extend the boundary layer.

In some softer rocks, especially limestones, algae may be partly (e.g. *Gongrosira*) or wholly endolithic (e.g. *Schizothrix*), living just beneath the stone surface. In some calcareous headwater streams, calcium carbonate is precipitated as travertine (also called 'tufa' or 'marl') as dissolved carbon dioxide in the spring water equilibrates with the atmosphere. The process is enhanced by the photosynthetic consumption of carbon dioxide by algae, such as *Oocardium* and *Rivularia*, and mosses, which then deposit delicate and elaborate 'casts' about themselves on the surfaces of riffles and waterfalls.

Epiphytes

In addition to furnishing further surfaces colonizable by algae, the blades and stems of submerged macrophytes (see Chapter 10) will, if sufficiently dense, create a local microenvironment of reduced velocity and turbulence (Hydraulics Research Laboratory 1985; Dawson 1989). Thus, while it is possible to recognize the analogous associations of epipelic diatoms (*Cocconeis*, *Navicula* and *Gomphonema* spp. on younger leaves; *Achnanthes*, *Meridion* and others are said to be more often associated with older or dying leaves; Margalef 1960), macrophyte beds in flowing water may provide (albeit temporary) refuges for larger unicells (e.g. *Closterium*), for the less robust filamentous and colonial forms (e.g. *Aulacoseira*, *Oncobyrsa*) and, indeed, to non-attached algae (including *Cryptomonas*) and free-swimming animals such as Daphniids. The presence of these more distinctive 'epiphytes' in the main flow depends greatly upon the extent of seasonal dieback and regrowth of the macrophyte beds and of the frequency of flushing events.

Epipsamms and epipels

Owing either to spatial variability in the velocity structure in the channel or to temporal fluctuation in the bulk discharge, many rivers will provide zones of low flow wherein water displacement is sufficiently weak for deposits of gravels,

sands and silts to acquire temporary stability. These locations furnish opportunities for the growth of algae which, while not apparently well-adapted to vigorous flow, are nevertheless sufficiently invasive to exploit them. Epipsammic algae attach to sandgrains. Epipelic algae are generally associated with fine sediments. The species concerned form loose mats or films made up of numerous gliding (*Oscillatoria*, *Phormidium*) or aggregating filaments (*Microcoleus*, *Mastigocladus*), nets (*Hydrodictyon*) or clusters of unicells (*Nitzschia*, *Caloneis*, *Surirella* and *Cymatopleura*). While suitable microhabitats persist, these algae collectively can represent a significant and expanding component of the fluvial microflora. At the same time, however, they remain vulnerable to increases in the discharge and the consequently enhanced shear forces and flushing rates.

Suspended algae

The live algae to be found in suspension in the open water may, to a greater or lesser extent, reflect the components of the various benthic associations, in proportion related to species-specific rates of dislodgement and flushing out by the flow. However, this benthic component is augmented, sometimes overwhelmingly, by habitually free-living species common in the plankton of lakes and ponds. Several long-standing tenets about this planktonic element, or 'potamoplankton' (Zacharias 1898), for instance that it originated primarily from standing waters in the catchment or that its substantial development could be manifest only in either very long or extremely sluggish rivers, are currently undergoing revision. Certainly, growth and replication of unicellular and simple coenobial species occurs whilst in fluvial suspension; if this criterion is used to characterize the potamoplankton (*cf.* Margalef 1960), then downstream development of populations may indeed be traced back to inocula from lentic waters in the catchment (Reynolds & Glaister 1992). Yet, equally, others may well originate in the benthos: Swale (1969) showed the plankton of the middle Severn, western Britain, to be frequently dominated by a species of *Navicula*.

The most prolific growths of planktonic algae in larger rivers appear not to be clearly derived from either source, whereas their distribution, in space as well as in time, would indicate that populations are effectively native to rivers. The most frequently encountered of these are the smaller species of centric (*Cyclotella* and, especially, *Stephanodiscus*) and pennate diatoms (*Nitzschia*, *Asterionella*), as well as a variety of green algae, such as *Chlorella*, *Ankistrodesmus*, *Scenedesmus* and *Pediastrum*, and euglenoids (*Trachelomonas*, *Phacus*); the rarer reports of dominance by other limnetic species, including *Dinobryon*, *Cryptomonas*, *Microcystis* and *Oscillatoria*, may be confined to just those long, sluggish pool-like rivers or to occasions of severe drought, once thought to apply to all potamoplankton.

Although there is often an evident trend for planktonic algae to be more abundant downstream (Greenberg 1964; Hynes 1970), neither the scale of downstream increase (Reynolds & Glaister 1992) nor the synchrony of maxima at different locations along the length of the river (Lack *et al* 1978) can be explained simply by reference to the downstream passage of a growing inoculum of cells. Reynolds' (1988) preliminary discussion of this paradox deduced that the presence of non-flowing water—whether in slow-flowing, intermediate reaches (e.g. the Sudanese White Nile; Prowse & Talling 1958), in impounded stretches of the water course (Talling & Rzóska 1967; Décamps *et al* 1984), in side arms (Wawrik 1962) and cuts (Moss & Balls 1989) or in a series of retentive in-channel 'dead-zones', as envisaged and defined by Valentine and Wood (1977; see also Chapter 5)—was crucial to the maintenance and dynamics of plankton in rivers. The results of subsequent investigations (Reynolds & Glaister 1989; Reynolds *et al* 1991) are beginning to amplify these contentions: potamoplankton is confined neither to particularly large nor particularly long rivers.

9.3 CONTROLLING FACTORS

As in lakes, where the understanding of the ecological factors regulating the dynamics of algal production has substantially evolved, the growth of autotrophs in running water is essentially dependent upon an adequate supply of light energy and dissolved inorganic nutrients, while species selection, abundance and dominance is influenced by competitive interactions and herbivorous consumers. In addition, river plankton is reliant upon maintenance in suspension if growth is to offset its seemingly inevitable consignment to the sea, while for the algae attached to the surface of gravel, stones and higher plants, size stability and orientation of the substratum, together with the frequency of its movement, contribute to the complex of microhabitats available. Many of the critical variables in streams and rivers are therefore related to, or are determined by, the physical properties of channel flow. It is expedient, therefore, to determine the ambient capacity factors—those 'regional features' of localities that govern the attachable quantities and composition of flowing-water algal communities—and then to consider the regulatory roles of seasonal and increasingly stochastic variability in channels.

Persistent 'regional' factors

The basic morphology of channels is ultimately governed at the catchment scale: the location with respect to latitude; the horizontal distance and vertical descent to the sea; the geology of bedrock and its tectonic history; its geochemical lability and resistance to erosion; and its climate, with particular reference to the amount of rainfall and the seasonality of its distribution. Scaling affects the characteristics of entire drainage basins (e.g. their areas, configurations, aspects, generated flows), through those of particular reaches (channel-form substrata, turbidities) to those distinguishing the diversity of microhabitats available for biological exploitation by algae (Fig. 9.1.).

Discharge

The linking property is the net discharge (*Q*), itself related to the balance between the precipitation input (*P*) less evapotranspirational loss (*E*) per unit of topographical catchment (*A*) upstream of the point of measurement:

$$Q = A(P - E) \tag{1}$$

and where

$$Q = Q_r - Q_t + Q_a \tag{2}$$

Q_r is the flow across the ground surface ('run-off'), Q_t is the fraction percolating laterally to the channel as groundwater ('throughflow') and Q_a is the fraction of percolating water penetrating to deeper aquifers and which does not necessarily re-enter the local surface drainage. The yield to channel flow (Q_c) is thus equivalent to $Q_r + Q_t$.

Fluvial chemistry

The capacity of the river to support algal production is initially related to the general ionic environment and, in particular, to the supply of nutrient resources to the water. When annual rainfall is distributed evenly through the seasons and throughflow dominates the drainage to the channels, the prolonged, intimate contact between the bedrock-derived mineral particles and the interstitial water ensures a steady, nutrient-enriched water flow. In contrast, where there is scant soil cover or the ground is already water-logged, the supply to channels is dominated by scarcely modified run-off and is very responsive to the periodicity of precipitation events.

In this way, headwater streams rising in hard-rock, mountainous regions of high precipitation yield weakly ionic, nutrient-poor waters in which the algal growth capacity may be severely limited. Such waters are particularly susceptible to acidification, carrying numerically sparse algal populations, perhaps characterized by the diatoms *Eunotia* and *Pinnularia*. Streams draining catchments based on soft or porous rocks (e.g. shales, limestone chalk, evaporites) or on depositional lowlands are generally of high ionic strength, alkaline and potentially rich in nutrients: at times, they may support heavy epipelic growths featuring (*inter alia*) *Navicula* spp., *Caloneis*, *Gyrosigma*. This potential may be further enhanced by agricultural activities in the catchment and by the products of sewage arising from human settlement.

Within these extremes, the total load of a solute (Λ) is generally a function of discharge, of the form:

$$\Lambda = k\, Q_c^f \tag{3}$$

where k is a local constant depending upon the availability of a given chemical and its solubility in water; f is generally less than 1, the concentration being diluted by higher discharges (curve 1 in Fig. 9.3). Although this applies to most major ions in well-mineralized waters, the pattern is not unique. More vigorous displacement of deep soil-water constituents by throughflow flushing can raise the concentration of (e.g.) N-species or organic matter in streams (curve 2 in Fig. 9.3), or be ultimately diluted by sustained run-off (curve 3), as in the elution of H^+ ions from peat soils.

Channel form

The depth of water in the river, its gradient and the 'roughness' of the channel play a leading role in delimiting algal microhabitats. From its first definable rills to its coastal outfall, the typical drainage basin is based upon the progressive confluence of streams of increasing rank (order) and increasing discharge capacity. Following the direction of slope, the shape and pattern are influenced by the gradient (Leopold *et al* 1964). Ideally, the long profile of a river tends toward a

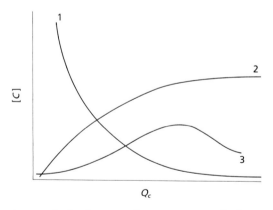

Fig. 9.3 Idealized curves of concentrations of solute concentration ([*C*]) as a function of fluctuations in channel flow (Q_c). Most catchment-derived determinants conform to curve 1 but the displacement of deep soil-water solutes conforms more closely to curve 2, while curve 3 describes the elution and ultimate dilution of H^+ ions from acid soils. Redrawn from Meybeck *et al* (1989).

catena; much of the work of downcutting, removal of eroded material and downstream deposition can be interpreted as the dynamic progression towards the ideal. On a smaller scale, natural channels follow a series of steps, manifest as cascades in rocky mountain streams and as alternating deeper pools and faster flowing shallow areas ('riffles') on gentler slopes. Lengthwise variations in the profile and sinuosity of the channel, as well as the implied sequence of erosion and deposition, contribute to a downstream alteration in the regional blend of local microenvironmental characters, associated with the progressive increase in ambient discharge. The algal habitats generated in channels are as varied as the geographical distribution in annual run-off generated by rainfall (P), between virtually zero (Atacama Desert, Chile and the eastern Sahara) and the 11.5 m frequently experienced at Kauai, Hawaii ($Q_c/A = 0.35$ m^3 km^2 s^{-1}), and as the range in drainage basin areas. The largest, the Amazon Basin (7.05×10^6 km^2) delivers to the sea some 6.7×10^{12} m^3 annually, at an average rate of 212×10^3 m^3 s^{-1} ($Q_c/A \sim 0.03$ m^3 km^{-2} s^{-1}). Algal communities have been studied in headwater steams where the reported discharges were eight to nine orders of magnitude less (Reynolds 1988). Throughout this range, the form and dimensions of the channel vary continuously to contain the flow of water, so the predominant habitat structure and, hence, the composition of the species assemblages present, respond to related variations in width (b), depth (h) and mean velocity ($\bar{u}$) in terms of the discharge,

$$b = k_1 (Q_c)^\chi$$

$$h = k_2 (Q_c)^\psi$$

$$\bar{u} = k_3 (Q_c)^\omega$$

Moreover, because at any point Q_c approximates to the product $bh\bar{u}$, the coefficients and exponents are cumulatively related:

$$k_1 k_2 k_3 = 1$$

$$\chi + \psi + \omega = 1$$

In other words, width, depth and velocity increase downstream, although less rapidly than does Q_c. Neither are the proportions constant; the ratio of $b{:}h$ is greatest in low-gradient channels through

unconsolidated materials (e.g. fluvoglacial deposits), while in adjacent pool and riffle sections carrying instantaneously similar discharge, changes in cross-sectional area, bh, are reciprocated by changes in mean velocity therethrough:

$$\bar{u} = Q_c/bh \qquad (4)$$

Velocity

One further generalization concerning mean velocities through a channel section is that expressed by the familiar Manning equation (5):

$$\bar{u} = {}^1/n \, (a/p)^{2/3} \, s^{1/2} \qquad (5)$$

where a is the cross-sectional area of the channel and p is its wetted perimeter (see Fig. 9.1; the quotient a/p is also known as the *hydraulic radius* and, in wide, smooth channels, approaches a value close to $a/b = h$); s is the slope in the sense $1/x$, where x is the horizontal distance per unit vertical descent; $1/n$ is a factor correcting for frictional resistance to flow, due, for example, to a rough or shallow bottom, or to abundant macrophytes in the channel. Ven Te Chow (1981) has recently updated estimates of n between ~ 0.03 in large, open channels, $0.04-0.06$ in smaller ones and up to ~ 0.15 in channels heavily beset with macrophyte growth. Except in cascading reaches confined within rocky gorges, mean velocity rarely exceeds the ~ 2 m s^{-1} at which most rivers accommodate the flow by enlarging their channels.

Velocity structure

Nevertheless, the range $0 < u < 2$ m s^{-1} embraces a spectrum of environments from weedy, silted backwaters to boulder-strewn rapids and deep, pool-like reaches. This variability characterizes rivers from source to outfall and by no means in a regular progression along its long profile.

It is equally important to emphasize that much the same range of velocities is encountered within individual reaches, at scales in the order of a few metres to a few millimetres. The classic vertical profile (i.e. in the z direction) of instantaneous velocities (u_z) in the downstream (x direction) reveals flow to be more rapid away from the frictionable bounding surfaces, adjacent to which the velocity approaches zero (see Fig. 9.2(a)). As the channel section alters through adjacent pools and riffles, so the velocity profile is altered downstream (see Figs 9.2(c) and (d)). Moreover, considerable lateral differences in velocity (i.e. in the y direction) are evident on bends where flow is centrifugally displaced (see Fig. 9.2(e)).

This variability in velocity structure is of profound significance to the ecology of fluvial algae. The presence of a near-bed boundary layer, perhaps a few millimetres in thickness, at the surfaces of stone and macrophytes furnishes a microhabitat suitable for the settlement and development of epipelic algae of lesser dimensions (generally <0.1 mm), yet which is sufficiently close to the main flow and its supply of dissolved gases and nutrients for favourable diffusion gradients to be maintained. Pool-like reaches may be better tolerated than shallow riffles by macrophytic algae, provided suitable sites for attachment are available; yet again, they benefit from the advective replenishment of mainflow solutes. Moreover, it is the lateral 'dead-zones' of reduced flow that retain water — progressively renewed by fluid exchange with the mainflow (Young & Wallis 1987) — that provide the enhanced opportunities for the growth and otherwise paradoxical downstream increase of planktonic algae.

Turbulence and shear

The condition for the persistence of planktonic cells in the open channel, as elsewhere, is that the motion of the water is sufficiently turbulent (Margalef 1960; see Chapter 5) constantly to redisperse suspended particles throughout the water depth. In flowing waters, turbulence is generated between adjacent water layers travelling at different velocities. At low velocities, viscosity overcomes the turbulence: the layers slide over one another in laminar flows. The ratio of the inertial to the viscous forces is expressed by the dimensionless Reynolds' number, Re:

$$Re = \rho_w \, ul/v \qquad (6)$$

where ρ_w is the density of the water, v is its viscosity and l is the appropriate length dimension; in the present consideration, it is

equated with the depth of water, h. As Re increases from 500 towards 2000, laminar flow breaks rapidly into turbulence (e.g. Smith 1975). This transition is inserted in the plot of h versus u in Fig. 9.4(a) to illustrate the conditions for the onset of turbulence in channels.

The intensity of the turbulence is quantifiable from velocity fluctuations in each plane, $\pm u'$, $\pm v'$ and $\pm w'$ about the mean downstream velocity, u, in the x, y and z directions respectively. To overcome the rapid tendency to zero of these series, it is customary to refer each quantity to its non-vanishing root mean square value. For example:

$$u' = [(\pm u')^2]^{1/2} \qquad (7)$$

The scale of turbulent intensity is the time-averaged product of the change of momentum imposed upon the main motion in the x direction; this is also known as the friction velocity, $u^\star$. Considering the vertical component:

$$u^\star = (u' \, w')^{1/2} \qquad (8)$$

The turbulence associated with flow applies a horizontal stress, called shear (τ), which is transmitted through adjacent vertical layers:

$$\tau_{xz} = \rho_w \, (u' \, w') \qquad (9)$$

whence:

$$u^\star = (\tau_{xz}/\rho_w)^{1/2} \qquad (10)$$

While turbulence is generally proportional to the velocity and the vertical distance above the bed (see Chapter 5), the stress on the bottom can be derived from its transmission through the velocity profile (du/dz); assuming the mean velocity, $\bar{u}$, to be located at $0.4h$ above the bed (see Fig. 9.2(a)),

$$u^\star = (\tau_0/\rho_w)^{1/2} = 0.4h \, (du/dz) \qquad (11)$$

Rearranging:

$$du/dz = u^\star/0.4h \qquad (12)$$

and integrating:

$$u = 2.5 \, u^\star \, \ln(h/c) \qquad (13)$$

c is an integration constant with a finite value related to the roughness of the bed: r_p corresponds to the height an object projects from the bed and:

$$c \sim r_p/30 \qquad (14)$$

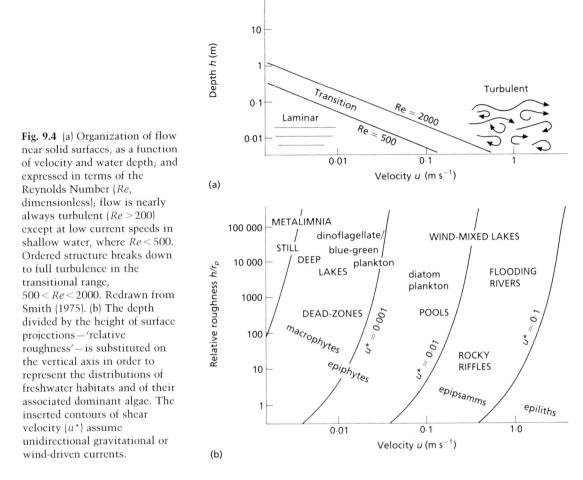

Fig. 9.4 (a) Organization of flow near solid surfaces, as a function of velocity and water depth; and expressed in terms of the Reynolds Number (Re, dimensionless); flow is nearly always turbulent ($Re > 200$) except at low current speeds in shallow water, where $Re < 500$. Ordered structure breaks down to full turbulence in the transitional range, $500 < Re < 2000$. Redrawn from Smith (1975). (b) The depth divided by the height of surface projections—'relative roughness'—is substituted on the vertical axis in order to represent the distributions of freshwater habitats and of their associated dominant algae. The inserted contours of shear velocity (u^*) assume unidirectional gravitational or wind-driven currents.

Then:

$$u_z = 2.5\, u^* \ln (30h/r_p) \tag{15}$$

so at $0.4h$,

$$u = 2.5\, u^* \ln (12h/r_p) \tag{16}$$

Depending on the ratio h/r_p, from approaching 1 for very stony streams to >5000 for deeper flows over fine silts, the value of u^* can range between $1/6$ and $1/30\, u$.

The relationships are ecologically instructive. For a given change in u, the stress applied to the bed increases geometrically and the thickness of the non-turbulent boundary layers adjacent to solid surfaces is correspondingly compressed (see Fig. 9.2(b)). If the bottom is 'rough', there will be many areas, chiefly on the downstream side of large boulders, where flow is weaker (see Fig. 9.2(d)). Local differences in shear will often influence the microscale distribution and abundance of attached algae and, equally, render some areas more susceptible to sediment deposition than others. This principle is extended in the construction of Fig. 9.4(b), which depicts the general range of aquatic algal habitats in terms of their relative roughness (h/r_p) and mean velocity (u). Isopleths of friction velocity (u^*) are superimposed. The plot distinguishes among the approximate characteristics of lakes, rivers, pools and riffles and suggests the distributions of the major algal life forms.

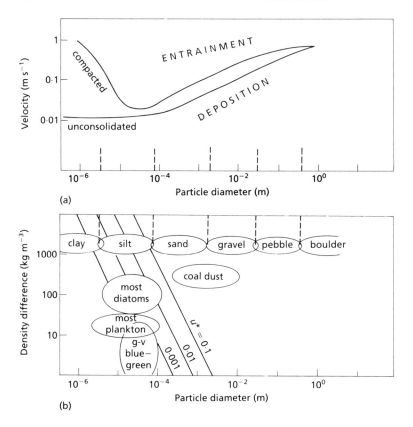

(a)

(b)

Fig. 9.5 (a) The entrainment and deposition of non-living particles of different sizes (clay to boulders) as a function of velocity (note that compacted beds of fine particles can present a collective, erosion-resistant, boundary layer, which may protect deposits from resuspension). (b) The differences in the densities of the particles and the surrounding water; a range of shear velocities, critical to the maintenance in suspension of fine silt and clay particles and of certain categories of planktonic algae including non-motile diatoms and gas-vacuolate blue-greens, is inserted.

Entrainment and settlement

Generally, the greater the velocity, the greater is the size of the largest particles moved and of the smallest unmoved (Fig. 9.5(a); for further discussion, see Chapter 5). The uniformly low local roughness coefficients offered by compacted fine sediments in large, gently graded, lowland rivers protects them from entrainment: $r_p <$ boundary layer thickness. The threshold for resuspension may depend upon the mean velocity accelerating to the point of channel enlargement! So far as benthic algae are concerned, the less frequently disturbed surfaces — large stones and boulders — should be more suitable habitats for large or slower-growing species. On the other hand, the significant presence of diatom growths on fine sediment surfaces will occur only in slow-flowing periods when the boundary layer is least compressed.

The criterion for entrainment of particles, including planktonic algae, whether introduced into

the river from external sources of resuspended from the bottom by turbulent shear, is that turbulent velocity fluctuations substantially exceed the intrinsic settling rate (w_s) of the particle. Humphries and Imberger (1982) proposed:

$$[(w')]^{1/2} > 15w_s \tag{17}$$

where w_s is the intrinsic settling rate of the particle which, subject to the criterion that the flow of water over the moving particle remains laminar, is described by the Stokes equation (18):

$$- w_s = d^2 \, g(\rho_a - \rho_w) \cdot (18_v \, \phi) \tag{18}$$

d being the diameter of the particle (or the diameter of a sphere of identical volume), ρ_a is its density, ρ_w is the density of the water, and η is its viscosity; ϕ is a 'form-resistance factor' compensating for distortions from the spherical form; g is gravitational acceleration. In the plot of particles in terms of their test-density difference and diameter (Fig. 9.5(b)), contours of critical u^*

values are inserted, assuming that $2u^*$ is adequate to meet the entrainment criterion (equation 17). It is evident that, under the turbulence conditions encountered in the majority of larger streams and rivers, clay and fine silt particles, as well as the cells of most phytoplankton, will be substantially entrained in the flow. Moreover, while the entrainment criterion continues to be satisfied but a basal boundary layer is maintained, the rate of population change due to loss from suspension (σ) by settlement is largely a function of water depth, not of velocity as might be supposed (Reynolds *et al* 1990):

$$N_t/N_0 = e^{-w_s t/h} \qquad (19)$$

where N_0 and N_t are the populations of particles separated by an interval t. Nevertheless, the faster is the rate of flow, the further downstream material will be transported in the same unit time before loss from suspension is complete.

Turbidity

Fine, non-living particles in suspension interfere with the penetration of underwater light and so constitute an important environmental factor in the growth and distribution of both attached and planktonic algae. Put at its simplest, the incident light penetrates beyond the reflective, uneven surface of the water (I_0) is reduced exponentially by absorption and scatter so that at depth z, the remaining light (I_z) is approximately given by:

$$I_z = I_0 e^{-\varepsilon z} \qquad (20)$$

where ε is the coefficient of monochromatic light extinction. This coefficient comprises two separate effects (absorbance in particular wavebands and backscatter) of several components, including the pure water itself, together with dissolved colour (ε_w), the suspended algae (ε_a, which is a product of the algae present and their specific areal absorbance; $\varepsilon_a = n.\varepsilon_s$) and the load of non-living clay and silt particles in suspension (ε_p). In rivers, where turbidity is often mainly due to non-living suspensoids, it is useful to express the suspended load as a light-extinction coefficient comparable with that for the algae. From a series of measurements of suspended solids in the Bristol Channel (Hydraulics Research Laboratory 1981) and the depth–time distribution of planktonic photosynthesis (Joint & Pomroy 1981) it has been possible to approximate ε_p as 20 m^{-1} (kg m^{-3})$^{-1}$. The impact on the penetration of a given light income, say up to 2 mmol photon m^{-2} s^{-1}, of suspended-solid loads in the observed range (1 g to 2 kg m^{-3}) is represented in Fig. 9(a). In many rivers, loads >0.1 kg m^{-3} may begin seriously to limit the amount of light reaching the bottom. For many bottom-dwelling attached algae which are unable to adapt to extreme shade (e.g. by raising cell-specific chlorophyll sufficiently to enhance their photosynthetic efficiencies), turbidity reduces the opportunities for survivorship of the population. For planktonic cells in a turbid river and exposed to extreme fluctuations in light, from $I_0 \rightarrow I_z$ and back within intervals of minutes, it is important to exploit the available energy as effectively as possible. One way to achieve this is to increase the light-harvesting capacity of the cells by increasing pigment content/unit biomass. As a result, low-light adapted cells have shortened response times in which to raise the photosynthetic rate, but have a widened absorption spectrum and an increased risk of photoinhibition near the surface.

As both the amount and longevity of particles in suspension as well as the distance that they are transported downstream are all likely to be properties of the flow through the individual reach, these quantities will vary with discharge. Figure 9.6(b) is redrawn from a derivation in Reynolds (1988) of ε_p ($+\varepsilon_w$) based on plankton chlorophyll, photosynthesis and inferred photic depth, at differing discharges in a section on the River Thames near Reading, as measured by Lack (1971) and Kowalczewski and Lack (1971); its inclusion here is to illustrate the sensitivity of the phytoplankton-carrying capacity of a lowland river to discharge-led variability in the suspended-solid load.

Seasonal factors

The principal factors regulating algal production in lakes are day length, solar energy income and, in turn, water temperature, especially in the higher latitudes (Lund 1965). This may be expected to hold for rivers, although seasonal

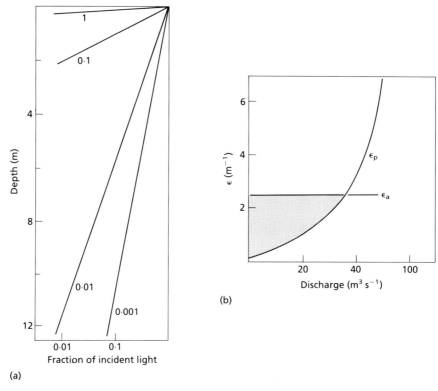

Fig. 9.6 Water clarity as a function of flow. (a) The fraction of the incident light penetrating the surface reaching into the water column at the selected concentrations of clay particles (in kg per m^3 of suspending water). (b) The turbidity response in the River Thames, due to non-living material (mostly clay), as a function of discharge (ε_p) and the extinction (ε_a) of a self-shaded concentration of planktonic diatoms in the River Thames at Reading; the stippled area therefore denotes capacity for net growth increment of fluvial phytoplankton. Revised representation of data derived by Reynolds (1988) from data presented in Lack (1971).

variations in discharge may have a more immediate impact upon the survival and growth of algae in given reaches. As with other aspects of fluvial phycology, there is a dearth of studies in which the effects of the various factors have been satisfactorily separated.

Light

In clear headwater streams, the incident light penetrating to the bed of the stream is barely modified either in intensity, spectral composition or duration. Many of the benthic chlorophytes are known, or are supposed, to tolerate high light intensities or, alternatively, are less adaptable to low average light intensities than representatives of algal groups with a greater capacity for chromatic adaptation. The filamentous green algae, *Ulothrix* and *Cladophora*, are among the genera said to be influenced by increasing day length; *Hildenbrandia* and *Batrachospermum* are genera of red algae said to be tolerant of a shading by overhanging trees and herbs (Whitton 1975) and which, under temperate deciduous forests, will be of increasing relative importance through the summer. Algae may be excluded from deeper surfaces during episodes of high turbidity which may well regulate the growth of phytoplankton independently of other factors. Williams (1964) attributed the winter growth of planktonic diatoms in central North American rivers, despite sustained flows and near-freezing temperatures, to the fact that little silt was washed in while the ground itself remained frozen.

Temperature

In rivers, temperature is assumed to respond more rapidly to increased solar radiation fluxes and, conversely, to lose heat to a cooler atmosphere more quickly than do lakes, by virtue of their relative shallowness and more vigorous, more complete vertical mixing. On the other hand, the general weakness of density stratification, save where it is supported by salinity gradients, means that the higher and lower temperature extremes at the surfaces of lakes ($\theta 35°C$, $<0°C$) are rarely reached in rivers. Mean temperatures are generally supposed to increase from source to mouth. On a smaller scale, there is also provisional evidence that the horizontal temperature differences can develop between the more retentive deadzones in lowland river reaches and the main channel flow, under appropriate warming or cooling episodes (Reynolds & Glaister 1989).

So far as the responses of algae are concerned, few data are available to verify the supposition that algae grow faster at higher temperatures. If river algae conform to patterns of growth of their limnetic counterparts (Reynolds 1989; Nielsen & Sand-Jensen 1990), then it would be expected that the algae of larger unit surface area:volume ratio ($>10^6$ m^2 m^{-3}) are the more rapidly growing at 20°C and only about half as fast at 10°C ($Q_{10} \sim 2$); species with relatively lower unit area:volume ratio ($<10^5$ m^2 m^{-3}) are both slower growing and relatively more sensitive to lowered temperatures ($Q_{10} \geq 3$). Most rivers appear to support attached and suspended algae belonging to the first category, so it is not easy to attribute distributions of species, either in time or in space, to water temperature alone. However, the optimal seasonal production of planktonic algae (e.g. Lack 1971) and of benthic assemblages (e.g. Phinney & McIntyre 1965), on whose data Fig. 9.7 is based, is clearly responsive to the interactions of lengthening days and rising temperatures.

Grazing

Various gastropods, certain mayflies, hydroptilid caddisflies, mites and specialized cyprinid fish feed on epilithic and epiphytic algae. Grazing tends to vary quasi-seasonally, reflecting the proliferation of algae, to some extent, the life-

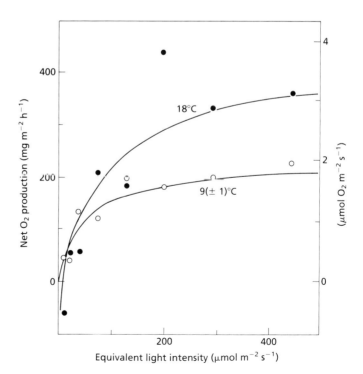

Fig. 9.7 Curves of net oxygen production by photosynthesizing benthic epiphytic algal communities, within two temperature ranges, as a function of light availability. Such data are relatively scarce, these examples being derived from the data of Phinney and McIntyre (1965).

cycles of the consumers they sustain and, perhaps, the ability of the algal community to 'recover' after episodes of heavy grazing. Algal production may also sustain, in part, a variety of benthic filter feeders, from simuliid larvae to bivalve molluscs, but this presence will rarely exert a direct effect upon the algal production. There have been many studies on the food requirements, feeding rates of a selection of common invertebrates and, in general terms, on the impact of grazing on the food resource (for reviews see Gregory 1983; Lamberti & Moore 1984). Gregory (1983) has also provided one of the few critical evaluations of the plant–herbivore relationship in lotic communities, wherein the influence of grazing on the structure of the algal assemblage was investigated. Attached diatoms are a frequent food source for many grazing animals, whereas filamentous forms and blue-green algal species are often supposed to be rejected. Nevertheless, their faster growth rates and their higher productivity (*sensu* production per unit biomass) at relatively low biomass can still enable films of diatoms to dominate heavily grazed benthic communities. On the other hand, the availability of suitable algae may regulate the growth rates, abundance and distribution of direct herbivores and, in turn, the extent to which the lotic ecosystem is supported by autochthonous-based production as opposed to catchment-derived, allochthonous components.

Stochastic factors

Both plants and animals in running waters are nevertheless governed to a large extent by external disturbances, notably the occurrence of high water flows. In headwater streams, episodes of enhanced surface run-off generally have the effect of reducing friction to permit faster velocities of streamflow; as discussed above, the principal effects on algae are expressed through the movement and, perhaps, overturning of larger stones and through the compression of the boundary 'logarithmic' (see Chapter 5; Fig. 9.2(b)) layer which exposes more prominent algae to mechanical stress. In deeper, low-gradient lowland reaches, where channel flow is already well equilibrated to fluvial geometry, the response to increased discharge may be relatively minor in terms of the channel section but, nevertheless, of a magnitude to effect the resuspension and transport of fine materials, with all the consequential impacts on turbidity, light penetration and redeposition rates. Such effects are generically similar from river to river but their characteristics will be unique to each system, according to river size, channel section, nature of the bed and seasonality of floods. Neither will there be, necessarily, in any given river, a direct relationship between turbidity load and discharge; indeed, a hysteretic effect of turbidity would be anticipated, first increasing abruptly and then falling back, following an abrupt change from one velocity state to another.

Except in regions where rainfall is itself markedly seasonal in distribution, it seems reasonable to suppose that whereas the annual alternation of the seasons remains a primary driving factor in the ecology of stream algae, the further the distance downstream, the more the seasonality is suppressed by the responses to frequent discharge variations. In both cases, the community structure is liable to rapid reversal and re-initiation. In the headwaters, the 'reset mechanism' (Gregory 1983) operates every autumn; in lowland rivers, the incidence of elevated discharges may be prompted several times per year. The impact is particularly evident in the comparison (Fig. 9.8) of the responses of phytoplankton in the lower River Thames during a dry and a much wetter summer, as observed and subsequently simulated by Whitehead and Hornberger (1984).

9.4 ECOLOGICAL PROCESSES

It was implied at the outset that survival prospects of algae in rivers are confronted by numerous problems. The subsequent sections have shown, however, that fluvial algae are apparently well adapted to exploit particular niches within the vast spectrum of potential microhabitats consequential upon the interactions of water chemistry, hydrology, channel-form and gradient, substratum, suspended-load and, ultimately, velocity parameters. Moreover, variability among these properties often occurs over short distances

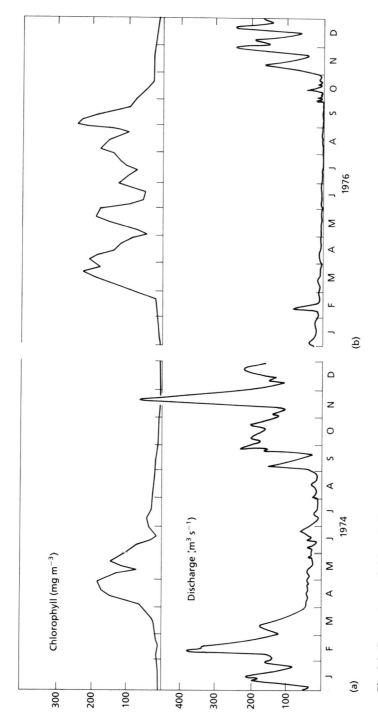

Fig. 9.8 Comparisons of chlorophyll *a* concentrations in (a) the River Thames between Staines and Teddington in relation to contemporaneous discharge fluctuations during 1974 (having a cool wet summer) and (b) 1976 (an exceptional drought year). Redrawn from data plotted in Whitehead and Hornberger (1984).

in space and over short intervals of time. Velocity and turbulence relate the spatial and temporal dimensions.

As with the understanding of other ecosystems (Wiens 1989), recognition of the importance of scaling effects is essential to defining key processes regulating ecological structure and function. Thus, although the longevity of a given water system is likely to be comparable to the timescale of the geomorphological evolution of its drainage basin, perhaps measurable in millions of years ($>10^{14}$ s), the mean residence time of flowing water may be a matter of a few hours to little more than a year or two (10^4–10^8 s). At the same time, the flowing water is constantly mixed, by turbulence, in the lateral and vertical directions (at scales of 10^1–10^3 s; Reynolds 1988), bringing positive benefits to both attached and suspended algae in enhanced dispersal, nutrient renewal and distribution, the removal of wastes and the exchange of respiratory gases. Other, negative, aspects include added turbidity and wash-out of suspended algal populations, although for plankton and, indeed, other species,

survival is assisted by the presence of refuges ('dead-zones') along the channel.

In Fig. 9.9, these scales of 'environmental variability' are compared with the timescales of the various algal responses. Whereas the organisms' physiological responses at scales $<10^4$ s will determine their fitness to function successfully in flowing water, the rate of cell replication (or generation time) is the final expression of the ability of a species to survive and recruit new cells to, and perhaps dominate, particular communities. Because, in many natural rivers, cell division times generally exceed 10^5 s, the key scales of variability are those operating at comparable frequency. This may bias the selective advantage among potentially competing planktonic algae towards species tolerant of high-frequency disturbance ('ruderal' R-species in the terminology of Grime 1979) or, briefly, towards those fast-growing opportunists (C-species) that quickly take advantage of intervals of more favourable growth (declining velocity, reduced turbidity, elevated nutrient concentration) associated with low discharge. Selective pressures are

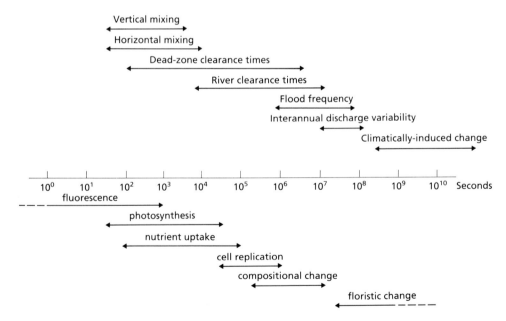

Fig. 9.9 Temporal dimensions of environmental fluctuations (against a common logarithmic scale, in seconds) and the hierarchy of algal responses, from the biochemistry and physiology of individual cells to their relative growth rates and the consequent adjustments in relative abundance and dominance of the community.

generally not in favour of the predominance of the relatively slow-growing but more persistent dinoflagellates and colonial cyanobacteria, save in those long or seasonally sluggish rivers or in extensively regulated catchments dominated by standing waters (Prowse & Talling 1958; Décamps *et al* 1984; Krogmann *et al* 1986; Mohammed *et al* 1989). Given a sufficient period of downstream travel, it is not unusual to find such limnetic forms as *Microcystis* developing in, and sometimes dominating, the lower reaches of slow-moving rivers. Thus, where disturbance is substantially damped out, species describable as *S*-strategists in Grime's (1979) terminology may be selected.

While not subject to the same continuous tendencies towards downstream elimination of the reproductive stock, attached algae are nevertheless sensitive to discharge variability. This is consequential not only upon the increased turbidity during flood, but also upon the compression of the boundary layers, and upon direct mechanical stress on protruding or pedunculate epiliths. Encrusting growths may be more resistant and possibly benefit from frequent disturbances to improve periodically their access to light and nutrient resources. The ability of accelerating velocities to move and turn over increasingly larger (heavier) stones constitutes a disturbance-frequency scale on which, at one extreme, algal colonization is effectively prevented and, at the other, it can proceed unchecked by movement of the substratum.

This counterposition of opportunities and destructive perturbations plays a major part in regulating the spatial and temporal distributions of the main species. It also helps to explain sharp seasonal and interannual contrasts in the abundance of certain species which are sometimes, although usually erroneously, attributed to pollution or enrichment of the system concerned.

The dynamic approach to the structure of lotic algal communities was prominently advanced by Margalef (1960). He envisaged a series of comparable ecological successions, each moving towards a climactic stage but each susceptible to the abrupt intervention of destructive mechanical forcing, 'resetting' (Gregory 1983) the system to an earlier successional condition: in short, what

is now referred to as Connell's (1978) Intermediate Disturbance Hypothesis. Under this scheme, the sequence in which species dominate is not necessarily a true ecological succession in the sense of Odum (1969). Moreover, different successional stages can be maintained simultaneously in adjacent zones.

Epilithic successions may be initiated on appropriate 'young' surfaces, such as stones freshly exposed to the light during a recent spate. Depending upon the water flow and the chemical nature (acidity, nutrient content), the pioneer community may be dominated by encrusting diatoms, in which *Ceratoneis* species are frequently represented or, as recently shown experimentally by Elber and Schanz (1990), by the chrysophyte *Hydrurus foetidus*. Communities often featuring species of *Achananthes* and *Meridion* maintain the succession, yielding place to larger diatom species of *Synedra* and *Aulacoseira*, or to red or blue-green algae in particular streams, before the eventual overgrowth by macrophytic species, including mosses, if the river bed is not disturbed or turned over once again. It is of interest that, for instance, Arnold and Macan (1969) observed dense growths of moss (*Amblystegium eurhynchioides*) to be associated only with large boulders and bands of solid rock exposed in the bed of a cascading hill stream, whereas the smallest stones carried the least growth of any kind; intermediate-sized stones and small boulders sometimes supported dense growths of epilithic diatoms, notably *Meridion* spp. and, later in spring, of *Lemanea*, and, in summer, of *Ulothrix* or *Nostoc*. On each occasion that the total biomass of attached algae increased, it is probable that the number of species represented diminished or an increasing fraction of the total biomass was invested in a diminishing number of species (Hynes 1970).

The precise species representation and dominance through this successional stage is highly variable form river to river, even from location to location within a given river or at the same point from year to year. Again, stochastic events and circumstances contribute to this variability but certain species-associations may occur preferentially in certain kinds of rivers, with identifiable 'regional' characteristics. For instance, according to Margalef's (1960) scheme, an *Aulacoseira–Navicula–Rhoicosphenia–Cymbella–Diploneis* association is frequently encountered in limestone mountains. *Phormidium* and *Lyngbya* are often well represented, as an encrusting growth habit, while *Amphipleura*, and other genera, e.g. *Chaetophora*, *Chara* and *Zygnema*, may be prominent in the quieter pools and backwaters. Mosses such as *Fontinalis* and *Fissidens* may quickly tend to dominance. In streams running over rather less alkaline granites, sandstones and slates, the analogous associations feature *Eunotia*, *Fragilaria*, *Melosira*, *Nitzschia* and *Pinnularia* spp., with *Oedogonium* and *Tribonema* often dominating in pool-like conditions, where the succession proceeds to liverworts and mosses and such flowering plants as *Callitriche* and *Ranunculus*. In nutrient-enriched, hard-water streams, the prevalent *Cocconeis–Melosira–Navicula–Surirella* assemblages often become overgrown by *Cladophora*, *Stigeoclonium* or, in some instances, by *Enteromorpha*. Similarly, humic and saline flows each support distinctive and successive algal assemblages. In every case, however, the deeper, more turbid, pools of downstream reaches become increasingly hostile to the growth of benthic algae, usually because of light deficiency, although a lack of suitably firm substrata, save for such constructions as piers or piling, may often be as important. In such downstream reaches, however, the principal survival opportunities for autotrophs pass increasingly to the phytoplankton.

The principal assemblages represented in the potamoplankton (namely the *Stephanodiscus–Nitzschia–Asterionella* group, prominent in moderate-to-large rivers, and the *Chlorella–Ankistrodesmus–Scenedesmus*–euglenoid group of more general occurrence in smaller, nutrient-rich streams), together with those more typically limnetic associations of Reynolds (1987) recognized to survive in slow-flowing reaches of rivers (notably the mesotrophic *Dinobryon–Sphaerocystis* grouping, the *Oscillatoria agardhii–Oscillatoria redekei* of shallow, well-mixed basins and the eutrophic summer grouping that includes *Microcystis*), appear to be selected by locally obtaining conditions of water depth, light availability and fluid displacement, rather

than in accord with a strict successional progression. The most familiar transition, that between the diatoms and green-algal/euglenoid associations, cannot be regarded as being autogenic; it is most commonly mediated in time by a change in discharge (and especially in depth), whereas spatial transitions from greens (upstream) to diatoms (downstream) may move down or up the river with seasonal changes in discharge (Sabater & Muñoz 1990).

In both the benthos and the plankton, successional sequences are generally brief, being subject to frequent truncation or reversal by abrupt changes in the ambient environment. Indeed, successions in flowing water can scarcely culminate in a climactic, internally regulated equilibrium, analogous to (say) the forest climax of the pond hydrosere (Symoens *et al* 1988); while water continues to flow through a defined channel and remains subject to externally driven fluctuations in discharge, the succession cannot move to its climax without the intervention of large-scale factors, such as climate change-mediated differences in hydrology or geomorphological alterations to the drainage basin. Rather, successions are initiated as pioneer species develop after a recent flood and they progress with the establishment of other plants, yet they fail to evolve beyond the flow-imposed plagioclimax condition, dominated by macrophytic algae, aquatic bryophytes or flowering plants.

Subclimactic communities are noted for their species diversity, which, in combination with the stochastic nature of the initial colonization and the subsequent development, contributes to the maintenance of the variety of lotic habitats necessary to the survival of a wide range of species.

It is the frequency of the environmental fluctuations and the periodicity and intensity of the major events which govern the ecology of many elements of the fluvial biota (see for example Hildrew & Townsend 1987) and of the fluvial algae in particular. In the representation of aquatic habitats against axes of resource availability and frequency of disturbance (Fig. 9.10(a)), most rivers will tend towards the upper right-hand corner. Following the logic of Grime (1979), exclusion of the biologically untenable contingency of low

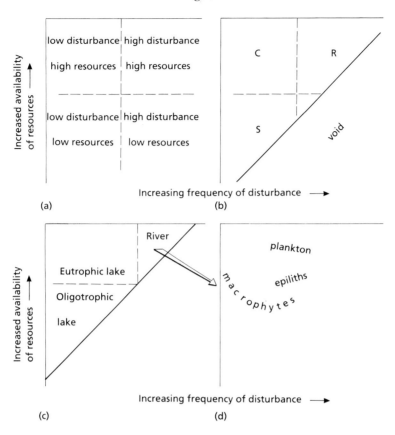

Fig. 9.10 (a) Theoretical contingencies of the incidence of the distinctive properties of the aquatic habitats of algae, as defined by axes representing available resources and disturbance frequency. (b) The primary growth and life-history strategies (colonist, stress-tolerant or ruderal — *C*, *S* or *R*) required by algae best adapted to the particular habitats (based on Grime 1979). (c) A proposed subdivision of aquatic habitats, with rivers generally occupying highly disturbed, resource-rich category; even so, within the *'RIVER'* apex, the relative success of particular algal types may be discussed (d).

resources and high disturbance leaves a triangular representation of available ecological range (Fig. 9.10(b)). The location of river algae in the right-hand apex (Figs 9.10(c) and (d)) serves to emphasize the overriding constraint placed on their ecologies by disturbance factors as opposed to, say, the chronic nutrient limitations of algae in oligo-trophic lakes. The deduction is important to the understanding of fluvial ecosystems, their management and their conservation.

REFERENCES

Arnold F, Macan TT. (1969) Studies on the fauna of a Shropshire hill stream. *Field Studies* 3: 159–84. [9.4]

Blum JL. (1956) The ecology of river algae. *Botanical Reviews* 22: 291–341. [9.1]

Blum JL. (1960) Algal populations in flowing waters. *Special Publications of the Pymatuning Laboratory for Field Biology* 2: 11–21. [9.1]

Boudou A, Ribeyre F. (1989) *Aquatic Exotoxicology: Fundamental Concepts and Methodologies*, Vol. I. CRC, Boca Raton. [9.1]

Connell JH. (1978) Diversity in tropical rain forests and coral reefs. *Science* 199: 1302–10. [9.4]

Dawson FH. (1989) Ecology and management of water plants in lowland streams. *Report, Freshwater Biological Association* 57: 43–60. [9.2]

Décamps H, Capblancq J, Tourenq JN. (1984) Lot. In: Whitton BA (ed.) *Ecology of European Rivers*, pp 207–35. Blackwell Scientific Publications, Oxford. [9.2, 9.4]

Elber F, Schanz F. (1990) Algae, other than diatoms affecting the density, species richness and diversity of diatom communities in rivers. *Archive für Hydrobiologie* **119**: 1–14. [9.4]

Greenberg AE. (1964) Plankton of the Sacramento River. *Ecology* **45**: 40–9. [9.2]

Gregory SV. (1983) Plant–herbivore interactions in stream systems. In: Barnes JR, Minshall GW (eds) *Stream Ecology*, pp 157–89. Plenum Press, New York. [9.3, 9.4]

Grime JP. (1979) *Plant Strategies and Vegetation Processes*. Wiley-Interscience, Chichester. [9.4]

Hildrew AG, Townsend CR. (1987) Organization in freshwater benthic communities In: Gee JHR, Giller PS (eds) *The Organization of Communities, Past and Present*, pp 347–71. Blackwell Scientific Publications, Oxford. [9.4]

Humphries SE, Imberger J. (1982) *The influence of the internal structure and dynamics of Burrinjuck Reservoir on phytoplankton blooms*. Centre for Water Research. Report ED. 82–023, University of Western Australia, Nedlands. [9.3]

Hydraulics Research Laboratory (1981) *The Servern Estuary:observation of tidal currents, salinities and suspended solids concentrations*. Report EX 966. Hydraulics Research Ltd., Wallingford. [9.3]

Hydraulics Research Laboratory (1985) *The hydraulic roughness of vegetation in open channels*. Report IT 281, Hydraulics Research Ltd., Wallingford. [9.2]

Hynes HBN. (1970) *The Ecology of Running Waters*. University of Liverpool Press, Liverpool. [9.1, 9.2, 9.4]

Joint IR, Pomroy AJ. (1981) Primary production in a turbid estuary. *Estuarine, Coastal and Shelf Science* **13**: 303–16. [9.3]

Kowalczewski A, Lack TJ. (1971) Primary production and respiration of the phytoplankton of the rivers Thames and Kennet at Reading. *Freshwater Biology* **1**: 197–212. [9.3]

Krogmann DW, Butalla K, Sprinkle J. (1986) Blooms of cyanobacteria in the Potomac River. *Plant Physiology* **80**: 667–71. [9.4]

Lack TJ. (1971) Quantitative studies on the phytoplankton of the rivers Thames and Kennet at Reading. *Freshwater Biology* **1**: 213–24. [9.3]

Lack TJ, Youngman RE, Collingwood RW. (1978) Observations on a spring diatom bloom in the River Thames. *Verhandlungen des internationale Vereinigung für theoretische und angewandte Limnologie* **20**: 1435–9. [9.2]

Lamberti GA, Moore JW. (1984) Aquatic insects as primary consumers. In: Resh VH, Rosenberg DM (eds) *The Ecology of Aquatic Insects*, pp 164–95. Praeger Scientific, New York. [9.3]

Leopold LB, Wolman MG, Miller JP. (1964) *Fluvial Processes in Geomorphology*. Freeman, San Francisco. [9.3]

Lund JWG. (1965) The ecology of the freshwater phytoplankton. *Biological Reviews of the Cambridge Philosophical Society* **40**: 231–93. [9.3]

Margalef R. (1960) Ideas for a synthetic approach to the ecology of running waters. *Internationale Revue des gesamten Hydrobiologie* **45**: 133–53. [9.2, 9.3, 9.4]

Meybeck M, Chapman DV, Helmer R. (1989). *Global Freshwater Ecology*. Basil Blackwell, Oxford. [9.3]

Mohammed AA, Ahmed AM, El-Otify AM. (1989) Field and laboratory studies on Nile phytoplankton in Egypt. IV. Phytoplankton of Aswan High Dam Lake (Lake Nasser). *Internationale Revue des gesamten Hydrobiologie* **74**: 549–78. [9.4]

Moss B, Balls H. (1989) Phytoplankton distribution in a floodplain lake and river system. II. Seasonal changes in the phytoplankton communities and their control by hydrology and nutrient availability. *Journal of Plankton Research* **11**: 839–67. [9.2]

Nielsen SL, Sand-Jensen K. (1990) Allometric scaling of maximal photosynthetic growth rate to surface/volume ratio. *Limnology and Oceanography* **35**: 177–81. [9.3]

Odum EP. (1969) The strategy of ecosystem development. *Science* **164**: 262–70. [9.4]

Phinney HK, McIntyre CD. (1965) Effect of temperature on metabolism of periphyton communities developed in laboratory streams. *Limnology and Oceanography* **10**: 341–4. [9.3]

Prowse GA, Talling JF. (1958). The seasonal growth and succession of plankton algae in the White Nile. *Limnology and Oceanography* **3**: 223–8. [9.2, 9.4]

Reynolds CS. (1987) The response of phytoplankton communities to changing lake environments. *Schweizerische Zeitschrift für Hydrologie* **49**: 220–36. [9.4]

Reynolds CS. (1988) Potamoplankton: paradigms, paradoxes and prognoses. In: Round FE (ed.) *Algae and the Aquatic Environment*, pp 285–311. Biopress, Bristol. [9.2, 9.5, 9.4]

Reynolds CS. (1989) Physical determinants of phytoplankton succession. In: Sommer U (ed.) *Plankton Ecology*, pp 9–56. Science-Tech Publishers, Madison. [9.3]

Reynolds CS, Glaister MS. (1989) Remote sensing of phytoplankton concentrations in a UK river. In: White SJ (ed.) *Proceedings of the NERC Workshop on Airborne Remote Sensing, 1989*, pp 131–40. Natural Environment Research Council, Swindon. [9.2, 9.3]

Reynolds CS, Glaister MS. (1992) Spatial and temporal changes in phytoplankton abundance in the upper and middle reaches of the River Severn. *Large Rivers* (in press). [9.2]

Reynolds CS, Carling PA, Beven K. (1991) Flow in river channels: new insights into hydraulic retention. *Archiv für Hydrobiologie* **121**: 171–79. [9.2]

Reynolds CS, White ML, Clarke RT, Marker AF. (1991)

Suspension and settlement of particles in flowing water: comparison of the effects of varying water depth and velocity in circulating channels. *Freshwater Biology* **24**: 23–24. [9.3]

Round FE. (1981) *The Ecology of the Algae*. Cambridge University Press, Cambridge. [9.1]

Sabater S, Muñoz I. (1990) Successional dynamics of the phytoplankton in the lower part of the River Ebro. *Journal of Plankton Research* **12**: 573–92. [9.4]

Smith IR. (1975) *Turbulence in Lakes and Rivers*. Scientific Publication No. 29. Freshwater Biological Association, Ambleside. [9.2, 9.3]

Swale EMF. (1969) Phytoplankton in two English rivers. *Journal of Ecology* **57**: 1–23. [9.2]

Symoens J-J, Kusel-Fetzmann E, Descy J-P. (1988) Algal communities in continental waters. In: Symoens JJ (ed.) *Vegetation of Inland Waters*, pp 183–221. Kluwer Academic Publishers, Dordrecht. [9.4]

Talling JF, Rzóska J. (1967) The development of plankton in relation to hydrological régime in The Blue Nile. *Journal of Ecology* **55**: 637–62. [9.2]

Valentine EM, Wood JR. (1977) Longitudinal dispersion with dead zones. *Journal of the Hydraulics Division, The American Society of Civil Engineers* **103**: 975–80. [9.2]

Ven Te Chow. (1981) *Open-channel Hydraulics*, international edition. McGraw-Hill Kogakusha, Tokyo. [9.3]

Wawrik F. (1962) Zur Frage: Fuhrt der Donaustrom autochtones Plankton? *Archiv für Hydrobiologie Supplementband* **27**: 28–35. [9.2]

Whitehead PG, Hornberger GM. (1984) Modelling algal behaviour in the River Thames. *Water Research* **18**: 945–53. [9.3]

Whitton BA. (1975) Algae. In: Whitton BA (ed.) *River Ecology*, pp 81–105. Blackwell Scientific Publications, Oxford. [9.1, 9.2, 9.3]

Wiens JA. (1989) Spatial scaling in ecology. *Functional Ecology* **3**: 385–97. [9.4]

Williams LG. (1964) Possible relationships between plankton-diatom species numbers and water-quality estimates. *Ecology* **45**: 809–23. [9.3]

Young PC, Wallis SG. (1987) The aggregated dead-zone model for dispersion in rivers. *Proceedings of the Conference on Water Quality Modelling in the Inland Natural Environment*, pp 421–33. BHRA, Cranfield. [9.3]

Zacharias O. (1898) Das Potamoplankton. *Zoologische Anzeiger* **21**: 41–8. [9.2]

10: Macrophytes

A.M.FOX

10.1 INTRODUCTION

The aquatic macrophytes comprise a diverse assemblage of plants that have become adapted from terrestrial species to life wholly, or partially, in fresh water. Their roles in aquatic environments have evoked increasing interest in recent years as multipurpose utilization of aquatic habitats has intensified. While much of the research in this field has concentrated on the nuisance species that interfere with water use, concerns over the restoration and protection of natural systems have prompted further studies relating to all aquatic plants. Many aquatic habitats can be effectively managed or protected only by applying a knowledge of the biology and ecology of macrophytes. However, more emphasis has been placed on the vegetation of lakes and canals than on that of rivers.

Both terms 'aquatic' and 'macrophytes' are open to interpretation but definitions and examples will be chosen here that describe the majority of plants found in rivers. 'Aquatic' plants will be defined according to Cook (1974) as those whose photosynthetically active parts are permanently or, at least, for several months each year submerged in, or floating on, fresh water. This definition allows differentiation between the truly aquatic riparian species and those that tolerate only occasional high flood stages or, as with many riverbank trees, are rooted in saturated substrata.

'Macrophytes' are limited to the macroscopic flora including aquatic spermatophytes (seed-bearing plants), pteridophytes (ferns and fern allies) and bryophytes (mosses and liverworts). Algae, such as the filamentous *Cladophora* spp.

and macroscopic charophytes (e.g. *Chara* and *Nitella*) are often included in this definition but they have been discussed in Chapter 9.

Macrophytes are often classified by their growth habit rather than taxonomically. Various subdivisions and terminologies have been proposed but a simple four-group system is widely accepted (Fig. 10.1: Sculthorpe 1967) *Emergent macrophytes* are rooted plants with most of their leaves and stem tissue above the water surface (e.g. *Phragmites australis*, *Typha latifolia*, *Sagittaria* spp). *Floating-leaved macrophytes* are rooted plants with most of their leaf tissue at the water surface (e.g. *Nymphaea alba*, *Nymphoides peltata*). *Free-floating macrophytes* are plants not rooted to the substratum but living unattached within or upon the water (e.g. *Ceratophyllum demersum*, *Eichhornia crassipes*, *Lemna* spp.). *Submerged macrophytes* are rooted plants with most of their vegetative tissue beneath the water surface (e.g. *Ranunculus penicillatus* var. *calcareus*, *Hydrilla verticillata*).

Although this system generally works well, depending upon environmental conditions or growth stage some species are not restricted to a single category. Amphibious plants (Hutchinson 1975) at maturity may have quite distinct (heterophyllous; defined by Sculthorpe 1967) forms of submerged and emersed leaves, the proportions of which vary with water depth (Spence *et al* 1987). Other heterophyllous species show temporal segregation of growth forms. Some (e.g. *Nuphar lutea*) have submerged leaves in the juvenile stage followed later by large floating or emergent leaves.

To appreciate the specialized adaptations that are exhibited by macrophytes in rivers, it is first

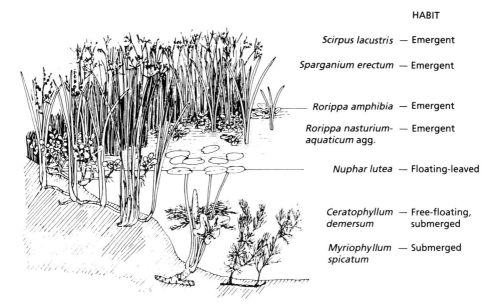

HABIT

Scirpus lacustris — Emergent

Sparganium erectum — Emergent

Rorippa amphibia — Emergent

Rorippa nasturium- — Emergent
aquaticum agg.

Nuphar lutea — Floating-leaved

Ceratophyllum — Free-floating,
demersum   submerged

Myriophyllum — Submerged
spicatum

Fig. 10.1 Examples of species of the four macrophyte habits found in a lowland, clay river (modified from Haslam 1978).

necessary to recognize general adaptations to the aquatic environment. All freshwater macrophytes have evolved from terrestrial species, either by reduction of terrestrial characteristics or by evolving secondary adaptations. The most significant differences between the terrestrial and aquatic environments are the much slower (10 000 times) rates of gas diffusion in water compared with air (e.g. movement of carbon dioxide to leaves for photosynthesis and oxygen to roots for respiration) and the attenuation of light entering water, by reflection, absorption and scatter.

A major adaptation to these constraints is the formation of aerial leaves, as seen in the emergent, floating-leaved and surface free-floating species. More specialized adaptations required for the growth of submerged tissues include thin, dissected leaves lacking an epidermis, internal gas transfer, and specialized photosynthetic physiology (Bowes & Salvucci 1989; Boston *et al* 1989).

In this chapter, the principles involved in species distribution will be discussed first. Next, aspects of quantitative macrophyte ecology that consider growth patterns and productivity will be introduced, with a detailed treatment reserved for Chapter 16 on primary production. Finally, these principles will be incorporated into a discussion of the interactions between macrophytes and human activities, emphasizing the need to understand such principles if rational river management is to be achieved.

10.2 DISTRIBUTIONAL ECOLOGY OF MACROPHYTES

The objective of distributional ecology is to examine the factors that determine the presence or absence of a species at a particular site. Three major factors are readily identified: dispersal, tolerance of abiotic environment, and interactions with the biota. The sequence in which they should be considered if the absence of a species from a particular site is to be determined is indicated by the series of questions in Fig. 10.2.

Dispersal

It is not difficult to envisage that the rate of downstream dispersal of macrophytes within a river system is determined by the size and weight

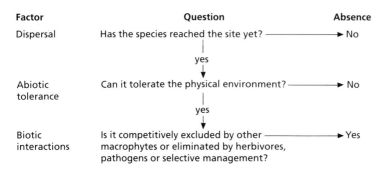

Factor	Question	Absence

Dispersal — Has the species reached the site yet? ——————→ No

yes ↓

Abiotic tolerance — Can it tolerate the physical environment? ——————→ No

yes ↓

Biotic interactions — Is it competitively excluded by other macrophytes or eliminated by herbivores, pathogens or selective management? ——————→ Yes

Fig. 10.2 The sequence of factors and questions to be addressed to determine why a species is absent from a particular site.

of the propagules and the force of water movement. Dispersal between river catchments or to upstream sites is less obvious and is dependent on a variety of means other than water movement.

Organs of dispersive propagation

Aquatic macrophytes in general, and submerged river plants in particular, show a tendency towards replacing sexual with vegetative reproduction (Sculthorpe 1967). For submerged macrophytes this may be related in all but a few species to the difficulty of raising the flowers above the water surface for aerial fertilization.

Organs of vegetative dispersal may include whole plants (e.g. *Lemna* spp., *Eichhornia crassipes*), shoot fragments including buds or nodes (e.g. *Ceratophyllum demersum*) or specialized organs, such as tubers (swollen stem or root sections, e.g. *Hydrilla verticillata*) and turions (dormant apices, e.g. *Hydrilla*, *Potamogeton* spp.). Spread of existing stands of macrophytes may be accomplished by other vegetative mechanisms (e.g. rhizomes, stolons) but they do not usually contribute to the colonization of new sites except under flood conditions with severe channel scouring.

Seeds are important organs of dispersal in emergent macrophytes which are less likely than other aquatic species to fragment or be wholly displaced. Flowers of emergent and surface free-floating species do not need to be modified from the terrestrial habit and are usually wind or insect pollinated (e.g. *Glyceria maxima*, *Hydrocharis morsus-ranae*). Floating-leaved species are usually fertilized in the same manner, with their

chief adaptation to the aquatic environment being the production of long peduncles (flower stalks) capable of lifting the flower above the water (e.g. *Nymphaea alba*). In flowing conditions these stems need to be longer than the depth of water to accommodate changes in water level and bending as a result of water velocity.

Ironically, the evolution of the angiosperm flower as a mechanism that liberated plants from a dependence on water for gamete dispersal has been only secondarily adapted to allow pollination in water in a very few submerged species (e.g. *Vallisneria spiralis*, *Zannichellia palustris*; Sculthorpe 1967).

Mechanisms of macrophyte dispersal

In addition to the downstream movement of reproductive propagules within a river, flooding may carry propagules to adjacent lakes, canals or ditches. However, flooding usually cannot carry propagules from one watershed to another.

Wind dispersal of small seeds is common in the aquatic grasses (e.g. *Phragmites*, *Phalaris* spp.) and other emergent species (e.g. *Typha*). Transport of specialized propagules (seeds, turions or tubers) or stem fragments (e.g. *Lemna* thalli, sections of flexuous *Elodea canadensis* stem) by birds and other animals, either within the digestive tract or by external attachment, can disperse plants between river catchments. Unlike the specialized reproductive propagules that can usually tolerate some degree of digestion or desiccation (Basiouny *et al* 1978), the distance that fragments of submerged species may be transported would be limited.

Whole macrophytes or their propagules are

often transported over long distances as a result of intentional (e.g. *Eichhornia crassipes* into the USA) or accidental carriage by humans. Accidental introductions to river catchments or countries could result from transport of water, sediments, crops (e.g. rice, cotton), animal fur or wool, or with desirable macrophytes, such as aquarium plants (Schmitz *et al* 1991). The often disastrous consequences of introducing exotic weedy macrophyte species to water bodies will be discussed in the last section of this chapter.

Geographical distribution of macrophytes

The geographical distribution of a species can give an indication of how effectively it is dispersed. *Phragmites australis* and *Typha* spp., which have small windborne seeds, are widely distributed around the world (Sculthorpe 1967); however, members of the Podostemaceae family are typically endemic to small geographical areas of tropical Africa or even to single river basins. The lack of dispersion in the latter group relates to the poor dispersal of their minute seeds to other suitable habitats (Willis 1914). Aquatic macrophytes are remarkable for their large proportions of ubiquitous and endemic species compared to terrestrial vegetation (Sculthorpe 1967).

Dispersal is not the only factor involved in determining geographical distribution. A narrow distribution may not be caused by poor dispersal but may result from limited tolerance of a species to certain climatic and/or geological conditions. A species on the verge of extinction or recently evolved will have a limited geographical range, regardless of its dispersive capabilities.

Tolerance of abiotic environment

The question of whether a species' incidence is determined by its dispersal (Fig. 10.2) will tend to be answered by a simple 'Yes' (the species has had the opportunity to reach the site) or 'No' (there has not been such an opportunity). The two other factors affecting species incidence (abiotic tolerance and biotic interactions), however, will also influence species abundance. A macrophyte species will be excluded from a site if it cannot

tolerate the abiotic conditions to which it would be exposed throughout its normal lifecycle. For annual species (e.g. many *Najas* spp.) this would mean that even if the mature plant could tolerate a particular environment, survival would be temporary if the seedling would perish under the same conditions. Many macrophytes which reproduce vegetatively do not exhibit such susceptible stages and so may tolerate wider environmental ranges.

The absence of a species will be determined by intolerance of extremes of abiotic conditions. The abundance of a species will be related to how near the conditions are to those optimal for maximum growth rates. Limits of tolerance and growth optima may be determined for a particular species and environmental variable by correlation of species presence or abundance with measured levels of environmental variables in the field. This method requires large numbers of field samples, and direct causal relationships cannot be assumed from such data. Growth under a range of controlled environmental conditions in artificial habitats or transplants between differing natural sites may also be used to determine tolerance and growth ranges for certain variables. However, several series of experiments with multiple variables are required to determine interactions between abiotic factors (e.g. Maberly 1985).

Although it is necessary to discuss each environmental variable separately, macrophytes do not respond to each variable independently. For example, macrophyte growth may be directly influenced by substratum particle size. However, such influence cannot be easily distinguished from the effects of water velocity, which not only affects macrophytes directly but also determines substratum particle size. The effects of environmental variables may also interact if suboptimal conditions of one variable alter the tolerance of a species to another variable.

A further complication of this dynamic system is that all variables do not remain constant over time within a habitat. Short-term, seasonal, or longer-term changes (flood, drought, turbidity) can improve or reduce the suitability of a habitat for a particular species. Temporary periods of adverse conditions may be survived by the production of specialized organs of perennation (e.g. seeds, tubers, turions, rhizomes). The frequency, duration and amplitude of environmental variations will influence the tolerance of a species to an affected habitat (Westlake 1981).

Water movement

Water movement is one of the most important and specific abiotic variables influencing species composition and location of plant communities in rivers. There can be direct and indirect effects on the vegetation related to aspects of water movement such as velocity, turbulence and erosive force. Direct effects of water movement include influences on photosynthetic rates and exposure to nutrients and carbon dioxide, mechanical damage and propagule establishment. Indirect effects occur through the influence of water movement on substrata and fauna.

Metabolic rates of macrophytes are generally higher in moving water than in still water (Westlake 1967; Madsen & Søndergaard 1983). At flow rates typically found within plant beds, photosynthesis increases with water velocity in riverine *Ranunculus* species. This may result from the reduction in thickness of the boundary layers surrounding the macrophyte leaves that occurs as water velocity increases. The thinner the boundary layer, the less limiting is the rate of diffusion of dissolved carbon sources across this layer, from the main body of water to the leaf surface (Westlake 1967).

Faster flowing water will tend to be more turbulent (especially as the substrata will tend to be coarser) and this may also improve macrophyte photosynthesis by increasing the aeration of the water with atmospheric carbon dioxide. The relationship of water turbulence and concentrations of dissolved carbon dioxide explains why macrophytes unable to utilize bicarbonate carbon sources, such as many mosses, are limited in calcareous waters to areas of turbulent flow as found in highland streams, riffles, or on weirs in lowland rivers (Bain & Procter 1980).

Water movement in rivers will also expose macrophytes to a constantly replenished source of nutrients, less likely to become diminished and limiting than in static water. However,

despite the metabolic stimulation resulting from the improved availability of carbon dioxide and nutrients in flowing water, there is little evidence that growth is faster in rivers than in lakes (Westlake 1975). This is probably because of increased organic metabolite losses (Wetzel 1969) resulting from the reduced boundary layer and, more significantly, the continual loss of plant material by mechanical damage.

Although the metabolic rates of some species may rise with increased water movement more than others, such factors tend to have more of an effect on total macrophyte abundance than on species selection. Some riverine species (e.g. *Ranunculus fluitans*) may only tolerate, or be competitive in, the improved metabolic conditions of flowing water. However, the tolerance of different species to the mechanical damage associated with various flow regimes probably has a greater impact on river community structure.

The most obvious mechanical influence of water movement on macrophyte communities is that free-floating species will be limited to areas, or periods, of reduced flow. This does not mean that such species cannot have major impacts on some rivers (e.g. infestations of *Eichhornia crassipes* in tropical rivers), but their role in most temperate rivers is minor compared with the rooted species.

The various types of mechanical damage that rooted river plants must tolerate — uprooting, battering, tangling, abrasion, erosion and sedimentation — have been described by Haslam (1978). The hydraulic resistance of individual plants depends on their dimensions relative to the direction of flow and their leaf, stem and branching shapes. Broad, bushy plants (e.g. *Myriophyllum spicatum, Rorripa* spp.) will create more resistance and be more susceptible to uprooting and battering than low, streamlined ones (e.g. *Ranunculus, Vallisneria* spp.). Increased anchoring strength reduces the susceptibility of a plant to uprooting and sediment erosion, and is greatest in plants with well-developed root and rhizome systems (e.g. *Sparganium erectum*). Tissue robustness determines susceptibility to abrasive damage and whether a plant will tear apart as water velocity increases. Battering damage is also influenced by the turbulence of flow and, when severe, stems of long flexuous plants may become tangled.

Laboratory experiments have been conducted on individual, and pieces of, macrophytes to try to determine hydraulic resistance and the susceptibility of different species to these various types of damage (Haslam 1978). In the field, however, the shape of plant beds may be more important than of individual stems (Pitlo & Dawson 1990). Water velocity is significantly reduced within beds of many riverine macrophyte species (Losee & Wetzel 1988; Getsinger *et al* 1990) with inner stems being protected from sudden increased water velocities and turbulence. Thus, species that have individual stems with high hydraulic resistance (e.g. *Elodea canadensis*) may form beds with an overall streamlined shape. Such species could not colonize sites with constantly high water velocity or turbulence but might survive short periods of such spate conditions once beds were established. This indicates how important the periodicity of extreme flow conditions may be and, hence, how the stability of flow regimes (as determined by the catchment geology) can influence macrophyte community structure.

Water velocity and turbulence also influence the establishment of macrophyte propagules. Unless they are sheltered from flow by being carried into backwaters, deep depressions or plant beds, buoyant or delicate propagules will be swept downstream or mechanically damaged before they can root and grow in most rivers. Very small propagules may become trapped in the boundary layers and dead water downstream of suitable obstacles, such as rocks and plant beds (*cf.* invertebrate habitats; Hynes 1970). All the dangers of mechanical damage discussed above must then be tolerated by the new plant as it grows out of the sheltered area. Although the hydraulic resistance of small shoots is less than of mature ones, their tissues are usually more delicate and their roots less secure.

Association of particular macrophyte species with certain ranges of water velocity has frequently been based on qualitative observations. The observations of Butcher (1933), for example, have often been cited (e.g. Hynes 1970; Westlake 1975). More recently, quantitative data have been

collected in large-scale field surveys of Britain and Europe (Haslam 1978, 1987; Holmes 1983) with correlations made between the presence or abundance of species and flow regime.

Substratum

The geology underlying a watershed has a major influence on the physical and chemical characteristics of its river channels and lake basins, and, hence, on macrophyte colonization. Catchment geology affects patterns of rainfall runoff and the stability of flow regimes. For example, chalk rivers have little seasonal fluctuation in discharge rates and provide stable conditions for macrophyte growth. Rivers rising on erosion-resistant rocks have spatey flow regimes, dependent upon seasonal rainfall distribution and support only those macrophytes tolerant of mechanical damage or able to rapidly recolonize a flood-scoured site.

On a more local scale, variations in the resistance of underlying rocks will determine erosional channel features such as riffles and pools. Variable water depth and flow rates between such features can influence the communities of macrophytes over distances of just a few metres.

Geology also determines the general type of substratum in the river channel, whether resistant rocks, the firm gravel of limestone rivers or the easily-eroded and silty deposits of clay catchments. Secondarily, local water velocity defines the characteristics of river substrata, with fine sediments and organic matter being eroded or deposited accordingly.

Pearsall (1920) first documented the associations between certain macrophytes and substratum-type in the Cumbrian lakes, particularly with reference to silting rates at the mouths of inflowing streams. Subsequent investigations of substratum–macrophyte relations have concentrated on lacustrine species, and on nutrient availability rather than physical characteristics of the sediments (Barko & Smart 1981b).

River surveys (Butcher 1933; Haslam 1978; Holmes 1983) have indicated that many macrophyte species are associated with sediment of particular particle size. Do these species have a requirement for certain substratum characteristics, or is that relationship coincidental to their selection by water movement and the causal relationship between flow rates and sediment type? Carefully controlled experiments (e.g. Barko & Smart 1986) would be needed to answer such a question.

Substratum particle size influences macrophyte colonization with respect to root depth and stability. Deep rooting patterns (e.g. *Sparganium erectum*) associated with thick layers of fine sediments stabilize plants during spate conditions when the upper soil is disturbed but are susceptible to uprooting in sustained increases in flow. Species associated with coarser particles (e.g. *Ranunculus* spp.) tend to have networks of curling roots that can securely anchor to individual stones and consolidate the sediment for greater stability in faster flowing rivers (Haslam 1978).

The effects of macrophytes on the substratum, other than rooting consolidation, include the production of organic material and sufficient slowing of water movement within plant beds to allow the sedimentation of suspended materials. This sediment accumulation may permit the colonization of species indicative of finer sediments and lesser flow. For example, in shallow chalk streams the accumulation of silt in *Ranunculus penicillatus* var. *calcareus* beds permits the colonization of the emergent *Rorripa nasturtium-aquaticum* (Dawson *et al* 1978).

Light

The attenuation of light in clear water is a major constraint to submerged macrophyte growth. Light availability principally affects the abundance of submerged vegetation, but since some species are better adapted to low light intensities than others (e.g. *Hydrilla*; Bowes *et al* 1977) it also has some influence on community structure. Factors that increase light attenuation in water proportionately reduce the depth of plant colonization.

The presence of suspended solids (turbidity) in moving water has the most influence on light attenuation. Prolonged periods of increased water velocity and factors in the catchment that elevate sediment loading can reduce plant growth, particularly in deep waters. Water colour is also

determined by characteristics of the river catchment. Heavily forested catchments and rivers with a high proportion of runoff from bogs and swamps may have brown, tannin-stained water that significantly reduces light penetration even if there is no turbidity.

In slow-moving water, shading by floating or floating-leaved macrophytes or phytoplankton, may limit the growth of submerged species. Epiphytic algae can significantly reduce photosynthetic rates of macrophytes that are susceptible to their colonization (Sand-Jensen & Borum 1984). Shading by riparian trees reduces the abundance of all types of macrophytes in narrow river channels (Canfield & Hoyer 1988).

Temperature

While some macrophyte species with a worldwide distribution do not appear to have any special temperature requirements (e.g. *Phragmites australis*, *Lemna minor*), other species are limited to certain climatic zones (e.g. cool: *Sparganium emersum*; warm: *Hydrilla verticillata*; Sculthorpe 1967; Barko & Smart 1981a). River plants do not appear to be affected by this any differently from lacustrine macrophytes but the source of a river's water may have a local influence on temperature. Spring-fed rivers will tend to be of more constant temperature throughout the year compared with rivers that are more dependent upon catchment runoff. Near the springs species may occur that require warmer (in the higher latitudes) or cooler (in the tropics) conditions than the vegetation occurring further downstream or in nearby rivers that are not spring-fed.

Water chemistry

The aspect of water chemistry that has most influence on macrophyte community structure is the complex relationship between pH, the predominant form of dissolved inorganic carbon (DIC), water hardness and calcium concentration (Butcher 1933; Spence 1967). Some species of aquatic plants are found only in acidic waters (e.g. *Potamogeton polygonifolius*) where carbon dioxide is the predominant source of DIC. Such species may be intolerant of more alkaline habitats because of an inability to use bicarbonate ions (Spence & Maberly 1985). The occurrence of certain species has been correlated with water calcium concentrations (Haslam 1978) or has been associated with catchment geologies that have a significant influence on water hardness and pH (Holmes 1983).

The correlation of macrophytes with nutrient (particularly phosphorus and nitrogen) concentrations is complicated by the ability of many species to obtain minerals either by foliar uptake from the water or from the substratum via their roots (Denny 1980). Many lowland rivers (especially temperate chalkstreams) have been shown to have an excess of available nutrients so that other factors must be responsible for limiting plant growth in these habitats (e.g. Ladle & Casey 1971; Canfield & Hoyer 1988).

Other aspects of water chemistry may have localized influences on macrophyte community composition. For example, elevated salinity may exclude intolerant species (e.g. *Hydrilla verticillata*; Haller *et al* 1974) from river/estuary boundaries and the outflow of saltwater springs or mining operations.

Water level fluctuations

The effect of catchment geology on the stability of river discharges has been mentioned in relation to macrophyte colonization. Some rivers that are not spring-fed may be susceptible to summer droughts near their sources. This limits all but the emergent vegetation, to species that produce propagules capable of tolerating periodic drying (e.g. seeds and tubers). Rivers that undergo continuous gradual changes in water levels tend to have a broader fringe of emergent vegetation than rivers that are either very stable or that are subject to sudden and brief water level fluctuations and storm erosion.

Interactions with biota

Once a species has reached a site that it can tolerate, its survival there depends upon its interactions with other biota. Severe competition between macrophyte species may result in the exclusion of one species from a site by another.

Under natural conditions, herbivores and plant pathogens, while reducing macrophyte abundance, will not eliminate their plant hosts. However, a reduction in vigour of macrophytes susceptible to grazing or pathogens might lead to competitive exclusion by less susceptible plant species.

Competition

Grime (1979) identified four resources for which plants compete: light, space, nutrients and water. By definition, aquatic macrophytes do not need to compete for water resources and in lowland rivers, at least, nutrients are often available in excess of requirements (Ladle and Casey 1971). Competition for light is probably the dominating factor in the relationships between river macrophytes.

Morphological features that allow one plant to modify the environment to the detriment of others provide greatest competitive advantages. For example, the shading of low-lying submerged plants by tall, canopy-forming species or by macrophytes of other habits. The production of allelopathic substances that inhibit the growth of other species may be regarded as a competitive characteristic. Most evidence for the release of such compounds has been produced under laboratory conditions (Leather & Einhellig 1986) and conclusive examples of this phenomenon have not been identified in rivers.

Grime (1979) listed 18 competitive characteristics for terrestrial plants, and those appropriate for aquatic macrophytes have been identified (Van *et al* 1978; Murphy *et al* 1990). Characteristics relevant to river plants include: canopy formation; use of bicarbonate for DIC; use of carbon dioxide in the air; early seasonal and daily growth resulting from a low light compensation point; low root/shoot ratio; and high litter production.

Most competition studies with macrophytes have been between major weed species and the native plants that they displace (e.g. *Hydrilla* and *Vallisneria*; Haller & Sutton 1975). De Wit replacement series have been carried out with floating and lacustrine submerged species (e.g. McCreary *et al* 1983; Agami & Reddy 1989). Emergent plants have been transplanted along gradients of changing species (Grace & Wetzel

1981) but few competition experiments have been carried out in flowing water.

Some species (usually those regarded as 'weedy') do appear to be able to dominate rivers to the exclusion of others. As predicted by Grime's triangular strategy theory, competitive exclusion to one or two species usually occurs in the least disturbed and physically stressed environments, such as nutrient-enriched lowland rivers (e.g. *Potamogeton pectinatus*; Holmes 1983).

River macrophytes play an important role in relation to other biota by providing habitat diversity (Westlake *et al* 1972). The large surface area of plants and local reductions in flow provide a substratum and shelter for epiphytic algae, invertebrates, fish and their eggs. Since some of these organisms are found in varying densities on different plant species (Soszka 1975; Miller *et al* 1989), a diversity of river macrophytes is usually desirable to promote a variety of other biota. Maintenance of macrophyte diversity by the prevention of competitive exclusion by species such as *Potamogeton pectinatus* is often a major objective of river vegetation management.

Herbivory and plant pathogens

Herbivory of submerged river plants is not extensive. Some snails (e.g. Sphaeridae), insects (e.g. Elmidae) and fish (e.g. cyprinid species) will scrape algal epiphytes off macrophytes, and other invert-

ebrates and fish feed on detritus composed largely of dead plant material (Hynes 1970). Often crayfish and temperate fish, such as *Rutilus* sp., will eat fresh vegetation in addition to their diet of invertebrates but there are few species that are exclusively herbivorous (e.g. *Tilapia* spp. and Grass Carp, *Ctenopharyngodon idella*).

As with the tropical herbivorous fish, some plant pathogens have been investigated with respect to their potential as biocontrol agents of weedy macrophytes (Charudattan 1990). However, relatively little is known about the incidence of pathogens affecting temperate riverine macrophytes.

Synthesis of the distributional ecology of macrophytes

By following the series of questions in Fig. 10.2 and by examining the environmental requirements and competitive characteristics of a macrophyte species, it should be possible both to determine why that species is absent from a particular site and to predict suitable habitats for its growth. Trying to explain or anticipate detailed spatial and temporal changes in species composition and relative abundances is more complex because of the many interrelationships between macrophytes and their environment. Detailed studies of some lowland rivers have resulted in models (Fig. 10.3) that attempt to predict some

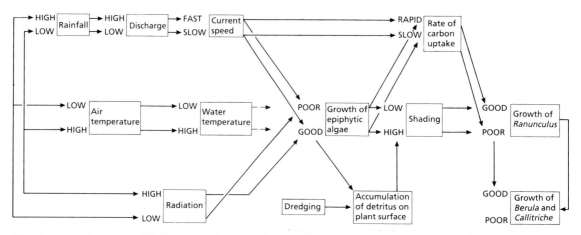

Fig. 10.3 A qualitative model of factors influencing the growth of *Ranunculus* at Bangor, UK (after Ham *et al* 1981).

simple changes in species dominance (Ham *et al* 1981). However, the production of more quantitative models for describing the dynamics of river communities is a far more daunting task requiring the interaction of researchers from a variety of physical and biological disciplines.

The dynamic interrelationships of plants, flow and sediments result in a mosaic of macrophyte species that changes by the month (Fig. 10.4), season or year (Butcher 1933; Ladle & Casey 1971). The unstable physical characteristics of river channels preclude the succession leading to climax vegetation often observed in small lakes and deltas. Rivers maintain vegetation typical of early succession due to periodic scouring. These may be seasonal events, washing out the summer's accumulation of vegetation, or the infrequent major spates that change the whole morphology of the channel. Only in the isolation of ox-bow lakes can the traditional concept of hydroseral succession proceed.

10.3 QUANTITATIVE ECOLOGY OF MACROPHYTES

In discussing the general quantitative ecology of macrophytes, the abundance and productivity of aquatic species and habitats are considered in relation to variable environmental factors. Temporal variations are considered with regard to seasonal development and growth patterns, and spatial variations are discussed in comparisons of the productivity of different types of aquatic vegetation and river sites. The difficulties of quantifying vegetation in flowing water provide a major constraint on this aspect of macrophyte ecology

Seasonal growth patterns

In temperate and subtropical climates, seasonal changes in day length and temperature undoubtedly play an important role in regulating macrophyte growth patterns, just as they do for terrestrial plants. Rainfall patterns, and hence river discharge and erosive forces, may be just as significant. Many of the examples mentioned will assume the typical temperate pattern of low summer rainfall and higher rainfall in winter and/or spring and autumn.

Late May

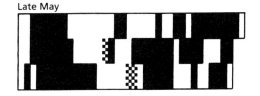

Late June

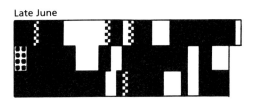

Late July

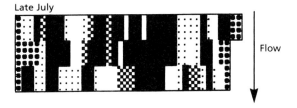

Flow

Early October

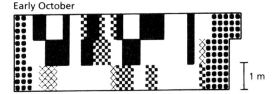

1 m

■ *Ranunculus penicillatus* var. *calcareus*.

▨ *Potamogeton pectinatus*.

▦ Other submerged species,
e.g. *Myriophyllum spicatum*.

▤ Filamentous algae, e.g. *Cladophora glomerata*.

⊞ Emergent species, e.g. *Phalaris arundinacae*.

Fig. 10.4 Maps of permanent transects in the lowland, clay/chalk stream River Windrush, UK, showing changes in species dominance throughout the summer of 1984. The dominant macrophytes in each 0.25 × 1.0-m rectangle based on surface coverage are indicated.

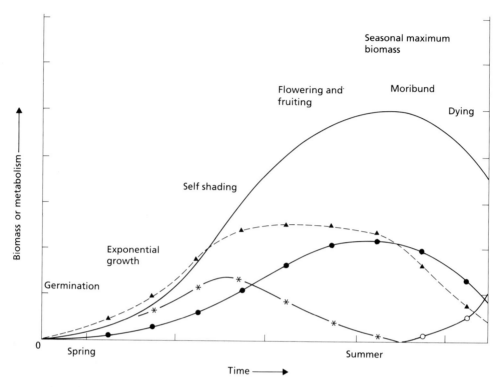

Fig. 10.5 Hypothetical growth and metabolism curves for an annual plant. − Biomass; ▲ current gross productivity; current net productivity; ● current respiration rate; ○ death losses (after Westlake 1965).

Westlake (1965) described the typical growth pattern for an annual plant assuming negligible overwintering biomass (Fig. 10.5). The timing of the summer maximum biomass will vary with species or even with biotype within a species (e.g. *Ranunculus penicillatus* var. *calcareus*; Dawson 1980a). The timing and severity of the first autumn storms will influence the postflowering wash-out.

This pattern is appropriate for many perennial submerged species with the exception that growth in the spring starts from an overwintering biomass which may vary annually, depending upon the previous summer's production and winter discharges. Detailed studies of the production of *Ranunculus penicillatus* var. *calcareus* (Dawson 1976) revealed such a pattern, with self-shading within large beds slowing the exponential growth rates of early summer. To raise the insect-pollinated flowers above the water surface, this species forms thick buoyant flowering stems that are more brittle than the vegetative stems. Losses of these fragile stems after flowering initiates the late summer reduction in biomass, with 75% of the shoots being fragmented and lost within 3 months.

Such growth patterns are evident from biomass samples of most submerged river plants because underground biomass is small (0.5% of maximum *Ranunculus* biomass; Dawson 1976) and moribund tissues are immediately removed. This is not the case for many emergent and floating-leaved species that possess substantial rhizome systems (e.g. *Typha, Phragmites, Nuphar*) or other perennating organs (e.g. bulb-like root stock of *Sagittaria*) which may comprise much of the overwintering biomass. Emergent species are not as affected by seasonal increases in water discharge as submerged species but are more susceptible to winter frost damage.

Some of the problems of estimating the abundance and productivity of macrophytes, such as whether to sample biomass from both above and below ground, are not unique to quantitative studies of aquatic plants (Westlake 1963). Biomass sampling in rivers does require some specialized techniques to prevent the downstream loss of harvested material. Dredges or grab samplers operated from boats or by divers may be necessary in deep water (Sliger *et al* 1990). In shallow rivers, material removed from quadrats can be collected in surrounding boxes or nets (e.g. Lambourn sampler; Hiley *et al* 1981). Dry weights should be measured because it is difficult to ensure a uniform removal of surface water. Also the water content of different macrophytes varies considerably, with ranges of 85−92% of tissue weight being water for submerged species and 70−85% for emergent species (Wetzel 1983).

Estimates of vegetation cover either by remote sensing (e.g. aeroplane or balloon photography) or ground mapping tend to be less time consuming than biomass sampling. Random or systematic point sampling within the channel, or by use of grapnels from the bank, provides estimates of relative species abundances. Use of permanent transects to map plant beds can indicate changes in the placement of species over time. The division of transects into rectangles, from which the dominant species is recorded (see Fig. 10.4; Wright *et al* 1981) can speed up mapping procedures without great loss of information. The size of sampling units should reflect the level of diversity at a site. The patchy and dynamic distribution of submerged species encountered in diverse river communities (Butcher 1933) is best indicated by mapping techniques, and such patterns have to be assessed in the choice of size, number and placement of biomass samples.

Dawson (1976) showed that the percentage cover of *Ranunculus* was not a good indicator of its biomass in a shallow stream. While the *Ranunculus* biomass was halved over a 4-year period, the cover of the stream bed did not change significantly. Volume estimates of bed size rather than just those expressed in terms of area might improve the relationship with biomass but would be almost as time consuming to collect (e.g. using a recording fathometer; Maceina *et al* 1984).

Productivity of macrophyte communities

The continual losses of tissue from macrophytes in the river channel (Dawson 1980b) and accumulation of litter by emergent species complicate the estimation of productivity using repeated biomass measurements. Estimates have been achieved by tagging individual plants and leaves, and measuring changes in their size (Odum 1957; Dawson 1976), but such methods are highly intensive and small-scale. Indirect methods have included measurement of photosynthetic substrates and products (Odum 1956) but these studies are often restricted to specialized areas (e.g. near a spring mouth or weir) and are usually short term.

As a result, the productivity of river macrophytes has not received the same attention as terrestrial, wetland or lacustrine vegetation. However, a knowledge of some of the environmental factors influencing macrophyte growth can help in predicting the relative productivities of various lotic sites.

Submerged species are less productive than emergent and surface-floating plants (Westlake 1963; Wetzel 1983) because of the slow diffusion of photosynthetic gases and the reduced intensity of incident light. By virtue of warmer and longer growing seasons, macrophytes in tropical rivers are more productive than temperate ones. Within a climatic zone, differences between rivers or reaches of a river depend upon all the abiotic features outlined above, as well as growth characteristics of the composite species.

10.4 INTERACTIONS BETWEEN MACROPHYTES AND HUMAN ACTIVITIES

As with many biological disciplines, the most intensively studied species of macrophytes are the aquatic weeds; by definition, those that interfere with human activities. Murphy (1988) reviewed the major aquatic weed species of the world, and many of the problems that they cause for humans.

Conversely, human activities related to the use of natural resources, power production and flood control often affect rivers and associated macrophytes. The important roles that macrophytes play in lotic habitats can be summarized as: primary production; photosynthetic production of oxygen; substratum for algae; habitat for invertebrates and fish eggs; nutrient cycling to and from sediments; stabilization of river beds and banks. In view of these roles, activities that reduce or drastically alter the composition of macrophyte communities must be acknowledged.

Macrophytes affecting human activities

Flooding risks

The ways in which macrophytes influence water flow in rivers tend to be related to the abundance and habit of the vegetation rather than to species composition. As the biomass of macrophytes in a channel increases so does the hydraulic roughness (Dawson 1978; see Chapter 5) which results in a decrease of mean water velocity. If the river discharge is constant, a reduction in velocity will cause an increase in water depth which may overflow the river banks, or raise the water table and flood or waterlog what is commonly valuable agricultural land.

The deposition of suspended sediments within macrophyte beds reduces the effective cross-section of the channel available for discharge. This can result in channelization of rapid flow between plant beds which will locally scour the substratum and may undercut river banks. Under spate conditions, these narrow channels cannot carry all the additional water, causing flooding.

Interference with fishing and navigation

Tall emergent vegetation along river banks can prevent access for shoreline fishing. Dense growths of submerged plants, particularly those that have reached the water surface, interfere both with fishing itself and the use of boats to get to areas inaccessible from the shore. Perennial submerged species can spoil the gravel spawning beds of salmonid fish (e.g. *Potamogeton pectinatus*; Caffrey 1990). High densities of photosynthesizing macrophytes can cause large diel fluctuations in oxygen which can stress fish. In non-turbulent waters, under prolonged hot and cloudy conditions, when photosynthesis does not exceed respiration, oxygen depletion can result in fish mortality (Brooker *et al* 1977).

Some of the greatest obstacles to commercial fishing and navigation have occurred on tropical rivers as a result of floating vegetation, such as *Eichhornia crassipes* and *Salvinia molesta*. Sudd formation (floating islands) is a major problem to navigation in parts of Africa and India, and results from dense mats of floating vegetation on top of which emergent and even woody species can grow (Sculthorpe 1967).

Reduction in water supply

The continuous loss of fragments from submerged vegetation near irrigation or drinking water intakes can interfere with water supply by the clogging of pumps and filters. In tropical climates the loss of water by the transpiration of emergent and some floating vegetation exceeds the evaporative loss from an equivalent area of unvegetated water surface (Rao 1988), and in arid environments such losses may significantly reduce water supply.

Habitat for disease organisms

Several human diseases are carried by intermediate hosts that favour aquatic habitats vegetated by macrophytes. Examples include: schistosomiasis (bilharzia) carried by aquatic snails; filariasis carried by *Mansonia* mosquitoes; and malaria carried by *Anopheles* mosquitoes. The intermediate hosts are either dependent upon certain macrophytes for completion of their life-cycle (e.g. *Mansonia* attaches to roots of *Pistia stratiotes*) or inhabit stagnant water resulting from the obstruction of water-courses by vegetation. The damming of many rivers and construction of irrigation canals has particularly increased the incidence of such diseases in many tropical areas of the world (Brown & Deom 1973).

Human activities affecting macrophytes

Weed management

To reduce the problems created by aquatic weeds,

various forms of plant management are employed throughout the world. Usually the objectives of responsible weed management are to reduce the biomass of problem species rather than the total elimination of vegetation. Management of some fisheries or conservation areas may not require reductions in total vegetation so much as increasing species diversity, or disturbing the environment to retain a colonizing flora.

Chemical control is less likely to be used for submerged weed management in rivers than in lentic systems because of the difficulties of maintaining a period of contact, sufficient for a phytotoxic effect, between the herbicide and plant tissues. Herbicides used in drainage and irrigation canals (e.g. acrolein, xylene) are not selective and so are not suitable for use in rivers where fauna and some vegetation are desired. Fewer herbicides are permitted to be applied to aquatic sites than terrestrial ones because of the sensitivity of the habitats and the concerns of residues reaching irrigation and potable water supplies.

Selective placement of the contact herbicide diquat has been achieved in rivers when a gel-like formulation with alginate is applied to submerged weed beds (Barrett & Murphy 1982; Fox & Murphy 1990), but this product is less satisfactory for large-scale management. Contact herbicides that do not damage underground organs tend to cause temporary reductions in vegetation. Prolonged (10–12 weeks) applications of fluridone in Florida, USA have resulted in the selective removal of the submerged weed *Hydrilla* from several kilometres of river (Haller *et al* 1990). Being a systemic herbicide such management should have a longer-lasting effect than contact herbicides. In the long term, appropriate herbicide applications to rivers, that reduce the biomass of a dominant species such as *Hydrilla*, are unlikely to have detrimental effects on the habitat. Systemic herbicides, such as glyphosate and dalapon, may be used extensively on emergent species. Repeated use here does have potential for permanent community changes although there is evidence that overall diversity is not reduced (Wade 1980).

Physical removal of weeds, while appearing to have an advantage over herbicides in removing nutrients from the system, is not very efficient in this objective because of the high proportion of water that must be moved. Whether weed removal is manual or mechanical, results tend to be short-lived with multiple harvests needed in a growing season, particularly of species that have basal meristems (e.g. grasses) or underground perennating organs. Removal of all plant material by dredging will have a more drastic short-term effect on the habitat if repeated regularly but, in the long term, communities may remain fairly stable (Wade 1978). It has been suggested that the characteristic submerged macrophyte communities of chalkstreams in the south of England have evolved as a result of their regular, physical management over several centuries (Westlake 1979).

The use of herbivorous fish (e.g. *Ctenopharyngodon idella*) for biological control of weeds in rivers has been limited because of the problems of confining the fish to reaches where management is needed. In tropical and subtropical areas, insects have been used with varying degrees of efficacy to control floating plants on rivers. Two notable successes have been *Agasicles* beetles released on alligatorweed (*Alternanthera philoxeroides*) in the south-eastern USA (Maddox *et al* 1971) and *Cyrtobagous* weevils on *Salvinia molesta* in Australia (Room *et al* 1981). Grazing of domestic animals on emergent vegetation commonly occurs, but over prolonged periods this may select for unpalatable species and trampling can accelerate bank erosion.

Macrophyte management by environmental manipulation has been proposed in the planting of trees on river banks to shade narrow river channels (Dawson & Haslam 1983). Regulation of water levels may reduce macrophyte growth but that is not usually the objective of such manipulation in rivers.

Pollution

Macrophytes are not used as commonly as invertebrate species as indicators of pollution in rivers. While reductions in the abundance of macrophytes that occasionally occur downstream of point sources of pollution can sometimes be ascribed to heavy metal or detergent toxicity, increased turbidity and smothering of plants by silt and mining spoils are more common causes (Hynes 1960). General increases in turbidity may result from catchment deforestation, poor agricultural practices and intensive boat usage. Macrophytes may also be shaded out by dense populations of sewage fungus, filamentous or epiphytic algae downstream of organic pollution.

Acidification of poorly buffered rivers either from point sources, such as coal mines, or acid rain deposition is likely to result in changes in the species composition of vegetation rather than rapid changes in productivity (Stokes 1986; Weatherley & Ormerod 1989). Heat pollution and changes in salinity also alter species composition (e.g. *Egeria densa* in British power-station outfall; Sculthorpe 1967).

Eutrophication is unlikely to increase macrophyte productivity in many lowland rivers that already have sufficient phosphorus and nitrogen for optimum growth rates. Nutrient inputs may increase macrophyte productivity in oligotrophic rivers but evidence of such changes is not always easy to find as it depends upon the availability of pre-enrichment data. Eutrophic conditions are usually associated with a loss of macrophyte diversity as highly competitive species predominate (e.g. *Potamogeton pectinatus*).

River engineering

Engineering projects, such as the building of bridges or channel straightening, will temporarily reduce macrophytes in the disturbed and silted areas. The eventual return of the original flora will depend upon any permanent changes in depth, substratum and flow, and upon species recruitment from upstream.

A change from riverine to lacustrine flora will occur immediately upstream of dam construction. As upstream water movement is reduced, floating species often dominate in hot climates, frequently disrupting the purpose of the dam (e.g. *Salvinia molesta* behind the Kariba irrigation dam in Zimbabwe; Sculthorpe 1967). The regulation of flow downstream may have an effect on the composition of submerged species and can reduce the width of the emergent fringe.

Introduction of exotic macrophytes

In the tropics the introduction of exotic macrophytes has probably been the human activity most responsible for severe weed problems in natural water bodies. Released from the constraints of natural competitors and herbivores that kept them in balance in their native countries, exotic species can rapidly dominate the flora, excluding native plants and reducing diversity. Floating (e.g. *Eichhornia crassipes*, *Salvinia molesta*) and canopy-forming submerged species (e.g. *Hydrilla verticillata*) have created problems in rivers throughout the world (Sculthorpe 1967; Haller 1978). In temperate climates exotic species have more often become established in the slower flows of lakes and canals (e.g. *Elodea canadensis* in British canals; Murphy et al 1982; *Crassula helmsii* potentially spreading from lakes to streams; Dawson 1983). However, some serious river infestations have occurred (e.g. *Myriophyllum spicatum* in Tennessee Valley, USA; Bates *et al* 1985).

Efforts to control existing exotic weed problems already strain the river management resources in many tropical countries. The prevention of spread between water bodies and prevention of the deliberate or accidental importation of new species must become an important objective of regulators who are concerned about river management.

ACKNOWLEDGEMENTS

The author is most grateful to Bill Haller and George Bowes for their helpful comments. Published with the approval of the Department of Agronomy and Florida Agricultural Experiment Station as Journal Series Number R-01196.

REFERENCES

Agami M, Reddy KR. (1989) Inter-relationships between *Salvinia rotundifolia* and *Spirodela polyrhiza* at various interaction stages. *Journal of Aquatic Plant Management* 27: 96–102. [10.2]

Bain JT, Procter CF. (1980) The requirement of aquatic bryophytes for free CO_2 as an inorganic carbon source: some experimental evidence. *New Phytologist* 86: 393–400. [10.2]

Barko JW, Smart RM. (1981a) Comparative influences of light and temperature on the growth and metabolism of selected submersed freshwater macrophytes. *Ecological Monographs* **51**: 219–35. [10.2]

Barko JW, Smart RM. (1981b) Sediment-based nutrition of submersed macrophytes. *Aquatic Botany* **10**: 339–52. [10.2]

Barko JW, Smart RM. (1986) Sediment-related mechanisms or growth limitation in submerged macrophytes. *Ecology* **67**: 1328–40. [10.2]

Barrett PRF, Murphy KJ. (1982) The use of diquat-alginate for weed control in flowing waters. *Proceedings of European Weed Research Society, 6th Symposium on Aquatic Weeds, 1982*, 200–8. European Weed Research Society Wageningen. [10.4]

Basiouny FM, Haller WT, Garrard LA. (1978) Survival of hydrilla (*Hydrilla verticillata*) plants and propagules after removal from the aquatic habitat. *Weed Science* **26**: 502–4. [10.2]

Bates AL, Burns ER, Webb DH. (1985) Eurasian water-milfoil (*Myriophyllum spicatum*) in the Tennessee Valley: an update on biology and control. *Proceedings of 1st International Symposium on Watermilfoil (Myriophyllum spicatum) and Related Halagoraceae spp.*, pp 104–15. Aquatic Plant Management Society, Washington, DC. [10.4]

Boston HL, Adams MS, Madsen JD. (1989) Photosynthetic strategies and productivity in aquatic systems. *Aquatic Botany* **34**: 27–57. [10.1]

Bowes G, Salvucci ME. (1989) Plasticity in the photosynthetic carbon metabolism of submersed aquatic macrophytes. *Aquatic Botany* **34**: 232–66. [10.1]

Bowes G, Van TK, Garrard LA, Haller WT. (1977) Adaptation to low light levels by hydrilla. *Journal of Aquatic Plant Management* **15**: 32–5. [10.2]

Brooker MP, Morris DL, Hemsworth RJ. (1977) Mass mortalities of adult salmon (*Salmo salar*) in the river Wye, 1976. *Journal of Applied Ecology* **14**: 409–17. [10.4]

Brown AWA, Deom JO. (1973) Summary: health aspects of man-made lakes. In: Ackerman W, White GR, Worthington GB (eds) *Man-made Lakes: Their Problems and Environmental Effects*, pp 755–64. William Byrd Press, Richmond, Virginia. [10.4]

Butcher RW. (1933) Studies on the ecology of rivers. I. On the distribution of macrophytic vegetation in the rivers of Britain. *Journal of Ecology* **21**: 58–91. [10.2, 10.3]

Caffrey JM. (1990) Problems relating to the management of *Potamogeton pectinatus* L. in Irish rivers. *Proceedings of the European Weed Research Society, 8th Symposium on Aquatic Weeds, 1990*, pp 61–8. European Weed Research Society, Wageningen. [10.4]

Canfield DE Jr, Hoyer MV. (1988) Influence of nutrient enrichment and light availability on the abundance of aquatic macrophytes in Florida streams. *Canadian Journal of Fisheries and Aquatic Sciences* **45**: 1467–72. [10.2]

Charudattan R. (1990) Biological control of aquatic weeds by means of fungi. In: Pieterse AH, Murphy KJ (eds) *Aquatic Weeds*, pp 186–201. Oxford University Press, Oxford. [10.2]

Cook CDK. (1974) *Water Plants of the World*. Dr W. Junk b.v., Publishers, The Hague. [10.1]

Dawson FH. (1976) The annual production of the aquatic macrophyte *Ranunculus penicillatus* var. *calcareus* (R.W. Butcher) CDK Cook. *Aquatic Botany* **2**: 51–73. [10.3]

Dawson FH. (1978) The seasonal effects of aquatic plant growth on the flow of water in a stream. *Proceedings of the European Weed Research Society, 5th Symposium on Aquatic Weeds, 1978*, pp 71–8. [10.4]

Dawson FH. (1980a) Flowering of *Ranunculus penicillatus* (Dun) Bab. var. *calcareus* (R.W. Butcher) C.D.K. Cook in the River Piddle (Dorset, England). *Aquatic Botany* **9**: 145–57. [10.3]

Dawson FH. (1980b) The origin, composition and downstream transport of plant material in a small chalk stream. *Freshwater Biology* **10**: 419–35. [10.3]

Dawson FH. (1983) *Crassula helmsii* (T. Kirk) Cockayne: is it an aggressive alien aquatic plant in Britain? *Biological Conservation* **42**: 247–72. [10.4]

Dawson FH, Haslam SM. (1983) The management of river vegetation with particular reference to shading effects of marginal vegetation. *Landscape Planning* **10**: 147–69. [10.4]

Dawson FH, Castellano E, Ladle M. (1978) Concept of species succession in relation to river vegetation and management. *Verhandlungen der Internationalen Vereinigung für Theoretische und Angewandte Limnologie* **20**: 1429–34. [10.2]

Denny P. (1980) Solute movement in submerged angiosperms. *Biological Reviews* **55**: 65–92. [10.2]

Fox AM, Murphy KJ. (1990) The efficacy and ecological impacts of herbicides and cutting regimes on the submerged plant communities of four British rivers. I. A comparison of management efficacies. *Journal of Applied Ecology* **27**: 520–40. [10.4]

Getsinger KD, Green WR, Westerdahl HE. (1990) Characterization of water movement in submersed plant stands. *Miscellaneous Paper A-90-5*. US Army Engineer Waterways Experiment Station, Vicksburg, Mississippi. [10.2]

Grace JB, Wetzel RG. (1981) Habitat partitioning and competitive displacement in cattails (*Typha*): experimental field studies. *The American Naturalist* **118**: 463–74. [10.2]

Grime JP. (1979) *Plant Strategies and Vegetation Processes*. John Wiley & Sons, New York. [10.2]

Haller WT. (1978) Hydrilla: a new and rapidly spreading aquatic weed problem. *Extension Circular S-245*. Agricultural Experiment Station/IFAS/Gainesville,

Florida/Agronomy Department, University of Florida. [10.4]

Haller WT, Sutton DL. (1975) Community structure and competition between hydrilla and vallisneria. *Hyacinth Control Journal* 13: 48–50. [10.2]

Haller WT, Sutton DL, Barlowe WC. (1974) Effects of salinity on growth of several aquatic macrophytes. *Ecology* 55: 891–4. [10.2]

Haller WT, Fox AM, Shilling DG. (1990) Hydrilla control program in the upper St. Johns River, Florida, USA. *Proceedings of the European Weed Research Society, 8th Symposium on Aquatic Weeds, 1990*, pp 111–16. European Weed Research Society, Wageningen. [10.4]

Ham SF, Wright JF, Berrie AD. (1981) Growth and recession of aquatic macrophytes on an unshaded section of the River Lambourn, England, from 1971 to 1976. *Freshwater Biology* 11: 381–90. [10.2]

Haslam SM. (1978) *River Plants*. Cambridge University Press, Cambridge. [10.1, 10.2]

Haslam SM. (1987) *River Plants of Western Europe*. Cambridge University Press, Cambridge. [10.2]

Hiley PD, Wright JF, Berrie AD. (1981) A new sampler for stream benthos, epiphytic macrofauna and aquatic macrophytes. *Freshwater Biology* 11: 79–85. [10.3]

Holmes NTH. (1983) Typing British rivers according to their flora. *Focus on Nature Conservation*, 4. Nature Conservancy Council, Peterborough. [10.2]

Hutchinson GE. (1975) *A Treatise on Limnology, Volume III, Limnological Botany*. John Wiley & Sons, New York. [10.1]

Hynes HBN. (1960) *The Biology of Polluted Waters*. Liverpool University Press, Liverpool. [10.4]

Hynes HBN. (1970) *The Ecology of Running Waters*. University of Liverpool Press, Liverpool. [10.2]

Ladle M, Casey H. (1971) Growth and nutrient relationships of *Ranunculus penicillatus* var. *calcareus* in a small chalk stream. *Proceedings of the European Weed Research Council, 3rd Symposium on Aquatic Weeds, 1971*, pp 53–64. European Weed Research Society, Wageningen. [10.2]

Leather GR, Einhellig FA. (1986) Bioassays in the study of allelopathy. In: Putnam AR, Tang C (eds) *The Science of Allelopathy*, pp 133–45. John Wiley & Sons, New York. [10.2]

Losee RF, Wetzel RG. (1988) Water movement within submersed littoral vegetation. *Verhandlungen der Internationalen Vereinigung für Theoretische und Angewandte Limnologie* 23: 62–6. [10.2]

Maberly SC. (1985) Photosynthesis by *Fontinalis antipyretica*. II. Assessment of environmental factors limiting photosynthesis and production. *New Phytologist* 100: 141–55. [10.2]

McCreary NJ, Carpenter SR, Chaney JE. (1983) Coexistence and interference in two submersed freshwater perennial plants. *Oecologia* 59: 393–6. [10.2]

Maceina MJ, Shireman JV, Langeland KA, Canfield DE Jr. (1984) Prediction of submersed plant biomass by use of a recording fathometer. *Journal of Aquatic Plant Management* 22: 35–8. [10.3]

Maddox DM, Andres LA, Hennessey RD, Blackburn RD, Spencer NR. (1971) Insects to control alligatorweed. An invader of aquatic ecosystems in the United States. *Bioscience* 21: 985–91. [10.4]

Madsen TV, Søndergaard M. (1983) The effects of current velocity on the photosynthesis of *Callitriche stagnalis* Scop. *Aquatic Botany* 15: 187–93. [10.2]

Miller AC, Beckett DC, Bacon EJ. (1989) The habitat value of aquatic macrophytes for macroinvertebrates: benthic studies in Eau Galle Reservoir, Wisconsin. *Proceedings of the 23rd Annual Meeting, Aquatic Plant Control Research Program Miscellaneous Paper A-89-1*, pp 190–201. US Army Engineer Waterways Experiment Station, Vicksburg, Mississippi. [10.2]

Murphy KJ. (1988) Aquatic weed problems and their management: a review. I. The worldwide scale of the aquatic weed problem. *Crop Protection* 7: 232–48. [10.4]

Murphy KJ, Eaton JW, Hyde TM. (1982) The management of aquatic plants in navigable canal systems used for amenity and recreation. *Proceedings of the European Weed Research Society, 6th Symposium on Aquatic Weeds, 1982*, pp 141–51. European Weed Research Society, Wageningen. [10.4]

Murphy KJ, Rørslett B, Springuel I. (1990) Strategy analysis of submerged lake macrophyte communities: an international example. *Aquatic Botany* 36: 303–23. [10.2]

Odum HT. (1956) Primary production in flowing waters. *Limnology and Oceanography* 1: 102–17. [10.3]

Odum HT. (1957) Trophic structure and productivity of Silver Springs, Florida. *Ecological Monographs* 27: 55–112. [10.3]

Pearsall WH. (1920) The aquatic vegetation of the English lakes. *Journal of Ecology* 8: 164–201. [10.2]

Pitlo RH, Dawson FH. (1990) Flow resistance by aquatic weeds. In: Pieterse AH, Murphy KJ (eds) *Aquatic Weeds*, pp 74–84. Oxford University Press, Oxford. [10.2]

Rao AS. (1988) Evapotranspiration rates of *Eichhornia crassipes* (Mart.) Solms, *Salvinia molesta* DS Mitchell and *Nymphaea lotus* (L.) Willd. Linn. in a humid tropical climate. *Aquatic Botany* 30: 215–22. [10.4]

Room PM, Harley KLS, Forno IW, Sands DPA. (1981) Successful biological control of the floating weed salvinia. *Nature* 294: 78–80. [10.4]

Sand-Jensen K, Borum J. (1984) Epiphyte shading and its effects on photosynthesis and diel metabolism of *Lobelia dortmanna* L. during the spring bloom in a Danish lake. *Aquatic Botany* 20: 109–19. [10.2]

Schmitz DC, Nelson BV, Nall L, Schardt JD. (1991) Exotic aquatic plants in Florida: a historical perspective and review of the present aquatic plant regulation program. In: Center TD, Doren RF, Hofstetter RL, Myers RL, Whiteaker LD (eds) *Proceedings of the Symposium on Exotic Pest Plants, Miami, Florida, 1988*, pp 303–26. US Dept. of Interior, National Park Service Natural Resources Publication Office, Denver, Colorado. [10.2]

Sculthorpe CD. (1967) *The Biology of Aquatic Vascular Plants*. Edward Arnold, London. [10.1, 10.2, 10.4]

Sliger WA, Henson JW, Shadden RC. (1990) A quantitative sampler for biomass estimates of aquatic macrophytes. *Journal of Aquatic Plant Management* **28**: 100–2. [10.3]

Soszka GJ. (1975) The invertebrates on submersed macrophytes in three Masurian lakes. *Ekologia Polska* **23**: 371–91. [10.2]

Spence DHN. (1967) Factors controlling the distribution of freshwater macrophytes with particular reference to the Lochs of Scotland. *Journal of Ecology* **55**: 147–70. [10.2]

Spence DHN, Maberly SC. (1985) Occurrence and ecological importance of HCO_3^- use among aquatic higher plants. In: Lucas WJ, Berry JA (eds) *Inorganic Carbon Uptake by Aquatic Photosynthetic Organisms*, pp 125–43. American Society of Plant Physiologists, Rockville, Maryland. [10.2]

Spence DHN, Bartley MR, Child R. (1987) Photomorphogenic processes in freshwater angiosperms. In: Crawford RMM (ed.) *Plant Life in Aquatic and Amphibious Habitats*, pp 153–66. Blackwell Scientific Publications, Oxford. [10.1]

Stokes PM. (1986) Ecological effects of acidification on primary producers in aquatic systems. *Water, Air and Soil Pollution* **30**: 421–38. [10.4]

Van TK, Haller WT, Bowes G. (1978) Some aspects of the competitive biology of *Hydrilla*. *Proceedings of the European Weed Research Society, 5th Symposium on Aquatic Weeds, 1978*, pp 117–26. European Weed Research Society, Wageningen. [10.2]

Wade PM. (1978) The effect of mechanical excavators on the drainage channel habitat. *Proceedings of the European Weed Research Society, 5th Symposium on Aquatic Weeds, 1978*, pp 333–42. European Weed Research Society, Wageningen. [10.4]

Wade PM. (1980) The effects of channel maintenance on the aquatic macrophytes of the drainage channels of the Monmouthshire Levels, South Wales,

1840–1976. *Aquatic Botany* **8**: 307–22. [10.4]

Weatherley NS, Ormerod SJ. (1989) Modelling ecological impacts of the acidification of Welsh streams: temporal changes in the occurrence of macroflora and macroinvertebrates. *Hydrobiologia* **185**: 163–74. [10.4]

Westlake DF. (1963) Comparisons of plant productivity. *Biological Reviews* **38**: 385–425. [10.3]

Westlake DF. (1965) Some basic data for investigations of the productivity of aquatic macrophytes. *Memorie dell'Istituto Italiano di Idrobiologia Dott Marco de Marchi* **18** (Suppl.): 229–48. [10.3]

Westlake DF. (1967) Some effects of low-velocity currents on the metabolism of aquatic macrophytes. *Journal of Experimental Botany* **18**: 187–205. [10.2]

Westlake DF. (1975) Macrophytes. In: Whitton BA (ed.) *River Ecology*, pp 106–28. Blackwell Scientific Publications, Oxford. [10.2]

Westlake DF. (1979) The ecology of chalk streams. *Watsonia* **12**: 387. [10.4]

Westlake DF. (1981) Temporal changes in aquatic macrophytes and their environment. In: Hoestlandt H (ed.) *Symposium Dynamique des Populations et Qualité des Eaux*. Acts Symp. Institute d'Ecologie du Bassin de la Somme, Chantilly, 5–8 November 1979, pp 111–38. [10.2]

Westlake DF, Casey H, Dawson FH, Ladle M, Mann RHK, Marker AFH. (1972) The chalk-stream ecosystem. In: Kajak Z, Hillbricht-Ilkowska A (eds) *Proceedings of the IBP-UNESCO Symposium on Productivity Problems of Freshwaters, Kazimierz, Dolny, 1970*, pp 616–35. Polish Scientific Publishers, Warszawa-Krakow, Poland. [10.2]

Wetzel RG. (1969) Factors influencing photosynthesis and excretion of dissolved organic matter by aquatic macrophytes in hard-water lakes. *Verhandlungen der Internationalen Vereinigung für Theoretische und Angewandte Limnologie* **17**: 72–85. [10.2]

Wetzel RG. (1983) *Limnology*, 2nd edn. Saunders College Publishing, Philadelphia. [10.3]

Willis JC. (1914) On the lack of adaptation in the Tristichaceae and Podostemaceae. *Proceedings of the Royal Society, London, Series B* **87**: 532–50. [10.2]

Wright JF, Hiley PD, Ham SF, Berrie AD. (1981) Comparison of three mapping procedures developed for river macrophytes. *Freshwater Biology* **11**: 369–79. [10.3]

11: Invertebrates

K.W.CUMMINS

11.1 INTRODUCTION

Historically the invertebrates as a group have received major attention in the study of running water ecosystems. The reasons for this are clear. The macroinvertebrates (approximately >0.5 mm; Cummins 1975a) stand as the link between algae and micro-organisms, which serve as their primary food resources, and the fish (and other vertebrates), for which they are prey. Macro-invertebrates are also intermediate in turnover rates. They have replacement times greater than the very small, rapid turnover micro-organisms, and faster replacement rates than the generally larger fish with their characteristic slower turn-over times. Thus, the macroinvertebrates are large enough to be observed with the unaided eye, abundant enough to be readily collected, and have lifecycles of suitable length (several weeks to 1 or 2 years) for short-term seasonal or annual field investigations. With these credentials, it is not surprising that the macroinvertebrates have been widely used to assess the prey base available to support fish populations (e.g. Waters 1988) and to evaluate water quality (e.g. Hilsenhoff 1987; Yount & Niemi 1990; Karr 1991).

Aside from the immense number of taxonomic studies on running-water invertebrates, especially on the aquatic insects (e.g. see review in Merritt & Cummins 1984), research on lotic invertebrates over the last century has focused on the patterns of distribution and abundance, and the environmental factors controlling these (e.g. Hynes 1970; Macan & Worthington 1951, Macan 1962, 1974). The study of the physical, chemical and biological (especially riparian vegetation, e.g. Corkum 1989) factors controlling the distribution and abundance

of lotic invertebrates has evolved to the use of clustering techniques or principal component analyses to identify the key parameters involved (e.g. Moss *et al* 1987).

Since the mid-1970s, emphasis has shifted from the primary focus on structure, towards studies on process and function (e.g. Cummins 1974, 1988; Fig. 11.1). Examples of the areas that have received particular attention are: rates of biomass production, life-history strategies, resource partitioning, and population parameters such as competition and predator–prey interactions. (e.g. Cummins 1973, 1974, 1975a, 1988; Hynes 1975; Macan 1977; Townsend 1980; Cummins *et al* 1984; Resh & Rosenberg 1984; Minshall *et al* 1985). In these and many other areas, there has been extensive debate concerning appropriate methodology for estimating the dimensions of the various parameters involved; an example being the debate about the most suitable technique for quantifying gross and net invertebrate production (e.g. Waters 1979a, 1979b; Benke 1984; Iverson & Dall 1989).

The microinvertebrates, or meiofauna (generally <0.5 mm in size), of running waters have received much less attention than the macroinvertebrates, for most of the same reasons cited above relative to micro-organisms. In recent years, some of the long-standing interest in the benthic meiofauna of marine systems has shifted to running waters (e.g. Ward & Voelz 1990). In particular, studies are now being conducted on benthic microcrustacea in lotic systems. Of course, the two groups overlap in size, and the distinction as to what is studied is largely a matter of the mesh size selected for sorting (see below).

The review presented in this chapter is not

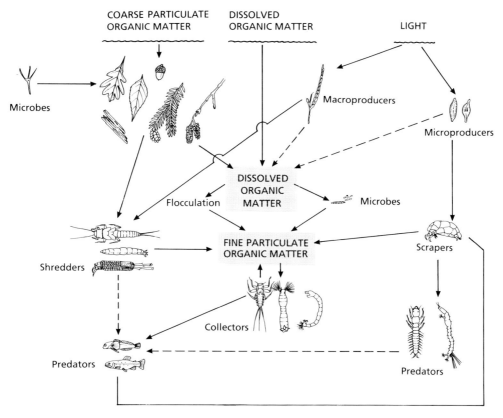

Fig. 11.1 A conceptual model of stream ecosystem structure with emphasis on invertebrate functional groups (shredders, collectors, scrapers and predators) and their food resource pools (coarse and fine particulate organic matter (CPOM and FPOM), periphyton (attached, non-filamentous algae and associated material) and prey) (after Cummins 1974, with permission).

exhaustive. Rather it is focused on selected aspects of running-water invertebrates, bearing on their roles in lotic ecosystem functioning and on their use in monitoring and predicting changing conditions in streams and rivers.

11.2 TAXONOMY, THE OPEN-ENDED AGENDA

The major accomplishments in defining the lotic invertebrate fauna have been in Europe and North America (e.g. Thienemann 1950; Edmondson 1959; Pennak 1978; Merritt & Cummins 1984). The task stands at a more complete state for Europe than North America because holarctic groups had much more restricted refugia in Europe

during the last glacial period, although the longer period of anthropogenic environmental disturbance and related extinctions have also been significant (Hynes 1960, 1970). In fact, because of the present global scale of environmental alteration of running waters, many species are being, and certainly will be, lost before they are known to science. Running-water invertebrates are an important component of the general concern over loss of global biodiversity (e.g. Wilson 1988). Taxonomic knowledge is clearly related to accessibility of environments: by the time the lotic invertebrates of a geographical region are reasonably well known, the majority of the habitats have been altered, and undetermined numbers of species lost. For example, in Europe very little 'virgin' habitat remains. In North America, the major ecoregions of the eastern (eastern deciduous forest) and central (tall and short grass prairies) portions have been severely altered, save for a few small virgin preserves, and only 10% of the virgin forests of the Pacific Northwest remain. The tropics, subtropics and major island systems are obviously on the same course, currently representing the areas of most accelerated loss of biodiversity (e.g. Wilson 1988). Therefore, even if we never know what the taxonomic identity of the world's lotic invertebrate fauna *is*, it is certain we shall never know what it *was*.

The level of taxonomic uncertainty for lotic invertebrates, notably the holometabolous insects, is exacerbated by the fact that the aquatic immature stages are often the most difficult to distinguish (a usual feature of neotenous similarity) and, classically, the systematics are based on the reproductive morphology of adult males that are rarely encountered during aquatic sampling. Often, special efforts are required to sample the adult males, because of their brief longevity (usually days to weeks) and reclusive (e.g. night active) or inaccessible (e.g. swarming at great height) terrestrial habits (e.g. Wessenberg-Lund 1943; Merritt & Cummins 1984). The necessary light trap and sweep collections in the riparian zone along running-water habitats need to be adopted as standard procedure in all lotic invertebrate studies concerned with species diversity (Merritt *et al* 1984).

11.3 SAMPLING AND DATA ANALYSIS IN LOTIC HABITATS

These activities have been referred to as the 'four S's' of lotic invertebrate ecology: Sampling, Sorting, Systematics and Statistics (Cummins 1966, 1972, 1975a). The standard reference on sampling design and statistical evaluation of the resulting freshwater benthic samples is Elliott (1977). Since the 1977 revision of this excellent treatment, there is little to add save the debate about 'pseudoreplication' (Hurlbert 1984), that is the failure to adequately replicate observations or experiments in space and/or time (see also Chapter 13). Invertebrate ecologists have long been aware that true replication in running-water studies is often difficult or impossible (e.g. the problem of a replicate for the Amazon River). An example is the problem of comparing the invertebrate standing stocks in 'paired' streams, in which each member of the 'pair' needs to be replicated as well as each time period, such as a season or year. Of course, there is the problem as to whether the assumptions of similarity of the 'pairs' are valid. Related problems arise concerning the need for replicate sampling within a single example of a given habitat type, such as a riffle, that does not serve as appropriate replication to generalize about all riffles in the stream being sampled. As described in detail in Elliott (1977), the number of samples required to arrive at a selected level of precision (defined as variation as a percentage of the mean) can be determined by the analysis of a preliminary set of data (see also Merritt & Cummins 1984; Chapter 13).

Where and how often to sample also remain perennial problems. It is clear from essentially all studies of benthic lotic invertebrates that their distributions are clumped, fitting a negative binomial distribution, and sampling design must account for these habitat-related patchy distributions (e.g. Macan 1958; Cummins 1966; Elliott 1977; Merritt *et al* 1984). Such non-continuous distributions have led many ecologists to employ stratified random sampling, which starts with the assumption that distributions are fundamentally different between major habitats such as riffles or pools, or to sample using a transect method that runs along known gradients from bank to channel thalweg (path of maximum current). The type of sampler is clearly dictated by the nature of the substrate, depth of the water, and current velocity (e.g. Merritt *et al* 1984). For example, larger, deeper rivers have been sampled with scuba gear (e.g. Minshall *et al* 1983a–b, 1985). The depth into the substrate to which the sample is taken should correspond generally to the depth of oxygenation, given the obligate aerobe status, at least in the long term, of the invertebrates. In some systems with good interstitial flow, organisms can be found to significant depth and into the bank as well (e.g. Williams 1989). It seems clear that the extent of this hyporeic zone should be evaluated in each system, and the sampling strategy adjusted accordingly. If half-metre sections of clean wood dowling are driven into the stream bottom in a transect across the channel with the position of the sediment water interface marked, removal of the rods after about a week (50–100 degree-days) will easily reveal the depth of the anoxic zone by the line of discoloration. This type of information can be used to establish approximate sampling depths in a given running-water reach.

Sorting of samples has been a major, time honoured and immensely time-consuming task in the study of running-water invertebrates. A variety of methods, including differential density solutions, various agitation and settling procedures, and specific stains combined with special wavelength lighting, all have been used. Unfortunately, even the best of these is only partially successful in shortening the task and is never universally applicable to all types of lotic samples. Thus, for the forseeable future, there will continue to be armies of sorters. The mesh size used to prepare the sediment and debris samples containing the invertebrates, in the field and/or the laboratory, strongly influences both the dimensions of the task and the accuracy of the estimates. The larger the sample and the finer the mesh the more likely it will be necessary to utilize some subsampling method (e.g. Merritt *et al* 1984). The still prevalent use of mesh sizes in the 0.5–1.0-mm range ensures that the early life stages of most invertebrates, as well as the majority of the meiofauna, will be lost. Historically, this has greatly reduced the number of midge larvae in stream and river benthic samples. Or, as the attitude could be summarized, 'If it's a group you despise, use a coarser mesh size'! Most macroinvertebrate studies now use a mesh size in the range of 250 μm which retains most insect first instars. The use of a 125-μm mesh retains essentially all insect larvae and a 50-μm mesh is reasonably efficient at retaining the meiofauna.

11.4 IN SEARCH OF PRODUCTION ESTIMATES

The methods used to calculate production of invertebrates vary depending on the portion of the community over which the estimate is intended to extend and the resolution of the data available relative to size (age) classes and seasonal patterns (e.g. Waters 1977, 1979a, 1979b; Benke 1984; see also Chapter 17).

The usual objective in studies of lotic invertebrate production is to estimate the elaboration of new biomass per square metre over some specified time span, normally a year. The calculation of total, or gross, production of the aquatic growth stage is independent of the fate of the biomass, for example loss to predation or emergence. Net production of a generation can be measured as the biomass of surviving adults. The field techniques require data on invertebrate size-specific biomass over the growth period. Most frequently the biomass is calculated by using length versus weight regressions (e.g. Meyer 1989) or by volume displacement coupled with the assumption of a specific gravity slightly greater than 1 (e.g. Stanford 1972) to convert size or volume to dry (or ash-free dry) mass; this is the size-frequency method. Although direct mass determinations should be made on dried (or ashed) fresh specimens, as a last resort some workers have weighed preserved material. These results will be highly variable owing to the amount of leaching of soluble fractions from the specimens which is dependent on the type of preservative, length of exposure, and thickness and extent of cuticle cover of the invertebrates.

Many estimates of lotic invertebrate production have taken a more limited taxonomic approach (e.g. Huryn & Wallace 1986; Benke & Parsons 1990; Huryn 1990; Lugthart *et al* 1991) as opposed to entire invertebrate community assessments (e.g. Benke *et al* 1984; Smock *et al* 1985). The size-frequency method of production estimation can be used at the level of a single species population, or some combination of species ranging up to the entire community. The method is very sensitive to differences in lifecycles. For example, including non-growing or diapausing stages in the time interval over which the calculations are applied yields an underestimate of production. Benke (1979) proposed that a cohort production interval (CPI) correction be made. That is, on an annual basis, the production estimate should be multiplied by 365 days/CPI so as to include only the growth period.

Because the growth of most lotic invertebrates

approaches a logarithmic function, individual instantaneous growth rate (IGR) can be calculated as the natural log of the final or maximum biomass minus the initial or minimum biomass divided by the time interval (Cummins & Merritt 1984). The relative growth rate (RGR), which is the change in biomass over a time interval divided by the median mass during that interval, multiplied by 100 gives the expression of growth as a percentage of bodyweight per time interval (e.g. Waldbauer 1968). Over short time spans, the RGR approximates the IGR very closely (e.g. Cummins *et al* 1973). The IGR calculated over an entire lifecycle approximates the ratio of annual or co-hort production divided by the mean annual or cohort standing stock biomass. This ratio, P/B, is an expression of turnover; that is, the rate at which biomass is replaced (the units of biomass produced per unit of standing stock biomass per unit area over the time span). A variety of studies have yielded P/B ratios of 3–6, with a mean of 5, for most univoltine species (e.g. Waters & Crawford 1973; Waters & Hokenstom 1980; Benke 1984). Bi- and multivoltine species would be expected to have ratios >6, and for those with very rapid turnover populations the ratio might be as high as 100 (Hauer & Benke 1991). Turnover rates of this magnitude would probably resolve the so-called 'Allen's paradox' which points to the difficulty of measuring an adequate invertebrate prey base to support lotic fish populations (Waters 1988).

The ratio of P/B or the turnover ratio (TR), which is often in the range of 4–5 but may be much higher in polyvoltine species, can be used to estimate the production (P) portion of the P/B relationship. A knowledge of average generation time of a given species allows an estimate of TR. For example, a typical univoltine species in the temperate zone would have a TR of 5. With an estimate of TR, a measurement of average standing stock biomass (B) can in turn yield a production estimate. The other most commonly used method of production estimation is to utilize independent laboratory (e.g. Cummins *et al* 1973) or field enclosure (e.g. Hauer & Benke 1991) measurements of average cohort instantaneous growth. That is, $G = \ln B_{\mathrm{f}}/B_{\mathrm{i}} \times t^{-1}$, where B_{i} and B_{f} are average initial and final biomasses of the cohort respectively, and t is the time interval. Multiplying this growth rate (G) by the numerical standing stock provides the production estimate for the species in question. Knowledge of generation times and measurements of G for selected species were used, together with laboratory data on ingestion and respiration rates, to arrive at the system level production values, summed by functional group (Fig. 11.1), that are given in Fig. 11.2 (where production = assimilation − respiration (CO_2) and assimilation = ingestion − egestion).

It is quite useful to express the growth rate of individuals, or the production of the population or community as a whole, in a form normalized for temperature, that is, degree-days (Andewartha & Birch 1954). This summation, which is the daily temperature (usually taken as the median temperature) cumulated over the time interval, allows comparisons or rates measured in lotic systems that differ in temperature. This allows factors regulating growth other than temperature to be exposed, or when the normalized rates are the same indicates the primary control by temperature.

The thermal regulation of growth and reproductive success, the latter being interpreted as 'fitness', in lotic invertebrate populations has been embodied in the Thermal Equilibrium Hypothesis as proposed by Vannote and Sweeney (1980). The basic feature of this hypothesis is that fecundity, which is proportional to female body size, will be maximized at the optimal temperature regime for the species along any gradient (latitude, altitude). This results in reduced fitness of populations living at temperatures above or below the optimum. In a test of the hypothesis by Rader and Ward (1990), mean body size and fecundity were predicted better than population density and biomass. Various sources of mortality and influences on growth other than temperature, such as food quality, are clearly also at work. For example, because of the critical importance of micro-organisms as a detrital component, and the direct relation between microbial growth and temperature, increases in temperature are likely to increase food quality for shredders and collectors that feed on particulate organic matter. This means that food quality would be, on average, greater at temperatures above the optimum for

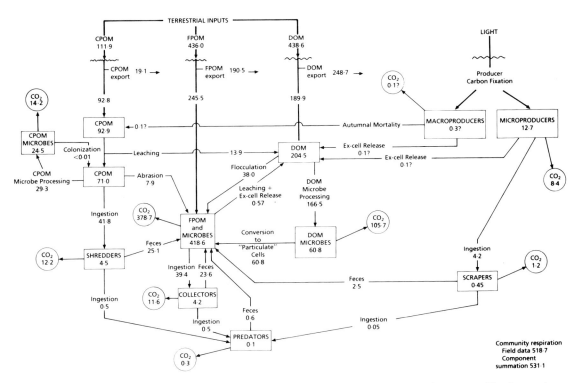

Fig. 11.2 The stream ecosystem model shown in Fig. 11.1 with actual measurements of state variables (invertebrate functional groups and their food resource pools), transfer functions, and carbon dioxide losses. Units are dry mass m^{-2} year^{-1} (after Boling *et al* 1975; Cummins *et al* 1973, 1981, 1983; Petersen *et al* 1989; KW Cummins & RC Petersen unpublished data).

many species. Further, the availability of high quality food has been shown to override the regulation of temperature on body size (e.g. Ward & Cummins 1979).

11.5 FUNCTIONAL ANALYSIS OF SPECIES ASSEMBLAGES

Habitats

There has been little substantive modification of the separation of running-water habitats into erosional and depositional, usually referred to as riffles and pools, since the classic paper of Moon (1939; see also Shelford 1914; Chapter 6). The subdivision of these two basic habitat types on geomorphic and hydrological bases has received significant attention (e.g. Davis & Barmuta 1989). Fast-flowing areas have been variously separated into rapids runs and glides, in addition to riffles,

which still stands as a general descriptor for all erosional habitats. In addition, pools as a general descriptor, backwaters and side-channels have been identified as specific depositional lotic habitats.

Habits

The adaptations of invertebrates to the unidirectional flow, modified by turbulence, of running water habitats has served as a major attraction to ecologists for some time (e.g. Shelford 1914; Wesenberg-Lund 1943; Hynes 1970; Macan 1974; Anderson & Wallace 1984). Morphological and behavioural adaptations of invertebrates to the hydraulic characteristics of their running-water environment serve as important examples in the comprehensive and delightful treatments by Vogel (1981, 1988).

The adaptations can be classified into three categories: (1) those involving position of the organism, such as locomotion, attachment (short or long term), and concealment (e.g. burrowing); (2) those associated with feeding (see functional feeding group classification below); and (3) those pertaining to reproduction (e.g. male egg 'protection' by adult male belostomatid hemipterans; e.g. Merritt & Cummins 1984).

Functional feeding groups

The technique of functional analysis of invertebrate feeding was first described 18 years ago (Cummins 1973) and has been modified in some details since then (e.g. Cummins 1974; Cummins & Klug 1979; Merritt & Cummins 1984; Cummins & Wilzbach 1985). The macroinvertebrate functional feeding group method is based on the association between a limited set of feeding adaptations found in freshwater invertebrates and their basic nutritional resource categories (Fig. 11.1; Cummins & Wilzbach 1985). These food resource categories are (Cummins 1974):

1 Detritus: coarse (CPOM) or fine (FPOM) particulate organic matter and the associated microbiota.
2 Periphyton: attached algae and associated entrained material.
3 Live macrophytes.
4 Prey.

The level of morphological and behavioural adaptation of the invertebrates that allows them to exploit these resource categories can be obligate or facultative (Cummins & Klug 1979). The obligate specialist forms are more readily displaced and the facultative generalists are more tolerant under conditions of disturbance. The presence and abundance of the various functional feeding groups, and the dominance of obligate or facultative representatives, is a direct reflection of the availability of the required food resources (both quantity and quality) and the condition of the related environmental parameters. An example here would be a change in the species and timing of the inputs of leaf litter to a headwater stream from the adjacent riparian zone (e.g. Cummins *et al* 1989).

The invertebrate functional groups are (see Fig. 11.1) (Merritt & Cummins 1984; Cummins & Wilzbach 1985):

1 Shredders feeding on CPOM (primarily litter of terrestrial origin from the riparian zone) or live macrophytes.
2 Collectors feeding on FPOM either by filtering from the water column (filtering collectors) or by 'mining' the sediments or browsing surface deposits (gathering collectors).
3 Scrapers feeding on periphyton.
4 Piercers feeding on macroalgae by piercing individual cells.
5 Predators feeding on prey.

The functional group analysis is made on a hierarchical basis of increasing levels of resolution. The first (lowest) level of resolution allows separation of live invertebrate collections in the field at an efficiency of 80–85%. The second level of resolution increases the efficiency by another 5–10%. When comparisons are to be made between sites on a regional basis, the level of resolution must be set so that all workers involved in the assessment can accomplish the task, with levels of greater resolution allowing groups with the appropriate expertise to produce more detailed analyses. Assignments to functional group of most of the North American genera of aquatic insects can be found in the ecological tables in Merritt and Cummins (1984). The feeding activity of shredders, primarily on plant litter of riparian origin, has a significant impact on the overall dynamics of organic matter fluxes in headwater lotic ecosystems. For example, at least 30% of the processing of coarse litter (CPOM) to finer particles (FPOM) can be attributed to shredder activity (e.g. Petersen & Cummins 1974; Cuffney *et al* 1990). In addition, the generation of FPOM can affect the growth of collectors that feed on the fine particles (e.g. Short & Maslin 1977) and shredder feeding can enhance the release of dissolved organic matter (DOM; Meyer & O'Hop 1983).

Invertebrate functional group analysis is sensitive to ecosystem properties of running waters. It is sensitive to both the normal pattern of geomorphic and concomitant biological changes that occur along river systems from headwaters to lower reaches (see Fig. 11.3) as well as to alterations in these patterns resulting from human impact. In headwaters, or extensively braided channels in the mid-reaches of river systems, the influence of riparian vegetation, through shading and litter inputs, is expressed in the general heterotrophic nature of such areas. Litter of terrestrial origin favours shredders, serving as their major food supply once it is appropriately conditioned by aquatic micro-organisms. As the litter is converted to finer organic particles (FPOM), it supports populations of collectors. The exclusion of light by riparian vegetation restricts in-stream primary production and consequently also limits the periphyton-grazing scrapers. The ratio of gross primary production (P) to community respiration (R) has been shown to be a useful index (P/R) of the balance between autotrophy and heterotrophy in running-water ecosystems (e.g. Vannote *et al* 1980; Table 11.1). The ratio of scrapers to shredders or to shredders + total collectors (SC/(SH + COL)) can serve as an index of P/R (Table 11.1).

In addition, the apportionment of coarse particulate (CPOM) to fine particulate (FPOM) organic matter shifts from relatively more to relatively less CPOM as the direct influence of the riparian vegetation is reduced — either as a natural consequence of increasing river size (i.e. reduced width of the riparian zone relative to river width) or because of human alterations (e.g. forest harvest or clearing for agriculture). For this relationship, the ratio of shredders to total collectors (SH/COL) is an index of the degree of the direct riparian influence as a litter source (Table 11.1). A further example can be found in the relative importance of FPOM in suspension (transport) compared with the depositional (benthic storage) component of total FPOM. If the input of FPOM is continuous, as it often is under disturbed conditions (e.g. organic effluents, severe erosion), then the nutritional resource base for filtering collectors (FC) is abundant and continuous. Usually this is in contrast to natural systems in which the input and movement of FPOM tracks the major pattern of the hydrograph, and the major reservoirs of particulate organic matter (POM) are in benthic storage for most of the year. This balance between storage and transport of POM can be reflected in the ratio of filtering to gathering collectors (FC/GC).

Table 11.1 Comparison of stream ecosystem parameters (P/R, CPOM/FPOM, storage POM/transport POM) and invertebrate functional feeding group ratios (SH/COL, SC/SH, SC/(SH + COL)) for the Kalamazoo River basin in southern Michigan, USA

Parameter	Stream order			
	1	2	3	5
Stream width (m)	1	5	10	45
Trophic status	Heterotrophic	Autotrophic	Heterotrophic	Autotrophic
P/R	0.47	1.13	0.90	1.23
Transport, CPOM/FPOM	0.022	0.016	0.019	0.022
Storage, CPOM/FPOM	0.36	0.11	0.15	0.10
POM, storage/transport	0.10	0.16	0.23	0.16
Mean annual invertebrates m$^{-2} \times 10^3$	19.6	15.0	63.6	41.7
SH/COL	0.22	0.003	0.002	0.001
FC/GC	0.67	0.42	0.45	1.50
SC/SH	0.18	12.23	3.99	16.91
SC/(SH + COL)	0.08	0.24	0.11	0.05

P, gross annual primary production; R, annual community respiration; POM, particulate organic matter; CPOM, coarse POM; FPOM, fine POM; SH, shredders; COL, total collectors; FC, filtering COL; GC, gathering COL. All invertebrate data used were means of fall–winter and spring–summer densities m^{-2} of individuals >0.5 mm (After Cummins *et al* 1981).

In the example shown in Table 11.1 for the Kalamazoo River Basin in southern Michigan, USA, the functional group ratios reflect the auto-trophic/heterotrophic nature of the four stream orders that were studied. The sites on orders 1 and 3 were heavily wooded; order 1 was canopy closed and order 3 was nearly so. Order 2 had been altered by timber removal, and the replace-ment grass and shrub riparian zone provided re-duced shading and litter inputs (Cummins *et al* 1981). This is reflected by the ecosystem ratios, *P/R* and CPOM/FPOM. The ratio of SC to SH (or SC/(SH + COL)) follows the autotrophic/hetero-trophic status as measured by *P/R*. The reduced ratio of SC/SH (or SC/(SH + COL)) at the order 5 site, despite the maximum *P/R* value, is an indi-cation of the dominance of vascular hydrophyte and filamentous macroalgal primary producers which are not suitable food resources for scrapers. However, the sloughed tissue and cells from these macrophytes probably yield significant FPOM of high quality (Cummins & Klug 1979) supporting large populations of filtering collectors as indi-cated by the FC/FG ratio in Table 11.1.

A generalized model of the shifts in the relative abundances of invertebrate functional groups along a river tributary system from headwaters to mouth as predicted by the river continuum concept (RCC; Vannote *et al* 1980; Minshall *et al* 1983b, 1985) is given in Fig. 11.3. The general pattern reflects: (1) the importance of litter inputs, normally maximized in the headwaters, influ-encing the relative density of shredders; (2) in-creases in scrapers where light and nutrients favour increased microphyte production, nor-mally in the wide, shallow mid-reaches; (3) the link between the abundance of collectors and FPOM, either in the headwaters related to litter processing, or in the lower reaches as the result of import from upstream tributaries and floodplain capture; and (4) the fairly constant relative abun-dance (approximately 10%) of predators in all reaches.

Microbial–invertebrate relationships

A major relationship between the invertebrates and the micro-organisms that co-occur with them in running-water environments concerns the general area of feeding biology. For example, the detrital feeding functional groups (shredders and collectors) have resident microflora, primarily in the hind-gut (Klug & Kotarski 1980), able to digest resistant plant compounds such as cellulose and lignin derivatives (e.g. Martin *et al* 1980, 1981a, 1981b; Lawson & Klug 1989). Regular resident microbial components in most shredders and collector hind-guts are large filamentous spore-forming bacteria. These are notably absent in the algal-feeding scrapers. Absorption occurs through the hind-gut wall of organic molecules from resident microbes that are useful to the invert-ebrate (Lawson & Klug 1989), and is enhanced by the heavy concentration of mitochondria in the area. However, materials in the hind-gut resulting from digestion by resident microbes also can be refluxed forward into the mid-gut, the normal site for absorption.

The invertebrate shredder–hyphomycete fungus relationship in streams is an example of a specific invertebrate–microbial association (see Chapter 15). Feeding by shredders can be shown to be correlated with the presence of the fungi on the leaf substrates, including selection over the presence of bacteria. Further, the key biochemical components involved in the selection have been shown to be specific 16- and 18-carbon 2 and 4 polyunsaturated lipids (Hanson *et al* 1983, 1985; Cargill *et al* 1985a, 1985b).

Drift

Invertebrate drift remains one of the most in-triguing phenomena in running waters (e.g. Waters 1972), and the mechanisms involved are still widely debated (e.g. Allan & Russek 1985; Wilzbach *et al* 1988). In particular, the role of directed ('behavioural') versus non-directed ('acci-dental') drift is a major point of interest (e.g. Allan *et al* 1986; Wilzbach 1990). Wilzbach *et al* (1988) have proposed an integrated behavioural classifi-cation for lotic invertebrates that combines drift pattern, habit and functional feeding group characteristics. For example, invertebrate scrapers (functional group) tend to be clingers (habit) and usually show non-directed drift (accidental drift), while many gathering collectors are swimmers or sprawlers and exhibit directed drift behaviour.

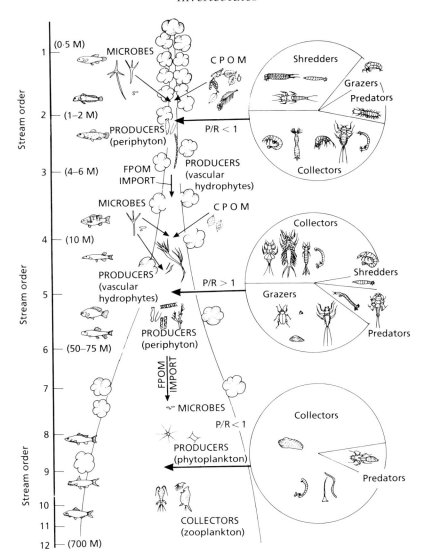

Fig. 11.3 A generalized model of the shifts in the relative abundances of invertebrate functional groups along a river tributary system from headwaters to mouth as predicted by the river continuum concept (RCC; e.g. Vannote *et al* 1980). The river system is shown as a single stem of increasing stream order and width. The headwaters (orders 1−3) are depicted as dominated by riparian shading and litter inputs resulting in a heterotrophic $P/R < 1$. The invertebrates are dominated by shredders which utilize riparian litter as their food resource once it has been appropriately conditioned by aquatic micro-organisms (especially aquatic hyphomycete fungi) and collectors that feed on fine particulate organic matter (FPOM). The mid-reaches (orders 4−6) are less dependent upon direct riparian litter input and with increased width and reduced canopy shading they are autotrophic with a $P/R > 1$. The shredders are reduced and the scrapers are relatively more important as attached microalgae become more abundant. The larger rivers are dominated by FPOM (and therefore collectors), and the increased transport load of this material together with increased depth results in reduced light penetration, and the system is again characterized by a $P/R < 1$ (after Cummins 1975b).

Drift has usually been viewed as the major mechanism of dispersal leading to population expansion through colonization of new habitats. However, the significantly higher mortality rate among drifting individuals relative to their non-drifting benthic counterparts (Cummins & Wilzbach 1988) and the more than adequate compensation for drift by local turnover (e.g. Waters 1961; Wilzbach & Cummins 1989) brings this view into question. The roles of food limitation and risk of predation remain the most often invoked drivers of the drift phenomenon, but the data are conflicting, suggesting that a generalized paradigm applying to all lotic invertebrate species may not be possible (Cummins & Wilzbach 1988; Wilzbach & Cummins 1989).

11.6 HABITAT QUALITY ASSESSMENT AND MANAGEMENT

As the demand for competing water uses continues to grow, so does the need for accurate and rapid biological monitoring of river water quality. Although chemical measurements continue to be critical for water quality assessment (e.g. Averett & McKnight 1987), biological measurements are often preferred because of their time-integrative nature — effects can be detected long after a short-term pulsed chemical input has dispersed. As the analysis of stresses to running-water ecosystems progressed from instantaneous measurements of selected chemical contaminants to biological measurements, such as biochemical oxygen demand (e.g. Calow *et al* 1990), the invertebrates emerged most frequently as the aquatic group of choice for water quality evaluations (Hynes 1970). The invertebrates are universally abundant in running waters, easily collected, large enough to be examined in the field with the unaided eye or a simple hand lens, and most have lifecycle lengths compatible with seasonal or annual sampling regimes. As indicated above, from an ecosystem perspective, macroinvertebrates are ideal integrators between the micro-organisms that dominate nutrient cycling and the fish that are often the product of interest in aquatic systems.

Both the field and laboratory phases of lotic studies, in which invertebrates are used for en-

vironmental evaluation, regardless of the specific approach applied, should be conducted on a hierarchical basis. If local (intra- or inter-site), regional, or global comparisons are to be made, a minimum level of resolution must be set so that all groups conducting the analyses can accomplish the task. The hierarchical approach, described above for functional group analysis, allows increased levels of resolution by those sampling who have the appropriate expertise, but allows all studies to be compared at the basic level agreed upon. This was the approach used in the river continuum studies in North America (e.g. Minshall *et al* 1983b, 1985).

The concept of index species (e.g. Hilsenhoff 1987; Lenat 1988) should certainly be modified to some form of index associations. A currently widely used method is the index of biotic integrity (e.g. Karr 1987, 1991). The described functional feeding group approach is another example. The index approach using stream or river invertebrates has been hindered by the incomplete state of taxonomic knowledge referred to above. That is, a methodology that seeks to measure system redundancy, namely many individuals but few species, usually finds redundancy in the taxonomy through an inability to separate species. This is a more significant problem in North America and in 'underdeveloped' countries than in Europe. For example, in one North American third-order woodland stream, the midge (Chironomidae) fauna is composed of more than 175 species on an annual basis (Coffman 1973, 1974). Often this dipteran family is the most diverse one represented in streams, and yet it is most frequently identified and counted only to the family, or possibly subfamily, tribe or (rarely) genus level.

Thus, although structural comparative analysis of running waters with respect to water quality or general environmental conditions (e.g. status of the riparian zone) can be assessed taxonomically, the incomplete state of knowledge about most invertebrate groups in most river systems results in poor levels of resolution unless immense effort is expended. Approaches that focus on functional aspects of invertebrate associations, such as the feeding groups described above, provide insight that is based on combinations of diverse taxa that require less time in categorization. This does not replace the taxonomically based structural analysis, but allows for evaluation of sytem functional status until such time as invertebrate systematics has progressed to the point that reliable alternative approaches are possible.

Large-scale regional inventories of running-water habitats usually include some assessment of invertebrate diversity. An important component of such inventories is the concept of reference sites and evaluations organized by ecological regions, or ecoregions (e.g. Hughes *et al* 1986; Hughes & Larsen 1988). In general, the reference sites are viewed as those with the highest diversity in a given ecoregion or subregion. In the USA, for example, the Environmental Protection Agency (EPA) Environmental Monitoring Assessment Program (EMAP) effort is organized according to ecoregions, and the US Geological Survey's National Water Quality Assessment (NAQWA) Program has established long-term reference sites.

Another area of important new initiatives for management involves the methodology of remote sensing. There has been greatly expanded attention focused on the riparian zone, as a terrestrial system inseparably linked to the biology of the associated running waters, especially the invertebrates (e.g. Cummins 1988). Inclusion of riparian corridors in management strategies for the biota, including the invertebrates, of a particular watercourse is taken as given in many cases. The analysis on a regional scale of the terrestrial biomes through which rivers and their tributary streams flow provides considerable insight into the specific nature of the linkage between invertebrate life histories and these terrestrial vegetation systems (e.g. Cummins 1988; Cummins *et al* 1989). Macro-scale remote sensing from satellite or low-level overflight by helicopter or fixed-wing aircraft can provide the necessary imagery for such analyses using appropriate geographical information system (GIS) software (e.g. White & MacKenzie 1986).

11.7 SOME RESEARCH FRONTIERS

Perhaps the most promising 'frontier' of research on running-water invertebrates involves their relationships with micro-organisms. Among the promising topics are: (1) the coevolution of invertebrates and micro-organisms resulting in controlled habitat (gut) advantages to the microbes and nutrient uptake advantages to the host invertebrates; (2) feeding strategies directed toward specific microbial populations (e.g. the hyphomycete–shredder linkage) or that result in the accumulation by the invertebrate of microbial enzymes for digestion of its gut contents; and (3) microbially mediated pathogenic mortality in natural lotic invertebrate populations (Cummins & Wilzbach 1988).

With regard to the first point, the coevolution of lotic invertebrates and micro-organisms can likely be viewed as a fundamental organizing principle in running-water environments worldwide. Invertebrate shredder feeding on hyphomycete fungi, which was discussed above is a good case in point. The aquatic insect taxa that dominate the shredder functional group evolved from terrestrial stock, as did the hyphomycetes that are probably aquatic vegetative forms of terrestrial Basidiomycetes. The coevolution of these two groups has been the key to terrestrial leaf litter turnover in headwater streams. The hyphomycetes, and the products of their metabolism of leaves in streams, constitute the nutritional base for shredders. In turn, shredder feeding is instrumental in opening up new colonization sites for these obligate aerobic fungi (e.g. Cummins 1974; Cummins & Klug 1979).

The matter of the natural regulation of lotic invertebrate populations by pathogens (e.g. May 1983; Cummins & Wilzbach 1988) will undoubtedly receive increasing attention because of the relationship to basic ecological population theory and questions related to the control of running-water pest and disease species such as the blackflies (Simuliidae).

As the release of genetically engineered organisms becomes a more common reality in the next decade, the relationship between the diversity of species as reproductively isolated entities and the genetic diversity residing within each species will receive increasing attention (e.g. Futuyma & Petersenk 1985; West-Eberhard 1989). In lotic ecosystems it is probable that there is some sort of general inverse relationship between interspecific taxonomic diversity and intraspecific genetic diversity. This means that a lotic habitat

The Biota

with few species, most of which might be generalists, could persist in time at least partly owing to the large genetic diversity embodied in most of the species. The alternative would be diverse invertebrate communities made up largely from genetically more narrow specialists. From the standpoint of lotic ecosystem persistence, the maintenance of functional diversity, or a particular functional balance, would be the critical issue. This functional balance could be achieved through either species or intraspecific genetic diversity.

A high priority in running-water studies should be the integration of the paradigms that are related to the understanding and prediction of the distribution and abundance of invertebrate populations. The integration of such paradigms as the river continuum concept (RCC; e.g. Vannote *et al* 1980) and the concepts related to patch dynamics and disturbance (e.g. Townsend 1989) would constitute a significant contribution to our approach to invertebrate studies and lotic ecology as a whole. Pringle *et al* (1988) concluded that the concepts of patch dynamics could be usefully integrated into the major running-water paradigms, the RCC and nutrient spiralling (Elwood *et al* 1983). Townsend has suggested that the RCC is not generally applicable, but patterns of patch dynamics might serve as a general organizing principle for running waters. Frid and Townsend (1989) concluded that patch dynamics theory, which was developed for terrestrial forest and marine intertidal systems, needed expansion before appropriate applications to other systems, such as lotic communities, could be made. Townsend (1989) has presented a convincing treatment of such an application of the theory to streams and rivers. However, most of the aspects of the RCC were rejected out of hand and no serious attempt at integration was made. Some useful points for studies of lotic invertebrates relative to patch dynamics and disturbance will be the more complete incorporation of island biogeography theory (e.g. Minshall *et al* 1983a) and the significance of the finding that drifting invertebrates have a higher pathogen load and represent likely mortality rather than highly successful colonists, even if settling space is available (e.g. Cummins & Wilzbach 1988).

REFERENCES

Allan JD, Russek E. (1985) The quantification of stream drift. *Canadian Journal of Fisheries and Aquatic Science* **42**: 210–15. [11.5]

Allan JD, Flecker AS, McClintock NL. (1986) Diel epibenthic activity of mayfly nymphs, and its non-concordance with behavioral drift. *Limnology and Oceanography* **31**: 1057–65. [11.5]

Anderson NH, Wallace JB. (1984) Habitat, life history, and behavioral adaptations of aquatic insects. In: Merritt RW, Cummins KW (eds) *An Introduction to the Aquatic Insects*, 2nd edn, pp 38–58. Kendall/Hunt, Dubuque, Iowa. [11.5]

Andewartha HG, Birch LC. (1954) *The Distribution and Abundance of Animals*. University of Chicago Press, Chicago. [11.4]

Averett RC, McKnight DM (eds) (1987) *Chemical Quality of Water and the Hydrologic Cycle*. Lewis Publishers, Michigan. [11.6]

Benke AC. (1979) A modification of the Hynes method for estimating secondary production with particular significance for multivoltine populations. *Limnology and Oceanography* **24**: 168–71. [11.4]

Benke AC. (1984) Secondary production of aquatic insects. In: Resh VH, Rosenberg DM (eds) *The Ecology of Aquatic Insects*, pp 289–322. Praeger, New York. [11.1, 11.4]

Benke AC, Parsons KA. (1990) Modelling black fly production dynamics in blackwater streams. *Freshwater Biology* **24**: 167–80. [11.4]

Benke AC, van Arsdall TC, Gillespie DH, Parish FK. (1984) Invertebrate productivity in a subtropical blackwater river: the importance of habitat and life history. *Ecological Monographs* **54**: 25–63. [11.4]

Boling RH Jr, Petersen RC, Cummins KW. (1975). Ecosystem Modeling for Small Woodland Streams. In: Patten BH (ed.) *Systems Analysis and Simulation in Ecology*, Vol. 3, pp 183–204. Academic Press, New York. [11.4]

Calow P, Armitage P, Boon P, Chave P, Cox E, Hildrew A. *et al* (1990) River water quality. *Ecology Issues* 1. British Ecology Society, London. [11.6]

Cargill AS II, Cummins KW, Hanson BJ, Lowry RR. (1985a) The role of lipids, fungi, and temperature in the nutrition of a shredder caddisfly, *Clistoronia magnifica*. *Freshwater Invertebrate Biology* **4**: 64–78. [11.5]

Cargill AS II, Cummins KW, Hanson BJ, Lowry RR. (1985b) The role of lipids as feeding stimulants for shredding aquatic insects. *Freshwater Biology* **15**: 455–64. [11.5]

Coffman WP. (1973) Energy flow in a woodland stream ecosystem: II. The taxonomic composition and phenology of the Chironomidae as determined by the collection of pupal exuviae. *Archives in Hydrobiology*

71: 281–322. [11.6]

Coffman WP. (1974) Seasonal differences in the diel emergence of a lotic chironomid community. *Ent Tidskr* **95**: 42–8. [11.6]

Corkum LD. (1989) Patterns of benthic invertebrate assemblages in rivers of northwestern North America. *Freshwater Biology* **21**: 191–205. [11.1]

Cuffney TF, Wallace JB, Lugthart GJ. (1990) Experimental evidence quantifying the role of benthic invertebrates in organic matter dynamics of headwater streams. *Freshwater Biology* **23**: 281–99. [11.5]

Cummins KW. (1966) A review of stream ecology with special emphasis on organism–substrate relationships. *Special Publications of the Pymatuning Laboratory of Ecology, University of Pittsburgh* **4**: 2–51. [11.3]

Cummins KW. (1972) What is a river? –Zoological description. In: Oglesby RT, Carlson CA, McCann JA. (eds) *River Ecology and Man*, pp 35–52. Academic Press, New York. [11.3]

Cummins KW. (1973). Trophic relations of aquatic insects. *Annual Review of Entomology* **18**: 183–206. [11.1, 11.5]

Cummins, KW. (1974) Structure and function of stream ecosystems. *BioScience* **24**: 631–41. [11.1, 11.5, 11.7]

Cummins KW. (1975a) Macroinvertebrates. In: Whitton BA (ed.) *River Ecology*, pp 170–98. Blackwell Scientific Publications, New York. [11.1, 11.3]

Cummins KW. (1975b) The ecology of running waters; theory and practice. In: Baker DB, Jackson WB, Prater BL (eds) *Proceedings of the Sandusky River Basin Symposium, International Joint Commission, Great Lakes Pollution*, pp 277–93. Environmental Protection Agency, Washington, DC. [11.5]

Cummins KW. (1988) The study of stream ecosystems: a functional view. In: Pomeroy LR, Alberts JJ (eds) *Concepts of Ecosystem Ecology*, pp 247–62. Springer-Verlag New York. [11.1, 11.6]

Cummins KW, Klug MJ. (1979) Feeding ecology of stream invertebrates. *Annual Review of Ecology and Systematics* **10**: 147–72. [11.5, 11.7]

Cummins KW, Merritt RW. (1984) Ecology and distribution of aquatic insects. In: Merritt RW, Cummins KW (eds) *An Introduction to the Aquatic Insects of North America*, pp 59–65. Kendall/Hunt, Dubuque, Iowa. [11.4]

Cummins KW, Wilzbach MA. (1985) *Field procedures for the analysis of functional feeding groups in stream ecosystems*. Pymatuning Laboratory of Ecology, University of Pittsburgh, Linesville, Pennsylvania. [11.5]

Cummins KW, Wilzbach MA. (1988) Do pathogens regulate stream invertebrate populations? *Verhandlungen Internationale Vereinigung für Theoretische und Angewandt Limnologie* **23**: 1232–43. [11.5, 11.7]

Cummins KW, Petersen RC, Howard FO, Wuycheck JC, Holt VI. (1973) The utilization of leaf litter by stream detritivores. *Ecology* **54**: 336–45. [11.4]

Cummins KW, Klug MJ, Ward GM, Spengler GL, Speaker RW, Ovink RW *et al* (1981) Trends in particulate organic matter fluxes, community processes, and macroinvertebrate functional groups along a Great Lakes Drainage Basin river continuum. *Verhandlungen Internationale Vereinigung für Theoretische und Angewandt Limnologie* **21**: 841–9. [11.4, 11.5]

Cummins KW, Sedell JR, Swanson FJ, Minshall GW, Fisher SG, Cushing CE *et al* (1983) Organic matter budgets for stream ecosystems: problems in their evaluation. In: Barnes JR, Minshall GW (eds) *Stream Ecology Application and Testing of General Ecological Theory*, pp 299–353. Plenum Press, New York. [11.4]

Cummins KW, Minshall GW, Sedell JR, Petersen RC. (1984) Stream ecosystem theory. *Verhandlungen Internationale Vereinigung für Theoretische und Angewandt Limnologie* **22**: 1818–27. [11.1]

Cummins KW, Wilzbach MA. Gates DM, Perry JB, Taliaferro WB. (1989) Shredders and riparian vegetation. *Bioscience* **39**: 24–30. [11.5, 11.6]

Davis JA, Barmuta LA. (1989) An ecologically useful classification of mean and near-bed flows in streams and rivers. *Freshwater Biology* **21**: 271–82. [11.5]

Edmondson WT (ed.) (1959) *Freshwater Biology*, 2nd edn. John Wiley & Sons, New York. [11.2]

Elliott JM. (1977) *Some Methods for the Statistical Analysis of Samples of Benthic Invertebrates*, 2nd edn. Scientific Publication of the Freshwater Biology Association UK, 25. [11.3]

Elwood JW, Newbold JD, O'Neill RV, VanWinkle W. (1983) Resource spiralling: an operational paradigm for analyzing lotic ecosystems. In: Fontaine TD, Bartell SM (eds) *The Dynamics of Lotic Ecosystems*, pp 3–27. Ann Arbor Science. Ann Arbor, Michigan. [11.7]

Frid CLJ, Townsend CR. (1989) An appraisal of the patch dynamics concept in stream and marine benthic communities whose members are highly mobile. *Oikos* **56**: 137–41. [11.7]

Futuyma DJ, Petersenk SC. (1985) Genetic variation in the use of resources by insects. *Annual Review of Entomology* **30**: 217–38. [11.7]

Hanson BJ, Cummins KW, Cargill AS, Lowry RR. (1983) Dietary effects on lipid and fatty acid composition of *Clistoronia magnifica* (Trichoptera: Limnephilidae). *Freshwater Invertebrate Biology* **2**: 2–15. [11.5]

Hanson BJ, Cummins KW, Cargill AS, Lowry RR. (1985) Lipid content, fatty acid composition, and the effect of diet on fats of aquatic insects. *Comparative Biochemistry and Physiology* **80B**: 257–76. [11.5]

Hauer FR, Benke AC. (1991) Rapid growth rates of snag-dwelling chironomids in a black water river: the influence of temperature and discharge. *Journal of the North American Benthological Society* **10**: (in press). [11.4]

Hilsenhoff WL. (1987) An improved biotic index of organic stream pollution. *Great Lakes Entomology* **20**: 31–9. [11.1, 11.6]

Hughes RM, Larsen DP. (1988) Ecoregions: an approach to surface water protection. *Journal of the Water Pollution Control Federation* **60**: 486–93. [11.6]

Hughes RM, Larsen DP, Omernik JM. (1986) Regional reference sites: a method for assessing stream pollution. *Environmental Management* **10**: 629–35. [11.6]

Hurlbert SH. (1984) Pseudoreplication and the design of ecological field experiments. *Ecological Monographs* **54**: 187–211. [11.3]

Huryn AD. (1990) Growth and voltinism of lotic midge larvae: patterns across an Appalachian mountain landscape. *Limnology and Oceanography* **35**: 339–51. [11.4]

Huryn AD, Wallace JB. (1986) A method for obtaining *in situ* growth rates of larval Chironomidae (Diptera) and its application to studies of secondary production. *Limnology and Oceanography* **31**: 216–32. [11.4]

Hynes HBN. (1960) *The Biology of Polluted Waters*. Liverpool University Press, Liverpool.

Hynes HBN. (1970) *The Ecology of Running Waters*. University of Toronto Press, Toronto. [11.1, 11.2, 11.5, 11.6]

Hynes HBN. (1975) The stream and its valley. *Verhandlungen Internationale Vereinigung für Theoretische und Angewandt Limnologie* **19**: 1–15. [11.1]

Iverson TB, Dall P. (1989) The effect of growth pattern, sampling interval and number of size classes on benthic invertebrate production estimated by the size-frequency method. *Freshwater Biology* **22**: 232–331. [11.1]

Karr JR. (1987) Biological monitoring and environmental assessment: a conceptual framework. *Environmental Management* **11**: 249–56. [11.6]

Karr JR. (1991) Biological integrity: a long neglected aspect of water resource management. Ecological Applications **1**: 26–35. [11.1, 11.6]

Klug, MJ, Kotarski S. (1980) Bacteria associated with the gut tract of larval stages of the aquatic cranefly *Tipula abdominalis* (Diptera: Tipulidae). *Applied Environmental Microbiology* **40**: 408–16. [11.5]

Lawson DL, Klug MJ. (1989) Microbial fermentation in the hind guts of two stream detritivores. *Journal of the North American Bethological Society* **8**: 85–91. [11.5]

Lenat DR. (1988) Water quality assessment of streams using a qualitative collection method for benthic macroinvertebrates. *Journal of the North American Bethological Society* **7**: 222–53. [11.6]

Lugthart GJ, Wallace JB, Huryn AD. (1991) Secondary production of chironomid communities in insecticide-treated and untreated headwater streams. *Freshwater Biology* 143–56. [11.4]

Macan TT. (1958) Methods of sampling the bottom fauna in stony streams. *Mitteilungen Internationale Vereinigung für Theoretische und Angewandte Limnologie* **8**: 1–21. [11.3]

Macan TT. (1962) The ecology of aquatic insects. *Annual Review of Entomology* **7**: 261–88. [11.1]

Macan TT. (1977) The influence of predation on the composition of freshwater animal communities. *Biological Reviews of the Cambridge Philosophical Society* **52**: 45–70. [11.1]

Macan TT. (1974) *Freshwater Ecology*, 2nd edn. John Wiley & Sons, New York. [11.1, 11.5]

Macan TT, Worthington EB. (1951) *Life in Lakes and Rivers*. Collins, London.

Martin MM, Martin JS, Kukor JJ, Merritt RW. (1980) The digestion of protein and carbohydrate by the stream detritivore, *Tipula abdominalis* (Diptera, Tipulidae). *Oecologia* **46**: 360–4. [11.5]

Martin MM, Kukor JJ, Martin JS, Merritt RW. (1981a) Digestive enzymes of larvae of three species of caddisflies (Trichoptera). *Insect Biochemistry* **11**: 501–5. [11.5]

Martin MM, Martin JS, Kukor JJ, Merritt RW. (1981b) The digestive enzymes of detritus-feeding stonefly nymphs (Plecoptera) Pteronarcyidae. *Canadian Journal of Zoology* **59**: 1947–51. [11.5]

May RM. (1983) Parasitic infections as regulators of animal populations. *American Scientist* **71**: 36–45. [11.7]

Merritt RW, Cummins KW. (eds) (1984) *An Introduction to the Aquatic Insects of North America*. Kendall/Hunt, Dubuque, Iowa. [11.1, 11.2, 11.3, 11.5]

Merritt RW, Cummins KW, Resh VH. (1984) Collecting, sampling, and rearing methods for aquatic insects. In: Merritt RW, Cummins KW. (eds) *An Introduction to the Aquatic Insects of North America*, pp 11–26. Kendall/Hunt, Dubuque, Iowa. [11.3]

Meyer JL, O'Hop J. (1983) Leaf-shredder insects as a source of dissolved organic carbon in headwater streams. *American Midland Naturalist* **109** 175–83. [11.5]

Minshall GW, Andrews DA, Manuel-Faler CY. (1983a) Application of island biogeographic theory to streams: macroinvertebrate recolonization in the Teton River, Idaho. In: Barnes JR, Minshall GW. (eds) *Stream Ecology*, pp 279–97. Plenum Press, New York. [11.3, 11.7]

Minshall GW, Petersen RC, Cummins KW, Bott TL, Sedell JR, Cushing CE, Vannote RL. (1983b) Inter-biome comparison of stream ecosystem dynamics. *Ecological Monographs* **53**: 1–25. [11.3, 11.5, 11.6]

Minshall GW, Cummins KW, Petersen RC, Cushing CE, Bruins DA, Sedell JR, Vannote RL. (1985)

Developments in stream ecology. *Canadian Journal of Fisheries and Aquatic Sciences* **42**: 1045–55. [11.1, 11.5, 11.6]

Moss D, Furse MT, Wright JF, Armitage PD. (1987) The prediction of the macro-invertebrate fauna of unpolluted running-water sites in Great Britain using environmental data. *Freshwater Biology* **17**: 41–52. [11.1]

Meyer E. (1989) The relationship between body length parameters and dry mass in running water invertebrates. *Archives of Hydrobiology* **117**: 191–203. [11.4]

Meyer JL, O'Hop J. (1983) Leaf-shredding insects as a source of dissolved organic carbon in headwater streams. *American Midland Naturalist* **109**: 175–83. [11.5]

Moon HP. (1939) Aspects of the ecology of aquatic insects. *Transactions of the British Entomological Society* **6**: 39–49. [11.5]

Pennak RW. (1978) *Freshwater Invertebrates of the United States*, 2nd edn. John Wiley & Sons, New York. [11.2]

Petersen RC, Cummins KW. (1974) Leaf processing in a woodland stream. *Freshwater Biology* **4**: 343–68. [11.5]

Petersen RC, Cummins KW, Ward GM. (1989) Microbial and animal processing of detritus in a woodland stream. *Ecological Monographs* **59**: 21–39. [11.4]

Pringle CM, Naiman RJ, Bretschko G, Karr JR, Oswood MW, Welcomme RL, Winterbourn MJ. (1988) Patch dynamics in lotic systems. *Journal of the North American Bethological Society* **7**: 503–24. [11.7]

Rader RB, Ward JV. (1990) Mayfly growth and population density in constant and variable temperature regimes. *Great Basin Naturalist* **50**: 97–106. [11.4]

Resh VH, Rosenberg DM. (eds) (1984) *The Ecology of Aquatic Insects*. Praeger Publishers, New York. [11.1]

Shelford VE. (1914) An experimental study of the behavior agreement among animals of an animal community. *Biological Bulletin* **26**: 294–315. [11.5]

Short RA, Maslin PE. (1977) Processing of leaf litter by a stream detritivore: effect on nutrient availability to collectors. *Ecology* **58**: 935–8. [11.5]

Smock LA, Gilinsky E, Stonebrunner DL. (1985) *Macroinvertebrate production in a southeastern United States blackwater stream. Ecology* **66**: 1491–503. [11.4]

Stanford JA. (1972) A centrifuge method for determining live weights of aquatic insect larvae, with a note on weight loss in preservative. *Ecology* **54**: 499–551. [11.4]

Thienemann A. (1950) *Verbreitungsgeschicte der susswassertierwelt Europas*. Die Binnengewasser, Stuttgart 18. [11.2]

Townsend CR. (1980) *The Ecology of Streams and Rivers*. Edward Arnold, London. [11.1]

Townsend CR. (1989) The patch dynamics concept of stream community ecology *Journal of the North American Bethological Society* **8**: 36–50. [11.7]

Vannote RL, Sweeney BW. (1980) Geographic analysis of thermal equilibria: a conceptual model for evaluating the effect of natural and modified thermal regimes on aquatic insect communities. *American Naturalist* **115**: 667–95. [11.4]

Vannote RL, Minshall GW, Cummins KW, Sedell JR, Cushing CE. (1980) The river continuum concept. *Canadian Journal of Fisheries and Aquatic Sciences* **37**: 130–7. [11.5, 11.7]

Vogel S. (1981) *Life in Moving Fluids*. Princeton University Press, Princeton, New Jersey. [11.5]

Vogell S. (1988) *Life's Devices*. Princeton University Press, Princeton, New Jersey. [11.5]

Waldbauer G. (1968) The consumption and utilization of food by insects. *Advances in Insect Physiology* **5**: 229–88. [11.4]

Ward GM, Cummins KW. (1979) Food quality on growth of a stream detritivore. *Paratendipes albimanus (meigen)* (Diptera: Chironomidae). *Ecology* **60**: 57–64. [11.4]

Ward JV, Voelz NJ. (1990) Gradient analysis of interstitial meiofauna along a longitudinal stream profile. *Stygologia* **5**: 93–9. [11.1]

Waters TF. (1961) Standing crop and drift of stream bottom organisms. *Ecology* **42**: 352–7. [11.5]

Waters TF. (1972) The drift of stream insects. *Annual Review of Entomology* **17**: 253–72. [11.5]

Waters TF. (1977) Secondary production in inland waters. *Advances in Ecological Research* **10**: 91–164. [11.4]

Waters TF. (1979a) Influence of benthos life history upon the estimation of secondary production. *Journal of the Fisheries Research Board of Canada* **36**: 1425–30. [11.1, 11.4]

Waters TF. (1979b) Benthic life histories: summary and future needs. *Journal of the Fisheries Research Board of Canada* **36**: 342–5. [11.1, 11.4]

Waters TF. (1988) Fish production – benthos production relationships in trout streams. *Polskie Archiwum Hydrobiologii* **35**: 548–61. [11.1, 11.4]

Waters TF, Crawford GW. (1973) Annual production of a stream mayfly population: a comparison of methods. *Limnology and Oceanography* **18**: 286–96. [11.4]

Waters TF, Hokenstrom JC. (1980) Annual production and drift of the stream amphipod *Gammarus pseudolimnaeus* in Valley Creek, Minnesota. *Limnology and Oceanography* **25**: 700–10. [11.4]

Wesenberg-Lund C. (1943) *Biologie der Susswasserinsekten*. Springer, Berlin. [11.2, 11.5]

West-Eberhard MJ. (1989) Morphological variation and width of ecological niche. *Annual Review of Ecology and Systematics* **20**: 249–78. [11.7]

White PS, MacKenzie MD. (1986) Remote sensing and

landscape pattern in Great Smokey Mountains biosphere reserve, North Carolina and Tennessee. In: Dyer MI, Crossley DA (eds) *Coupling of ecological studies with remote sensing: potentials at four biosphere reserves in the United States*, pp 52–70. Bureau of Oceans and International Environmental Affairs, US Dept. State Publ. 9504, Washington, DC. [11.6]

Williams DD. (1989) Towards a biological and chemical definition of the hyporeic zone of two Canadian rivers. *Freshwater Biology* 22: 189–208. [11.3]

Wilson EO. (ed.) (1988) *Biodiversity*. National Academic Press, Washington, DC. [11.2]

Wilzbach MA. (1990) Non-concordance of drift and benthic activity in Baetis. *Limnology and Oceanography* 35: 945–52. [11.5]

Wilzbach MA, Cummins KW. (1989) An assessment of short-term depletion of stream macroinvertebrate benthos by drift. *Hydrobiologia* 185: 29–39. [11.5]

Wilzbach MA, Cummins KW, Knapp R. (1988) Towards a functional classification of stream invertebrate drift. *Verhandlungen Internationale Vereinigung für Theoretische und Angewandt Limnologie* 23: 1244–64. [11.5]

Yount JD, Niemi GJ (eds) (1990) Recovery of lotic communities and ecosystems following disturbance: theory and applications. *Environmental Management* (special issue) 4: 515–762. [11.1]

12: Riverine Fishes

P.B. BAYLEY AND H.W. LI

12.1 INTRODUCTION

This chapter puts the challenges of understanding and managing riverine fish and fisheries in an ecological and evolutionary context. An appreciation of the adaptations of fish and the reasons for characteristic assemblages will lead to a more strategic view of riverine fisheries management than the tactical ones often employed, such as maximizing the yield of a single species while ignoring environmental variation. We propose that most serious fisheries management problems result from actions that have changed the hydrological regime, habitats and/or fish fauna, thereby disrupting the long-term, dynamic patterns to which the indigenous fishes are adapted. This has resulted in the need for restoration in many systems. We discuss the options available to research and management agencies in the light of current limitations on our ability to sample fish quantitatively and our knowledge of their spatial and temporal dependence on their environment.

Apart from their aesthetic value, riverine fishes are important because they can be harvested for human consumption, caught for recreation, are useful as indicators of the well-being of the environment, or serve as appropriate subjects for testing principles of population or community ecology. Considerable knowledge has resulted from the independent pursuit of these interests. However, this knowledge is dwarfed by information of which we are ignorant, which includes many of the concepts and tools necessary to make inferences about fish populations or communities based on the few systems we can afford to study intensively.

In order that projects pursuing any of these interests are directed towards conservation or recovery of the natural system, have long-term economic viability, and have general application in fish resource management, we propose four requirements. First, the project must recognize (or investigate) the constraints imposed by the evolutionary adaptations (section 12.3) and interactions (section 12.5) of the taxa concerned. Second, the biases and variance of the sampling process must be known with sufficient accuracy (sections 12.4, 12.7; Chapter 13), and the long-term management costs of monitoring the resource with that accuracy must be included. Third, the investigator should be aware of the feasibility and cost of restoring a damaged system (section 12.6). Last, but not least, the project must be designed and reported to enable generalizations across systems on appropriate scales and classifications (sections 12.2, 12.7). Considering that many publications still account for few, if any, of these factors, there is much room for improvement.

12.2 CLASSIFICATIONS AND UNIFYING CONCEPTS FOR RIVERINE FISHES

There are about 8500 freshwater fish species (Lowe-McConnell 1987), most of which occur in rivers or connected floodplains. Current technology and resources (section 12.4) are probably insufficient to complete ecological studies by species and to predict population trends by stock before some of these species become extinct naturally. In view of the unnatural changes occurring in systems since the industrial revolution and the subsequent human population explosion, it is clearly impossible to achieve a moderate level of predictive capability for each stock of each species of interest with respect to their exploitation or conservation.

Therefore, more studies must focus on comparisons among systems or their components, so that information from intensive, localized studies can be used to manage, conserve and restore fish populations and communities across many systems. This requires classifications of ecologically equivalent units that comprise functionally similar species and/or life stages so that generalizations can be tested. Classification of units, gradients of key variables within units, and the scale adopted depend on the problem and the information available. The pluralistic approach (Schoener 1986, 1987) emphasizes differences between ecological communities based on organismic and environmental axes. Elements of these axes, such as body size (section 12.3) and stream discharge values (section 12.2), may jointly indicate appropriate boundaries for working definitions of classification units. Inadequacies in data are common, and generalization can be more limited by appropriate survey information across systems than by results from localized studies.

Classifications at different spatial scales (sections 12.2–12.5) and unifying concepts within and across scales (section 12.6) are presented as heuristic tools to understand how fishes are organized in river systems within the hierarchy of spatial and temporal scales available.

Spatial and temporal scales and hierarchies

The spatial and temporal scales of environmental units available for studying river fishes are correlated (Fig. 12.1(a)). Unifying concepts and classifications of fish assemblages need to recognize this correlation as well as the hierarchical structure of these units, whose physical characteristics persist on scales of 10^5–10^6 to 10^{-1}–10^0 years from landscapes to microhabitats, respectively (Frissell *et al* 1986), and are extended to evolutionary scales at the zoogeographical level in Fig. 12.1(a). Hierarchical scaling promotes the most effective solution of ecological problems (Allen & Starr 1982). The hierarchy implies

that the larger, more stable, environment imposes limits on the smaller, more variable, environmental units. Habitats, for example, can be classified within broader units, and thus lend themselves to statistically nested designs for the testing of differences in fish assemblages or other attributes.

Johnson (1980) suggested that resource selection by species follows a hierarchy from the zoogeographical range (first order), through microhabitat scales, with resource selection in each order being conditional on a lower order. Although we question his unidirectional dependence of selection (e.g. the home range can depend on the selection of habitat as well as vice versa) and the separation of feeding and habitat usage into different orders, it is important to understand resource selection in the context of spatial and temporal scales. Resource selection by a fish depends on a series of conditions: (1) ability to disperse among fluvial systems on a zoogeographical scale; (2) seasonal migrations of some species limited by basin extent, geomorphology and habitat availability; (3) home range limited by physicochemical factors (habitat distribution) on the reach or stream scale; and (4) activity under the constraints of biological interactions which include the probability of being killed, availability of prey and reproductive requirements at the microhabitat scale (Fig. 12.1(b)).

The time scales are complex, because those relating to persistence of environmental units (Fig. 12.1(a)) extend to evolutionary scales, and are two to three orders of magnitude greater (at similar spatial scales) than the ecological scales corresponding to the response times of individual fish (Fig. 12.1(b)). The formation of species assemblages depends on zoogeographical limits derived in evolutionary time scales (Fig. 12.1(a)), morphological and physiological preadaptations constraining distributional limits in ecological time scales (Fig. 12.1(b)), and interactions among species which fine-tune assemblage structure in ecological and evolutionary time scales. The following four subsections discuss the usefulness of classifying and predicting properties of fish and environments at decreasing spatial scales.

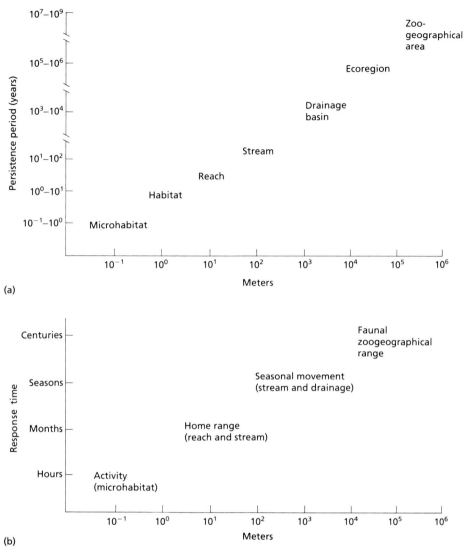

Fig. 12.1 (a) Persistence period versus spatial scale of environmental units (after Frissell *et al* 1986). (b) Response time versus spatial scale of fish movements.

Zoogeographical scales

What can be learned from comparative zoogeography that is useful to the manager or the ecologist who is not concerned with historical reasons for current fish assemblages? Can one generalize among similar systems of different zoogeographical regions? Only with great care; for instance, physically similar drainage systems in Poland and Ontario, Canada differ in species richness and body-size distributions among species (Mahon 1984). In contrast, Moyle and Herbold (1987) found great similarity in assemblage structure among cold headwater streams of Europe and eastern and western North America, and found common structural features among the warmwater fishes of Europe and western North America. These patterns may have resulted from the degree of similarity of conditions during Pleistocene glaciation events. In a comparison of external morphological measurements among stream fish assemblages from North and southern South America, Strauss (1987) found that correlations were high among North American zoogeographical areas and low between the two subcontinents, and reflected the phylogenetic constraints on morphology. To the extent that morphologies reflect ecological attributes, this evidence suggests that it is unwise to infer ecological similarities among assemblages that are distantly related phylogenetically. Likewise, evolutionary convergence of body form is not inevitable because there is not a unique morphological solution to a specific environment (Mayden 1987; Strauss 1987).

This phylogenetic limitation applies to comparisons of ecologies at the species level. At the opposite extreme of comparing gross energy transfers, such as fishery yield from a large proportion of the fish taxocene (section 12.4), similarities across zoogeographical areas reflect similarities in environment rather than in phylogenies.

Regional scales: ecoregions and hydrology

Ecoregional classifications such as those pioneered by Omernik (1987) and his co-workers (e.g. Larsen *et al* 1986; Hughes *et al* 1987; Whittier *et al* 1988; Hughes *et al* 1990) describe fish distributions at the landscape level by relating species presence to geomorphic/land-use patterns. Its use in anticipating where species may become endangered or in formulating broad management policies is not disputed, but in Wisconsin the classification was found to be fairly imprecise and fish assemblages were better classified by general habitat variables (Lyons 1989).

Ecoregions can divide basins and combine adjacent basins, resulting in a departure from the zoogeographical > basin > stream > habitat hierarchy of spatial units. Departing from this hierarchy and devising a network of scales is possible technically, but migratory fish can occupy two or more ecoregions. A classification that predicts and explains fish assemblages should have mechanistic links to finer scales; an ecoregional approach will not successfully predict or explain the presence of a migratory species which also depends on environments in adjoining ecoregions,

including ocean habitats for anadromous species.

In addition, there is a cumulative effect of landscape that can also affect non-migratory fishes. A point in a stream downstream of an ecoregional boundary may reflect properties, such as hydrology, temperature and chemistry (e.g. Elwood *et al* 1983), that are controlled by the upstream ecoregion rather than by the ecoregion in which the point lies. Landscape elements from an ecoregional classification may be useful if they reflect environmental features affecting fish populations; the problem lies in assigning those elements within boundaries independently of watersheds and in implying that their effect is independent of their distance to the lotic area of interest (see Osborne & Wiley 1988 for an alternative approach).

Poff and Ward (1989) developed a regional classification of stream communities based upon variation in streamflow patterns of 78 streams across the USA. In Fig. 12.2(a) we have attempted to summarize the essential features of Poff and Ward's nine stream types plus large river-floodplains, in terms of four key hydrological attributes that are independent of spatial scale. Although this condensed representation assumes interdependence among four attributes and cannot represent all hydrological types precisely, these attributes were defined and positioned to account for Poff and Ward's significant correlations. Thus, data from most individual streams would occupy relatively small areas within the corresponding triangle. All types show some overlap with neighbors. Data for snowmelt, snow and rain, and winter rain showed considerable overlap (Poff & Ward 1989) and they, in turn, form a continuum with surrounding perennial types in the right triangle and with intermittent types in the left triangle (Fig. 12.2(a)). The left triangle is drawn incomplete because there is not a complementary relationship between intermittency and low annual flow variability, and examples do not exist along that axis.

Figure 12.2(b) shows some of the expected trends in fish population and community properties across different combinations of these hydrological attributes. In addition to these, Poff and Ward (1989) have suggested other plausible properties. Most published work has failed to put fish studies in an adequate hydrological frame-

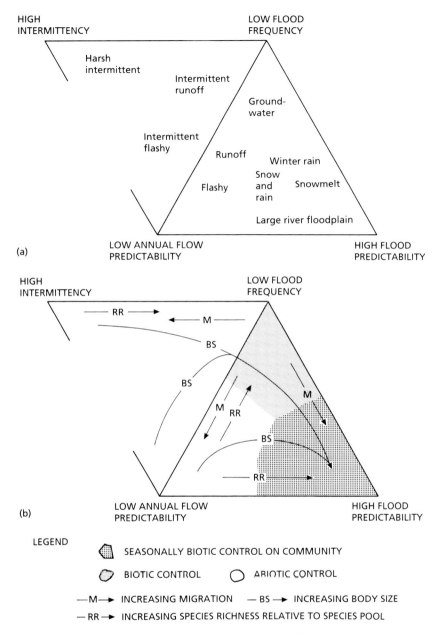

Fig. 12.2 (a) Stream types arranged according to approximate proportions of hydrological attributes (estimated partly from data in Poff & Ward 1989, Table 3). (b) Ecological attributes of fish species and assemblages corresponding to stream types in (a). Right-hand triangle in (a) and (b) includes all perennial stream types.

work, but there are exceptions. For instance, fish inhabiting streams subject to regular desiccation and flash floods were found to respond differently to streamflow changes than those restricted to small, clear creeks with permanent flow, resulting in different assemblages (Rohm *et al* 1987; Matthews 1988; Matthews *et al* 1988).

We believe that a hydrological approach, possibly in conjunction with different temperature regimes and landscape properties, may provide a superior regional template for understanding community function, life-history patterns and making inferences than an ecoregional classification, because the hydrology is more directly associated with physical constraints on fish habitats while reflecting some geomorphological features of the basin. Also, one could monitor ecologically significant departures from the natural regime due to anthropomorphic disturbance.

Basin and reach scales: zonation

The basis of zonation is to find large gradients (generic sense) to which the fauna must respond. Zonation schemes have used stream order (Sheldon 1968; Lotrich 1973; Horwitz 1978), hydraulic stress and power (Statzner 1987), temperature (Gard & Flittner 1974), habitat heterogeneity (Gorman & Karr 1978; Gorman 1988) and physico-chemical gradients (Echelle & Schnell 1976; Matthews & Styron 1981). Huet's (1947, 1959) longitudinal zonation used a combination of gradient and stream width in European rivers to relate reaches to fish communities characterized by individual species. This approach is more difficult to apply to rich faunal assemblages covering various climatic and geomorphic zones (Allen 1969), such as in North America and the tropics, but it can be useful when empirically derived for a particular area (e.g. Angermeier & Karr 1983; Moyle & Senanayake 1984; Matthews 1986b). However, longitudinal zonation does not explain how stream reaches influence assemblages, does not account for potadromy, and does not explain the distribution of fishes in rivers with significant floodplains.

Habitat and microhabitat scales

It is surprising that the distribution of habitat types within reaches has not received much attention because the measurement of riffle: pool ratios to characterize stream reaches is a cherished tradition. However, Bisson and his associates have refocused awareness on habitats as channel units (e.g. side-scour pools, step pools, riffles, glides), first by creating a typology based on hydrological features (Bisson et al 1982), followed by an examination of fish distributions among these habitats

(Bisson et al 1988). Building on this theme, Hicks (1990) found that the physical characteristics and distribution of habitats and faunal assemblages were different for sandstone versus basaltic drainages within the same ecoregion. Furthermore, habitats in sandstone drainages were more sensitive to changes in low summer discharge following logging than basaltic ones (Hicks et al 1991).

The Instream Flow Incremental Methodology (IFIM) is based primarily on microhabitat use patterns of fishes, e.g. current, depth and substrate (Bovee 1982). Some workers have claimed that IFIM shows promise for cold, headwater salmonid stream assemblages (Newcombe 1981; Moyle & Baltz 1985), where conditions and assemblages are quite consistent worldwide (Moyle & Herbold 1987), and for a few obligate warmwater stream fishes (Orth & Maughan 1982). IFIM does not work well when fishes do not have stereotyped behaviours on a limited spatial scale. Many stream fishes have extensive home ranges, select habitats at the channel unit level, or are relatively non-selective in microhabitat choice or are selective only under severe conditions (Angermeier 1987; Felley & Felley 1987; Ross et al 1987; Scarnecchia 1988). IFIM depends on fish biomass density being linearly related to the area of each available habitat type, which has been shown to be invalid (Mathur et al 1985; Conder & Annear 1987). In addition, IFIM does not work well when fishes are influenced by behavioural trade-offs such as risk avoidance or competitive spatial partitioning (Baltz et al 1982) (section 12.5). Finally, IFIM relates ephemeral, short-term behaviours to microhabitats, and such small-scale relationships will not necessarily maintain their predictive quality when extrapolated to larger scales (see Fig. 12.1(b)) that are more appropriate for management. The same general criticisms hold for the related Habitat Suitability Index (HSI) models (US Fish and Wildlife Service 1981; Terrell et al 1982).

River continuum and flood pulse concepts

Two unifying concepts, the river continuum (Vannote et al 1980) and the flood pulse (Junk et al 1989), provide ecological templates which can be used to compare and contrast fish communities or guilds within and among systems (Fig. 12.3). Both concepts have the potential to guide the derivation of sets of hypotheses to identify dominant mechanisms, in particular those operating between adjacent spatiotemporal scales, and either should improve the derivation of classifications at various scales to provide predictions useful for management in appropriate systems.

Although both concepts are designed to work up to a drainage basin scale, they are mutually exclusive in low-gradient potamon reaches, because the flood pulse concept recognizes the periodic nature of the interaction between the flood pulse and the floodplain which influences the adaptations of fish species. In contrast, the phytoplankton-dominated description of low-gradient reaches in the river continuum concept has more in common with heavily regulated rivers that have been denied access to floodplains, such as the Thames (Mann *et al* 1970). Such rivers, which are widespread in the temperate zone,

present a dilemma in formulating a unifying ecological concept. Can we produce useful classifications of systems which are manipulated to the extent that adaptive and coevolutionary features of the fish species are no longer relevant, or do we have to undertake detailed investigations in each unique system?

The flood pulse concept has, at most, peripheral importance in the higher gradient rhithron, where many of the longitudinal processes of the continuum concept provide a more appropriate description. However, the original continuum concept needs to be adapted to, or excepted from, the following: differences in upstream riparian vegetation (Barmuta & Lake 1982; Wiley *et al* 1990), discontinuities (Statzner 1987; Naiman *et al* 1988; Pringle *et al* 1988) and upstream transport of nutrients and biomass through migrations of temperate (Hall 1972; Li *et al* 1987) and tropical fishes (Petrere 1985; Welcomme 1985).

In conclusion, the development of unifying con-

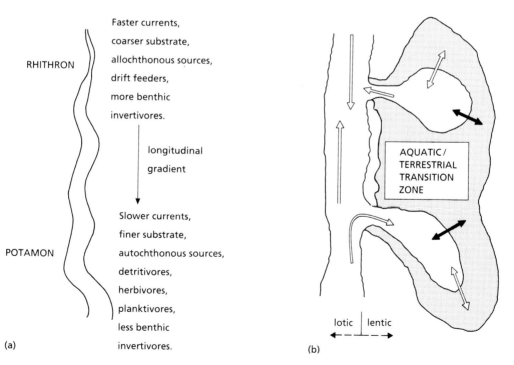

Fig. 12.3 Some contrasts between (a) the river continuum and (b) flood pulse concepts as applied to fishes. (⇨'white' fishes; ➤ 'black' fishes; arrows indicate migrations.)

cepts is still in its infancy and will remain so unless more studies and surveys follow guidelines such as those outlined in section 12.7. Nevertheless, the concepts are currently helpful in understanding life-history adaptations and in classifying functional groups to act as templates for testable hypotheses.

12.3 LIFE-HISTORY ADAPTATIONS

A Gleasonian view of fish species adaptations based on physiological and morphological responses to the environment is presented in this section. This is a complementary explanation of longitudinal zonation to that presented by (Horwitz 1978) with lateral effects of floodplains incorporated. We propose that, at least in the headwaters, constraints of habitat size, habitat variety (see also Gorman & Karr 1978) and hydraulic forces limit diversification of body form and limit resource partitioning according to the principle of limiting similarity (MacArthur & Levins 1967). We describe life-history adaptations to different classes of the environment in terms of body morphology and activity, P/B ratio and body size, trophic properties, reproductive strategies, and trade-offs and polymorphisms.

The rhithron and the potamon impose different requirements on fish life (Fig. 12.3). The classic, high-gradient rhithron is usually more restrictive: it is smaller, cooler, more highly oxygenated, and comprised primarily of fastwater habitats: rapids, riffles, cascades and step-pools. The low-gradient rhithron typical of surface-drained, low-lying areas is characterized by more variable temperature, oxygen and discharge (Wiley *et al* 1990). The potamon comprises habitats with a greater variety of size, depth and flow: large river channels and pools, braided stream channels, oxbows and sloughs, and habitats of the floodplain. The gradient is usually minimal, temperatures are higher, and some habitats become hypoxic. These features, and local variants, have a profound effect on the adaptations discussed below.

Morphology and activity

The attributes of rhithron fishes are more constrained by habitat size (Schlosser 1982) and hydraulic power and variability (Statzner 1987; Wiley *et al* 1990) than the fishes of the potamon. This is especially noticeable in terms of their size, shape and metabolism (Table 12.1). Smaller body size is not related just to the scale of available in-stream cover (Schlosser 1987); the demands for acceleration and agility in a turbulent environment also favour a smaller body mass (Webb & de Buffrénil 1990). A larger body increases speed, but at the expense of agility and manoeuvrability (Webb & de Buffrénil 1990).

Body shapes of non-benthic rhithron fishes should be closely distributed around an optimal fineness ratio (FR = ratio of body length to depth) of 4.5 (Webb 1975), in contrast to greater variation in the potamon. Scarnecchia (1988) found greater variation in FR for fish assemblages from reaches with complex flow than from those inhabiting fast, channelized reaches.

The greater muscular force needed for agility in fast currents of the rhithron suggests that those fishes should have higher metabolic rates than potamonic fishes of similar size. Clausen's (1936) initial evidence for this argument was flawed because allometric relationships were ignored. However, our survey of the metabolic literature suggests that Clausen's intuition may have been correct (Table 12.2). Haemoglobin of fishes inhabiting fast, cold stream reaches delivers greater amounts of oxygen to active tissues than in those fish found in typical potamonic habitats (Powers 1972; Cech *et al* 1979; Powers *et al* 1979). These generalizations apply to fishes swimming in the water column. Substrate-oriented fishes, including those that inhabit the hyporheos, have metabolic rates and morphologies reflecting low-flow and benthic environments (Facey & Grossman 1990).

P/B ratios and body size versus habitats

Even though metabolic expenditures are higher in the rhithron, fish communities in smaller streams tend to have higher P/B ratios (production/biomass or turnover rate on an annual basis; see section 12.4) than in larger streams (Lotrich 1973). We found a significant correlation between body size (as mean weight) and P/B ratios for 32 freshwater fish species ($r = -0.47$,

Table 12.1 Typical distributions of fish attributes along the stream gradient

| Attribute | Rhithron | Stream gradient | |
		Intermediate	Potamon
Temperature guild	Stenotherms (high gradient) Eurytherms (temperate-continental, except springfed)	Mesotherms or eurytherms (temperate)	Eurytherms (temperate)
Metabolism	High standard metabolism		Low standard metabolism (except floodplain habitats)
Fitness ratio*	FR near 4.5 (optimal score)		FR variable
Oxygen-binding affinity or haemoglobin	Low (high gradient)		High
Fish size	Smaller		Variable
Life span	Relatively short		Greater longevity
P/B ratio (index of r_{max})	Relatively high		Relatively low (except floodplain habitats)
Reproductive guild (Balon 1975)	Lithophils and phytophils	Phytolithophils and psammophils and pelagophils	Lithopelagophils

* FR, fitness ratio (body length/depth).

$P < 0.01$), which confirms an earlier analysis of diverse groups (Banse & Mosher 1980) and recent analyses of cohorts within species (Boudreau & Dickie 1989).

Small body size is encountered in the rhithron in the form of small species and young of larger species which move upstream to spawn. Small body size is also dominant in seasonally inundated habitats in floodplains where high growth rates and production occur (Bayley 1983, 1988b; Junk *et al* 1989). Both environments have two significant factors in common: they are shallow and are subjected to flooding and dewatering to a much greater extent than the main channel of the potamon. The P/B ratio is strongly related to the intrinsic rate of increase, r_{max}, of the population. Small species and young stages of larger species with high r_{max} values and other *r*-traits (*sensu* Pianka 1970) are expected to dominate habitats in the rhithron (Hall 1972) and floodplains (Junk *et al* 1989). An exception to this may be small

streams of unusual constancy of flow (groundwater type; see Fig. 12.2(a)). In the lotic component of the potamon, which is normally characterized by lower productivity, larger species with low P/B ratios and r_{max} values, and possibly species with relatively more *K*-traits (Pianka 1970), are expected to be relatively common, although large *r*-trait species whose young occupy the rhithron or floodplain may dominate.

Trophic adaptations

One morphological design often affects the rest of the body plan (Thompson 1942). Body size places limits on trophic specialization, life span, and reproductive capacity. Average food particle size is approximately 0.07 of fish length (L) (Kerr 1974), ranges up to 0.33 L for obligate piscivores (Popova 1978), but generally decreases as L increases (Webb & de Buffrénil 1990). Typical rhithron fishes are small and are primarily adapted to

Table 12.2 Standard metabolic rates of stenotherms and eurytherms among temperate fishes

Type	Fresh weight (g)	Temperature (°C)	Standard metabolic rate (mg kg^{-1}h^{-1})	Reference
Stenotherms				
Salmo trutta	100	9.5	73	Elliot (1979)
Oncorhynchus clarki	83–93	10	7	Dwyer & Kramer (1975)
		20	129	
Oncorhynchus mykiss	100	15	11	Rao (1968)
Oncorhynchus nerka	100	15	76	Brett (1965)
Salvelinus fontinalis	100	20	147	Beamish & Mookherjii (1964)
Eurytherms				
Gila atraria	100	18–20	42	Rajagopal & Kramer (1974)
Rhinichthys cataractae	100	18–22	40	Rajagopal & Kramer (1974)
Carassius auratus	87	20	50	Smit (1965)
Cyprinus carpio	100	10	16	Beamish & Mookherjii (1964)
		20	48	
Mylopharodon conocephalus	100	20	51	Alley (1977)
Catostomus catostomus	100	10	35	Beamish (1964)
		20	110	
Lepomis macrochirus	100	10	31	Wohlschlag & Juliano (1959)
		20	116	
Ictalurus nebulosus	100	10	20	Beamish (1964)
		20	66	

consume small aquatic and terrestrial invertebrates, especially terrestrial drift.

More diversification is possible in the potamon because greater habitat and food diversity confers advantages to large and small fishes (Welcomme 1985). A wide range of trophic adaptations results from the enormous productivity and variety of food in the tropical floodplain (Junk *et al* 1989). Seasonal changes result in oppportunistic behaviour. There are many generalists that consume invertebrates and plant matter of aquatic and terrestrial origin (Bayley 1988b).

Detritus and associated microflora appear to be the main repository of organic matter in river-floodplains. Subsequently many fishes have evolved to specialize on fine detritus (e.g. Bowen *et al* 1984). Such fish in the tropics start feeding on detritus when very small (Bayley 1988b) but become large (2–5 kg) compared with the small particle sizes they depend on.

The combination of seasonal food availability and low-water periods has resulted in many species building up large fat reserves to survive a season without feeding and to provide additional energy for migrations (Junk 1985). This applies to tropical and temperate rivers; the season of fasting for most species is the dry season in the tropics and the winter in the temperate zone (Cunjak & Power 1987; Cunjak *et al* 1987). Seasonal fluctuation in food supply demands that most fishes be flexible in food selection and/or energy storage within the limits of their size.

Reproductive strategies

Changes in substrates along the river continuum influence the mode of reproduction. Substrate size is much larger in high-gradient rhithron zones, corresponding to the greater hydraulic power in the steeper gradient (Richards 1982). Therefore, reproductive guilds (*sensu* Balon 1975) tend to be ordered from lithophils in high-gradient rhithron to lithopelagophils and psammophils in the potamon (see Table 12.1). Further generalization is possible by coupling Balon's reproductive guilds with Hokanson's (1977) concepts of temperature guilds. Hokanson (1977) recognized that physiological adaptations to seasonal temperatures affected patterns of growth, gonadal development and reproductive timing of temperate fishes (Table 12.3). In general, temperate fishes in the rhithron spawn primarily from the autumn to the spring, where oxygen conditions permit longer development or dormant periods, whereas most of the fishes remaining in the potamon spawn during the spring, summer or autumn months.

Large floodplains are still connected to rivers in many tropical regions and a few temperate ones, resulting in a full spectrum of lotic to lentic habitats and strong seasonal effects due to the flood pulse (Fig. 12.3) which may be independent of temperature. All spawning strategies seem to point towards giving as many young as possible access to the very productive, newly flooded, shallow areas created during the flood pulse (Junk *et al* 1989). However, the means to obtain this goal vary. The South-East Asian classification of 'black' fish and 'white' fish communities is a useful first-order differentiation of river-floodplain species (Welcomme 1985), and is partly based on reproductive strategy. Black fish prefer lentic habitats and undertake local migrations between floodplain habitats only in response to water level changes (Fig. 12.3). Many black fish species are multiple spawners and practice parental care. White fish undertake longitudinal migrations in the river, tend to be annual spawners influenced by the flood cycle, but also use the floodplain for feeding in nursery areas and, in some species, for spawning (Welcomme 1985; Junk *et al* 1989). Many white species are pelagophils, spawning in the main channel just before, or on the rise of, the flood pulse (de Godoy 1954;

Bayley 1973; Schwassmann 1978). This results in the eggs and larvae being dispersed widely over the floodplain during the flood pulse.

There are equivalents to this classification in temperate systems. For example, in North America the bluegill, *Lepomis macrochirus*, and the largemouth bass, *Micropterus salmoides*, are nestbuilding, floodplain lake species that correspond to black fish. The white basses, *Morone* spp., undergo longitudinal migrations when permitted and produce semibuoyant eggs in open water, and correspond to white fish.

Trade-offs in life-history strategies and polymorphism

Trade-offs in life histories and body designs are caused by conflicting selective pressures, particularly with reproductive strategies. Darwinian fitness of fishes is related to survival and fecundity, which are both size related (Calow 1985). The critical energy trade-off is between investment in gametes or in somatic tissues. Fish subjected to physically taxing spawning migrations tend to be semalparous and invest more energy in egg production; iteroparous fish do not exhaust body stores during the migration and tend to allocate energy to ensure post-spawning survival (Schaffer & Elson 1975; Glebe & Leggett 1981a, 1981b).

Iteroparous and semalparous life histories can occur in a single species; e.g. steelhead trout, *Oncorhynchus mykiss* (Ward & Slaney 1988); Atlantic salmon, *Salmo salar* (Schaffer & Elson 1975); and American shad, *Alosa sapidissima* (Glebe & Leggett 1981a, 1981b). A species possessing this polymorphic attribute is buffered against environmental uncertainty through increased specialization without losing fitness from con-

Table 12.3 Hokanson's (1977) temperature guilds for freshwater fishes in temperate zones

Guild	Optimal temperature (°C)	UILT (°C)	Gonadal growth phase	Spawning phenology
Stenotherm	20	<26	Summer (<20°C)	Autumn to spring (<15°C)
Mesotherm	0–28	<28–34	Autumn and winter (<28–34°C)	Spring (3–23°C)
Eurytherm	<28	<34	Increasing photoperiod (12°C)	Spring to autumn (15–32°C)

UILT, upper incipient lethal temperatures.

flicting gene recombinations. The species survives as a metapopulation, some populations of which are in the process of expanding their range and compensating for contractions by others in response to a changing environment.

Polymorphism may reflect adaptive responses to selective forces (Seghers 1974; Suzumoto *et al* 1977; Endler 1980; Taylor & McPhail 1985a, 1985b; Wade 1986), genetic drift (Allendorf & Phelps 1980) or a combination of both processes (Hatch 1990). Polymorphism can reflect fine tuning to local conditions (e.g. Zimmerman & Wooten 1981; Calhoun *et al* 1982; Matthews 1986b; Hulett 1991) or responses to large environmental gradients (Riddell & Leggett 1981; Riddell *et al* 1981; Schreck *et al* 1986; Hatch 1990). Some traits exhibit high degrees of heritability as demonstrated in breeding experiments focusing on the genetics of disease resistance (Wade 1986; Withler & Evelyn 1990). Polymorphism has been attributed to phenotypic plasticity responding to environmental change (Gee 1972; Stearns 1980; Metcalfe 1989).

Sedentary and anadromous phenotypes among salmonids have been attributed to phenotypic plasticity with arctic charr (Nordeng 1983) and to genetic differentiation with rainbow trout (Currens *et al* 1990). Various stocks exhibit polymorphic life-history traits, including migratory/sedentary tendencies among some white fishes of the river-floodplains (Welcomme 1985) and the diverse potamodromous life histories exhibited by Yellowstone cut-throat trout, *Oncorhynchus clarki bouvieri* (Varley & Gresswell 1988).

How should these stocks be managed? The most conservative, but recommended, approach is to assume that polymorphic traits are under genetic control and to adopt a policy that conserves the diversity of the metapopulation. This relates to problems with hatchery-reared fish.

Augmenting stocks from hatcheries

Hatchery fish account for the production of approximately 80% of the salmonids in the Columbia River. However, mitigation using current hatchery techniques will, at best, result in a pyrrhic victory. Although it is theoretically possible to manage hatcheries to conserve genetic

resources (Nyman & Ring 1989) and create conditions for heterozygous populations (Schreck *et al* 1986), much genetic diversity is lost through genetic drift and founder's effect (Allendorf & Phelps 1980; Ryman & Stahl 1980; Quattro & Vrijenhoek 1989) and a recent study has shown differences in selection between local hatchery and wild populations of chinook salmon (Hulett 1991). Hatchery-raised fishes tend to be more aggressive and dominate wild cohorts (Nickelson *et al* 1986; Noble 1991). The degree to which aggressive dominance is influenced by early exposure to hatchery conditions compared with genetic influences is still unknown. In any case, if hatcheries are to be used, genetic and environmental factors must be accounted for.

12.4 QUANTITATIVE MEASURES OF FISH POPULATIONS AND YIELD PREDICTION

Measures of the quantities by number or weight of fish in populations, guilds or communities is important for the management of fish for human exploitation and for the testing of effects of environmental change. Such changes can be caused by natural events (section 12.5) or human impacts (section 12.6). Details of the biases and variance associated with specific sampling methods are discussed in Chapter 13.

Rational exploitation for food or recreation requires prediction of the yield or the proportion of production that is available for human use on a sustained basis. In principle, this can be estimated through traditional dynamic pool models, from production estimates, or from comparative approaches. There are distinctions between the approaches to problems of exploitation and environmental change, but there are common practical limitations that will become apparent below.

Fish population dynamics

Fish population models (Schaefer 1954; Beverton & Holt 1957; Gulland 1969; Ricker 1975) have evolved to predict trends in intensive fisheries of significant value as measured by market economists, such as those exploiting offshore marine and anadromous populations. They explicitly or implicitly incorporate birth, death and growth rates on a given number of individuals. The simplest, the logistic or surplus-yield model (Schaefer 1954), requires the definition of a stock or viable population, yield estimates, and reliable indices of fishing mortality rate (fishing effort) and of population abundance (catch per unit effort) for a series of years. Typically, the most critical assumptions are that yield estimates be at equilibrium with the current exploitation rate, catch per unit effort be proportional to abundance density, and that the environment limiting the population is constant.

The yield-per-recruit model (Beverton & Holt 1957) requires knowledge of the age structure and natural mortality rate of the population in addition to the logistic model requirements. It provides an estimate of the fishing mortality rate that maximizes yield per fish recruited to the fishery. So far no models described have required population size estimates, but to estimate yield from the yield-per-recruit model the number of recruits obviously needs to be estimated each year (e.g. Pope 1972). Predicting recruits from abundance of parents has a good theoretical foundation (Ricker 1975), but with rare exceptions (Elliott 1985) these models are very noisy and have no predictive capability.

Can these models help us with river fisheries? They are sound theoretically, but their application is limited by the quality of data, the cost of obtaining data, and changes in the environment. Elliott's experience suggests that stream fish are more appropriate for the testing of population regulation mechanisms. More accurate estimates are possible in streams, although with more effort than most biologists realize (Bayley 1985), and fisheries in large rivers are generally more difficult to sample than benthic fish in many continental shelf environments. Fishery studies have tended to underrate environmental effects in all systems, but streams and smaller rivers are typically the most physically variable of them all.

A principal physical variable is discharge. Because fish recruitment and production are generally considered to be most affected by events in the first year of life, the variability of abundance of 0+ and 1+ fish was estimated from the literature (Table 12.4). Coefficients of variation of annual fish population measures were compared with those of annual discharge measures (Fig. 12.4). Despite the variety of streams, species, methods and lengths of stream sampled there is a clear dependence between year-to-year variability of population size and discharge. There are probably various mechanisms that connect discharge to population size, and critical periods of flow within systems may provide clues. Considering that for a given accuracy and time scale discharge is cheaper to measure than population size. It is disappointing that discharge is so often inadequately monitored.

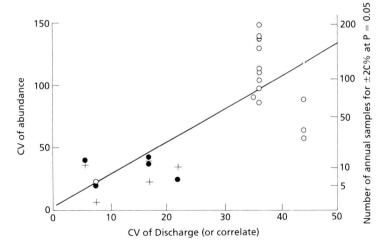

Fig. 12.4 Coefficient of variation (CV) of autumn fish abundance versus CV of annual discharge or correlate (see Table 12.4 for further explanation of data and sources). Number of samples (=years) required for a precision of $\pm 20\%$ at $P = 0.05$, based on central limit theory applied to CV of abundance predicted by regression, is shown at right. ($\bullet$, 0+ fish; $+$, 1+ fish; $\circ$, mostly 0+ and 1+ fish.

Table 12.4 Coefficients of variation for fish abundances estimated near the end of the growing season and for discharges

Stream	Country	Years of study	Species	Mean dimensions of stream segment (m)		Age class	No. of years	% Coefficient of variation of annual	
				Length	Width			Abundance	Discharge
Black Brows Beck	UK	1967–83[1]	*Salmo trutta*	75	0.8	0+	17	24	26
			Salmo trutta	75	0.8	1+	16	34	26
Hunt Creek	USA	1949–62[2]	*Salvelinus fontinalis*	2816	5.6	0+	14	20	7.4
			Salvelinus fontinalis	2816	5.6	1+	14	6.3	7.4
Hunt Creek	USA	1968–86[3]	*Salvelinus fontinalis*	1610	4.0	0/1+	20	22	7.4
Laurence Creek	USA	1953–70[4]	*Salvelinus fontinalis*	5630	—	0+	18	39	5.5
			Salvelinus fontinalis	5630	—	1+	18	36	5.5
Shelligan Burn	UK	1966–75[5]	*Salmo trutta*	167	3.6	0+	10	41	17
			Salmo trutta	167	3.6	1+	10	22	17
			Salmo salar	167	3.6	0+	10	37	17
			Salmo salar	167	3.6	1+	10	39	17
Maquoketa River	USA	1978–82[6]	*Micropterus dolomieui*	3540	27.0	0/1+	5	90	35
Middle Fork Salt River	USA	1985–89[7]	*Lepomis cyanellus*	250	1.1	0/1+	5	86	36
			Pimephales promelas	250	1.1	0/1+	5	104	36
			Lepomis humilis	250	1.1	0/1+	5	113	36
			Notemigonus crysoleucas	250	1.1	0/1+	5	97	36
			Cyprinella lutrensis	250	1.1	0/1+	5	129	36

Titus Creek	USA	1985–89[7]	Hybopsis dorsalis	250	2.9	0/1+	5	148	36
			Lepomis cyanellus	250	2.9	0/1+	5	137	36
			Etheostoma nigrum	250	2.9	0/1+	5	139	36
			Semotilus atromaculatus	250	2.9	0/1+	5	138	36
			Pimephales promelas	250	2.9	0/1+	5	133	36
Jordan Creek	USA	1950–53[8]	Pimephales notatus	1450	6.7	0/1+	4	88	44
			Campostoma anomalum	1450	6.7	0/1+	4	63	44
			Ericymba buccata	1450	6.7	0/1+	4	57	44

[1] August/September abundance data from Elliott (1985); discharge estimate uses annual rainfall variation from June to August (Elliott 1984).

[2] September abundance data from McFadden *et al* (1967); CV of flow estimated from range of 21–27 cubic feet per second.

[3] Autumn abundance data of 2–4.9-inch fish in control stretch in Alexander and Hansen (1988), CV discharge assumed same as during previous period on Hunt Creek.

[4] September abundance data from R. Hunt, Wisconsin Department of Natural Resources, personal communication. Discharge CV estimated from 1961–6 data only.

[5] Autumn abundance data from Egglishaw and Shackley (1977); discharge estimate uses annual rainfall variation as a proxy.

[6] Autumn abundance data of fish <20 cm (mostly 0+ fish) from Paragamian (1987); discharge data May–September from Paragamian and Wiley (1987).

[7] Autumn abundance data of Smale and Rabeni (1990); the five most abundant species from each stream are shown; discharge estimate uses annual rainfall variation as a proxy.

[8] August abundance data from Larimore (1955); discharge estimate uses US Geological Survey data from River Vermillion downstream.

It is sobering to consider the number of samples, which in this case is years, required to obtain estimates of cohort sizes within ±20% of the mean with 95% confidence (see right ordinate, Fig. 12.4). Although in some cases the methods could be improved to reduce bias, reduction in measurement variance usually requires more sampling, which may itself influence the population in a small stream. The streams with lowest discharge variability (Table 12.4), situated in northern Michigan and Wisconsin, are fed by groundwater. The other examples of salmonid streams, although they are surface run-off, are unusually stable. In contrast, the examples of higher discharge variability, which happen to be surface run-off midwestern streams, show what to some would be an alarming variability, although the examples from Missouri include two drought years. While much of this variability might be explained by discharge or other factors within streams, this does not help prediction if next year's weather has to be predicted first. Thus, there is an important distinction between a *post-hoc* investigation, which for example tests the effect of an environmental change given a discharge pattern, and one that needs to predict the future trend of a population for management of a fishery. Problems of assessing effects of environmental change on salmonid populations have been assessed by Hall and Knight (1981).

Although high recruitment variability is not restricted to stream fish (Dennis *et al* 1985), the resources available to adequately sample stream fish, define each stock, and account for environmental change are quite inadequate to consider using traditional models on the majority of stream or river populations. An exception is the valuable sockeye salmon mixed-stock fishery of the River Fraser, Canada, but even with 34 years of spawner-recruit data, extreme harvest rate reductions on some stocks over two generations (8 years) were deemed necessary to give >50% chance of detecting an increase in spawner abundance (Collie *et al* 1990).

Fish production

Production is defined as the total addition of biomass or equivalent potential energy to a population, including that from individuals not surviving the time period concerned. Therefore, production can represent resources available for other trophic levels. The International Biological Program (IBP) prompted many biologists to undertake fish production studies since the 1960s. These efforts have been useful in providing first-order estimates of trophic efficiency and energy flow, but have limited utility for estimating the proportion of production available to humans or to predict changes in the community due to an environmental change. These limitations result as much from the lack of accuracy in methods as from the lack of appropriate comparisons among systems.

Unfortunately, many biologists have pursued fish production estimates, which are expensive endeavours, without a clear purpose. In addition, some bemoan the lack of production estimates for the first year of life, implying that only estimates for the whole population are useful. Sometimes the contrary is true. For instance, the production of exploited year-classes furnishes an upper limit to the yield potential. The production of a particular size-range of fish may better reflect the influence of a critical habitat type. The implication that the production of a fish population represents a single trophic level is usually false. Young fish can be trophically more similar to (or even consumed by) some invertebrates than older conspecifics. Designs for studies requiring knowledge of annual production of young fish should be aware that such estimates in streams may be as variable as those for 0+ abundance levels (Fig. 12.4) and be subject to similar biases, because both estimates depend on the same limited options for estimating abundance density (Bayley 1985). A similar argument applies to the use of bioenergetics to estimate production or yield.

It is useful to estimate the production of the whole fish taxocene to compare the broad, upper-trophic-level productivity of different systems whose fauna are dominated by fish (e.g. Table 6.14 in Welcomme 1985) or to estimate the primary production sources required to support the taxocene (Bayley 1989). This approach could be refined by stratifying the production along criteria of individual body size and trophic group (or guild within zoogeographical areas), but should preferably include all significant fauna and not just fish. Conversely, comparing the production of fish populations at the species level or between species from different environments has no utility, unless differences in competitive or predatory environments and in population structure are accounted for. (For more discussion, see Chapter 17.)

Multispecies yields

The development and application of dynamic models, such as the surplus-yield model, to multispecies fisheries is in its infancy, and is severely limited by empirical data (Pauly & Murphy 1982). However, there is surprising consistency among yields of tropical river-floodplain fisheries (Welcomme 1985) that can largely be explained by fishing effort using a function related to the surplus-yield model (Bayley 1988a). This consistency across fisheries may be a result of common socioeconomic development of fishing communities in response to high species diversity (yields of species groups tend to be more stable than individual species), the dominance of native, coevolved fish fauna in tropical systems, and/or the relatively natural flood pulses and floodplain habitats in the tropics.

However, residual variability is significant, and the explanation of 74% of the variance in yield per unit floodplain area by fishing intensity (Bayley 1988a) should not hide the fact that the prediction of an individual system is still too imprecise for most management needs in individual river fisheries. Other factors at the comparative level, such as indices of system productivity or, at the individual system level, the hydrological regime in previous years (Welcomme 1985) should be explored to refine these 'top-down' approaches. However, the current models are adequate to alert politicians and economists of the current or potential animal protein yields that may be lost, and are currently being lost, by many basin development projects.

Summary

Three principal problems limit our ability to measure and predict fish populations: deriving sampling methods of known accuracy (bias and precision), accounting for temporal and spatial variability, and accounting for environmental effects, particularly discharge (see Chapter 13). We consider age determination to be a secondary issue with most populations. Available models are not considered to be as limiting as good quality empirical data. Simulation models (e.g. DeAngelis *et al* 1991) can suggest plausible sets of mechanisms and driving forces, and reasons for the high variance found empirically. Such models indicate probable areas requiring better empirical data, which can be obtained to generate another cycle of improved simulations. In the absence of a large injection of empirical data, however, it is unlikely that simulation modelling will increase the accuracy of annual prediction that managers often require, even if such predictions are conditional on future weather patterns. More empirical data based on parallel time series of independent systems is one approach to developing more efficient syntheses and predictions (section 12.7).

12.5 COMMUNITY ECOLOGY AND MANAGEMENT

Fish resources have not traditionally been managed at the assemblage or ecosystem level. At best, we have monitored the welfare of a small fraction of the assemblage and concentrated management activities within the confines of the river channel. The difficulties of estimating populations (section 12.4), the need to account for environmental constraints and adaptations (section 12.3), and the effects of changes in single components on the assemblage (this section) all indicate the necessity to understand community- or assemblage-level processes at the habitat, stream and ecosystem levels, and to develop fisheries management approaches at those levels. But do we know enough to provide viable alternatives to current management methods?

Abiotic or biotic control in response to disturbance

Fisheries managers need to know the extent to which a change in one species affects the responses of the other species in the assemblage. If biotic interactions such as predation or competition are strong, one presumes that the loss or addition of a single species would change the structure of the remaining species of the assemblage. If a physical change, such as an environmental disturbance or intense exploitation, principally affected one species, would this result in more changes in the assemblage than would be expected from only the autecologies of the remaining species? If physical forces are highly dynamic, perhaps biotic factors do not have sufficiently short response times to intervene (Hutchinson 1961; Wiens 1984). These regulatory issues are at the heart of the controversy over whether abiotic or biotic effects govern assemblage structure of stream fishes (Moyle & Li 1979; Grossman *et al* 1982; Herbold 1984; Rahel *et al* 1984; Yant *et al* 1984; Grossman *et al* 1985).

Considerations of scale and generally acceptable definitions of terms (Schoener 1987) can increase the odds of solving this difficult problem. First, regional influences on hydrology (section 12.2; Fig. 12.2) should serve as the largest scale in which to assess the role of abiotic factors in structuring fish assemblages. Second, it is important to distinguish between contingency and magnitude as factors of disturbance (Colwell 1974). High flushing flows during normal periods of peak discharge, within typical bounds, do not disturb fauna coadapted to seasonal contingencies (Resh *et al* 1988). However, such flows during the base flow period can severely impact fauna. In systems with variable flood frequency and poor predictability, such as the prairie streams described by Matthews (1987), disturbances such as drought and flash floods are frequent events and when measured at small scales (habitat units and days) result in apparently randomly varying assemblages; but at larger scales (reaches and years) assemblages appear to be relatively stable (Ross *et al* 1985; Matthews 1986a; Matthews *et al* 1988). (Comparisons between scales should ensure that a difference is not an artefact of different measurement precision; for instance, a smaller standard error can result from a larger statistical unit associated with the larger scale.)

Intermittent disturbances such as floods of low predictability can act as resetting mechanisms

(Starrett 1951; Fisher 1983; Matthews 1986a) or mediate competitive and predatory interactions (Meffe 1984). Conversely, the lack of a flood in a river-floodplain where predictability is high (see Fig. 12.2(a)) comprises a disturbance because the fauna are adapted to a regular flood pulse (Junk *et al* 1989). Variable and stable assemblage components were observed following a severe drought over a 40-month period (Freeman *et al* 1988). Some fishes remained persistent, characterized by relatively stable populations, whereas others were less persistent and highly variable. The effect on the assemblage depended on timing, especially in relation to recruitment dynamics, frequency and magnitude of the event, and regional adaptation (Freeman *et al* 1988).

Predation and competition

Biological interactions are apparent at smaller spatial scales, such as at the stream reach or channel unit level. Studies indicating competition or predation suggest that assemblage structure may be controlled by non-random factors because observations of behavioural patterns, especially in experiments, appear to be predictable and repeatable.

In general, examples of competitive displacement from large sections of stream result from species introductions (Fausch & White 1981, 1986; Cunjak & Green 1983, 1984; Meffe 1984; Castleberry & Cech 1986; Fausch 1988). Competition among native assemblages generally evokes resource partitioning through selective and interactive segregation (Baker & Ross 1981; Baltz & Moyle 1984; Moyle & Senanayake 1984; Baltz *et al* 1987; Gorman 1987; Ross *et al* 1987; Wikramanayake & Moyle 1989; Wikramanayake 1990). For instance, the innate hierarchy of similar-sized salmonids in Cascadian streams is coho salmon > steelhead trout > cutthroat trout > chinook salmon in order of dominance (Li *et al* 1987). This social hierarchy has a direct effect on selection and use of microhabitats and habitat units among sympatric salmonids. Coho salmon force other salmonids to less desirable locations.

A comparison of stream sites with and without piscivores by Bowlby and Roff (1986) indicated that the biomass of non-piscivorous fish was lower in the presence of piscivores. Piscivores can shape species composition, morphology, behaviour and size structure of their prey, as demonstrated when guppies of Trinidad were exposed to the piscivores *Rivulus harti, Crenicichla alta* and *Hoplias malabaricus* (Seghers 1974; Endler 1980). Predators can influence patterns of habitat use, activity budgets and foraging patterns of the prey through intimidation (Fraser & Cerri 1982; Gilliam & Fraser 1987).

Fish diseases and parasites

The effects of fish diseases on fish assemblage structure have been underestimated. Parasites can have more varied direct and indirect impacts on communities than predators (Holmes 1982; Holmes & Price 1986). Although spectacular epizootics have killed millions of fishes (Rohovec & Fryer 1979; Wurtsbaugh & Tapia 1988), the impacts of disease can be subtle. Patterns of abundance and distribution may appear to be governed by random physical processes when in fact a subset of species may be susceptible to a microparasite while others are not. As stated by Price *et al* (1986, p. 499) '... many interactions that appear to be between two species actually involve a third. Unless this is recognized, models will either fail to match field reality or will match it spuriously.'

Microparasitic infection has probably been a major selective force on fish assemblages of the Pacific Northwest (Li *et al* 1987). Subpopulations of steelhead trout, coho salmon and chinook salmon differ in genetic resistance to various microparasites (Suzumoto *et al* 1977; Winter *et al* 1979; Wade 1986). The degree of immunity reflects the historical distributions in space and time of the microparasite and host (Buchanan *et al* 1983). Microparasites, such as *Ceratomyxa shasta* and *Flexibacter columnaris*, can consequently act as zoogeographical barriers to specific populations. For instance, the epizootic of *F. columnaris* in the Columbia River during 1973 was responsible for the loss of approximately 80% of the spawning run of Snake River steelhead in eastern Oregon and Idaho (Becker & Fugihara 1978). Steelhead stocks that ascend early in the year in typically cold water, when columnaris

disease is neither prevalent nor virulent, were especially vulnerable. Columnaris disease became rampant when the Columbia River warmed early in the drought year of 1973, but not all size classes and species of fish were susceptible. In general, larger cypriniform fishes, a group which is tolerant of higher temperatures, were more immune (Becker & Fugihara 1978).

Monitoring changes in fish assemblages

It is well known that sufficient stress or 'press disturbance' on a community will reduce the number of species (species richness). We know that the number of species is related to the size of the basin (Welcomme 1985) and that this relationship changes among regions (Fausch *et al* 1984; Moyle & Herbold 1987) and continents (Welcomme 1985; Moyle & Herbold 1987). To the degree that species richness is predictable in natural systems, departures from the norm at a comparable scale can serve as an indicator of stress upon stream communities. However, even if non-native species are excluded and natural conditions are well defined, relative species richness is a crude measure of stress because it does not reflect different stress tolerances among species, is prone to measurement error because some species are always rare (Sheldon 1987) and suffers from measurement bias when gear selectivity cannot be corrected (Bayley *et al* 1989).

The Index of Biological Integrity (IBI) is an attempt to provide a more robust measure of stress or stream degradation than relative richness, diversity or methods using an indicator species (Karr 1981; Fausch *et al* 1984, 1990; Karr *et al* 1986, 1987), and has received wide publicity and use in the USA. The IBI is a sum of 12 metrics, each scored as odd numbers up to 5, that are estimated to represent the degree to which a particular stream locality is degraded from its natural state. The metrics, which are applied to fish that are not young-of-the-year, include species richness, proportions of trophic groups, total abundance, proportions of stress-tolerant species specific to the region concerned, abundance of all species, extent of hybridization and parasitism. Subsets of these metrics have varying degrees of correlation. Like relative species richness or diversity, the IBI is only as good as the information on the natural fish community for the reach being assessed; it is sensitive (to an unknown degree) to sampling bias; it does not relate directly to known ecological relationships among the species and with their environment; and it does not identify types of stress or disturbance. We are not suggesting that any single number could represent all these factors (multivariate approaches are called for) or that the pertinent ecological information is available (it usually is not). However, much information already exists that quantifies species/habitat relationships (sections 12.2 and 12.3). Although the IBI does not link species composition to physical characteristics, it could be modified to detect disruptions of longitudinal zonation patterns (Fausch *et al* 1984). If physical factors influence fish species distributions, then one should be able to infer 'habitat integrity' from species composition. Conversely, the IBI was insensitive to the environmental impacts of massive military exercises on a small prairie watershed (Bramblett & Fausch 1991), where hardy fishes are naturally subjected to flash floods and droughts, and whose presence and structure depend on their rate of colonization rather than habitat changes.

The IBI has definite limits and its assumptions should be carefully examined, as there are other more quantitative alternatives (Fausch *et al* 1990; Bramblett & Fausch 1991). However, the index is a laudable attempt to provide a yardstick for stream managers who have a good knowledge of local natural history (Karr *et al* 1986), but who lack cause-specific ecological information to identify the mechanism(s) and source(s) of degradation. Therefore, in common with any composite index composed of arbitrary elements, the IBI should be regarded as a useful management tool for preliminary diagnosis rather than as an index of ecological or heuristic value.

12.6 ANTHROPOGENIC DISTURBANCES, MITIGATION, AND RESTORATION

We have presented evidence that different species and life stages of fishes are adapted to particular components of the riverscape, and are sensitive to

the spatial scales and dynamics in each system. Disruption of riverscape processes causes great damage. River fisheries management operates on the river channel, yet economic and social activities outside of the lotic habitat have profound cumulative impacts on fishes.

The effects of and recovery from disturbance are scale dependent with respect to fishes (see Fig. 12.1(b)) and environmental units (see Fig. 12.1(a)). Microhabitat units are more sensitive to anthropogenic disturbances but are also the quickest to recover. In contrast, damages to the river from the surrounding landscape may be measured in hundreds of years, if one takes into account geomorphic processes and key structural organisms. For example, it is estimated that at least 500 years are required to restore drainages of the Pacific Northwest where ancient forests of Douglas firs have been clearcut (Li *et al* 1987).

Major impacts on regional scales have occurred in the plains streams of Kansas and Missouri (Cross & Moss 1987; Pflieger & Grace 1987). The combination of dams, revetments and jetties created deeper, clearer, more channelized streams of high velocity. The native fishes are morphologically and behaviourally adapted to shallow, silty, highly braided rivers. This native fauna is disappearing and introduced species are becoming dominant. Further west, two-thirds of the fish species in the Colorado basin are introduced and 17 of 54 native species are threatened, endangered or extinct (Carlson & Muth 1989) owing to similar effects of mainstream dams. Dams impede the distribution of material and energy transfer through the drainage basin, obstruct spawning migrations (Bonetto *et al* 1989; Barthem *et al* 1991), inhibit reproduction by altering thermal regimes (Baxter 1977), alter faunal structure through habitat change (Bain *et al* 1988), are centers for disease transmission (Becker & Fugihara 1978) and create unstable fish assemblages (Gelwick & Matthews 1990). Impoundments in tropical river systems, where extant floodplains have demonstrated high fish yields (Welcomme 1985; Bayley 1991), can devastate fish production, in particular downstream where the flood pulse is affected (Bonetto *et al* 1989; Junk *et al* 1989).

The temptation is to mitigate for damage using

complicated, expensive, engineering solutions because these are politically attractive; but this has unpleasant consequences that are expensive in the long run. Gore (1985) correctly stated that river restoration should be equated to 'recovery enhancement' because it should accelerate the natural process of recovery. In general, the most cost-effective mitigation procedures mimic natural processes. Such procedures will also enable relatively painless restoration in the long term.

Mitigation to restore native fishes in the intermountain western USA will require burning and logging stands of newly invaded juniper (J. Sedell, personal communication, Forest Sciences Laboratory, Oregon State University). Native bunch grasses have virtually disappeared in just over a century. They did not coevolve with large grazing ungulates and declined when livestock was introduced. Exotic plants soon became established and artificial suppression of the natural fire cycle led to poorer water infiltration, greater siltation and runoff, and the massive invasion of junipers. Each mature juniper evapotranspirates 15–35 gallons (68–160 litres) of water daily, which resulted in many streams becoming ephemeral. Restoration of natural fire cycles will provide short-term mitigation, but full restoration of the watershed will require decades.

Habitat restoration has been very successful at reach, channel unit and microhabitat levels in the central USA (White & Brynildson 1967; Hunt 1976). However, the success of enhancement projects from Alaska and the Pacific Northwest has, with few exceptions (House & Boehne 1985, 1986), not been determined (Hall & Baker 1982). Blatant failures occurred when mitigation was attempted without a regional context; what succeeds in a low gradient stream in Wisconsin will not necessarily apply to high gradient streams along the Pacific coast. Recovery times of successful projects are only in terms of years at the reach scale. Hunt (1976) found maximum response to improvements of channel structures in 5 years.

Bayley (1991) has argued that the best short-term approach to restore the function of river-floodplains impacted by navigation impoundments is to control discharge to mimic the pattern of the natural hydrograph, providing that artificial levees are also removed. However, any extensive restoration of large rivers will require watershed restoration which reduces the drainage rate, and dam removal which will restore the natural downstream transport of sediment and permit the natural flood pulse and associated vegetation to return. This is not a restoration that is just in the interest of fish and other native biota, but will provide the most cost-effective flood control (see Belt 1977).

Mitigation using exotic fishes is a great temptation. However, it is one of the poorest tools because exotic fishes can cause more problems than solutions, and the treatment is irreversible. Exotic species have contributed to the defaunation of many areas, especially areas that are disturbed (e.g. Moyle *et al* 1986; Carlson & Muth 1989; Moyle & Williams 1990). Predation and competition have been implicated in many instances (Moyle *et al* 1986), but the role of fish diseases can be insidious. Transfers of exotic fish diseases are occurring at unprecedented rates (Ganzhorn *et al* in press) and coadapted (micro) parasite–host populations are more stable than novel ones. As in germ warfare, parasites carried by resistant hosts can enhance their invasion by infecting native competitors (Holmes 1982; Freeland 1983).

12.7 CONCLUSIONS AND FUTURE RESEARCH DIRECTIONS

We argue that in most parts of the world it is now more important to spend resources on the restoration and conservation of riverine environments than to promote and maximize exploitation of species of current interest. Exceptions are relatively pristine river-floodplains in some developing countries, where sustained food fisheries are a human necessity and are advisable to provide a defence against destructive river-basin development schemes (Bayley & Petrere 1989).

Given an environment to which they are adapted, riverine fish are remarkably persistent because of the intensity and frequency of natural disturbances in evolutionary time, such as extreme floods and droughts. Therefore, in the long term the cost of correcting mistakes in management of the exploitation process is negligible compared to that of correcting or compensating for the effects of permanent changes to the environment, such as dams, water extraction, floodplain habitat removal, chronic pollution or introduction of exotics.

Research on riverine fish is controlled to a large extent by cost which is governed by sampling limitations (Fig. 12.5). There is a high variance in population size (see Fig. 12.4) which may be accounted for only partly by measurement variance, so increased sampling intensity is not necessarily a solution. As is typical with all ecosystems, the population level is the noisiest, with many measures of individual organisms and communities increasing in stability on a comparable temporal scale (Fig. 12.5). Therefore, how can we obtain the information to understand the environmental requirements to which the fish have adapted, estimate an acceptable management protocol, and apply such findings to problems of conservation, restoration, or exploitation in other systems?

We believe that a co-ordinated approach that combines extensive, protracted, empirical data collection with experiments and modelling is required. The co-operation of fishery managers is important so that some experimental control over the exploitation process is achieved. However, fisheries managers have only limited control over most aspects of the environment, and the co-operation of agencies and the private sector that are associated with the hydrology, siltation, chemical loads and in-stream structures is essential. This degree of integration is rare because of the different philosophies and individuals involved, which results in educational and institutional divisions.

Long time-series are advocated as a partial solution to improved empirical data needs. Such series are important in assessing responses by fish to unpredictable disturbances or long-term effects such as global warming. However, when seeking 'natural experiments' from the data, confounding

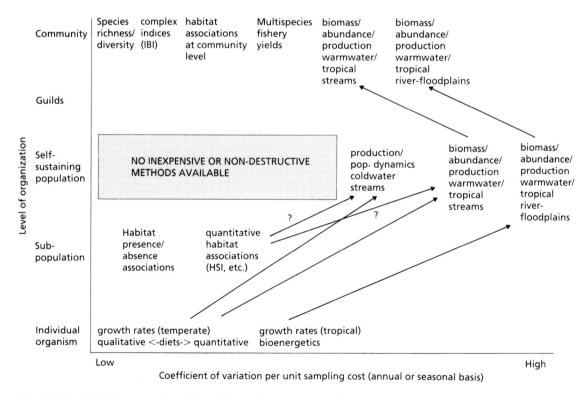

Fig. 12.5 Lotic fish investigations by level of organization versus adequate precision per unit cost. (Arrows indicate which higher level studies typically utilize information from a lower level.)

among time-correlated variables can produce too many alternative explanations. Concurrent time-series among 'ecologically comparable' systems (section 12.2) increases our ability to unravel time-correlated effects, such as when different climatic events affect fish in subsets of the systems being monitored. By ecologically comparable, we mean that the systems have comparable habitats and communities and have environments that are not drastically altered from important pristine conditions, particularly with respect to the hydrological regime. Comparisons among stressed systems are an alternative and are sometimes unavoidable, but the variety of stresses among systems may result in too many variables to control. In some river systems of developed countries this line of investigation will not be useful until some degree of restoration is attained.

Concurrent time-series of sufficient observation frequency and density will facilitate scale-dependent analyses of fish and habitat interactions that will permit more reliable inferences in other systems. One could obtain time-series by spending money on population estimates for a limited number of systems, or by spending a similar amount on community/guild/habitat/organism measures for a greater number of systems. In the majority of situations we advocate the latter approach, provided that every attempt is made to estimate fishery yield and effort. Such analyses should at least indicate the most economical paths towards the goal of restoration in many rivers and provide estimates of yield trends across sets of rivers.

Although extensive empirical data and natural experiments will narrow down the number of possible factors needed to improve prediction and understanding, the generality of results will be limited to those systems or very similar ones. Determination of dominant sets of mechanisms through experimentation and modelling within the context of a sound empirical database will broaden the application to more systems and reduce the future costs of empirical data collection. Reductionist approaches in the absence of preliminary analyses of empirical field data can result in the proposal of biologically plausible but ecologically insignificant mechanisms.

The cost and difficulty of analysing extensive field and experimental data have been reduced by improved computational facilities and methods. However, the cost of obtaining data has at least kept pace with inflation, and inadequate data have often been collected and published. If we are going to overcome the challenges of managing riverine fishes and their environment, researchers must strive for better quality data by investing more time to understand the sampling properties of their methods and to obtain the co-operation of managers and even the public to provide more extensive data. The ideal of cost-effective fishery management in rivers will not be achieved by a dependence on, or subjugation to, engineering approaches or byproducts, such as the development of unnatural hydrological regimes and attempts to utilize artificial habitats with hatchery-reared or exotic fishes.

REFERENCES

Alexander GR, Hansen EA. (1988) *Decline and recovery of a brook trout stream following an experimental addition of sand sediment.* Fisheries Research Report, July 11 1943. Michigan Department of Natural Resources. [12.4]

Allen KR. (1969) Distinctive aspects of the ecology of stream fishes: a review. *Journal of the Fisheries Research Board of Canada* **26**: 1429–38. [12.2]

Allen TFH, Starr TB. (1982) *Hierarchy: Perspectives for Ecological Complexity.* University of Chicago Press, Chicago. [12.2]

Allendorf FW, Phelps SR. (1980) Loss of genetic variation in a hatchery stock of cutthroat trout. *Transactions of the American Fisheries Society* **109**: 537–43. [12.3]

Alley DW Jr. (1977) *The energetic significance of microhabitat selection by fishes in the foothill Sierra stream.* M.S. Dissertation, University of California, Davis, California. [12.3]

Angermeier PL. (1987) Spatiotemporal variation in habitat selection by fishes in small Illinois streams. In: Matthews WJ, Heins DC (eds) *Community and Evolutionary Ecology of North American Stream Fishes*, pp 52–60. University of Oklahoma Press, Norman, Oklahoma. [12.2]

Angermeier PL, Karr JR. (1983) Fish communities along environmental gradients in a system of tropical streams. *Environmental Biology of Fishes* **9**(2): 117–35. [12.2]

Bain MB, Finn JT, Booke HE. (1988) Streamflow regulation and fish community structure. *Ecology* **69**: 382–92. [12.6]

Baker JA, Ross ST. (1981) Spatial and temporal resource

utilization by southeastern cyprinids. *Copeia* **1981**: 178–89. [12.5]

Balon EK. (1975) Reproductive guilds of fishes: a proposal and definition. *Journal of the Fisheries Research Board of Canada* **32**(6): 821–64. [12.3]

Baltz DM, Moyle PB. (1984) Segregation by species and size classes of rainbow trout, *Salmo gairdneri*, and Sacramento sucker, *Catostomus occidentalis*, in three California streams. *Environmental Biology of Fishes* **10**: 101–10. [12.5]

Baltz DM, Moyle PB, Knight NJ. (1982) Competitive interactions between benthic stream fishes, riffle sculpin, *Cottus gulosus*, and speckled dace, *Rhinichthys osculus*. *Canadian Journal of Fisheries and Aquatic Sciences* **39**: 1502–11. [12.2]

Baltz DM, Vondracek B, Brown LR, Moyle PB. (1987) Influence of temperature on microhabitat choice by fishes in a California stream. *Transactions of the American Fisheries Society* **116**: 12–20. [12.5]

Banse K, Mosher S. (1980) Adult body mass and annual production biomass relationships of field populations. *Ecological Monographs* **50**: 355–79. [12.3]

Barmuta LA, Lake PS. (1982) On the value of the river continuum concept. *New Zealand Journal of Marine and Freshwater Research* **16**: 227–31. [12.2]

Barthem RB, Ribeiro MLB, Petrere M Jr. (1991) Life strategies of some long-distance migratory catfish in relation to hydroelectric dams in the Amazon Basin. *Biological Conservation* **55**: 339–45. [12.6]

Baxter RM. (1977) Environmental effects of dams and impoundments. *Annual Review of Ecology and Systematics* **8**: 255–84. [12.6]

Bayley PB. (1973) Studies on the migratory characin, *Prochilodus platensis* Holmberg 1889 (Pisces, Characoidei) in the Rio Pilcomayo, South America. *Journal of Fish Biology* **5**: 25–40. [12.3]

Bayley PB. (1983) *Central Amazon fish populations: biomass, production and some dynamic characteristics.* PhD Dissertation, Dalhousie University, Halifax, Nova Scotia. [12.3]

Bayley PB. (1985) Sampling problems in freshwater fisheries. In: O'Hara K, Aprahamian C, Leah RT (eds) *Proceedings of the Fourth British Freshwater Fisheries Conference, 1–3 April*, pp 3–11. University of Liverpool, Liverpool. [12.4]

Bayley PB. (1988a) Accounting for effort when comparing tropical fisheries in lakes, river-floodplains, and lagoons. *Limnology and Oceanography* **33**: 963–72. [12.4]

Bayley PB. (1988b) Factors affecting growth rates of young tropical fishes: seasonality and density-dependence. *Environmental Biology of Fishes* **21**: 127–42. [12.3]

Bayley PB. (1989) Aquatic environments in the Amazon Basin, with an analysis of carbon sources, fish production, and yield. *Special Publication of the Canadian Journal of Fisheries and Aquatic Sciences* **106**: 399–408. [12.4]

Bayley PB. (1991) The flood pulse advantage and the restoration of river-floodplain systems. *Regulated Rivers: Research and Management* **6**: 75–86. [12.4, 12.6]

Bayley PB, Petrere M Jr. (1989) Amazon fisheries: assessment methods, current status, and management options. *Special Publication of the Canadian Journal of Fisheries and Aquatic Sciences* **106**: 385–98. [12.7]

Bayley PB, Larimore RW, Dowling DC. (1989) Electric seine as a fish-sampling gear in streams. *Transactions of the American Fisheries Society* **118**: 447–53.

Beamish FWH. (1964) Respiration of fishes with special emphasis on standard oxygen consumption. II. Influence of weight and temperature on respiration of several species. *Canadian Journal of Zoology* **42**: 178–88. [12.3]

Beamish FWH, Mookherjii PS. (1964) Respiration of fishes with special emphasis on standard oxygen consumption. I. Influence of weight and temperature on respiration of goldfish, *Carassius auratus* L. *Canadian Journal of Zoology* **42**: 161–75. [12.3]

Becker CD, Fugihara MP. (1978) *The bacterial pathogen, Flexibacter columnaris, and its epizootiology among Columbia River fish: a review and synthesis.* American Fisheries Society Monographs No. 2, Bethesda, Maryland. [12.5, 12.6]

Belt CB Jr. (1977) The 1973 flood and man's constriction of the Mississippi River. *Science* **189**: 681–4. [12.6]

Beverton RJH, Holt SJ. (1957) *On the Dynamics of Exploited Fish Populations.* Fisheries Investigations Series 2, Vol. 19. UK Ministry of Agriculture and Fisheries, London. [12.4]

Bisson PA, Nielsen JL, Palmason RA, Grove LE. (1982) A system of naming habitat types in small streams, with examples of habitat utilization by salmonids during low stream flow. In: Armandtrout NB (ed.) *Acquisition and Utilization of Aquatic Habitat Information*, pp 62–73. Western Division of the American Fisheries Society, Portland, Oregon. [12.2]

Bisson PA, Sullivan K, Nielsen JL. (1988) Channel hydraulics, habitat use and body form of juvenile coho salmon, steelhead, and cutthroat trout in streams. *Transactions of the American Fisheries Society* **117**: 262–73. [12.2]

Bonetto AA, Wais JR, Castello HP. (1989) The increasing damming of the Paraná Basin and its effects on the lower reaches. *Regulated Rivers: Research and Management* **4**: 333–46. [12.6]

Boudreau PR, Dickie LM. (1989) Biological model of fisheries production based on physiological and ecological scalings of body size. *Canadian Journal of Fisheries and Aquatic Sciences* **46**: 614–23. [12.3]

Bovee KD. (1982) *A guide to stream habitat analysis using the Instream Flow Incremental Methodology.*

Instream Flow Information Paper No. 12 FWS/OBS-82–26. US Fish and Wildlife Service Biological Services Program, Cooperative Instream Flow Service Group. Fort Collins, Co. [12.2]

Bowen SH, Bonetto AA, Ahgren MO. (1984) Microorganisms and detritus in the diet of a typical neotropical riverine detritivore, *Prochilodus platensis* (Piscies: Prochilodontidae). *Limnology and Oceanography* **29**(5): 1120–2. [12.3]

Bowlby JN, Roff JC. (1986) Trophic structure in southern Ontario streams. *Ecology* **67**: 1670–9. [12.5]

Bramblett RG, Fausch KD. (1991) Variable fish communities and the Index of Biological Integrity in a western Great Plains river. *Transactions of the American Fisheries Society* **120**: 752–69. [12.5]

Brett JR. (1965) The relation of size to rate of oxygen consumption and sustained swimming speed of sockeye salmon (*Oncorhynchus nerka*). *Journal of the Fisheries Research Board of Canada* **22**: 1491–501. [12.3]

Buchanan DV, Sanders JE, Zinn JL, Fryer JL. (1983) Relative susceptibility of four strains of summer steelhead to infection by *Ceratomyxa shasta. Transactions of the American Fisheries Society* **112**: 541–3. [12.3]

Calhoun SW, Zimmerman EG, Beitinger TL. (1982) Stream regulation alters acute temperature preferenda of red shiners, *Notropis lutrensis. Canadian Journal of Fisheries and Aquatic Sciences* **39**: 360–3. [12.3]

Calow P. (1985) Adaptive aspects of energy allocation. In: Tytler P, Calow P (eds) *Fish Energetics: New Perspectives*, pp 13–31. The Johns Hopkins University Press, Baltimore, Maryland. [12.3]

Carlson CA, Muth RT. (1989) The Colorado River: lifeline of the American Southwest. *Special Publication of the Canadian Journal of Fisheries and Aquatic Sciences* **106**: 220–39. [12.6]

Castleberry DT, Cech JJ Jr. (1986) Physiological responses of a native and introduced desert fish to environmental stressors. *Ecology* **67**: 912–18. [12.5]

Cech JJ Jr, Mitchell SJ, Massingill MJ. (1979) Respiratory adaptations of Sacramento blackfish, *Orthodon microlepidotus* (Ayers), for hypoxia. *Comparative Biochemistry and Physiology* **63A**: 411–15. [12.3]

Clausen RG. (1936) Oxygen consumption in fresh water fishes. *Ecology* **17**: 216–26. [12.3]

Collie JS, Peterman RM, Walters CJ. (1990) Experimental harvest policies for a mixed-stock fishery: Fraser River sockeye salmon, *Oncorhynchus nerka. Canadian Journal of Fisheries and Aquatic Sciences* **47**: 145–55. [12.4]

Colwell RK. (1974) Predictability, constancy, and contingency of periodic phenomena. *Ecology* **55**: 1148–53. [12.5]

Conder AL, Annear TC. (1987) Test of weighted usable area estimates derived from a PHABSIM model for instream flow studies on trout streams. *North American Journal of Fisheries Management* **7**: 339–50. [12.2]

Cross FB, Moss RE. (1987) Historic changes in fish communities and aquatic habitats in plains streams of Kansas. In: Matthews WJ, Heins DC (eds) *Community and Evolutionary Ecology of North American Stream Fishes*, pp 155–65. University of Oklahoma Press, Norman, Oklahoma. [12.6]

Cunjak RA, Green JM. (1983) Habitat utilization by brook char (*Salvelinus fontinalis*) and rainbow trout (*Salmo gairdneri*) in Newfoundland streams. *Canadian Journal of Zoology* **61**: 1214–19. [12.5]

Cunjak RA, Green JM. (1984) Species dominance by brook trout and rainbow trout in a simulated stream environment. *Transactions of the American Fisheries Society* **113**: 737–43. [12.5]

Cunjak RA, Power G. (1987) The feeding and energetics of stream-resident trout in winter. *Journal of Fish Biology* **31**: 493–511.

Cunjak RA, Curry RA, Power G. (1987) Seasonal energy budget of brook trout in streams: implications of a possible deficit in early winter. *Transactions of the American Fisheries Society* **116**: 817–28. [12.3]

Currens KP, Schreck CB, Li HW. (1990) Allozyme and morphological divergence of rainbow trout (*Oncorhynchus mykiss*) above and below waterfalls in the Deschutes River, Oregon. *Copeia* **1990**: 730–46. [12.3]

de Godoy MP. (1954) Locais de desovas de peixes num trecho do Rio Mogi Guaçu, Estado de São Paulo, Brasil. *Revista Brasileira de Biologia* **14**(4): 375–96. [12.3]

DeAngelis DL, Godbout L, Shuter BJ. (1991) An individual-based approach to predicting density-dependent dynamics in smallmouth bass populations. *Ecological Modelling* **57**: 91–116. [12.4]

Dennis B, Brown BE, Stage AR, Burkhart HE, Clark S. (1985) Problems of modeling growth and yield of renewable resources. *The American Statistician* **39**(4): 374–83. [12.4]

Dwyer WP, Kramer RH (1975) The influence of temperature on scope for activity in cutthroat trout, *Salmo clarki. Transactions of the American Fisheries Society* **104**: 552–4. [12.3]

Echelle AA, Schnell GD. (1976) Factor analysis of species association among fishes of the Kiamichi River, Oklahoma. *Transactions of the American Fisheries Society* **105**: 17–31. [12.2]

Egglishaw HJ, Shackley PE. (1977) Growth, survival and production of juvenile salmon and trout in a Scottish stream, 1966–75. *Journal of Fish Biology* **11**: 647–72. [12.4]

Elliott JM. (1979) Energetics of freshwater teleosts. *Symposia of the Zoological Society of London* **44**: 29–61. [12.3]

Elliott JM. (1984) Numerical changes and population regulation in young migratory trout *Salmo trutta* in a

Lake District stream, 1966–83. *Journal of Animal Ecology* **53**: 327–50. [12.4]

Elliott JM. (1985) Population regulation for different life-stages of migratory trout *Salmo trutta* in a Lake District stream, 1966–83. *Journal of Animal Ecology* **54**: 617–38. [12.4]

Elwood JW, Newbold JD, O'Neill RV, van Winkle W. (1983) Resource spiraling: an operational paradigm for analyzing lotic ecosystems. In: Fontaine TD, Bartell SM (eds) *Dynamics of Lotic Ecosystems*, pp 3–28. Ann Arbor Science, Ann Arbor, Michigan. [12.2]

Endler J. (1980) Natural selection on color patterns in *Poecilia reticulata. Evolution* **34**: 76–91. [12.3, 12.5]

Facey DE, Grossman GD. (1990) The metabolic cost of maintaining position for four North American stream fishes. *Physiological Zoology* **63**: 757–76. [12.3]

Fausch KD. (1988) Tests of competition between native and introduced salmonids in streams: what have we learned? *Canadian Journal of Fisheries and Aquatic Sciences* **45**: 2238–46. [12.5]

Fausch KD, White RJ. (1981) Competition between brook trout (*Salvelinus fontinalis*) and brown trout (*Salmo trutta*) for positions in a Michigan stream. *Canadian Journal of Fisheries and Aquatic Sciences* **38**: 1220–7. [12.5]

Fausch KD, White RJ. (1986) Competition among juveniles of coho salmon, brook trout, and brown trout in a laboratory stream, and implications for Great Lakes tributaries. *Transactions of the American Fisheries Society* **115**: 363–81. [12.5]

Fausch KD, Karr JR, Yant PR. (1984) Regional application of an index of biotic integrity based on stream fish communities. *Transactions of the American Fisheries Society* **113**: 39–55. [12.5]

Fausch KD, Lyons J, Karr JR, Angermeier PL. (1990) Fish communities as indicators of environmental degradation. *American Fisheries Society Symposium* **8**: 123–44. [12.5]

Felley JD, Felley SM. (1987) Relationship between habitat selection by individuals of a species and patterns of habitat segregation among species: fishes of the Calcasieu drainage. In: Matthews WJ, Heins DC. (eds) *Community and Evolutionary Ecology of North American Stream Fishes*, pp 61–8. University of Oklahoma Press, Norman, Oklahoma. [12.2]

Fisher SG. (1983) *Succession in Streams*. Plenum Press, New York. [12.5]

Fraser DF, Cerri RD. (1982) Experimental evaluation of predator–prey relationships in a patchy environment: consequences for habitat use patterns in minnows. *Ecology* **63**: 307–13. [12.5]

Freeland WJ. (1983) Parasites and the coexistence of animal host species. *American Naturalist* **121**: 223–36. [12.6]

Freeman MC, Crawford M, Barrett J, Facey DE, Flood M, Stouder D, Grossman GD. (1988) Fish assemblage stability in a southern Appalachian stream. *Canadian Journal of Fisheries and Aquatic Sciences* **45**: 1949–58. [12.5]

Frissell CA, Liss WJ, Warren CE, Hurley MD. (1986) A hierarchical framework for stream habitat classification: viewing streams in a watershed context. *Environmental Management* **10**: 199–214. [12.2]

Ganzhorn J, Rohovec JS, Fryer JL. (in press) Dissemination of microbial pathogens through introductions and transfers of finfish. In: Rosenfield A, Mann R (eds) *Dispersal of Living Organisms into Aquatic Ecosystems*, Maryland Sea Grant Program, College Park, Maryland. [12.6]

Gard R, Flittner GA. (1974) Distribution and abundance of fishes in Sagehen Creek, California. *Journal of Wildlife Management* **38**: 347–58.

Gee JH. (1972) Adaptive variation in swimbladder length and volume in dace, genus *Rhinichthys. Journal of the Fisheries Research Board of Canada* **29**: 119–27. [12.3]

Gelwick FP, Matthews WJ. (1990) Temporal and spatial patterns in littoral-zone fish assemblages of a reservoir (Lake Texoma, Oklahoma-Texas, U.S.A.). *Environmental Biology of Fishes* **27**: 107–20. [12.6]

Gilliam JF, Fraser DF. (1987) Habitat selection under predation hazard: test of a model with foraging minnows. *Ecology* **68**: 1856–62. [12.5]

Glebe BD, Leggett WC. (1981a) Latitudinal differences in energy allocation and use during the freshwater migrations of American shad (*Alosa sapidissima*) and their life history consequences. *Canadian Journal of Fisheries and Aquatic Sciences* **38**: 806–20. [12.3]

Glebe BD, Leggett WC. (1981b) Temporal, intra-population differences in energy allocation and use by American Shad (*Alosa sapidissima*) during the spawning migration. *Canadian Journal of Fisheries and Aquatic Sciences* **38**: 795–805. [12.3]

Gore JA. (1985) Introduction. In: Gore JA (ed.) *The Restoration of Rivers and Streams*, pp vii–xii. Butterworth, Boston. [12.6]

Gorman OT. (1987) Habitat segregation in an assemblage of minnows in an Ozark stream. In: Matthews WJ, Heins DC (eds) *Community and Evolutionary Ecology of North American Stream Fishes*, pp 33–41. University of Oklahoma Press, Norman, Oklahoma. [12.5]

Gorman OT. (1988) The dynamics of habitat use in a guild of Ozark minnows. *Ecological Monographs* **58**: 1–18. [12.2]

Gorman OT, Karr JR. (1978) Habitat structure and stream fish communities. *Ecology* **59**: 507–15. [12.2, 12.3]

Grossman GD, Moyle PB, Whitaker JO Jr. (1982) Stochasticity in structural and functional characteristics of an Indiana stream fish assemblage: a test of community theory *American Naturalist* **120**: 423–54. [12.5]

Grossman GD, Freeman MC, Moyle PB, Whitaker JO Jr. (1985) Stochasticity and assemblage organization in an Indiana stream fish assemblage. *American Naturalist* **126**: 275–85. [12.5]

Gulland JA. (1969) *Manual of methods for fish stock assessment: Part 1. Fish population analysis.* FAO Manuals in Fisheries Science, No. 4. Food and Agriculture Organization, Rome. [12.4]

Hall CAS. (1972) Migration and metabolism in a temperate stream ecosystem. *Ecology* **53**: 586–604. [12.2, 12.3]

Hall JD, Baker CD. (1982) Rehabilitating and enhancing stream habitat: 1. Review and evaluation. In: Meehan WR (ed.) *Influences of forest and rangeland management on anadromous fish habitat in western North America.* USDAS Forest Service, General Technical Report PNW-138, Portland, Oregon. p. 29. [12.6]

Hall JD, Knight NJ. (1981) *Natural variation in abundance of salmonid populations in streams and its implications for design of impact studies.* EPA-600/S3-81–021. July 1981. US Environmental Protection Agency report. Corvallis, Oregon. [12.4]

Hatch KM. (1990) *Phenotypic comparison of thirty-eight steelhead* (Oncorhynchus mykiss) *populations from coastal Oregon.* M.S. Dissertation, Oregon State University, Corvallis, Oregon. [12.3]

Herbold B. (1984) Structure of an Indiana stream fish association: choosing an appropriate model. *American Naturalist* **124**: 561–72. [12.5]

Hicks BJ. (1990) *The Influence of geology and timber harvest on channel morphology and salmonid populations in Oregon Coast Range streams.* PhD Dissertation, Oregon State University, Corvallis, Oregon. [12.2]

Hicks BJ, Hall JD, Bisson PA, Sedell JR. (1991) Response of salmonid populations to habitat changes caused by timber harvest. In: Meehan WR (ed.) *The Influence of Forest and Rangeland Management on Salmonids and their Habitat,* pp 484–518. Special Publication of the American Fisheries Society, Bethesda, Maryland. [12.2]

Hokanson KEF. (1977) Temperature requirements of some percids and adaptations to the seasonal temperature cycle. *Journal of the Fisheries Board of Canada* **34**: 1524–50. [12.3]

Holmes JC. (1982) Impact of infectious disease agents on the population growth and geographical distribution of animals. In: Anderson RM, May RM (eds) *Population Biology of Infectious Diseases,* pp 37–51. Springer-Verlag, Berlin. [12.5, 12.6]

Holmes JC, Price PW. (1986) Communities of parasites. In: Kikkikawa J, Anderson DJ (eds) *Community Ecology: Pattern and Process,* pp 187–213. Blackwell Scientific Publications, Melbourne. [12.5]

Horwitz RJ. (1978) Temporal variability patterns and the distributional patterns of stream fishes. *Ecological Monographs* **48**: 307–321. [12.2, 12.3]

House RA, Boehne PL. (1985) Evaluation of instream structures for salmonid spawning and rearing in a coastal Oregon stream. *North American Journal of Fisheries Management* **5**: 283–95. [12.6]

House RA, Boehne PL. (1986) *Effects of instream structure on salmonid habitat and populations in Tobe Creek, Oregon. North American Journal of Fisheries Management* **6**: 38–46. [12.6]

Huet M. (1947) Aperçu des relations entre la pente et les populations piscicoles des eaux courantes. *Schweizerische Zeitschrift für Hydrologie* **11**: 333–51. [12.2]

Huet M. (1959) Profiles and biology of Western European streams as related to fish management. *Transactions of the American Fisheries Society* **88**: 155–63. [12.2]

Hughes RM, Rexstad E, Bond CE. (1987) The relationship of aquatic ecoregions, river basins and physiographic provinces to the ichthyogeographic regions of Oregon. *Copeia* **1987**: 423–52. [12.2]

Hughes RM, Whittier TR, Rohm CM, Larsen DP. (1990) A regional framework for establishing recovery criteria. *Environmental Management* **14**: 673–83. [12.2]

Hulett PL. (1991) *Patterns of genetic inheritance and variation through ontogeny for hatchery and wild stocks of chinook salmon.* M.S. Dissertation, Oregon State University, Corvallis, Oregon. [12.3]

Hunt RL. (1976) A long-term evaluation of trout habitat development and its relation to improving management-related research. *Transactions of the American Fisheries Society* **105**: 361–4. [12.6]

Hutchinson GE. (1961) Paradox of the plankton. *American Naturalist* **95**: 137–46. [12.5]

Johnson DH. (1980) The comparison of usage and availability measurements for evaluating resource preference. *Ecology* **61**: 65–71. [12.2]

Junk WJ. (1985) Temporary fat storage, an adaptation of some fish species to the waterlevel fluctuations and related environmental changes of the Amazon system. *Amazoniana* **9**: 315–51. [12.3]

Junk WJ, Bayley PB, Sparks RE. (1989) The flood pulse concept in river-floodplain systems. *Special publication of the Canadian Journal of Fisheries and Aquatic Sciences* **106**: 110–27. [12.2, 12.3, 12.5, 12.6]

Karr JR. (1981) Assessment of biotic integrity using fish communities. *Fisheries (Bethesda)* **6**(6): 21–7. [12.5]

Karr JR, Fausch KD, Angermeier PL, Yant PR, Schlosser IJ. (1986) *Assessing biological integrity in running waters. A method and its rationale* Special publication No. 5, September 5. Illinois Natural History Survey. Champaign, Illinois. [12.5]

Karr JR, Yant PR, Fausch KD, Schlosser IJ. (1987) Spatial and temporal variability of the index of biotic integrity in three midwestern streams. *Transactions of the American Fisheries Society* **116**: 1–11. [12.5]

Kerr SR. (1974) Theory of size distribution in ecological communities. *Journal of the Fisheries Research Board of Canada* **31**: 1859–62. [12.3]

Larimore RW. (1955) Minnow productivity in a small Illinois stream. *Transactions of the American Fisheries Society* **84**: 110–16. [12.4]

Larsen DP, Omernik JM, Hughes RM, Rohm CM, Whittier TR, Kinney AJ et al (1986) Correspondence between spatial patterns in fish assemblages in Ohio, USA, streams and aquatic ecoregions. *Environmental Management* **10**: 815–28. [12.2]

Li HW, Schreck CB, Bond CE, Rexstad E. (1987) Factors influencing changes in fish assemblages of Pacific Northwest streams. In: Matthews WJ, Heins DC (eds) *Community and Evolutionary Ecology of North American Stream Fishes*, pp 193–202. University of Oklahoma Press, Norman, Oklahoma. [12.2, 12.5, 12.6]

Lotrich VA. (1973) Growth, production, and community composition of fishes inhabiting a first-, second-, and third-order stream of eastern Kentucky. *Ecological Monographs* **43**: 377–97. [12.2, 12.3]

Lowe-McConnell RH. (1987) *Ecological Studies in Tropical Fish Communities.* Cambridge Tropical Biology Series. Cambridge University Press, Cambridge. [12.2]

Lyons J. (1989) Correspondence between the distribution of fish assemblages in Wisconsin streams and Omernik's ecoregions. *American Midland Naturalist* **122**: 163–82. [12.2]

MacArthur RH, Levins R. (1967) The limiting similarity, convergence and divergence of coexisting species. *American Naturalist* **101**: 377–85. [12.3]

McFadden JT, Alexander GR, Shetter DS. (1967) Numerical changes and population regulation in brook trout *Salvelinus fontinalis. Journal of the Fisheries Research Board of Canada* **24**: 1425–59. [12.4]

Mahon R. (1984) Divergent structure in fish taxocenes of north temperate streams. *Canadian Journal of Fisheries and Aquatic Sciences* **41**: 330–50. [12.2]

Mann KH, Britton RH, Kowalczewski A, Lack TJ, Mathews CP, McDonald I. (1970) Productivity and energy flow at all trophic levels in the River Thames, England. In: Kajak Z, Hillbricht-Ilkowska A (eds) *Productivity Problems of Freshwaters*, pp 579–96. Proceedings of the IBP-UNESCO Symposium on Productivity Problems of Freshwaters, Kazimierz Dolny, Poland. [12.2]

Mathur D, Bason WH, Purdy EJ Jr, Silver CA. (1985) A critique of the Instream Flow Incremental Methodology. *Canadian Journal of Fisheries and Aquatic Sciences* **42**: 825–31. [12.2]

Matthews WJ. (1986a) Fish faunal structure in an Ozark stream: stability, persistence, and a catastrophic flood. *Copeia* **1986**: 388–97. [12.5]

Matthews WJ. (1986b) Geographic variation in thermal tolerance of a widespread minnow (*Notropis lutrensis*) of the North American mid-west. *Journal of Fish Biology* **28**: 407–17. [12.2, 12.3]

Matthews WJ. (1988) North American prairie streams as systems for ecological study. *Journal of the North American Benthological Society* **7**: 387–409. [12.2]

Matthews WJ, Styron JT Jr. (1981) Tolerance of headwater vs. mainstream fishes for abrupt physicochemical changes. *American Midland Naturalist* **105**: 149–58. [12.2]

Matthews WJ, Cashner RC, Gelwick FP. (1988) Stability and persistence of fish faunas and assemblages in three midwestern streams. *Copeia* **1988**: 945–55. [12.2, 12.5]

Matthews WM. (1987) Physicochemical tolerance and selectivity of stream fishes as related to their geographic ranges and local distributions. In: Matthews WJ, Heins DC (eds) *Community and Evolutionary Ecology of North American Stream Fishes*, pp 111–20. University of Oklahoma Press, Norman, Oklahoma. [12.5]

Mayden RL. (1987) Historical ecology and North American highland fishes: a research program in community ecology. In: Matthews WJ, Heins DC (eds) *Community and Evolutionary Ecology of North American Stream Fishes*, pp 210–22. University of Oklahoma Press, Norman, Oklahoma. [12.2]

Meffe GK. (1984) Effects of abiotic disturbance on coexistence of predatory–prey fish species. *Ecology* **65**: 1525–34. [12.5]

Metcalfe NB. (1989) Differential response to a competitor by Atlantic salmon adopting alternative life-history strategies. *Proceedings of the Royal Society of London B* **236**: 21–7. [12.3]

Moyle PB, Baltz DM. (1985) Microhabitat use by an assemblage of California stream fishes: developing criteria for instream flow determinations. *Transactions of the American Fisheries Society* **114**: 695–704. [12.2]

Moyle PB, Herbold B. (1987) Life-history patterns and community structure in stream fishes of western North America: comparisons with eastern North America and Europe. In: Matthews WJ, Heins CD (eds) *Community and Evolutionary Ecology of North American Stream Fishes*, pp 25–32. University of Oklahoma Press, Norman, Oklahoma. [12.2, 12.5]

Moyle PB, Li HW. (1979) Community ecology and predator-prey systems in warmwater streams. In: Stroud RE, Clepper HE (eds) *Predator–Prey Systems in Fisheries Management*, pp 171–81. Sports Fisheries Institute, Washington, DC. [12.5]

Moyle PB, Senanayake FR. (1984) Resource partitioning among the fishes of rainforest streams in Sri Lanka. *Journal of Zoology, London* **202**: 195–223. [12.2, 12.5]

Moyle PB, Williams JE. (1990) Biodiversity loss in the temperate zone: decline of the native fish fauna of

California. *Conservation Biology* **4**: 275–84. [12.6]

Moyle PB, Daniels RA, Herbold BL, Baltz DM. (1986) Patterns in distribution and abundance of a noncoevolved assemblage of estuarine fishes in California. *Fishery Bulletin (NOAA)* **84**: 105–17. [12.6]

Naiman RJ, Décamps H, Pastor J, Johnston CA. (1988) The potential importance of boundaries to fluvial ecosystems. *Journal of the North American Benthological Society* **7**(4): 289–306. [12.2]

Newcombe C. (1981) A procedure to estimate changes in fish populations caused by changes in stream discharge. *Transactions of the American Fisheries Society* **110**: 382–390. [12.2]

Nickelson TE, Solazzi MF, Johnson SL. (1986) Use of hatchery coho salmon (*Oncorhynchus kisutch*) presmolts to rebuild wild populations in Oregon coastal streams. *Canadian Journal of Fisheries and Aquatic Sciences* **43**: 2443–9. [12.3]

Noble SL. (1991) *Impacts of earlier emerging steelhead fry of hatchery origin on the social structure, distribution, and growth of wild steelhead fry.* M.S. Dissertation, Oregon State University, Corvallis, Oregon. [12.3]

Nordeng H. (1983) Solution to the 'char problem' based arctic char (*Salvelinus alpinus*) in Norway. *Canadian Journal of Fisheries and Aquatic Sciences* **40**: 1372–87. [12.3]

Nyman L, Ring O. (1989) Effects of hatchery environment on three polymorphic loci in arctic char (*Salvelinus alpinus* species complex). *Nordic Journal of Freshwater Research* **65**: 34–43. [12.3]

Omernik JM. (1987) Ecoregions of the conterminous United States (with map supplement). *Annals of the Association of American Geographers* **77**: 118–25. [12.2]

Orth DJ, Maughan OE. (1982) Evaluation of the incremental methodology for recommending instream flows for fishes. *Transactions of the American Fisheries Society* **110**: 1–13. [12.2]

Osborne LL, Wiley MJ. (1988) Empirical relationships between land use/cover and stream quality in an agricultural watershed. *Journal of Environmental Management* **26**: 9–27. [12.2]

Paragamian VL. (1987) Production of smallmouth bass in the Maquoketa River, Iowa. *Journal of Freshwater Ecology* **4**: 141–8. [12.4]

Paragamian VL, Wiley MJ. (1987) Effects of variable streamflows on growth of smallmouth bass in the Maquoketa River, Iowa. *North American Journal of Fisheries Management* **7**: 357–62. [12.4]

Pauly D, Murphy GI. (eds) (1982) *Theory and Management of Tropical Fisheries.* International Center for Living Aquatic Resources Management, Manila, Conference Proceedings 9. Cronulla, Australia. [12.4]

Petrere M Jr. (1985) *Migraciones de peces de agua dulce en America Latina: algunos comentarios.*

COPESCAL Doc. Ocas. No. 1. FAO Organizacion de las Naciones Unidas para la Agricultura y la Alimentacion. COPESCAL/OP1. p. 17. Rome, Italy. [12.2]

Pflieger WL, Grace TB. (1987) Changes in the fish fauna of the lower Missouri River, 1940–1983. In: Matthews WJ, Heins DC (eds) *Community and Evolutionary Ecology of North American Stream Fishes,* pp 166–77. University of Oklahoma Press, Norman, Oklahoma. [12.6]

Pianka ER. (1970) On *r*- and *K*-selection. *American Naturalist* **104**: 592–7. [12.3]

Poff LN, Ward JV. (1989) Implications of streamflow variability and predictability for lotic community structure: a regional analysis of streamflow patterns. *Canadian Journal of Fisheries and Aquatic Sciences* **46**: 1805–18. [12.2]

Pope JG. (1972) An investigation of the accuracy of virtual population analysis using cohort analysis. *International Commission for the Northwest Atlantic Fisheries Research Bulletin* **9**: 65–74. [12.4]

Popova OA. (1978) The role of predacious fish in ecosystems. In: Gerking SD (ed.) *Ecology of Freshwater Fish Production,* pp 215–49. Blackwell Scientific Publications, Oxford. [12.3]

Powers DA. (1972) Hemoglobin adaptation for fast and slow water habitats in sympatric catostomid fishes. *Science* **177**: 360–2. [12.3]

Powers DA, Fyhn HJ, Fyhn UEH, Marin JP, Garlick RL, Wood SC. (1979) A comparative study of the oxygen equilibria of blood from 40 genera of Amazonian fishes. *Comparative Biochemistry and Physiology* **62A**: 67–85. [12.3]

Price PW, Westoby M, Rice B, Atsatt PR, Fritz RS, Thompson JN, Mobley K. (1986) Parasite mediations in ecological interactions. *Annual Review of Ecology and Systematics* **17**: 487–506. [12.5]

Pringle CM, Naiman RJ, Bretschko G, Karr JR, Oswood MW, Webster JR *et al.* (1988) Patch dynamics in lotic systems: the stream as a mosaic. *Journal of the North American Benthological Society* **7**(4): 503–24. [12.2]

Quattro JM, Vrijenhoek RC. (1989) Fitness differences among remnant populations of the endangered Sonoran topminnow. *Science* **245**: 976–78. [12.3]

Rahel FJ, Lyons JD, Cochran PA. (1984) Stochastic or deterministic regulation of assemblage structure? It may depend on how the assemblage is defined. *American Naturalist* **124**: 583–9. [12.5]

Rajagopal PK, Kramer RH. (1974) Respiratory metabolism of Utah chub, Gila atraria (Girard) and speckled dace, *Rhinichthys osculus* (Girard). *Journal of Fish Biology* **6**: 215–22. [12.3]

Rao GMM. (1968) Oxygen consumption of rainbow trout (*Salmo gairdneri*) in relation to activity and salinity. *Canadian Journal of Zoology* **62**: 866–87. [12.3]

Resh VH, Brown AV, Covich AP, Gurtz ME, Li HW,

Minshall GW *et al.* (1988) The role of disturbance in stream ecology. *Journal of the North American Benthological Society* **7**: 433–55. [12.5]

Richards K. (1982) *Rivers: Form and Process in Alluvial Channels.* Methuen, London. [12.3]

Ricker WE. (1975) Computation and interpretation of biological statistics of fish populations. *Bulletin of the Fisheries Research Board of Canada* **191**: 382. [12.4]

Riddell BE, Leggett WC. (1981) Evidence of an adaptive basis for geographic variation in body morphology and time of downstream migration of juvenile Atlantic salmon (*Salmo salar*). *Canadian Journal of Fisheries and Aquatic Sciences* **38**: 308–20. [12.3]

Riddell BE, Leggett WC, Saunders RL. (1981) Evidence of adaptive polygenic variation between two populations of Atlantic salmon (*Salmo salar*) native to tributaries of the S.W. Miramichi River, N.B. *Canadian Journal of Fisheries and Aquatic Sciences* **38**: 321–33. [12.3]

Rohm CM, Giese JW, Bennett CC. (1987) Evaluation of an aquatic ecoregion classification of streams in Arkansas. *Journal of Freshwater Ecology* **4**: 127–40. [12.2]

Rohovec JS, Fryer JL. (1979) Fish health management in aquaculture. In: Klingeman PC (ed.) *Aquaculture: A Modern Fish Tale*, pp 15–36. Water Resource Research Institute, Oregon State University, Seminar Series, SEMIN WR 026–79, Corvallis, Oregon. [12.5]

Ross ST, Matthews WJ, Echelle AA. (1985) Persistence of stream fish assemblages: effects of environmental change. *American Naturalist* **126**: 24–40.

Ross ST, Baker JA, Clark KE. (1987) Microhabitat partitioning of southeastern stream fishes: temporal and spatial predictability. In: Matthews WJ, Heins DC (eds) *Community and Evolutionary Ecology of North American Stream Fishes*, pp 42–51. University of Oklahoma Press, Norman, Oklahoma. [12.2, 12.5]

Ryman N, Stahl G. (1980) Genetic changes in hatchery stocks of brown trout (*Salmo trutta*). *Canadian Journal of Fisheries and Aquatic Sciences* **37**: 82–7. [12.3]

Scarnecchia DL. (1988) The importance of streamlining in influencing fish community structure in channelized and unchannelized reaches of a prairie stream. *Regulated Rivers: Research and Management* **2**: 155–66. [12.2, 12.3]

Schaefer MB. (1954) Some aspects of the dynamics of populations important to the management of the commercial marine fisheries. *Bulletin of the Inter-American Tropical Tuna Commission* **1**: 27–56. [12.4]

Schaffer WM, Elson PF. (1975) The adaptive significance of variation in life history among local populations of Atlantic salmon in North America. *Ecology* **56**: 577–90. [12.3]

Schlosser IJ. (1982) Fish community structure and function along two habitat gradients in a headwater stream. *Ecological Monographs* **52**(4): 395–414. [12.3]

Schlosser IJ. (1987) A conceptual framework for fish communities in small warmwater streams. In: Matthews WJ, Heins DC (eds) *Community and Evolutionary Ecology of North American Stream Fishes*, pp 17–24. University of Oklahoma Press, Norman, Oklahoma. [12.3]

Schoener TW. (1986) Overview: kinds of ecological communities — ecology becomes pluralistic. In: Diamond J, Case TJ (eds) *Community Ecology*, pp 467–79. Harper and Row, New York. [12.2]

Schoener TW. (1987) Axes of controversy in community ecology. In: Matthews WJ, Heins DC (eds) *Community and Evolutionary Ecology of North American Stream Fishes*, pp 8–16. University of Oklahoma Press, Norman, Oklahoma. [12.2, 12.5]

Schreck CB, Li HW, Hjort RC, Sharpe CS, Currens KP, Hulett PL *et al.* (1986) *Stock identification of the Columbia River chinook salmon and steelhead trout.* Bonneville Power Administration Final Report, Portland, Oregon. [12.3]

Schwassmann HO. (1978) Times of annual spawning and reproductive strategies in Amazonian fishes. In: Thorp JE (ed.) *Rhythmic Activity of Fish*, pp 186–200. Academic Press, London. [12.3]

Seghers BH. (1974) Schooling behavior in the guppy (*Poecilia reticulata*): an evolutionary response to predation. *Evolution* **28**: 486–9. [12.3, 12.5]

Sheldon AL. (1968) Species diversity and longitudinal succession in stream fishes. *Ecology* **49**: 193–8. [12.2]

Sheldon AL. (1987) Rarity: patterns and consequences for stream fishes. In: Matthews WJ, Heins DC (eds) *Community and Evolutionary Ecology of North American Stream Fishes*, pp 203–9. University of Oklahoma Press, Norman, Oklahoma. [12.5]

Smale MA, Rabeni CF. (1990) *Stream fish communities and stream health in headwater streams draining agricultural watersheds in Missouri.* Progress Report to Missouri Department of Natural Resources, Water Quality Division, July 1990. Cooperative Fish and Wildlife Research Unit, University of Missouri, Missouri. [12.4]

Smith H. (1965) Some experiments on the oxygen consumption of goldfish (*Carassius auratus* L.) in relation to swimming speed. *Canadian Journal of Zoology* **43**: 623–33. [12.3]

Starrett WC. (1951) Some factors affecting the abundance of minnows in the Des Moines River, Iowa. *Ecology* **32**(1): 13–27. [12.5]

Statzner B. (1987) Characteristics of lotic ecosystems and consequences for future research directions. *Ecological Studies* **61**: 365–90. [12.2, 12.3]

Stearns SC. (1980) A new view of life-history evolution. *Oikos* **35**: 266–81. [12.3]

Strauss RE. (1987) The importance of phylogenetic constraints in comparisons of morphological structure among fish assemblages. In: Matthews WJ, Heins DC (eds) *Community and Evolutionary Ecology of North American Stream Fishes*, pp 136–43. University of Oklahoma Press, Norman, Oklahoma. [12.2]

Suzumoto BK, Schreck CB, McIntyre JD. (1977) Relative resistances of three transferrin genotypes of coho salmon (*Oncorhynchus kisutch*) and their hematological responses to bacterial kidney disease. *Canadian Journal of Fisheries and Aquatic Sciences* **34**: 1–8. [12.3, 12.5]

Taylor EB, McPhail JD. (1985a) Variation in body morphology among British Columbia populations of coho salmon (*Oncorhynchus kisutch*). *Canadian Journal of Fisheries and Aquatic Sciences* **42**: 2020–8. [12.3]

Taylor EB, McPhail JD. (1985b) Variation in burst and prolonged swimming performance among British Columbia populations of coho salmon, *Oncorhynchus kisutch. Canadian Journal of Fisheries and Aquatic Sciences* **42**: 2029–33. [12.3]

Terrell JW, McMahon TE, Inskip PD, Raleigh RF, Williamson KL. (1982) *Habitat suitability index models: Appendix A. Guidelines for riverine and lacustrine applications of fish HSI models with the habitat evaluation procedures.* FWS/OBS-82/10A. US Fish and Wildlife Service. p. 54. Washington, DC. [12.2]

Thompson DW. (1942) *On Growth and Form.* McMillan, New York. [12.3]

US Fish and Wildlife Service. (1981) *Standards for the development of habitat suitability index models.* 103 ESM. US Fish and Wildlife Service, Division of Ecological Services. Washington, DC. [12.2]

Vannote RL, Minshall GW, Cummins KW, Sedell JR, Cushing CE. (1980) The river continuum concept. *Canadian Journal of Fisheries and Aquatic Sciences* **37**: 130–7. [12.2]

Varley JD, Gresswell RE. (1988) Ecology, status and management of the Yellowstone cutthroat trout. *American Fisheries Society Symposium* **4**: 13–24. [12.3]

Wade MG. (1986) *The relative effects of* Ceratomyxa shasta *on crosses of resistant and susceptible stocks of summer steelhead.* M.S. Dissertation, Oregon State University, Corvallis, Oregon. [12.3, 12.5]

Ward BR, Slaney PA. (1988) Life history and smolt-to-adult survival of Keogh River steelhead trout (*Salmo gairdneri*) and the relationship to smolt size. *Canadian Journal of Fisheries and Aquatic Sciences* **45**: 1110–22. [12.3]

Webb PW. (1975) Hydrodynamics and energetics of fish propulsion. *Canadian Bulletin of Fisheries and Aquatic Sciences* **190**: 1–159. [12.3]

Webb PW, de Buffrénil V. (1990) Locomotion in the biology of large aquatic vertebrates. *Transactions of the American Fisheries Society* **119**: 629–41. [12.3]

Welcomme RL. (1985) *River Fisheries.* FAO Fisheries Technical Paper No. 262. Food and Agriculture Organization, Rome, Italy. 330. [12.2, 12.3, 12.4, 12.5, 12.6]

White RJ, Brynildson OM. (1967) *Guidelines for management of trout stream habitat in Wisconsin.* Technical Bulletin 39. Department of Natural Resources, Wisconsin. [12.6]

Whittier TR, Hughes RM, Larsen DP. (1988) Correspondence between ecoregions and spatial patterns in stream ecosystems in Oregon, USA. *Canadian Journal of Fisheries and Aquatic Sciences* **45**: 1264–78. [12.2]

Wiens JA. (1984) On understanding a non-equilibrium world: myth and reality in community patterns and processes. In: Strong DR Jr, Simberloff D, Abele LG, Thistle AB (eds) *Ecological Communities: Conceptual Issues and the Evidence*, pp 439–57. Princeton University Press, Princeton, New Jersey. [12.5]

Wikramanayake ED. (1990) Ecomorphology and biogeography of a tropical stream fish assemblage. *Ecology* **71**: 1756–64. [12.5]

Wikramanayake ED, Moyle PB. (1989) Ecological structure of tropical fish assemblages in wet-zone streams of Sri Lanka. *Journal of Zoology (London)* **281**: 503–26. [12.5]

Wiley MJ, Osborne LL, Larimore RW. (1990) Longitudinal structure of an agricultural prairie river system and its relationship to current stream ecosystem theory. *Canadian Journal of Fisheries and Aquatic Sciences* **47**: 373–84. [12.2, 12.3]

Winter GW, Schreck CB, McIntyre JD. (1979) Resistance of different stocks and transferrin genotyes of coho salmon and steelhead trout to bacterial kidney disease and vibriosis. *Fishery Bulletin (NOAA)* **77**: 795–802. [12.5]

Withler RE, Evelyn TPT. (1990) Genetic variation in resistance to bacterial kidney disease within and between two strains of coho salmon from British columbia. *Transactions of the American Fisheries Society* **119**: 1003–9. [12.3]

Wohlschlag DE, Juliano RO. (1959) Seasonal changes in bluegill metabolism. *Limnology and Oceanography* **4**: 195–205. [12.3]

Wurtsbaugh WA, Tapia RA. (1988) Mass mortality of fishes in Lake Titicaca (Peru–Bolivia) associated with the protozoan parasite *Ichthyophthirius multifiliis. Transactions of the American Fisheries Society* **117**: 213–17. [12.5]

Yant PR, Karr JR, Angermeier PL. (1984) Stochasticity in stream fish communities: an alternative interpretation. *American Naturalist* **124**: 573–82. [12.5]

Zimmerman EG, Wooten MC. (1981) Increased heterozygosity at the MDH-B locus in fish inhabiting a rapidly fluctuating thermal environment. *Transactions of the American Fisheries Society* **110**: 410–16. [12.3]

13: The Sampling Problem

R.H.NORRIS, E.P.McELRAVY
AND V.H.RESH

13.1 INTRODUCTION

Aims of this chapter

Decisions about sampling requirements are generally acknowledged as an important component of river and stream studies. Sampling decisions, however, are often based on tradition rather than on optimizing the data obtained per unit effort. To us, this is the essence of the sampling problem.

In this chapter we shall review general principles and considerations in sampling river and stream environments and emphasize some essential topics that have not been previously examined or about which confusion still exists. These will cover: sampling requirements for determining values of estimates (e.g. densities) compared with detecting differences among sites or times; the use of transformations; and case histories analysing factors influencing sampling design.

Much has been written on optimizing study designs and readers are referred to Cochran (1963), Eberhardt (1976, 1978), Green (1979), Resh (1979), Waters and Resh (1979), Bernstein and Zalinski (1983), Millard and Lettenmaier (1985), Millard et al (1985), Norris and Georges (1986), and Waters and Erman (1990). We have also provided a summary of selected references on methods to study algae, macrophytes, bacteria and fungi, macroinvertebrates and fish to direct the reader into these subject areas (Table 13.1).

Setting objectives and the role of sampling

Statements on the need for good study design with clearly stated objectives have been repeated so often as to be tedious (e.g. Eberhardt 1963, 1976, 1978; Green 1979; Waters & Resh 1979; Ellis & Lacey 1980; Rosenberg et al 1981; Cairns & Pratt 1986; Norris & Georges 1986; Cooper & Barmuta 1991). No doubt much of this repetition is seen as necessary because poorly stated objectives are still commonly encountered. Although there are always logistic or economic constraints on what or how much can be measured, the relationships among variables being measured and the problem of interest must be clear; little is gained from sampling irrelevant variables. A discussion of how different types of objectives may control sampling is provided by Ellis and Lacey (1980).

There are some questions that sampling cannot hope to cover: it cannot be used to provide judgements on 'good' or 'bad' scales, provide direct measures of aesthetic values or environmental ambience, or directly decide on what is, for example, 'wild and scenic'. Sampling can be used to: detect environmental change through space and time; make comparisons with pre-established standards; establish relationships between variables by statistical inference, experimentation or manipulation; develop predictions of effects or on relationships; and provide estimates of environmental variables such as population size or community structure. Commonly, the questions to be answered by sampling fall into two groups: (1) estimates of variables of interest such as the population density at one site; and (2) comparisons among sites or times. This important distinction has been ignored in discussions on the number of replicate sample units needed in a study. For example, estimates of fish production for calculation of harvestable yield may require estimates of totals or means, but often problems

Table 13.1 Sources of references for sampling structural and functional components of rivers that may be affected by disturbance for five groups of organisms

Measure	Taxonomic category				
	Algae	Macrophytes	Bacteria/fungi	Invertebrates	Fish
Standing stock, biomass	Aloi 1990 Trotter & Hendricks 1979 Blum 1956	Dennis & Isom 1984 Downing & Anderson 1986	Kogure et al 1984 Bott 1973 Karl 1985	Elliott 1977 Merritt et al 1984	Nielsen & Johnson 1983 Lagler 1978 Robson & Spangler 1978
Transport, drift	Power 1990 Lamberti & Resh 1987 Stevenson 1983	Dawson 1980 Dawson 1988	Webster et al 1987	Waters 1972 Brittain & Eikeland 1988	Harvey 1987 Gale & Mohr 1978
Production	Aloi 1990 Karl 1985	Dennis & Isom 1984 Hill et al 1984	Karl 1985 Moriarty 1986	Benke 1984 Rigler & Downing 1984	Chapman 1978
Taxonomic richness	Aloi 1990 Holder-Franklin 1986 Blum 1956	Dennis & Isom 1984	De Long et al 1989 Karl 1985 Bohlool & Schmidt 1980 Holder-Franklin 1986	Resh & McElravy 1992	Haedrich 1983 Lowe-McConnell 1978 Lundberg & McDade 1990 Strauss & Bond 1990.
Structural, functional, trophic diversity	Steinman & McIntire 1986	den Hartog 1982 den Hartog & van der Velde 1988 Best 1988	Karl 1985 Atlas 1985 Niels & Bell 1986	Cummins 1973 Merritt & Cummins 1984	Bowen 1983 Windell & Bowen 1978
Nutrient cycling	Mulholland et al 1990	Carpenter & Lodge 1986 Denny 1980	Webster & Benfield 1986 Atlas 1985 Boylen & Soracco 1986	Elwood et al 1983 Webster & Benfield 1986	Cederholm & Peterson 1985 Richey et al 1975
Life-history patterns	Bold & Wynne 1985	Dennis & Isom 1984 Kautsky 1988	Andrews 1986 Andrews 1991	Butler 1984	Baltz 1990
Size spectra	Steinman & McIntire 1986 Malone 1980	Wright et al 1981	Fry 1988 Andrews 1991	Mundie 1971 Clifford & Zelt 1972	Lagler 1978
Biotic interactions (within category)	Bothwell 1988 Peterson et al 1983	Carpenter & Lodge 1986	Atlas 1985 Boylen & Soracco 1986	Peckarsky 1984 Power et al 1988	Crowder 1990

of environmental concern will be comparisons of differences between (or among) sites or times. Parts of the sampling design may be similar for both types of questions (e.g. devices used), but the hypotheses tested, the number of sample unit replicates (i.e. the sample size), and the data and analysis needs may be quite different.

There will always be some differences between

the information obtained from measurements of the small part of the environment that has been 'sampled' compared with what actually exists in the total environment. This is the error associated with the data collected, and some information is needed about this error to estimate the risk of drawing wrong conclusions after examining the data obtained from sampling. 'The risk of being wrong' forms the basis for hypothesis testing, and it is needed whether data are being tested in the formal statistical sense or simply using scientific judgement. The validity of using scientific judgement, often based on years of experience and training, and possibly on extensive historical information, is often undervalued in favour of 'more rigorous' statistical approaches (see Berry 1987; Fryer 1987; Miller 1989). Statistical hypothesis testing usually is a more impartial approach (provided that the study has been properly designed) than using scientific judgement alone, particularly in situations requiring unequivocal results that are necessary for some management practices.

Sampling in rivers differs from many other environments because the unidirectional flow creates problems that are otherwise uncommon, e.g. difficulties in finding controls and replicating samples units. Commonly, upstream sites are used as the control against which changes at downstream sites are tested, and the problems with this design have received some attention (Eberhardt 1978; Hurlbert 1984; Cooper & Barmuta 1992; Norris & Georges 1992). Environmental conditions may be quite different at upstream and downstream sites, and upstream conditions may affect results downstream; in addition, the treatments themselves often are not properly replicated (Hurlbert 1984; Cooper & Barmuta 1992; Norris & Georges 1992). The following are also important: catchments usually will be more different from each other than sites within catchments; the same sites vary over time; and once a parcel of water has passed a site its effects will not be repeated. Because of these difficulties, interpretation based on scientific judgement alone may carry more weight with regard to rivers than other habitat types.

13.2 GENERAL PRINCIPLES

Study types

Studies in lotic systems are generally performed in one of two ways: (1) as mensurative studies, where space and time are the main experimental variables or treatments; and (2) manipulative studies, usually involving some intervention by the experimenter to control one or more external factors relative to the experimental units (Hurlbert 1984).

Interpretation of mensurative studies is mostly by inference, often (but not always) based on statistical analysis. Data from these studies will be from what Eberhardt (1963) terms 'unconfined populations'; if analysed by ANOVA, a random-effects model is usually used because of many uncontrolled factors. In these situations, homogeneity of variances is a critical factor (often achieved only after transformation), and sampling will need to be random or stratified random where only features of interest are sampled randomly. Any inferences made about the populations can be supported with probability statements, and thus be less affected by a poor choice of mathematical or ecological model (Eberhardt 1963). Functional relationships may not be apparent from this approach, and interpretation will depend on ingenuity and foresight in designing the study as well as expert knowledge of the field (Eberhardt 1963). This latter point has been criticized by some authors (e.g. Hurlbert 1984), but knowledge and expertise are usually prerequisites for adequate interpretation, regardless of sampling design. Indeed, where statistical testing is not done, it is the only basis for interpretation.

Analysis of data from manipulative studies is often based on a fixed model ANOVA where heterogeneity of variances may be of little concern (Eberhardt 1963). Manipulative studies of disturbance effects in streams are constrained by ethical and practical difficulties; nevertheless, many have been performed. The issues involved have been discussed in detail by Resh *et al* (1988), who concluded that studies involving severe degradation should meet high standards in statistical design and measurement procedures to minimize the likelihood of inconclusive results and the need for repetition. The same high standards are also needed for mensurative studies, and thus design requirements are similar in both cases. Data needs and methods of analysis have been covered in detail by Cooper and Barmuta (1992), who discussed manipulative experimental designs, and by Norris and Georges (1992), who considered analysis options more generally.

Design of many studies on rivers is difficult because of temporal problems that occur independent of disturbance, and because sites cannot be replicated easily. A commonly used design for assessing disturbance is to locate sites upstream and downstream of the disturbance, and then to replicate sampling units (or measurements) on each sampling occasion before and after the disturbance occurs. This would appear to be a simple two-way ANOVA with two factors, each with two levels: area (control and disturbed) by time (before and after) (Barmuta 1987). However, treatments within the design are not properly replicated because the replicate sampling units are only nested within each of the cells (Barmuta 1987). Eberhardt (1976) called this 'pseudodesign', and Hurlbert (1984) popularized it as 'pseudoreplication' because it is possible that some factor other than that being tested may affect the downstream impact area but not the control. Thus, the results of tests of significance need to be qualified in that they depend on the validity of using upstream sites as a control for downstream sites (Eberhardt 1978).

When sampling units at each site and time have not been replicated, it is possible to use samples through time as replicates (Eberhardt 1976; Stewart-Oaten et al 1986). However, this practice should be used with caution. Although it is valid in the sense that significant results can be believed, statistical testing in ANOVA that uses samples through time as replicates will be inefficient unless true trends in time are identical in magnitude and direction for all sites (Norris & Georges 1992). Normally, this would not be expected in most studies; to ensure the best opportunity for detecting true differences between sites (avoiding type II errors, see below), it is necessary to replicate sampling units on each visit to each site (Norris & Georges 1992).

Habitat types and stratification

Habitat stratification is implicit in the design of any stream or river study that describes sampling as being done in a riffle, pool, or any kind of 'uniform' habitat. Random sampling in rivers is inevitably stratified random sampling. Why do we stratify our sampling efforts? Ideally it is to make the sampling universe reflect the population universe of the organisms under question so that precision of estimates is maximized for minimal sampling effort. Sampling should include both a spatial and a temporal component; migratory species, such as diadromus fish, emphasize this point.

In practice, stratification is done to reduce variability or to facilitate intersite comparisons; both of these approaches improve the efficiency of sampling. Reductions in sampling variability from stratification by depth, current, substratum composition, or individual and combined hydraulic features have been repeatedly demonstrated for benthos (Resh 1979) and other freshwater fauna and flora including aquatic mammals (Statzner et al 1988). Stratification may result from choice of sampling devices (e.g. glass slides, allowing preferential colonization of some algal taxa), but usually it is done by spatial delineation. Although hundreds of publications have discussed stream zones and distribution of benthos, a hierarchical framework, such as suggested by Frissel et al (1986), is generally used in practice. In their approach, a stream system (say 10^3 m) is subdivided into successively lower levels: stream segment (10^2 m), reach (10^1 m), pool/riffle (10^0 m) and microhabitat (10^{-1} m) are the most widely used levels.

Restriction of the sampling universe to a level below that of the population universe may have a logistical advantage (and a corresponding statistical justification) of maximizing precision of estimates, but there are risks and uncertainties associated with stratification. For example, how representative is the restricted sample? Does it comprise a consistent proportion of the population (with different size classes or stages, or in successive years) in study sites under comparison? Moreover, the temporal and spatial variability inherent in unstratified designs may clearly provide an explanation of factors explaining observed patterns. This latter reason explains why pilot studies most often involve transect or other designs that maximize variability, allowing strata to be selected for follow-up analyses.

Measurement types

Physical and chemical measures

Many physical and chemical methods have been highly standardized and often are tested with interlaboratory comparisons. Possibly the most widely accepted reference for standardization of physical, chemical and many biological methods is that from the American Public Health Association (1989). Because of the high degree of standardization, especially for commonly used physical and chemical measures, no attempt has been made to summarize the literature further. Neither have we attempted to deal with hydrometric measurements, although hydraulic regimes, in terms of frequency, intensity and predictability of extreme flows, are important when interpreting data from rivers (Resh *et al* 1988).

Many of the standard methods for physical and chemical data (e.g. American Public Health Association 1989) are seen as being inexpensive, fast and easy to perform; consequently they have found their way into legislation protecting water quality, with regulatory standards usually being set based on toxicity testing. Caution is urged in selecting these physical and chemical measures, because most are surrogates for measuring biological changes of interest, and many, such as pesticide determination, are neither inexpensive nor quick. This is not an argument in favour of one form of sampling over another, but rather that the variables and methods chosen should be relevant to the study problem.

Biological measures

Structural and functional components of rivers that may be affected by disturbance can be summarized in ten categories (Resh *et al* 1988), and we have presented methods of measurement relevant to each of these for major categories of organisms (Table 13.1). Vertebrates other than fish have not been included because we felt that these came under the umbrella of wildlife, rather than river, management; references to these can be found in more specific texts (e.g. Seber 1982).

13.3 VARIABILITY AND TRANSFORMATIONS

Variability

Variability in measurements is probably the most significant factor to be considered when implementing a sampling programme, and it has received attention from several authors, for example Eberhardt (1978) for studies on animal populations in general; Downing (1979), Resh (1979), Allan (1984), Schwenneker and Hellenthal (1984), Morin (1985), Norris and Georges (1985, 1986), and Canton and Chadwick (1988) for invertebrates; Downing and Anderson (1986) for macrophytes; Ridley-Thomas *et al* (1989) for periphyton; Kosinski (1984) on oxygen productivity; Morin *et al* (1987) on primary and secondary production estimates; Mahon (1980), and Raleigh and Short (1981) on fish; Clarke (1990) for suspended sediments; Crisp (1990) for stream temperature; Jordan (1989) for pH; and Kerekes and Freedman (1989) on seasonal variations in water chemistry.

Variability of measurements has two components to be considered when selecting methods and determining required levels of replication: precision and accuracy. Precision is a measure of the similarity of repeated measurements, and is most important when comparisons are being made. Accuracy refers to the closeness of a measurement to its true value, and is most important when estimates are sought. Whether estimating means (for population numbers, or the legal limit of a chemical concentration) or determining whether differences between sites or times exist, confidence in conclusions will depend on a knowledge of data variability (measures of precision). Adequate replication (at least at some stage in the study) in both sample collection and analysis can provide a measure of precision. Failure to include adequate replication can lead to estimates that are so imprecise as to be little better than guesses, or in an inability to decide whether differences exist and thus be a costly waste of resources. Likewise, too much replication of sampling units will provide more data than are needed to demonstrate the points of interest, and again wastes resources. Much work on variability has considered only the variance of an estimator (e.g. total numbers), rather than consideration of the variance in maximizing the power of a study design to detect change (Millard & Lettenmaier 1985). Decisions on precision and the power of designs to detect change are both issues that we will discuss in detail (section 13.4).

Alternatively, variability may be viewed as potentially valuable information rather than just as noise. Resh (1979) has given a detailed presentation of the lifecycle and behavioural features of a net spinning caddisfly that would contribute to sampling variability (or worse, bias) in a randomized design, but provide information in a restricted design. Thus, variability in space (e.g. habitat selection) or time (e.g. reproductive phases) may sometimes need to be maximized as being of direct interest. Likewise, variability may be important in elucidating mechanisms (Resh & Rosenberg 1989). Prior knowledge of variability may be also needed, such as that just described, so that sampling can avoid or minimize its impacts (as discussed by Eberhardt 1963, 1976).

Organisms are rarely distributed evenly or randomly through the environment. Clumping usually occurs because many environmental features to which organisms respond are unevenly distributed. Behavioural, reproductive and dispersal features of some species may also produce aggregation without the influence of environmental factors. Clumping is also dependent on the scale at which sampling is performed. Clumps of high densities may be unevenly distributed through an environment of otherwise low density, but distribution within clumps may be regular because of competition for resources. Aquatic organisms most commonly have a negative binomial distribution where the variance is greater than the mean (Elliott 1977; Eberhardt 1978). Likewise physical and chemical variables may be unevenly distributed in the environment, but less attention seems to be paid to the problem of sampling for them because streams are gener-

ally considered to be well mixed. An example would be if downstream transport of phosphorus is of interest. Most of the phosphorus may move in particulate form during high flow. How, for example, should one cope with particles that are too large for the necks of commonly used sampling bottles. Dead fish, birds and mammals (and other large particles) represent clumps of high concentrations of phosphorus commonly transported under such conditions but are usually ignored. The statistical distributions of physical and chemical data have not been considered in nearly as much detail as biological ones because usually only single collections are made, or if multiple collections are made they are combined before analysis, thus depriving the investigator of any measure of variability.

Need for transformation

A further problem is encountered when statistical tests are applied to data representing environmental attributes that are contagiously distributed: the possible need for transformation to meet the tests' assumptions. These include: (1) linearity between the variance of the means of replicated samples and the average variance of the values that make up each sample; (2) homogeneity of variances; and (3) the normal distribution of data. Other assumptions of randomness in sampling and independence of observations are generally achieved by sound sampling design (Green 1979), although independence may be a greater problem in rivers because of unidirectional flow. The assumptions that replicated values of a measure are normally distributed and variances are equal, are often beyond the control of the investigator at the time of data collection and must be dealt with during preliminary analysis. Violations of these assumptions are generally expressed as a relationship between the variance and the mean, which should be independent for normally distributed data. Such a relationship is common for counts of benthic macroinvertebrates (Downing 1979; Morin 1985; Norris & Georges 1986) and is generally in the form $S^2 = a\bar{Y}^b$ (Taylor 1961). If sampling units have been replicated at each site or time, a plot of the sample variances against the sample means, followed by a test of the correlation between the two, will indicate whether a transformation is necessary (Table 13.2).

Traditionally, the square-root transformation ($Y' = SQRT (Y + \frac{1}{2})$) is applied if the variance and mean are approximately equal (corresponding to a random spatial or temporal distribution), and a log transformation ($Y' = \log_{10} (Y + 1)$) is applied when the variance is consistently larger than the mean (corresponding to a clumped distribution) (Elliott 1977). In extreme cases no transformation will normalize such data. After the data have been transformed, their means and variances can be calculated again and plotted against each other to see if the transformation has been successful (Norris & Georges 1991).

Transformation — case studies

To demonstrate the effects of collections with low numbers of organisms (or many zeros) on the effectiveness of transformation, two data-sets have been used: counts of the bug *Aphelocheirus aestivalis* from two sites, the upper and lower Schierenseebach, Germany (a full description of sites and methods can be found in Statzner (1978) and Statzner *et al* (1988)), and counts of total numbers of invertebrates and the mayfly *Baetis tricaudatus* from three sites in a northern Californian coast range stream, Hunting Creek (see Resh & Jackson 1991 for site description and methods). Several transformations ($\log_{10}$, square root, fourth root, and Taylor's power (Taylor 1961)) have been applied to the above data. The Shapiro–Wilk statistic (SAS Institute 1987) was used to test for normality on each transformation, as well as the raw data. *Apelocheirus aestivalis* data were then grouped into twos to double the sample unit size, removing most zero counts from the new data-set, and the transformations and tests for normality were reapplied (Table 13.2). For the purposes of this example, and because the collections of *A. aestivalis* were made randomly, these data were arbitrarily grouped into lots of five for calculation of means and variances. Five replicates from each site and time were used for the Hunting Creek data. $\log_{10}$ of variances were regressed against their corresponding means (also transformed $\log_{10}$) for the raw data. Significant relationships between means and variances were

used to determine power transformations using Taylor's relationship as discussed below (Table 13.2). If the data were to be used for comparisons between sites, Taylor's power relationship would need to be calculated on the combined sites and a single transformation used, because all data would need to be handled consistently.

The presence of many zeros in a data-set results in the transformation's inability to normalize the data (Table 13.2). This problem is overcome in large degree by doubling the sample unit size, thereby eliminating zeros and increasing the count per collection. Similar findings are reported by Allan (1984), who also cited Anderson (1965) as stating that variance-stabilizing transformations could not be found for collections with a mean ($\overline{Y}$) < 3. Square root transformations ($(Y + \frac{1}{2})^{0.5}$) were frequently not strong enough to normalize the data (Table 13.2). Fourth root transformations (recommended by Downing 1979) and the commonly used $\log_{10}$ were most effective, with the latter slightly more so (Table 13.2). The highly recommended transformations based on Taylor's power relationship performed poorly except where sample sizes and/or means were large (e.g. total numbers in Hunting Creek; Table 13.2).

Allan (1984) suggested that at higher mean densities the fourth root and power transformations were favoured over the log, but with small means the log transformation was most efficient, even if it did not always eliminate the dependency of the variance on the mean. However, the findings reported here indicate that the logarithmic transformation was as least as good or better than the other transformations used in all examples (Table 13.2). The degree to which transformation alters data distribution must also be considered; for example the effect of not properly normalizing data with a transformation is to reduce the sensitivity of ANOVA, and increase the likelihood of concluding no effect exists when in fact one does (type II error) (Cochran 1947). In cases where the log transformation produced distributions still significantly different from normal (Table 13.2), observation of frequency plots of the data indicated that departure from normality was not severe. Thus, in many cases where a transformation is applied but the data distribution is still

Table 13.2 Effectiveness of transformations on normalizing distributions of *Aphelocheirus aestivalis* in collections from the Upper (site 1) and Lower (site 2) Schierenseebach, Germany, in samples of single-unit size (1X) or double-unit size (2X), and for three sites in Hunting Creek, northern California, based on five replicates from each site collected using a Surber sampler in mid-April from 1984 to 1990

Data-set	n	$\bar{Y}$	S^2	Raw[a]		$\log_{10}(Y+1)$		$(Y+\tfrac{1}{2})^{0.5}$		$Y^{0.25}$		$Y^{1-0.5b}$	
				R^2	W	R^2	W	R^2	W	R^2	W	R^2	W
A. aestivalis													
Both sites 1X	135	9.5	244	0.852	0.561	0.007	0.938	0.558	0.858	0.069	0.884		b
				0.000	0.000	0.684	0.570	0.000	0.000	0.019	0.000		
Both sites 2X	66	19.3	502	0.872	0.674	0.684	0.979	0.809	0.891	0.373	0.967		b
				0.000	0.000	0.511	0.629	0.000	0.000	0.035	0.187		
Site 1 1X	67	5.0	29	0.628	0.776	0.122	0.927	0.204	0.934	0.592	0.813	0.069	0.693
				0.001	0.000	0.241	0.000	0.121	0.002	0.002	0.000	0.365	0.000
Site 1 2X[c]	33	10.1	55		0.849		0.963		0.958		0.970		0.064
					0.000		0.367		0.272		0.575		0.409
Site 2 1X	68	13.8	420	0.896	0.626	0.038	0.926	0.508	0.882	0.154	0.889	0.670	0.596
				0.000	0.000	0.525	0.000	0.006	0.000	0.185	0.000	0.000	0.000
Site 2 2X[c]	33	28.5	789		0.745		0.965		0.923		0.971		0.962
					0.000		0.427		0.026		0.568		0.364
Hunting Creek													
All sites and all years													
Total numbers	99	497.6	277006	0.562	0.767	0.096	0.977	0.129	0.024	0.032	0.994	0.015	0.978
				0.000	0.000	0.185	0.383	0.120	0.000	0.305	0.246	0.605	0.430
B. tricaudatus	99	49.3	4810	0.928	0.711	0.045	0.954	0.459	0.905	0.032	0.961	0.879	0.530
				0.000	0.000	0.368	0.006	0.001	0.000	0.453	0.026	0.000	0.000

[a] R^2 on raw data has been calculated on variances and means in groups of five for *A. aestivalis* and on sets of five replicates for Hunting Creek, which were then logged and regressed.

[b] Power transformations were not performed for the total *A. aestivalis* data-set because the exponent was close to 2 indicating $\log_{10}$ to be appropriate (see text).

[c] *Aphelocheirus aestivalis* regressions were not done for individual sites because the number of samples was too small to form enough groups for a meaningful analysis.

R^2 for regression of means against variances, W is the Shapiro–Wilk test of normality. Upper lines are R^2 and W values, lower lines are levels of significance, n = number of observations, $\bar{Y}$ = the arithmetic mean and s^2 = variance. Transformations: Raw = raw data, $\log_{10}$ = $\log_{10}(Y+1)$, $(Y+\tfrac{1}{2})^{0.5}$ = square root, $Y^{0.25}$ = fourth root, $Y^{1-0.5b}$ = power transformation from the variance to mean relationship, *Aphelocheirus aestivalis* site 1 $Y^{0.15}$, site 2 $Y^{0.06}$, Hunting Creek, total numbers $Y^{0.21}$, *Baetis tricaudatus* $Y^{0.05}$.

significantly different from normal the likelihood of a type II error is reduced, and the transformation may be judged acceptable.

Sampler unit size

It has often been stated that larger numbers of smaller (in area or volume) sampling units will provide estimates with better precision than if fewer numbers of larger ones had been collected (Elliott 1977; Downing 1979). The advantages may not be as great as first thought (Riddle 1989) because of the additional problem of the intractability of zero counts to transformation. This point is demonstrated when the sample unit size is doubled for *A. aestivalis*; this resulted in better normalization of the data (Table 13.2). Where costs have been considered it has been recommended that small sample unit areas be used when mean densities are high, and larger ones where they are low (Morin 1985). This also coincides with data needs for transformation that have been demonstrated here; in cases of low densities, larger sample unit sizes will tend to eliminate zeros. One way of determining minimum sample unit area is by selecting the size required to eliminate most zero counts.

A more profound problem exits for physical and chemical data where values are reported as below the detection limits (usually called censored data) of the methods or apparatus used for measurement, but not zero. Several methods have been proposed for dealing with this problem after the data have been collected (Tiku 1967; Helsel 1986; El-Shaarawi 1989; El-Shaarawi & Dolan 1989; Norris & Georges 1991). If transformations are required on such data, their effectiveness will depend on their distribution and the way in which the below-detection-limit results are handled. In pollution studies where high concentrations are usually of concern, below-detection-limit readings are unlikely to be a problem. A study designed to investigate the relationships of low concentrations of essential chemicals might need to consider sampling alternatives such as evaporation, chelation or some other method of concentration to overcome this difficulty.

Selecting a transformation

Taylor (1961, 1980) found that for many types of aggregated data (e.g. abundance data), the following power relationship exists between the variance and the mean:

$$S^2 = a\overline{Y}^b$$

and that $Y' = Y^{1-0.5b}$ or $Y' = (Y + c)^{1-0.5b}$ (c is a constant such as 0.5, as used for the square root transformation) is the appropriate transformation for breaking the relationship between the mean and the variance. The parameter b is obtained by linearizing the above equation with a log–log transformation and applying linear regression. When $b = 0$, the distribution is considered normal and no transformation is necessary (for a discussion see Taylor 1961; Elliott 1977). When $b = 2$, a log transformation is appropriate (Elliott 1977; Green 1979).

In all cases of the data analysed here, variances were much greater than their corresponding means (Table 13.2), indicating aggregated distributions; a negative binomial distribution is commonly found in this type of data (Taylor 1961; Elliott 1977). In all cases there were significant relationships between means and variances (Table 13.2) and these were used to calculate the appropriate transformation based on Taylor's power relationship.

Poor performance of transformations based on Taylor's power relationship (Table 13.2) may have occurred because a log–log transformation may yield a grossly biased estimate of b (Zar 1968; Sprugel 1983), and the degree of bias will depend in part on the amount of scatter about the power relationship. Caution should also be exercised when calculating this transformation because it is optimized for the sample at hand; in fact, what is required is a transformation that will correct the population from which the sample was drawn (Norris & Georges 1991). These problems, its poor performance (Table 13.2), and the difficulty of calculating the power relationship to obtain the transformation, suggest that its use should not be recommended.

One of the most widely used statistical analyses is ANOVA, which is based on a linear relationship between the variance of means of replicated samples and the average variance of the values that make up each sample. The observed variance among the means for different sites or times can be compared to that expected to arise from this relationship; the ratio of the two variances is then compared using an *F*-test. ANOVA involving two or more factors assumes additivity, which Eberhardt (1978) considers more important than homogeneity of variances (normality). Many of the factors affecting environmental variables may be multiplicative so that a logarithmic transformation will make them additive, thus conforming to the model for ANOVA. Thus, the efficiency of the logarithmic transformation demonstrated above, its ease of computation, and its attributes of creating additivity for the ANOVA model from multiplicative environmental relationships lead us to suggest that it is the most favoured transformation. However, a single transformation will probably not be appropriate at all densities, and care should be taken to check the effects of the transformation used (Morin 1985).

13.4 THE REPLICATION PROBLEM

Estimates and comparisons

One of the most widely discussed topics in sampling is: 'how many replicates should be taken'? This can vary greatly depending on such simple considerations as whether collections are made to provide estimates of numbers of organisms or concentrations (Elliott 1977; Downing 1979; Green 1979; Norris & Georges 1986), or to detect a desired difference among means (Sokal & Rohlf 1981; Schwenneker & Hellenthal 1984; Zar 1984; Norris & Georges 1986). Estimates often need to provide a tightly defined measure of the real number or amount present, using data that are untransformed. However, because changes between sampling occasions may often be great and comparisons are usually tested using data that have been transformed to meet the assumptions of statistical tests, comparisons may require less sampling effort than estimates. This section examines the consequences of choices made rather than specifics of study design (which are determined by study objectives).

Studies of an organism's demography, biomass or abundance, or the assessment of chemical concentrations with regard to standards, often require calculations of sample size using formulae that provide the number of replicates necessary for a desired precision around the mean (formulae 1 and 2, Table 13.3). Studies to determine the sample size required to detect a difference or a change of a desired magnitude usually make use of a second type of formulae (formulae 3, 4 and 5, Table 13.3). This distinction has often been neglected, but it is important because it is usually unacceptable to transform data when an estimate of a mean is needed, whereas transformation is often appropriate and required when data from sites or times are being compared with statistical procedures. Some authors have suggested that estimate data can be transformed and then back-transformed, e.g. Allan (1984) who also provides worked examples of resulting asymmetrical confidence limits. A geometric mean (from back-transformation of logged data) will always be lower than or equal to (but never more than) the arithmetic mean. It will give an indication of where the number of organisms or concentration was most of the time or in most of the replicates most of the time. This may be appropriate for some studies (e.g. long-term resources needed to support a population). The investigator needs to make this decision on whether transformed data can be used before initiating the study because less replication is generally needed for transformed data (Norris & Georges 1986). Statistical properties of data are altered by transformation, and investigators need to be sure of the meaning of back-transformed results. The following discussion proceeds on the premise that estimates will usually not be transformed.

Additionally, all the formulae require some prior knowledge of the variability expected in the data. This should present little difficulty if the well accepted view on the need for a pilot study before committing full resources to a programme is adhered to. All formulae also present the problem of deciding on the desired level of precision, or the magnitude of the difference that the programme is desired to detect.

Table 13.3 Formulae for calculating sample sizes (n) for a required precision around the mean, and for testing for differences between means

(1) $$n = \frac{S^2}{d^2\,\overline{Y}^2} \quad \text{(Elliott 1977, p. 129)}$$

n = number of replications
S^2 = sample variance
d = the ratio of the standard error to the mean, e.g. $d = 0.2$ for SE $\pm 20\%$ of $\overline{Y}$
$\overline{Y}$ = arithmetic mean

(2) $$n = \frac{t^2\,S^2}{d^2\,\overline{Y}^2} \quad \text{(Elliott 1977, p. 130)}$$

d = relative error as percentage CL of $\overline{Y}$, e.g. $d = 0.4$ for CL $\pm 40\%$ $\overline{Y}$
t = value from Student's t distribution with n degrees of freedom

(3) $$n \geq 2\left(\frac{\sigma}{\delta}\right)^2\left\{t_{\alpha[\nu]} + t_{2(1-p)[\nu]}\right\}^2 \quad \text{(Sokal \& Rohlf 1981, p. 263)}$$

σ = true standard deviation
δ = difference between means expressed as a percent of $\overline{Y}$, e.g. $= 20$ for a 20% difference between means
P = desired probability that a difference will be found to be significant
ν = degrees of freedom of the sample standard deviation with a groups and n replications per group
$t_{\alpha[\nu]}$ and $t_{2(1-P)[\nu]}$ = values from a t table with ν degrees of freedom and corresponding to probabilities of α and $2(1-P)$, respectively. Note if $P = \frac{1}{2}$ then $t_{1[\nu]} = 0$

(4) $$n = \frac{2S_P^2\,(t_{\alpha(2),2(n-1)})^2\,F_{\beta(1),2(n-1),\nu}}{d^2} \quad \text{(Zar 1984, p. 133)}$$

d = relative error in terms of CL of $\overline{Y}$ expressed as an absolute value
S_P^2 = pooled variance with ν degrees of freedom
$1-\alpha$ = the confidence level of the desired confidence interval
$1-\beta$ = is the probability that the half-width of the confidence interval will not exceed d
F = critical values from an F table for numerator degrees of freedom of $2(n-1)$ and denominator degrees of freedom of ν
t = critical values from a t table with degrees of freedom $2(n-1)$

(5) $$n = \frac{S^2\,(q_{\alpha,DF,k})^2\,F_{\beta(1),DF,\nu}}{d^2} \quad \text{(Zar 1984, p. 193)}$$

d = half-width of the $1-\alpha$ confidence interval
$1-\beta$ = the probability that the half-width of the confidence interval will not exceed d
S^2 = estimated error variance with ν degrees of freedom
k = total number of means
DF = error degrees of freedom that the experiment would have with the estimated n (i.e. $DF = k(n-1)$) for a single factor ANOVA
q = critical values from a q table

Sample size for estimates

Precision for estimates and sample size required

The level of precision around a mean value indicates the degree of confidence that can be placed in it. Fortunately, decisions can be made on the desired precision *before* embarking on a sampling programme, and the number of replicates needed to achieve it can be easily calculated (formulae 1 and 2, Table 13.3).

The level of precision accepted for a study is

important from the point of view of available resources because the number of replicates needed to achieve a desired precision increase with its inverse square (see Elliott 1977). Thus, four times as many replicates would be needed to achieve a precision of ±20% of the mean as for ±40% of the mean. At large sample sizes ($n > 30$), the value of t included in formula 2 is close to two and the results it yields will be almost identical to those from formula 1 (Table 13.3). At smaller sample sizes, formula 2 should be used and solved iteratively. The number of replicates needed for a given level of precision is also affected by two other features: the size of the mean and the degree of aggregation (Resh 1979). Larger means need fewer replicates for a given degree of aggregation and level of precision (Morin 1985; Norris & Georges 1986). Thus, if larger sample unit (area or volume) sizes are used, fewer replicates are needed for the same precision (Downing 1979). Higher levels of aggregation in variables of interest (e.g. indicated by b in Taylor's power relationship, section 13.2) need more replicates to attain the desired precision.

Replication to achieve a desired precision has been the subject of much discussion with respect to benthic macroinvertebrates. Chutter and Noble (1966) corrected analyses carried out earlier by Needham and Usinger (1956) for benthic invertebrate data from Prosser Creek, California. They concluded that hundreds of replicates would be needed to achieve estimates of biomass and abundance with 95% CL of the mean that would be acceptable. Allan (1984) suggested that 10 to 20 replicates would be needed for moderate precision (95% CL ±30−55% of the mean) for counts of benthic macroinvertebrates, and 50 replicates for high precision (95% CL ±10−25% of the mean). Elliott's (1977) popular book has been used extensively by stream ecologists, and his worked example for calculating required sample sizes using a precision of CL ±40% has resulted in this figure generally being accepted as a reasonable precision. The question to be answered is not how many replicates are needed for a given level of precision, but rather what level of precision is needed to meet the study objectives? Subsequent to this decision, calculation of the required number of replicates is straightforward.

Levels of precision in stream data have been discussed more generally by Norris and Georges (1986) who also included typical levels of accuracy (closeness to the real value) commonly accepted for some physical and chemical variables. These range from ±10% for biochemical oxygen demand and total Kjeldahl nitrogen to ±2% for electrical conductivity. Acceptable precisions on chemical determinations (for repeated field measurements) may be much tighter than they are for biological measurements. If this is so, several replicate collections also may be needed for estimation of physical and chemical variables, a practice usually not done. Note that the above precision is based on repeated collections from the field, *not* repeated measurements in the laboratory on the same water sample; the latter gives a precision only for the method and apparatus, and not for the environment.

Choosing levels of precision for estimates

Levels of precision needed should not be blindly accepted from past work (e.g. 95% CL ±40% of the mean), but must be determined with scientific expertise to constitute a reasonable estimate of abundance or concentration, relative to the study objectives. Several factors may be considered when making this decision. First, some understanding will be needed of the natural variability of environmental features of interest. This may be obtained from past work, a pilot study, or drawn from the expertise of the investigator. For example, electrical conductivity is usually conservative in terms of variability, perhaps slowly increasing downstream, but it usually varies little through time or within sites. Alternatively, total nitrogen may be highly variable, being affected by nitrification and de-nitrification processes, temperature, oxygen concentrations and biological activity. Organism numbers may vary by orders of magnitude depending on lifecycle status, behavioural responses and natural environmental influences such as drought or flood, which may be independent of the effect of pollutants. Experienced scientists will have knowledge of what ranges can be expected for the variables to be measured; levels of precision can be set relative to these ranges. For example, in an unpolluted river where phosphorus concentrations are known to limit algal growth, estimates of a mean of 30 μg l^{-1} total P and 95% CL ±10 μg (33%) would probably be acceptable because a range of 10–40 μg l^{-1} might be considered usual, and unlikely to effect substantial changes in algal growth. If the threshold for algal growth in the same river was known to be near 40 μg l^{-1} P and a phosphorus discharge was to begin, replicates to achieve estimation of a mean of 40 μg l^{-1} 95% CL ±5 μg (12%) might be needed to provide data for stream-protection decisions. Alternatively, it may be deemed that algal growth will be controlled if the phosphorus concentrations remain below 50 μg l^{-1} for 90% of the time. In this case, the number of replicates needed would be calculated to yield a mean with the upper boundary for the 90% confidence interval at 50 μg l^{-1}. In the last two examples, as the mean approaches the concentrations considered important, tighter confidence limits may be needed and consequently larger numbers of replicates must be collected. Experienced scientists usually will be able to decide on how much variation will be important.

Precision of estimates — case studies

The number of replicates needed for precisions based on confidence limits of 20, 40, 80 and 100% have been calculated for the Schierenseebach and Hunting Creek data-sets described above. Calculations have been done across all sites, and because site differences may increase variability, the number of replicates needed may be somewhat inflated. Most analyses lose power if sample sizes are unequal and if assumptions are violated to some degree; other difficulties such as number of species being related to total numbers of organisms collected will also make interpretation difficult. Thus, calculations for sample size to yield a minimum level of precision are done with data from the most variable site, or across all sites as presented here. Because of large changes in t values, the calculation is inefficient for sample sizes <3; thus a value of three replicates has been reported as the minimum.

Levels of precision selected must meet study objectives as well as be realistic. A sample size of 265 for CL ±20% for the rare and endangered *A. aestivalis* (Statzner 1978) may exceed available resources, whereas 68 for CL ±40% may be reasonable (in fact equal to the number collected in the original study) (Table 13.4). To understand what this means we must consider the level of precision relative to the real values of the mean. At CL ±20% there will be a 95% probability that the real mean will lie somewhere between 8 and 12 animals per sample, and at CL ±40% it will lie somewhere between 5 and 14 animals per sample. Familiarity with the nature of the animal's habitat and its survival characteristics relative to density,

Table 13.4 Sample sizes needed to estimate population numbers of *Aphelocheirus aestivalis* in the Upper and Lower (combined) Schierenseebach, Germany, for single (1X) and double (2X) sample unit sizes, and *Baetis tricaudatus* and total numbers of benthic invertebrates in Hunting Creek, California. Calculations using formula 2 (Elliott 1977) and iterated, for raw and log_{10}-transformed data

Variable	n	Mean	Variance	Replicates for CLs ±20%	±40%	±80%	±100%
A. aestivalis							
1X raw	135	9.452	244.2	265	68	19	13
1X log_{10}	135	0.748	0.238	43	13	5	5
2X raw	68	18.765	497.84	139	37	11	8
2X log_{10}	68	1.099	0.185	18	6	3	3
Hunting Creek							
Total number raw	99	497.58	277006	110	30	9	7
Total number log_{10}	99	2.48	0.21	6	2	3	3
B. tricaudatus raw	99	49.32	4810	193	50	15	10
B. tricaudatus log_{10}	99	1.29	0.44	28	9	4	3

together with the study objectives, should enable a decision to be made easily. The effect of doubling the sample unit size is approximately to halve the required number of replicates (Table 13.4). For example, 40 replicates per site (a sample unit size double that used in the original study) would be sufficient to yield mean estimates of abundance with 95% CL ±40% and would also avoid the transformation problems found with frequent zero counts (section 13.2). In the original study performed on this rare animal (Statzner *et al* 1988), the replication proposed here would have been appropriate.

Required numbers of replicates are quite high even at the CL ±40% level for Hunting Creek data, although estimates of total numbers of macroinvertebrates may need fewer replicates than required for individual taxa (Table 13.4). Numbers of replicates required to yield CL ±80% would have been reasonable (nine for total numbers and 15 for *Baetis tricaudatus*; Table 13.4) and likely to be affordable for an ongoing sampling programme. Hunting Creek sites are unimpacted and variability between sites, as well as between years (especially for *B. tricaudatus*), often exceeds 80% of individual means, and can be considered as 'natural variability' (see p. 300). If density estimates were made relative to natural variability, CL ±80% of the means might be sufficient to provide reasonable confidence in the data for each sampling occasion. In this case, precision could be selected relative to expected ranges of natural variability obtained either from a pilot study or from the experience of the investigators. Levels of precision suggested here are generally lower than other authors have considered (see above) but these are probably realistic relative to the magnitude of expected differences. If population numbers vary by an order of magnitude, there is little sense in sampling for precisions of CL ±5% or even CL ±40%.

Precision of estimates and transformed data

In many cases, an index of abundance or concentration may be all that is needed, and data may be transformed to provide this information. In such cases where the number of replicates needed for a required level of precision is calcu-

lated on the transformed data, many fewer replicates are needed for the same precision (Norris & Georges 1986; Table 13.4). Eberhardt (1978) suggested that it matters little what the units of an index are, and he argued against the need to back-transform. If data can be used when transformed, high levels of precision might be obtained with little effort (Table 13.4), a consideration that has been neglected in discussions of determining sample size for estimates (Elliott 1977; Downing 1979; Allan 1984).

Improving precision for estimates

There are several ways of improving precision without substantially changing allocated resources. Major improvements may be made by decreasing the sample unit size and collecting more replicates (Downing 1979, 1989). This will be so, provided that counts in each collection are not too small (Riddle 1989) and that frequent zeros are avoided (section 13.2). Sampling stratification also may reduce variability (section 13.2), and so improve precision. Such improvements may be more difficult for chemical sampling where the most straightforward way of improving precision is to increase the number of replicates collected.

There will be cases where acceptable levels of precision are difficult to achieve, for example obtaining estimates of a rare species; this is also true for highly toxic substances when their critical levels are below the detection limits of current methods or technology. These are problem areas that are necessarily expensive.

Determination of levels of precision for a sampling programme for estimates will include the following considerations.

1 *Determination of the ranges for the variables of interest from a pilot study, previous work, or the experience of the investigator.* Wider ranges may need less precise estimates.

2 *Determination of the biological or chemical consequences of decisions made based on the estimates provided.* If consequences are not expected even within wide ranges, or are unlikely to be serious, low precisions can be selected. Conversely, if consequences may occur in a nar-

row range and/or they are serious, corresponding high levels of precision are required.

3 *Determine how close estimates are likely to be to levels where an effect will occur.* Small differences require higher levels of precision than would otherwise be needed.

4 *Determination of the characteristics of the organisms or chemicals.* High reproductive potentials or high rates of gain/loss generally need lower levels of precision.

5 *Determination of whether the data can be used transformed.* High levels of precision may be obtained with lower numbers of replicates when using transformed data, but care should be exercised using back-transformed data.

Sample size for comparisons

Sample size comparison decisions

The objectives of many sampling programmes in rivers involve the determination of change, such as the differences between two or more sites or times. The problem of determining the number of replicates required to detect a given difference between means with a known chance of error is related to precision, but it has received less attention than the question of estimating mean concentrations or densities. Often means are compared using t-tests or ANOVA with data meeting the assumptions of these tests (section 13.3). Usually the only concern with comparisons are whether or not the means being compared are different, i.e. in which direction they differ (see below), and not with the absolute values. This is where confusion arises when calculating sample sizes required for comparisons, as compared with those done for estimates. Most discussions are concerned with calculating sample size to detect differences between means using a test such as ANOVA (e.g. Schwenneker & Hellenthal 1984) and most texts (e.g. Snedecor & Cochran 1967, Sokal & Rohlf 1981, Zar 1984) suggest doing this with untransformed data. However, with most environmental data, comparisons to test for significant differences between means are made with transformed data. Calculations for required sample size should therefore be done on the data in the form in which they will be analysed (Norris & Georges 1986).

Three decisions need to be made before calculating sample size for comparing means once the difference to be detected is known:

1 *The formula to be used.* There are two types of formulae, closely related, but with important differences in their calculation (formula 3, and formulae 4 and 5, Table 13.3): formula 3 (Sokal & Rohlf 1981) uses the expected percentage difference between two means; formulae 4 and 5 (Zar 1984) use the actual difference expected between means (which could be converted to a percentage). Formula 4 is for the two-sample case, and formula 5 is for three or more samples.

2 *The level of significance.* Both types of formulae include the level of significance (α), or the probability at which means will be accepted as being different. This is also the probability of drawing a conclusion that the means being compared are different when in fact they are not (type I error). Conventionally, environmental studies accept α as 0.05, but some cases may call for different levels (see Schwenneker & Hellenthal 1984).

3 *The power of the test.* Both formulae also include β ($1 - P$ in formula 3), which is the power of the test, or the probability of drawing a conclusion that the means being compared are not different when in fact they are (type II error). Unfortunately, little discussion is available to guide selection of the level for β.

The following discussion will consider points (1) and (3) using analyses on the Schierenseebach and Hunting Creek data-sets. All analyses on *A. aestivalis* have been done with the individual collections grouped into two to eliminate the problem of zeros with transformation. Logarithmic transformations were chosen based on previous arguments (section 13.2).

Choosing a formula to calculate sample size for comparisons

The initial choice of formula type (3, 4, or 5, Table 13.3) is determined by whether the investigator is able to express the difference between means that the test is to detect in absolute terms (formula 4 or 5, Table 13.3) or merely as a percentage (formula 3, Table 13.3). Generally, for specification of a difference in absolute terms, some estimate of at least two means is needed, using the methods we employed in the following study. Alternatively, if only one mean is available, the investigator needs to specify the departure to be detected. In our example of *A. aestivalis*, the means were 9.8 at site 1 upstream of the lake and 27.7 at site 2 downstream (see Table 13.2) giving a difference of 17.9. A number smaller than this could be included in the formula, with the degree of conservatism being determined by the investigator. If only the mean from site 2 was available, a decision would need to be made on how much difference would be important; perhaps this difference could be as low as 7 or 8. If only the mean from site 1 was available, the difference selected might have been lower (e.g. 2 or 3), unless the expected direction of change was an increase, in which case 7 or 8 might still be accepted. Clearly the size of the mean used, the expected direction of change, and the characteristics of the organism all affect decisions on the expected level of difference for use in the formula.

If a percentage difference is to be used (formula 3, Table 13.3), it could be estimated from both or either mean. Here we have a dilemma. The usual convention would be to calculate it as a percentage of the largest mean as follows:

$$\text{Percentage difference} = 100 - \left(\frac{\overline{Y}_{min}}{\overline{Y}_{max}} \times 100 \right)$$

In our example the percentage difference, or decrease, would be about 65%. If an increase from the smallest mean was of interest, the calculation would be:

$$\text{Percentage increase} = \left(\frac{\overline{Y}_{max}}{\overline{Y}_{min}} \times 100 \right) - 100$$

or about 182%. When considering difference and change as a percentage, the expected direction is critical and the two equate as follows:

Percentage increase on smallest mean	Percentage difference between means, or percentage decrease
10	9
20	17
50	33
100	50
200	67
300	75

Where small changes are considered important (20% or less), the difference between percentage increase and percentage decrease is trivial. However, the percentage difference (the decrease calculated using the formula above) is limited to 100% while the percentage increase is unlimited. The percentage to be included in formula 3 (Table 13.3) will depend on the expected direction of change and individual situations. For example, if changes in concentrations of toxic substances such as trace metals are of concern, an increase will usually be important. Alternatively, if low levels of dissolved oxygen are problematic, a decrease will be important. Most impacts on rivers cause a decrease in biota but some, such as the effects of nutrients on algal growth, effect increases. Calculations on the percentage difference should be done on control or reference sites for the expected direction of change. Of course, in some cases the direction of change cannot be specified, such as in baseline or general ecological studies where differences may occur in either direction. In these cases, a general estimate may be obtained across all sites and then the percentage changes or absolute differences determined relative to expected ranges. The smallest changes of interest should be included in the formula to determine sample size.

An important point following from this discussion is that tabulated values of *t* and *F* used in the formulae should be from a one-tailed test if the direction of change can be specified, and two-tailed if not. The test will be more powerful if the direction of change can be specified and consequently fewer replicates will be needed (e.g. *Aphelocheirus aestivalis* (see Table 13.6)).

Choosing the power of the test

Having accepted the significance level of α to be 0.05 (which is in general use), the next decision to be made is the level of β, which is the likelihood of making a type II error (Sokal & Rohlf 1981) and which is included in the formula. The level of β, or the power of the test, is related to α but is also dependent on the distributions of the data-sets being compared (Sokal & Rohlf 1981). The latter will affect it in three ways. First, where overlap of the tails of the distributions being compared is large, β will also be large; where it is small, β will also be small. Second, where the means of the data-sets being compared are far apart, β will be small; where they are closer it will be larger. Third, β is inversely proportional to sample size, being small at large sample sizes and large at small ones. Therefore levels of β may change depending on the nature of the data-sets being compared and the sample size, even though α remains constant at 0.05. Unfortunately, there is little discussion on how to determine what level of β should be included in the formulae to calculate sample size (Sokal & Rohlf 1981; Zar 1984), on what levels of β equate with levels of α for different types of data, or even, what levels are acceptable in biomonitoring work.

The most straightforward way to increase the power of a statistical test, such as ANOVA, when comparing means is to increase replication (sample size). This has the effect of reducing β relative to a fixed level of α and a fixed difference between means (Sokal & Rohlf 1981). Non-normally distributed data may also reduce the power of the test (increasing β) (Cochran 1947), but this interaction with the data distribution can be changed with transformation (section 13.2).

To assess the relationship of α and β with benthic data we drew 50 sets of collections at random from the *A. aestivalis* data-set for each of 10, 15, 17 and 20 replicates and performed ANOVA to test for significant differences (Table 13.5). The threefold difference between means in the full data-set (see Table 13.2) could be considered significant and this was verified by ANOVA for both raw and log-transformed data (sample size 33, Table 13.5). Therefore, where ANOVA failed to show significant differences for smaller sample sizes the results were considered to be type II errors. As expected, type II errors decreased as sample size increased for both raw and log-transformed data (Table 13.5). This analysis also demonstrates empirically that about 20 replicates are needed for raw data and about 10–15 for logged data to avoid a type II error and provide sufficient statistical power to detect the difference (Table 13.5). This is substantially lower than the estimates of required replicates calculated and presented above. Usually, fewer replicates are needed for transformed data (demonstrated empirically in Table 13.5) because homogeneity of variances, achieved through transformation, may increase the power of ANOVA (see below). The relationship between transformation and sample-size calculations are discussed below.

This type of analysis is not feasible when designing a sampling programme where the replication needed will be decided on a continuum from guesswork to use of sample-size formulae

Table 13.5 Tests for significant differences between means for a number of sample sizes of counts of *Aphelocheirus aestivalis*, Schierenseebach, Germany

Data type	n	F	P	$\overline{Y}_2$	$\overline{Y}_1$	Type II errors
Raw	5	3.33	0.103	30.6	9.06	5
	10	5.53	0.052	28.45	9.32	2
	15	7.34	0.025	29.86	10.10	2
	20	8.56	0.015	29.17	10.32	1
	33	13.26	0.0005	28.54	10.12	–
Log_{10}	5	3.18	0.095	1.341	0.947	5
	10	5.84	0.073	1.289	1.912	3
	15	11.24	0.015	1.355	0.946	0
	20	9.75	0.012	1.305	0.961	0
	33	16.21	0.0002	1.307	0.958	–

Values are means of 50 analyses, except for full sample of 33 which was one analysis. n = number of observations; F = F-value from ANOVA; P = probability level from ANOVA; $\overline{Y}_1$, $\overline{Y}_2$ = arithmetic means from sites 1 and 2, respectively.

based on some prior knowledge of data variability. Levels of β to include in the formulae were not available to us, and no convention has been established. To estimate levels of β we performed the *A. aestivalis* ANOVAs just mentioned (Table 13.5); the power (level of β) of each of these tests was determined using the approach of Zar (1984, p. 174), using an *a priori* level of α of 0.05. This was done separately for raw and log-transformed data because the relationships would be expected to be different for different data distributions (Sokal & Rohlf 1981). The values of β were then averaged (observations >0.7 were set to 0.7 for this calibration) yielding 0.30 for raw data and 0.22 for logged data. These were then used in formulae 3 and 4 for calculating sample size. We saw this as being a useful way of estimating levels of β at which the ANOVAs were operating. When these values were substituted into formulae 3 and 4 (Table 13.3) and the actual differences between means in the existing data also included (as percentages or absolute numbers), calculated sample sizes were equivalent to those that demonstrated differences at the 0.05 level of significance (i.e. 15–20, Table 13.6). This indicates that both formulae produce accurate estimates of sample size when levels of β are equivalent to those operating in the tests performed on the data used here. Average levels of β used here were larger than might have been suggested by worked examples in texts (Sokal & Rohlf 1981; Zar 1984), or those used in other work (e.g. Schwenneker & Hellenthal 1984). If those smaller levels had been used, calculated sample sizes would have been larger, possibly resulting in an inefficient sampling programme. In contrast, when sample sizes are small (e.g. $n = 5$ for Hunting Creek), β may be high. Comparisons across all three sites for total numbers of individuals and number of *B. tricaudatus* in the Hunting Creek data-set (Table 13.6) showed that about three-quarters of the tests had values of β > 0.7. The importance of determining, *a priori*, the levels of β as well as α during the design phase of a biomonitoring study is clearly demonstrated here.

Levels of β estimated here are likely to be different for data collected from different places and for different data types. Further testing would be needed to establish whether the mean values of β are general for benthic invertebrates. Similar analyses should also be performed for physical and chemical data to establish values of β to be used in formulae to calculate sample sizes for them.

Higher levels of β indicate greater chances of making a type II error, concluding that an effect did not exist when in fact it did. The levels we have used provide sample sizes with 30% (raw) and 22% (logged) chances of making a type II error. Risks as high as these of drawing an incorrect conclusion may be unacceptable in some cases, especially for assessment of environmental impact where it is usually important to detect an impact if one exists. In these cases smaller values of β may be included in the sample size formulae to reduce the risk to acceptable levels.

Variance, transformation and sample size for comparisons

Formulae 3, 4 and 5 (Table 13.3) all use an estimate of the variance in the numerator to calculate a required sample size. Thus, larger variances result in larger sample sizes if the differences to be detected remain equivalent. The best estimate of variance for ANOVA (and for use in the formulae) is the pooled variance (Snedecor & Cochran 1967 p. 101) across sites or times. Generally this will be smaller than a variance calculated for all sites because among-site differences will contribute in addition to within-site differences. There will be little difference between the pooled variance and total variance if the ANOVA assumption of homogeneity of variances is met. One of the reasons for transforming data is to meet this assumption (Section 13.3). Thus, data meeting the assumption of homogeneity of variances have a lower pooled variance than those not doing so. Consequently, when data are properly transformed, sample-size requirements are smaller than for the same data untransformed (Table 13.6), and calculations of required sample size should be done on the transformed data. Heterogeneity of variances will increase the likelihood of a type II error in ANOVA (Cochran 1947), but calculated sample-size requirements on such data will be larger, thus offsetting the problem (Table 13.6).

The Biota

Table 13.6 Sample sizes needed to detect differences between means from three sites in Hunting Creek, northern California, and two sites in the lower Schierenseebach, Germany, using the formulae of Sokal and Rohlf (1981, p. 263) (S&R) and Zar (1984). The formula from Zar (1984, p. 133) was used for the two-sample case for *Aphelocheirus aestivalis* and the formula from Zar (1984, p. 193) was used for the Hunting Creek multisample case. Percentage differences are the actual differences between the smallest and largest means, α has been set at 0.05 and β at 0.30 for raw data and 0.22 for log-transformed data. Results of ANOVA (F) and corresponding significance levels (P) show the comparisons for which significant differences were found (underlined). $\overline{Y}$ = arithmetic mean, S_P^2 = pooled sample variance, CV = coefficient of variation.

Data type	Year	$\overline{Y}$	S_P^2	CV	Percentage difference	F	P	Sample size S&R	Sample size Zar
HUNTING CREEK, CALIFORNIA									
Total *n* (raw)	1985	289	39441	68.7	52.5	1.40	0.284	22	27
	1986	184	29240	92.9	68.6	1.15	0.351	24	29
	1987	251	14217	47.5	45.3	1.98	0.181	15	18
	1988	689	86126	42.6	69.7	8.71	0.005	6	7
	1989	963	561947	77.9	75.2	2.69	0.108	14	18
	1990	853	245831	58.2	36.7	0.96	0.409	32	39
Total *n* (log₁₀)	1985	2.378	0.071	11.2	10.6	1.26	0.319	18	21
	1986	2.081	0.173	20.0	23.0	1.83	0.206	12	14
	1987	2.355	0.033	7.8	10.2	2.74	0.105	10	11
	1988	2.764	0.029	6.2	16.5	10.66	0.002	3	4
	1989	2.841	0.081	10.0	19.5	5.85	0.017	5	6
	1990	2.850	0.074	9.6	10.5	1.89	0.193	14	16
Baetis tricaudatus (raw)	1985	52	6073	151.0	93.6	1.35	0.295	33	40
	1986	32	1490	119.8	90.7	2.71	0.111	23	28
	1987	18	514	127.4	89.1	1.89	0.194	26	32
	1988	81	3691	75.1	69.6	2.31	0.141	16	19
	1989	90	6155	87.1	98.0	5.34	0.022	11	13
	1990	47	3871	132.8	95.2	3.45	0.066	25	31
Baetis tricaudatus (log₁₀)	1985	1.327	0.195	33.3	60.9	8.76	0.004	6	7
	1986	1.228	0.136	30.0	56.8	8.34	0.006	5	7
	1987	0.995	0.158	40.0	67.6	7.56	0.008	6	8
	1988	1.625	0.305	34.0	52.1	5.46	0.021	8	9
	1989	1.520	0.155	25.9	72.6	22.58	0.000	3	4
	1990	1.347	0.092	22.5	61.6	18.60	0.000	3	4
LOWER SCHIERENSEEBACH, GERMANY									
Aphelocheirus aestivalis Total *n* (raw)		28.5	422.3	72.0	64.5	13.26	0.000		
Two-sided values of *t*								17	13
One-sided values of *t*								8	10
Total *n* (log₁₀)		1.307	0.124	26.9	26.7	16.21	0.000		
Two-sided values of *t*								17	12
One-sided values of *t*								8	9

Smaller sample sizes and tighter confidence limits were obtained with transformed data (Table 13.6). Transformed data, plus the large differences between means in the test data, yielded the relatively small sample sizes needed to demonstrate differences between means using ANOVA (Table 13.6). Variability in the test data are well within the ranges of variability considered normal for benthic invertebrates (see Eberhardt (1978) for examples). The sample sizes suggested here are much smaller than those indicated by others (e.g. Schwenneker & Hellenthal 1984), but their use of raw data and stringent levels of β would have yielded much higher replicate numbers than might be necessary to demonstrate differences that are of interest.

Determination of required sample sizes for comparing means will include the following considerations:

1 *Determination of the difference or change to be detected.*

2 *Determination of the variance.* This should be a pooled variance across all sites, or times, calculated using data in the form in which they will be analysed. Where it is not possible to calculate a pooled variance and some other estimate of variance is used, conservative sample size estimates are likely to be generated.

3 *Determine if the direction of change can be specified.* This may affect calculation of the percentage difference to be accepted. Also, if the direction can be specified, values for α and β should be for one-tailed rather than two-tailed tests.

4 *Selection of the formula to be used.* This is based largely on whether differences between means are to be specified as a percentage or in absolute terms, and whether two or several means are to be compared. The percentage differences may not be affected as much as the absolute differences by subsequent changes in sampling methods.

5 *Selection of the significance or probability of a type I error at which the test will operate.* Conventionally, this is 0.05 for environmental data. Management and scientific needs may dictate other levels in special circumstances.

6 *Selection of the probability level that the test will find a difference when one exists (avoiding a type II error).* Different levels may be accepted for different data types: transformed, raw, chemical or biological. Results presented here suggest β = 0.30 for raw data on benthic invertebrates, and β = 0.22 for log-transformed data.

ACKNOWLEDGEMENTS

B. Statzner and J.A. Gore provided *Aphelocheirus aestivalis* data. Information for Table 13.1 was provided by T. Bott, D. Erman, G. Lamberti, F. Ligon, B. Orr, C. Dahm, T. Arsuffi and M. Power. A. Georges commented on the manuscript.

Allan JD. (1984) Hypothesis testing in ecological studies of aquatic insects. In: Resh VH, Rosenberg DM (eds) *The Ecology of Aquatic Insects*, pp 485–507. Praeger Publishers, New York. [13.3, 13.4]

Aloi JE. (1990) A critical review of recent freshwater periphyton field methods. *Canadian Journal of Fisheries and Aquatic Sciences* **47**: 656–70. [13.1]

American Public Health Association, American Water Works Association, Water Pollution Control Federation (1989) *Standard Methods for the Examination of Water and Wastewater* 17th edn. American Public Health Association, Washington, DC. [13.2]

Anderson FS. (1965) The negative binomial distribution and sampling of insect populations. In: *Proceedings of the XIIth International Congress of Entomology*, p. 395. London. [13.3]

Andrews JH. (1986) *r*- and *K*-selection and microbial ecology. *Advances in Microbial Ecology* **9**: 99–147. [13.1]

Andrews JH. (1991) *Comparative Ecology of Microorganisms and Macroorganisms*. Brock/Springer Series in Contemporary Bioscience. Springer Verlag, New York. [13.1]

Atlas RM. (1985) Applicability of general ecological principles to microbial ecology. In: Poindexter JS, Leadbetter ER (eds) *Bacteria in Nature, Vol 2. Methods and Special Applications in Bacterial Ecology*, pp 339–70. Plenum, New York. [13.1]

Baltz DM. (1990) Autecology. In: Schreck CB, Moyle PB (eds) *Methods for Fish Biology*, pp 585–607. American Fisheries Society, Bethesda, Maryland. [13.1]

Barmuta LA. (1987) Polemics, aquatic insects and biomonitoring: an appraisal. In: Majer JD (ed.) *The Role of Invertebrates in Conservation and Biological Survey*, pp 65–72. Western Australian Department of Land Management Report, Perth, Western Australia. [13.2]

Benke AC. (1984) Secondary production of aquatic insects. In: Resh VH, Rosenberg DM (eds), *The*

Ecology of Aquatic Insects, pp 289–322. Praeger Publishers, New York. [13.1]

Bernstein BB, Zalinski J. (1983) An optimum sampling design and power tests for environmental biologists. *Journal of Environmental Management* **16**: 35–43. [13.1]

Berry RJ. (1987) Scientific natural history: a key base to ecology. *Biological Journal of the Linnaean Society* **32**: 17–29. [13.1]

Best EPH. (1988) The phytosociological approach to the description and classification of aquatic macrophyte vegetation. In: Symoens JJ (ed.) *Vegetation of Inland Waters. Handbook of Vegetation Science 15/1*, pp 155–82. Kluwer Academic Publishers, Dordrecht. [13.1]

Blum JL. (1956) The ecology of river algae. *Botanical Reviews* **22**: 291–341. [13.1]

Bohlool BB, Schmidt EL. (1980) The immuno-fluorescence approach in microbial ecology. *Advances in Microbial Ecology* **4**: 203–41. [13.1]

Bold HC, Wynne MJ. (1985) *Introduction to the Algae*. Prentice-Hall, Englewood Cliffs, New Jersey. [13.1]

Bothwell ML. (1988) Growth rate responses of lotic periphyton diatoms to experimental phosphorus enrichment: the influence of temperature and light. *Canadian Journal of Fisheries and Aquatic Sciences* **45**: 261–70. [13.1]

Bott TL. (1973) Bacteria and the assessment of water quality. In: Cairns J Jr, Dickson KL (eds) *Biological Methods for the Assessment of Water Quality*, pp 61–75. A symposium presented at the 75th annual meeting, American Society for Testing and Materials. American Society for Testing and Materials Special Technical Publication 528. American Society for Testing and Materials, Philadelphia. [13.1]

Bowen SH. (1983) Quantitative description of the diet. In: Nielsen LA, Johnson DL (eds) *Fisheries Techniques*, pp 325–36. The American Fisheries Society, Bethesda, Maryland. [13.1]

Boylen CW, Soracco RJ. (1986) Autecological studies in microbial limnology. In: Tate RL III (ed.) *Microbial Autecology*, pp 183–231. John Wiley and Sons, New York. [13.1]

Brittain JE, Eikeland TJ. (1988) Invertebrate drift — a review. *Hydrobiologia* **166**: 77–93. [13.1]

Butler MC. (1984) Life histories of aquatic insects. In: Resh VH, Rosenberg DM (eds) *The Ecology of Aquatic Insects*, pp 24–55. Praeger Publishers, New York. [13.1]

Cairns J Jr, Pratt JR. (1986) Developing a sampling strategy. In: Isom BG (ed.) *Rationale for Sampling and Interpretation of Ecological Data in the Assessment of Freshwater Ecosystems*, pp 168–86. American Society for Testing and Materials, Philadelphia. [13.1]

Canton SP, Chadwick JW. (1988) Variability in benthic invertebrate density estimates from stream samples. *Journal of Freshwater Ecology* **4**: 291–7. [13.3]

Carpenter SR, Lodge DM. (1986) Effects of submersed macrophytes on ecosystems processes. *Aquatic Botany* **26**: 291–341. [13.1]

Cederholm CJ, Peterson NP. (1985) The retention of coho salmon (*Oncorynchus kisutch*) carcasses by organic debris in small streams. *Canadian Journal of Fisheries and Aquatic Sciences* **42**: 1222–5. [13.1]

Chapman DW. (1978) Production. In: Bagenal T (ed.) *Methods for Assessment of Fish Production in Fresh Waters* 3rd edn, pp 202–18. IBP Handbook 3, Blackwell Scientific Publications, Oxford. [13.1]

Chutter FM, Noble RG. (1966) The reliability of a method of sampling stream invertebrates. *Archiv für Hydrobiologie* **62**: 95–103. [13.4]

Clarke RT. (1990) Bias and variance of some estimators of suspended sediment load. *Journal des Sciences Hydrologiques* **35**: 253–61. [13.3]

Clifford HF, Zelt KA. (1972) Assessment of two mesh sizes for interpreting life cycles, standing crop, and percentage composition of stream insects. *Freshwater Biology* **2**: 259–69. [13.1]

Cochran WG. (1947) Some consequences when the assumptions for analysis of variance are not satisfied. *Biometrics* **3**: 22–38. [13.3, 13.4]

Cochran WG. (1963) *Sampling Techniques* 2nd edn, pp 413. John Wiley and Sons, New York. [13.1]

Cooper SD, Barmuta LA. (1992) Field experiments in biomonitoring. In: Rosenberg DM, Resh VH (eds) *Freshwater Biomonitoring Using Macroinvertebrates*. pp 395–437. Chapman and Hall, New York. [13.1, 13.2]

Crisp DT. (1990) Simplified methods of estimating daily mean stream water temperature. *Freshwater Biology* **23**: 457–62. [13.3]

Crowder LB. (1990) Community ecology. In: Schreck CB, Moyle PB (eds) *Methods for Fish Biology*, pp 609–32. American Fisheries Society, Bethesda, Maryland. [13.1]

Cummins KW. (1973) Trophic relations of aquatic insects. *Annual Review of Entomology* **18**: 183–206. [13.1]

Dawson FH. (1980) The origin, composition and downstream transport of plant material in a small chalk stream. *Freshwater Biology* **10**: 419–35. [13.1]

Dawson FH. (1988) Water flow and the vegetation of running waters. In: Symoens JJ (ed.) *Vegetation of Inland Waters. Handbook of Vegetation Science 15/1*, pp 283–309. Kluwer Academic Publishers, Dordrecht. [13.1]

Dennis WM, Isom BG (eds). (1984) *Ecological Assessment of Macrophyton: Collection, Use, and Meaning of Data*. American Society for Testing and Materials Special Technical Publication 843. American Society for Testing and Materials, Philadelphia. [13.1]

Denny FH. (1980) Solute movement in submerged angiosperms. *Biological Reviews* **55**: 65–92. [13.1]

Downing JA. (1979) Aggregation, transformation, and the design of benthos sampling programs. *Journal of the Fisheries Research Board of Canada* **36**: 1454–63. [13.3, 13.4]

Downing JA. (1989) Precision of the mean and the design of benthos sampling programmes: caution revised. *Marine Biology* **103**: 231–4. [13.4]

Downing JA, Anderson MR. (1986) Estimating the standing biomass of aquatic macrophytes. *Canadian Journal of Fisheries and Aquatic Sciences* **42**: 1094–104. [13.1, 13.3]

Eberhardt LL. (1963) Problems in ecological sampling. *Northwest Science* **37**: 144–54. [13.1, 13.2, 13.3]

Eberhardt LL. (1976) Quantitative Ecology and Impact Assessment. *Journal of Environmental Management* **4**: 213–17. [13.1, 13.2, 13.3]

Eberhardt LL. (1978) Appraising variability in population studies. *Journal of Wildlife Management* **42**: 207–38. [13.1, 13.2, 13.3, 13.4]

Elliott JM. (1977) *Some Methods for the Statistical Analysis of Samples of Benthic Invertebrates* 2nd edn. Freshwater Biological Association Scientific Publication No. 25, Cumbria, UK. [13.1, 13.3, 13.4]

Ellis JC, Lacey RF. (1980) Sampling: defining the task and planning the scheme. *Water Pollution Control* **79**: 699–718. [13.1]

El-Shaarawi AH. (1989) Inferences about the mean from censored water quality data. *Water Resources Research* **25**: 685–90. [13.3]

El-Shaarawi AH, Dolan DM. (1989) Maximum likelihood estimation of water quality concentrations from censored data. *Canadian Journal of Fisheries and Aquatic Sciences* **46**: 1033–9. [13.3]

Elwood JW, Newbold JD, O'Neill RV, van Winkle W. (1983) Resource spiraling: an operational paradigm for analyzing lotic ecosystems. In: Fontaine TD III, Bartell SM (eds) *Dynamics of Lotic Ecosystems*, pp 3–27. Ann Arbor Science Publishers, Ann Arbor. [13.1]

Frissell CA, Liss WJ, Warren CE, Hurley MD. (1986) A hierarchical framework for stream habitat classification: viewing streams in a watershed context. *Environmental Management* **10**: 199–214. [13.2]

Fry JC. (1988) Determination of biomass. In: Austin B (ed.) *Methods in Aquatic Bacteriology*, pp 27–82. John Wiley and Sons, New York. [13.1]

Fryer G. (1987) Quantitative and qualitative: numbers and reality in the study of living organisms. *Freshwater Biology* **17**: 447–55. [13.1]

Gale WF, Mohr HW Jr. (1978) Larval fish drift in a large river with a comparison of sampling methods. *Transactions of the American Fisheries Society* **107**: 46–55. [13.1]

Green RH. (1979) *Sampling Design and Statistical Methods for Environmental Biologists*. John Wiley and Sons, New York. [13.1, 13.3, 13.4]

Haedrich RL. (1983) Reference collections and faunal surveys. In: Nielsen LA, Johnson DL (eds) *Fisheries Techniques*, pp 275–82. The American Fisheries Society, Bethesda, Maryland. [13.1]

den Hartog C. (1982) Architecture of macrophyte-dominated aquatic communities. In: Symoens JJ, Hooper SS, Compere P (eds) *Studies on Aquatic Vascular Plants*, pp 222–34. Royal Botanical Society of Belgium, Brussels. [13.1]

den Hartog C, van der Velde G. (1988) Structural aspects of aquatic plant communities. In: Symoens JJ (ed.) *Vegetation of Inland Waters. Handbook of Vegetation Science 15/1*, pp 113–53. Kluwer Academic Publishers, Dordrecht. [13.1]

Harvey BC. (1987) Susceptibility of young-of-the-year fishes to downstream displacement by flooding. *Transactions of the American Fisheries Society* **116**: 851–5. [13.1]

Helsel DR. (1986) Estimation of distributional parameters for censored water quality data. In: El-Shaarawi AH, Kwiatkowski RE (eds) *Developments in Water Science*, pp 137–57. Elsevier, Amsterdam. [13.3]

Hill BH, Webster JR, Linkins AE. (1984) Problems in the use of closed chambers for measuring photosynthesis by a lotic macrophyte. In: Dennis WM, Isom BG (eds) *Ecological Assessment of Macrophyton: Collection, Use, and Meaning of Data*, pp 69–75. American Society for Testing and Materials Special Technical Publication 843. American Society for Testing and Materials, Philadelphia. [13.1]

Holder-Franklin MA. (1986) Ecological relationships of microbiota in water and soil as revealed by diversity measurements. In: Tate RL III (ed.) *Microbial Autecology*, pp 93–132. John Wiley and Sons, New York. [13.1]

Hurlbert SH. (1984) Pseudoreplication and the design of ecological field experiments. *Ecological Monographs* **45**: 1922–7. [13.1, 13.2]

Jordan C. (1989) The mean pH of mixed fresh waters. *Water Research* **23**: 1331–4. [13.3]

Karl DM. (1985) Determination of *in situ* microbial biomass, viability, metabolism and growth. In: Poindexter JS, Leadbetter ER (eds) *Bacteria in Nature, Vol 2. Methods and Special Applications in Bacterial Ecology*, pp 45–176. Plenum, New York. [13.1]

Kautsky L. (1988) Life stages of aquatic soft bottom macrophytes. *Oikos* **53**: 126–35. [13.1]

Kerekes J, Freedman B. (1989) Seasonal variations of water chemistry in oligotrophic streams and rivers in Kejimkujik-National-Park Novia Scotia. *Water Air and Soil Pollution* **46**: 131–44. [13.3]

Kogure K, Simidu U, Taga N. (1984) An improved direct viable count method for aquatic bacteria. *Archiv für Hydrobiologie* **102**: 117–22. [13.1]

Kosinski RJ. (1984) A comparison of the accuracy and precision of several openwater oxygen productivity techniques. *Hydrobiologia* **119**: 139–48. [13.3]

Lagler KF. (1978) Capture, sampling and examination of fishes. In: Bagenal T (ed.) *Methods for Assessment of Fish Production in Fresh Waters* 3rd edn, pp 7–47. IBP Handbook No. 3, Blackwell Scientific Publications, Oxford. [13.1]

Lamberti GA, Resh VH. (1987) Seasonal patterns of suspended bacteria and algae in two northern California streams. *Archiv für Hydrobiologie* **110**: 45–57. [13.1]

De Long EF, Wickham GS, Pace NR. (1989) Phylogenetic stairs: ribosomal RNA-based probes for identification of single cells. *Science* **243**: 1360–3. [13.1]

Lowe-McConnell RH. (1978) Identification of freshwater fishes. In: Bagenal T (ed.) *Methods for Assessment of Fish Production in Fresh Waters* 3rd edn, pp 48–83. IBP Handbook No. 3, Blackwell Scientific Publications, Oxford. [13.1]

Lundberg JG, McDade LA. (1990) Systematics. In: Schreck CB, Moyle PB (eds) *Methods for Fish Biology*, pp 65–108. American Fisheries Society, Bethesda, Maryland. [13.1]

Mahon R. (1980) Accuracy of catch effort methods for estimating fish density and biomass in streams. *Environmental Biology of Fishes* **5**: 343–64. [13.3]

Malone TC. (1980) Algal size. In: Morris T (ed.) *The Physiological Ecology of Phytoplankton*, pp 433–63. University of California Press, Berkeley. [13.1]

Merritt RW, Cummins KW (eds). (1984) *An Introduction to the Aquatic Insects of North America* 2nd edn. Kendall/Hunt Publishing Company, Dubuque. [13.1]

Merritt RW, Cummins KW, Resh VH. (1984) Collecting, sampling, and rearing methods for aquatic insects. In: Merritt RW, Cummins KW (eds) *An Introduction to the Aquatic Insects of North America* 2nd edn. pp 11–26. Kendall/Hunt Publishing Company, Dubuque. [13.1]

Millard SP, Lettenmaier DP. (1985) Optimal design of biological sampling programs using the analysis of variance. *Estuarine Coastal Shelf Science* **42**: 637–45. [13.1, 13.3]

Millard SP, Yearsley JR, Lettenmaier DP. (1985) Space–time correlation and its effects on methods for detecting aquatic ecological change. *Canadian Journal of Fisheries and Aquatic Sciences* **42**: 1391–400. [13.1]

Miller C. (1989) Down with numbers. *Biocycle* **30**: 20. [13.1]

Moriarty DJW. (1986) Measurement of bacterial growth rates in aquatic systems from rates of nucleic acid synthesis. *Advances in Microbial Ecology* **6**: 245–92. [13.1]

Morin A. (1985) Variability of density estimates and the optimization of sampling programs for stream benthos. *Canadian Journal of Fisheries and Aquatic Sciences* **42**: 1530–40. [13.3, 13.4]

Morin A, Mousseau TA, Roff DA. (1987) Accuracy and precision of secondary production estimates. *Limnology and Oceanography* **32**: 1342–52. [13.3]

Mulholland PJ, Steinman AD, Elwood JW. (1990) Measurement of phosphorus uptake length in streams: Comparison of radiotracer and stable PO_4 releases. *Canadian Journal of Fisheries and Aquatic Sciences* **32**: 817–20. [13.1]

Mundie JH. (1971) Sampling benthos and substrate materials, down to 50 microns in size, in shallow streams. *Journal of the Fisheries Research Board of Canada* **28**: 849–60. [13.1]

Needham PR, Usinger RL. (1956) Variability in the macrofauna of a single riffle in Prosser Creek, California, as indicated by the Surber sampler. *Hilgardia* **24**: 383–409. [13.4]

Niels AL, Bell PE. (1986) Determination of individual organisms and their activities *in situ*. In: Tate RL III (ed.) *Microbial Autecology*, pp 27–60. John Wiley and Sons, New York. [13.1]

Nielsen LA, Johnson DL (eds). (1983) *Fisheries Techniques*. The American Fisheries Society, Bethesda, Maryland. [13.1]

Norris RH, Georges A. (1985) *Importance of sample size for indicators of water quality*. International Symposium on Biological Monitoring of the State of the Environment (Bioindicators), New Delhi, 11–13 October 1984, pp 37–49. Indian Academy of Science. [13.3]

Norris RH, Georges A. (1986) Design and analysis for assessment of water quality. In: Deckker P, Williams WD (eds) *Limnology in Australia*, pp 555–72. Commonwealth Scientific and Industrial Research Organization, Melbourne, Australia and Dr W.Junk, Dordrecht, the Netherlands. [13.1, 13.3, 13.4]

Norris RH, Georges A. (1991) Analysis and interpretation of benthic macroinvertebrate surveys. In: Rosenberg DM, Resh VH (eds) *Freshwater Biomonitoring Using Macroinvertebrates*, pp 230–82. Chapman and Hall, New York [13.1, 13.3]

Peckarsky BL. (1984) Sampling the stream benthos. In: Downing JA, Rigler FH (eds) *A Manual on Methods for the Assessment of Secondary Productivity in Fresh Waters*, pp 131–60. IBP Handbook 17, Blackwell Scientific Publishers, Oxford. [13.1]

Peterson BJ, Hobbie JE, Corliss TL, Kriet K. (1983) A continuous-flow periphyton bioassay: tests of nutrient limitation in a tundra stream. *Limnology and Oceanography* **28**: 583–91. [13.1]

Power ME. (1990) Benthic turfs vs floating mats of algae in river food webs. *Oikos* **58**: 67–79. [13.1]

Power ME, Stewart AJ, Matthews WJ. (1988) Grazer control of algae in an Ozark mountain stream: effects of a short-term exclusion. *Ecology* **69**: 1894–8. [13.1]

Raleigh RF, Short C. (1981) Depletion sampling in stream ecosystems: assumptions and techniques. *Progress in Fish Culture* **43**: 115–20. [13.3]

Resh VH. (1979) Sampling variability and life history features: basic considerations in the design of aquatic insect studies. *Journal of the Fisheries Research Board of Canada* **36**: 290–311. [13.1, 13.2, 13.3, 13.4]

Resh VH. Jackson JK. (1992) Rapid assessment approaches to biomonitoring using benthic macroinvertebrates. In: Rosenberg DM, Resh VH (eds) *Freshwater Biomonitoring and Benthic Macroinvertebrates*, pp 192–229. Chapman and Hall, New York. [13.3]

Resh VH, McElravy EP. (1992) Contemporary quantitative approaches to biomonitoring using benthic macroinvertebrates. In: Rosenberg DM, Resh VH (eds) *Freshwater Biomonitoring and Benthic Macroinvertebrates*, pp 157–91. Chapman and Hall, New York. [13.1]

Resh VH, Rosenberg DM. (1989) Spatial–temporal variability and the study of aquatic insects. *Canadian Entomologist* **121**: 941–63. [13.3]

Resh VH, Brown AV, Covich AP, Gurtz ME, Li HW, Minshall GW *et al.* (1988) The role of disturbance in stream ecology. *Journal of the North American Benthological Society* 7: 433–55. [13.2]

Richey JE, Perkins MA, Goldman CR. (1975) Effects of kokanee salmon (*Oncorhynchus nerka*) decomposition of the ecology of a subalpine stream. *Canadian Journal of Fisheries and Aquatic Sciences* **32**: 817–20. [13.1]

Riddle MJ. (1989) Precision of the mean and the design of benthos sampling programmes: caution advised. *Marine Biology* **103**: 225–30. [13.3, 13.4]

Ridley-Thomas CJ, Austin A, Lucey WP, Clark MJR. (1989) Variability in the determination of ash free dry weight for periohyton communities—a call for a standard method. *Water Research* **23**: 3–8. [13.3]

Rigler FH, Downing JA. (1984) The calculation of secondary productivity. In: Downing JA, Rigler FH (eds) *A Manual on Methods for the Assessment of Secondary Productivity in Fresh Waters*, pp 19–58. IBP Handbook No. 17. 2nd edn. Blackwell Scientific Publications, Oxford. [13.1]

Robson DS, Spangler GR. (1978) Estimation of population abundance and survival. In: Gerking SD (ed.) *Ecology of Freshwater Fish Production*, pp 26–51. John Wiley & Sons, New York. [13.1]

Rosenberg DM, Resh VH, Balling SS, Barnby MA, Collins JN, Durbin DV *et al.* (1981) Recent trends in environmental impact assessment. *Canadian Journal of Fisheries and Aquatic Sciences* **38**: 591–624. [13.1]

SAS Institute. (1987) *SAS/STAT Guide for Personal Computers*. Version 6. SAS Institute, Cary, North Carolina. [13.3]

Schwenneker BW, Hellenthal RA. (1984) Sampling considerations in using stream insects for monitoring water quality. *Environmental Entomology* **13**: 741–50. [13.3, 13.4]

Seber GAF. (1982) *Estimation of Animal Abundance.* Griffin, London. [13.2]

Snedecor GW, Cochran WG. (1967) *Statistical Methods* 6th edn. The Iowa State University Press, Ames, Iowa. [13.4]

Sokal RR, Rohlf FJ. (1981) *Biometry* 2nd edn. W.H. Freeman Company, London. [13.4]

Sprugel DG. (1983) Correcting for bias in log-transformed allometric equations. *Ecology* **64**: 209–10. [13.3]

Statzner B. (1978) The effects of flight behaviour on the larval abundance of Trichoptera in the Schierenseebrooks (North Germany). In: *Proceedings of the 2nd International Symposium on Trichoptera*, pp 121–34. Junk, The Hague. [13.3, 13.4]

Statzner B, Gore JA, Resh VH. (1988) Hydraulic stream ecology: observed patterns and potential applications. *Journal of the North American Benthological Society* 7: 307–60. [13.2, 13.3, 13.4]

Steinman AD, McIntire CD. (1986) Effects of current velocity and light energy on the structure of periphyton assemblages in laboratory streams. *Journal of Phycology* **22**: 352–61. [13.1]

Stevenson RJ. (1983) Effects of current and conditions simulating autogenically changing microhabitats on benthic diatom immigration. *Ecology* **64**: 1514–24. [13.1]

Stewart-Oaten A, Murdoch WW, Parker KR. (1986) Environmental impact assessment: 'Pseudoreplication: in time?'. *Ecology* **67**: 929–40. [13.2]

Strauss RE, Bond CE. (1990) Taxonomic methods: morphology. In: Schreck CB, Moyle PB (eds) *Methods for Fish Biology*, pp 109–140. American Fisheries Society, Bethesda, Maryland. [13.1]

Taylor LR. (1961) Aggregation, variance and the mean. *Nature (London)* **189**: 732–5. [13.3]

Taylor LR. (1980) New light on the variance/mean view of aggregation and transformation: comment. *Canadian Journal of Fisheries and Aquatic Sciences* **37**: 1330–2. [13.3]

Tiku ML. (1967) Estimating the mean and standard deviation from a censored normal sample. *Biometrika* **54**: 155–8. [13.3]

Trotter DM, Hendricks AC. (1979) Attached, filamentous algal communities. In: Weitzel RL (ed.) *Methods and Measurements of Periphyton Communities: a Review*, pp 58–60. American Society for Testing and Materials Special Technical Publication 690. American Society for Testing and Materials, Philadelphia. [13.1]

Waters TF. (1972) The drift of stream insects. *Annual Review of Entomology* **17**: 253–72. [13.1]

Waters WE, Erman DC. (1990) Research methods: concept and design. In: Schreck CB, Moyle PB (eds) *Methods for Fish Biology*, pp 1–34. American Fisheries Society, Bethesda, Maryland. [13.1]

Waters WE, Resh VH. [1979] Ecological and statistical features of sampling insect populations in forest and aquatic environments. In: Patil GP, Rosenzweig ML (eds) *Contemporary Quantitative Ecology and Related Econometrics*, pp 569–617. International Co-operative Publishing House, Fairland, Maryland. [13.1]

Webster JR, Benfield EF. (1986) Vascular plant breakdown in freshwater ecosystems. *Annual Review of Ecology and Systematics* **17**: 567–94. [13.1]

Webster JR, Benfield EF, Golladay SW, Hill BH, Hornick LE, Kazmierczak RF, Berry WB. (1987) Experimental studies of physical factors affecting seston transport in streams. *Limnology and Oceanography* **32**: 848–63. [13.1]

Windell JT, Bowen SH. (1978) Methods for study of fish diets based on analysis of stomach contents. In: Bagenal T (ed.) *Methods for Assessment of Fish Production in Fresh Waters* 3rd edn, pp 219–26. IBP Handbook No. 3, Blackwell Scientific Publications, Oxford. [13.1]

Wright JF, Hiley PD, Ham SF, Berrie AD. (1981) Comparison of three mapping procedures developed for river macrophytes. *Freshwater Biology* **11**: 369–79. [13.1]

Zar JH. (1968) Calculation and miscalculation of the allometric equation as a model in biological data. *Bioscience* **18**: 1118–20. [13.3].

Zar JH. (1984) *Biostatistical Analysis* 2nd edn. Prentice Hall, Englewood Cliffs, New Jersey. [13.4]

Part 3: Inputs and Pathways of Matter and Energy

Part 2 was arranged trophically, but was essentially concerned with the structural organization of river communities of microbes and multicellular plants and animals. This Part is concerned with the functioning of ecosystems. It begins with an analysis of functional links—food webs—and the rules governing them (Chapter 14). The functional biology of the primary sources of energy and matter is then dealt with in Chapters 15 (decomposer systems) and 16 (primary producers). The ways that energy flows through river communities and matter spirals within them are treated respectively in Chapters 17 and 18. Again, general principles rather than specific details are emphasized, and this leads to a further consideration of general patterns of function and structure embodied in, for example, the river continuum concept.

14: Food Webs and
Species Interactions

A.G.HILDREW

14.1 PATTERNS IN FOOD WEBS

The idea that the nutrition of living things links them together into a network of trophic interactions — known as food webs — is perhaps the most familiar in ecology. It is also deceptively simple. The study of food webs involves investigations of both their structure and function. Structure may be revealed when the species in a community are written down systematically and an audit made of the feeding links between them — 'joining up the dots'. As we shall see, there appear to be patterns in the structure of such webs that emerge from the examination of data from many communities. The most important question then is whether these patterns in food-web structure result from the nature of the biological interactions, embodied in the web, for example predation and competition, or from some extrinsic feature(s) of the environment. This is the motivation for juxtaposing in this section food web 'function', the processes by which species interact, with observed patterns in food web structure.

While much of this section may dwell on quite basic aspects of river and stream ecology, a better understanding of food webs is rich in promise for the applied science of river management. Progress in this area could provide a real understanding of river communities, which are important both as targets for environmental management and as tools in environmental assessment. The really remarkable feature of this field is how little we know of food webs in general, and how little effort is going into the study of this unifying feature of natural ecosystems.

Apparent patterns in food webs have emerged by examination of patchy and sometimes inadequate published information, for instance the 113 webs compiled by Briand and Cohen (1987). Some of the best-known patterns follow:

1 Food chains are short, usually five species or less.

2 Connectance (the proportion of possible feeding links between species which are actually realized) declines with increasing species richness.

3 The proportions of 'top', 'intermediate' and 'basal' species in food webs are constant.

4 Omnivory (species feeding at more than one level in the web) is rare.

5 Food webs from 'constant' environments have relatively more omnivory and higher connectance than those from 'fluctuating' environments.

6 Food webs vary with environmental structure and food chains are longer in three- than in two-dimensional habitats.

The reader should refer to Lawton (1989) for a fuller list, references and explanation. Briefly, some of these patterns may be real, some artefacts, but several of them are amenable to explanation by a number of competing hypotheses. The textbook explanation that food chains are short because of the low efficiency of energy transfer along them is now controversial, for food chain length may not vary along gradients from very high to very low productivity (Briand & Cohen 1987). Most of the food-web patterns can be predicted by models based on Lotka–Volterra dynamical interactions. Upper limits to species richness and connectance are then constrained by stability and, further, food-web structure will relate to environmental variability (because complex webs are dynamically 'fragile' and cannot persist in more variable environments). A non-dynamic hypothesis, the cascade model, offers an alternative explanation (Cohen & Newman 1985). It assumes a constant number of trophic links per species and that there is a hierarchy such that species feed only on those below them in this trophic 'cascade'. Warren and Lawton (1987) add the idea that the hierarchy could be based on body size, predators feeding only on species smaller than themselves.

But what of the specifics of freshwater food-webs? Only 21 of Briand and Cohen's (1987) compilation are from freshwaters and of these only nine are from rivers or streams. This is unsurprising because there are very few studies of running waters indeed that provide food-web descriptions sufficiently detailed to compare with the observed patterns. Jeffries and Lawton (1985) showed that there is a constant ratio of predator to prey species in a broad range of freshwaters, including some data from streams and rivers. This constant ratio is widespread in a variety of habitats (Pimm *et al* 1991). Briand (1985) compared food webs in a variety of freshwater habitats with others. He found the shortest food webs in streams. They were also complex (highly connected and wide) compared with lake and river webs which were thinner, longer and exhibited fewer connections.

In Table 14.1, data have been added from the food web of Broadstone Stream (Hildrew *et al* 1985), which is far more complete than any in Briand's (1985) compilation. In only one respect does Broadstone Stream conform markedly to Briand's stream characteristics: it has high linkage complexity (SC_{max}, 10.54, Table 14.1).

Connectance is a parameter of central importance in both static and dynamic theories of community structure (Warren 1990). Whereas dynamic theories predict that connectance should decline with increasing species richness, the static cascade model simply assumes an equivalent proposition: that the number of feeding links per species is constant. The latter theory thus needs an independent explanation for patterns of connectance. Pimm (1982) proposed that species may simply be restricted in the range of others they can feed upon because of limitations of behaviour or morphology. Warren (1990), however, suggested that as total species richness increases, prey would become increasingly closely packed in the predator's niche space and the number of feeding links per species would thus also increase. He showed that the number of links will in fact increase in proportion to S^2, and that the proportionality constant will be that fraction of total

Table 14.1 Food-web variables for Broadstone Stream (Hildrew *et al* 1985) and group means from a compilation of webs (consisting of 'trophic' species, see text) by Briand (1985)

	Broadstone Stream		Briand (1985)		
	Taxonomic species	Trophic species	Streams ($n = 7$)	Rivers ($n = 2$)	Others (non-freshwater) ($n = 20$)
S	24	17	11–19	?	?
C_{min}	0.35	0.29			
C_{max}	0.76	0.62			
SC_{max}	18.2	10.54	9.28	5.86	6.38
Max chain	5	5	4.00	5.00	5.25
Mean chain	4.1	3.7	2.29	2.91	2.99
Fraction sp.					
Basal	0.17	0.23	0.19	0.26	0.20
Intermediate	0.71	0.59	0.47	0.61	0.57
Top	0.12	0.18	0.33	0.13	0.23

S, species richness; C, connectance (fraction of all possible links realized; min excludes competitive interactions, max includes them); max and mean chain, longest and average chain lengths in the web; a basal species has no prey, intermediate species have both prey and predators, top species have no predators.

niche space exploited by predators (large for webs of generalists, small for webs of specialists) (Fig. 14.1(a)).

Data from the benthic invertebrates of two fishless, acidic freshwaters, Skipwith pond (Warren 1990) and Broadstone stream (Hildrew *et al* 1985), place them clearly among the highly connected webs of generalists (Fig. 14.1(b)). It may be premature, however, to categorize all freshwater food webs in this fashion; there are simply too few good data. Indeed, Paine (1988) argues that patterns of connectance in food webs in general may be artefacts, generated by the increasing difficulty of tallying feeding interactions in species-rich communities. Thus, the decline in connectance with species richness in Fig. 14.1(b) may be more apparent than real.

The poverty of the stream data in the Briand (1985) compilation (Table 14.1) is illustrated by the fact that the species richness of the nine stream communities ranged from 11 to 19. Broadstone stream has an extremely impoverished fauna, yet even that has a list of nearly 30 macro-invertebrate taxonomic species and about 20 'trophic' species (those taxonomic species with identical predators and prey are lumped to form a trophic species). To this number must now be added the ten or so species of microarthropods (harpacticoid and cyclopoid copepods, benthic Cladocera and mites) found by Rundle and Hildrew (1990). These are still aggregated in the Broadstone web because their trophic interactions are not known. Many other groups remain unstudied in this 'simple' stream community.

We have so far dwelt mainly on patterns in the structure of food webs, based on whether interactions are present or absent. But these interactions also vary in strength, and some at least may be dynamically trivial. If we can identify strong and weak interactions between 'elements' (often aggregates of species appearing to act in the food web in a similar way) perhaps this will reveal the main features of real webs and the most important restrictions on their structure (Menge & Sutherland 1987). In addition to better descriptions of food webs, we therefore need

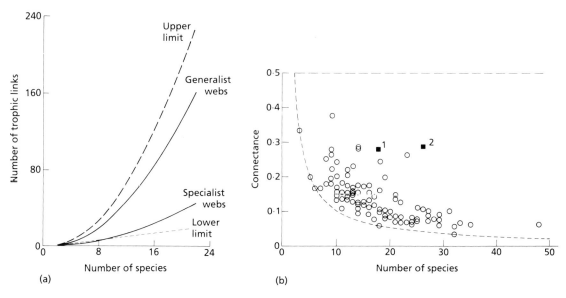

Fig. 14.1 (a) Upper and lower limits (broken lines) to the relationship between the total number of trophic links in a food web and species richness. Webs of generalists and specialists fall close to these upper and lower limits respectively. (b) Connectance declines with species richness in the 113-web compilation of Briand and Cohen (1987) (the broken lines are theoretical upper and lower limits to connectance derived from the limits to trophic links shown in (a)). The solid square symbols add data from a fishless, acidic stream (1, Hildrew *et al* 1985) and a similar pond (2, Warren 1990). (a) and (b) are both after Warren (1990).

experimental evidence on the role and strength of biotic interactions before we can distinguish between the theoretical options.

Lawton (1989) has seen two major difficulties with models of food-web structure based on Lotka–Voltera dynamics. The first is the assumption that trophic links involve closely coupled enemy–victim interactions, and the second is that all interactions take place in a spatially homogeneous world. Both will raise eyebrows among many empirical ecologists, for modern emphasis in the ecology of running waters has been on disturbance and on patchy, stochastic environments (Hildrew & Townsend 1987; Pringle *et al* 1988; Resh *et al* 1988). However, Pimm *et al* (1991) have pointed out that many of the predictions of dynamic models are actually quite robust to these detailed assumptions.

14.2 BIOTIC INTERACTIONS IN RUNNING-WATER FOOD WEBS

The best-known categories of interactions involve either positive/negative effects on the two interactors or negative/negative effects. In running waters the first category is commonly exemplified by herbivory, consumption by detritivores of microbes attached to dead organic matter and by conventional predation. There are then cases of competitive interactions, which are not strictly direct trophic links but may be important in the dynamics of food webs through indirect effects. There are also some indications that mutualisms (positive/positive interactions) may be widespread in freshwater communities (see below).

Herbivory

There are four main sources of autochthonous primary production in streams and rivers: rooted aquatic angiosperms and bryophytes, their algal epiphytes, and the algal elements of biofilms on mineral particles. The reasons for the apparently negligible grazing of living angiosperm tissue in freshwaters are essentially unknown (but see Newman 1991). By far the best known herbivore systems are the invertebrate and fish grazers of algae in stony streams.

There has been a flurry of experimental work on invertebrate/algal interactions in streams in the last 10 years (reviews in Gregory 1983; Lamberti & Moore 1984: recent examples by Feminella *et al* 1989; Lamberti *et al* 1989). This field has undoubtedly been one of the big success stories of stream ecology and has revealed widespread, although not ubiquitous, strong interactions between grazers and grazed. Features of the algal community often altered by grazing include algal biomass, physical structure, species composition, productivity per unit biomass and total primary production. The determinants of the apparent strength of the interaction are: (1) the productive capacity of the algae (usually set by nutrient and/or light supply); (2) the type of grazer tested (caddis and snails are often particularly effective, mayflies sometimes not); (3) the onset of grazing in relation to the development of algal growth (algae may 'escape' control if the onset of grazing is too late); and (4) hydraulic disturbance in the stream (which may overwhelm biotic interactions).

Hill and Knight (1987) established a gradient of mayfly (*Ameletus validus*) densities in a number of *in-stream* perspex channels in a northern California stream. Periphyton biomass declined with increasing grazer density in the channels (Fig. 14.2(a)). They also found increased chlorophyll *a* per unit biomass and a decreased contribution of the loosely attached upper layer of periphyton to total algal biomass. Feminella *et al* (1989) compared the algal biomass accrued on tiles placed either on the streambed or on raised platforms (this treatment excluding crawling grazers, mainly cased caddis and snails) (Fig. 14.2(b)). Raised tiles accumulated more algal biomass than controls in two of the three study streams. Crawling grazers also maintained low algal biomass across a wide range of tree canopy cover in one of the streams (Fig. 14.2(c)).

In a search for indirect trophic effects in river food webs, Hill and Harvey (1991) recently manipulated insectivorous fish and grazing snails in a series of channels in a shaded headwater stream. Snail (*Elimia clavaeformes*) grazing substantially reduced the biomass of loosely attached periphyton but had no significant effect on the tightly attached layer. Snail grazing had opposing effects

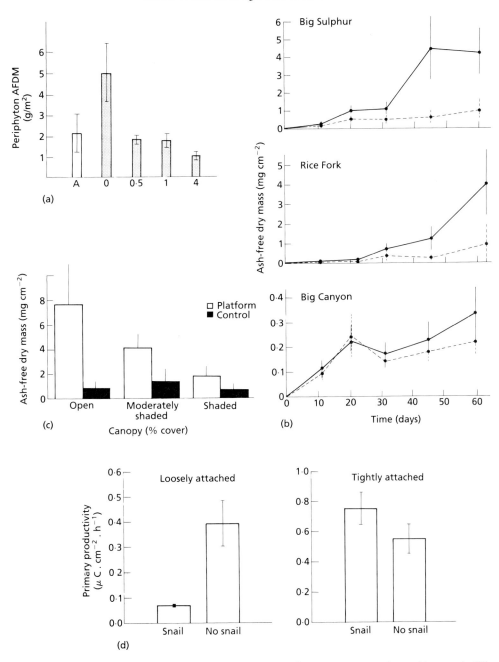

Fig. 14.2 Grazer/algal interactions. (a) The effects of grazing mayflies on mean periphyton biomass (±SE);
A, ambient grazer density; 0, 0.5, 1, 4 are multiples of ambient (Hill & Knight 1987). (b) Feminella *et al* (1989)
measured the accrual of mean periphyton biomass (±95% CL) on raised tiles (solid lines), from which crawling
grazers were excluded, and control tiles (broken lines) in three Californian streams. (c) They found that periphyton
biomass responded to light on raised (platform, □) but not on control (■) tiles. (d) Snail grazing in a shaded, headwater
US stream reduced mean (±SE) primary production in the loosely attached periphytic layer but increased it in the
tightly attached layer (Hill & Harvey 1991).

on primary production in the loosely and tightly attached layers, apparently stimulating it in the latter (Fig. 14.2(d)). Light, however, also varied generally among the channels and it was light, rather than the presence of snails or fish, which accounted for most of the variation in total primary production.

Although grazing clearly affects the species composition of periphyton, its influence on species diversity is less clear. Slight, intermediate and intense grazing by *Baetis* mayflies, the snail *Juga* and the caddis *Dicosmoecus* produced no effect, an increase and a strong decrease in algal diversity, respectively (DeNicola *et al* 1990). Vulnerable species, such as the large, stalked diatoms of the genera *Gomphonema* and *Cymbella*, are often enormously reduced in abundance even by low densities of grazers (McCormick & Stevenson 1989). In terms of these true *species* interactions, however, the analysis of food webs in these grazing systems is in its infancy. Periphytic communities in streams are species rich and no published empirical web offers anything like a complete description.

Community level, structural features of periphyton are thus clearly affected by grazing in streams. The loss of loosely attached, large, or otherwise vulnerable species often leads to a reduction in biomass. However, at least biomass-specific primary production may actually be stimulated by grazers (Fig. 14.2(d)). The mechanism is presumably by enhancing a resource, such as light or nutrients, through the removal of senescent cells, detritus or silt. Overall primary production is commonly then limited by abiotic factors (light in shaded streams, nutrients in open streams) or by disturbance. But what of the reciprocal half of the interaction?

Are the grazers food limited?

There has been much less work on the 'bottom-up' effects of algal food on grazers in streams but several significant results, mainly with caddis larvae, have been reported. In McAuliffe's (1984a) experiments *Glossosoma* reduced periphyton abundance on experimental bricks. He then exposed bricks with differing resource abundance and *Glossosoma* densities to short-term coloniz-

ation by other mobile grazers. The experiments seem to show quite clearly the effect of exploitative, interspecific competition on colonization (Fig. 14.3(a)). Other experimenters have demonstrated food limitation and intraspecific competition for food. For example, Lamberti *et al* (1987) manipulated densities of *Helicopsyche* in enclosures. The final weight of larvae declined with

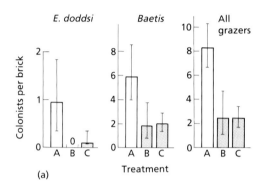

(a)

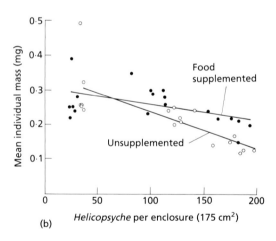

(b)

Fig. 14.3 (a) Exploitative competition among grazers. Colonization by the mayflies *Ephemerella doddsi* and *Baetis*, and by total grazers, of experimental bricks in treatments A (high algal biomass and no interference competition from the grazing caddis *Glossosoma*), B (low algal density and no interference competition) or C (low algal density and *Glossosoma* present) (McAuliffe 1984a). (b) Weight attained by the grazing caddis *Helicopsyche* kept for 60 days at various densities in enclosures in a Californian stream. ●, Food supplementation; ○, no supplementation (Lamberti *et al* 1987). Lines are least square regressions.

increasing larval density, but when food was supplemented this effect was slight (Fig. 14.3(b)). Essentially similar results were obtained by Hart (1987) and Lamberti *et al* (1989).

In summary, there is evidence both of strong top-down regulation of algal community structure and of bottom-up resource limitation of grazers. In such systems supplementation of limiting resources to primary producers might lead not to an increased biomass of algae but to that of their grazers or both. We must temper this by remembering that most experimental approaches so far have been at small spatial and temporal scales. Also, experimenters tend to emphasize positive results. Frequent resetting of algal succession by spates or droughts may weaken, or render intermittent, this 'enemy–victim' interaction. Biggs and Close (1989) thus found that the hydrological regime in nine gravel-bed rivers in New Zealand explained more of the variation in algal biomass and chlorophyll *a* than did nutrient concentrations. In particular, percentage time in flood was an important variable, and flow disturbances may determine the extent to which grazing interactions are important.

The mysteries of detritus

Most streams, with the particular exception of desert systems, are heterotrophic, with total respiration exceeding gross primary production (Busch & Fisher 1981; Winterbourn 1986). The detrital subsidy comes from the catchment in the form of particulate plant litter and dissolved organic matter. Even autochthonous primary production, particularly that due to aquatic vascular macrophytes, is often incorporated in stream food webs as detritus, rather than being grazed as living material. Detritus is, therefore, quantitatively the most important fuel for running-water food webs.

There is a good deal of circumstantial evidence that detritivorous animals feeding on tree leaves are food limited (e.g. Gee 1988). Recent experiments have tested this contention. Smock *et al* (1989) manipulated the number of debris dams in coastal plain streams of the south eastern USA and found that shredder abundance increased with dam abundance. Richardson (1991) imposed three rates of detritus supply to replicated, stream-side

channels. In treatments with extra detritus there were increases in the density of several common detritivorous insects and in the biomass of detritivores overall (Fig. 14.4). It is the retention characteristics of streams which often seems to determine the importance of quantitative food limitation to detritus feeders (Hildrew *et al* 1991). However, there may also be important limitations imposed on detritus feeders by the quality of their food (Groom & Hildrew 1989).

Detritus is a complex mixture of non-living and microbial components, and the precise relationship between detritivores and these two fractions of their food is still a matter of controversy (see Chapter 15). Microbially unconditioned leaf litter may be more or less indigestible by animals, and colonization by microbes and partial decomposition yields a more palatable food source which may support higher individual growth (Groom & Hildrew 1989; Fig. 14.5). In terms of species interactions and food webs there are potentially five separate but not exclusive modes of interaction between decomposer microbes and detritivorous animals (Fig. 14.6):

1 A simple 'predator/prey' interaction in which animals assimilate microbial carbon or nutrients but little of the detrital material itself (Fig. 14.6(a)). Animals certainly do feed on the microbial element, although controversy abounds over whether microbial biomass is sufficient for their

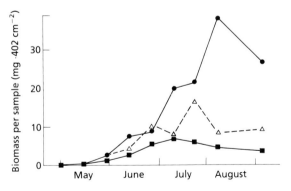

Fig. 14.4 Richardson (1991) supplemented leafy detrital food at two levels above ambient in streamside channels on a Canadian stream. This shows seasonal change in the biomass of a total of nine detritivorous taxa. Each point is a mean of two replicate channels. ●, High; △, intermediate; ■, 'natural'.

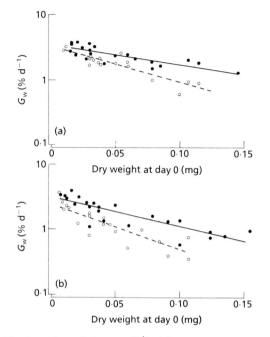

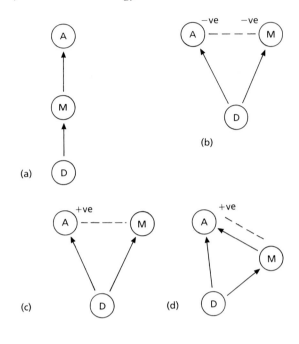

Fig. 14.5 Growth $(G_w, \% \ d^{-1})$ of the common detritivorous stonefly *Nemurella pictetii* as a function of initial weight when fed on (a) alder and (b) beech leaves which have been conditioned in acid (○) or neutral (●) stream water. Growth is significantly faster on alder than on beech and on neutral- than on acid-conditioned litter. Both indicate an enhancement of growth on microbially well-conditioned litter (Groom & Hildrew 1989).

needs (Findlay *et al* 1986; Arsuffi & Suberkropp 1988).

2 A competitive interaction between microbiota and animals for dead organic matter — microbes are essentially a sink for detrital carbon, which is mainly mineralized and not passed on to animals (Fig. 14.6(b)).

3 A commensal interaction in which animals benefit from the presence of microbial populations, although not principally by the ingestion of microbial biomass *per se* (Fig. 14.6(c)). Microbes soften and partially digest leaf tissue, rendering it more nutritious for animals (Bärlocher & Kendrick 1975).

4 A combination between predation and commensal interactions (Fig. 14.6(d)) is possible if, as well as gaining nutrition from microbial biomass, microbial enzymes can be sequestered, remain

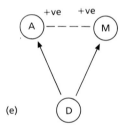

Fig. 14.6 Hypothetical food-web interactions between the dead fraction of detritus (D), micro-organisms (M) and detritivorous animals (A). Trophic interactions (solid lines) show arrows from 'prey' to 'predator'. Non-trophic interactions (broken lines) have positive or negative signs to indicate possible competition, commensalism or mutualism. (a) Simple predator/prey interaction. (b) Animals and micro-organisms as competitors for detritus. (c) Commensal interaction in which animals benefit from partial digestion and softening of leaves by micro-organisms. (d) Commensal/predator–prey interaction in which animals benefit from sequestered enzymes. (e) Mutualism in which animals benefit from microbial conditioning and micro-organisms benefit, for example, from comminution of particles.

active in animal guts and enhance the digestion by the animal of the dead fraction of the detritus (Sinsabaugh *et al* 1985).

5 A mutualistic interaction in which both microbes and animals benefit (Fig. 14.6(e)), for instance the animals by the partial digestion of detritus by microbes, and the microbes by the increased fragmentation of the substrate by animal feeding.

These subtle interactions lie at the very base of running-water food webs, yet we know little of their quantitative relative importance in any whole web. Although good progress has been made in several respects, we need to expand our efforts in microbial ecology perhaps more urgently than in any other area. The incorporation of dissolved organic matter into river food webs, centred on the formation of particles and the growth of slimy biofilms on stones, is understood mainly in 'process terms' (see Chapter 15) and not as part of a taxonomically discriminated food web. It seems inevitable that this component of the river community will remain as an aggregated 'black box' for the forseeable future.

In conclusion, there is evidence of bottom-up control of populations of detritivorous animals, both through food quantity and quality. Whether their grazing on micro-organisms decomposing detritus ever limits microbial biomass or production remains a matter largely of conjecture. It may be that detritus-based communities are examples of 'donor control', in which the consumed influences the consumer but the effect is not reciprocated. At the limit it is clear that running-water animals have little influence on the gross supply of allochthonous organic matter, but they might affect the rate at which it is made available by microbes. Donor-controlled food webs can theoretically have a different structure from those dominated by closely coupled interactions (Pimm 1982; Lawton 1989), with greater connectance and omnivory.

Predation

Predators are numerous in streams and rivers and include a variety of invertebrate taxa, most fish, amphibia such as salamander larvae, and streamside birds and mammals. The role of predation by fish and invertebrates in the population dynamics of their prey and in structuring benthic communities, particularly in stony trout streams, is a matter of enduring controversy (see Cooper *et al* 1990). Many features of the morphology, physiology and behaviour of prey animals seem related to defence against fish and invertebrate predators, indicating that predation is a powerful selective force in their evolution (e.g. Peckarsky 1984; Peckarsky & Penton 1988; Soluk 1990). Estimates of food consumption, both by fish and invertebrate predators, continue to yield estimates greater than the quantities thought to be produced in the benthos (Allan 1983; Waters 1988). These indirect lines of evidence indicate a heavy exploitation of prey populations and possibly a case of 'closely coupled enemy–victim interactions'.

Many ecologists have been tempted to test this idea in field experiments, often using various kinds of enclosures or exclosures in which prey density attained is compared between replicates with and without predators. In sharp contrast to the obvious effects of similar manipulations in many grazer/herbivore systems, the results have rarely been spectacular and often insignificant (Cooper *et al* 1990). The inference is that either predation really is dynamically trivial in running waters or that the experiments are unable to detect the effects through some details of scale, timing or design.

A number of authors have speculated on possible confounding processes in the detection of predator effects (e.g. Gilliam *et al* 1989). Perhaps the most important is the extremely patchy distribution of prey which masks the impact of predators in enclosures. One obvious implication is that a large number of replicates is needed to detect differences between treatments. More interesting is the possibility that prey exchange with the surrounding benthos rapidly replaces prey consumed. One determinant of prey exchange is the mesh size of the enclosure walls, often chosen as that which will just retain predators. In a recent analysis of 52 published experiments, Cooper and colleagues (1990) found a significant relationship between predator impact and mesh size of enclosures. They also found a tendency in their own field experiments for predator impacts to be more easily detected when using fine mesh enclosures. The effects of trout in California stream pools were also greater when prey turnover (drift out of each pool per day divided by the number of invertebrates in the pool) was low. Experiments by Lancaster *et al* (1991) are consistent with the notion that a high rate of exchange of prey with benthos may mask effects; when they restricted exchange by reducing enclosure mesh sizes, predator impact was significant. Both mathematical (Cooper *et al* 1990) and graphical (Lancaster *et al* 1991) models indicate the clear effect of exchange rate on the detection of predator impact at the patch (=enclosure) scale.

Intriguingly, predation on the stream benthos could therefore be a case of a process which is often not apparent at the small scale but which is important at broader spatial scales. Predation on mobile and patchily distributed prey can be intense overall whilst remaining elusive in patch scale experiments. The comparison with benthic algae and their herbivores is well made; algae are sessile and local depredations on them easy to detect. What has slowly been learned of predation in streams has made field manipulations more sophisticated and consistently rewarding. Recent experiments by Lancaster (1991) with predatory stoneflies involved replicate streamside channels rather than mesh enclosures and enabled careful accounting of arrival and departure of prey by drift. The results are rather clearcut and indicate a strong effect of a stonefly predator on the accumulation of two prey taxa (Fig. 14.7(a)).

We now seem close to a general conclusion, based both on circumstantial and direct experimental evidence, that predators in streams can exert definite top-down control of their prey. The limitations to this conclusion include the following:

1 Fluctuations in discharge or other disturbances can reset benthic densities and overwhelm the effects of predation (Lancaster *et al* 1990; Lancaster 1990).

2 Predator impact may be limited to those hydraulic conditions which favour the predators. Some recent experiments by Peckarsky *et al* (1990) have illustrated this nicely (Fig. 14.7(b)). Hansen *et al* (1991) showed also that predation by a triclad on blackfly larvae was mediated by flow

in quite subtle ways. Not only did flow preferences differ between predator and prey (thus imposing partially separate microdistributions), but the ability of triclads successfully to complete prey capture declined at higher velocities. They explicitly linked this to the idea that there are refugia for prey in streams and rivers, provided by spatial variations in the forces of flow. Flow refugia are only one example of a whole range of circumstances, including substrate heterogeneity and temporal variations, which may act similarly (e.g. Hildrew & Townsend 1977). Refugia strongly stabilize predator/prey models (Begon *et al* 1990) and could thus be of dynamic significance in food webs.

3 Predators vary in their effectiveness. For instance, among the fish, salmonids seem to affect only rather large or otherwise vulnerable prey (e.g. Schofield *et al* 1988), while benthic feeders (Cottidae, Cyprinidae) have more widespread effects (e.g. Gilliam *et al* 1989; Schlosser & Ebel 1989). Salmonidae may also be relatively effective in slow rather than in fast-flowing streams (Cooper *et al* 1990).

4 Prey differ in their vulnerability through a miscellany of devices such as shells, cases and retreats. Selection pressures imposed by predators on how, where and when species feed, and upon the evolution of antipredator devices, probably play an important role in structuring freshwater communities (Jeffries & Lawton 1984).

Finally, it is clear that many predators are markedly polyphagous in running waters and take prey roughly in proportion to their abundance. In these circumstances, while predators influence prey, it is possible that the population dynamics of individual prey taxa have little effect on predators. Such a circumstance will influence stability of food-web models in ways which have yet to be explored (Lawton 1989).

Interference, competition for space and 'self-damping' in river food webs

Interference competition, including pseudointerference (Free *et al* 1977), is strongly stabilizing in population and food-web models (Lawton 1989). How widespread is it in river food webs? Some well-studied fractions of communities seem

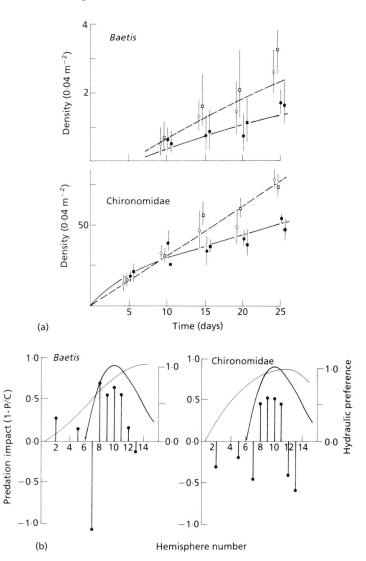

Fig. 14.7 (a) Lancaster (1990) shows the colonization by two prey taxa of four streamside channels with (solid symbols) or without (open symbols) predatory stoneflies (*Doroneuria baumanni*). Colonization curves are for with-predator (solid lines) or no-predator (broken lines) treatments. (b) In enclosure experiments, Peckarsky *et al* (1990) found that the impact of the predatory stonefly *Dinocras cephalotes* on two prey taxa varied along a hydraulic gradient (measured using the standard hemispheres of Statzner and Müller (1989), higher hemisphere numbers indicating greater forces of flow). Curves show the hydraulic preferences of predator (solid) and prey (broken). Predator impact (solid points) is measured as a relationship between mean prey density in cages with (P) or without (C) predators.

completely free of signs of structuring by competition for time, space or food (Tokeshi 1986; Tokeshi & Townsend 1987; Hildrew & Townsend 1987). Yet interference competition seems widespread among 'sessile' species of the stony benthos and there are examples of both interspecific and intraspecific competition for space among filter feeders, algal grazers and predators.

Various kinds of caddis larvae build semipermanent retreats and defend territories around them (Edington & Hildrew 1982). These include the filter-feeding Hydropsychidae, which are aggressive and stridulate (Matczak & Mackay 1990), the predatory Polycentropodidae (Hildrew & Townsend 1980), and the herbivorous Psychomyiidae and Hydroptilidae (Hart 1985). These families all contain dominant, competitive species that are important 'players' in many stream communities. The predatory larva of the polycentropodid *Plectrocnemia conspersa* is numerous in some fishless, acidic English streams, where it is an important predator on the remaining benthos. There is some evidence that its population density is limited by competition for space, since supplementation of net-spinning sites led to clear local increases in density (Lancaster *et al* 1988; Fig. 14.8(a)). Territoriality and interference competition is not limited to caddis larvae, and has also been described in grazing, tube-dwelling chironomids and in odonate larvae (Wiley & Kohler 1984; Johnson 1991).

Such species may pre-empt space and compete strongly with other sessile or mobile species (McAuliffe 1984b; Dudley *et al* 1990). Coexistence and relative abundance may then be determined by disturbance that opens up gaps which can be exploited by the less competitive species. This seems to be the case with *Hydropsyche*, and the more opportunistic *Simulium*, in some Californian streams (Hemphill & Cooper 1983; Fig. 14.8(b)). Hildrew & Townsend (1987) speculated that the sessile, territorial lifestyle is sustainable only in streams in which such disturbance is limited in its severity; for instance, where there is stable substrate and moderate or infrequent fluctuations in discharge and water level. Sessile species are absent from many streams and are abundant where stable substrate is combined with abundant renewable food (e.g. Valett & Stanford 1987).

14.3 INTERACTION WEBS IN RUNNING WATERS

The experiments and processes described so far are all limited to some extent to small spatial or temporal scales and to relatively simple, pairwise interactions. However, there is great interest in community ecology generally in multilevel, complex or indirect interactions (Kerfoot & Sih 1987; Huang & Sih 1990), and their importance in the regulation of trophic structure. Some of these complex interactions are plainly demonstrated in lakes (Carpenter 1988) but have attracted less attention in streams and rivers. The 'interaction' webs (partial networks of strong and weak biotic interactions between species or groups of similar species in communities) which often result from such studies (e.g. Menge & Sutherland 1987) form a 'half-way house' between the study of pairwise interactions of neighbouring compartments in

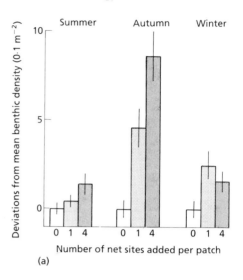

(a)

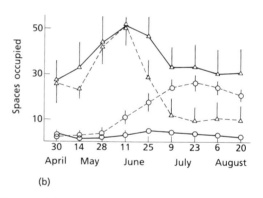

(b)

Fig. 14.8 (a) Lancaster *et al* (1988) supplemented net-spinning sites in small patches of streambed for the predatory caddis *Plectrocnemia conspersa*. Local density (measured as the deviation from that in the benthos) responded to the supplementation in each of three seasonal experiments. (b) Hemphill and Cooper (1983) regularly disturbed, by scrubbing, quadrats of bedrock in a Californian stream occupied by both *Simulium* and *Hydropsyche* larvae. Control quadrats were undisturbed. Counts are the mean (±SE) number of small spaces occupied by either insect between April and August. Notice that *Simulium* (△) showed a decline in the controls (broken line) whereas the larger, aggressive *Hydropsyche* (○) took up more spaces in these undisturbed areas.

the web, and the practically intractable problem of identifying every species and assessing every interaction.

Some authors have attempted large-scale, whole-river manipulations (despite the virtual impossibility of replication). Fertilization with phosphorus of the Kuparek River, a tundra stream in northern Alaska, was associated with an increase in algal biomass and productivity, an increase in bacterial activity in biofilms and an apparent increase in growth of blackfly larvae (Peterson *et al* 1985). Hershey and Hiltner (1988) subsequently found that the density of a filter feeding caddis, *Brachycentrus*, increased in the fertilized reach while that of blackflies declined, perhaps because of interference competition with the caddis. These results point to the possibility of impacts across several levels in the food web and of indirect interactions among animals.

Others have performed smaller scale, replicated experiments with several trophic levels using separate pools (Power *et al* 1985; McCormick 1990), enclosures (Hill & Harvey 1991) or replicated in-stream channels (Hart & Robinson 1990). Strongly positive results, in terms both of 'bottom-up' and 'top-down' control, have been obtained most often where the prevailing conditions are markedly pond-like. For instance, Power (1990) experimented with the food web around boulders in the Eel River, California at summer base flow. The web (Fig. 14.9(a)) consists essentially of four compartments: large fish, small predators (juvenile fish and invertebrates), tuft-weaving chironomids and the fast-growing macroalga Cladophora and its epiphytes.

Power (1990) found that the large fish depress the densities of the smaller predators; this releases the herbivorous chironomids from predation and, in turn, these reduce the Cladophora to a low prostrate form with few epiphytes. Removal of large fish results in high densities of smaller predators, fewer chironomids and tall, upright tufts of Cladophora with many epiphytes. These manipulations provide the first support in rivers for the Hairston *et al* (1960) and Fretwell (1977) model of community organization, in which alternate trophic levels are controlled by predation and competition (Fig. 14.9(b) and (c)). The main biological reason for this is the fact that the

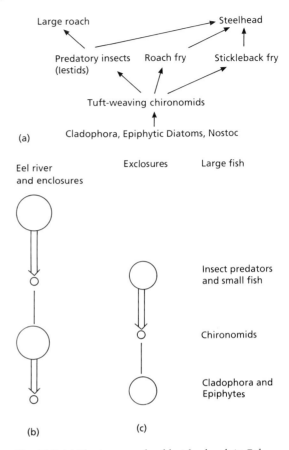

Fig. 14.9 (a) The 'summer boulder' food web in Eel River, California (Power 1990). (b) Power's exclusion of large fish changed the interaction web from a four-level to (c) a three-level system and altered the relative abundance of the elements (as indicated by the size of the circles).

herbivore is invulnerable to the top carnivore, in this case the large fish which rarely eat the chironomids. Again, Power (1990) is careful to acknowledge that such interactions are likely to be detectable only in rivers where the frequency of disturbance does not exceed the capacity of the system to re-establish itself.

In Fig. 14.10(a) a speculative interaction web for a hypothetical stony trout stream has been constructed. This draws on instances of strong interactions in the literature, many of them mentioned above, but in no single site have all these interactions been demonstrated. It is oversimplified in inumerable respects, not least because it represents only salmonid fish and ignores terrestrial linkages with birds and piscivorous mammals. The main features of this system are marked omnivory, self-damping by salmonid populations, strong interactions between algae and their grazers, and control of detritivores by their food and of large, vulnerable invertebrates (often predatory) by fish. Such a system will not be marked by obvious alternate level regulation by competition and predation, as is the Eel River. This is because it does not include a level of herbivorous/detritivorous consumers which are invulnerable to the dominant, but not to the subdominant, predator. It remains to be seen whether such an interaction web is realistic, but it obviously bears a close resemblance to the Menge and Sutherland (1987) model of community control. The main difficulty with this model comes with its assumption that physical disturbance affects sessile organisms least. These are rather less well represented in streams and, at least as far as animals go, seem rather vulnerable to physical disturbance.

Physiological stress, in the form of acidification, removes vulnerable species and whole levels in the food web of trout streams (Sutcliffe & Hildrew 1989). Fish are lost, as are many algal grazers. The web shown in Fig. 14.10(b) is derived mainly from the work of myself and collaborators in some southern English streams (e.g. Schofield *et al* 1988; Groom & Hildrew 1989; Lancaster *et al* 1991). The main features are marked omnivory, strong top-down control of prey by abundant, polyphagous invertebrate predators (which are themselves self-damped) and overall bottom-up control of productivity by detrital food quality.

Almost all stream ecologists are aware of the vulnerability of their systems to physical disturbance. Yet there is no very well-developed theoretical or empirical approach to the nature of disturbance and its effect on running waters: this is a very obvious growth area for research (Resh *et al* 1988). There is not much evidence for the wholesale loss of complete interaction elements in trout streams which are just fairly torrential and 'flashy'. It seems most likely that these kinds of hydraulic physical disturbances simply weaken species interactions, as shown in Fig. 14.10(c) and

(d), while not altering the basic shape of the web overall. It should be kept in mind that all streams will be subject to reset by disturbance (Fisher *et al* 1982) and to periods when conditions are less extreme. Systems differ from each other in frequency of disturbance, predictability and mean conditions. The strong interactions illustrated will 'flicker' on and off in concert with the environment and, as Statzner (1987) has pointed out, different components of the community will have different recovery rates after reset. This makes the point that we need studies of interactions and food webs throughout the year, to take some account of seasonal and other temporal changes.

14.4 GAPS IN KNOWLEDGE AND FUTURE DEVELOPMENTS

Undoubtedly the most important lacuna in food-web research is in the area of the 'microbial food loop', in which stream research lags behind its marine equivalent by perhaps 10 years (Carlough & Meyer 1990). Research on the energetic base of streams has stressed three main sources and modes of transfer into food webs. These are: (i) terrestrial leaf litter and the detritus/microbe/shredder pathway; (ii) benthic algae grazed by macroinvertebrates; and (iii) dissolved organic matter, taken up in biofilms and grazed by invertebrates (see Winterbourn 1986 for a review). Groundwater, discharging up through bed sediments, is a possible source of the latter material (Hynes 1983). Recent research shows that biotic immobilization of dissolved free amino acids by sediments cores is very effective (Fiebig & Lock 1991).

A fourth pathway is for dissolved organics, either autochthonous or allochthonous in origin, to be taken up by suspended bacteria in suspension which could then be consumed by Protozoa. Carlough and Meyer (1990) estimated that suspended Protozoa cleared an average of 47% of the water column of the Ogeechee River per day. These Protozoa were then available to micro-filterers such as *Simulium*. More generally there are some estimates of the densities both of suspended and sedimentary bacteria and Protozoa, the meiofauna, and of bacterivory by Protozoa

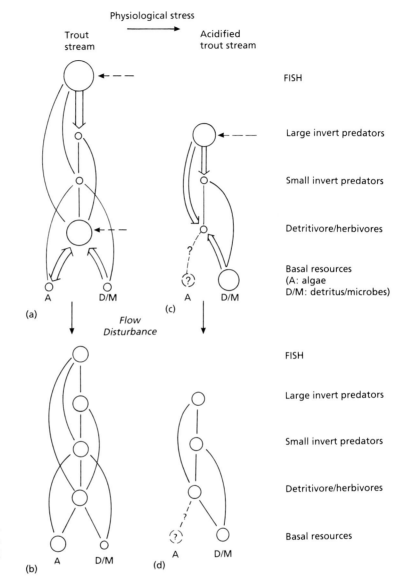

Fig. 14.10 Hypothetical interaction webs in (a) a stony trout stream, (b) a trout stream subject to frequent flow disturbance, (c) an acidified stony stream and (d) a similarly disturbed acid stream. Narrow solid lines represent trophic interactions; thick lines represent strong interactions with direction of main effect indicated by arrows; horizontal broken lines represent incidences of 'self-damping'. Size of circles crudely represents abundance of elements. The role of algae and their grazers in acid streams remains uncertain.

(e.g. Baldock & Sleigh 1988; Bott & Kaplan 1989, 1990; Carlough & Meyer 1989, 1990). The incorporation of dissolved organic matter into river food webs, probably through a microbial link, is a subject of the highest priority for research.

A food source that appears not often to be directly exploited in streams and rivers is that of living aquatic macrophytes (see Newman 1991). This is possibly due to chemical defences

by macrophytes (Ostrofsky & Zettler 1986), although this subject is poorly explored in freshwaters. Aquatic angiosperms are a taxonomically scattered assemblage, consisting of a restricted number of representatives from a wide range of terrestrial groups (Sculthorpe 1967). They invaded freshwaters after the primarily aquatic insect orders did, and may be chemically well defended against the generalist detritivore/herbivores predominant amongst the really common aquatic insects. Newman (1991) points out that 63% of the freshwater phytophagous insect species are from primarily terrestrial insect orders (Lepidoptera, Coleoptera and Orthoptera). While aquatic macrophytes have clearly taken some of their terrestrial herbivores with them, species richness is restricted: about 45% of terrestrial insects are phytophagous compared with 11% of aquatic species. Perhaps this restriction can be ascribed to the patchy taxonomic representation of angiosperms in freshwaters. In effect there are large 'voids' of niche space separating 'islands' of plants.

Others argue that there is a coevolved mutualism between grazers, especially aquatic snails, and macrophytes (Bronmark 1985; Thomas 1987). Plants gain from the grazing of epiphytic algae from their surfaces, thus reducing shading, and snails gain food, shelter and access to dissolved organic molecules. A further possible role for mutualisms in freshwater food webs is now suggested between grazers and the epiphytic algae themselves (Underwood & Thomas 1990). Many algae seem to survive passage through grazer guts and actually benefit from grazing through access to mineral nutrients. The grazers would gain from the release of dissolved organics from algae in the gut. McCormick (1990) recently found that moderate herbivory enhanced algal growth in isolated stream pools under conditions of nitrogen limitation.

The role of parasites and diseases in stream and river food webs is often discounted (e.g. Statzner 1987). However, Cummins and Wilzbach (1988) have argued that pathogenic infections represent the primary regulator of stream invertebrate populations. They were able to muster rather little evidence but this is an area that surely merits further investigation. In the context of food webs, it would be particularly interesting if webs in running waters turned out generally to be composed of mixtures of conventional predators with microparasites and diseases. Such webs 'violate the assumptions of the (size) cascade model' (Lawton 1989).

Finally, one of the reasons river food webs may have appeared rather 'short' is that a potentially important link is almost always ignored — predation by streamside birds and mammals. For instance, one family of passerine birds, the Cinclidae or dippers, are particularly adapted to fast-flowing rivers and feed exclusively on small fish and stream invertebrates. The five species of dippers cover extremely large areas, including the Andes (two species), the Rocky Mountains (one), the Himalayas and western Pacific Islands (one), and the Palaearctic from western Europe to the Himalayas, north Africa and Persia (one) (Ormerod & Tyler 1991). Ormerod and Tyler estimated that the Eurasian dipper could account for almost the entire production of Cottidae in some small Welsh streams. These fish provide the birds with an essential source of calcium and are taken particularly in late summer. Similarly, hydropsychid caddis larvae are heavily preyed upon, making up between 12% and 26% of the diet of dippers on circumneutral streams, which could account for about 11–24% of hydropsychid production. We have already seen the dangers and discrepancies inherent in these kinds of calculations but this example surely demonstrates that land–water food web linkages should not be ignored.

More generally, the evident extension of river food webs into the riparian zone and beyond holds the possibility that, although there may be some compartmentalization of webs at habitat boundaries, isolation is by no means complete. River food webs, representing mainly interactions within the channel, are firmly 'nested' into the primarily terrestrial webs of river catchments. This is a community-level expression of the idea that running waters are part of catchment scale ecosystems.

14.5 FOOD WEBS AND RIVER MANAGEMENT

In areas where most river ecologists work, almost all running waters are profoundly affected by humans, and it seems that human impacts invariably alter the scope and influence of species interactions in food webs. The application of ideas in food-web ecology to river management and conservation, however, is in its infancy. Among the most obvious effects of humans are those on water quality (Calow *et al* 1990). The only substantial applications of community level phenomena to problems with water quality are in establishing methods of assessment, most obviously using various indices based on macroinvertebrates, algae or other groups. Community structure can also be used to establish 'targets' for environmental management and to judge conservation value. There has been little attempt, however, to establish the *processes* by which communities respond to pollution and thus underly those 'structures' (e.g. species composition and richness, relative abundance) which are taken to indicate various kinds of chemical stress. Emphasis has been strongly on the direct toxicity of pollutants to individual species while the possible ramifying effects through food webs have largely been ignored. I have tried to point out these effects, in the context of acidification, elsewhere (Sutcliffe & Hildrew 1989). Species introductions are another impact of humans which can be understood in the context of food webs. Many introductions appear to have little further ecological effect; a few have had profound impacts several links away in the web.

Perhaps the most interesting and pervasive of all are the ecological effects of a number of physical changes commonly wrought by engineering works on channel form and confinement (e.g. Brookes 1988). Contact and exchange between land and water is perhaps the most basic property of river systems, and its influence pervades almost every section of this book. The importance of land/water interactions, including those between the surface flow and groundwater, have been recognized particularly by Hynes (1975, 1983) and demonstrated most fruitfully at the ecosystem level of organization by Likens (1985). The extension of running-water food webs on to the land has been briefly mentioned above and finds its most spectacular examples in the great tropical floodplain rivers. In the Amazon, many adult fish, their eggs or fry migrate, or are swept by seasonal spates, out on to enormous inland floodplains where they may feed on prey, fruit, seeds or detritus of terrestrial origin (Welcomme 1979). Aquatic invertebrates may also share in this periodic use of floodplains (Gladden & Smock 1990). But most north temperate rivers, especially in Europe, have lost their floodplains to agriculture

and urban development, and with them probably a large number of extra food web linkages. Many important land/water interactions remain, particularly where riparian vegetation is intact, and the importance of these edges or 'ecotones' is increasingly recognized (Naiman & Décamps 1990).

River engineering essentially confines rivers within their channels and tries to isolate them from the land. Ecological processes within the channels themselves are also altered by the engineers. Their task is essentially to prevent flooding and to remove water from catchments as soon as possible. Channels are straightened and deepened, and weedbeds and snags are removed to lessen hydraulic resistance (Brookes 1988). This reduces the surface area and physical complexity of the habitat and removes refugia from predation for vulnerable species and will alter the strength and outcome of species interactions.

Conventional river engineering, therefore, reduces the retention time of water and organic matter. By comparison, flow dead-zones in natural river channels, essentially unmixed parcels of non-flowing water, increase the average retention time of the river. Reynolds (1988) recognized that these features of flow, associated with channel features such as meanders, snags and holes in the bed, could at last provide an explanation for the persistence of phytoplankton in rivers. They could similarly account for the rich populations of water-column Protozoa in some rivers (Carlough & Meyer 1990). Flow dead-zones could also provide temporary refugia for invertebrates and fish during flow events, reducing catastrophic mortality and speeding up the return time of the system after disturbance (Hildrew *et al* 1991). In summary, most human impacts on water quality, channel form and flow must have simplified river food webs and increased the influence of disturbance.

14.6 CONCLUSIONS

Based on this brief survey of river food webs and the biotic interactions embedded in them, it would be difficult to conclude that dynamic constraints alone determine what trophic struc-

tures occur. While interactions are very important in certain circumstances, species dynamics
are patchy in space and time and seem not to
constitute 'closely coupled enemy—victim interactions'. Very speculatively, most interactions
are probably weak most of the time. The static,
trophic cascade model of Cohen and Newman
(1985) therefore seems a much more profitable
starting point. Recall that this proposes a trophic
hierarchy such that species feed only on others
below them. Warren and Lawton (1987) stress the
role of body size in determining the order of
vulnerability.

What seems to determine the trophic structure
of Eel River (Power 1990) is the ecological
disparity between top and bottom — the smallest
prey are not vulnerable to the top predators. With
strong interactions among elements, this produces a trophic structure close to the model of
Hairston *et al* (1960) and Fretwell (1977). This
does not seem obviously to be the case in many
other stream and river food webs, where omnivory
(species feeding on two or more levels in the web)
is prominent. Perhaps river food webs are more
generally short and broad because of a limited
'ecological disparity' between top and bottom,
most obviously but not always based on size.
Habitat structure, including refugia provided by
complexity on vertical and horizontal spatial
scales (Ward 1989), must also play a part in this
'disparity' and thus to the chain length and the
degree of omnivory in food webs. Since so much
of the anthropogenic change wrought on rivers
removes physical complexity, food-web shapes
and structures characteristic of 'pristine' systems
could have been early casualties.

Finally, it is clear that this is an area where
fundamental and applied ecology have related
aims. We presently have a restricted ecological
understanding of the basis of community structures in rivers. The need is to move from empiri-
cism and rule of thumb to understanding and
prediction within the constraints imposed by a
very variable sytem. Both better food-web descriptions and studies on strong and weak interactions
are badly needed.

ACKNOWLEDGEMENTS

I am very grateful to the colleagues who responded
to my requests for help and for reprints and
preprints of their work: Dave Allen, Paul L.
Angermeier, Randi A. Hansen, David Hart, Walter
Hill, Gary A. Lamberti, Jonathan T. Morales,
Raymond M. Newman, Christopher G. Peterson,
John Richardson, Amy D. Rosemond, Lane C.
Smith, Dan Soluk and Phil Warren. I particularly
thank Mary Power for the Eel River food web, and
Jill Lancaster, John Lawton, Mike Winterbourn
and an anonymous referee for their helpful
reviews.

REFERENCES

Allan JD. (1983) Predator—prey relationships in streams.
In: Barnes JR, Minshall GW (eds) *Stream Ecology:
Application and Testing of General Ecological
Theory*, pp 191—229. Plenum Press, New York. [14.2]

Arsuffi TL, Suberkropp K. (1988) Effects of fungal mycelia and enzymatically degraded leaves on feeding
and performance of caddisfly (Trichoptera) larvae.
Journal of the North American Benthological Society
7: 205—11. [14.2]

Baldock BM, Sleigh MA. (1988) The ecology of benthic
Protozoa in rivers: seasonal variation in numerical
abundance in fine sediments. *Archiv für Hydrobiologie* 111: 409—21. [14.4]

Bärlocher F, Kendrick B. (1975) Leaf-conditioning by
microorganisms. *Oecologia* 20: 359—62. [14.2]

Begon M, Harper JL, Townsend CR. (1990) *Ecology:
Individuals, Populations and Communities* 2nd edn.
Blackwell Scientific Publications, Oxford. [14.2]

Biggs BJF, Close ME. (1989) Periphyton biomass dynamics in gravel bed rivers: the relative effects of
flows and nutrients. *Freshwater Biology* 22: 209—31.
[14.2]

Bott TL, Kaplan LA. (1989) Densities of benthic Protozoa
and nematodes in a Piedmont stream. *Journal of the
North American Benthological Society* 8: 187—96.
[14.4]

Bott TL, Kaplan LA. (1990) Potential for protozoan
grazing of bacteria in streambed sediments. *Journal
of the North American Benthological Society* 9:
336—45. [14.4]

Briand F. (1985) Structural singularities of freshwater
food webs. *Verhandlungen der Internationalen
Vereinigung für Theoretische und Angewandte
Limnologie* 22: 3356—64. [14.1]

Briand F, Cohen JE. (1987) Environmental correlates of
food chain length. *Science* 238: 956—60. [14.4]

Bronmark C. (1985) Interactions between macrophytes, epiphytes and herbivores: an experimental approach. *Oikos* **45**: 26–30. [14.4]

Brookes A. (1988) *Channelized Rivers: Perspectives for Environmental Management*. John Wiley & Sons, Chichester. [14.5]

Busch DE, Fisher SG. (1981) Metabolism of a desert stream. *Freshwater Biology* **11**: 301–7. [14.2]

Calow P, Armitage P, Boon P, Chave P, Cox E, Hildrew A et al (1990) *River Water Quality*. The Ecological Issues No. 1, British Ecological Society: Field Studies Council Publishers, Shrewsbury. [14.5]

Carlough LA, Meyer JL. (1989) Protozoans in two southeastern blackwater rivers and their importance to trophic transfer. *Limnology and Oceanography* **34**: 163–77. [14.4]

Carlough LA, Meyer JL. (1990) Rates of protozoan bacterivory in three habitats of a southeastern blackwater river. *Journal of the North American Benthological Society* **9**: 45–53. [14.4, 14.5]

Carpenter SR (ed.) (1988) *Complex Interactions in Lake Communities*. Springer-Verlag, New York. [14.3]

Cohen JE, Newman CM. (1985) A stochastic theory of community food webs. 1. Models and aggregated data. *Proceedings of the Royal Society of London, B* **224**: 421–48. [14.1, 14.6]

Cooper SD, Walde SJ, Peckarsky BL. (1990) Prey exchange rates and the impact of predators on prey populations in streams. *Ecology* **71**: 1503–14. [14.2]

Cummins KW, Wilzbach MA. (1988) Do pathogens regulate stream invertebrate populations? *Verhandlungen der Internationalen Vereinigung für Theoretische und Angewandte Limnologie* **23**, 637–41. [14.4]

DeNicola DM, McIntire CD, Lamberti GA, Gregory SV, Ashkenas LR. (1990) Temporal patterns of grazer–periphyton interactions in laboratory streams. *Freshwater Biology* **23**: 475–89. [14.2]

Dudley TL, D'Antonio CM, Cooper SC. (1990) Mechanisms and consequences of interspecific competition between two stream insects. *Journal of Animal Ecology* **59**. 849–66. [14.2]

Edington JM, Hildrew AG. (1982) *Caseless Caddis Larvae of the British Isles*. Scientific Publications of the Freshwater Biological Association, Ambleside. [14.2]

Feminella JW, Power ME, Resh VH. (1989) Periphyton responses to invertebrate grazing and riparian canopy in three northern California coastal streams. *Freshwater Biology* **22**: 445–57. [14.2]

Fiebig DM, Lock MA. (1991) Immobilization of dissolved organic matter from groundwater discharging through the stream bed. *Freshwater Biology* **26**: 45–55. [14.4]

Findlay S, Meyer JL, Smith PJ. (1986) Incorporation of microbial biomass by *Peltoperla* sp. (Plecoptera) and *Tipula* sp (Diptera). *Journal of the North American Bethological Society* **5**: 306–10. [14.2]

Fisher SG, Gray LJ, Grimm NB, Busch DE. (1982) Temporal succession in a desert stream ecosystem following flash flooding. *Ecological Monographs* **52**: 93–110. [14.3]

Free CA, Beddington JR, Lawton JH (1977) On the inadequacy of simple models of mutual interference for predation and parasitism. *Journal of Animal Ecology* **46**: 543–54. [14.2]

Fretwell SD. (1977) The regulation of plant communities by food chains exploiting them. *Perspectives in Biology and Medicine* **20**: 169–85. [14.3, 14.6]

Gee JRD. (1988) Population dynamics and morphometrics of *Gammarus pulex* L.: evidence of seasonal food limitation in a freshwater detritivore. *Freshwater Biology* **19**: 333–43. [14.2]

Gilliam JF, Fraser DF, Sabat AM. (1989) Strong effects of foraging minnows on a stream benthic invertebrate community. *Ecology* **70**: 445–52. [14.2]

Gladden JE, Smock LA. (1990) Macroinvertebrate distribution and production on the floodplains of two lowland headwater streams. *Freshwater Biology* **24**: 533–45. [14.5]

Gregory SV. (1983) Plant–herbivore interactions in stream systems. In: Barnes JR, Minshall GW (eds) *Stream Ecology: Application and Testing of General Ecological Theory*, pp 157–89. Plenum Press, New York. [14.2]

Groom AP, Hildrew AG. (1989) Food quality for detritivores in streams of contrasting pH. *Journal of Animal Ecology* **58**: 863–81. [14.2, 14.3]

Hairston NG, Smith FE, Slobodkin LB. (1960) Community structure, population control and competition. *American Naturalist* **94**: 421–5. [14.3, 14.6]

Hansen RA, Hart DD, Merz RA. (1991) Flow mediates predator–prey interactions between triclad flatworms and larval blackflies. *Oikos* **60**: 187–96. [14.2]

Hart DD. (1985) Causes and consequences of territoriality in a grazing stream insect. *Ecology* **60**: 404–14. [14.2]

Hart DD. (1987) Experimental studies of exploitative competition in a grazing stream insect. *Oecologia* **73**: 41–7. [14.2]

Hart DD, Robinson CT. (1990) Resource limitation in a stream community: phosphorus enrichment effects on periphyton and grazing. *Ecology* **71**: 1494–1502. [14.3]

Hemphill N, Cooper SD. (1983) The effect of physical disturbance on the relative abundance of two filter-feeding insects in a small stream. *Oecologia* **58**: 378–82. [14.2]

Hershey AE, Hiltner AL. (1988) Effect of a caddisfly on blackfly density: interspecific interactions limit blackflies in an arctic river. *Journal of the North American Benthological Society* **7**: 188–96. [14.3]

Hildrew AG, Townsend CR. (1977) The influence of substrate on the functional response of *Plectrocnemia conspersa* (Curtis) larvae (Trichoptera: Polycentropodidae). *Oecologia* **31**: 21–6. [14.2]

Hildrew AG, Townsend CR. (1980) Aggregation, interference and foraging by larvae of *Plectrocnemia conspersa* (Trichoptera: Polycentropodidae). *Animal Behaviour* **28**: 553–60. [14.2]

Hildrew AG, Townsend CR. (1987) Organization in freshwater benthic communities. In: Gee JHR, Giller PS (eds) *Organization of Communities: Past and Present*, pp 347–71. Symposia of the British Ecological Society, Blackwell Scientific Publications, Oxford. [14.1, 14.2]

Hildrew AG, Townsend CR, Hasham A. (1985) The predatory Chironomidae of an iron-rich stream: feeding ecology and food web structure. *Ecological Entomology* **10**: 403–13. [14.1]

Hildrew AG, Dobson MK, Groom A, Ibbotson A, Lancaster J, Rundle SD. (1991) Flow and retention in the ecology of stream invertebrates. *Verhandlungen der Internationalen Vereinigung für Theoretische und Angewandte Limnologie* **24**: 1742–47. [14.2, 14.5]

Hill WR, Harvey BC. (1991) Periphyton responses to higher trophic levels and light in a shaded stream. *Canadian Journal of Fisheries and Aquatic Sciences* **47**: 2307–14. [14.2, 14.3].

Hill WR, Knight AW. (1987) Experimental analysis of the grazing interaction between a mayfly and stream algae. *Ecology* **68**: 1955–65. [14.2]

Huang C, Sih A. (1990) Experimental studies on behaviorally mediated, indirect interactions through a shared predator. *Ecology* **71**: 1515–22. [14.3]

Hynes HBN. (1975) The stream and its valley. *Verhandlungen der Internationalen Vereinigung für Theoretische und Angewandte Limnologie* **19**: 1–15. [14.5]

Hynes HBN. (1983) Groundwater and stream ecology. *Hydrobiologia* **100**: 93–9. [14.4, 14.5]

Jeffries MJ, Lawton JH. (1984) Enemy free space and the structure of ecological communities. *Biological Journal of the Linnean Society* **23**: 269–86. [14.2]

Jeffries MJ, Lawton JH. (1985) Predator–prey ratios in communities of freshwater invertebrates: the role of enemy free space. *Freshwater Biology* **15**: 105–12. [14.1]

Johnson DM. (1991) Behavioral ecology of larval dragonflies and damselflies. *Trends in Ecology and Evolution* **6**(1): 8–13. [14.2]

Kerfoot WC, Sih A. (1987) *Predation: Direct and Indirect Impacts on Aquatic Communities*. University Press of New England, Hanover, New Hampshire. [14.3]

Lamberti GA, Moore JW. (1984) Aquatic insects as primary consumers. In: Resh VH, Rosenberg DM (eds) *The Ecology of Aquatic Insects*, pp 164–95. Praeger, New York. [14.2]

Lamberti GA, Feminella JW, Resh VH. (1987) Herbivory and intraspecific competition in a stream caddisfly population. *Oecologia* **73**: 75–81. [14.2]

Lamberti GA, Gregory SV, Ashkenas LR, Steinman AD, McIntire CD. (1989) Productive capacity of periphyton as a determinant of plant herbivore interactions in streams. *Ecology* **70**: 1840–56. [14.2]

Lancaster J. (1990) Predation and drift of lotic macroinvertebrates during colonization. *Oecologia* **85**: 457–63. [14.2]

Lancaster J, Hildrew AG, Townsend CR. (1988) Competition for space by predators in streams: field experiments in a net-spinning caddisfly. *Freshwater Biology* **20**: 185–93. [14.2]

Lancaster J, Hildrew AG, Townsend CR. (1990) Stream flow and predation effects on the spatial dynamics of benthic invertebrates. *Hydrobiologia* **203**: 177–90. [14.2]

Lancaster J, Hildrew AG, Townsend CR. (1991) Invertebrate predation on patchy and mobile prey in streams. *Journal of Animal Ecology* **60**: 625–41. [14.2, 14.3]

Lawton JH. (1989) Food webs. In: Cherrett JM (ed.) *Ecological Concepts*, pp 43–78. Symposia of the British Ecological Society, Blackwell Scientific Publications, Oxford. [14.1, 14.2, 14.4]

Likens GE (ed.). (1985) *An Ecosystem Approach to Aquatic Ecology: Mirror Lake and its Environment*. Springer-Verlag, New York. [14.5]

McAuliffe JR. (1984a) Resource depression by a stream herbivore: effects on distributions and abundances of other grazers. *Oikos* **42**: 327–33. [14.2]

McAuliffe JR. (1984b) Competition for space, disturbance, and the structure of a benthic stream community. *Ecology* **65**: 894–908. [14.2]

McCormick PV. (1990) Direct and indirect effects of consumers on benthic algae in isolated pools of an ephemeral stream. *Canadian Journal of Fisheries and Aquatic Sciences* **47**: 2057–67. [14.3, 14.4]

McCormick PV, Stevenson RJ. (1989) Effects of snail grazing on benthic algal community structure in different nutrient environments. *Journal of the North American Benthological Society* **8**: 162–72. [14.2]

Matczak TZ, Mackay RJ. (1990) Territoriality in filter-feeding caddisfly larvae: laboratory experiments. *Journal of the North American Benthological Society* **9**: 26–34. [14.2]

Menge BA, Sutherland JP. (1987) Community regulation: variation in disturbance, competition, and predation in relation to environmental stress and recruitment. *American Naturalist* **130**: 730–57. [14.1, 14.3]

Naiman RJ, Décamps H (eds). (1990) *The Ecology and Management of Aquatic–Terrestrial Ecotones*. Man and the Biosphere Series, Vol. 4, UNESCO, Parthenon Publishing Group, Paris. [14.5].

Newman RM. (1991) Herbivory and detritivory on freshwater vascular macrophytes by aquatic invertebrates: a review. *Journal of the North American Benthological Society* **10**: 89–114. [14.2, 14.4]

Ormerod SJ, Tyler SJ. (1991) Predatory exploitation by a river bird, the dipper *Cinclus cinclus* (L.), along acidic and circumneutral streams in upland Wales. *Freshwater Biology* **25**: 105–116. [14.4]

Ostrofsky ML, Zettler ER. (1986) Chemical defenses in aquatic plants. *Journal of Ecology* **74**: 279–87. [14.4]

Paine RT. (1988) On food webs: road maps of interactions or grist for theoretical development? *Ecology* **69**: 1648–54. [14.1]

Peckarsky BL. (1984) Predator–prey interactions among aquatic insects. In: Resh VH, Rosenberg DM (eds) *The Ecology of Aquatic Insects*, pp 196–254. Praeger, New York. [14.2]

Peckarsky BL, Penton MA. (1988) Why do *Ephemerella* nymphs scorpion posture: a 'ghost of predation past'? *Oikos* **53**: 185–93. [14.2]

Peckarsky BL, Horn SC, Statzner B. (1990) Stonefly predation along a hydraulic gradient: a field test of the harsh–benign hypothesis. *Freshwater Biology* **24**: 181–91. [14.2]

Peterson BJ, Hobbie JE, Hershey AE, Lock MA, Ford TE, Vestal JR *et al.* (1985) Transformation of a tundra river from heterotrophy to autotrophy by addition of phosphorus. *Science* **229**: 1383–6. [14.3]

Pimm SL. (1982) *Food Webs*. Chapman and Hall, London. [14.1, 14.2]

Pimm SL, Lawton JH, Cohen JE. (1991) Food web patterns, their causes and their consequences. *Nature* **350**: 669–74. [14.1]

Power ME. (1990) Effects of fish in river food webs. *Science* **250**: 811–14. [14.3, 14.6]

Power ME, Mathews WJ, Stewart AJ. (1985) Grazing minnows, piscivorous bass, and stream algae: dynamics of a strong interaction. *Ecology* **66**: 1448–56. [14.3]

Pringle CM, Naiman RJ, Bretschko G, Karr JR, Oswood MW, Webster JR *et al.* (1988) Patch dynamics in lotic systems: the stream as a mosaic. *Journal of the North American Benthological Society* **7**: 503–24. [14.1]

Resh VH, Brown AV, Covich AP, Gurtz ME, Li HW, Minshall GW *et al.* (1988) The role of disturbance in stream ecology. *Journal of the North American Benthological Society* **7**: 433–55. [14.1, 14.3]

Reynolds CS. (1988) Potamoplankton: paradigms, paradoxes and prognoses. In: Round FE (ed.) *Algae and the Aquatic Environment*, pp 285–311. Biopress, Bristol. [14.5]

Richardson JS. (1991) Seasonal food limitation of detritivores in a montane stream: an experimental test. *Ecology* **72**: 873–87. [14.2]

Rundle SD, Hildrew AG. (1990) The distribution of micro-arthropods in some southern English streams: the influence of physicochemistry. *Freshwater Biology* **23**: 411–31. [14.1].

Schlosser IJ, Ebel KK. (1989) Effects of flow regime and cyprinid predation on a headwater stream. *Ecological Monographs* **59**: 41–57. [14.2]

Schofield K, Townsend CR, Hildrew AG. (1988) Predation and the prey community of a headwater stream. *Freshwater Biology* **20**: 85–95. [14.2, 14.3]

Sculthorpe CD. (1967) *The Biology of Aquatic Vascular Plants*. St. Martin's Press, New York. [14.4]

Sinsabaugh RL, Linkins AE, Benfield EF. (1985) Cellulose digestion and assimilation by three leaf-shredding aquatic insects. *Ecology* **66**: 1464–71. [14.2]

Smock LA, Metzler GM, Gladden JE. (1989) Role of debris dams in the structure and functioning of low-gradient headwater streams. *Ecology* **70**: 764–75. [14.2]

Soluk DA. (1990) Postmolt susceptibility of *Ephemerella* larvae to predatory stoneflies: constraints on defensive armour. *Oikos* [14.2]

Statzner B. (1987) Characteristics of lotic ecosystems and consequences of future research directions. In: Schulze ED, Zwolfer H (eds) *Potentials and Limitations of Ecosystem Analysis*, pp 365–90. Ecological Studies 61, Springer, Berlin. [14.3, 14.4].

Statzner B, Müller R. (1989) Standard hemispheres as indicators of flow characteristics in lotic benthic research. *Freshwater Biology* **21**: 445–9. [14.2]

Sutcliffe DW, Hildrew AG. (1989) Invertebrate communities in acid streams. In: Morris R Brown DJA, Brown JA (eds) *Acid Toxicity and Aquatic Animals*, pp 13–29. Seminar Series of The Society for Experimental Biology, Cambridge University Press, Cambridge. [14.3, 14.5]

Thomas JD. (1987) An evaluation of the interactions between freshwater pulmonate snail hosts of human schistosomes and macrophytes. *Philosophical Transactions of the Royal Society of London, B* **315**: 75–125. [14.4]

Tokeshi MW. (1986) Resource utilization, overlap and temporal community dynamics: a null model analysis of an epiphytic chironomid community. *Journal of Animal Ecology* **55**: 491–506. [14.2]

Tokeshi MW, Townsend CR. (1987) Random patch formation and weak competition: coexistence in an epiphytic chironomid community. *Journal of Animal Ecology* **56**: 833–45. [14.2]

Underwood GJC, Thomas JD. (1990) Grazing interactions between pulmonate snails and epiphytic algae and bacteria. *Freshwater Biology* **23**: 505–22. [14.4]

Valett HM, Stanford JA. (1987) Food quality and hydropsychid caddisfly density in a lake outlet stream in Glacier National Park, Montana, USA. *Canadian Journal of Fisheries and Aquatic Sciences* **44**: 77–82.

[14.2]

Ward JV. (1989) The four-dimensional nature of lotic ecosystems. *Journal of the North American Benthological Society* **8**: 2–8. [14.6]

Warren PH. (1990) Variation in food web structure: the determinants of connectance. *American Naturalist* **136**: 689–700. [14.1]

Warren PH, Lawton JH. (1987) Invertebrate predator–prey body size relationships: an explanation for upper triangular food webs and patterns in food web structure? *Oecologia* **74**: 231–5. [14.1, 14.6]

Waters TF. (1988) Fish production–benthos production relationships in trout streams. *Polskie Archiwum Hydrobiologii* **35**: 545–61. [14.2]

Welcomme RL. (1979) *Fisheries Ecology of Floodplain Rivers*. Longman, London. [14.5]

Wiley MJ, Kohler SL. (1984) Behavioral adaptations of aquatic insects. In: Resh VH, Rosenberg DM (eds) *The Ecology of Aquatic Insects*, pp 101–33. Praeger, New York. [14.2]

Winterbourn MJ. (1986) Recent advances in our understanding of stream ecosystems. In: Polunin N (ed.) *Ecosystem Theory and Application*, John Wiley & Sons, New York, pp 240–68. [14.2, 14.4]

15: Detritus Processing

L.MALTBY

15.1 INTRODUCTION

Detritus processing, the incorporation of non-living organic material into living biomass, plays a central role in the functioning of many flowing water systems. This account of detritus processing begins with a description of detritus (section 15.2), the organisms (detritivores and micro-organisms) involved in its processing (section 15.3) and the interactions between them (section 15.4). There is then a brief consideration of the enzymatic capabilities of invertebrate detritivores (section 15.5) before the relative importance of micro-organisms and detritivores in detritus processing is discussed (section 15.6). Finally, the implications of changes in water quality (section 15.7) and riparian vegetation (section 15.8) for detritus processing are considered.

15.2 FORMS AND SOURCES OF DETRITUS

Detritus has been described as an assemblage of living and dead material incorporating fungi, bacteria, microinvertebrates, algal cells and the substrate itself (Anderson & Cargill 1987). Although properties such as source, associated microbes, energy content, percentage ash and C:N ratios have been used to classify detritus (Cummins & Klug 1979), the most commonly used scheme is based on particle size. Detritus is divided into three major classes within which there are a number of subgroups: coarse particulate organic matter (CPOM, >1 mm), fine particulate organic matter (FPOM, >0.45 μm <1 mm) and dissolved organic matter (DOM, <0.45 μm) (Cummins 1974).

Detritus in streams may be either allochthonous, i.e. imported from the surrounding catchment (e.g. leaves, twigs, branches), or autochthonous, i.e. produced by the system itself (e.g. algae, dead aquatic animals and plants). The relative importance of allochthonous and autochthonous detritus will vary both spatially and temporally (Webster & Benfield 1986).

In wooded headwater streams, shading restricts primary production and CPOM is a major energy source. Inputs of allochthonous detritus into small streams flowing through mature forests may account for more than 95% of the total organic matter inputs (Triska et al 1982). In larger streams, where channels are wider and there is less shading, primary production is greater and allochthonous CPOM inputs are less important (Conners & Naiman 1984). Detrital material in middle- and high-order streams consists of FPOM transported from upstream regions as well as decomposing algae and macrophytes (Anderson & Cargill 1987). Consequently, whereas in small woodland streams the food of detritivores would be mostly allochthonous, in more open systems it would be primarily autochthonous. Evidence in support of this comes from studies of gut contents and stable isotope analysis (e.g. Rounick et al 1982; Ross & Wallace 1983; Winterbourn et al 1986).

The composition of the riparian vegetation influences both the quality and quantity of organic matter entering the aquatic environment. The quantity of organic detritus in a stream at any specific time is a balance between input and output processes (Swanson et al 1982). Small forested streams are extremely retentive and primarily export material in the form of FPOM or

DOM (Cummins *et al* 1983; Speaker *et al* 1984). Only in larger streams, where the formation of retention structures (e.g. log jams) is restricted or removed by the water flow, is CPOM such as leaf litter effectively transported (Triska *et al* 1982).

Coarse particulate organic matter

Particulate detrital inputs range from rapidly processed leaves, needles and twigs to large, slowly processed, woody debris. Although woody material has a lower food value than non-woody material, it may sustain organisms during periods when few leaves or needles are available. It is also important in that it increases the stability of the channel and retards the loss of other more palatable detritus (Bilby & Ward 1989; Winkler 1991).

The majority of studies on decomposition have concentrated on leaf material, in particular from deciduous trees. Breakdown rate of leaf litter varies between species such that leaves from non-woody plants break down significantly faster than leaves from woody plants. Within non-woody plants, leaves of submerged and floating macrophytes break down significantly faster than those from emergent macrophytes, terrestrial grasses or ferns (Webster & Benfield 1986).

Factors that affect breakdown rate include the level of nutrients, fibre and inhibitory chemicals present in the plant tissue. For example, leaves of woody plants with a high nitrogen and/or low lignin content break down faster than those with a low nitrogen and/or high lignin content (Kaushik & Hynes 1971; Sedell *et al* 1975; Suberkropp *et al* 1976). The breakdown rate of conifer needles is slower than that for deciduous leaves, due partly to their thick cuticle and the presence of inhibitory substances (Bärlocher & Oertli 1978a, 1978b).

Anderson and Cummins (1979) suggested that food items can be organized from low to high quality as follows: wood, terrestrial leaf litter, FPOM, decomposing vascular macrophytes, filamentous algae, diatoms and animal tissue. Such schemes define food quality in terms of the chemical properties of the organic material, i.e. nutrient content (nitrogen, protein) or abundance of refractory materials (lignin, cellulose). The

observation that nitrogen and lignin content correlate well with breakdown rates has resulted in the suggested use of the lignin:nitrogen ratio as an index of leaf quality (Melillo *et al* 1982). However, although these may be convenient measures with which to categorize food items it must be remembered that the only true measure of food quality is the performance (i.e. growth, reproduction, survival) of the organism feeding upon it. As detritivores vary in their digestive capabilities (section 15.5), any particular type of organic material may be of high food quality for one organism but of low food quality for another.

The quality of the input of CPOM, defined in gross chemical terms, varies seasonally. In spring and summer it consists of high-nutrient pollen, flowers and insect excretory products, whereas in autumn it consists principally of leaves, and in winter and spring wood is the major input (Anderson & Sedell 1979). Most allochthonous input enters woodland streams during the autumn although leaves that fall on to the floodplain may not enter until major flooding occurs. Such flooding often occurs in spring and results in an input of particulate organic matter to the stream at a time when the material that entered in the autumn is greatly depleted. Under such circumstances floodplains may act as a store of organic material for the stream (Cummins *et al* 1983). For example, Cuffney (1988) calculated that 5.5 kg AFDW m^{-2} (ash-free dry weight) of particulate organic matter entered the Ogeechee River (USA) channel from the floodplain.

Fine particulate organic matter

FPOM can be produced by physical abrasion, microbial action, feeding by invertebrates, scouring of algae, bacteria and fungal spores from surfaces, flocculation of DOM or soil erosion. It is a mixture of shredded vascular plant material, faeces, algae and micro-organisms (Short & Maslin 1977; Anderson & Sedell 1979; Bowen 1984; Findlay & Arsuffi 1989). Models of detritus processing in streams (e.g. Boling *et al* 1975) suggest that small particles have undergone considerable processing and decomposition and should therefore contain the largest amount of refractory material such as lignin and cellulose

(Suberkropp *et al* 1976; Cummins & Klug 1979). However, contrary to these predictions, the percentage of lignin and cellulose and the C:N ratio decreases as particle size decreases (Ward 1986; Sinsabaugh & Linkins 1990). Possible explanations for this include: (i) naturally occurring FPOM comes from a variety of sources with different processing rates and therefore predicted patterns may be obscured (Ward 1986); (ii) as particle size decreases, surface area increases, thereby providing a larger surface for immobilizing bacteria and nutrients. Sinsabaugh and Linkins (1990) measured the microbial respiration of particles of differing sizes and found that below 2.5 mm, microbial oxygen consumption was inversely proportional to particle size.

Dissolved organic matter

DOM arises from a number of sources, both detrital and non-detrital (see Fig. 8.5). It may be 'leaked' by living algae and macrophytes, released during lysis of dead cells, or arise from microbial activities, groundwater seepages, throughfall from the canopy, and surface runoff. Miller (1987) emphasized the potential importance of non-detrital DOM as an energy source for microheterotrophs. He compared the biomass of heterotrophic micro-organisms on leaves incubated in control and darkened troughs in a stream. There were significantly more bacteria and fungi colonizing leaf litter incubated in the open troughs (Fig. 15.1) and this was interpreted as evidence in support of the hypothesis that non-detrital DOM, released by microalgae, was utilized by bacteria and

fungi colonizing dead leaves. Compared with leaf leachate, which is released rapidly once the leaf enters the water (Kaushik & Hynes 1971), non-detrital DOM is less seasonal, being released throughout the lifespan of aquatic algae and macrophytes (Wetzel 1975; Bott & Ritter 1981), and consists mainly of low molecular weight labile compounds (Kaplan & Bott 1985). However, although there is little doubt that algae, when physiologically stressed, release DOM, figures of 40–50% of photosynthetically fixed carbon being released as DOC (e.g. Haack & McFeters 1982) may be overestimates. Berman (1990) suggested that a more realistic figure for actively growing, healthy algae, is less than 10%.

Some fungi (Bengtsson 1982) and invertebrates (Hipp *et al* 1986; Salonen & Hammer 1986; Thomas *et al* 1990) can assimilate DOM, although the main routes by which it is processed are: conversion to a particulate form by physicochemical processes (Bowen 1984; McDowell 1985); adsorption on to surfaces (Armstrong & Bärlocher 1989); or assimilation by bacteria (Mann 1988). Bacteria may be either suspended in the water column or associated with organic layers on stones. These organic layers, which consist of a matrix of polysaccharide slime containing fungi, bacteria, algae and FPOM, are an important site for the processing of organic material (Lock & Hynes 1976; Kaplan & Bott 1983, 1985; Lock *et al* 1984). They also provide a potential food source for detritivores (Rounick *et al* 1982; Rounick & Winterbourn 1983; Lock *et al* 1984; Bärlocher & Murdoch 1989). For example, the mayfly *Stenonema* acquires 47% of its daily carbon needs

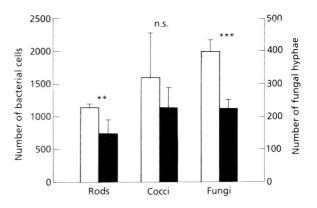

Fig. 15.1 Relative abundance of bacterial cells (median number/ten microscope fields) and fungal hyphae (median number/four 12-mm transects) accumulating *in situ* on leaf litter in troughs open to natural illumination (□) and from which natural illumination was excluded (■). Error bars represent 95% CL. ** significant difference at $P < 0.01$; *** significant difference at $P < 0.001$; n.s., no significant difference (after Miller 1987).

from bacteria associated with stone surface layers (Edwards & Meyer 1990) and the gut contents of *Tipula caloptera* and *Gammarus tigrinus* can release amino acids from organic layers (Armstrong & Bärlocher 1989; Bärlocher & Murdoch 1989).

DOM provides a continuous input of energy which can be utilized by invertebrates either after flocculation to FPOM or conversion to microbial biomass (Merritt *et al* 1984). In marine systems, considerable attention has been paid to the incorporation of DOM into bacterial biomass which is then grazed by protozoans (Porter *et al* 1985; Azam & Cho 1987). This 'microbial loop' has been less intensively studied in flowing freshwater systems even though recent studies have demonstrated its potential importance in the trophic dynamics and energy budgets of such systems (Bott & Kaplan 1990; Carlough & Meyer 1990). Whereas in marine systems free-living bacteria are mainly responsible for DOM uptake from the water column (Azam *et al* 1990), in rivers and streams attached bacteria are probably more important (Edwards & Meyer 1986). Bacteria suspended in the water column may be either

consumed directly by macroinvertebrates or consumed by protozoans which in turn become prey for macroinvertebrate filter-feeders (Wallace *et al* 1987; Carlough & Meyer 1989). For example, 20–67% of the daily growth of simuliids in the Ogeechee River is due to incorporation of bacterial carbon (Edwards & Meyer 1987).

Interactions between DOM, bacteria, protozoans and macroinvertebrates are illustrated in Fig. 15.2.

15.3 ORGANISMS INVOLVED IN DETRITUS PROCESSING

Detritivores

Detritivores include representatives of a wide number of taxonomic groups including arthropods, molluscs, annelids and fish. However, the majority of detritivores are insects and, to a lesser degree, crustaceans (Table 15.1). Detritivores do not necessarily feed exclusively on one particular type of food but may switch diet either in response to changes in the abundance of a particular cat-

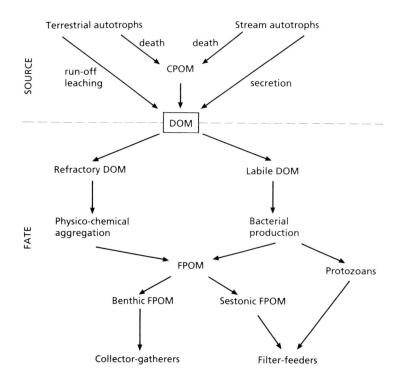

Fig. 15.2 Source and fate of dissolved organic matter (DOM) in streams. CPOM, coarse particulate organic matter; FPOM, fine particulate organic matter.

egory of detritus (Roeding & Smock 1989) or in response to changes in nutritional requirements during development (Fuller & Mackay 1981; Anderson & Cargill 1987). For example, in the few months before emergence, the diet of hydropsychids switches from being dominated by FPOM to being dominated by animal prey (Ross & Wallace 1983).

Freshwater animals have been categorized on the basis of the type of food they utilize and the way they acquire it (Chapter 11). The 'functional feeding groups' utilizing detritus are shredders, collector-filterers, collector-gatherers and gougers (Merritt & Cummins 1978; Anderson *et al* 1984). Shredders feed upon decomposing vascular plant material (CPOM). Their activity results in the production of FPOM either in the form of faeces or fragments. Invertebrates feeding on FPOM may either filter it out of the water column, in which case they are known as collector-filterers (e.g. simuliids, net-spinning caddisfly larvae), or feed on material that has been deposited on the streambed. These animals are known as collector-gatherers and include mayflies and chironomids. Gougers, which are sometimes incorporated

within the shredder category (Cummins *et al* 1989), feed on wood and include elmid beetles (e.g. *Lara avara*) and caddisfly larvae (e.g. *Heteroplectron californicum*).

Micro-organisms

The types of heterotrophic microorganisms associated with detritus were described in Chapter 8. In general it is assumed that fungi are more important in the breakdown of CPOM than bacteria. This has been investigated by exposing leaf discs to bactericides and/or fungicides and either measuring their respiration rate (Triska 1970) or weight loss (Kaushik & Hynes 1971). For example, Triska (1970) measured the respiration rate of leaf discs recovered from streams and incubated with or without bactericides. The results from these studies are presented in Fig. 15.3. Throughout the year, fungal respiration was greater than bacterial respiration and this difference was greatest during the early stages of decomposition (i.e. winter). The relative abundance of fungi and bacteria on CPOM is dependent both upon the vegetation type and stage of decomposition (Fig. 15.4).

Table 15.1 Detritivorous freshwater arthropods

Class	Order	Family
Insecta	Trichoptera	Brachycentridae, Calamatoceridae, Glossosomatidae, Helicopsychidae, Hydropsychidae, Hydroptilidae, Lepidostomatidae, Leptoceridae, Limnephilidae, Molannidae, Oeconesidae, Odontoceridae, Philopotamyidae, Phryganeiidae Polycentropodidae, Psychomyiidae, Seriostomatidae
	Diptera	Ceratopogonidae, Chironomidae, Culicidae, Dixidae, Empididae, Ephydridae, Psychodidae, Ptychopteridae, Simuliidae, Stratiomyidae, Syrphidae, Tipulidae
	Ephemeroptera	Baetidae, Caenidae, Ephemerellidae, Ephemeridae, Heptageniidae, Leptophlebeidae, Potamanthidae, Siphlonuridae, Tricorythidae
	Plecoptera	Capniidae, Chloroperlidae, Leuctridae, Nemouridae, Peltoperlidae, Pteronarcidae, Taeniopterygidae
	Coleoptera	Elmidae, Helodidae, Hydrophilidae, Psephenidae, Ptilodactylidae, Scirtidae
	Hemiptera	Corixidae
Crustacea	Isopoda	Asellidae
	Amphipoda	Gammaridae

Based on information from Cummins (1974), Petersen and Cummins (1974), Merritt and Cummins (1978), Anderson and Sedell (1979), and Anderson and Cargill (1987).

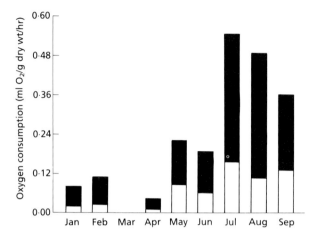

Fig. 15.3 Oxygen consumption (ml O_2/g dry weight/h) of alder leaf discs during decomposition. □, Bacterial respiration; ■, fungal respiration (after Triska 1970).

Bacteria may be more important in the latter stages of processing when particle size is reduced (Suberkropp & Klug 1976). Sinsabaugh and Linkins (1990) suggested that particles of less than 2–4 mm were too small to support the growth of fungal hyphae, and therefore microbial processing of these particles is dependent upon bacteria. Bacteria may also play an important role in the early decomposition of recalcitrant leaves, such as those of members of the Fagaceae (Iversen 1973; Davis & Winterbourn 1977), and are the major assimilators of DOM (section 15.2).

Within the fungi, the relative importance of 'terrestrial' over 'aquatic' groups is more difficult to assess owing to the different methodologies used by different workers (see Chapter 8). Many studies have identified fungal assemblages based on conidial production and have therefore immediately restricted their survey to aquatic hyphomycetes (e.g. Triska 1970; Padgett 1976). Other studies have used particle plating techniques but have incubated the leaf material on nutrient-rich agar at high temperatures, conditions well suited for 'terrestrial' hyphomycetes (e.g. Kaushik & Hynes 1971). Suberkropp and Klug (1976) performed a detailed study of the bacteria and fungi colonizing leaf material in Augusta Creek, Michigan. Several different techniques were used including direct observation, water incubation and particle plating using different agars and incubation temperatures. They concluded that fungi do dominate the microflora and that, although aquatic hyphomycetes are dominant, and therefore assumed most important, other groups such as Oomycetes and terrestrial hyphomycetes are also present. These results substantiated earlier conclusions drawn from a scanning electron microscopic study (Suberkropp & Klug 1974). The importance of different fungal species in decomposition depends on their enzymatic capability (Chapter 8, section 8.3) and their interaction with detritivores (section 15.4).

15.4 FUNGAL–INVERTEBRATE INTERACTIONS

Importance of microbes as mediators of food quality

When CPOM (e.g. leaves) enters a water body, soluble compounds such as carbohydrates, amino acids and phenolic substances are leached out leaving the structural molecules behind. The DOM is either assimilated by bacteria or precipitated by physicochemical processes, whereas the CPOM is either consumed by animals or colonized by micro-organisms (Fig. 15.5). Early studies of leaf decomposition concluded that leaching may remove about 5–30% of the initial dry weight of leaves in 1–2 days depending on the species involved (e.g. Petersen & Cummins 1974). However, these studies used pre-dried leaves and more recent studies have found that, for fresh leaves, both leaching and rate of colonization by aquatic hyphomycetes are reduced (Gessner & Schwoerbel 1989; Bärlocher 1991). The rate at

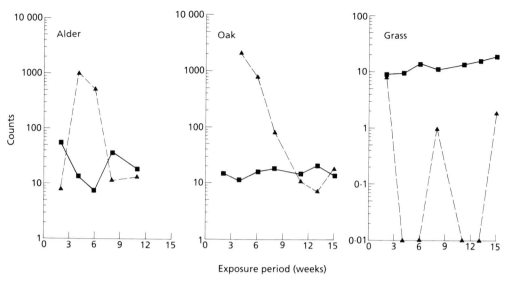

Fig. 15.4 Relative abundance of bacteria (■———■) and aquatic hyphomycete conidia (▲———▲) on different substrates. Data from Chamier (1987).

which energy flows through detritus-based communities is a function of the rate at which the refractory molecules remaining after leaching can be hydrolysed and assimilated by micro-organisms and detritivores. As mentioned in section 15.2, detritus consists not only of particulate organic matter but also contains micro-organisms. To understand energy flow in detritus-based systems it is important to know the relative importance of detrital structural molecules and microbial biomass to detritivore nutrition.

Many studies have confirmed that conditioned leaves (i.e. those colonized by micro-organisms) are more palatable to detritivores than unconditioned leaves (Kaushik & Hynes 1971; Kostalos & Seymour 1976; Rossi & Fano 1979; Golladay *et al* 1983; Bueler 1984) and that palatability is a function of: leaf type (Triska 1970; Kaushik & Hynes 1971, Bärlocher & Kendrick 1973a; Irons *et al* 1988), fungal species (Bärlocher & Kendrick 1973a; Marcus & Willoughby 1978; Fano *et al* 1982; Suberkropp *et al* 1983; Rossi *et al* 1983; Rossi 1985) and incubation time (Iversen 1973; Anderson & Grafius 1975; Arsuffi & Suberkropp 1984; Bueler 1984). When offered a choice, not only can detritivores discriminate between conditioned and sterile food, but they can also discriminate between leaf discs colonized by different fungal species, and can even discriminate between fungal patches on the same leaf (Arsuffi & Suberkropp 1985).

Detritivores exhibit preferences for different leaf types and tend to prefer those with the fastest rate of decay, although less preferred leaves can be made more palatable by inoculating them with a preferred fungus (Bärlocher & Kendrick 1973a). Therefore, not only are the leaf and micro-organisms important in themselves, but the interaction between them also influences food choice.

Several studies have attempted to correlate feeding preference with the physical (e.g. softness), chemical (e.g. nitrogen) or mycological (e.g. biomass, enzymatic capability) properties of inoculated leaf litter (Bärlocher & Kendrick 1973a; Suberkropp *et al* 1983; Bärlocher 1985). However, these studies have failed to provide a convincing explanation for the food choices observed, suggesting that other factors such as the accumulation of specific substances (e.g. lipids) and/or the breakdown of lignin or polyphenols may be important. Evidence that lipids may be important cues for detritivores comes from a study by Cargill and colleagues (1985) (see also Hanson *et al* 1983). Larvae of the caddisfly

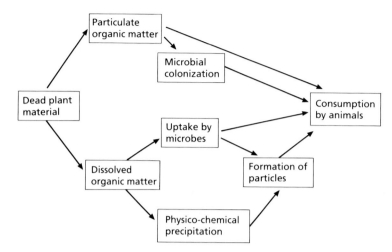

Fig. 15.5 Pathways of detritus processing in freshwater systems (after Mann 1988).

Clistoronia magnifica were reared on one of five diets: conditioned alder leaves, alder plus a mixture of fatty acids, alder and wheat grains, alder plus fungi, or fungi alone. Animals grew well and reproduced on a diet of alder plus wheat but did not complete their development on either alder alone or alder plus fatty acids. Larvae fed alder plus fungi or fungi alone developed successfully but failed to reproduce. The authors suggested that, in order to reproduce, *C. magnifica* needs to acquire large lipids stores during the last larval instar. These triglyceride stores could either be assimilated directly from the food or synthesized from carbohydrates. Wheat grains are a good quality food as they provide a rich source of soluble carbohydrates. This experiment highlights the fact that care must be taken when assessing food quality; just because a food promotes growth does not mean that the completion of the life cycle will be possible.

Preference rankings may also be the consequence of unpalatable species producing distasteful or toxic metabolites. There is some evidence that fungal preference changes through time, owing to the different processing rates of the fungi, and that leaves may become post-conditioned (*sensu* Boling *et al* 1975). For example, larvae of the caddisfly *Pschoglypha* were offered a choice of leaf discs inoculated with the fungus *Clavariopsis aquatica* and incubated for 7, 14 or 21 days. The consumption of food incubated for 7 days was three times greater than the consumption of leaf material incubated for 14 or 21 days, even though leaf material incubated for longer periods was softer and had a higher nitrogen content (Arsuffi & Suberkropp 1984). Whether such decreases in palatability with increasing incubation time are due to the accumulation of distasteful or toxic compounds or due to some other factor(s) is unknown.

Fungal preferences vary between taxa. For example, *Lemonniera aquatica* was not consumed by any of the three species of Trichoptera investigated by Arsuffi and Suberkropp (1984), although it was highly preferred by the mollusc *Helisoma trivolvis* (Arsuffi & Suberkropp 1989). The degree of selectivity also varies, with some detritivores (e.g. *Gammarus*) being much more selective than others (e.g. *Helisoma trivolvis, Pteronarcella badia*) (Arsuffi & Suberkropp 1989). Differences in food choice and degree of selectivity may be due to differences in digestive ability and mobility, more mobile species being more selective than relatively immobile ones (Bärlocher 1982). Other factors may also be important as food preferences vary both within (Christensen 1977) and between (Rossi *et al* 1983) populations of the same species.

Microbes as a food source

What is the relative importance of the microbes themselves as a food source, compared with the organic matter with which they are associated?

In feeding trials, the performance of detritivores fed fungal mycelia has been reported to be better than (Bärlocher & Kendrick 1973b; Rossi & Fano 1979; Cargill *et al* 1985), similar to (Kostalos & Seymour 1976; Marcus & Willoughby 1978; Arsuffi & Suberkropp 1988) or worse than (Bärlocher & Kendrick 1973b; Willoughby & Sutcliffe 1976; Marcus & Willoughby 1978; Sutcliffe *et al* 1981) that of animals fed decaying leaf material. However, it is difficult to compare the results from different studies as they have used different fungi, conditioning times and animals. For example, Sutcliffe *et al* (1981) found that the growth rate of the amphipod *Gammarus pulex* was lower when fed on a mycelial diet of *Lemonniera aquatica* or *Clavariopsis aquatica* compared with animals fed a diet of conditioned or unconditioned elm leaves. In contrast, Bärlocher and Kendrick (1973b) found that while some fungal species resulted in poor growth of *Gammarus pseudolimnaeus*, others supported growth rates two to five times greater than those for animals fed unconditioned leaves. When assessing the importance of micro-organisms in the diet of detritivores, differences in food quality must be taken into account. *Lemonniera aquatica* was one of the least preferred fungi by *Gammarus* sp. and *Gammarus pulex* (Arsuffi & Suberkropp 1989; Graça 1990) which may explain the poor performance of *G. pulex* fed *L. aquatica* mycelia (Sutcliffe *et al* 1981).

Not all fungal species increase the palatability of leaf material to the same extent and there is some evidence that animals show preferences for those diets affording the best survivorship and growth (Kostalos & Seymour 1976; Arsuffi & Suberkropp 1986), although this is not necessarily true in every case (Bärlocher & Kendrick 1973a, 1973b).

Although most of the work on fungal–invertebrate interactions has concentrated on aquatic hyphomycetes, other fungi are commonly found in freshwaters (Kaushik & Hynes 1971; Bärlocher & Kendrick 1974; Park 1974; Suberkropp & Klug 1976) and there is evidence that some of these (e.g. *Pythium*, *Cladosporium*, *Epicoccum*) can degrade plant material (Park & McKee 1978; Godfrey 1983). It has been argued that these so-called 'terrestrial' hyphomycetes are outcompeted by aquatic ones at the low temperatures prevalent in northern temperate streams during winter (see Chapter 8). However, 'terrestrial' hyphomycetes may be important in mediterranean and tropical regions where higher water temperatures would favour them. In fact, Rossi and his co-workers, studying the trophic biology of detritivores in Italian freshwaters, have concentrated on 'terrestrial' hyphomycetes (Fano *et al* 1982; Rossi 1985; Basset & Rossi 1987).

Can microbial biomass present on particulate organic matter support detritivore populations?

Many detritivores have a rapid gut throughput time (e.g. <2 h for *Gammarus pulex* (Willoughby & Earnshaw 1982; Welton *et al* 1983)) and this has been taken as evidence that the digestive mechanism of these organisms is mainly one of stripping microbial nutrients rather than digestion of the more refractory detritus itself. Although the quantitative importance of micro-organisms has been demonstrated for some detritivores (Calow 1974; Bärlocher & Kendrick 1975; Arsuffi & Suberkropp 1986), other evidence is less clearcut. For example, although it is possible to rear simuliids on pure cultures of bacteria in the laboratory (Fredeen 1964), under field conditions the concentrations of bacteria may be too low to meet their nutritional requirements (Baker & Bradnam 1976; Edwards & Meyer 1987).

Radioactive isotopes have been used to estimate the contribution of microbial biomass to invertebrate energy budgets (Findlay & Tenore 1982; Lawson *et al* 1984; Findlay *et al* 1984, 1986a, 1986b). Lawson *et al* (1984), studying the dipteran larva *Tipula abdominalis*, concluded that the biomass of microbes on detritus did not influence larval growth but did influence consumption rate. They concluded that *Tipula* did require bacteria and fungi in their diet but not as a carbon or nitrogen source because 73–89% of their growth was derived from the leaf matrix itself. Results obtained for *Tipula* cannot be viewed as being representative of detritivores in general as they are unusual in having an alkaline midgut (pH > 11) which is thought to be an adaptation for protein digestion (Martin *et al* 1980; Bärlocher 1982, 1985). Findlay, Meyer and Smith (1986a) fed the isopod *Lircus* either oak or willow leaves inoculated with either the zygomycete *Mucor* or the hyphomycete. *Triscelophorus*. They found that the contribution of fungal carbon to the total carbon respired varied between treatments from about 1% for animals fed willow plus *Triscelophorus* to 57% for animals fed oak plus *Mucor*. Therefore, although the authors concluded from their study that fungi alone could not meet the carbon requirements of *Lircus*, this is obviously dependent upon the fungal species present. If other, more preferred, fungi were used, the animals' carbon requirements may have been met. In contrast to the studies of Findlay *et al* (1986a, 1986b) and Lawson *et al* (1984), Arsuffi and Suberkropp (1986) concluded that fungal biomass could account for a significant proportion of the growth of trichopteran larvae.

The inability of fungal biomass to support detritivore populations has also been suggested by others workers who believe that, because microbial biomass of detritus is low, the main role of microbes is as modifiers of leaf material (Iversen 1973; Cummins & Klug 1979; Bärlocher & Kendrick 1981). Estimates of fungal biomass may vary depending on the technique used (Newell *et al* 1986). Moreover, Findlay and Arsuffi (1989) observed that the contribution of microbial biomass to the amount of organic carbon in the leaf–microbe complex varied both within and between leaf types (Fig. 15.6). However, it was

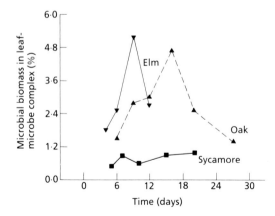

Fig. 15.6 Temporal variation in the contribution of microbial biomass to the amount of organic carbon in leaf–microbe complexes (after Findlay & Arsuffi 1989).

always less than 6% and, of this, fungi constituted between 82% and 96% of the microbial biomass on both oak and elm leaves. Although there is less microbial than leaf biomass in leaf–microbe complexes, micro-organisms are assimilated with a greater efficiency than the plant tissue itself. For example, although *Gammarus pseudolimnaeus* assimilates only 18% of elm and maple leaves, 67–93% of the energy in the fungi colonizing the leaf material is assimilated (Bärlocher & Kendrick 1975). Similarly, *Hyalella azteca* utilizes 60–90% of the bacterial biomass it ingests but assimilates only 5% of elm leaves (Hargrave 1970).

There is evidence, albeit equivocal, in support of the importance of microbes, especially fungi, both as modifiers of leaf material and as a food source. In an attempt to tease out whether food preferences exhibited by trichopteran larvae were due to the fungal biomass *per se* or due to the enzymatic action of the fungi on the leaves, Arsuffi and Suberkropp (1988) altered these two parameters by conditioning leaves in the presence or absence of glucose. The addition of glucose suppresses the production of extracellular polysaccharidases. Leaves that had been conditioned in the presence of glucose had a thick mycelial mat over their surface but they were not softened. In feeding trials, leaves conditioned in the presence of glucose were preferred to leaves conditioned without added glucose. This, and the observation that larvae feeding on the glucose-treated leaves fed almost exclusively on the fungal mat, led Arsuffi and Suberkropp (1988) to conclude that *Psychoglypha* larvae preferred fungal mycelium over the combination of fungal biomass and enzyme-modified leaf tissue.

There is no reason to assume that the relative importance of microbes as food will be the same for all detritivores. Evidence for the importance of microbes as modifiers of leaf material includes the observation that treating leaves with an extract of fungal enzymes increases their palatability to *Gammarus pseudolimnaeus* (Bärlocher & Kendrick 1973a, 1975) and, when offered a choice of conditioned leaves or pure cultures of *Tetrachaetum elegans*, *Gammarus minus* preferred conditioned leaves (Kostalos & Seymour 1976).

15.5 ENZYMATIC CAPABILITIES OF DETRITIVORES

As leaf material can be made more palatable by treatment with either fungal enzymes or hydrolysing agents (Bärlocher & Kendrick 1973a), it has been suggested that the importance of micro-organisms to detritivores lies in their ability to break down leaf material into subunits that the detritivores can utilize. If this is true, the degree to which any specific detritivore relies on microbial conditioning will be dependent upon its own enzymatic capabilities.

The major component of autumn-shed leaves is cell wall material consisting of cellulose and pectin. There are three possible sources of cellulose-degrading enzymes associated with detritivores: (i) tissue-level synthesis; (ii) microbial symbionts; and (iii) 'acquired' enzymes.

All aquatic detritivores examined to date can degrade cellobiose, the basic subunit of cellulose; some are able to digest soluble cellulose (carboxymethyl cellulose, CMC); but only a few show enzymatic activity towards native crystalline cellulose (Bjarnov 1972; Monk 1976, 1977). Tissue-level synthesis of cellulases has been recorded for molluscs, annelids, crustaceans and insects (Calow & Calow 1975; Monk 1976; Schoenberg *et al* 1984; Chamier & Willoughby 1986; Chamier 1991), although some of the evidence is confusing, especially for members of the genus *Gammarus*. Whereas, Chamier and Willoughby (1986) stated that *Gammarus pulex* produces cellulases, chitinases and β-1,3-glucanase, Bärlocher and Porter (1986) suggested that similar enzymes, found in the gut of *Gammarus tigrinus*, arise from ingested fungi. For a third species, *Gammarus fossarum*, Bärlocher (1983) states that the release of reducing substances from microbially conditioned leaves was due to a combination of microbial and gut enzymes. Moreover, if freshly shed leaves were pretreated with 70% alcohol to remove soluble phenols, reducing sugars could be released by gut extracts of this species (Bärlocher 1982). Although it is certainly not beyond the bounds of possibility that members of a single genus differ in their enzymatic capabilities, another possible reason for the different results could be related to the

methods used in the different studies. Chamier and Willoughby (1986) found it was necessary to incubate material at 37°C for 44 hours in order to detect enzymatic activity; no activity could be accurately measured at lower temperatures or over shorter time intervals. As the normal gut passage time of *Gammarus* at field temperatures (<20°C) is less than 2 hours, the ecological relevance of these results is questionable. Although the other studies (Bärlocher 1982; Bärlocher & Porter 1986) also used incubation times in excess of 2 hours (5 h and 14 h respectively), they did use realistic incubation temperatures. Martin *et al* (1981a), working on caddisfly larvae, could also only demonstrate the presence of enzymes active against CMC at elevated temperatures (i.e. 37°C).

Microbial symbiosis is thought to be important in the digestive physiology of the dipteran *Tipula abdominalis* (Klug & Kotarski 1980; Sinsabaugh *et al* 1985; Lawson & Klug 1989) whereas a range of other detritivores (Plectoptera, Trichoptera, Amphipoda) may utilize 'acquired' microbial enzymes (Martin *et al* 1981b; Bärlocher 1982; Sinsabaugh *et al* 1985; Bärlocher & Porter 1986).

Bärlocher (1982) has suggested that the digestive physiology of detritivores is related to their feeding behaviour and mobility. Shredders that specialize in digesting optimally conditioned leaf material (rich in microbes, high in enzymatic activity) must either be highly mobile or have access to other sources of food (e.g. *Gammarus* sp.). Relatively immobile species that are obligate detritivores must possess the digestive ability to break down the refractory material itself (e.g. *Tipula abdominalis*).

15.6 RELATIVE IMPORTANCE OF MICROBES AND INVERTEBRATES IN DETRITUS PROCESSING

Several studies have attempted to assess the relative importance of microbial and invertebrate activities in the processing of particulate organic material. As with other aspects of detritus processing, the majority of studies have concentrated on the breakdown of CPOM, in particular leaves. The results from these investigations are equivocal; whereas some conclude that invertebrates are not important in leaf breakdown (e.g. Mathews & Kowalczewski 1969; Kaushik & Hynes 1971; Reice 1978; Paul *et al* 1983; Leff & McArthur 1989), others have concluded that they are important (e.g. Petersen & Cummins 1974; Hart & Howmiller 1975; Iversen 1975; Minshall *et al* 1982; Kirby *et al* 1983; Oberndorfer *et al* 1984; Imbert & Pozo 1989).

Such differences may in part be due to the techniques and study sites used. Judgements as to the importance of invertebrates in detritus processing have been based on of two types of study: either (a) manipulation of the intensity of invertebrate feeding (Herbst 1982; Newman *et al* 1987; Newman 1990); or (b) correlation between rate of decomposition and invertebrate density (Reice 1978; Minshall *et al* 1982; Kirby *et al* 1983; Leff & McArthur 1989). Some of these correlative studies have related decomposition rates to total invertebrate density, rather than just shredder density (e.g. Leff & McArthur 1989). Such an approach may confound any effect shredders may be having on decomposition rates with the attractiveness of leaf packs as shelters or as traps for FPOM.

Invertebrate feeding has been manipulated by either manipulating predation pressure (e.g. Oberndorfer *et al* 1984) or by manipulating densities of shredders in enclosures (e.g. Newman 1990) or within artifical streams (e.g. Petersen & Cummins 1974). For example, Newman (1990) manipulated the density of shredders (*Gammarus pseudolimnaeus*) in mesh bags containing watercress (*Nasturtium officinale*). The rate of breakdown was positively correlated with initial shredder density although the relationship was less pronounced in the coarser mesh bags, probably due to loss of particles in the size range from 1 mm to 200 μm (Fig. 15.7).

The most frequently employed approach is to place leaves in enclosures of different mesh sizes, thereby restricting access to invertebrates (Mathews & Kowalczewski 1969; Kaushik & Hynes 1971; Hart & Howmiller 1975; Iversen 1975; Cummins *et al* 1980; Imbert & Pozo 1989). Placing leaves in mesh bags not only restricts invertebrate processing but it may also inhibit microbial processing by reducing the flow of water through the bag, resulting in the establishment of

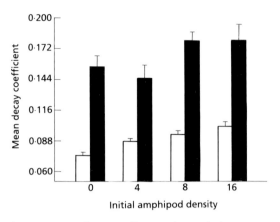

Fig. 15.7 Mean decay coefficients (+2 SE) of *Nasturtium officinale* in mesh bags stocked with *Gammarus pseudolimnaeus*. □, 200 μm-mesh; ■, 1 mm-mesh (after Newman 1990).

hypoxic or even anoxic conditions. Enclosing leaves in fine mesh bags will also influence loss rates due to fragmentation. In an attempt to distinguish between leaf weight loss due to invertebrate feeding or to fragmentation, Stewart and Davies (1989) devised a mesh bag, the body of which consisted of fine mesh (180 μm) and the top of coarse mesh (5 mm). This 'composite' mesh bag would therefore allow macroinvertebrates access but would retain fragments larger than 180 μm. The use of 'composite' mesh bags in combination with fine and coarse bags allows a distinction to be made between fragmentation and invertebrate feeding. In studies using *Cunonia capensis* and *Ilex mitis*, most of the breakdown of *Cunonia* was due to invertebrate feeding (no difference between composite and coarse mesh bags), whereas both invertebrate feeding and fragmentation was important for *Ilex* (Fig. 15.8).

Cummins and colleagues (1980) concluded that, whereas decomposition of leaf packs (groups of leaves loosely attached an anchor, e.g. house brick) was representative of the breakdown of leaves in riffle areas, decomposition of leaves in mesh bags was more similar to the slower processing rates of leaves in more stagnant pool areas. In contrast, Mutch *et al* (1983) concluded that the processing rate of willow leaves was always greater for leaves in mesh bags than in leaf

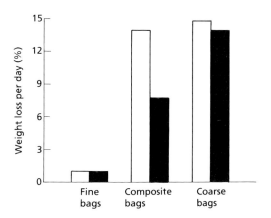

Fig. 15.8 Weight loss of *Cunonia capensis* (■) and *Ilex mitis* (□) leaf material from fine (180 μm), coarse (5 mm) or composite (180 μm and 5 mm) mesh bags (after Stewart & Davies 1989).

packs, irrespective of whether they were in riffles or pools.

The relative importance of macroinvertebrates in the breakdown of leaf litter varies between sites. Mathews and Kowalczewski (1969) concluded that macroinvertebrates were not important in litter breakdown in the River Thames (UK) and Reice (1978) could find no direct relationship between the number of individuals or species of macroinvertebrates in leaf packs and their rate of decomposition. However, both these studies were performed at sites where shredders were poorly presented in the natural community.

Therefore, although detritivores are known to feed selectively on conditioned leaf material, and some studies have calculated that 20–30% of litter processing is due to the activity of shredders (Petersen & Cummins 1974), results from field studies are far from conclusive.

15.7 EFFECT OF WATER QUALITY ON DECOMPOSITION

Many physicochemical factors have been shown to influence decomposition rates, including temperature, calcium, nitrogen, phosphorus and dissolved oxygen (Egglishaw 1968, 1972; Iversen 1975; Howarth & Fisher 1976; Reice 1977; Federle & Vestal 1980; Cummins *et al* 1983;

Burton *et al* 1985). Recently, several studies have been designed to investigate the effect of anthropogenic changes in water quality on processing rates. These include studies into the effects of pH (Traaen 1980; Hildrew *et al* 1984; Burton *et al* 1985; Mackay & Kersey 1985; Chamier 1987; Mulholland *et al* 1987), chlorine (Newman *et al* 1987), acid mine drainage (Carpenter *et al* 1983; Gray & Ward 1983), heavy metals (Giesy 1978; Leland & Carter 1985), coal ash effluent (Forbes & Magnuson 1980) and pesticides (Wallace *et al* 1982; Fairchild *et al* 1983; Cuffney *et al* 1984, 1990). All these pollutants cause a reduction in the rate of processing of CPOM. Such a reduction may be due to either low microbial activity or changes in detritivore abundance and/or feeding behaviour.

Insecticides

Insecticides have been used as a means of quantifying the role of benthic invertebrates in decomposition (Wallace *et al* 1982; Cuffney *et al* 1984, 1990). The insecticide methoxychlor was applied to a headwater stream seasonally for 2 years. It caused a large increase in invertebrate drift and a reduction in leaf litter processing rates. Although shredder density was reduced after application of the insecticide, microbial respiration was not affected (Cuffney *et al* 1990). This study provided strong evidence for the importance of invertebrates over microbes in the processing of leaf litter in such streams.

Acidity

The inhibition of decomposition by low pH appears to occur at around pH 5 (Schindler 1980; Traaen 1980; but see McGeorge *et al* 1991). For example, Burton *et al* (1985) observed a reduction in leaf decomposition rate in acidified artificial streams (pH 4–4.3) compared with control streams (pH 7.2–7.5). However, this reduction was significant only after 6–7 months' exposure (Fig. 15.9). The effect of low pH on fungal communities is equivocal. Whereas, McKinley and Vestal (1982) recorded a decrease in fungi with increasing acidity such that they were virtually absent at pH 4, experimental acidification to pH

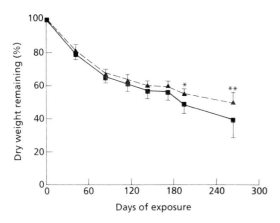

Fig. 15.9 Effect of acidification on the decomposition of *Acer sacchanum* leaves in artificial streams. Error bars represent 95% CL. ▲ − − −▲, pH 4.0–4.3; ■——■, pH 7.2−7.5 (after Burton *et al* 1985).

4.3 resulted in an increase in fungal colonization of leaf litter (van Frankenhuyzen *et al* 1985). Studies on aquatic hyphomycetes have found that their decompositional abilities are strongly reduced at pH 4 (Thompson & Bärlocher 1989), although they grow best on solid media at pH 4–5 and in liquid medium at pH 5−6 (Rosset & Bärlocher 1985).

Field studies into the effect of acidity on decomposition have correlated reduction in decomposition with reduction in microbial activity (Chamier 1987; Mulholland *et al* 1987; Palumbo *et al* 1987). As increasing acidity reduces microbial activity, and given that microbes are important to the nutrition of detritivores, either as modifiers of organic material or as a food source (Section 15.4), it may be predicted that the density of shredders would also be reduced in acid streams. However, several workers have observed an increased abundance of shredders in acid streams (Hildrew *et al* 1984; Mackay & Kersey 1985; Mulholland *et al* 1987) even though acidity does reduce food quality (Groom & Hildrew 1989). Particulate organic matter is a finite resource for which microbes and detritivores compete (Bärlocher 1980). The conditioning of organic matter by microbes has costs as well as benefits for detritivores. Although microbes increase the nutritional quality of the substrate, if they are too abundant or active they may considerably reduce the period over which it is available. A stress-induced reduction in microbial processing rate may therefore be advantageous to shredders by extending or increasing the availability of CPOM.

Heavy metals

Studies of the effect of metals on decomposition rates have been concerned either with individual metals (Giesy 1978; Leland & Carter 1985) or with acid mine drainage which contains heavy metals such as iron and zinc (Carpenter *et al* 1983; Gray & Ward 1983; Maltby & Booth 1991). Leland and Carter (1985) observed an inhibition in the rate of leaf litter decomposition at 2.5 μg copper l^{-1} and Giesy (1978) recorded a reduction in microbial colonization and leaf decomposition at 5 μg cadmium l^{-1}. Acid mine drainage also reduces decomposition rates and microbial activity (Carpenter *et al* 1983; Gray & Ward 1983); the fungal group most affected being the aquatic hyphomycetes (Maltby & Booth 1991). Downstream of the input of acid mine effluents, iron precipitates out as ferric hydroxide. The reduction in fungal abundance could therefore be due either to a direct toxic effect of the metals or to a physical change in the leaf surface which either smothers the fungi or prevents their spores from germinating.

Heavy metals reduce both the rate of growth and spore production of aquatic hyphomycetes, sporulation being the most sensitive parameter (Abel & Bärlocher 1984; S. Bermingham, unpublished data). The sensitivity of fungi to heavy metals varies between species; species of aquatic hyphomycetes found downstream of a mine effluent appear to be less sensitive to metals than those only found upstream (S. Bermingham, unpublished data).

Micro-organisms colonizing detritus accumulate toxins, including metals, which are then transferred along the food chain causing a reduction in the viability of detritivores and the organisms predating upon them (Patrick & Loutit 1976, 1978; Duddridge & Wainwright 1980; Pinkney *et al* 1985; Abel & Bärlocher 1988).

Chlorine

The effect of chlorine on the decomposition of *Potamogeton crispus* was investigated by Newman *et al* (1987) in the artificial stream system at Monticello (US EPA). Sites within streams were dosed with 10, 75 or 250 μg chlorine l^{-1} and weight loss of material from fine (210 μm) and coarse (1 mm) mesh bags was recorded after 4 and 11 days of exposure. Upstream sections of streams acted as controls. Microbial colonization and processing was reduced in both the 75-μg l^{-1} and 250-μg l^{-1} treatments during the first 4 days of exposure (Fig. 15.10). By 11 days there was no significant difference in either the rate of decomposition, bacterial density or microbial respiration rate between fine bags in either the control or dosed sites. There was, however, still a difference in the decomposition rates of leaves in coarse mesh bags, suggesting that feeding by macroinvertebrates was still inhibited. Chlorine appears to influence microbial decomposition by reducing colonization rates but not equilibrium densities. Reduced bacterial densities have been recorded below some chlorination sites (Maki *et al* 1986).

15.8 EFFECT OF CHANGES IN RIPARIAN VEGETATION ON DETRITUS PROCESSING

As stated in section 15.2, the riparian vegetation influences both the quantity and quality of allochthonous detritus entering streams and rivers. Given the high retention capacity of wooded streams and the distinct food preferences of detritivores, there is likely to be a strong link between the diversity and abundance of detritivores in streams and the riparian vegetation. In addition, the life cycle of detritivores, especially shredders, would be expected to be synchronized with the seasonal inputs of material. Therefore, changing the riparian vegetation may have considerable effects on both the structure and functioning of the adjacent aquatic community.

The structure or abundance of riparian vegetation may be altered as a result of many human activities. Stout and Coburn (1989) investigated the impact of road construction on leaf processing in a North American stream system. Two years after the construction work had been completed, leaf processing rates were still reduced downstream of the road compared with upstream reference sites. The authors concluded that the reduction in processing rates was due to a decrease in shredder abundance caused by the removal of riparian vegetation during road construction.

The principal human activity influencing riparian vegetation is forestry. Forestry practices not only alter the abundance and composition of the riparian vegetation but also affect water quality (Benfield *et al* 1991; Holopainan *et al* 1991). For example, streams are subjected to increased loadings of suspended solids, resulting in increased turbidity and sedimentation, as well as inputs of chemicals such as fertilizers and pesticides (Campbell & Doeg 1989).

Cummins *et al* (1989) have developed a model which predicts seasonal changes in shredder biomass based on information on the composition and percentage cover of riparian vegetation. These authors have suggested that the degree to which observations of shredder populations deviate from those predicted by the model could be used as a means of evaluating the impact of changes in riparian vegetation on stream invertebrates.

15.9 SUMMARY

When detritus enters a water body, soluble compounds are leached out, leaving the more recalcitrant structural molecules behind. The rate at which detritus is processed is therefore dependent upon the rate at which these structural polymers are assimilated into living biomass. The two groups of organisms involved in the processing of detritus are micro-organisms (bacteria and fungi) and detritivores (principally macroinvertebrates).

The relative importance of detritivores and micro-organisms in detritus processing is a matter for continuing debate. There is evidence both for the importance of detritivores over micro-organisms and *vice versa*. One possible reason for the differing results could be related to the sites used. Those studies that have concluded that detritivores are relatively unimportant in detritus

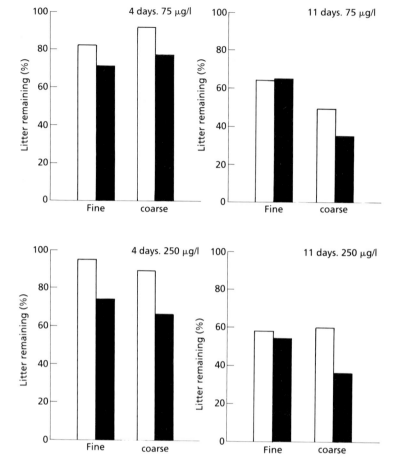

Fig. 15.10 Effect of chlorine on the decomposition of *Pomatogeton crispus* in coarse (1 mm) and fine (210 µm) mesh bags. □, Control sites; ■, dosed sites (after Newman *et al* 1987).

processing have tended to be conducted at sites where accumulations of CPOM and/or shredders are rare and therefore the invertebrate community would be ill adapted to cope with artificial inputs of CPOM.

Another controversial area concerns the interaction between micro-organisms and detritivores. Many studies have shown that detritivores exhibit clear preferences for microbially conditioned detritus; however, what is not clear, is whether microbes are important to detritivores as modifiers of detritus or as a food source. Evidence cited in favour of micro-organisms as modifiers of detritus include: low biomass of micro-organisms relative to detritus; fast gut passage times; and results from radiolabelling studies. Evidence cited in favour of micro-organisms as a food source

comes primarily from laboratory studies of the performance (growth, reproduction) of detritivores on different diets. It is likely that the relative importance of micro-organisms to the nutrition of detritivores is dependent upon both the species of detritivore studied and the types of micro-organisms used in the experiments.

Anthropogenic activities that affect detritus processing include those resulting in changes in water quality and/or riparian vegetation. The impact of pesticides, acidity and heavy metals on detritus processing has been investigated. In all cases processing rates were decreased, although whether such decreases were due to low microbial activity or changes in detritivore abundance and/or feeding behaviour is open to debate.

The majority of investigations into detritus

processing in freshwaters have concentrated on northern temperate, upland wooded streams where the main source of detritus is allochthonous CPOM, primarily autumn shed leaves and, to a lesser extent, woody debris. Most information is therefore available on the processing of leaf litter. Fungi are the major micro-organisms colonizing leaf material during the early stages of decomposition. Bacteria are more important during the latter stages of decomposition and may be important in the processing of highly recalcitrant leaf material. Detritivores feeding on leaf litter are referred to as shredders. Their activity results in fragmentation of the leaf litter, which, together with their faecal material, produces FPOM which can be utilized by other members of the detrital food web. FPOM is colonized by bacteria and the macroinvertebrates feeding upon it (i.e. collectors) either filter it from the water column or gather it from the river bed.

A third category of organic matter in freshwaters is DOM, which may be either detrital or non-detrital in origin (Miller 1987). Most studies of DOM utilization have concentrated on its incorporation into organic layers on stones or its flocculation into FPOM. Another possible pathway is the incorporation of DOM into suspended bacteria which are then grazed by protozoans. This so-called 'microbial-loop', important in the trophic dynamics of standing-water systems (marine and freshwater), has been virtually ignored in studies of running-water systems.

REFERENCES

Abel TH, Bärlocher F. (1984) Effects of cadmium on aquatic hyphomycetes *Applied and Environmental Microbiology* **48**: 245–51. [15.7]

Abel T, Bärlocher F. (1988) Uptake of cadmium by *Gammarus fossarum* (Amphipoda) from food and water. *Journal of Applied Ecology* **24**: 223–31. [15.7].

Anderson NH, Cargill AS. (1987) Nutritional ecology of aquatic detritivorous insects. In: Slansky F, Rodriguez JR (eds), *Nutritional Ecology of Insects, Mites, Spiders and Related Invertebrates*, pp 903–25. John Wiley & Sons, New York. [15.2, 15.3]

Anderson NH, Cummins KW. (1979) Influences of diet on the life-histories of aquatic insects. *Journal of the Fisheries Research Board of Canada* **36**: 335–42. [15.2]

Anderson NH, Grafius E. (1975) Utilization and pro-

cessing of allochthonous material by stream Trichoptera. *Verhandlungen Internationale Vereinigung für Theoretische und Angewandt Limnologie* **19**: 3083–8. [15.4]

Anderson NH, Sedell JR. (1979) Detritus processing by macroinvertebrates in stream ecosystems. *Annual Review of Entomology* **24**: 351–77. [15.2, 15.3]

Anderson NH, Steedman RJ, Dudley T. (1984) Patterns of exploitation by stream invertebrates of wood debris. *Verhandlungen Internationale Vereinigung für Theoretische und Angewandt Limnologie* **22**: 1847–52. [15.3]

Armstrong SM, Bärlocher F. (1989) Adsorption and release of amino acids from epilithic biofilms in streams. *Freshwater Biology* **22**: 153–9. [15.2]

Arsuffi TL, Suberkropp K. (1984) Leaf processing capabilities of aquatic hyphomycetes: interspecific differences and influence on shredder feeding preferences. *Oikos* **42**: 144–54. [15.4]

Arsuffi TL, Suberkropp K. (1985) Selective feeding by stream caddisfly (Trichoptera) detritivores on leaves with fungal-colonized patches. *Oikos* **45**: 50–8. [15.4].

Arsuffi TL, Suberkropp K. (1986) Growth of two stream caddisflies (Trichoptera) on leaves colonized by different fungal species. *Journal of the North American Benthological Society* **5**: 297–305. [15.4]

Arsuffi TL, Suberkropp K. (1988) Effects of fungal mycelia and enzymatically degraded leaves on feeding and performance of caddisfly (Trichoptera) larvae. *Journal of the North American Benthological Society* **7**: 205–11. [15.4]

Arsuffi TL, Suberkropp K. (1989) Selective feeding by shredders on leaf-colonizing stream fungi: comparison of macroinvertebrate taxa. *Oecologia* **79**: 30–7. [15.4]

Azam F, Cho BC. (1987) Bacterial utilization of organic matter in the sea. In: Fletcher M, Gray TRG, Jones JG (eds) *Ecology of Microbial Communities*, pp 261–81. Cambridge University Press, Cambridge. [15.2]

Azam F, Cho BC, Smith DC, Simon M. (1990) Bacterial cycling of matter in the pelagic zone of aquatic ecosystems. In: Tilzer MM, Serruya C (eds) *Large Lakes, Ecological Structure and Function*, pp 477–88. Springer-Verlag, Berlin. [15.2]

Baker JH, Bradnam LA. (1976) The role of bacteria in the nutrition of aquatic detritivores. *Oecologia* **24**: 95–104. [15.4]

Bärlocher F. (1980) Leaf-eating invertebrates as competitors of aquatic hyphomycetes. *Oecologia* **47**: 303–6. [15.7]

Bärlocher F. (1982) The contribution of fungal enzymes to the digestion of leaves by *Gammarus fossarum* Koch (Amphipoda). *Oecologia* **52**: 1–4. [15.4, 15.5]

Bärlocher F. (1983) Seasonal variation of standing crop and digestibility of CPOM in a Swiss Jura stream. *Ecology* **64**: 1266–1272. [15.5]

Bärlocher F. (1985) The role of fungi in the nutrition

of stream invertebrates. *Botanical Journal of the Linnean Society* **91**: 83–94. [15.4]

Bärlocher F. (1991) Fungal colonization of fresh and dried leaves in the River Teign (Devon, England). *Nova Hedwigia* **52**: 349–357. [15.4]

Bärlocher F, Kendrick B. (1973a) Fungi and food preferences of *Gammarus pseudolimnaeus*. *Archiv für Hydrobiologie* **72**: 501–16. [15.4, 15.5]

Bärlocher F, Kendrick B. (1973b) Fungi in the diet of *Gammarus pseudolimnaeus*. *Oikos* **24**: 295–300. [15.4]

Bärlocher F, Kendrick B. (1974) Dynamics of the fungal populations on leaves in a stream. *Journal of Ecology* **62**: 761–91. [15.4]

Bärlocher F, Kendrick B. (1975) Assimilation efficiency of *Gammarus pseudolimnaeus* (Amphipoda) feeding on fungal mycelium or autumn-shed leaves. *Oikos* **26**: 55–9. [15.4]

Bärlocher F, Kendrick B. (1981) Role of aquatic hyphomycetes in the trophic structure of streams. In: Wicklow DT, Carroll GC (eds) *The Fungal Community. Its Organization and Role in the Ecosystem*, pp 743–60. Marcel Dekker, New York. [15.4]

Bärlocher F, Murdoch JH. (1989) Hyporheic biofilms — a potential food source for interstitial animals. *Hydrobiologia* **184**: 61–7. [15.2]

Bärlocher F, Oertli JJ. (1978a) Colonization of conifer needles by aquatic hyphomycetes. *Canadian Journal of Botany* **56**: 57–62. [15.2]

Bärlocher F, Oertli JJ. (1978b) Inhibitors of aquatic hyphomycetes in dead conifer needles. *Mycologia* **70**: 964–74. [15.2]

Bärlocher F, Porter CW. (1986) Digestive enzymes and feeding strategies of three stream invertebrates. *Journal of the North American Benthological Society* **5**: 58–66. [15.5]

Bassett A, Rossi L. (1987) Relationship between trophic niche breadth and reproductive capabilities in a population of *Proasellus coxalis*. *Functional Ecology* **1**: 13–18. [15.4]

Benfield EF, Webster JR, Golladay SW, Peters GT, Stout BM. (1991) Effects of forest disturbance on leaf breakdown in southern Appalachian streams. *Verhandlungen Internationale Vereinigung für Theoretische und Angewandt Limnologie* **24**: 1687–1690. [15.8]

Bengtsson G. (1982) Patterns of amino acid utilization by aquatic hyphomycetes. *Oecologia* **55**: 355–63. [15.2]

Berman T. (1990) Microbial food-webs and nutrient cycling in lakes: changing perspectives. In: Tilzer MM, Serruya C (eds) *Large Lakes, Ecological Structure and Function*, pp 511–25. Springer-Verlag, Berlin. [15.2]

Bilby RE, Ward JW. (1989) Changes in characteristics and function of woody debris with increasing size of streams in Western Washington. *Transactions of the American Fisheries Society* **118**: 368–78. [15.2]

Bjarnov N. (1972) Carbohydrases in *Chironomus*, *Gammarus* and some Trichoptera larvae. *Oikos* **23**: 261–3. [15.5]

Boling RH, Goodman ED, van Sickle JA, Zimmer JO, Cummins KW, Petersen RC, Reice SR. (1975) Toward a model of detritus processing in a woodland stream. *Ecology* **56**: 141–51. [15.2, 15.4]

Bott TL, Kaplan LA. (1990) Potential for protozoan grazing of bacteria in streambed sediments. *Journal of the North American Benthological Society* **9**: 336–45. [15.2]

Bott TL, Ritter FP. (1981) Benthic algal production in a piedmont stream measured by ^{14}C and dissolved oxygen procedures. *Journal of Freshwater Ecology* **1**: 267–78. [15.2]

Bowen SH. (1984) Evidence of a detritus food chain based on consumption of organic precipitates. *Bulletin of Marine Science* **35**: 440–8. [15.2]

Bueler CM. (1984) Feeding preferences of *Pteronarcys pictetii* (Plecoptera: Insecta) from a small, acidic woodland stream. *Florida Entomologist* **67**: 393–401. [15.4]

Burton TM, Stanford RM, Allan JW. (1985) Acidification effects on stream biota and organic matter processing. *Canadian Journal of Fisheries and Aquatic Sciences* **42**: 669–75. [15.7]

Calow P. (1974) Evidence for bacterial feeding in *Planorbis contortus* Linn. (Gastropoda: Pulmonata). *Proceedings of the Malacological Society of London* **41**: 145–56. [15.4]

Calow P, Calow LJ. (1975) Cellulase activity and niche separation in freshwater gastropods. *Nature* **255**: 478–80. [15.5]

Campbell IC, Doeg TJ. (1989) Impact of timber harvesting and production on streams: a review. *Australian Journal of Marine and Freshwater Research* **40**: 519–39. [15.8]

Cargill AS, Cummins KW, Hanson BJ, Lowry RR. (1985) The role of lipids, fungi and temperature in the nutrition of a shredder caddisfly, *Clistoronia magnifica*. *Freshwater Invertebrate Biology* **4**: 64–78. [15.4]

Carlough LA, Meyer JL. (1989) Protozoans in two southeastern blackwater rivers and their importance to trophic transfer. *Limnology and Oceanography* **34**: 163–77. [15.2]

Carlough LA, Meyer JL. (1990) Rates of protozoan bactivory in three habitats of a southeastern blackwater river. *Journal of the North American Benthological Society* **9**: 45–53. [15.2]

Carpenter J, Odum WE, Mills A. (1983) Leaf litter decomposition in a reservoir affected by acid mine drainage. *Oikos* **41**: 165–72. [15.7]

Chamier A-C. (1987) Effect of pH on microbial degradation of leaf litter in seven streams of the English Lake District. *Oecologia* **71**: 491–500. [15.3, 15.7]

Chamier A-C. (1991) Cellulose digestion and metabolism in the freshwater amphipod *Gammarus pseudo-limnaeus* Bousfield. *Freshwater Biology* **25**: 33–40. [15.5]

Chamier A-C, Willoughby LG. (1986) The role of fungi in the diet of the amphipod *Gammarus pulex* (L): an enzymatic study. *Freshwater Biology* **16**: 197–208. [15.5]

Christensen B. (1977) Habitat preference among amylase genotypes in *Asellus aquaticus* (Isopoda, Crustacea). *Hereditas* **87**: 21–6. [15.4]

Conners ME, Naiman RJ (1984) Particulate alloch-thonous inputs: relationships with stream size in an undisturbed watershed. *Canadian Journal of Fisheries and Aquatic Sciences* **41**: 1473–84. [15.2]

Cuffney TF. (1988) Input, movement and exchange of organic matter within a subtropical coastal black-water river-floodplain system. *Freswater Biology* **19**: 305–20. [15.2]

Cuffney TF, Wallace JB, Lugthart GJ. (1990) Experimental evidence quantifying the role of benthic invertebrates in organic matter dynamics of headwater streams. *Freshwater Biology* **23**: 281–99. [15.7]

Cuffney TF, Wallace JB, Webster JR. (1984) Pesticide manipulation of a headwater stream: invertebrate responses and their significance for ecosystem processes. *Freshwater Invertebrate Biology* **3**: 153–71. [15.7]

Cummins KW. (1974) Structure and function of stream ecosystems. *BioScience* **24**: 631–41. [15.2, 15.3]

Cummins KW, Klug MJ. (1979) Feeding ecology of stream invertebrates. *Annual Review of Ecology and Systematics* **10**: 147–72. [15.2, 15.4]

Cummins KW, Spengler GL, Ward GM, Speaker RM, Ovink RM, Mahan DC, Mattingly RL. (1980) Processing of confined and naturally entrained leaf litter in a woodland stream ecosystem. *Limnology and Oceanography* **25**: 952–7. [15.6].

Cummins KW, Sedell JR, Swanson FJ, Minshall GW, Fisher SG, Cushing CE *et al.* (1983) Organic matter budgets for stream ecosystems: problems in their evaluation. In: Barnes JR, Minshall GW (eds) *Stream Ecology: Application and Testing of General Ecological Theory*, pp 299–353. Plenum Press, New York. [15.2, 15.7]

Cummins KW, Wilzbach MA, Gates DM, Perry JB, Taliaferro WB. (1989) Shredders and riparian vegetation. *BioScience* **39**: 24–30. [15.3, 15.8]

Davis SF, Winterbourn MJ. (1977) Breakdown and colonization of *Nothofagus* leaves in a New Zealand stream. *Oikos* **28**: 250–5. [15.3]

Duddridge JE, Wainwright M. (1980) Heavy metal accumulation by aquatic fungi and reduction in viability of *Gammarus pulex* fed Cd^{2+} contaminated mycelium. *Water Research* **14**: 1605–11. [15.7]

Edwards RT, Meyer JL. (1986) Production and turnover of planktonic bacteria in two southeastern blackwater rivers. *Applied and Environmental Microbiology* **52**: 1317–23. [15.2]

Edwards RT, Meyer JL. (1987) Bacteria as a food source for black fly larvae in a blackwater river. *Journal of the North American Benthological Society* **6**: 241–50. [15.2, 15.4]

Edwards RT, Meyer JL. (1990) Bactivory by deposit-feeding mayfly larvae (*Stenonema* spp.). *Freshwater Biology* **24**: 453–62. [15.2]

Egglishaw HJ. (1968) The quantitative relationship between bottom fauna and plant detritus in streams of different calcium concentrations. *Journal of Applied Ecology* **5**: 731–40. [15.7]

Egglishaw HJ. (1972) An experimental study of the breakdown of cellulose in fast-flowing streams. *Memorie Istituto Italiano di Idrobiologia* **29** (Suppl.): 405–28. [15.7]

Fairchild JF, Boyle TP, Robinson-Wilson E, Jones JR. (1983) Microbial action in detritus leaf processing and the effects of chemical perturbation. In: Fontaine TD, Bartell SM (eds) *Dynamics of Lotic Ecosystems*, pp 437–56. Ann Arbor Science, Michigan. [15.7]

Fano EA, Rossi L, Basset A. (1982) Fungi in the diet of three benthic invertebrate species. *Bollettino di Zoologia* **49**: 99–105. [15.4]

Federle TW, Vestal JR. (1980) Lignocellulose mineralization by arctic lake sediments in response to nutrient manipulation. *Applied and Environmental Microbiology* **40**: 32–9. [15.7]

Findlay SEG, Arsuffi TL. (1989) Microbial growth and detritus transformations during decomposition of leaf litter in a stream. *Freshwater Biology* **21**: 261–9. [15.2, 15.4]

Findlay S, Tenore K. (1982) Nitrogen source for a detritivore: detritus substrate versus associated microbes. *Science* **218**: 371–3. [15.4]

Findlay S, Meyer JL, Smith PJ. (1984) Significance of bacterial biomass in the nutrition of a freshwater isopod (*Lirceus* sp.). *Oecologia* **63**: 38–42. [15.4]

Findlay S, Meyer JL, Smith PJ. (1986a) Contribution of fungal biomass to the diet of a freshwater isopod (*Lirceus* sp.). *Freshwater Biology* **16**: 377–85. [15.4]

Findlay S, Meyer JL, Smith PJ. (1986b) Incorporation of microbial biomass by *Peltoperla* sp. (Plecoptera) and *Tipula* sp. (Diptera). *Journal of the North American Benthological Society* **5**: 306–10. [15.4]

Forbes AM, Magnuson JJ. (1980) Decomposition and microbial colonization of leaves in a stream modified by coal ash effluent. *Hydrobiologia* **76**: 263–7. [15.7]

Fredeen FJH. (1964) Bacteria as food for blackfly larvae (Diptera: Simuliidae) in laboratory cultures and in natural streams. *Canadian Journal of Zoology* **42**: 527–48. [15.4]

Fuller RL, Mackay RJ. (1981) Effects of food quality on the growth of three *Hydropsyche* species (Trichoptera:

Hydropsychidae). *Canadian Journal of Zoology* **59**: 1133–40. [15.3]

Gessner MO, Schwoerbel J. (1989) Leaching kinetics of fresh leaf-litter with implications for the current concept of leaf-processing in streams. *Archiv für Hydrobiologie* **115**: 81–90. [15.4]

Giesy JP. (1978) Cadmium inhibition of leaf decomposition in an aquatic microcosm. *Chemosphere* **6**: 467–75. [15.7]

Godfrey BES. (1983) Growth of two terrestrial microfungi on submerged alder leaves. *Transactions of the British Mycological Society* **81**: 418–21. [15.4]

Golladay SW, Webster JR, Benfield EF. (1983) Factors affecting food utilization by a leaf shredding aquatic insect: leaf species and conditioning time. *Holarctic Ecology* **6**: 157–62. [15.4]

Graça MAS. (1990) *Observations on the feeding biology of two stream-dwelling detritivores*: Gammarus pulex *(L)* and Asellus aquaticus *(L)*. Unpublished PhD thesis, University of Sheffield, Sheffield. [15.4]

Gray LJ, Ward JV. (1983) Leaf litter in streams receiving treated and untreated metal mine drainage. *Environment International* **9**: 135–8. [15.7]

Groom AP, Hildrew AG. (1989) Food quality for detritivores in streams of contrasting pH. *Journal of Animal Ecology* **58**: 863–81. [15.7]

Haack TK, McFeters GA. (1982) Nutritional relationships among microorganisms in an epilithic biofilm community. *Microbial Ecology* **8**: 115–26. [15.2]

Hanson BJ, Cummins KW, Cargill AS, Lowry RR. (1983) Dietary effects on lipid and fatty acid composition of *Clistoronia magnifica* (Trichoptera: Limniphilidae). *Freshwater Invertebrate Biology* **2**: 2–15. [15.4]

Hargrave BT. (1970) The utilization of benthic microflora by *Hyallela azteca* (Amphipoda). *Journal of Animal Ecology* **39**: 427–37. [15.4]

Hart SD, Howmiller RP. (1975) Studies on the decomposition of allochthonous detritus in two southern California streams. *Verhandlungen Internationale Vereinigung für Theoretische und Angewandt Limnologie* **19**: 1665–74. [15.6]

Herbst GN. (1982) Effects of leaf type on the consumption rates of aquatic detritivores. *Hydrobiologia* **89**: 77–87. [15.6]

Hildrew AG, Townsend CR, Francis J. (1984) Cellulolytic decomposition in streams of contrasting pH and its relationship with invertebrate community structure. *Freshwater Biology* **14**: 323–8. [15.7]

Hipp E, Mustafa T, Bickel U, Hoffman KH. (1986) Integumentary uptake of acetate and propionate (VFA) by *Tubifex* sp. a freshwater oligochaete. I. Uptake rate and transport kinetics. *Journal of Experimental Zoology* **240**: 289–97. [15.2]

Holopainen A-L, Huttunen P, Ahtiainen M. (1991) Effects of forestry practices on water quality and primary productivity in small forest brooks. *Verhandlungen*

Internationale Vereinigung für Theoretische und Angewandt Limnologie **24**: 1760–1766. [15.8]

Howarth RW, Fisher SG. (1976) Carbon, nitrogen and phosphorus dynamics during leaf decay in nutrient-enriched stream microecosystems. *Freshwater Biology* **6**: 221–8. [15.7]

Imbert JB, Pozo J. (1989) Breakdown of four leaf litter species and associated fauna in a Basque Country forested stream. *Hydrobiologia* **182**: 1–14. [15.6]

Irons JG, Oswood MW, Bryant JP. (1988) Consumption of leaf detritus by a stream shredder: influence of tree species and nutrient status. *Hydrobiologia* **160**: 53–61. [15.4]

Iversen TM. (1973) Decomposition of autumn-shed beech leaves in a springbrook and its significance for the fauna. *Archiv für Hydrobiologie* **72**: 305–12. [15.3, 15.4]

Iversen TM. (1975) Disappearance of autumn shed beech leaves placed in bags in small streams. *Verhandlungen Internationale Vereinigung für Theoretische und Angewandt Limnologie* **19**: 1687–92. [15.6, 15.7]

Kaplan LA, Bott TL. (1983) Microbial heterotrophic utilization of dissolved organic matter in a piedmont stream. *Freshwater Biology* **13**: 363–77. [15.2]

Kaplan LA, Bott TL. (1985) Acclimation of stream-bed heterotrophic microflora: metabolic responses to dissolved organic matter. *Freshwater Biology* **15**: 479–92. [15.2]

Kaushik NK, Hynes HBN. (1971) The fate of dead leaves that fall into streams. *Archiv für Hydrobiologie* **68**: 465–515. [15.2, 15.3, 15.4, 15.6]

Kirby JM, Webster JR, Benfield EF. (1983) The role of shredders in detrital dynamics of permanent and temporary streams. In: Fontaine TD, Bartell SM (eds) *Dynamics of Lotic Ecosystems*, pp 425–35. Ann Arbor Science, Michigan. [15.6]

Klug MJ, Kotarski S. (1980) Bacteria associated with the gut tract of larval stages of the aquatic cranefly *Tipula abdominalis* (Diptera; Tipulidae). *Applied Environmental Microbiology* **40**: 408–16. [15.5]

Kostalos M, Seymour RL. (1976) Role of microbial enriched detritus in the nutrition of *Gammarus minus* (Amphipoda). *Oikos* **27**: 512–16. [15.4]

Lawson DL, Klug MJ. (1989) Microbial fermentation in the hindguts of two stream detritivores. *Journal of the North American Benthological Society* **8**: 85–91. [15.5]

Lawson DL, Klug MJ, Merritt RW. (1984) The influence of the physical, chemical and microbiological characteristics of decomposing leaves on the growth of the detritivore *Tipula abdominalis* (Diptera: Tipulidae). *Canadian Journal of Zoology* **62**: 2339–43. [15.4]

Leff LG, McArthur JV. (1989) The effect of leaf pack composition on processing: a comparison of mixed and single species packs. *Hydrobiologia* **182**: 219–24. [15.6]

Leland HV, Carter JL. (1985) Effects of copper on production of periphyton, nitrogen fixation and processing of leaf litter in a Sierra Nevada, California, stream. *Freshwater Biology* **15**: 155–73. [15.7]

Lock MA, Hynes HBN. (1976) The fate of dissolved organic carbon derived from autumn-shed maple leaves (*Acer saccharum*) in a temperate hard water stream. *Limnology and Oceanography* **21**: 436–43. [15.2]

Lock MA, Wallace RR, Costerton JW, Ventullo RM, Charlton SE. (1984) River epilithon: towards a structural–functional model. *Oikos* **42**: 10–22. [15.2]

McDowell WH. (1985) Kinetics and mechanisms of dissolved organic carbon retention in a headwater stream. *Biogeochemistry* **2**: 329–53. [15.2]

McGeorge JE, Jagoe CH, Risley LS, Morgan MD. (1991) Litter decomposition in low pH streams in the New Jersey Pinelands. *Verhandlungen Internationale Vereinigung für Theoretische und Angewandt Limnologie* **24**: 1711–1714. [15.7]

McKinley VL, Vestal JR. (1982) Effects of acid on plant litter decomposition in an arctic lake. *Applied and Environmental Microbiology* **43**: 1188–95. [15.7]

Mackay RJ, Kersey KE. (1985) A preliminary study of aquatic insect communities and leaf decomposition in acid streams near Dorset, Ontario. *Hydrobiologia* **122**: 3–11. [15.7]

Maki JS, LaCroix SJ, Hopkins BS, Staley JT. (1986) Recovery and diversity of heterotrophic bacteria from chlorinated drinking waters. *Applied and Environmental Microbiology* **51**: 1047–55. [15.7]

Maltby L, Booth R. (1991) The effect of coal-mine effluent on fungal assemblages and leaf breakdown. *Water Research* **25**: 247–50. [15.7]

Mann KH. (1988) Production and use of detritus in various freshwater, marine and coastal marine ecosystems. *Limnology and Oceanography* **33**: 910–39. [15.2, 15.4]

Marcus JH, Willoughby LG. (1978) Fungi as food for the aquatic invertebrate *Asellus aquaticus*. *Transactions of the British Mycological Society* **70**: 143–6. [15.4]

Martin MM, Martin JS, Kukor JJ, Merritt RW. (1980) The digestion of protein and carbohydrate by the stream detritivore *Tipula abdominalis* (Diptera, Tipulidae). *Oecologia* **46**: 360–4. [15.4]

Martin MM, Kukor JJ, Martin JS, Lawson DL, Merritt RW. (1981a) Digestive enzymes of larvae of three species of caddisflies (Tricoptera). *Insect Biochemistry* **11**: 501–5. [15.5]

Martin MM, Martin JS, Kukor JJ, Merritt RW. (1981b) The digestive enyzmes of detritus-feeding stonefly nymphs (Plecoptera; Pteronarcyidae). *Canadian Journal of Zoology* **59**: 1947–51. [15.5]

Mathews CP, Kowalczewski A. (1969) The disappearance of leaf litter and its contribution to production in the River Thames. *Journal of Ecology* **57**: 543–52. [15.6]

Melillo JM, Aber JD, Mutatore JF. (1982) Nitrogen and lignin control of hardwood leaf litter decomposition dynamics. *Ecology* **63**: 621–6. [15.2]

Merritt RW, Cummins KW. (1978) *An Introduction to the Aquatic Insects of North America*. Kendall/Hunt Publishing, Iowa. [15.3]

Merritt RW, Cummins KW, Burton TM. (1984) The role of aquatic insects in the processing and cycling of nutrients. In: Resh VH, Rosenberg DM (eds) *The Ecology of Aquatic Insects*, pp 134–63. Praeger Publications, New York. [15.2]

Miller JC. (1987) Evidence for the use of non-detrital dissolved organic matter by microheterotrophs on plant detritus in a woodland stream. *Freshwater Biology* **18**: 483–94. [15.2, 15.9]

Minshall GW, Brock JT, LaPoint TW. (1982) Characterization and dynamics of benthic organic matter and invertebrate functional feeding group relationships in the Upper Salmon River, Idaho (USA). *Internationale Revue de Gesamten Hydrobiologie* **67**: 793–820. [15.6]

Monk DC. (1976) The distribution of cellulase in freshwater invertebrates of different feeding habits. *Freshwater Biology* **6**: 471–5. [15.5]

Monk DC. (1977) The digestion of cellulose and other dietary components and pH of the gut in the amphipod *Gammarus pulex* (L.) *Freshwater Biology* **7**: 431–40. [15.5]

Mulholland PJ, Palumbo AV, Elwood JW, Rosemond AD. (1987) Effects of acidification on leaf decomposition in streams. *Journal of the North American Benthological Society* **6**: 147–58. [15.7]

Mutch RA, Steedman RJ, Berté SB, Pritchard G. (1983) Leaf breakdown in a mountain stream: a comparison of methods. *Archiv für Hydrobiologie* **97**: 89–108. [15.6]

Newell SY, Fallon RD, Miller JD. (1986) Measuring fungal-biomass dynamics in standing-dead leaves of a salt-marsh vascular plant. In: Moss ST (ed.) *The Biology of Marine Fungi*, pp 19–25. Cambridge University Press, Cambridge. [15.4]

Newman RM. (1990) Effects of shredding amphipod density on watercress *Nasturtium officinale* breakdown. *Holarctic Ecology* **13**: 293–9. [15.6]

Newman RM, Perry JA, Tam E, Crawford RL. (1987) Effects of chronic chlorine exposure on litter processing in outdoor experimental streams. *Freshwater Biology* **18**: 415–28. [15.6, 15.7]

Oberndorfer RY, McArthur JV, Barnes JR, Dixon J. (1984) The effect of invertebrate predators on leaf litter processing in an alpine stream. *Ecology* **65**: 1325–31. [15.6]

Padgett DE. (1976) Leaf decomposition by fungi in a tropical rainforest stream. *Biotropica* **8**: 166–78. [15.3]

Palumbo AV, Bogle MA, Turner RR, Elwood JW, Mulholland PJ. (1987) Bacterial communities in acidic and circumneutral streams. *Applied and Environ-*

mental Microbiology **53**: 337–44. [15.7]

Park D. (1974) Accumulation of fungi by cellulose exposed in a river. *Transactions of the British Mycological Society* **63**: 437–47. [15.4]

Park D, McKee W. (1978) Cellulolytic *Pythium* as a component of the river mycoflora. *Transactions of the British Mycological Society* **71**: 251–9. [15.4]

Patrick FM, Loutit MW. (1976) Passage of metals in effluents, through bacteria to higher organisms. *Water Research* **10**: 333–5. [15.7]

Patrick FM, Loutit MW. (1978) Passage of metals to freshwater fish from their food. *Water Research* **12**: 395–8. [15.7]

Paul RW, Benfield EF, Cairns J. (1983) Dynamics of leaf processing in a medium-sized river In: Fontaine TD, Bartell SM (eds) *Dynamics of Lotic Ecosystems*, pp 403–23. Ann Arbor Science, Michigan. [15.6]

Petersen RC, Cummins KW. (1974) Leaf processing in a woodland stream. *Freshwater Biology* **4**: 343–68. [15.3, 15.4, 15.6]

Pinkney AE, Poje GV, Sansur RM, Lee CC, O'Connor JM. (1985) Uptake and retention of ^{14}C-Aroclor 1254 in the amphipod *Gammarus tigrinus*, fed contaminated fungus, *Fusarium oxysporum*. *Archives of Environmental Contamination and Toxicology* **14**: 59–64. [15.7]

Porter KG, Sherr EB, Sherr BF, Pace M, Sanders RW. (1985) Protozoa in planktonic food webs. *Journal of Protozoology* **32**: 409–15. [15.2]

Reice SR. (1977) The role of animal associations and current velocity in sediment-specific leaf litter decomposition. *Oikos* **29**: 357–65. [15.7]

Reice SR. (1978) Role of detritivore selectivity in species-specific litter decomposition in a woodland stream. *Verhandlungen Internationale Vereinigung für Theoretische und Angewandt Limnologie* **20**: 1396–1400. [15.6]

Roeding CE, Smock LA. (1989) Ecology of macroinvertebrate shredders in a low-gradient sandy-bottomed stream. *Journal of the North American Benthological Society* **8**: 149–61. [15.3]

Ross DH, Wallace JB. (1983) Longitudinal patterns of production, food consumption and seston utilization by net-spinning caddisflies (Trichoptera) in a south Appalachian stream (USA). *Holarctic Ecology* **6**: 270–84. [15.2, 15.3]

Rosset J, Bärlocher F. (1985) Aquatic hyphomycetes: influences of pH, Ca^{2+} and HCO_3^- on growth *in vitro*. *Transactions of the British Mycological Society* **84**: 137–45. [15.7]

Rossi L. (1985) Interactions between invertebrates and microfungi in freshwater ecosystems. *Oikos* **44**: 175–84. [15.4]

Rossi L, Fano AE. (1979) Role of fungi in the trophic niche of the congeneric detritivores *Asellus aquaticus* and *Asellus coxalis* (Isopoda). *Oikos* **32**: 380–5. [15.4]

Rossi L, Basset A, Nobile L. (1983) A coadapted trophic niche in two species of crustacea (Isopoda): *Asellus aquaticus* and *Proasellus coxalis*. *Evolution* **37**: 810–20. [15.4]

Rounick JS, Winterbourn MJ. (1983) The formation, structure and utilization of stone surface organic layers in two New Zealand streams. *Freshwater Biology* **13**: 57–72. [15.2]

Rounick JS, Winterbourn MJ, Lyon GL. (1982) Differential utilization and autochthonous inputs by aquatic invertebrates in some New Zealand streams: a stable carbon isotope study. *Oikos* **39**: 191–8. [15.2]

Salonen K, Hammer T. (1986) On the importance of dissolved organic matter in the nutrition of zooplankton in some lake waters. *Oecologia* **68**: 246–53. [15.2]

Schindler DW. (1980) Experimental acidification of a whole lake: a test of the oligotrophication hypothesis. In: Drabløs D, Tollan A (eds). *Proceedings International Conference on the Ecological Impact of Acid Precipitation, Norway 1980*, pp 370–4. SNSF Project. [15.7]

Schoenberg SA, Maccubbin AE, Hodson RE. (1984) Cellulose digestion by freshwater microcrustacea. *Limnology and Oceanography* **29**: 1132–6. [15.5]

Sedell JR, Triska FJ, Triska NS. (1975) The processing of conifer and hardwood leaves in two coniferous forest streams: I. weight loss and associated invertebrates. *Verhandlungen Internationale Vereinigung für Theoretische und Angewandt Limnologie* **19**: 1617–27. [15.2]

Short RA, Maslin PE. (1977) Processing of leaf litter by a stream detritivore: effect on nutrient availability to collectors. *Ecology* **58**: 935–8. [15.2]

Sinsabaugh RL, Linkins AE. (1990) Enzymic and chemical analysis of particulate organic matter from a boreal river. *Freshwater Biology* **23**: 301–9. [15.2, 15.3]

Sinsabaugh RL, Linkins AE, Benfield EF. (1985) Cellulose digestion and assimilation by three leaf-shredding aquatic insects. *Ecology* **66**: 1464–71. [15.5]

Speaker R, Moore K, Gregory S. (1984) Analysis of the process of retention of organic matter in stream ecosystems. *Verhandlungen Internationale Vereinigung für Theoretische und Angewandt Limnologie* **22**: 1835–41. [15.2]

Stewart BA, Davies BR. (1989) The influence of different litter bag designs on the breakdown of leaf material in a small mountain stream. *Hydrobiologia* **183**: 173–7. [15.6]

Stout BM III, Coburn CB Jr. (1989) Impact of highway construction on leaf processing in aquatic habitats of eastern Tennesse. *Hydrobiologia* **178**: 233–42. [15.8]

Suberkropp K, Klug MJ. (1974) Decomposition of deciduous leaf litter in a woodland stream. I. A scanning electron microscopic study. *Microbial Ecology* **1**: 96–103. [15.3]

Suberkropp K, Klug MJ. (1976) Fungi and bacteria associated with leaves during processing in a woodland stream. *Ecology* **57**: 707–19. [15.3, 15.4]

Suberkropp K, Godshalk GL, Klug MJ. (1976) Changes in the chemical composition of leaves during processing in a woodland stream. *Ecology* **57**: 720–7. [15.2]

Suberkropp K, Arsuffi TL, Anderson JP. (1983) Comparison of degradative ability, enzymatic activity and palatability of aquatic hyphomycetes grown on leaf litter. *Applied and Environmental Microbiology* **46**: 237–44. [15.4]

Sutcliffe DW, Carrick TR, Willoughby LG. (1981) Effects of diet, body size, age and temperature on growth rates in the amphipod *Gammarus pulex. Freshwater Biology* **11**: 183–214. [15.4]

Swanson FJ, Gregory SV, Sedell JR, Campbell GA. (1982) Land–water interaction: the riparian zone. In: Edmonds RL (ed.) *Analysis of Coniferous Forest Ecosystems in the Western United States*, pp 267–91. Hutchinson Ross, Stroudsburg, Pennsylvania. [15.2]

Thomas JD, Kowalczyk C, Somasundaram B. (1990) The biochemical ecology of *Biomaphalaria glabrata*, a freshwater pulmonate mollusc: the uptake and assimilation of exogenous glucose and maltose. *Comparative Biochemistry and Physiology* **95A**: 511–28. [15.2]

Thompson PL, Bärlocher F. (1989) Effect on pH on leaf breakdown in streams and in the laboratory. *Journal of the North American Benthological Society* **8**: 203–10. [15.7]

Traaen TS. (1980) Effects of acidity on decomposition of organic matter in aquatic environments. In: Drabløs D, Tollan A (eds) *Proceedings International Conference on the Ecological Impact of Acid Precipitation, Norway 1980*, pp 340–1. SNSF Project. [15.7]

Triska FJ. (1970) *Seasonal distribution of aquatic hyphomycetes in relation to the disappearance of leaf litter from a woodland stream.* Unpublished PhD thesis, University of Pittsburg, Pittsburg. [15.3, 15.4]

Triska FJ, Sedell JR, Gregory SV. (1982) Coniferous forest streams. In: Edmonds RL (ed.) *Analysis of Coniferous Forest Ecosystems in the Western United States*, pp 292–332. Hutchinson Ross, Stroudsburg, Pennsylvania. [15.2]

van Frankenhuyzen K, Geen GH, Koivisto C. (1985) Direct and indirect effects of low pH on the transformation of detrital energy by the shredding caddisfly, *Clistoronia magnifica* (Banks) (Limnephilidae. *Canadian Journal of Zoology* **63**: 2298–304. [15.7]

Wallace JB, Benke AC, Lingle AH, Parsons K. (1987) Trophic pathways of macroinvertebrate primary consumers in subtropical blackwater streams. *Archiv für Hydrobiologie* **74** (Suppl.): 423–51. [15.2]

Wallace JB, Webster JR, Cuffney TF. (1982) Stream detritus dynamics: regulation by invertebrate consumers. *Oecologia* **53**: 197–200. [15.7]

Ward GM. (1986) Lignin and cellulose content of benthic fine particulate organic matter (FPOM) in Oregon cascade mountain streams. *Journal of the North American Benthological Society* **5**: 127–39. [15.2]

Webster JR, Benfield EF. (1986) Vascular plant breakdown in freshwater ecosystems. *Annual Review of Ecology and Systematics* **17**: 567–94. [15.2]

Welton JS, Ladle M, Bass JAB, John IR. (1983) Estimation of gut throughput time in *Gammarus pulex* under laboratory and field conditions with a note on the feeding of young in the brood pouch. *Oikos* **41**: 133–8. [15.4]

Wetzel RG. (1975) *Limnology.* W.B. Saunders, Philadelphia. [15.2]

Willoughby LG, Earnshaw R. (1982) Gut passage times in *Gammarus pulex* (Crustacea, Amphipoda) and aspects of summer feeding in a stony stream. *Hydrobiologia* **97**: 105–17. [15.4]

Willoughby LG, Sutcliffe DW. (1976) Experiments on feeding and growth of the amphipod *Gammarus pulex* (L.) related to its distribution in the River Duddon. *Freshwater Biology* **6**: 577–86. [15.4]

Winterbourn MJ, Rounick JS, Hildrew AG. (1986) Patterns of carbon resource utilization by benthic invertebrates in two British river systems: a stable isotope study. *Archiv für Hydrobiologie* **3**: 349–61. [15.2]

Winkler G. (1991) Debris dams and retention in a low order stream (a backwater of Oberer Seebach – Ritrodat-Lunz study area, Austria). *Verhandlungen Internationale Vereinigung für Theoretische und Angewandt Limnologie* **24**: 1917–1920. [15.2]

16: Primary Production

R.G.WETZEL AND A.K.WARD

16.1 INTRODUCTION

The roles of photosynthetic primary producers within lotic ecosystems have been the subject of past controversy. Early evaluations of the magnitude of primary productivity in comparison with other sources of organic matter, particularly importations of organic matter from the drainage basin (e.g. Hynes 1970; Fisher & Likens 1973), suggested stream primary production was essentially irrelevant. Often the rationale for such evaluations focused on the amount of readily utilizable particulate organic matter available for consumption by herbivorous organisms, particularly benthic invertebrates. The reason for this evaluation was because many of the early ecosystem level studies focused on small streams that were heavily shaded by forest canopies during much of the active growing season and, therefore, were determined to have a negative net primary productivity. Autochthonous primary productivity within the stream by algae and macrophytes was often found to be less than respiratory losses of organic matter on an annual basis. As a result, streams in general were often referred to as heterotrophic; that is, greater amounts of organic matter were being decomposed in running waters than were being produced autochthonously within them. In reviews of the subject by Wetzel (1975) and subsequently by Minshall (1978), Bott (1983) and Bott et al (1985), among others, a more expansive view of the importance of primary production was adopted as a broader array of lotic ecosystems was examined and a more robust conceptual framework of the longitudinal aspects of river drainages and organic matter inputs was formulated (Vannote et al 1980). More recently,

even the importance of primary producers within heavily shaded, small, woodland streams has been recognized (e.g. Mayer & Likens 1987; Stock & Ward 1989).

Concurrent with the development of these concepts was an enlarged view of the 'boundaries' of stream ecosystems (Hynes 1974) and the importance of interfaces between the stream channel and other features of the stream catchment, specifically riparian (e.g. Cummins et al 1984), floodplain/wetland (e.g. Sedell & Froggatt 1984; Meyer 1990) and subsurface (hyporheic) regions (Grimm & Fisher 1984; Triska et al 1989a, 1989b). The existence and characteristics of these interfaces can clearly have a major impact on instream processes, including primary productivity. Therefore, a recent evolution in the conceptualization of the operation of stream and river ecosystems has concerned the coupled integration of dissolved and particulate nutrient/energy sources originating from outside the traditionally viewed stream channel with those produced and/or transformed within it. Much effort is now directed toward the sites of entry of this material, particularly from hyporheic groundwater through sediments to the overlying water and from riparian vegetation of adjacent land–water interface and floodplain regions.

Finally, much attention is correctly being directed to the fates of primary production within running waters. An evaluation of the factors important to the rates of primary production is inseparable from a discussion of decomposition and herbivory. As elaborated below, decomposition is the primary process of nutrient recycling that is supporting the primary production. Furthermore, herbivory can reduce as well as stimulate primary productivity. Although light availability, current, grinding action, and other physical factors remain as important regulators of the magnitude of primary productivity, the importance of microbial and higher-organism interactions are being recognized as integrating forces influencing both the primary productivity and the functional roles of primary producers in river ecosystems.

16.2 METHODS FOR MEASURING LOTIC PRIMARY PRODUCTIVITY AND THE MATTER OF SCALE

The methods used to estimate within-channel primary productivity have developed over many years in conjunction with the evolution and maturation of the total field of lotic ecology. Early estimates of lotic ecosystem metabolism, including primary production and respiration, used measurements of change in dissolved oxygen or carbon from pH in bulk streamwater measured at two stations over a diurnal period (e.g. Odum 1956; Hall & Moll 1975; reviewed by Wetzel & Likens 1991). Although this technique provides estimates of gross primary production in streams/rivers, it has been generally replaced by other methods because of problems in calculation of gas diffusion constants (particularly in small, rapidly flowing streams) and an inability to evaluate relative rates associated with subcomponents of the stream reach between the two stations (e.g. among habitats).

During the past 15 years, Plexiglas chambers equipped with submersible pumps to circulate streamwater over benthic substrata have been widely used as an *in-situ* technique to estimate various aspects of lotic, benthic metabolism (Bott *et al* 1978; Gregory 1980; Minshall *et al* 1983; Naiman 1983). The specific design and size of the chambers can be varied to accommodate the needs of the individual investigator, and changes in oxygen, pH, and total inorganic carbon or incorporation of $^{14}CO_2$ and $H^{14}CO_3$ into plant organic matter over a measured incubation period are used to estimate lotic primary productivity. The use of radiolabelled inorganic carbon is an extremely sensitive technique that probably best estimates net photosynthesis, as is assumed in other aquatic ecosystems, but does not provide

information on community respiration (Wetzel & Likens 1991). However, since benthic samples are radioassayed after the end of the incubation period, different plant morphological types and/or taxa from the same chamber can be radioassayed separately or examined by autoradiography and their primary production rates compared (Ward *et al* 1985; Stock *et al* 1987; Stock & Ward 1989). Overall, *in-situ* chamber methods provide a means of estimating primary productivity of within-channel benthic habitats under conditions which reasonably simulate stream conditions and at a convenient scale for the operator. One disadvantage is that benthic substrata and streamwater placed in the chamber are removed from the renewal effects of overlying streamwater and underlying sediments, a problem which can, however, be ameliorated by short incubation times (Stream Solute Workshop 1990). A frequent assumption of this method is that values based on metabolic rates of samples covering a square metre or less of stream bottom can be extrapolated to much larger areas.

Currently, all ecosystem researchers, including lotic ecologists, face similar technological problems in addressing the need to make estimates of primary productivity (and other metabolic processes) at larger scales than have previously been routinely measured (Risser *et al* 1988). Several new methods are emerging with strong potential in this area, although published research on streams/rivers is generally lacking. Combinations of satellite, low-altitude aerial photography, and infrared analyses have been used to estimate primary production over large areas of ocean and terrestrial landscapes and could be applicable to large rivers and associated lateral wetland areas as well (e.g. Brown *et al* 1985; Eppley *et al* 1985; Aumen *et al* 1991). Estimates of net ecosystem production based on measurements of diel atmospheric gas flux over large areas have been used in forest ecosystems (e.g. Ryan 1990). Also, recent application of long-path Fourier-Transform Infrared Spectroscopy (FTIR) to ecosystem analysis has resulted in the ability to measure the flux of multiple gases over landscapes at spatial scales up to 1 km (Gosz *et al* 1988).

At the other end of the spatial spectrum, there is an increasing recognition of the importance of microbial biofilms in lotic ecosystems, including their composition, complexity and community interactions. More sophisticated technology with resolution in the 5–100-μm scale range is needed to analyse and interpret the roles of photosynthetic and other components of these communities. Relevant technologies include the use of oxygen and pH microelectrodes as well as combinations of scanning electron microscopy and autoradiography (e.g. Revsbech 1983; Carlton & Wetzel 1987; Burkholder *et al* 1990).

16.3 COMPARATIVE PRIMARY PRODUCTIVITY

Comparative analyses of primary productivity among diverse ecosystems are difficult because of large differences in the analytical methods, conversion factors and sampling frequencies employed (*cf.*, for example, Dickerman *et al* 1986). On a global basis, the primary productivity of inland waters is considered to be minor, because of the relatively small surface area covered by

standing or running water (Woodwell *et al* 1978). In addition, the productivity of phytoplankton in any water is low; productivity cannot be increased because of light limitations, regardless of adequacy of nutrient and other growth factors, to levels approaching those of most terrestrial ecosystems (Wetzel 1983, 1990). The oft-cited comparative table of Woodwell *et al* (1978) is reiterated in Table 16.1; however, the values for inland waters are probably greatly underestimated. Among rivers, much of this underestimation has resulted from detailed estimations of the rates of primary productivity only by within-channel attached algae.

Productivity rates of attached algae In a detailed comparison of the rates of in-stream algal primary productivity from many geographical and geological regions of North America to those intensively studied sites in Table 16.2, rates were found to be similar (Bott *et al* 1985). Concentrations of chlorophyll *a* of benthic organisms spanned an order of magnitude ($10-100$ mg m^{-2})

Table 16.1 Net primary productivity of major plant communities of the earth based on biomass estimates

	Net primary productivity (g carbon m^{-2} year^{-1})
Aquatic ecosystems	
Swamp and marsh	1350.0
Marine, algal beds and reefs	1166.7
Marine, estuaries	714.3
Marine upwelling zones	250.0
Lakes and streams	200.0
Marine, continental shelf	161.7
Marine, open oceans	56.3
Terrestrial ecosystems	
Tropical rain forest	988.2
Tropical seasonal forest	720.0
Temperate evergreen forest	580.0
Temperate deciduous forest	542.9
Savanna grassland	406.7
Boreal forest	358.3
Woodland and shrubland	317.7
Cultivated land	292.9
Temperate grassland	266.7
Tundra and alpine meadow	62.5
Desert shrub	38.9
Rock, sand	1.3

After Woodwell *et al* (1978).

Table 16.2 Mean annual net primary productivity of in-stream community producers of North American streams

	Mean annual net community primary productivity (mg carbon m^{-2} day^{-1})
Eastern deciduous forest stream, coastal climate, Pennsylvania[a]	$-27.0-246.8$ ($\bar{x}_4$ 127.4)
Mesic hardwood forest stream, continental climate, Michigan[a]	$-55.9-486.4$ ($\bar{x}_4$ 207.3)
Cool, arid climate stream, much precipitation as snowfall, coniferous vegetation on north-facing slopes, sagebrush on south-facing slopes, Idaho[a]	$48.2-524.7$ ($\bar{x}_7$ 415.2)
Northern cool-desert stream open, no canopy, Idaho[b] Autochthonous	
Macrophytes	$47.2-147.6$ ($\bar{x}_3$ 93.4)
Periphyton	$465.6-1748.0$ ($\bar{x}_3$ 1004.0)
Allochthonous	$0.7-16.9$ ($\bar{x}_3$ 7.0)
Coniferous forest stream, coastal climate, Oregon[a]	$-21.0-93.8$ ($\bar{x}_4$ 45.1)

[a] Extracted from data of Bott (1983).
[b] Calculated from Minshall (1978) assuming 1 kcal = 4.6 g organic matter and 46.5% carbon in organic matter of aquatic plants (Westlake 1965).

with little geographical differentiation within a seasonal range of 1 mg m^{-2} to slightly in excess of 300 mg m^{-2}. Highest amounts were found in Great Plains or desert regions of high productivity.

Gross primary productivity (GPP) of within-stream benthic algae (Table 16.3), as well as community respiration (CR_{24}, consumption of dissolved oxygen in 24 h), commonly increases with downstream direction (e.g. Bott *et al* 1985). Photosynthetic efficiencies (percentage of ir-radiance (400–750 nm) utilized in existing rates of net or gross photosynthetic production) of benthic algae generally are below 4%, with averages near 1% and extremes of about 7% in regions of high nutrient enrichments. Photosynthetic efficiencies generally decline downstream with increasing light availability.

Accurate comparisons of the *annual* primary productivity of in-stream producers (periphyton, phytoplankton and macrophytes) simultaneously with the productivity of other producers of the saturated-soils of the floodplain and bank regions of river ecosystems do not exist. These data are important, however, to correct interpretations of composite ecosystem production and utilization of organic matter entering the ecosystem.

This underestimation has resulted, in part, from a too-narrow definition of the boundaries of standing and running-water ecosystems. That is, only the readily observed central or channel water was recognized as the water body 'proper', and the importance of adjacent interface zones and their interactions with the central water bodies were virtually ignored. In particular, recognition of the expansion of stream/river boundaries to include floodplain, riparian and other features in recent years has developed in parallel with a more integrated ecosystem paradigm of running waters, which gained great impetus from Hynes (1974). These concepts have roots in the earlier perceptions of Thienemann (1926) and Lindeman (1942) in which aquatic ecosystems were viewed as an integration of interdependent component parts and processes. Therefore, whether inland water bodies are labelled as lotic or lentic, all reside in and are closely coupled with their drainage basins through interface regions. As a result, addressing the primary productivity of stream/ river *ecosystems* requires complete analyses of

Table 16.3 Example rates of in-stream primary productivity

Site	Mean primary productivity (g carbon m^{-2} day^{-1})	Technique	Reference
Morgans Creek, Kentucky	0.004–0.0007	Biomass change	Minshall (1967)
Walker Branch, Tennessee	0.008–0.011	Biomass change	Elwood & Nelson (1972)
Glade Branch, Virginia	0.010	^{14}C	Hornick *et al* (1981)
Piney Branch, Virginia	0.011	^{14}C	Hornick *et al* (1981)
Guys Run, Virginia	0.018	^{14}C	Hornick *et al* (1981)
Berry Creek, Oregon	0.099	Oxygen exchange in recirculating chamber	Reese (1967)
Yellow Creek, Alabama	0.0493	^{14}C	Lay & Ward (1987)
Little Schultz Creek, Alabama	0.0912	^{14}C	Lay & Ward (1987)
New Hope Creek, North Carolina	0.33–1.43	Diurnal oxygen curve	Hall (1972)
Deep Creek, Idaho	1.18	Oxygen exchange, upstream/downstream	Minshall (1978)
Catahoula Creek, Mississippi	1.85	Oxygen exchange, upstream/downstream	de la Cruz & Post (1977)
White Clay Creek, Pennsylvania	2.22	O_2/CO_2 exchange	Bott *et al* (1978)

[a] Enriched by fertilizer and sewage.
[b] From Lay & Ward (1987).

the productivity both within and adjacent to the channel *per se*. Such comparative data from individual ecosystems are almost completely lacking. Further, although the expanded boundaries of lotic ecosystems can easily be recognized at the conceptual level, specific definitions of the exact physical dimensions in any given ecosystem can be more difficult. How should the boundaries of lotic ecosystems be defined, and how does that definition affect our perception of the importance and functions of primary production in streams and rivers?

16.4 FUNCTIONAL BOUNDARIES OF RUNNING-WATER ECOSYSTEMS

Most reported values of the primary productivity of streams/rivers have been restricted to those producers submersed within a traditionally viewed river channel, i.e. one that includes only plants associated with the visible, surface-flowing water and some of its benthic surfaces (Table 16.4). However, both small stream and large river ecosystems can have soils adjacent to the main channel that are water saturated some distance from the channel. In stream/river channels with gentle slopes this area can involve a region many times larger (>30 times) than the area of the open stream water. Therefore, contemporary concepts of the boundaries of stream ecosystems include lateral and subsurface regions around the main channel. Hence, stream ecosystems are 'bounded' by the subsurface, groundwater–stream-water interface rather than the main channel *per se* (Fig. 16.1). This region encompasses biologically active gradients between the two and is defined by the hydrological movement of water through the entire stream ecosystem (e.g. Triska *et al* 1989b). This perception of lotic boundaries has important implications for the functional aspects of metabolism of biota, particularly the primary producers, within the running waters as well as the dynamics

Table 16.4 Estimates of annual net primary productivity, estimated from seasonal maximum biomass, of aquatic macrophytes of river ecosystems

	g AFDW m^{-2} year^{-1}	g carbon m^{-2} year^{-1} [d]
Temperate rivers[a]		
Emergent	320–3712	149–1726
Submersed	8–395	4–184
Polluted Wisconsin stream[b]		
Submersed	1162	540
Amazon River[c]		
'Floating meadows'	2430–4050	1130–1883

[a] Extracted from the review of Rodgers *et al* (1983) from numerous sources.
[b] Calculated from Madsen & Adams (1988).
[c] Calculated from Junk (1970).
[d] Using an average value of 46.5% of AFDW (ash-free dry weight = organic carbon (*cf.* Westlake 1965).

of biogeochemical cycling. In the lateral regions, flood-tolerant vegetation adapted to saturated soils contributes to the primary production of lotic ecosystems. Some plants, such as emergent and floating-leaved rooted macrophytes, may be semi-immersed in the channel water. Other more woody shrubs and riparian species, such as willow and alder, can be rooted in the subsurface, hyporheic zone adjacent to the channel, but are included in the region hydrologically defined above as part of the stream ecosystem. The organic matter produced by these plants is in part decomposed at or near the sites of production. Part of the organic

matter produced is transported to the river water, both overland and within hyporheic groundwater.

Based on the above concept of the boundaries of lotic ecosystems, vegetation that should be included in comparative analyses of primary productivity of rivers includes all aquatic and amphibious higher and lower plants: (1) emergent wetland plants, riverine trees and shrubs, floating non-rooted macrophytes, floating-leaved rooted macrophytes, and submersed rooted macrophytes; (2) epiphytic algae and cyanobacteria associated with submersed aquatic macrophytes; (3) epipelic algae and cyanobacteria associated

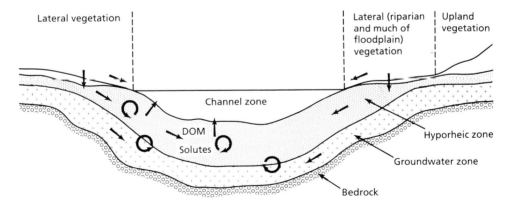

Fig. 16.1 Conceptualization of lateral and vertical boundaries of lotic ecosystems (modified extensively from Triska *et al* 1989b). The stream ecosystem boundary is defined as the hyporheic/groundwater interface and thereby includes a substantial volume beneath and lateral to the main channel. Vegetation rooted in the hyporheic zone is therefore part of stream ecosystem primary production. Arrows indicate flow pathways of dissolved organic matter and inorganic solutes derived from plant detritus within the stream ecosystem.

with particulate detritus and sediments in the understorey floodplain and bank wetlands; (4) epilithic algae and cyanobacteria associated with rock surfaces (from sand grains to boulders); and (5) phytoplankton and drift algae from epilithic and other attached sources.

The conceptual inclusion of all of these plants as part of lotic ecosystems is important because their productivity and metabolism have major effects on these systems. One of the most important effects of vegetation growing in lateral regions is the contribution of organic matter to the recipient stream, both in the particulate and dissolved form, a portion of which is metabolized microbially for potential utilization by higher trophic levels. Because of the importance of this organic matter to the metabolism, biogeochemical cycling of nutrients, and support of higher trophic levels within running-water systems, these organic matter sources drive subsequent metabolism and are functionally part of stream/river ecosystems. We contend that exclusion of this primary productivity by an arbitrary physical boundary and treatment as 'allochthonous organic matter loading' imposes an artificial demarcation that is incongruous with an integrated ecosystem perspective.

16.5 LONGITUDINAL AND LATERAL EFFECTS ON PRIMARY PRODUCTIVITY GRADIENTS

Longitudinal effects: light versus geomorphological characteristics

Much of the control of the observed primary productivity among running waters has been attributed to availability of solar insolation reaching the water and its attenuation within the water (e.g. Wetzel 1975; Minshall 1978; Bott 1983). As is the situation among standing waters, once nutrients are provided from both the water and the substrata in adequate supply to support maximum rates of photosynthesis, light availability is the dominant factor regulating photosynthesis. On the basis of several detailed comparative studies conducted in North America, a pattern of increased autotrophy with greater stream order

was found (Minshall *et al* 1983; Naiman 1983). Variance from this general correlation is wide, however. Some of these differences are related to geomorphological characteristics of hydraulic energy as well as the influence of human activities, particularly from deforestation and agriculture, which have led to high inorganic loadings, sustained turbidity and light reductions.

Although the correlation of river primary productivity to stream order is direct, that relationship applies among relatively non-disturbed ecosystems (Fig. 16.2(a)). Much of the increase in autotrophy is associated with increasing contributions of periphytic algae. In non-disturbed river

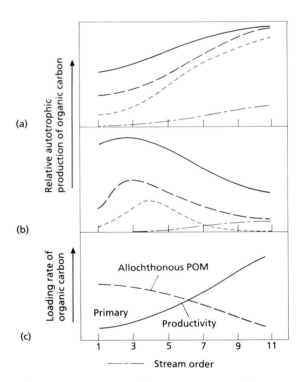

Fig. 16.2 Comparison of changes in patterns of autotrophic production of organic carbon with increasing stream order in (a) non-disturbed and (b) disturbed lotic ecosystems. (c) Longitudinal gradient of organic carbon loading from allochthonous (upland) versus lotic ecosystem primary productivity. ——, Phytosynthetically derived *dissolved* organic carbon from wetlands and floodplain; – – – –, macrophytes (emergent and submersed); ----, periphyton; –·–·–, phytoplankton.

ecosystems, as the slope decreases with increasing order, hydraulic energy dissipates to conditions where high water volumes are inadequate to transport much of the sediment loads. The result is widespread deposition with river braiding and meandering (e.g. Minshall *et al* 1985; Kellerhals & Church 1989; Sedell *et al* 1990; see also Part 1), which leads to exponential increases of shallow substrata for colonization by algae. In addition, the sediment deposits within the streams commonly augment large bank regions of saturated soils and floodplains, and provide habitats for both submersed and emergent macrophytes and wetland vegetation. Similarly, particulate detrital deposits from high vegetative productivity within these land–water margins provide large surface areas which are rapidly colonized by microbial producers and degraders. Biogeochemical cycling of essential nutrients is very rapid (see Chapter 18). Furthermore, with the reduction of hydraulic energy, grinding action is reduced, light availability increases somewhat with lowered particulate loads, and the development and productivity of phytoplankton increase.

Because light availability is a major regulator of primary productivity of river ecosystems, this variable can be examined in relation to stream gradients. Unfortunately marked alterations in the light quality of most river ecosystems have occurred as land clearance, agriculture and other disturbances have increased (e.g. Sedell & Froggatt 1984; Triska 1984; Yount & Nieme 1990). Light available for photosynthesis *within* river channels *per se* is regulated by both canopy cover and turbidity (Fig. 16.3). Light availability is commonly low in small streams and rivers, at least seasonally during the active growing season, because of extensive vegetative canopy in both non-disturbed as well as disturbed river ecosystems. Inorganic and organic turbidity was considerably lower before extensive land clearance and agriculture than is now commonly the situation. Undisturbed rivers usually have low loads of turbidity and maintain high water transparency as size increases down-gradient (Fig. 16.3). A few examples of transparent waters in large river ecosystems still exist (e.g. portions of South America and Siberian USSR), but they are becoming increasingly rare. In general, inorganic turbidity

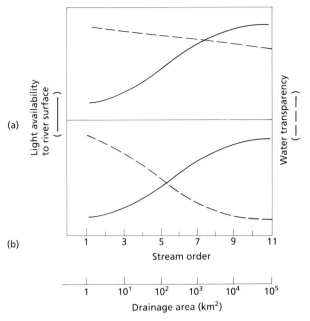

Fig. 16.3 Comparison of changes in patterns of light penetration to the stream water (water transparency) with increasing stream order in (a) preagricultural non-disturbed and (b) present/disturbed lotic ecosystems.

increases with increasing drainage area (*cf.* Slaymaker 1988) and stream order.

Longitudinal downstream gradients in resources result in a predictable structuring of the biota (Vannote *et al* 1980; Sedell *et al* 1989). Nutrient loadings usually increase as the sizes of the drainage basins and stream order increase, particularly among most river valleys that have been disturbed by agriculture. As a result, nutrient limitations to primary productivity of rivers generally decrease with increasing stream order. Primary productivity within the rivers, however, generally decreases with increasing stream order because of the light limitations (see Fig. 16.2(b)). In particular, the contributions by periphyton and submersed macrophytes decline with increasing size and turbidity.

Lateral contributions to autotrophic production of organic matter

The discussion above addresses contemporary syntheses that have emphasized the changes in photosynthetic metabolism observed along the length of stream gradients, i.e. with increasing stream order. Despite the reductions of *within-riverwater* primary productivity, we argue that, in general, the primary productivity of the river *ecosystem* increases with increasing stream order, particularly in non-disturbed but also in disturbed river systems (see Fig. 16.2(c)).

As the gradient slope decreases with increasing stream order and increased deposition of bedloads occurs, the percentage of the river valley containing saturated sediments and hydrosoils increases markedly. Colonization of these bank, wetland and floodplain areas by higher vegetation results in very high primary productivity (Table 16.5). Rates of primary productivity of these land–water interface areas are consistently high; in fact, they are among the highest of the biosphere. Decomposition of much of the particulate organic matter produced in these sites occurs at or near the sites of production. A very large portion of the organic matter synthesized by these high producers, however, is exported to the river water *per se* as dissolved organic matter.

Although there is a tendency to consider lateral development of stream ecosystems as increasing with increasing stream order, in fact, lateral regions of both small streams *and* relatively large rivers can be restricted by geomorphological and/ or human-induced features of the catchment. Reaches of lotic ecosystems are considered 'constrained' when the valley floor is narrower than two active stream channel widths (Sedell *et al*

1990). These river systems occur where natural geological or features constructed by humans constrict the valley floor and limit lateral mobility and adjacent plant communities. As a consequence, loading of particulate and dissolved organic matter from these lateral sources is relatively low, even though the total amount may constitute a significant portion of the total organic matter driving the ecosystem metabolism.

Unconstrained river reaches have valley floors that are wider than two active channel widths and lack major lateral geological or artificial constraints (Gregory *et al* 1991). These river ecosystems are characterized by active migrations of the channel that form extensive floodplains and often braided channels (Pinay *et al* 1990; Wissmar & Swanson 1990). The reduced erosive energy and depositional character results in widespread islands, peninsulas and inundated marginal environments with extensive development of fluvial and wetland vegetation (Fig. 16.4). The contemporary operational understanding of river ecosystems in three spatial dimensions and a fourth temporal dimension (Ward 1989) correctly incorporates the *dynamic scales* to account for the continual lateral expansion and contraction of river ecosystems. Clearly, the primary production through the floodplain to the valley walls loads the river channels with particulate and dissolved detrital organic carbon. Organic carbon produced in the floodplains is integrated with that produced within the channel *per se* (Table 16.6).

Table 16.5 Estimates of inputs of particulate organic matter to a 135-km reach of the New River (North Carolina, Virginia, West Virginia)

Source	Input (mT AFDW year^{-1})	Percentage of total input
Allochthonous		
Upstream and tributary	5893	53.8
Within study area	64	0.5
Autochthonous	3570	32.6
Periphyton	1435	13.1
Aquatic macrophytes		
Total particulate organic matter input	10962	

AFDW (ash-free dry weight). From Hill & Webster (1983).

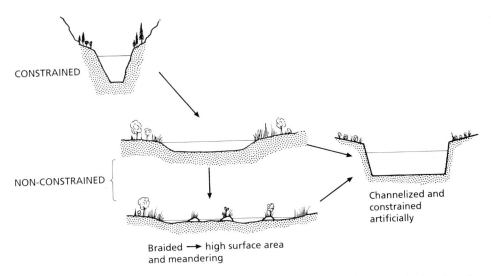

CONSTRAINED

NON-CONSTRAINED

Channelized and
constrained
artificially

Braided → high surface area
and meandering

Fig. 16.4 Conceptualization of changes in extent of lateral vegetation and hyporheic zones (subsurface, dotted) in constrained, non-constrained and channelized lotic ecosystems.

16.6 FATES OF LOTIC PRIMARY PRODUCTION

Grazer pathways

Much of the research regarding fate of lotic primary production over the past decade has been directed toward consumption of within-channel, benthic primary production by macroinvertebrates and fish (e.g. Gregory 1983; Power *et al* 1985; Steinman *et al* 1987; Lamberti *et al* 1987; Power 1990; see also Chapters 11 & 15). Many of these studies have utilized elegant laboratory experiments in which the effects of various grazers on benthic periphyton communities grown on tiles have been examined. In general, these studies have shown that grazers can decrease periphyton biomass, decrease areal primary productivity, increase biomass specific primary productivity, and alter algal species and chemical

Table 16.6 Annual mean concentrations of organic matter in transport in the Ogeechee River

	Mean mg AFDW/l	Percentage of total organic matter
Dissolved organic matter	25.4	96.32
Particulate organic matter		
Amorphous material (bacteria)	0.301	1.141
Amorphous material (protozoans)	0.039	0.148
Amorphous material (other)	0.521	1.976
Vascular plant detritus	0.028	0.106
Algae (mostly diatoms)	0.060	0.228
Fungi	0.014	0.053
Animals	0.002	0.008
Total particulate organic matter	0.97	3.68
Total organic matter in transport	26.37	100.00

AFDW (ash-free dry weight). Modified from Benke & Meyer (1988).

composition. Since grazer consumption of periphyton production represents a first step in transfer of this energy resource to higher trophic levels, it is not surprising that so much emphasis has been in this direction and that results have frequently been used to infer that the majority of lotic primary production is transferred to higher trophic levels. However, definitive research addressing the quantity of benthic periphyton production that is transferred to grazers, as opposed to that which is unconsumed and enters detrital pathways (including dissolved and particulate components), has never been completed under natural conditions in streams or rivers. In cases where comparative, *in-situ* epilithic (algal plus bacterial) and grazer (snail) productivity data exist and have been extrapolated to estimate this quantity in streams with heavily grazed surfaces in forested catchments, results suggest that periphyton production was conservatively three to ten times that consumed by grazers (Stock & Ward 1989).

Compared with studies on lotic grazer consumption of benthic periphyton, substantially less research has been directed toward grazer impact on within-channel macrophytes, lateral vegetation and associated attached communities. As with lake littoral and wetland vegetation, most of this lotic plant material is probably not readily consumed directly because of the reduced palatability of the structural components. However, even in studies which have shown that substantial invertebrate production can be supported by river vascular plants (water lilies), the majority (80–90%) of plant production still entered detrital pathways (Wallace & O'Hop 1985).

Of the different types of plant communities that contribute to lotic ecosystem production (see above), we would predict that benthic periphyton communities would be most susceptible to consumption and direct transfer to higher trophic levels. However, although this material is important to the nutritional needs of the animals, most periphyton production in streams enters detrital pathways directly or in the form of incompletely digested animal egesta. Only minor portions of within-channel macrophytic or lateral vegetation would be directly consumed. Therefore, the majority of lotic *ecosystem* production enters detrital pathways and is modified primarily through microbial metabolism.

Nutrient regeneration and detrital pathways

Lotic primary production that is not consumed directly by higher trophic levels contributes both particulate and dissolved organic matter to stream/river ecosystems. Most lotic *ecosystem* primary production undergoes direct microbial degradation. However, the exact pathways and mechanisms of transformation, as well as the ultimate fate of much of this material, are still not well understood. This is particularly true of the dissolved organic fraction. In general the fate of particulate organic matter which falls directly into the stream channel in relatively constrained, upland streams has been studied to a much greater extent (e.g. Petersen *et al* 1989; review by Ward *et al* 1990) than the fate of unconsumed microbial primary production (e.g. epilithic, epipelic, epiphytic) in both constrained and unconstrained streams or rivers or macrophytes in laterally expansive regions.

In all stream ecosystems, a portion of the riparian vegetation will fall directly in the channel where it will be transformed by both biotic and abiotic processes or buried and stored within the sediment. However, most of the lateral vegetation will fall to the ground beside the main channel and be incorporated into subsurface soil regions, including hyporheic zones. Both organic matter stored and/or produced within the main channel, as well as that which feeds into the hyporheic zone from lateral regions, can fuel microbially active subsurface regions, where both aerobic and anaerobic processes further transform the detrital organic material (see Fig. 16.1). Some of this dissolved organic and solute material eventually re-emerges in the main channel, where it can serve as nutrient sources to within-channel communities, including primary producers (e.g. Crocker & Meyer 1987; Dahm *et al* 1987; Triska *et al* 1989b; Coleman & Dahm 1990; DeAngelis *et al* 1990).

As lateral regions of lotic ecosystems increase relative to the main channel, retention and recycling mechanisms associated with the increased complexity of the regions become more intense and circuitous. Although budgets incorporating transformation of total detrital material in laterally expansive regions of lotic ecosystems are lacking, we hypothesize that: (1) a substantial portion of the particulate organic matter will be retained and decomposed within the lateral region; (2) much of the detrital material will be converted to gaseous end-products within the lateral regions; and (3) dissolved organic matter will be the primary form of detrital material exported to the main channel. In general, we would expect community response to detrital inputs to be metabolically coordinated and oriented toward conservation, retention and recycling (*cf.* Wetzel 1990).

The basis for these hypotheses stems largely from research on lake littoral and wetland ecosystems, where retention and recycling processes occur at several levels (Wetzel 1990). Similar retention and recycling pathways probably exist for the laterally expansive regions of lotic ecosystems. Firstly, gases produced metabolically in photosynthesis and respiration as well as acquired nutrients are retained within individual plants. As the environment becomes more restrictive, such as extremely slow aqueous diffusion rates of nutrients and gases of anoxia in nearly all sediment substrata, anatomical, morphological, and physiological plant adaptations for retention and re-utilization of essential resources increase in frequency and size. Examples include the retention, storage and re-utilization of inorganic carbon from respiration and photorespiration in internal gas spaces for subsequent recycling in photosynthesis. Limiting nutrients are largely translocated to and stored in perennating tissues during dormancy periods. Because much of the existing quantities of resources are retained and recycled, any of the external loadings of nutrients that are assimilated can be directed largely to net productivity.

Secondly, critical nutrients and organic carbon released by macrophytic primary producers are rapidly and efficiently assimilated and incorporated into the metabolism and productivity of microbiota associated with surfaces. Photosynthetic productivity of the attached algae can be extraordinarily high. In waters with submersed vegetation, the surface area for epiphytic microfloral colonization and development is increased and results in annual productivity values often far in excess of those of the supporting macrophytes. Again *recycling* of limiting essential gases, particularly oxygen and carbon dioxide, and dissolved nutrients within the attached community complex of algae, bacteria, fungi, protozoans, particulate detritus and inorganic precipitates, is the key to the commonly observed high sustained growth of attached microflora (Wetzel 1990). Intensive recycling of nutrients and gases in these microcommunities allows maintenance of photosynthetic biomass while minimizing losses and, as with the emergent macrophytes, allowing imported nutrients to be utilized primarily for net growth.

It should be noted that current velocities of water among these land–water gradients and over attached microbial communities can alter uptake rates of gases and nutrients. The diffusion rates across the boundary layer, however, can be enhanced only modestly by increased current velocities within the usual limnological range (e.g. Riber & Wetzel 1987). Even within dense macrophyte beds, flows are around and over the plants, and currents within the bed are largely slow and laminar (e.g. Losee & Wetzel 1992). As a result, most gaseous and nutrient fluxes within these attached communities are diffusion based and slow.

As a result of the above factors, much of the particulate organic matter in expansive lateral regions of lotic ecosystems is probably decomposed within the land–water interface regions to gaseous end-products. Nutrients released by the decomposition processes are usually rapidly assimilated by attached micro-organisms associated with particulate detritus, living plant tissues and inorganic substrata. Although irregular spates of variable magnitude can move particulate organic matter down-gradient, most exported organic matter is predominantly as dissolved organic substances. As the dissolved organic matter moves laterally towards the river channels, relatively energy-rich and chemically labile portions of the mixture of many organic compounds are utilized by attached microbiota selectively in conformance with kinetic models developed for polymer degradation (e.g. Cunningham & Wetzel 1989).

Inputs and Pathways of Matter and Energy

The more recalcitrant dissolved organic compounds can be partially degraded by ultraviolet photolysis as well as metabolism by the attached microflora (e.g. Mickle & Wetzel 1978, 1979; Wetzel 1991, 1992).

Dissolved organic compounds from peat drainage can affect nutrient availability and alter primary productivity. Humic complexes of aluminum and iron can bind phosphate ions, increase soil retention and reduce phosphorus availability for photosynthesis (e.g. Jackson & Schindler 1975). Phosphate sorbed to ferric iron–humic complexes may be released by ultraviolet-induced photoreduction of ferric iron to the ferrous state (Frankco & Heath 1982; Cotner & Heath 1990). In addition, colloidal humic–iron complexes can sequester iron and other micronutrients and possibly reduce primary productivity (Jackson & Hecky 1980). Dissolved humic compounds can also alter the reactivity of phosphate coprecipitation with calcium carbonate and other compounds (e.g. Otsuki & Wetzel 1972, 1973; Stewart & Wetzel 1981). These humic and fulvic acids can have variable effects on algal photosynthesis and productivity (Stewart & Wetzel 1982; Devol *et al* 1984). Interactive effects probably occur by complexation and restriction of extracellular enzyme activities and reduction of availability of essential nutrients from organic compounds (see discussion below). Where inorganic nutrient availability is relatively high, the effects of humic compounds would be minimal.

The dissolved organic compounds of wetland/ littoral and periphytic origins can function in two important ways in river ecosystems. Decomposition of the large reservoir of dissolved organic matter, although slow (1–2% per day), is collectively a major component of energy dissipation of the ecosystem (Wetzel 1983, 1984). Second, release of large amounts of dissolved organic compounds by wetland/littoral macrophytes and attached microbiota is seasonal and coupled to complex interactions among photosynthetic production, nutrient availability, phased senescence, plant tissue characteristics, and conditions, including hydrology, of decomposition (Wetzel 1990). Many of these relatively recalcitrant humic compounds complex chemically with extracellular enzymes of algae and bacteria (e.g.

Wetzel 1991, 1992). In this manner, enzymatic reactions and metabolic kinetics of river planktonic and periphytic organisms can be regulated in both positive and negative ways.

REFERENCES

Aumen NG, Miller DE, Crist CL. (1991) Contribution of mudflat primary production to a reservoir carbon budget. *Verh Internat Verein Limnol* **24**: 1304–08. [16.2]

Benke AC, Meyer JL. (1988) Structure and function of a blackwater river in the southeastern USA. *Verh Internat Verein Limnol* **23**: 1209–18. [16.5]

Bott TL. (1983) Primary productivity in streams. In: Barnes JR, Minshall GW (eds) *Stream Ecology. Application and Testing of General Ecological Theory*, pp 29–53. Plenum Press, New York. [16.1, 16.3, 16.5]

Bott TL, Brock JT, Cushing CE, Gregory SV, King D, Petersen RC. (1978) A comparison of methods for measuring primary productivity and community respiration in streams. *Hydrobiologia* **60**: 3–12. [16.2, 16.8]

Bott TL, Brock JT, Dunn CS, Naiman RJ, Ovink RW, Petersen RC. (1985) Benthic community metabolism in four temperate stream systems: an inter-biome comparison and evaluation of the river continuum concept. *Hydrobiologia* **123**: 3–45. [16.1, 16.3]

Brown O, Evans RH, Gordon HR, Smith RC, Baker KS. (1985) Blooming off the US coast: a satellite description. *Science* **229**: 163–7. [16.2]

Burkholder JM, Wetzel RG, Klomparens KL. (1990) Direct comparison of phosphate uptake by adnate and loosely attached microalgae within an intact biofilm matrix. *Applied and Environmental Microbiology* **56**: 2882–90. [16.2]

Carlton RG, Wetzel RG. (1987) Distribution and fates of oxygen in periphyton communities. *Canadian Journal of Botany* **65**: 1031–7. [16.2]

Coleman RL, Dahm CN. (1990) Stream geomorphology: effects on periphyton standing crop and primary production. *Journal of the North American Benthological Society* **9**: 293–302. [16.6]

Cotner JB Jr, Heath RT. (1990) Iron redox effects on photosensitive phosphorus release from dissolved humic materials. *Limnology and Oceanography* **35**: 1175–81. [16.6]

Crocker MT, Meyer JL. (1987) Interstitial dissolved organic carbon in sediments of a southern Appalachian headwater stream. *Journal of the North American Benthological Society* **6**: 159–67. [16.6]

de la Cruz A, Post AH. (1977) Production and transport of organic matter in a woodland stream. *Archives of Hydrobiology* **80**: 227–38. [16.3]

Cummins KW, Minshall GW, Sedell JR, Cushing CE,

Petersen RC. (1984) Stream ecosystem theory. *Verhandlungen Internationale Vereinigung für Theoretische und Angewandt Limnologie* **22**: 1818–27. [16.1]

Cunningham HW, Wetzel RG. (1989) Kinetic analysis of protein degradation by a freshwater wetland sediment community. *Applied and Environmental Microbiology* **55**: 1963–7. [16.6]

Dahm CN, Trotter EH, Sedell JR. (1987) Role of anaerobic zones and processes in stream ecosystem productivity. In: Averett RC, McKnight DM (eds) *Chemical Quality of Water and the Hydrologic Cycle*, pp 157–178. Lewis Publ., Chelsea, Michigan. [16.6]

DeAngelis DL, Mulholland PJ, Elwood JW, Palumbo AV, Steinman AD. (1990) Biogeochemical cycling constraints on stream ecosystem recovery. *Environmental Management* **14**: 685–97. [16.6]

Devol AH, Santos AD, Forsberg BR, Zaret TM. (1984) Nutrient addition experiments in Lago Jacaretinga, Central Amazon, Brazil: 2. The effect of humic and fulvic acids. *Hydrobiologia* **109**: 97–103. [16.6]

Dickerman JA, Stewart AJ, Wetzel RG. (1986) Estimates of net annual aboveground production: sensitivity to sampling frequency. *Ecology* **67**: 650–9. [16.3]

Elwood JW, Nelson DJ. (1972) Periphyton production and grazing rates in a stream measured with a P^{32} material balance method. *Oikos* **23**: 295–303. [16.3]

Eppley RW, Stewart E, Abbott MR, Heyman U. (1985) Estimating ocean primary production from satellite chlorophyll: introduction to regional differences and statistics from the southern California bight. *Journal of Plankton Research* **7**: 57–70. [16.2]

Fisher SG, Likens GE. (1973) Energy flow in Bear Brook, New Hampshire: an integrative approach to stream ecosystem metabolism. *Ecological Monographs* **43**: 421–39. [16.1]

Francko DA, Heath RT. (1982) UV-sensitive complex phosphorus: association with dissolved humic material and iron in a bog lake. *Limnology and Oceanography* **27**: 564–9. [16.6]

Gosz JR, Dahm CN, Risser PG. (1988) Long-path FTIR measurement of atmospheric trace gas concentrations. *Ecology* **69**: 1326–30. [16.2]

Gregory SV. (1980) *Effects of light, nutrients, and grazing on periphyton communities in streams.* Dissertation. Oregon State University, Corvallis, Oregon. [16.2]

Gregory SV. (1983) Plant–herbivore interactions in stream systems. In: Barnes JR, Minshall GW. (eds) *Stream Ecology: Application and Testing of General Ecological Theory*, pp 157–189. Plenum Press, New York. [16.6]

Gregory SV, Swanson FJ, McKee WA. (1991) An ecosystem perspective of riparian zones. *BioScience* **41**: 540–41.

Grimm NB, Fisher SG. (1984) Exchange between surface and interstitial water: Implications for stream metabolism and nutrient cycling. *Hydrobiologia* **111**: 219–228. [16.1]

Hall CAS. (1972) Migration and metabolism in a temperate stream ecosystem. *Ecology* **53**: 585–604. [16.3]

Hall CAS, Moll R. (1975) Methods of assessing aquatic primary productivity. In: Lieth H, Whittaker RH (eds) *Primary Productivity of the Biosphere*, pp 19–53. Springer-Verlag, New York. [16.2]

Hill BH, Webster JR. (1983) Aquatic macrophyte contribution to the New River organic matter budget. In: Fontaine TD, Bartell SM (eds) *Dynamics of Lotic Ecosystems*, pp 273–82. Ann Arbor Science, Ann Arbor, Michigan. [16.5]

Hornick LE, Webster JR, Benfield EF. (1981) Periphyton production in an Appalachian mountain trout stream. *American Midland Naturalist* **106**: 22–36. [16.3]

Hynes HBN. (1970) *The Ecology of Running Waters.* Liverpool University Press, Liverpool. [16.1]

Hynes HBN. (1974) The stream and its valley. *Verh Internat Verein Limnol* **19**: 1–15. [16.1, 16.3]

Jackson TA, Hecky RE. (1980) Depression of primary productivity by humic matter in lake and reservoir waters of the boreal forest zone. *Canadian Journal of Fisheries and Aquatic Sciences* **37**: 2300–17. [16.6]

Jackson TA, Schindler DW. (1975) The biogeochemistry of phosphorus in an experimental lake environment: evidence for the formation of humic–metal-phosphate complexes. *Verhandlungen Internationale Vereinigung für Theoretische und Angewandt Limnologie* **19**: 211–21. [16.6]

Junk W. (1970) Investigations on the ecology and production biology of the 'floating meadows' (Paspalo-Echinochloetum) on the middle Amazon. Part 1. The floating vegetation and its ecology. *Amazoniana, Kiel* **2(4)**: 449–95. [16.4]

Kellerhals R, Church M. (1989) The morphology of large rivers: characterization and management. In: Dodge DP (ed.) *Proceedings of the International Large River Symposium*, pp 31–48. Canadian Special Publication of Fisheries and Aquatic Sciences 106. Ottawa, Ontario. [16.5]

Lamberti GA, Ashkenas LR, Gregory SV, Steinman AD. (1987) Effects of three herbivores on periphyton communities in laboratory streams. *Journal of the North American Benthological Society* **6**: 92–104. [16.6]

Lay JA, Ward AK. (1987) Algal community dynamics in two streams associated with different geological regions in the southeastern United States. *Archives of Hydrobiology* **108**: 305–24. [16.3]

Lindeman RL. (1942) The trophic–dynamic aspect of ecology. *Ecology* **23**: 399–418. [16.3]

Losee RF, Wetzel RG. (1992) Littoral flow rates within and around submersed macrophyte communities.

Freshwater Biology (in press) [16.6]

Madsen JD, Adams MS. (1988) The seasonal biomass and productivity of the submerged macrophytes in a polluted Wisconsin stream. *Freshwater Biology* **20**: 41–50. [16.4]

Mayer MS, Likens GE. (1987) The importance of algae in a shaded headwater stream as food for an abundant caddisfly (Trichoptera) *Journal of the North American Benthological Society* **6**: 262–9. [16.1]

Meyer JL. (1990) A blackwater perspective on riverine ecosystems. *BioScience* **40**: 643–51. [16.1]

Mickle AM, Wetzel RG. (1978) Effectiveness of submersed angiosperm–epiphyte complexes on exchange of nutrients and organic carbon in littoral systems. II. Dissolved organic carbon. *Aquatic Botany* **4**: 317–29. [16.6]

Mickle AM, Wetzel RG. (1979) Effectiveness of submersed angiosperm–epiphyte complexes on exchange of nutrients and organic carbon in littoral systems. III. Refractory organic carbon. *Aquatic Botany* **6**: 339–55. [16.6]

Minshall GW. (1967) Role of allochthonous detritus in the trophic structure of a woodland springbrook community. *Ecology* **48**: 139–49. [16.3]

Minshall GW. (1978) Autotrophy in stream ecosystems. *BioScience* **28**: 767–71. [16.1, 16.3, 16.5]

Minshall GW, Petersen RC, Cummins KW, Bott TL, Sedell JR, Cushing CE, Vannote RL. (1983) Interbiome comparison of stream ecosystem dynamics. *Ecological Monographs* **53**: 1–25. [16.2, 16.5]

Minshall GW, Cummins KW, Petersen RC, Cushing CE, Bruns DA, Sedell JR, Vannote RL. (1985) Developments in stream ecosystem theory. *Canadian Journal of Fisheries and Aquatic Sciences* **42**: 1045–55. [16.5]

Naiman RJ. (1983) The annual pattern and spatial distribution of aquatic oxygen metabolism in boreal forest watersheds. *Ecological Monographs* **53**: 73–94. [16.2, 16.5]

Odum HT. (1956) Primary production of flowing waters. *Limnology and Oceanography* **2**: 85–97. [16.2]

Otsuki A, Wetzel RG. (1972) Coprecipitation of phosphate with carbonates in a marl lake. *Limnology and Oceanography* **17**: 763–7. [16.6]

Otsuki A, Wetzel RG. (1973) Interaction of yellow organic acids with calcium carbonate in fresh water. *Limnology and Oceanography* **18**: 490–4. [16.6]

Petersen RC Jr, Cummins KW, Ward GM. (1989) Microbial and animal processing of detritus in a woodland stream. *Ecological Monographs* **59**: 21–39. [16.6]

Pinay G, Décamps H, Chauvet E, Fustec E. (1990) Functions of ecotones in fluvial systems. In: Naiman RJ, Décamps H (eds) *The Ecology and Management of Aquatic–Terrestrial Ecotones*, pp 141–69. Parthenon Publishing Group, Carnforth, UK. [16.5]

Power ME. (1990) Resource enhancement by indirect effects of grazers: armored catfish, algae, and sedi-ment. *Ecology* **71**: 897–904. [16.6]

Power ME, Matthews WJ, Stewart AJ. (1985) Grazing minnows, piscivorous bass, and stream algae: dynamics of a strong interaction. *Ecology* **66**: 1448–56. [16.6]

Reese WH. (1967) *Physiological ecology and structure of benthic communities in a woodland stream*. PhD dissertation, Oregon State University, Corvallis, Oregon. [16.3]

Revsbech NP. (1983) *In situ* measurement of oxygen profiles of sediments by use of oxygen microelectrodes. In: Gnaiger E, Forstner H (eds) *Polarographic Oxygen Sensors*, pp 265–73. Springer-Verlag, Berlin. [16.2]

Riber HH, Wetzel RG. (1987) Boundary layer and internal diffusion effects on phosphorus fluxes in lake periphyton. *Limnology and Oceanography* **32**: 1181–94. [16.6]

Risser PG, Rosswall T, Woodmansee RG. (1988) Spatial and temporal variability of biospheric and geospheric processes: a summary. In: Rosswall R, Woodmansee RG, Risser PG (eds) *Scales and Global Change*, pp 1–10. John Wiley & Sons, New York. [16.2]

Rodgers JH Jr, McKivitt ME, Hammerlund DO, Dickson KL. (1983) Primary production and decomposition of submergent and emergent aquatic plants of two Appalachian rivers. In: Fontaine TD, Bartell SM (eds) *Dynamics of Lotic Ecosystems*, pp 283–301. Ann Arbor Science, Ann Arbor, Michigan. [16.4]

Ryan S. (1990) Diurnal CO_2 exchange and photosynthesis of the Samoa tropical forest. *Global Biogeochemical Cycles* **4**: 69–84. [16.2]

Sedell JR, Froggatt JL. (1984) Importance of streamside forests to large rivers: the isolation of the Willamette River, Oregon, USA from its floodplain by snagging and streamside forest removal. *Verhandlungen Internationale Vereinigung für Theoretische und Angewandt Limnologie* **22**: 1828–34. [16.1, 16.5]

Sedell JR, Richey JE, Swanson FJ. (1989) The river continuum concept: a basis for the expected ecosystem behavior of very large rivers? In: Dodge DP (ed.) *Proceedings of the International Large River Symposium*, Canadian Special Publication of Fisheries and Aquatic Sciences 106. [16.5]

Sedell JR, Reeves GH, Hauer FR, Stanford JA, Haukins CP. (1990) Role of refugia in recovery from disturbances: modern fragmented and disconnected river systems. *Environmental Management* **14**: 711–24. [16.5]

Slaymaker O. (1988) Slope erosion and mass movement in relation to weathering in geochemical cycles. In: Lerman A, Meybeck M (eds) *Physical and Chemical Weathering in Geochemical Cycles*, pp 83–111. Kluwer Academic Publishing, Dordrecht. [16.5]

Steinman AD, McIntire CD, Gregory SV, Lamberti GA, Ashkenas LR. (1987) Effects of herbivore type and density on taxonomic structure and physiognomy of algal assemblages in laboratory streams. *Journal of the North American Benthological Society* **6**: 175–88. [16.6]

Stewart AJ, Wetzel RG. (1981) Dissolved humic materials: photodegradation, sediment effects, and reactivity with phosphate and calcium carbonate precipitation. *Archives of Hydrobiology* **92**: 265–86. [16.6]

Stewart AJ, Wetzel RG. (1982) Influence of dissolved humic materials on carbon assimilation and alkaline phosphatase activity in natural algal–bacterial assemblages. *Freshwater Biology* **12**: 369–80. [16.6]

Stock MS, Ward AK. (1989) Establishment of a bedrock epilithic community in a small stream: microbial (algal and bacterial) metabolism and physical structure. *Canadian Journal of Fisheries and Aquatic Sciences* **46**: 1874–83. [16.1, 16.2, 16.6]

Stock MS, Richardson TD, Ward AK. (1987) Distribution and primary productivity of the epizoic macroalga *Boldia erythrosiphon* (Rhodophyta) in a small Alabama stream. *Journal of the North American Benthological Society* **6**: 168–74. [16.2]

Stream Solute Workshop (1990) Concepts and methods for assessing solute dynamics in stream ecosystems. *Journal of the North American Benthological Society* **9**: 95–119. [16.2]

Thienemann A. (1926) Der Nahrungskreislauf im Wasser. *Verhandlungen der Deutschen Zoologischen Gesellschaft* **31**: 29–79. [16.3]

Triska FJ. (1984) Role of wood debris in modifying channel geomorphology and riparian areas of a large lowland river under pristine conditions: a historical case study. *Verhandlungen Internationale Vereinigung für Theoretische und Angewandt Limnologie* **22**: 1876–92. [16.5]

Triska FJ, Kennedy VC, Avanzino RJ, Zellweger GW, Bencala KE. (1989a) Retention and transport of nutrients in a third-order stream: channel processes. *Ecology* **70**: 1877–92. [16.1]

Triska FJ, Kennedy VC, Avanzino RJ, Zellweger GW, Bencala KE. (1989b) Retention and transport of nutrients in a third-order stream in northwestern California: hyporheic processes. *Ecology* **70**: 1893–905. [16.1, 16.4, 16.6]

Vannote RL, Minshall GW, Cummins KW, Sedell JR, Cushing CE. (1980) The river continuum concept. *Canadian Journal of Fisheries and Aquatic Sciences* **37**: 130–7. [16.1, 16.5]

Wallace JB, O'Hop J. (1985) Life on a fast pad: waterlily leaf beetle impact on water lilies. *Ecology* **66**: 1534–44. [16.6]

Ward AK, Dahm CN, Cummins KW. (1985) *Nostoc* (Cyanophyta) productivity in Oregon stream ecosystems: invertebrate influences and differences between morphological types. *Journal of Phycology* **21**: 223–7. [16.2]

Ward GM, Ward AK, Dahm CN, Aumen NG. (1990) Origin and formation of organic and inorganic particles in aquatic systems. In: Wotton RS (ed.) *The Biology of Particles in Aquatic Systems*, pp 27–56. CRC Press, Boca Raton, Florida. [16.6]

Ward JV. (1989) The four-dimensional nature of river ecosystems. *Journal of the North American Benthological Society* **8**: 2–9. [16.5]

Westlake DF. (1965) Some basic data for investigations of the productivity of aquatic macrophytes. *Memorie dell'Istituto Italiano di Idrobiologia* **18**(Suppl): 229–48. [16.3, 16.4]

Wetzel RG. (1975) Primary production. In: Whitton BA (ed.) *River Ecology*, pp 230–47. Blackwell Scientific Publications, Oxford. [16.1, 16.5]

Wetzel RG. (1983) *Limnology* 2nd edn. Saunders College Publishing, Philadelphia. [16.3, 16.6]

Wetzel RG. (1984) Detrital dissolved and particulate organic carbon functions in aquatic ecosystems. *Bulletin of Marine Sciences* **35**: 503–9. [16.6]

Wetzel RG. (1990) Land–water interfaces: metabolic and limnological indicators. *Verhandlungen Internationale Vereinigung für Theoretische und Angewandt Limnologie* **24**: 6–24. [16.3, 16.6]

Wetzel RG. (1991a) Extracellular enzymatic interactions in aquatic ecosystems: storage, redistribution, and interspecific communication. In: Chróst RJ (ed.) *Microbial Enzymes in Aquatic Environments*, pp 6–28. Springer-Verlag, New York. [16.6]

Wetzel RG. (1992) Gradient-dominated ecosystems: sources and regulatory functions of dissolved organic matter in freshwater ecosystems. In: Kairesalo T, Jones RI (eds) *Dissolved Organic Matter in Lacustrine Ecosystems: Energy Source and System Regulator.* *Hydrobiologia* **229**: 181–98. [16.6]

Wetzel RG, Likens GE. (1991) *Limnological Analyses* 2nd edn. Springer-Verlag, New York. [16.2]

Wissmar RC, Swanson FJ. (1990) Landscape disturbances and lotic ecotones. In: Naiman RJ, Décamps H. (eds) *The Ecology and Management of Aquatic–Terrestrial Ecotones*, pp 65–89. Parthenon Publishing Group, Carnforth, UK. [16.5]

Woodwell GM, Whittaker RH, Reiners WA, Likens GE, Delwiche CC, Botkin DB. (1978) The biota and the world carbon budget. *Science* **199**: 141–6. [16.3]

Yount JD, Niemi GJ. (1990) Recovery of lotic communities and ecosystems from disturbance—a narrative review of case studies. *Environmental Management* **14**: 547–69. [16.5]

17: Energy Budgets

P.CALOW

17.1 INTRODUCTION

Since the classical work of Lindeman (1942) at Cedar Lake, one possible way of considering processes going on within ecosystems has been in terms of energy fluxes. There are a number of reasons why this has proved useful. First, processes involving different systems and different organisms can be expressed in common units, so comparison becomes possible. Second, energy yield from one population, trophic level or community can be represented as production that can be used by dependant systems, so enabling the quantification of interactions between ecological units.

The routes of energy flow and the rules governing them have been described in Chapter 14. Here we consider the rates of energy flow and some of the rules governing them. After a general descrip-

tion of ecosystem energetics, this chapter briefly describes some of the techniques needed for making these measurements before summarizing some results on individuals and populations of river organisms, and finally whole ecosystems. The flow of energy in river ecosystems is accompanied by the spiralling of matter and this is addressed in Chapter 18.

17.2 ENERGY FLOWS THROUGH ECOSYSTEMS

A basic diagram of energy flow through a general ecosystem is given in Fig. 17.1. Energy can flow into ecosystems from sunlight through photosynthesis; but whereas this *must* be the sole input for the biosphere, particular ecosystems might derive energy from other sources as already elaborated organic material. In freshwater systems

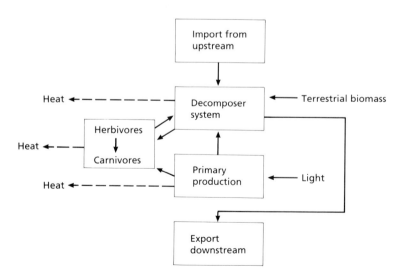

Fig. 17.1 Energy flow through a stretch of river (very general representation).

370

this can occur as either dissolved or particulate material. Energy, once in, passes from one trophic group (primary producers, decomposers, herbivores, carnivores, i.e. trophic levels) to another. As a basic requirement, what goes into each system — ecosystem, trophic levels, populations, individuals — must ultimately come out, albeit in transformed state. This is a requirement of the first law of thermodynamics. Moreover, the second law of thermodynamics requires degradation and increased entropy, and one manifestation of this is heat production and loss from each level leading to transfer efficiencies considerably lower than 100%. These ecological efficiencies were first compiled by Lindeman (see below) and are what is behind the pyramidical structuring of ecosystems in terms of energy flow and, to some extent, biomass and numbers (Phillipson 1966).

Currently, the most widely used unit of energy is the joule (J) and of energy flow (J s^{-1}) is the watt. However, in biological terms it is inconvenient to jump between these, and so energy budgets are generally expressed exclusively in joules (usually over a specified time). Moreover, many of the pre-1980s energy budgets were expressed in calories, and where appropriate this chapter refers to original units — but the conversion is straightforward (4.2 J cal^{-1}; 4.2 kJ kcal^{-1}).

Measuring the potential energy content of biological material is therefore of central importance for work on ecological energetics. It can be determined in at least one of three ways:

1 *Calorimetry*: The basis of this method involves combusting a sample, in the presence of excess oxygen under high pressure, in a chamber, the bomb, and sensing the heat that emanates as a result of this. A useful account of calorimetry is given by Grodzinski *et al* (1975).

2 *Wet oxidation*: The basis of this method consists of treating samples with a solution of a strong oxidant ($K_2Cr_2O_7$ or KIO_3). The difference between the initial quantity of oxidant and the amount left after exposure to the sample represents the amount of oxygen needed to oxidize the organic fraction of the sample. Despite considerable differences in chemical composition, the quantity of energy per unit of oxygen used varies little from one organic substance to

another, and a mean of 14.2 kJ g^{-1} oxygen is now generally used by hydrobiologists (Winberg 1971).

3 *Estimation from chemical composition*: Since lipids, proteins and carbohydrates have characteristic energy equivalents, it is straightforward to estimate the joule equivalent of a sample of biomass from a knowledge of the relative proportion of biochemical substances it contains (see Winberg 1971).

For a compilation of energy equivalents the reader is referred to Cummins and Wuychek (1971). In general, the following ranking of energy equivalents applies: animals > plants > detritus. Aquatic macrophytes predominantly contain cellulose (carbohydrate) and hence have relatively low energy equivalents of approximately 16 kJ g^{-1} (Boyd 1970). Allochthonous detritus, derived largely from plants, has a similarly low value (e.g. Kaushik & Hynes 1971). By contrast, values for microalgae are about 20 kJ g^{-1}.

17.3 INITIAL PRODUCTION

As noted above, for most ecosystems the predominant source of energy income is from sunlight and photosynthesis. This is classical primary production — the total synthesis of plant materials is referred to as gross primary production, and this less that lost to plant metabolism (i.e. that which is, in principle, available to the next trophic level) as net primary production.

The sources of primary production in flowing waters are algae (Chapter 9) and macrophytes (Chapter 10). Methods available for measuring primary production are touched on in these chapters and in more detail in Chapter 16. For more information the reader is referred to Vollenweider (1974) and Wetzel and Likens (1990).

The present understanding of the relationship between the supply of energy from primary production and the demand from herbivores in lotic systems is that the former is several times greater than the latter (Chapter 16), and this is particularly the case for macrophytes (Chapter 10). Hence, most of this primary production is yielded to the detritus. Moreover, with some variation between and within systems (as discussed below), biomass produced outside the system, and entering as particulate or dissolved organic material, is *the* major source of energy income.

Microbial decomposers play a key role in the yield of this dead organic material to consumers, either simply by conditioning the detritus or by converting it into microbial biomass that is more easily utilized by the metazoan consumers. The relative importance of these different roles is currently uncertain, but, clearly, measuring microbial production is of considerable importance. Methods for doing this either for the fungal and bacterial components separately or for the decomposer community in total are dealt with in Chapters 8 and 15.

The incorporation of dissolved organic matter, from the water column via a microbial/protozoan loop, has been speculated upon in Chapters 14 and 15. Such loops appear to be of considerable importance in the economy of marine and 'still' freshwater systems and involve water-column microbes. Likely to be of more importance in the utilization of dissolved organic matter in flowing waters are the 'microbial slimes' that adhere to submerged surfaces.

Whether the action of microbial decomposers is treated as primary or secondary production in lotic systems has been the subject of some debate (Wiegert & Owen 1971), but this is probably largely semantic. A summary of the main initial energy inputs to flowing water systems is given in Fig. 17.2.

17.4 SECONDARY PRODUCTION

Energy bound in plant biomass, detritus and microbial biomass is used by animals to produce their own biomass (secondary production). The processes involved, and how they are measured, are described in this section by particular reference to individuals (see also Chapters 11 and 12).

The energy budget of an individual animal can be summarized as follows:

$$I - F = A = R + G + U$$

where I is food ingested, F is food defaecated (egested), A is food absorbed, R is respiratory heat loss, G is energy accumulated in biomass, and U is energy loss in excretion (Ex) and secretion.

Of the food ingested (I) some is absorbed (A) and some lost as faeces (F). The efficiency by which food is digested and absorbed, absorption efficiency (AE), is given by:

$$AE = A/I = (I - F)/I$$

There are two main ways of measuring ingestion rates and absorption efficiencies:

1 *Gravimetric*: weighing food before and after exposure to a feeder, and quantitatively collecting faeces from a particular meal. Wet-to-dry weight and dry weight-to-joule conversion equations can enable these results to be expressed in energy terms.

2 *Markers*: To determine the ingestion rate of filter-feeding *Simulium* larvae in a stream, Ladle *et al* (1972) introduced activated charcoal powder upstream of the animals for a short time (as a pulse). This was taken in by the larvae, and the effect was to mark the gut contents with a black band. Samples of larvae were collected immediately and then at frequent intervals, the samples being preserved at once. From these, the average rate of passage of food through the gut

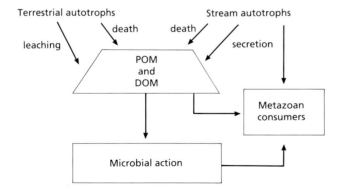

Fig. 17.2 Initial inputs of energy to lotic ecosystems.

could be calculated. This, multiplied by the average weight of gut contents, gives the rate of feeding. Similar techniques have been applied in the laboratory to fishes, where chromic oxide has frequently been used as a marker but also various radiotracers (e.g. ^{32}P, ^{51}Cr, ^{144}Ce) and X-ray techniques have been employed (Talbot 1983).

If the marker is not absorbed, then estimates of its concentration in the faeces as compared with the food can be used to estimate absorption efficiencies, i.e. AE = density of marker in food/density of marker in faeces. Thus chromium and cerium are not absorbed to any appreciable extent by animal membranes and are useful in this context. Clearly, this technique can be carried out on a small sample of faeces and so does not require complete, quantitative collection of faeces (Weeks & Rainbow 1990). If it is assumed that animals can digest only the organic fraction of their food then the ash content can also be used as an indigestible indicator. This is particularly useful for estimating AE in the field where non-quantitative samples of food and faeces need only be taken for an ash analysis (Conover 1966).

As indicated above, heat is lost from all energy conversions and ultimately from the bodies of organisms — and is therefore usually a major negative term in energy budgets.

There are two distinct ways of measuring metabolic heat loss.

1 *Direct calorimetry*: Gnaiger (1979) has used a system for measuring heat production directly in small, freshwater invertebrates. Sensitive thermopyles measure the temperature of water flowing into and out of a well-lagged chamber containing the animal.

2 *Indirect methods*. The amount of metabolic heat produced is proportional to metabolic rate, and that in turn is mainly dependent on oxidative processes and hence proportional to oxygen consumption. The oxyjoule equivalents (joules lost per unit oxygen consumed) depend upon substrates used but, in general, a figure of about 20 kJ l^{-1} oxygen inspired is usually appropriate (Lampert 1984).

There are other sources of energy loss that should be taken into account. Some energy is lost as a result of the loss of excretory products, i.e. breakdown products of excess resources, usually amino acids, from the food. Most of the energy is released in this catabolism, and the major end-product of freshwater animals is NH_3. Winberg (1956) erroneously assumed that the energy value of this is negligible, but the actual value is about 25 J mg^{-1} for ammonia nitrogen (Elliott & Davison 1975). Hence energy losses in the excretory products of fishes may be greater than 3–5% of the daily energy intake (Elliott 1976). These energy losses increase as both fish size and water temperature increase, and are also affected by ration size (Elliott 1979).

Many aquatic invertebrates also secrete large quantities of mucus. Similarly, fish secrete mucus. Potentially, these mucopolysaccharides can represent an appreciable loss of energy (e.g. Calow & Woollhead 1977), but probably make a significant contribution to 'slimes' on submerged surfaces that possibly play an important part in ecosystem energy flows.

17.5 INDIVIDUAL ENERGY BUDGETS WITH SPECIAL REFERENCE TO FISHES

Not surprisingly, in view of their commercial importance, the most detailed work has been carried out on fish energy budgets, pioneered by Ivlev (1945, 1961), but stimulated by the reviews of Brett and Groves (1979) and Elliott (1979) (see also Tytler & Calow 1983).

General summarizing energy budgets have been produced (Brett & Groves 1979) for carnivorous and herbivorous fishes (latter based on data from grass carp):

for carnivores, $100I = 44R + 29G + 27E$

for herbivores, $100I = 16R + 3G + 81E$

where E includes both faeces and excreta.

These have to be treated with considerable caution, but the lower absorption efficiency for herbivores is notable and expected, i.e. approximately $(100-80)/100$ for herbivores compared with approximately $(100-27)/100$ for carnivores. Because of this, it follows that herbivores should have lower gross growth efficiencies (3% versus 29%), but the lower net growth efficiency (approximately $3/(16+3)$ versus $29/(29+44)$, again ignoring excretion) is not typical of other animals (e.g. see Ricklefs 1980).

17.6 POPULATION PRODUCTION

Production (P) of a population of animals is defined as follows:

$$P = nA - nR$$

where A and R are appropriate individual energy budget terms and n is population size, or more usually population density (number per unit habitat). Hence P is very frequently expressed in kJ unit habitat^{-1} time^{-1}. In consequence, population production could be defined by summing individual energy budgets over time, and making due allowance for age and/or size structure. But this would be a very tedious way of measuring net production. Note also that n is not constant — some production will be lost to mortality and some might be gained from immigration (Wiegert 1968), and this complication has to be taken into account (see below).

Clearly, production, as defined above, must also be equivalent to the biomass accumulated by a population over a time interval (including that lost to mortality), so in principle this could be determined directly. Imagine a population of n_0 animals hatching simultaneously from eggs and with a starting size of m_0. The initial biomass is therefore $n_0 m_0$ and is attributed to the production of the preceding generation. Now, imagine the animals all growing at the same rate until eventually a number of animals die or are eaten. If this loss occurs instantaneously when the mass of individuals is m_1, the total production to that moment is $n_0 (m_1 - m_0)$. If the remaining animals (n_1) continue growing at the same rate until other animals die at mass m_2, the production over this second interval is $n_1 (m_2 - m_1)$, and generally over the whole lifecycle is:

$$P = n_0(m_1 - m_0) + n_1(m_2 - m_1) + \ldots + \\ n_i(m_i - m_{i-1})$$

This illustrates the basis of all methods for calculating secondary production from biomass changes. Of course, the real situation is rarely as simple as this; not all individuals in a cohort are born at the same time with the same dry mass,

nor do they die at the same time, and we can rarely measure the mass of individuals at the point of death. Due largely to the pioneering work of Allen (1951), working on trout in a New Zealand stream, various graphical techniques have been developed to address these problems. For a detailed review see Rigler and Downing (1984). There are also some short-cut methods of estimating production (see Chapter 11 and below).

17.7 USEFUL RATIOS

Between 1964 and 1974 there was a concerted international effort to gather information on the productivity of ecological communities. This was known as the International Biological Programme (IBP). A major aim of the IBP was to document, and understand, the relationship between production rate (P) and standing crop (B) in so-called P/B ratios. From all this work, the following generalizations emerged: P/B ratios for zooplankton are about $15-20$, for benthic invertebrates about $3-6$, and for fishes about $0.5-1$ (Waters 1977). These are also discussed by Cummins in Chapter 11. Clearly, if biomass is known and these P/B ratios can be relied on, then P can be estimated from B multiplied by the appropriate ratio.

The efficiency of transfer between trophic levels is often called an ecological efficiency, but sometimes a Lindeman efficiency after the person who first calculated it for a natural ecosystem. A Lindeman efficiency is a measure of the efficiency by which one population, or indeed the whole trophic level (=sum of all individuals of all populations exploiting a particular food type within a community), converts the energy that flows into it to energy that is used by the next trophic level. It is equivalent to P/NA only for steady-state populations, where standing crop or biomass remains constant, i.e. production has to be balanced by biomass in mortality lost to the next trophic level.

There are some flaws in the way that Lindeman calculated his efficiencies (Colinvaux & Barnett 1979), but the low efficiencies that he obtained of around 10% seem to apply generally (Colinvaux 1986). Some note has already been made of the consequences of these for energy flow and biomass relationships (see above), but it should also be noted that if the absolute quantity of biologically useful energy reduces from one trophic level to another, this then puts a constraint on the number of feasible links in food chains. This may provide part of the explanation for food chains rarely being longer than five links, but there are other explanations (Lawton 1989, and Chapter 14).

P/R is an ecosystem ratio of total primary production to total community respiration and is considered further in the next section.

17.8 ENERGY FLOW THROUGH FLOWING-WATER ECOSYSTEMS

An important feature of freshwater systems that makes them attractive for, and amenable to, ecological research is their discrete nature; lakes are bounded on all sides and rivers on two sides by land. This has been of particular importance for energy budget analysis. Yet even here energy transactions across the land/water interface (Chapter 16) can be of overriding importance and make it necessary to distinguish between autochthonous energy inputs (from primary production within the system) and allochthonous inputs (organic imports from terrestrial habitats). There can also be water to land flow of energy, for example during flooding, from the faeces of birds

and amphibious animals, and for the emergence of insects. These are considered further in Chapter 21. And, of course, within any stretch of channel there has to be upstream and downstream connections.

One of the first ecosystem energy budgets to be completed was that of Silver Springs (Fig. 17.3). This is a very large spring with an outflow about 100 m wide. Here, macrophytes make a large contribution to primary production, but periphyton on their surface are even more important. In desert streams, with little or no bank cover, in-stream primary production of periphyton is also the dominant form of energy input to the system (Busch & Fisher 1981). In most lotic systems, however, allochthonous organic material plays a dominant part in the input of energy. This is illustrated in Fig. 17.4 for Bear Brook, which receives almost all its input from the forest canopy, and the same was true for Augusta Creek (see Fig. 11.2). The River Thames has significant amounts of phytoplankton production (Fig. 17.5). Because of human navigation, this system is like a canal; dams or weirs occur at frequent intervals along its length, acting as series of impoundments.

A most important feature of lotic systems, though, is that they change in systematic fashion from source to sink (see Fig. 11.3, Fig. 16.2 & Fig. 16.3). Because of the likelihood of shading

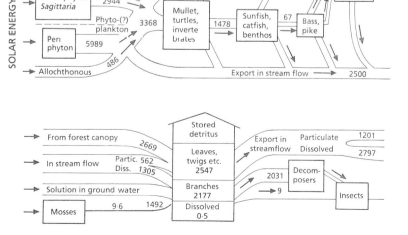

Fig. 17.3 Energy flow through Silver Springs (units are kcal m^{-2} year^{-1}) (after Mann 1980).

Fig. 17.4 Energy flow through Bear Brook (units are kcal m^{-2} year^{-1}) (after Mann 1980).

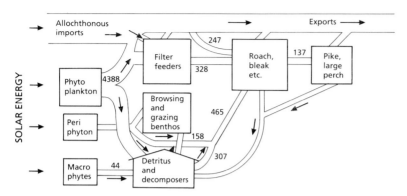

Fig. 17.5 Energy flow through the River Thames, UK (units are kcal m^{-2} year^{-1}) (after Mann 1980).

from surrounding vegetation, heterotrophic processes tend to dominate in upper reaches; primary production becomes more important in middle reaches; because of sediment loads, causing shading, heterotrophic processes are likely to become dominant again in lower courses. P/R ratios are often used as indicators of autotrophy and heterotrophy (Chapter 11). $P/R > 1$ indicates autotrophy whereas $P/R < 1$ suggests heterotrophy. The sequence of change in P/R along a river system is illustrated in Fig. 11.3, and is likely to follow the general rule of moving from <1 in upper reaches to >1 in middle reaches and then return to <1 in lower reaches. These, of course, are generalizations, and exceptions are likely to arise, particularly due to disturbances. An organic effluent input, for example, is likely to lower P/R ratios downstream of it and hence cause P/R to reduce to below 1 locally, even in middle reaches.

Clearly, for any energetically stable system, energy input should equal output, i.e. $P = R$. But where the system is open, as is the case for a stretch of river, imbalances can be maintained by export of excess $(P > R)$ or import to replace a deficit $(P < R)$. Hence, as general indicators of disturbance and stress P/R, R/B and P/B ratios should be used with caution. Papers by Odum (1985), Margalef (1975) and Calow (1991) deal with these issues in a very general way.

17.9 RIVER CONTINUUM CONCEPT

The shifts in major sources of energy income along the lengths of rivers, together with the way the energetics of the ecosystems in one stretch

are influenced by the energetics of upstream stretches (by import of production and washed-out detritus) and in turn influence the energetics of downstream stretches (by export), will lead to a longitudinal continuum of ecosystem processes. It is arguable that there are predictable changes in these processes, reflected, for example, by changes in P/R ratios (see above) along the continuum and that these drive trophodynamics which in turn determine trophic structure (Chapter 11). This is the basis of the river continuum concept (Chapters 11 & 23).

That there is a continuum of energetic processes in systems that are longitudinally extended is *almost* beyond doubt—apart from isolated and semi-isolated pockets that might persist in these continua (e.g. dead-zones; Chapters 5 & 9). But how general and predictable are the continua in both functional and structural terms through space and time?

A first general comment is that though function determines structure (e.g. food webs and community structure) it would appear that, ecologically, many variants of structure are compatible with a particular type of functioning dependent upon other conditions (Odum 1985).

Second, it has already been noted that general longitudinal patterns within systems can be disturbed by pollution, land management (e.g. forestation or deforestation), and also potentially by local discontinuities in natural features such as geology and gradients. Between systems, changes in the latter and also water chemistry are likely to cause variation in the longitudinal patterns in function and structure (Winterbourn 1986).

Finally, an obvious feature of flowing-water systems is their temporal variability — driven by variability in physical conditions (see Part 2). These, together with physical variability along the length of a river, are likely to lead to a patchiness in susceptibility to scouring and washing away of communities and hence to a patchiness, at any one time, in the development of community and ecosystem characteristics in particular places (Townsend 1989; Statzner & Higler 1986). The developmental stage in recolonization (succession) is known to have an effect on energetic function (Odum 1962) and community structure (Williams & Hynes 1976).

The extent to which this patchiness, dependent on temporal variability, dominates the longitudinal pattern may vary from one system to another. This is discussed further in Chapter 23.

REFERENCES

Allen KR. (1951) The Horokiwi Stream. A study of a trout population. *New Zealand Marine Department of Fisheries Bulletin* **10**: 1–231. [17.6]

Boyd CE (1970) Amino acid, protein and caloric content of vascular aquatic macrophytes. *Ecology* **51**: 902–6. [17.2]

Brett JR, Groves TDD. (1979) Physiological energetics. In: Hoar WS, Randall DH, Brett JR (eds) *Fish Physiology* **8**: pp 279–352. Academic Press, New York. [17.5]

Busch DE, Fisher SG. (1981) Metabolism of a desert stream. *Freshwater Biology* **11**: 301–7. [17.8]

Calow P. (1991) Physiological costs of combating chemical toxicants: ecological implications. *Comparative Biochemistry and Physiology* **100 C**: 3–6. [17.8]

Calow P, Woollhead AS. (1977) The relationship between ration, reproductive effort and the evolution of life-history strategies — some observations on freshwater triclads. *Journal of Animal Ecology* **46**: 765–81. [17.4]

Colinvaux PA. (1986) *Ecology*. John Wiley & Sons, New York. [17.7]

Colinvaux PA, Barnett BD. (1979) Lindeman and the ecological efficiency of wolves. *American Naturalist* **114**: 707–18. [17.7]

Conover RJ. (1966) Assimilation of organic matter by zooplankton. *Limnology and Oceanography* **11**: 338–45. [17.4]

Cummins KW, Wuychek JC. (1971) Caloric equivalents for investigation in ecological energetics. *International Association of Theoretical and Applied Limnology Communication* **18**: 1–158. [17.2]

Elliott JM. (1976) Energy losses in the waste products of brown trout (*Salmo trutta* L.). *Journal of Animal Ecology* **45**: 561–80. [17.4]

Elliott JM. (1979) Energetics of freshwater teleosts. *Symposium of the Zoological Society of London* **44**: 29–61. [17.4, 17.5]

Elliott JM, Davison W. (1975) Energy equivalents of oxygen consumption in animal energetics. *Oecologia* **19**: 195–201. [17.4]

Gnaiger E. (1979) Direct calorimetry in ecological energetics. Long-term monitoring of aquatic animals. *Experimentia* **57** (Suppl.): 155–65. [17.4]

Grodzinski W, Klekowski RZ, Duncan A. (eds) (1975) *Methods for Ecological Bioenergetics*. IBP Handbook No. 24, Blackwell Scientific Publications, Oxford. [17.2]

Ivlev VS. (1945) (*The biological productivity of waters*). *Uspekhi Sovremennoi Biologii* (Fisheries Research Board, Canada, Translation Ser. No. 394). [17.5]

Ivlev VS. (1961) *Experimental Ecology of the Feeding of Fishes*. Yale University Press, New Haven. [17.5]

Kaushik NK, Hynes HBN. (1971) The fate of dead leaves that fall into streams. *Archiv für Hydrobiologie* **68**: 465–515. [17.2]

Ladle M, Bass JAB, Jenkins WR. (1972) Studies on production and food consumption by the larval Simuliidae (Diptera) of a chalk stream. *Hydrobiologia* **39**: 429–48. [17.4]

Lampert W. (1984) The measurement of respiration. In: Downing JA, Rigler FH (eds) *A Manual on Methods for the Assessment of Secondary Productivity in Fresh Waters*. IBP Handbook No. 17, 2nd edn. Blackwell Scientific Publications, Oxford. [17.4]

Lawton JH. (1989) Food webs. In: Cherrett JM (ed.) *Ecological Concepts* pp 43–78. Symposia of the British Ecological Society, Blackwell Scientific Publications, Oxford. pp 413–68. [17.7]

Lindeman RL. (1942) The trophic dynamic aspects of ecology. *Ecology* **23**: 399–418. [17.1]

Mann KH. (1980) The total aquatic system. In: Barnes RSK, Mann KH (eds) *Fundamentals of Aquatic Ecosystems*, pp 185–220. Blackwell Scientific Publications, Oxford. [17.8]

Margalef R. (1975) Human impact on transportation and duration in ecosystems. How far is extrapolation valid? In: *Proceedings of the 1st International Congress of Ecology*, pp 237–41. Centre for Agricultural Publishing and Documents, Wageningen, the Netherlands. [17.8]

Odum EP. (1962) Relationships between structure and function in the ecosystem. *Japanese Journal of Ecology* **12**: 108–18. [17.9]

Odum EP. (1985) Trends expected in stressed ecosystems. *BioScience* **35**: 419–22. [17.8, 17.9]

Phillipson J. (1966) *Ecological Energetics*. Edward Arnold, London. [17.2]

Ricklefs RC. (1980) *Ecology* 2nd edn. Wilson & Sons Ltd., Chichester, New York. [17.5]

Rigler FH, Downing JA. (1984) The calculation of secondary productivity. In: Downing JA, Rigler FH (eds) *A Manual on Methods for the Assessment of Secondary Productivity in Fresh Waters.* IBP Handbook No. 17, 2nd edn. Blackwell Scientific Publications, Oxford. pp 19–58. [17.6]

Statzner B, Higler B. (1986) Stream hydraulics as a major determinant of invertebrate zonation patterns. *Freshwater Biology* **16**: 127–37. [17.9]

Talbot C. (1983) Laboratory methods in fish feeding and nutritional studies. In: Tytler P, Calow P (eds) *Fish Energetics. New Perspectives*, pp 125–34. Croom Helm, London. [17.4]

Townsend CR. (1989) The patch dynamics of stream community ecology. *Journal of the North American Benthological Society* **8**: 36–50. [17.9]

Tytler P, Calow P. (1983) *Fish Energetics. New Perspectives.* Croom Helm, London. [17.5]

Vollenweider RA. (1974) *A Manual on Methods for Measuring Primary Production in Aquatic Environments.* IBP Handbook No. 12. Blackwell Scientific Publications, Oxford. [17.3]

Waters TF. (1977) Secondary production in inland waters. *Advances in Ecological Research* **10**: 91–164. [17.7]

Weeks JM, Rainbow PS. (1990) A dual labelling tech- nique to measure the relative assimilation efficiencies of invertebrates taking up trace metals from food. *Functional Ecology* **4**: 711–17. [17.4]

Wetzel RG, Likens GE. (1990) *Limnological Analysis* 2nd edn. Spring & Verlas, New York. [17.3]

Wiegert RG. (1968) Population energetics of Meadow Spittlebugs, *Philaenus spumarius* L., as affected by migration and habitat. *Ecological Monographs* **34**: 217–41.

Wiegert RG, Owen DF. (1971) Trophic structure, available resources and population density in terrestrial vs. aquatic ecosystems. *Journal of Thermodynamic Biology* **30**: 69–81. [17.3]

Williams DD, Hynes BN. (1976) The recolonisation mechanisms of stream benthos. *Oikos* **27**: 265–77. [17.9]

Winberg CG. (1956) (*Rate of metabolism and food requirements of fishes*). Minsk Belorussion State University. (Fisheries Research Board, Canada Translation Series. No. 194; 1–253.) [17.4]

Winberg CG. (1971) *Methods for the Estimation of Production of Aquatic Animals.* Academic Press, London. [17.2]

Winterbourn MT. (1986) Recent advances in our understanding of stream ecosystems. In: Polunin N (ed.) *Ecosystem Theory and Applications*, pp 240–68. John Wiley & Sons, London. [17.9]

18: Cycles and Spirals of Nutrients

J.D. NEWBOLD

18.1 INTRODUCTION

Rivers play a major role in global biogeochemical cycling by transporting elements from terrestrial environments to the sea (Chapter 4). Many of these elements are essential nutrients and are utilized by river biota. A number of the major ions found in river water, such as Ca, Mg, K, Na, Si and Cl, are often present well in excess of any biological demands within the river, and may pass through the river system virtually unaffected. Other elements — notably, carbon, phosphorus and nitrogen, or particular chemical forms of these elements — may be in relatively short supply and undergo considerable utilization as they pass downstream.

Biota remove nutrients from river water, but they also regenerate nutrients to the water. This cycling of nutrients within the river may proceed intensively and yet produce small or negligible net effects on nutrient concentrations. In fact, for nutrients such as phosphorus, which does not exchange with the atmosphere, biota cannot alter the long-run total transport substantially. On the other hand, biota do influence the chemical and physical forms of nutrients, and the timing of nutrient transport. These effects may interact, in turn, with physical transport processes. For example, the biota might speed downstream transport by converting particulate-bound nutrients to dissolved forms or reducing detrital particles to smaller, more easily transported, sizes. Carbon, nitrogen and, to a limited extent, sulphur exchange with the atmosphere. For these elements, biota may strongly influence long-run total transport. Chapter 4 described variation in chemical composition *among* river systems. This chapter considers the dynamics of selected chemicals (nutrients) *within* river systems. For related reviews see Meyer *et al* (1988) and Stream Solute Workshop (1990).

Cycling is of interest not only because the biota may affect nutrient concentrations but because nutrient concentrations may affect the biota. These interactive influences, however, occur on a template of continual downstream transport, so that biotic processes in upstream reaches may influence those in downstream reaches. As a nutrient atom undergoes a series of transformations, completing a 'cycle' by returning to a previous state, it also traverses some distance downstream. This open, or longitudinally displaced, cycling has been termed 'spiralling' (Webster 1975; Wallace *et al* 1977; Webster & Patten 1979). Within the framework of the spiralling concept we can view cycling as involving exchanges and transformations that can be quantified on an areal or volumetric basis; it does not involve downstream transport (Fig. 18.1(a)). In this sense, cycling occurs at any point in the river, regardless of the fact that the atoms involved in the cycle represent an ever-changing population, and no individual atoms remain in place long enough to complete a cycle. Spiralling, on the other hand, involves measures of both cycling and downstream transport and refers explicitly to the longitudinal scale over which cycles occur (Fig. 18.1(b)).

18.2 OVERVIEW OF CYCLING AND SPIRALLING

Cycles

Figure 18.2 provides a simplified overview of the cycling of the three major elements we shall

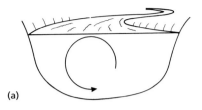

(a)

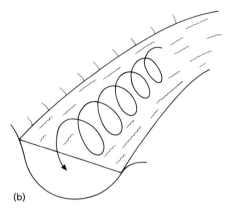

(b)

Fig. 18.1 (a) The nutrient cycle as viewed from a perspective that does not 'see' downstream transport. (b) The nutrient cycle, in conjunction with downstream transport, describes a spiral.

consider here: carbon, phosphorus and nitrogen. In this view of the river ecosystem, the metabolic activity within the river depends upon two major sources of organic carbon: primary production that occurs within the river (autochtonous carbon) and organic carbon supplied from the terrestrial environment (allochthonous carbon). Organic carbon passes through the food web (within which are significant subcycles not represented here) to an ultimate fate of being respired to dissolved inorganic carbon (DIC). Because DIC exchanges relatively freely with atmospheric carbon dioxide, we represent here only the organic-carbon 'half-cycle', which represents the primary pathway of energy flow in the ecosystem. The base of the carbon food web consists of both primary producers (algae, cyanobacteria, macrophytes, bryophytes; see Chapters 9, 10 & 16), and microbial consumers of the allochthonous carbon (bacteria and fungi; see Chapters 8 & 15). Both groups use inorganic phosphorus and nitrogen from the river water, and these elements flow through the food web in rough stoichiometric proportion to the flow of carbon and energy. As the carbon is respired to DIC, the phosphorus and nitrogen are regenerated as inorganic forms to be recycled to the algae and microbes.

In many ecosystems, a large fraction of the metabolism is supported by recycled nutrient (Pomeroy 1970), i.e. total nutrient utilization

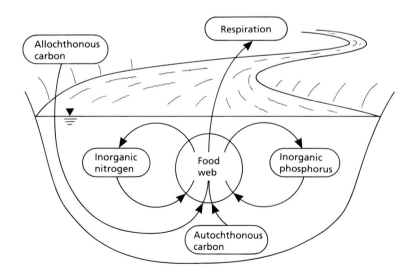

Fig. 18.2 Cycles of nitrogen and phosphorus are coupled to the 'half-cycle' of organic carbon.

within the ecosystem exceeds the supply of new nutrient from outside the ecosystem (Dugdale & Goering 1967). By delineating ecosystem boundaries and comparing imports and exports of nutrients to internal storage and utilization, much can be inferred about the significance of nutrient cycling to the productivity of the ecosystem. How does this logic apply to a stream or river? If the 'ecosystem' is a very short river reach, upstream imports and downstream exports may overwhelm internal utilization, suggesting that cycling is unimportant. Yet if the reach is very long, the opposite might be the case. To address these issues, we turn to the spiralling concept.

Spirals

A given nutrient atom, as it passes downstream, may be used again and again, the amount of utilization depending on the 'tightness' of the spirals (Webster & Patten 1979) or the downstream displacement from one cycle to the next. This in turn depends not only on how quickly cycling occurs, but also on the retentiveness of the ecosystem, or the degree to which the downstream transport of nutrient is retarded relative to that of water. That is, if it takes an average time, t_C, for the average nutrient atom to complete a cycle, while it moves downstream at an average velocity, v_T, then the cycle is completed over a downstream distance of:

$$S = v_T t_C \tag{1}$$

where S is the spiralling length (Elwood *et al* 1983). The velocity, v_T, may be near that of water, v_W, in large rivers, but very much lower in streams and rivers where nutrients reside in the sediments for a high proportion of the time.

The cycle of a nutrient such as phosphorus or nitrogen can be thought of as consisting of two components: (1) the biological assimilation ('uptake') of dissolved inorganic nutrients from the water column; and (2) the subsequent biological processing and movement through the food web leading eventually to regeneration to the inorganic form (Fig. 18.3). Spiralling length consists of the average downstream distance travelled by a dissolved nutrient atom until uptake (the 'uptake length') plus the downstream distance travelled within the biota until regeneration (the 'turnover length'). To illustrate the relationship between spiralling length, nutrient fluxes and nutrient retentiveness, we shall simplify the river ecosystem to two compartments: water (W), by which we mean the inorganic nutrient dissolved in the water, and biota (B), by which we mean nutrient in living tissue.

Suppose that the biotic utilization of nutrients from the water compartment is U, expressed as mass per unit area per unit time ($M L^{-2} T^{-1}$), and the downstream flux of dissolved nutrient is F_W, as mass per unit width of river per unit time ($M L^{-1} T^{-1}$), which can also be expressed as $F_W = C_W v_W d$, where C_W is the dissolved nutrient concentration, v_W is the water velocity, and d is river depth. Thus, in each unit distance of river, a

Fig. 18.3 Spiralling in a simplified river ecosystem consisting of two compartments: water (W) and biota (B). The spiralling length, S, is the average distance a nutrient atom travels downstream during one cycle through the water and biotic compartments. It is the sum of the uptake length (S_w) and the turnover length (S_n), and can be calculated from the downstream nutrient fluxes (F_w and F_n) and the exchange fluxes (U and R), as described in the text.

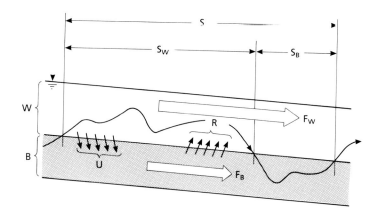

proportion, k_L $[L^{-1}] = U/F_w$, of the downstream flux is taken up by the biota. Depending on the level of nutrient regeneration and on external sources of nutrients, F_w may decline, remain uniform, or increase in the downstream direction. Provided, however, that the ratio U/F_w (i.e. k_L) remains uniform, then the particular atoms, F_w^*, passing any point in the river will disappear from the water column exponentially with distance, x:

$$F_w^* = F_{w0}^* \exp(-k_L x) \qquad (2)$$

where $F_{w0}^* = F_w$ at $x = 0$. For example, if a tracer (such as $^{32}PO_4$ for phosphorus) is injected into a river, equation (2) describes the longitudinal disappearance of the tracer and can be used to estimate k_L (Ball & Hooper 1963; Newbold *et al* 1981). The average travel distance, or uptake length, is given by $S_w = 1/k_L$, which can be verified by a simple integration of equation (2). Thus, since $k_L = U/F_w$:

$$S_w = F_w/U \qquad (3)$$

If followed downstream at the water velocity, v_w, the nutrient atoms disappear exponentially with time (as well as distance), at the rate $k_w = k_L v_w$. Substituting for k_L, we can write $S_w = 1/k_L = v_w/k_w$, or, given that the residence time in the water is $t_w = 1/k_w$, $S = v_w T_w$. The stock of inorganic nutrient, on a unit-area basis, is $X_w = C_w d$, so that the uptake can be written as $U = k_w X_w$.

The turnover length, or distance of travel in the biotic compartment, is analogous to the uptake length. Suppose that the standing stock of nutrient in the biota is X_B $[M L^{-2}]$, and that a fraction, k_B, is regenerated per unit time giving a residence time for nutrient atom of $1/k_B$, and a regeneration flux of R $[M L^{-2} T^{-1}] = k_B X_B$. The biotic compartment is distributed between that in the water column X_{sus} (moving downstream at v_w) and that in the sediments, X_{sed} (with zero velocity). That is, $X_B = X_{sus} + X_{sed}$. The weighted average velocity at which the compartment as a whole moves downstream, then is $v_B = (X_{sus}/X_B)v_w$. Therefore, during its residence time of $1/k_B$ in the biotic compartment, a nutrient atom will travel an average distance:

$$S_B = v_B/k_B \qquad (4)$$

which is the turnover length.

The downstream flux (per unit width) of the biotic compartment is $F_B [M L^{-1} T^{-1}] = v_w X_{sus} = v_B X_B$. From this, and the definition of R above, it can be seen that equation (4) is equivalent to:

$$S_B = F_B/R \qquad (5)$$

The total downstream displacement during one cycle, or spiralling length, then is:

$$S = S_w + S_B = v_w/k_w + v_B/k_B = F_w/U + F_B/R \qquad (6)$$

In the idealized case in which the river is at steady state and is longitudinally uniform on the scale of the spiralling length, the regeneration flux equals the uptake flux $(R = U)$ so that:

$$S = F_T/U \qquad (7)$$

where F_T is the total downstream nutrient flux, or $F_w + F_B$. Thus, the spiralling length represents the distance of river over which utilization is equal to the downstream flux of nutrients or, alternatively, over which the downstream flux is cycled, on average, only once.

Equation (7) quantifies the intuitive idea that more intensive nutrient utilization involves cycling over shorter distances, or 'tighter' spiralling. The total time for completing a cycle is $t_C = 1/k_w + 1/k_B$, while the average downstream velocity for the entire nutrient stock is, $v_T = F_T/X_T$, where $X_T = X_w + X_B$. By combining these definitions with equation (6) and assuming $R = U$, one can obtain equation (1), i.e. $S = v_T t_C$.

An actual analysis might involve several ecosystem compartments (see section 18.6). Each compartment has a turnover rate and a downstream velocity, and hence a turnover length. The total spiralling length can be determined from the individual turnover lengths by weighting each compartmental turnover length by the proportion of the total uptake flux that passes through it, and summing these weighted values (Newbold *et al* 1983a). The biotic compartment presented here, and its turnover length, may be thought of as an aggregation of the entire portion of the cycle in which the nutrient is not in the dissolved inorganic form, and hence is temporarily unavailable for plant and microbial uptake. This may include nutrient in dead tissue and nutrient in less available organic molecules. However, if the biotic compartment actually consists of several compartments with differing turnover rates and transport velocities, the equation $S_B = F_B/R$ is only an approximation unless the ecosystem is at steady state and without longitudinal gradients in standing stock.

18.3 NUTRIENT LIMITATION AND NUTRIENT RETENTION

Primary productivity in most of the world's lakes is limited by phosphorus, nitrogen, or both, even when these nutrients are supplied at relatively high levels of anthropogenic enrichment (Dillon & Rigler 1974; Vollenweider 1976; Elser *et al* 1990). Nutrient limitation of primary productivity and microbial activity in streams and smaller rivers, however, appears to be far less widespread and largely restricted to near-pristine conditions (see Chapter 16). Phosphorus limitation of algal growth or production has been reported in several streams (Stockner & Shortreed 1978; Elwood *et al* 1981b; Horner & Welch 1981; Pringle & Bowers 1984; Bothwell 1985, 1988; Peterson *et al* 1985; Freeman 1986; Knorr & Fairchild 1987; Pringle 1987; Keithan *et al* 1988; Biggs & Close 1989; Hart & Robinson 1990; Horner *et al* 1990; Lock *et al* 1990), and there have been a few reports of phosphorus limitation of heterotrophic microbial activity (Elwood *et al* 1981b; Klotz 1985; Lock *et al* 1990). In all cases concentrations of soluble reactive phosphorus (SRP; see section 18.6) were ≤ 15 μg l^{-1} and frequently <5 μg l^{-1}. Mixed results have been reported in the range of $3-50$ μg l^{-1} (Wurhmann & Eichenberger 1975; Wong & Clark 1976; Horner & Welch 1981; Horner *et al* 1983; Bothwell 1985, 1989; Klotz 1985; Hullar & Vestal 1989), while higher concentrations have not revealed limitation (Pringle *et al* 1986; Knorr & Fairchild 1987; Kiethan *et al* 1988; Munn *et al* 1989).

Periphyton growth has responded to nitrogen enrichment in several streams where inorganic nitrogen concentrations were less than 60 μg l^{-1} (Grimm & Fisher 1986; Hill & Knight 1988; Triska *et al* 1989a), and has failed to respond in streams where inorganic nitrogen exceeded 50 μg l^{-1} (Wurhmann & Eichenberger 1975; Pringle *et al* 1986; Knorr & Fairchild 1987; Kiethan *et al* 1988; Munn *et al* 1989). The apparently consistent transition at $50-60$ μg l^{-1} should be viewed with caution because of the small number of studies involved, and because inorganic nitrogen includes both ammonium and nitrate, which are assimilated with differing efficiencies (see section 18.7). Meyer and Johnson (1983) inferred that microbial activity in decomposing leaves was nitrogen limited at 8 μg l^{-1}, whereas Newbold *et al* (1983b) saw no effect at 30 μg l^{-1}.

One explanation for the lower incidence of nutrient limitation in streams than in lakes is that water velocity and associated turbulence have an enriching effect (Ruttner 1963). Uptake of nutrients by periphyton is enhanced at higher velocities (Whitford & Schumacher 1961, 1964), presumably because diffusion barriers to nutrient transport are reduced (see Chapter 9). Yet this explanation is not sufficient because: (1) planktonic algae regularly deplete nutrients to levels far below those at which nutrient limitation has been observed in streams; and (2) algal mats or biofilms present a diffusion barrier that at least partially offsets the enriching effect of velocity (Bothwell 1989).

An alternative explanation for the lower incidence of nutrient limitation in streams is that they are open systems with a large capacity to retain nutrients. Lakes are vulnerable to nutrient limitation in large part because they act, in the short run, as closed systems. The total nutrient in the water column remains relatively constant but, as plankton populations increase, dissolved available nutrient is incorporated into biomass. Net incorporation of nutrient necessarily ceases when nearly all of the nutrient is sequestered. It is this very low residual concentration, rather than the potentially high initial concentration, that actually limits algal growth.

In a stream with a continual supply of nutrient from the watershed, benthic organisms may deplete the available nutrient supply in three ways: (1) by temporarily reducing downstream flux (F_T) during nutrient accumulation (Grimm 1987); this depletion ceases at steady state; (2) diverting nutrients to the atmosphere or terrestrial environment; this is rarely significant except for denitrification (section 18.7); and (3) detaching and carrying the nutrient downstream in unavailable form, i.e. as the flux, F_B. In the latter case, the flux of dissolved inorganic nutrient at steady state is $S_W = F_T - F_B$. Recalling from the previous section that the total biomass X_B migrates downstream at the velocity v_B, and that $F_B = v_B X_B$, we see that nutrient depletion increases as the retentiveness of the stream or river decreases. A retentive stream is characterized by a short turnover length relative to the uptake length since, if $R = U$, equation (6) yields:

$$S_W/S_B = F_W/F_B \qquad (8)$$

Many streams and smaller rivers are highly retentive $(v_B \ll v_W)$, supporting large standing stocks with little nutrient depletion (Newbold *et al* 1982b; Newbold 1987; section 18.6). Thus, we would expect nutrient limitation to occur only as a transient phenomenon during rapid growth, or when nutrient is supplied from the watershed at concentrations already low enough to be limiting.

Growth rates of many algae and bacteria are saturated by nutrient concentrations substantially below the limiting concentrations observed in the field studies reviewed above. To explain this discrepancy, Bothwell (1989) suggested and experimentally supported the hypothesis that nutrient concentrations can be locally depleted to limiting levels within algal biofilms. A second possible explanation for the discrepancy — applicable to phosphorus limitation — is that soluble reactive phosphorus measurements may overestimate PO_4-P concentrations (section 18.6).

In larger, less retentive, rivers we might expect nutrient depletion and nutrient limitation to be more prevalent. In an entirely planktonic river, where $v_B = v_W$, nutrient in biomass depletes the dissolved inorganic nutrient on a one-to-one basis, as in a closed algal culture or the epilimnion of a lake. River phytoplankton has been reported to deplete phosphorus from municipal discharges to limiting levels (Décamps *et al* 1984). But in several rivers, phytoplankton has been observed not to be nutrient limited (Burkholder-Crecco & Bachmann 1979; Megard 1981; Elser & Kimmel 1985; Krogstad & Løvstad 1989; Wiley *et al* 1990). Three factors — non-algal turbidity, extremely high nutrient concentrations and water residence time — seem to account for the absence of nutrient limitation. Søballe and Kimmel (1987), in a comparison of 126 rivers with 149 natural lakes in the USA, found that in rivers with residence times (i.e. time-of-travel from the river source) greater than about 50 days, the relationship between total phosphorus and phytoplankton populations approximated that of lakes (strongly implying nutrient limitation), whereas in rivers with shorter residence times, phytoplankton populations were lower than in lakes with equivalent phosphorus concentrations.

18.4 MODELLING SPATIAL AND TEMPORAL VARIATION

The spiralling concept as presented so far does not account for spatial and temporal variation in nutrient fluxes and stocks. Mathematical simulations are widely used to describe such variations, as well as other aspects of ecosystem dynamics in rivers. Many river models express the behaviour of the concentration, C, of a solute or suspensoid in the water column as a one-dimensional advection–dispersion equation, one form of which is:

$$\underbrace{\frac{\partial C}{\partial t} = \frac{-Q}{A}\frac{\partial C}{\partial x}}_{\text{Advection}} + \underbrace{\frac{1}{A}\frac{\partial}{\partial x}\left[AD\frac{\partial C}{\partial x}\right]}_{\text{Dispersion}}$$

$$\underbrace{+ \frac{\partial Q}{\partial x}(C_L - C)}_{\text{Inflow and dilution}} + \underbrace{N}_{\text{Net of sources and sinks}} \qquad (9)$$

where x is downstream distance, Q is stream flow, A is cross-sectional area, D is the coefficient of longitudinal dispersion $[L^2\ T^{-1}]$, C_L is the concentration of influent water, and $N[M\ L^{-3}\ T^{-1}]$ is the net of gains from all losses to other ecosystem compartments, the atmosphere, or the terrestrial environment.

Differences among models include the forms of the source and sink terms, whether the channel and flow characteristics can vary with space and time, the number of solutes and suspensoids that are modelled simultaneously, and the number of additional, non-transporting compartments that are coupled to the source and sink terms. Only the simplest forms have analytical solutions; most require numerical solution by computer.

Figure 18.4 presents a model for the simple

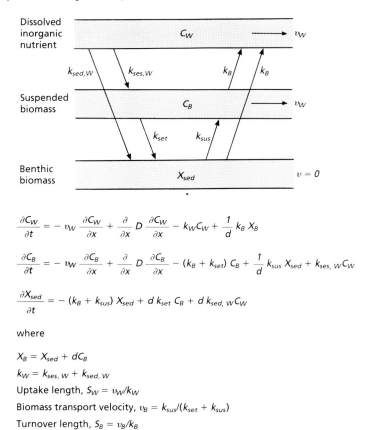

$$\frac{\partial C_W}{\partial t} = - v_W \frac{\partial C_W}{\partial x} + \frac{\partial}{\partial x} D \frac{\partial C_W}{\partial x} - k_W C_W + \frac{1}{d} k_B X_B$$

$$\frac{\partial C_B}{\partial t} = - v_W \frac{\partial C_B}{\partial x} + \frac{\partial}{\partial x} D \frac{\partial C_B}{\partial x} - (k_B + k_{set}) C_B + \frac{1}{d} k_{sus} X_{sed} + k_{ses, W} C_W$$

$$\frac{\partial X_{sed}}{\partial t} = - (k_B + k_{sus}) X_{sed} + d k_{set} C_B + d k_{sed, W} C_W$$

where

$X_B = X_{sed} + dC_B$

$k_W = k_{ses, W} + k_{sed, W}$

Uptake length, $S_W = v_W/k_W$

Biomass transport velocity, $v_B = k_{sus}/(k_{set} + k_{sus})$

Turnover length, $S_B = v_B/k_B$

Fig. 18.4 Linear, donor-controlled (or tracer) model for a two-compartment, longitudinally uniform, stream ecosystem.

two-compartment ecosystem used above to introduce spiralling. The biotic compartment is subdivided into a suspended portion and a sedimentary portion. The advection–diffusion equations are simplified, relative to equation (9), by assuming uniform flow and cross-section. The equation for the sediment compartment contains no transport terms and is expressed in terms of mass per unit area. As illustrated in Fig. 18.4, spiralling length can be computed from the model parameters.

An important aspect of this model is that all transfers among compartments vary linearly with the nutrient content of the donor compartment. A linear, donor-controlled model cannot incorporate many mechanisms known to control exchanges among compartments. It does not, for example, include an influence of biomass on nutrient uptake, or any means for saturating uptake at high nutrient concentrations. A model of this type is, however, appropriate for simulating the

dynamics of a tracer (such as a radioisotope) added to a stream or river in which other aspects of the ecosystem, including the cycling of the naturally occurring nutrient, are relatively constant (O'Neill 1979). This is because a tracer, by definition, is added at a level too low to affect the dynamics of naturally occurring nutrient or any other aspect of the ecosystem. The model, therefore, simulates only the simple mixing of the tracer through compartments at rates determined by the steady fluxes of the naturally occurring nutrient. Model coefficients estimated by simulating a tracer injection, therefore, quantify the cycling of the ecosystem under the conditions of the experiment, but may poorly predict responses to altered conditions, such as addition of polluting levels of a nutrient.

Where the objective is to describe or investigate the dynamics of nutrient concentration over a range of concentrations, such as might be produced by an experimental nutrient addition, non-linear functions may be used. For example, to model uptake by attached algae of nitrate added to experimental flumes, Kim *et al* (1990) used an additional water compartment with no down-stream velocity representing an 'exchange zone' near the streambed. Nutrient transfer into the exchange zone from the water column was governed by first-order kinetics, while nutrient uptake occurred from the exchange zone according to Michaelis–Menten kinetics, i.e.:

$$U = X_{\text{B}} \, V_{\text{max}} \left(\frac{C_{\text{ex}}}{K_{\text{s}} + C_{\text{ex}}} \right) \tag{10}$$

where C_{ex} is the NO_3-N concentration in the exchange zone, K_{s} is the half-saturation constant at which uptake is half its maximum rate, V_{max} [M T^{-1} per mass of X_{B}]. Note that equation (10) predicts that uptake length will increase with increasing concentrations, as has been observed experimentally by Mulholland *et al* (1990). In non-biological applications, other types of exchanges, such as sorption isotherms (Bencala 1984), and chemical submodels (Chapman 1982), have been coupled to equation (9). Functions used in modelling trace contaminant behaviour have been reviewed by Thomann (1984), Kuwabara and Helliker (1988), and O'Connor (1988a, 1988b, 1988c).

The dynamics of an added nutrient, unlike those of a tracer, may interact with other aspects of the ecosystem. For example, using equation (10) to describe an increase in uptake due to added nutrients raises the question of whether the biomass, X_{B}, increases in response. This, in turn, may depend on whether there is a secondary effect on consumer populations that might keep X_{B} in check (section 18.8). Considerations such as these lead to increasingly complex ecosystem models in which the dynamics of a single nutrient are only a small part.

Such complexity appears, to varying degrees, in water quality simulation models, designed to predict effects of potential waste discharges, or of reducing existing discharges (Chapra & Reckhow 1983; Orlob 1983; Bowie *et al* 1985; Thomann & Mueller 1987; McCutcheon 1989). Several water quality models simulate phosphorus and nitrogen dynamics, including release of these nutrients

from discharged organic matter, uptake by and influence on the growth of algae, releases from decaying algae, and exchanges with sediments. Often, however, the focus is on the net effects on water-column concentrations, rather than on cycling *per se*, and in some models the sediments are regarded as potentially infinite sources or sinks without mass balance constraints. Water quality simulation models are typically constructed by using functional relationships and parameter values available in the scientific literature (see Bowie *et al* 1985) and then adjusted to bring model simulations into agreement with data from a particular river. These data are often limited, either in quantity or in dynamic range, so that unequivocal parameter estimations and rigorous testing of the model are rarely feasible. None the less, the models are undeniably effective in simulating major water quality parameters, and they serve to identify areas in which scientific understanding of ecosystem processes is limited.

18.5 ORGANIC CARBON SPIRALLING

The carbon cycle in rivers differs from that of phosphorus and nitrogen in that: (1) the inorganic phase of the cycle (CO_2, HCO_3^-, and CO_3^{2-}) exchanges freely with atmospheric carbon dioxide; and (2) a variable, but sometimes very large, proportion of the organic phase begins not with uptake of inorganic carbon (i.e. photosynthesis), but with lateral accrual of organic carbon from the terrestrial environment. In describing spiralling of carbon, therefore, we focus on the portion of the cycle, or half-cycle, involving organic carbon, beginning with its entry into the river from terrestrial sources or formation by photosynthesis and ending in respiration to carbon dioxide. The organic matter turnover length, S_{C}, is the expected downstream travel distance associated with this half-cycle (Newbold *et al* 1982a). As we shall see below, both computation and meaningful interpretation of S_{C} are made problematic by the very great diversity of organic carbon forms in a river.

For the moment, however, we assume that all organic carbon in the river is of similar quality, degrading at the rate k_{C}, and migrating downstream at the velocity, v_c. The turnover length, then (as derived above), is $S_c = v_c/k_c$. If X_c is the areal standing stock of organic carbon (hereafter 'carbon'), then the areal rate of carbon loss to respiration is $R = k_c X_c$, and the downstream flux of carbon per unit width is $F_T = v_c X_c$. Simple substitutions yield:

$$S_c = F_T/R_w \qquad (11)$$

This allows turnover length to be estimated from measurements of downstream transport and areal respiration in the river. S_c, like spiralling length, measures the combined effects of retentiveness (v_c) and rate of biological processing (k_c) in determining how effectively a unit of river bottom processes nutrient supplied from upstream. Moreover, under our simplifying assumption of uniform carbon quality, the turnover length describes how carbon entering the river at a particular location affects downstream metabolism. For example, if I^* is a steady input (per unit area) to a unit length of river at location $x = 0$, then the respiration $R^*(x)$ at downstream distance x attributable to I^* is given by $R^* = (I_0^*/S_c) \exp(-x/S_c)$. This is essentially the same model that Streeter and Phelps (1925) used to describe the downstream effects of organic matter in sewage effluent. Streeter and Phelps assumed that all organic matter remained in suspension ($v_c = v_w$), expressed R in terms of oxygen utilization, and coupled this model with one describing oxygen depletion and re-aeration. The Streeter–Phelps approach, although sometimes highly elaborate, remains the basis for many currently used water quality models (Gromiec *et al* 1983; *cf.* section 18.4).

Estimates of organic carbon turnover length, based on equation (11), have been made for several streams and rivers (Newbold *et al* 1982a; Minshall *et al* 1983, 1992; Edwards & Meyer 1987; Naiman *et al* 1987; Meyer & Edwards 1990; Richey *et al* 1990). Values ranged from 2.9 km for a small wooded stream in New Hampshire, USA (data from Fisher & Likens 1973), to 4000 km for the Amazon River (Richey *et al* 1990). Minshall *et al* (1983) estimated the turnover lengths of only particulate carbon (i.e. excluding dissolved organic matter). Their estimates ranged from 1.0 to 10 km in headwater North American streams

to 250 km in the McKenzie River, Oregon, with an average flow of 55 m^3 s^{-1} (Minshall *et al* 1983). The major influence on turnover length is stream size. Areal respiration often, but not always, increases in the downstream direction, and the variation usually remains within one order of magnitude (Minshall *et al* 1983, 1992; Bott *et al* 1985; Naiman *et al* 1987; Meyer & Edwards 1990). Water-column organic matter concentration may increase somewhat with stream size but on the whole remains remarkably constant (Schlessinger & Melack 1981; Mulholland & Watts 1982). It is, therefore, the simple variation in depth and velocity that accounts for most of the size effect on turnover length (since $F_T = v_w d C_c$, where C_c is total organic carbon concentration).

Carbon turnover lengths are generally longer than the stream or river in which they were measured. That is, most of the carbon entering a river is transported, either to a larger downstream river or to an estuary or sea. Much of the carbon reaching the sea is highly refractory (Ittekot 1988), and probably undergoes very little degradation within the river system. Clearly, much of the metabolism in a river is of more labile forms, such as algae and fresh leaf litter, which contribute little to transport. This diversity of forms affects the estimation of turnover length. Equation (11), in effect, yields an average of the individual turnover of the various forms of carbon, weighted by their relative contribution to the measured respiration. Respiration of forms whose turnover length is short relative to river length can be expected to be about equal to the rate of lateral input (on an annual basis), while respiration of forms with very long turnover lengths must be far less than lateral inputs. Thus, a turnover length weighted by carbon inputs rather than by respiration would be much longer than estimated by equation (11).

The promise of analysing organic carbon dynamics from the perspective of spiralling is that it might provide a measure of the upstream–downstream interdependence of the river ecosystem. Such a measure would be useful from a practical standpoint, for example in evaluating the importance of protecting headwater stream reaches to maintain a downstream fishery. It

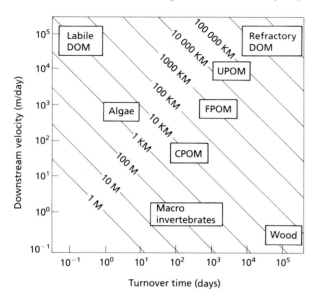

Fig. 18.5 Postulated turnover lengths (diagonal isoclines) for carbon of various forms in a medium-size river (e.g. with a flow of 3–30 m³ s⁻¹). Values represent rough averages from various sources and some speculation. Data sources include: Waters 1961, 1965; Fisher 1977; Marker & Gunn 1977; Minshall *et al* 1983, 1992; Bott *et al* 1985; Webster & Benfield 1986. CPOM, course particulate organic matter (>1 mm); FPOM, fine particulate organic matter (50 µm to 1 mm); UPOM, ultrafine particulate organic matter (<50 µm); DOM, dissolved organic matter.

would also provide an approach to analysing longitudinal variations in ecosystem structure, such as those postulated by the river continuum concept (Vannote *et al* 1980). If natural carbon were uniform in character, the turnover length, in conjunction with the simple exponential decay model presented above, would provide such a measure as it does for the case of organic pollution. But the diversity of organic carbon prevents this interpretation. If, for example, a turnover length of 100 km is an average of two kinds of carbon with turnover lengths of 1 km and 199 km, respectively, the turnover length tells us little about the longitudinal interdependence of the river ecosystem. What is necessary, is to describe carbon spiralling in terms of specific carbon forms, accounting not only for the individual turnover lengths of these forms, but for the transformation of carbon among them.

Such an analysis has not been conducted, but sufficient information exists to illustrate its potential. Figure 18.5 presents rough turnover times $(1/k_c)$ and downstream transport velocities (v_c) for a small river (e.g. baseflow 3–30 m³ s⁻¹), representing a composite of several studies. Turnover lengths for the individual forms range over approximately eight orders of magnitude from ~100 m for macroinvertebrates to ~10⁶ km or more for refractory dissolved organic matter. Note

that labile DOM and woody debris, which occupy opposite extremes on the turnover time and velocity axes, both have intermediate turnover lengths in the range of 10^2–10^4 m. The downstream influence of carbon of a given form depends not only on the turnover length shown here, but also on its conversion to other forms. For example, carbon initially in algae and coarse particulate organic matter (CPOM) may remain local if assimilated by macroinvertebrates, but pass downstream if it is leached as labile DOM or converted to fine particles (FPOM). In rivers of larger or smaller size, we would expect turnover times to remain roughly as in Fig. 18.5, but downstream velocities would differ dramatically. Transport velocities of CPOM, and FPOM in first-order streams, for example, are of the order of 0.1 and 1–10 m day⁻¹, respectively.

Figure 18.5 suggests that longitudinal linkages — or influences of upstream ecosystems on downstream ecosystems — may be transmitted primarily through labile DOM, algae, CPOM, FPOM, and wood; that is, the forms whose turnover lengths are on the scale of the river length. In contrast, most of the downstream transport of carbon in unpolluted streams and rivers occurs as refractory DOM and particles smaller than 50 µm (UPOM) (Wallace *et al* 1982; Minshall *et al* 1983), which are forms with extremely long turnover lengths. A complete analysis will require coupling the turnover-length concept with budgetary studies of individual reaches (e.g. Fisher & Likens 1973; Fisher 1977; Mulholland 1981; Cummins *et al* 1983; Richey *et al* 1990), and with geomorphic approaches which consider the role of all channels, of all sizes, within a river network (e.g. Cummins *et al* 1983; Naiman *et al* 1987; Meyer & Edwards 1990).

18.6 PHOSPHORUS

Forms and concentrations

Phosphorus is an essential nutrient centrally involved in energy transformations within organisms, making up roughly 0.1–1% of organic matter. Figure 18.6 illustrates the major aspects of the phosphorus cycle in rivers. Phosphorus in water is normally categorized as being either dissolved or particulate, depending on whether it passes 0.45-μm membrane filter. 'Dissolved' phosphorus, therefore, may include a substantial colloidal component. Within the dissolved fraction, inorganic P (DIP) occurs as orthophosphate (PO_4), which is usually estimated by variations of the molybdenum blue method (Strickland & Parsons 1972). Because this method may overestimate the orthophosphate by hydrolysing organic and colloidally bound forms (Rigler 1966; Stainton 1980; Tarapchak 1983), measurements based on the molybdenum blue method are often referred to as 'soluble reactive phosphorus' (SRP). 'Total dissolved phosphorus' (TDP) represents a molybdenum blue assay for phosphorus following an acid digestion that releases the dissolved phosphorus in organic forms. The organic fraction (DOP) is not well characterized, but in lake water some consists of inositol hexaphosphate (Herbes *et al* 1975; Eisenreich & Armstrong 1977) and DNA fragments (Minear 1972).

In unpolluted rivers, SRP averages about 10 μg l^{-1} (Meybeck 1982) on a worldwide basis, while total dissolved phosphorus averages about 25 μg l^{-1}. Dissolved phosphorus concentrations may increase with discharge, but rarely by a factor of more than 2–4 during peak flows (Kunishi *et al* 1972; Leonard *et al* 1979; Meyer & Likens 1979; Saunders & Lewis 1988). Agricultural activities may increase dissolved phosphorus to the range of 50–100 μg l^{-1} (Omernik 1977; Smart *et al* 1985; Mason *et al* 1990), and to >500 μg l^{-1} during snowmelt (Rekolainen 1989). Municipal effluents, however, may increase concentrations to the range of 1000 μg l^{-1} (Meybeck 1982).

Particulate phosphorus includes phosphorus

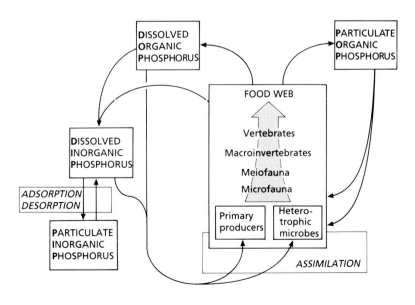

Fig. 18.6 Schematic of the phosphorus cycle in streams and rivers.

incorporated into mineral structures, adsorbed on to surfaces, primarily of clays, and incorporated into organic matter, and averages worldwide about 500 µg l^{-1} (Meybeck 1982). Concentrations of suspended particulates vary greatly with land use and erodibility of the watershed (Cosser 1989; Karlsson & Löwgren 1990) and increase dramatically with stormflows (Kunishi *et al* 1972; Verhoff & Melfi 1978; Meyer & Likens 1979; Munn & Prepas 1986; Prairie & Kalff 1988; Krogstad & Løvstad 1989).

Dissolved-to-particulate transfers

Movement from the dissolved inorganic (DIP) compartment to other ecosystem compartments is perhaps best observed by adding a radio-isotope tracer, ^{33}P-PO$_4$, directly to a stream river. Unfortunately, this is not generally feasible, and such experiments have been conducted in only a few streams and one relatively small river. The uptake length can be estimated from equation (2) and the longitudinal disappearance of ^{32}P relative to a hydrological tracer such as ^{3}H-H$_2$O or chloride. In Walker Branch, a first-order woodland stream in Tennessee, USA (baseflow 3–8 l s^{-1}), ^{32}P-estimated uptake lengths ranged from 21 to 165 m, varying roughly inversely with the quantity of leaf detritus on the stream bottom (Mulholland *et al* 1985b). Corresponding rates of phosphorus uptake were 1.8–22 mg^{-2} day^{-1}. In the Sturgeon River, Michigan, USA, with a flow increasing downstream from 1.1 to 1.4 m^3 s^{-1}, Ball and Hooper (1963) estimated uptake lengths between 1100 and 1700 m in successive reaches (with no downstream trend). From their data, areal uptake of phosphorus averaged about 10 mg^{-2} day, or near the middle of the range of estimates from Walker Branch.

Chambers, or microcosms, offer a promising but underexploited alternative method for using ^{32}P to estimate phosphorus uptake (Corning *et al* 1989; Paul *et al* 1989; Paul & Duthie 1989; Stream Solute Workshop 1990). Corning *et al* (1989) incubated rocks from Canadian streams and rivers under *in situ* conditions of temperature and flow in closed 6-litre recirculating chambers. Initial loss of ^{32}P from solution is first order until regeneration of ^{32}P becomes significant. Corning

et al reported uptake rates by the epilithon ranging from 0.5 to 4.3 mg m^{-2} day^{-1}.

Estimates of phosphorus removal based on tracer data may be erroneously high if the true levels of PO$_4$ are below the value estimated as SRP. This problem has proved particularly vexing in lakes, where SRP may overestimate PO$_4$ by more than an order of magnitude (Rigler 1966; White *et al* 1981). Peters (1981), using a radiological bioassay, found that PO$_4$ constituted between 6% and 44% of SRP in water from various river waters, but the radiobiological method (based on Rigler 1966) has been criticized by Tarapchak and Herche (1988). Jordan and Dinsmore (1985), also using radiobioassay, concluded that in the River Main in Northern Ireland the readily available phosphorus (presumably PO$_4$) accounted for 76% of the SRP. Newbold *et al* (1983a) inferred from exchange rates with periphyton of known pool size, that true PO$_4$ concentrations in Walker Branch were at least 70% of the average SRP.

Several studies have observed the uptake of stable phosphorus (^{31}P) experimentally injected into small streams (McColl 1974; Meyer 1979; Aumen *et al* 1990; Mulholland *et al* 1990; Munn & Meyer 1990; D'Angelo *et al* 1991). In a few cases, the observed net uptake of phosphorus has been negligible, indicating that whatever uptake occurs in the stream is saturated at ambient phosphorus concentrations. In most cases, however, considerable retention of the added phosphorus occurred, with uptake lengths in the range of 5–200 m. An uptake length measured by adding stable phosphorus is equivalent to one measured using ^{32}P only if the uptake flux of phosphorus increases in direct proportion to the increase in concentration (i.e. *F/U* remains constant). Mulholland *et al* (1990) found that an addition which increased PO$_4$ concentration from a background of 2.9 µg l^{-1} to 7.4 µg l^{-1} yielded an uptake length 55% longer than that measured by ^{33}P at the ambient concentration (2.9 µg l^{-1}). Estimates made using this approach, therefore, provide only an upper bound for uptake length.

A number of mass balance studies have shown phosphorus removal from solution downstream from municipal and agricultural sources of phosphorus (Keup 1968; Taylor & Kunishi 1971; Johnson *et al* 1976; Harms *et al* 1978; Hill 1982).

In these studies, DIP concentrations declined longitudinally from peaks sometimes exceeding 1000 µg l^{-1} (Harms *et al* 1978), with net uptake rates in the range of 10 to >100 mg m^{-2} day^{-2} (in most cases, these uptake rates are inferred because few studies reported net uptake directly). Phosphorus apparently accumulates in the sediments by physical adsorption (as discussed below) during periods of steady flow, and is then transported downstream as particulate load during storms (Cahill *et al* 1974; Harms *et al* 1978; Verhoff & Melfi 1978; Verhoff *et al* 1979).

As phosphorus leaves the dissolved form in the water column, it may be transferred either to the sediments or to suspended solids (seston) within the water column. Uptake by seston may be negligible in very small streams (e.g. 1.3% of total uptake; Newbold *et al* 1983a), but increases in importance with stream size (Ball & Hooper 1963; Paul *et al* 1989) and may account for nearly all of the uptake in large rivers with true phytoplankton populations (Décamps *et al* 1984). As in marine and lacustrine environments, most of the sestonic uptake appears to occur within the smallest size fractions (e.g. <1 µm; Corning *et al* 1989). Phosphorus transferred to seston may subsequently settle to the sediments. Simmons and Cheng (1985) concluded that this was the major pathway for removal of phosphorus from the water column in a river receiving sewage effluent.

Biological uptake

Phosphorus is removed from the water by algae (including cyanophyte bacteria), heterotrophic microbes, macrophytes, bryophytes and riparian plants. The kinetics of phosphorus uptake by planktonic algae have received extensive attention (see Cembella *et al* 1984a, 1984b), and the fundamentals undoubtedly apply to attached algae. Uptake generally follows Michaelis–Menten kinetics (see equation (10)). Bothwell (1985) found that K_s for periphytic river diatoms ranged from 0.5 to 7.2 µg l^{-1}, or below that for most phytoplankton populations. Algae (and bacteria) can take up phosphorus at a much greater rate than they can utilize it for growth so that V_{max} can vary greatly, declining as cells accumulate phosphorus (Rhee 1974). As a result, steady-state growth is saturated by concentrations very much lower than K_s (Droop 1973; Rhee 1974). Uptake of phosphorus by attached algae is also influenced by transport of phosphorus to and into the algal biofilms (section 18.3).

Few data exist to evaluate the relative importance of ecosystem components in taking up phosphorus from the water column. In Walker Branch, which is well shaded, 60% of the phosphorus uptake in the summer was accounted for by large detritus (>1 mm), 35% by fine particles (<1 mm) and 5% by the epilithon, consisting primarily of diatoms (Newbold *et al* 1983a). The phosphorus content of leaf detritus increases during decomposition (Kaushik & Hynes 1968; Meyer 1980), and the rate of phosphorus uptake correlates with measures of metabolic activity (Gregory 1978; Mulholland *et al* 1984; Elwood *et al* 1988). In less shaded streams and rivers, the epilithon and associated algae are probably the dominant fate of phosphorus (Ball & Hooper 1963; Peterson *et al* 1985). Ball and Hooper (1963) did not make a mass accounting for the phosphorus taken up, but they did find that weight-specific uptake by macrophytes (*Chara*, *Fontinalis* and *Potomogeton*) was far slower than by periphyton.

Adsorption

Adsorption and desorption of phosphorus on to organic and inorganic surfaces are continual kinetic processes. Thus, some portion of the phosphorus cycle (e.g. as observed via ^{32}P injection) involves entirely abiotic processes. Unfortunately, few studies have addressed phosphorus adsorption from a kinetic standpoint (except see Pomeroy *et al* 1965; Li *et al* 1972, Furumai *et al* 1989), so that it is difficult to evaluate the magnitude of this cycling. Studies of ^{32}P uptake by epilithon and by particulate organic matter have found that uptake is small or negligible under sterile conditions (Gregory 1978; Elwood *et al* 1981a; Paul *et al* 1989). These studies, however, involved very low phosphorus concentrations and did not include large quantities of inorganic fine particles.

Most studies of phosphorus adsorption in rivers have focused not on the kinetic exchange of phosphorus, but on the role of sediments in controlling or 'buffering' dissolved phosphorus concentrations. Adsorption of phosphorus on to natural river sediments and suspensoids generally corresponds to a Langmuir isotherm (Green *et al* 1978; McCallister & Logan 1978; Stabel & Geiger 1985; Furumai *et al* 1989), which has the form:

$$X = bkC/(1 + kC) \qquad (12)$$

in which X is the quantity of adsorbed phosphorus per mass of sediment, C is the dissolved phosphorus concentration, b is the 'adsorption maximum' in units of X, and k is the 'adsorption energy' in units of $1/C$. At phosphorus concentrations below $0.1–1.0$ mg l^{-1}, the isotherms are approximately linear (i.e. $kC \ll 1$). For fine particles (<0.1 mm), which account for nearly all of the sorption capacity (Meyer 1979), the slope (bk) of the isotherms ranges from about 0.1 to 1.0 µg P g^{-1} sediment per µg P l^{-1} water.

Sediment particles are often characterized in terms of their 'equilibrium phosphorus concentration' (EPC), which is the water concentration at which phosphorus is neither adsorbed nor desorbed (White & Beckett 1964; Klotz 1988). The EPC reflects the quantity, X, of adsorbed phosphorus and, in terms of the Langmuir isotherm, represents the value of C that satisfies equation (12). Sediments that enter rivers may either desorb phosphorus to the water (Mayer & Gloss 1980; Gloss *et al* 1981; Grobbelaar 1983; Viner 1988), or adsorb phosphorus from the water (Kunishi *et al* 1972) depending on whether the initial EPC exceeds the water phosphorus concentration. The EPC of stream sediments is often near the stream water phosphorus concentration (Taylor & Kunishi 1971; Meyer 1979; Mayer & Gloss 1980; Hill 1982; Klotz 1985, 1988; Munn & Meyer 1990), indicating rapid equilibration. This equilibration has frequently been interpreted as evidence that the sediments control or 'buffer' stream water concentrations. Such buffering, however, implies a net uptake or release of phosphorus by the sediments which, under steady conditions, will diminish as the sediment EPC adjusts to the concentration of the source water. Thus, it would seem premature to conclude that the sediment EPC controls the stream water concentration (rather than the converse) without additional information about source water concentrations, the actual quantity of sediments in the streambed, or both.

In general, we would expect sediments to influence phosphorus concentrations when: (1) there is an abundance of fine inorganic particles (e.g. several hundred kg m^{-2}); and (2) a large discrepancy exists between the EPC of sediments entering the river and the concentration of phosphorus entering the river. These conditions would explain the prolonged periods of high net phosphorus removal downstream from pollution sources discussed above. In streams and rivers with few fine inorganic sediments and low dissolved phosphorus inputs, however, the role of adsorption in controlling phosphorus concentrations may be minimal, and perhaps exceeded by the influence of biological processes. Mulholland *et al* (1990), for example, observed an apparent transition from biotically dominated uptake to sorption-dominated uptake at about 5 µg l^{-1}, a concentration that might have saturated biological uptake capacity.

Regeneration

Where field injections of ^{32}P have been made, the loss of ^{32}P from the water has not been accompanied by commensurate depletion of stable phosphorus (Ball & Hooper 1963; Newbold *et al* 1983a; Mulholland *et al* 1985b), implying that regeneration of phosphorus approximately balances gross uptake. Other evidence for regeneration comes from direct observation of ^{32}P in stream water for several weeks following a ^{32}P release (Elwood & Nelson 1972), and from ^{32}P accumulation in detritus and periphyton placed into a stream after the initial labelling (Ball & Hooper 1963; Elwood & Nelson 1972). Similar results have been obtained in laboratory streams (Short & Maslin 1977; Paul & Duthie 1989). Studies with ^{32}P have shown that phosphorus associated with epilithon and detritus returns to the water with a turnover time of a week or less, whereas turnover in fine sediments may be an order of magnitude slower (Ball & Hooper 1963; Elwood & Nelson 1972; Newbold *et al* 1983a).

Regeneration of phosphorus from the biota may occur via: (1) excretion of phosphorus (either as DIP or DOP) from living algae and bacteria;

(2) release of phosphorus upon death and lysis of cells; and (3) ingestion followed by egestion, excretion, and death of animal consumers. Although the first two pathways may in some cases be significant (Barsdate *et al* 1974; Lean & Nalewajko 1976; DePinto 1979), there is a growing consensus from marine and lacustrine studies that animal consumers, particularly Protozoa and other microscopic animals, represent the major pathway. Regeneration by consumers is discussed in section 18.8.

A portion of the phosphorus lost from organisms may be in the form of dissolved organic phosphorus (DOP). Much of the DOP released may be hydrolysed rapidly to DIP by alkaline phosphatase and ultraviolet degradation (Francko 1986). Alkaline phosphatase, which is capable of hydrolysing a range of organic phosphorus compounds, has been detected in the sediments of some streams (Sayler *et al* 1979; Klotz 1985). However, the bulk of the DOP that is found in natural waters is not hydrolysed by alkaline phosphatase (Herbes *et al* 1975; Jordan & Dinsmore 1985; Hino 1989). Mulholland *et al* (1988) found that about 12% of ^{32}P released from decomposing leaf detritus was organic with molecular size >5000 daltons. This organic fraction appeared to be far less available for utilization than PO_4, with only about 10% utilized within 24 hours of incubation. High molecular weight organic phosphorus has been identified in other river and lake waters (Lean 1973; Downes & Paerl 1978; Peters 1978, 1981; White *et al* 1981) where it has similarly been shown to be biologically available, but at a slower rate than PO_4 (Pearl & Downes 1978; Peters 1981). It remains unclear whether utilization of DOP can occur directly (Smith *et al* 1985), possibly involving enzymatic activity on cell surfaces (Ammerman & Azam 1985), or whether the DOP is first hydrolysed within the water. Whatever the ecological significance of this distinction, it is clearly important to the interpretation of ^{32}P kinetics. The potential role of DOP in cycling has special significance for rivers because DOP is highly transportable yet unavailable for rapid uptake and thus could contribute substantially to spiralling length.

Phosphorus spiralling in a woodland stream

An extensive quantification of phosphorus spiralling has been attempted for only one stream, Walker Branch, Tennessee (Newbold *et al* 1983a). This description was obtained by fitting a multi-compartmental model to the ^{32}P data to estimate transfer coefficients among all compartments (Fig. 18.7). From the model, spiralling within each compartment is described with three parameters: the residence time in the compartment, the fraction of the cycling phosphorus flux passing through the compartment, and the downstream velocity of the compartment (Table 18.1). The spiralling length was 190 m, consisting of 165 m of travel in the water (as described above), and 25 m of travel in the particulate compartments (CPOM, FPOM and epilithon). Phosphorus taken up by consumers had a downstream travel distance of 2 m, but since only 2.8% of the flux passed through this compartment, the contribution to total spiralling length was only 0.06 m. Although dissolved phosphorus accounted for most of the spiralling length, its turnover time of 75 min was a small fraction of the 18 days required to complete a cycle.

Walker Branch is highly retentive of phosphorus. In effect, the biotic phase 'slows' the average movement of phosphorus downstream by a factor of about 300 relative to the water velocity. Stated another way, the standing stock of phosphorus is 300 times greater than in the absence of retention. The snail *Elimia* (=*Goniobasis*) *claeviformis* plays a large role in this retention, accounting for 23% of the standing stock exchangeable phosphorus, and drifting downstream at a velocity of <1 cm day^{-1}.

18.7 NITROGEN

Nitrogen cycle

Nitrogen is a fundamental constituent of protein and, like phosphorus, its availability can frequently limit algal and microbial productivity. The major pathways of the nitrogen cycle in aquatic systems are shown in Fig. 18.8 (see also Kaushik *et al* 1981; Sprent 1987), although not all of these pathways have been studied extensively

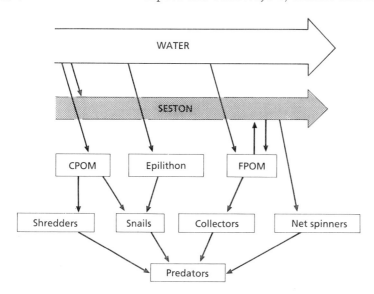

Fig. 18.7 Ecosystem compartments and flows used to analyse spiralling of phosphorus in a small woodland stream (from Newbold *et al* 1983a). In addition to the phosphorus fluxes indicated by arrows on the diagram, every compartment releases phosphorus directly back to the water compartment. These arrows are omitted for simplicity. Dynamics of the water and seston compartments were described by partial differential equations similar to equation (9) and coupled to other compartments as illustrated in Fig. 18.3.

in rivers. The nitrogen cycle in rivers is more complex than the phosphorus cycle in several important respects. First, the processes of nitrogen fixation and denitrification involve exchanges with atmosphere, so that cycling of nitrogen cannot be considered closed with respect to the air–water interface. (Atmospheric exchanges of ammonia may also be significant under some

Table 18.1 Calculation of spiralling indices for phosphorus in Walker Branch (from Newbold *et al* 1983a)

Compartment	t_i (days)	v_i (m day^{-1})	s_i (m)	b_i	S_i (m)
Water	0.052	3200	165	1.0	165
Particulates					
CPOM[†]	6.9	0.06	0.40	0.60	0.24
FPOM (fast)[‡]	6.9	7.4	51	0.27	13
FPOM (slow)[‡]	99.0	1.4	141	0.080	11
Epilithon	5.6	0.0	0.0	0.054	0.0
Total particulates	14.0*	1.8*	25*	1.0	25
Consumers					
Snails	150	0.005	0.77	0.024	0.019
Shredders	76	0.13	9.8	0.0002	0.0030
Collectors	105	0.12	13	0.003	0.03
Net spinners	220	0.0	0.0	0.007	0.0
Predators	14	0.007	0.10	0.008	0.0008
Total consumers	150*	0.013	2.0*	0.028	0.056
Total	18*	10	190	1.0	190

* Weighted average; [†] coarse particulate organic matter (>1 mm); [‡] fine particulate organic matter (<1 mm). Two turnover rates, 'fast' and 'slow', were resolved from the ^{32}P dynamics of the FPOM compartment.
t_i is average residence time in compartment i; v_i is average downstream velocity of compartment i. Average travel distance in a compartment, $s_i = v_i t_i$. The contribution of compartment i to spiralling length is $S_i = b_i s_i$, where b_i is the probability of passing through compartment i. Particulate and consumer turnover lengths are the sums of S_i values within these categories.

circumstances.) Second, in addition to its role as a fundamental cellular constituent in all organisms, nitrogen is also involved in biologically mediated redox reactions, such as nitrification which yields energy for metabolism and consumes oxygen, and denitrification, in which nitrate serves as a terminal electron acceptor. These processes involve specialized organisms and occur with very different stoichiometric coupling to carbon flow than does cellular assimilation of nitrogen. Finally, there are two major sources of inorganic nitrogen for algal and microbial assimilation (NH_4 and NO_3), rather than one (PO_4) in the phosphorus cycle. Understanding of the nitrogen cycle in rivers is limited not only by the complexity of the cycle, but by the fact that very few isotopic tracer studies have been conducted in rivers or in river-simulating microcosms. The available isotope,

^{15}N, is not radioactive, but is otherwise more difficult and expensive to work with than are the radioisotopes of phosphorus.

Estimates by Meybeck (1982) of worldwide average concentrations of dissolved nitrogen in unpolluted rivers are ($\mu g\ l^{-1}$): dissolved organic nitrogen (DON) 260; nitrate 100; ammonium 15; nitrite 1. In agricultural watersheds in North America, DON averages about 1000 $\mu g\ l^{-1}$, and dissolved inorganic nitrogen typically ranges from ~700 to 5000 $\mu g\ l^{-1}$ (Omernik 1977). Algae and other micro-organisms tend to use ammonium in preference to nitrate, which must be reduced before it can be assimilated (Sprent 1987). Based on results from lacustrine and marine ecosystems (Eppley *et al* 1979; Axler *et al* 1982) it is reasonable to conjecture that in many streams and rivers, ammonium supplies a substantial part, if not the

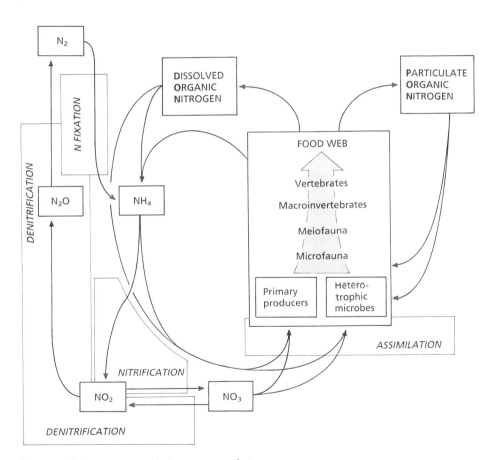

Fig. 18.8 Schematic of the nitrogen cycle in streams and rivers.

majority, of the nitrogen assimilation by algae and heterotrophic micro-organisms. Much of this ammonium may be supplied by excretion from consumers and cycle rapidly (and over short distances) back into the microbial community. Stanley and Hobbie (1981), using ^{15}N, found that uptake by river plankton of ammonium exceeded that of nitrate by a factor of three, while the ammonium concentration averaged only about half that of nitrate. Thus, the ammonium pool turned over approximately six times faster than the nitrate pool. The uptake of dissolved inorganic nitrogen (DIN, consisting of nitrate plus ammonia) averaged about 5 mg m^{-2} h^{-1}, but peaked at near 70 mg m^{-2} h^{-1} in the summer at maximum photoplankton concentrations. On an annual basis, the utilization of DIN within the 60-km study reach exceeded upstream inputs to the reach by a factor of three, corresponding to an uptake length of 20 km, with an average turnover time of the DIN pool of about 10 days. In the summer, under high demand and low flows, turnover times fell to 3–10 h, corresponding to uptake lengths of <1 km. Stanley and Hobbie also concluded that bacteria may have played an important role, in addition to phytoplankton, in taking up the DIN, and that approximately one-third of the DON entering the reach was converted to DIN, presumably through microbial assimilation and regeneration.

Biological uptake

Little is known about the magnitude of ammonium utilization by benthic organisms. Several studies have documented rapid uptake by the benthos of ammonium introduced at enriching levels (McColl 1974; Newbold et al 1983b; Richey et al 1985; Hill & Warwick 1987). Ammonium, however, is similar to phosphorus in that it is strongly subject to physical sorption and cells can incorporate ammonia temporally in excess of growth requirements (Eppley & Renger 1974; Conway & Harrison 1977). Thus it is difficult to draw inferences from these studies about the utilization of ammonium under natural concentrations.

As with ammonium there are no studies other than that of Stanley and Hobbie (1981) of nitrate

gross uptake under undisturbed conditions. However, there have been a number of mass balance studies observing disappearance of nitrate, which has in some cases been equated to a net incorporation of nitrate by benthic algae and heterotrophic micro-organisms. Fisher et al (1982) and Grimm (1987) observed longitudinal declines in nitrate concentrations in a desert stream associated with rapid growth of benthic algae following scouring storms. Maximum calculated uptake rates were 240 mg m^{-2} day^{-1}. Fisher et al (1982) noted that comparing net primary production to calculated uptake rates gave a C:N ratio of 4:1, or lower than the Redfield ratio of 6.6:1 (Redfield et al 1963), suggesting that some of the net nitrate uptake went to other fates, such as denitrification.

Uptake of experimentally injected nitrate has been observed in several streams (McColl 1974; Sebetich et al 1984; Triska et al 1989a; Aumen et al 1990; Munn & Meyer 1990; Webster et al 1991) in which background DIN concentrations were <150 µg l^{-1}. In other streams, all with background concentrations >100 µg l^{-1}, no net uptake was observed (McColl 1974; Richey et al 1985; Aumen et al 1990). The relation to DIN, however, is not entirely consistent, and background ammonium concentrations (not always reported) may be an important factor. In the most detailed study, where substantial net uptake of added nitrate occurred over 9 days of injection (Triska et al 1989a, 1989b), background ammonium concentrations were undetectable, while background nitrate was 25–50 µg l^{-1}.

Kim et al (1990) analysed uptake kinetics of nitrate added to artificial channels in which periphyton grew on Plexiglas slides. They obtained half-saturation constants for uptake (K_s) in the range of 50–80 µg l^{-1} from laboratory batch experiments, and then fitted an uptake model (section 18.4) to the dynamics of nitrate injected into their channels. From this, they obtained values for maximum uptake in the range of 0.15–0.65 µg N s^{-1} g^{-1} ash-free-dry-mass (AFDM) of periphyton.

Nitrification

Nitrification, the biological oxidation of ammonium to nitrite and then to nitrate, seems to occur in streams and rivers whenever both ammonium and oxygen are available. It is of particular interest as a water quality issue because it can reduce potentially toxic levels of ammonium discharges. Nitrification, however, consumes dissolved oxygen, and so exacerbates the oxygen depletion associated with oxidation of organic wastes. Much of the dissolved and particulate organic nitrogen in waste discharges may also ultimately be nitrified, after degradation to ammonium. Rates of nitrification in rivers, therefore, are sometimes reported as the rate at which both ammonium and organic nitrogen are converted to nitrate. These rates vary approximately in the range of $0.1-0.5$ day^{-1} depending on temperature, the proportion of the nitrogen source that is ammonium, and other factors (McCutcheon 1987). Nitrification may play a large role in maintaining ammonium levels in unpolluted streams at low levels. Triska *et al* (1990) found that nitrification occurred in the sediments, reducing relatively high ammonium levels in influent groundwater to low levels observed in the stream. Nitrification appears to be associated largely with the sediments in streams and smaller rivers (Matulewich & Finstein 1978; Cooper 1983; Cooper 1984) but may occur largely in suspension in larger rivers (McCutcheon 1987). Estimated benthic rates of nitrification in streams range from 29 mg m^{-2} day^{-1} (Chatarpaul *et al* 1980) to 2.5 g m^{-2} day^{-1} (Cooper 1984).

Denitrification

Denitrification, the reduction of nitrate to dinitrogen, is carried out by bacteria using the nitrate as the terminal electron acceptor for oxidative metabolism in the absence of oxygen. Denitrification is included in the general category of dissimilatory nitrate reduction, which can also include reduction of nitrate to ammonium. Denitrification can be a significant pathway for loss of nitrogen from the river and has been implicated in explaining downstream declines in nitrogen concentrations in several watersheds (Kaushik *et al* 1975; Hill 1979, 1988; Swank & Caskey 1982; Cooper & Cooke 1984). Most estimates of denitrification in streams fall into the range of $10-200$ mg m^{-2} day^{-1} (Chatarpaul &

Robinson 1979; Cooper & Cooke 1984; Duff *et al* 1984b; Hill & Sanmugadas 1985; Christensen & Sørensen 1988; Seitzinger 1988; Christensen *et al* 1989). Nearly all of these measurements, however, have been conducted in streams with relatively high nitrate concentrations (i.e. $0.5-10$ mg l^{-1}), and in streams that are not nitrate enriched normal rates of denitrification may be much lower (Duff *et al* 1984b; S.P. Seitzinger, personal communication).

Although denitrification requires anoxic conditions, it has been observed in aerated sediments and in relatively thin epilithic films (Nakajima 1979; Duff *et al* 1984b; Duff & Triska 1990). Evidently, denitrification occurs in microzones of anoxia within the sediments and biofilms. Oxygen produced by benthic algae may inhibit denitrification (Duff *et al* 1984b) resulting in diel variations in denitrification rates (Christensen *et al* 1990).

Denitrification requires an organic carbon source and proceeds faster where more carbon is available in the water and in the sediments (Hill & Sanmugadas 1985; Duff & Triska 1990). Typical denitrification rates of $10-200$ mg m^{-2} day^{-1} are equivalent to the respiration of $0.03-0.7$ g O$_2$ m^{-2} day^{-1}. By comparison, oxygen consumption in rivers generally ranges from 0.2 to 10 g O$_2$ m^{-2} day^{-1} (Bott *et al* 1985). Thus, denitrification may contribute a significant portion of the oxidative metabolism in streams where nitrate supply is high and appropriate habitat and carbon resources are available. Denitrification may occur simultaneously with nitrification (Duff *et al* 1984a; Cooke & White 1987). In Little Lost Man Creek, California, USA, nitrogen enters the deep sediments (hyporheos) as ammonium, mixes with oxygenated interstitial water and nitrifies (Triska *et al* 1990), yielding a net downstream increase in nitrogen concentrations. Yet significant denitrification also occurs within the hyporheos, presumably in anoxic microzones (Duff & Triska 1990). Chatarpaul *et al* (1980) found that tubificid worms in sediments enhanced both nitrification and denitrification.

Most of the measurements reported above estimated denitrification by the acetylene block method (Balderston *et al* 1976; Yoshinari *et al* 1976), although ^{15}N (e.g. van Kessel 1977;

Chatarpaul & Robinson 1979) and direct measurement of nitrogen evolution (Seitzinger 1988) have also been employed. The acetylene block method is subject to various uncertainties, the most serious of which is that it interferes with nitrification. As a result, the method can greatly underestimate denitrification when it is coupled to nitrification of an ammonia source (SP Seitzinger, CP Nielson, J Caffrey & PB Christenson, unpublished data).

Nitrogen fixation

Blue-green algae and micro-organisms fix nitrogen in streams and rivers, but few measurements have been made. Horne and Carmiggelt (1975) reported nitrogen fixation by the blue-green alga *Nostoc* of $42-360$ mg N m^{-2} year^{-1}, while Francis *et al* (1985) estimated nitrogen fixation in pool sediments of 5.1 g N m^{-2} year^{-1}. Even these higher values, however, are small in comparison with other nitrogen fluxes as noted above.

18.8 THE ROLE OF CONSUMERS

Animals in the food web (see Chapters 11 & 14) may account for a considerable portion of the nutrient cycle, but their influence on nutrient cycling involves, in addition: (1) direct influences on prey populations, which may 'cascade' through several trophic levels; (2) indirect influences of regenerated nutrients; and (3) indirect influences of physical transformations and translocations of nutrients and of the physical habitat (Kitchell *et al* 1979).

Algal grazers have been reported to assimilate $30-98\%$ of algal nitrogen uptake (Grimm 1988) and roughly 80% of algal phosphorus uptake (Mulholland *et al* 1983). This is consistent with many studies showing that grazing macroinvertebrates and fish maintain biomass of epilithic algae substantially below levels that would occur in the absence of grazers (e.g. McAuliffe 1984; Jacoby 1985; Stewart 1987; Power *et al* 1988; Feminella *et al* 1989; Hart & Robinson 1990). Indirect, or cascading, control of algal biomass by predators of grazers has also been demonstrated (Power *et al* 1985; Gilliam *et al* 1989; Power 1990). The weight-specific productivity of grazed

algae normally exceeds that of ungrazed algae (Kehde & Wilhm 1972; Summer & McIntire 1982; Gregory 1983; Lamberti & Resh 1983; Lamberti *et al* 1987, 1989; Hill & Harvey 1990), and a similar effect has been observed on phosphorus turnover rate (Mulholland *et al* 1983). However, the net effect of depressing biomass and increasing biomass turnover is normally to decrease productivity (and nutrient uptake) on an areal basis, and only slight, or statistically non-significant, cases of an actual stimulation of areal productivity have been observed.

The possibility that some moderate level of grazing might stimulate, or maximize, areal productivity is of interest because the phenomenon has been observed in lentic systems and microcosms (Cooper 1973; Flint & Goldman 1975; Bergquist & Carpenter 1986; Elser & Goldman 1991), and is usually attributed to the effect of consumers in regenerating nutrients, as predicted by nutrient cycling models (Lane & Levins 1977; Carpenter & Kitchell 1984). Stimulation may have gone undetected in streams because experiments have employed flumes that are too short to observe recycling effects (i.e. much shorter than the spiralling length). However, analysis of a spiralling model (Newbold *et al* 1982b) indicated that the stimulation effect (which would produce a shorter spiralling length) is possible in streams, but unlikely to occur because: (1) streams do not suffer depletion of the limiting nutrient in the same manner as lakes or closed systems (section 18.3); and (2) a simultaneous effect of grazers, which is to dislodge and suspend particles, tends to nullify the benefits of regeneration.

In woodland streams, heterotrophic microbes may take up far more inorganic nutrients from the water than do autotrophs (e.g. Table 18.1; see also Chapters 8 & 15). It appears, however, that relatively little of this uptake is assimilated or regenerated by macroinvertebrate consumers. In Walker Branch, Tennessee, for example, where detritus accounts for 95% of the phosphorus uptake, <3% of this is eventually assimilated by macroinvertebrates (Table 18.1). Studies of detritus processing have similarly shown that carbon assimilation by macroinvertebrates is only a small percentage of carbon lost to microbial respiration (Cummins *et al* 1973; Fisher & Likens

1973; Webster 1983; Smock & Roeding 1986; Petersen *et al* 1989; see Chapter 17). In marine and lacustrine environments, the protozoans and other very small animals (i.e. the meiofauna) assimilate much of the microbial production and play a major role in nutrient regeneration (e.g. Johannes 1965; Andersen *et al* 1986; Caron & Goldman 1990). Work in riverine environments, although still limited, suggests an equally strong role for meiofauna (Findlay *et al* 1986; Bott & Kaplan 1990; Crosby *et al* 1990; see also Chapter 15). In streams and rivers, the meiofauna may include early-instar insect larvae, as well as protozoans and animals that remain microscopic throughout their life history. Whether energy and nutrients from the meiofauna can then pass to higher trophic levels via predation remains essentially uninvestigated.

Macroinvertebrate consumers exert a substantially larger effect on nutrient cycling than their energy consumption would suggest (*cf* Chapter 17). Macroinvertebrate shredders increase significantly the speed the conversion of large detrital particles to small particles (e.g. Petersen & Cummins 1974; Wallace *et al* 1982; Cuffney *et al* 1990) with consequent effects on particle transport. Cuffney *et al* (1990) found that removal of macroinvertebrate populations from a small stream reduced annual transport of seston by 40%. Webster (1983) estimated that macroinvertebrate activity accounted for 27% of annual seston transport, and as much as 83% of transport during low summer flow. From the standpoint of phosphorus and nitrogen transport, these effects are probably not important since, as we have seen, turnover lengths in headwater streams are short. The transport does, however, increase the organic carbon turnover length, depleting carbon stocks in headwater reaches, and so depresses the available substrate for energy and nutrient utilization.

Newbold *et al* (1982b) hypothesized that, by increasing the ratio of surface to volume of detrital particles, leaf shredders might enhance microbial activity and nutrient uptake. Experimental tests, however, showed little evidence of this effect. Shredders instead decrease phosphorus utilization both through direct assimilation of the detrital pool and through increasing downstream particulate losses (Mulholland *et al* 1985a).

Net-spinning caddisfly (Trichoptera) and blackfly (Diptera: Simuliidae) larvae filter particles from the water column and therefore actually reduce transport. Although downstream declines in seston transport have been observed below reservoirs and lake outfalls where filter feeders are abundant (e.g. Maciolek & Tunzi 1968) and Ladle *et al* (1972) inferred a substantial effect of *Simulium* on particle transport in a chalk stream, it appears that filter feeders in most streams and rivers exert a small or insignificant effect on total particle transport (e.g. McCullough *et al* 1979; Newbold *et al* 1983a). However, they may substantially reduce the transport of particles in specific high-quality food classes (Georgian & Thorp 1992).

REFERENCES

Ammerman JW, Azam F. (1985) Bacterial 5'-nucleotidase in aquatic ecosystems: a novel mechanism of phosphorus regeneration. *Science* **227**: 1338–40. [18.6]

Andersen OK, Goldman JC, Caron DA, Dennett MR. (1986) Nutrient cycling in a microflagellate food chain: III. Phosphorus dynamics. *Marine Ecology Progress Series* **31**: 47–55. [18.8]

Aumen NG, Hawkins CP, Gregory SV. (1990) Influence of woody debris on nutrient retention in catastrophically disturbed streams. *Hydrobiologia* **190**: 183–92. [18.6, 18.7]

Axler RP, Gersberg RM, Goldman CR. (1982) Inorganic nitrogen assimilation in a subalpine lake. *Limnology and Oceanography* **27**: 53–65. [18.7]

Balderston WL, Sherr B, Payne WJ. (1976) Blockage by acetylene of nitrous oxide reduction in *Pseudomonas perfectomarinus*. *Applied and Environmental Microbiology* **31**: 504–8. [18.7]

Ball RC, Hooper FF. (1963) Translocation of phosphorus in a trout stream ecosystem. In: Schultz V, Klement AW Jr (eds) *Radioecology*, pp 217–28. Reinhold Publishing, New York. [18.2, 18.6]

Barsdate RJ, Prentki RT, Fenchel T. (1974) Phosphorus cycle of model ecosystems: significance for decomposer food chains and effect of bacterial grazers. *Oikos* **25**: 239–51. [18.6]

Bencala K. (1984) Interactions of solutes and streambed sediment. 3. A dynamic analysis of coupled hydrologic and chemical processes that determine solute transport. *Water Resources Research* **20**: 1804–14. [18.4]

Bergquist AM, Carpenter SR. (1986) Limnetic herbivory: effects on phytoplankton populations and primary production. *Ecology* **67**: 1351–60. [18.8]

Biggs BJF, Close ME. (1989) Periphyton biomass dynamics in gravel bed rivers: the relative effects of flows and nutrients. *Freshwater Biology* **22**: 209–31. [18.3]

Bothwell ML. (1985) Phosphorus limitation of lotic periphyton growth rates: an intersite comparison using continuous-flow troughs (Thompson River system, British Columbia). *Limnology and Oceanography* **30**: 527–42. [18.3, 18.6]

Bothwell ML. (1988) Growth rate responses of lotic periphytic diatoms to experimental phosphorus enrichment: the influence of temperature and light. *Canadian Journal of Fisheries and Aquatic Sciences* **45**: 261–70. [18.3]

Bothwell ML. (1989) Phosphorus-limited growth dynamics of lotic periphytic diatom communities: areal biomass and cellular growth rate responses. *Canadian Journal of Fisheries and Aquatic Sciences* **46**: 1293–301. [18.3]

Bott TL, Kaplan LA. (1990) Potential for protozoan grazing of bacteria in streambed sediments. *Journal of the North American Benthological Society* **9**: 336–45. [18.8]

Bott TL, Brock JT, Dunn CS, Naiman RJ, Ovink RW, Petersen RC. (1985) Benthic community metabolism in four temperate stream systems: an inter-biome comparison and evaluation of the river continuum concept. *Hydrobiologia* **123**: 3–45. [18.5, 18.7]

Bowie GL, Mills WB, Porcella DP, Campbell CL, Pagenkopf JR, Rupp GL *et al* (1985) *Rates, constants, and kinetics formulations in surface water quality modeling* 2nd edn. EPA 600/3-85/040. EPA Environmental Research Laboratory, Athens, Georgia. [18.4]

Burkholder-Crecco JM, Bachmann RW. (1979) Potential phytoplankton productivity of three Iowa streams. *Proceedings of the Iowa Academy of Sciences* **86**: 22–5. [18.3]

Cahill TH, Imperato P, Verhoff FH. (1974) Phosphorus dynamics in a watershed. *Journal of the Environmental Engineering Division, Proceedings of the American Society of Civil Engineers* **100**: 439–58. [18.6]

Caron DA, Goldman JC. (1990) Protozoan nutrient regeneration. In: Capriuto GM (ed.) *Ecology of Marine Protozoa*, pp 283–306. Oxford University Press, New York. [18.8]

Carpenter SR, Kitchell JF. (1984) Plankton community structure and limnetic primary production. *American Naturalist* **124**: 159–72. [18.8]

Cembella AD, Antia NJ, Harrison PJ. (1984a) The utilization of inorganic and organic phosphorus compounds as nutrients by eukaryotic microalgae: a multidisciplinary perspective. Part 1. *Critical Reviews in Microbiology* **10**: 317–91. [18.6]

Cembella AD, Antia NJ, Harrison PJ. (1984b) The utiliz-

ation of inorganic and organic phosphorus compounds as nutrients by eukaryotic microalgae: a multidisciplinary perspective. Part 2. *Critical Reviews in Microbiology* **11**: 13–81. [18.6]

Chapman BM. (1982) Numerical simulation of the transport and speciation of nonconservative chemical reactants in rivers. *Water Resources Research* **18**: 155–67. [18.4]

Chapra SC, Reckhow KH. (1983) *Engineering Approaches for Lake Management. Vol. 2: Mechanistic Modeling.* Butterworth Publishers, Boston. [18.4]

Chatarpaul L, Robinson JB. (1979) Nitrogen transformations in stream sediments: ^{15}N studies. In: Litchfield CD, Seyfried PL (eds) *Methodology for Biomass Determinations and Microbial Activities in Sediments*, pp 119–27. ASTM STP 673. American Society for Testing and Materials. Philadelphia, USA. [18.7]

Chatarpaul L, Robinson JB, Kaushik NK. (1980) Effects of tubificid worms on denitrification and nitrification in stream sediment. *Canadian Journal of Fisheries and Aquatic Sciences* **37**: 656–63. [18.7]

Christensen PB, Sørensen J. (1988) Denitrification in sediments of lowland streams: Regional and seasonal variation in Gelbæck and Rabis Bæk, Denmark. *FEMS Microbial Ecology* **53**: 335–44. [18.7]

Christensen PB, Nielsen LP, Revsbech NP, Sørensen J. (1989) Microzonation of denitrification activity in stream sediments as studied with a combined oxygen and nitrous oxide microsensor. *Applied and Environmental Microbiology* **55**: 1234–41. [18.7]

Christensen PB, Nielsen LP, Sørensen J, Revsbech NP. (1990) Denitrification in nitrate-rich streams: Diurnal and seasonal variation related to benthic oxygen metabolism. *Limnology and Oceanography* **35**: 640–51. [18.7]

Conway HL, Harrison PJ. (1977) Marine diatoms grown in chemostats under silicate or ammonium limitation. IV. Transient response of *Chaetoceros debilis*, *Skeletonoma costatum*, and *Thalassiosira gravida* to a single addition of the limiting nutrient. *Marine Biology* **43**: 33–43. [18.7]

Cooke JG, White RE. (1987) The effect of nitrate in stream water on the relationship between denitrification and nitrification in a stream-sediment microcosm. *Freshwater Biology* **18**: 213–26. [18.7]

Cooper AB. (1983) Effect of storm events on benthic nitrifying activity. *Applied and Environmental Microbiology* **46**: 957–60. [18.7]

Cooper AB. (1984) Activities of benthic nitrifiers in streams and their role in oxygen consumption. *Microbiol Ecology* **10**: 317–34. [18.7]

Cooper AB, Cooke JG. (1984) Nitrate loss and transformation in 2 vegetated headwater streams. *New Zealand Journal of Marine and Freshwater Research* **18**: 441–50. [18.7]

Cooper DC. (1973) Enhancement of net primary productivity by herbivore grazing in aquatic laboratory microcosms. *Limnology and Oceanography* **18**: 31–7. [18.8]

Corning KE, Duthie HC, Paul BJ. (1989) Phosphorus and glucose uptake by seston and epilithon in boreal forest streams. *Journal of the North American Benthological Society* **8**: 123–33. [18.6]

Cosser PR. (1989) Nutrient concentration–flow relationships and loads in the South Pine River, south-eastern Queensland. 1. Phosphorus loads. *Australian Journal of Marine and Freshwater Research* **40**: 613–30. [18.6]

Crosby MP, Newell RIE, Langdon CJ. (1990) Bacterial mediation in the utilization of carbon and nitrogen from detrital complexes by *Crassostrea virginica*. *Limnology and Oceanography* **35**: 625–39. [18.8]

Cuffney TF, Wallace JB, Lugthart GJ. (1990) Experimental evidence quantifying the role of benthic invertebrates in organic matter dynamics of headwater streams. *Freshwater Biology* **23**: 281–99. [18.8]

Cummins KW, Petersen RC, Howard FO, Wuycheck JC, Holt VI. (1973) The utilization of leaf litter by stream detritivores. *Ecology* **54**: 336–45. [18.8]

Cummins KW, Sedell JR, Swanson FJ, Minshall GW, Fisher SG, Cushing CE *et al.* (1983) Organic matter budgets for stream ecosystems: problems in their evaluation. In: Barnes JR, Minshall GW (eds) *Stream Ecology. Application and Testing of General Ecological Theory* pp 299–353. Plenum Press, New York. [18.5]

D'Angelo DJ, Webster JR, Benfield EF. (1991) Mechanisms of stream phosphorus retention: an experimental study. *Journal of the North American Benthological Society* **10**: 225–37. [18.6]

Décamps H, Capblancb J, Tourenq JN. (1984) Lot. In: Whitton BA (ed.) *Ecology of European Rivers*. pp 207–35. Blackwell Scientific Publications. [18.3, 18.6]

DePinto JV. (1979) Water column death and decomposition of phytoplankton: an experimental and modeling review. In: Scavia D, Robertson A (eds) *Perspectives on Lake Ecosystem Modeling*, pp 25–52. Ann Arbor Science Publishers, Ann Arbor, Michigan. [18.6]

Dillon PJ, Rigler FH. (1974) The phosphorus–chlorophyll relationship in lakes. *Limnology and Oceanography* **19**: 767–73. [18.3]

Downes MT, Paerl HW. (1978) Separation of two dissolved reactive phosphorus fractions in lakewater. *Journal of the Fisheries Research Board of Canada* **35**: 1635–9. [18.6]

Droop MR. (1973) Some thoughts on nutrient limitation in algae. *Journal of Phycology* **9**: 264–72. [18.6]

Duff JH, Triska FJ. (1990) Denitrification in sediments from the hyporheic zone adjacent to a small forested

stream. *Canadian Journal of Fisheries and Aquatic Sciences* **47**: 1140–7. [18.7]

Duff JH, Stanley KC, Triska FJ, Avanzino RJ. (1984a) The use of photosynthesis-respiration chambers to measure nitrogen flux in epilithic algal communities. *Verhandlungen Internationale Vereinigung für Theoretische und Angewandt Limnologie* **22**: 1436–43. [18.7]

Duff JH, Triska FJ, Oremland RS. (1984b) Denitrification associated with stream periphyton: chamber estimates from undisrupted communities. *Journal of Environmental Quality* **13**: 514–18. [18.7]

Dugdale RC, Goering JJ. (1967) Uptake of new and regenerated nitrogen in primary productivity. *Limnology and Oceanography* **12**: 196–206. [18.2]

Edwards RT, Meyer JL. (1987) Metabolism of a subtropical low gradient blackwater river. *Freshwater Biology* **17**: 251–64. [18.5]

Eisenreich SJ, Armstrong DE. (1977) Chromatographic investigation of inositol phosphate esters in lake waters. *Environmental Science and Technology* **11**: 497–501. [18.6]

Elser JJ, Goldman CR. (1991) Zooplankton effects on phytoplankton in lakes of contrasting trophic status. *Limnology and Oceanography* **36**: 64–90. [18.8]

Elser JJ, Kimmel BL. (1985) Nutrient availability for phytoplankton production in a multiple-impoundment series. *Canadian Journal of Fisheries and Aquatic Sciences* **42**: 1359–70. [18.3]

Elser JJ, Marzolf ER, Goldman CR. (1990) Phosphorus and nitrogen limitation of phytoplankton growth in the freshwaters of North-America: a review and critique of experimental enrichments. *Canadian Journal of Fisheries and Aquatic Sciences* **47**: 1468–77. [18.3]

Elwood JW, Nelson DJ. (1972) Periphyton production and grazing rates in a stream measured with a ^{32}P material balance method. *Oikos* **23**: 295–303. [18.6]

Elwood JW, Newbold JD, O'Neill RV, Stark RW, Singley PT. (1981a) The role of microbes associated with organic and inorganic substrates in phosphorus spiralling in a woodland stream. *Verhandlungen Internationale Vereinigung für Theoretische und Angewandt Limnologie* **21**: 850–6. [18.6]

Elwood JW, Newbold JD, Trimble AF, Stark RW. (1981b) The limiting role of phosphorus in a woodland stream ecosystem: effects of P enrichment on leaf decomposition and primary producers. *Ecology* **62**: 146–58. [18.3]

Elwood JW, Newbold JD, O'Neill RV, Van Winkle W. (1983) Resource spiralling: an operational paradigm for analyzing lotic ecosystems. In: Fontaine TD III, Bartell SM (eds) *The Dynamics of Lotic Ecosystems*, pp 3–27. Ann Arbor Science, Ann Arbor, Michigan. [18.2]

Elwood JW, Mulholland PJ, Newbold JD. (1988)

402 Inputs and Pathways of Matter and Energy

Microbial activity and phosphorus uptake on decomposing leaf detritus in a heterotrophic stream.
*Verhandlungen Internationale Vereinigung für
Theoretische und Angewandt Limnologie* **23**:
1198–208. [18.6]

Eppley RW, Renger EH. (1974) Nitrogen assimilation of
an oceanic diatom in nitrogen-limited continuous
culture. *Journal of Phycology* **10**: 15–23. [18.7]

Eppley RW, Renger EH, Harrison WG, Cullen JJ. (1979)
Ammonium distribution in southern California
coastal water and its role in the growth of phytoplankton. *Limnology and Oceanography* **24**: 495–509.
[18.7]

Feminella JW, Power ME, Resh VH. (1989) Periphyton
responses to invertebrate grazing and riparian canopy
in three northern California coastal streams. *Freshwater Biology* **22**: 445–57. [18.8]

Findlay S, Carlough L, Crocker MT, Gill HK, Meyer JL,
Smith PJ. (1986) Bacterial growth on macrophyte
leachate and fate of bacterial production. *Limnology
and Oceanography* **31**: 1335–41. [18.8]

Fisher SG. (1977) Organic matter processing by a stream-
segment ecosystem: Fort River, Massachusetts, USA.
Internationale Revue der Gesamten Hydrobiologie
62: 701–27.

Fisher SG, Likens GE. (1973) Energy flow in Bear Brook,
New Hampshire: an integrative approach to stream
ecosystem metabolism. *Ecological Monographs* **43**:
421–39. [18.5, 18.8]

Fisher SG, Gray LJ, Grimm NB, Busch DE. (1982)
Temporal succession in a desert stream ecosystem
following flash flooding. *Ecological Monographs* **52**:
93–110. [18.7]

Flint RW, Goldman CR. (1975) The effects of a benthic
grazer on the primary productivity of the littoral zone
of Lake Tahoe. *Limnology and Oceanography* **20**:
935–44. [18.8]

Francis MM, Naiman RJ, Melillo JM. (1985) Nitrogen
fixation in subarctic streams influenced by beaver
(*Castor canadensis*). *Hydrobiologia* **121**: 193–202.
[18.7]

Francko DA. (1986) Epilimnetic phosphorus cycling:
influence of humic materials and iron on coexisting
major mechanisms. *Canadian Journal of Fisheries
and Aquatic Sciences* **43**: 302–10. [18.6]

Freeman MC. (1986) The role of nitrogen and phos-
phorus in the development of *Cladophora glomerata*
(L.) Kutzing in the Manawatu River, Zew Zealand.
Hydrobiologia **131**: 23–30. [18.3]

Furumai H, Kondo T, Ohgaki S. (1989) Phosphorus
exchange kinetics and exchangeable phosphorus
forms in sediments. *Water Research* **23**: 685–91.
[18.6]

Georgian T, Thorp JH. (1992) Effects of microhabitat
selection on feeding rates of net-spinning caddisfly
larvae. *Ecology* **73**: 229–40. [18.8]

Gilliam JF, Fraser DF, Sabat AM. (1989) Strong effects of
foraging minnows on a stream benthic invertebrate
community. *Ecology* **70**: 445–52. [18.8]

Gloss SP, Reynolds RC Jr, Mayer LM, Kidd DE. (1981)
Reservoir influences on salinity and nutrient fluxes
in the arid Colorado River basin. In: Stefan HG (ed.)
Symposium on Surface Water Impoundments, 2–5
June 1980, Minneapolis, pp 1618–29. *American
Society of Civil Engineers*, New York. [18.6]

Green DB, Logan TJ, Smeck NE. (1978) Phosphate
adsorption–desorption characteristics of suspended
sediments in the Maumee River Basin of Ohio.
Journal of Environmental Quality **7**: 208–12. [18.6]

Gregory SV. (1978) Phosphorus dynamics on organic
and inorganic substrates in streams. *Verhandlungen
Internationale Vereinigung für Theoretische und
Angewandt Limnologie* **20**: 1340–46. [18.6]

Gregory SV. (1983) Plant–herbivore interactions in
stream systems. In: Minshall GW, Barnes JR (eds)
*Stream Ecology. Application and Testing of General
Ecological Theory*, pp 157–89. Plenum Press, New
York. [18.8]

Grimm NB. (1987) Nitrogen dynamics during succession in a desert stream. *Ecology* **68**: 1157–70.
[18.3, 18.7]

Grimm NB. (1988) Role of macroinvertebrates in
nitrogen dynamics of a desert stream. *Ecology* **69**:
1884–93. [18.8]

Grimm NB, Fisher SG. (1986) Nitrogen limitation in
a Sonoran Desert stream. *Journal of the North
American Benthological Society* **5**: 2–15. [18.3]

Grobbelaar JU. (1983) Availability to algae of N and P
adsorbed on suspended solids in turbid waters of the
Amazon River. *Archiv für Hydrobiologie* **96**: 302–16.
[18.6]

Gromiec MJ, Loucks DP, Orlob GT. (1983) Stream
quality modeling. In: Orlob GT (ed.) *Mathematical
Modeling of Water Quality: Streams, Lakes, and
Reservoirs*, pp 176–226. John Wiley and Sons, New
York. [18.5]

Harms LL, Vidal PH, McDermott TE. (1978) Phosphorus
interactions with stream-bed sediments. *Journal of
the Environmental Engineering Division of the
American Society of Civil Engineers* **104**: 271–88.
[18.6]

Hart DD, Robinson CT. (1990) Resource limitation in a
stream community: phosphorus enrichment effects
on periphyton and grazers. *Ecology* **71**: 1494–502.
[18.3, 18.8]

Herbes SE, Allen HE, Mancy KH. (1975) Enzymatic
characterization of soluble organic phosphorus in lake
water. *Science* **187**: 432–34. [18.6]

Hill AR. (1979) Denitrification in the nitrogen budget of
a river ecosystem. *Nature* **281**: 291–2. [18.7]

Hill AR. (1982) Phosphorus and major cation mass
balances for two rivers during low summer flows.

Freshwater Biology **12**: 293–304. [18.6]

Hill AR. (1988) Factors influencing nitrate depletion in a rural stream. *Hydrobiologia* **160**: 111–22. [18.7]

Hill AR, Sanmugadas K. (1985) Denitrification rates in relation to stream sediment characteristics. *Water Research* **19**: 1579–86. [18.7]

Hill AR, Warwick J. (1987) Ammonium transformations in springwater within the riparian zone of a small woodland stream. *Canadian Journal of Fisheries and Aquatic Sciences* **44**: 1948–56. [18.7]

Hill WR, Harvey BC. (1990) Periphyton responses to higher trophic levels and light in a shaded stream. *Canadian Journal of Fisheries and Aquatic Sciences* **47**: 2307–14. [18.8]

Hill WR, Knight AW. (1988) Nutrient and light limitation of algae in two northern California streams. *Journal of Phycology* **24**: 125–32. [18.3]

Hino S. (1989) Characterization of orthophosphate release from dissolved organic phosphorus by gel filtration and several hydrolytic enzymes. *Hydrobiologia* **174**: 49–55. [18.6]

Horne AJ, Carmiggelt CJW. (1975) Algal nitrogen fixation in California streams: seasonal cycles. *Freshwater Biology* **5**: 461–70. [18.7]

Horner RR, Welch EB. (1981) Stream periphyton development in relation to current velocity and nutrients. *Canadian Journal of Fisheries and Aquatic Sciences* **38**: 449–57. [18.3]

Horner RR, Welch EB, Veenstra RB. (1983) Development of nuisance periphytic algae in laboratory streams in relation to enrichment and velocity. In: Wetzel RG (ed.) *Periphyton of Freshwater Ecosystems*, pp 121–34. Junk Publishers, The Hague. [18.3]

Horner RR, Welch EB, Seeley MR, Jacoby JM. (1990) Responses of periphyton to changes in current velocity, suspended sediment and phosphorus concentration. *Freshwater Biology* **24**: 215–32. [18.3]

Hullar MA, Vestal JR. (1989) The effects of nutrient limitation and stream discharge on the epilithic microbial community in an oligotrophic Arctic stream. *Hydrobiologia* **172**: 19–26. [18.3]

Ittekot V. (1988) Global trends in the nature of organic matter in river suspensions. *Nature* **332**: 436–8. [18.5]

Jacoby JM. (1985) Grazing effects on periphyton by *Theodoxus fluviatilis* (Gastropoda) in a lowland stream. *Journal of Freshwater Ecology* **3**: 265–74. [18.8]

Johannes RE. (1965) Influence of marine protozoa on nutrient regeneration. *Limnology and Oceanography* **10**: 434–42. [18.8]

Johnson AH, Bouldin DR, Goyette EA, Hedges AM. (1976) Phosphorus loss by stream transport from a rural watershed: quantities, processes, and sources. *Journal of Environmental Quality* **5**: 148–57. [18.6]

Jordan C, Dinsmore P. (1985) Determination of biologically available phosphorus using a radiobioassay technique. *Freshwater Biology* **15**: 597–603. [18.6]

Karlsson G, Löwgren M. (1990) River transport of phosphorus as controlled by large scale land use changes. *Acta Agriculturae Scandinavica* **40**: 149–62. [18.6]

Kaushik NK, Hynes HBN. (1968) Experimental study on the role of autumn-shed leaves in aquatic environments. *Journal of Ecology* **56**: 229–43. [18.6]

Kaushik NK, Robinson JB, Sain P, Whiteley HR, Stammers WN. (1975) A quantitative study of nitrogen loss from water of a small spring-fed stream. In: *Water Pollution Research in Canada*, pp 110–17. Proceedings of the 10th Canadian Symposium, Toronto, February 1974. Institute for Environmental Studies, University of Toronto, Toronto. [18.7]

Kaushik NK, Robinson JB, Stammers WN, Whiteley HR. (1981) Aspects of nitrogen transport and transformation in headwater streams. In: Lock MA, Williams DD (eds) *Perspectives in Running Water Ecology*, pp 113–39. Plenum Press, New York. [18.7]

Kehde PM, Wilhm JL. (1972) The effects of grazing by snails on community structure of periphyton in laboratory streams. *The American Midland Naturalist* **87**: 8–24. [18.8]

Keithan ED, Lowe RL, Deyoe HR. (1988) Benthic diatom distribution in a Pennsylvania stream: role of pH and nutrients. *Journal of Phycology* **24**: 581–5. [18.3]

Keup LE. (1968) Phosphorus in flowing waters. *Water Research* **2**: 373–86. [18.6]

Kim BK, Jackman AP, Triska FJ. (1990) Modeling transient storage and nitrate uptake kinetics in a flume containing a natural periphyton community. *Water Resources Research* **26**: 505–15. [18.4, 18.7]

Kitchell JF, O'Neill RV, Webb D, Gallepp GW, Bartell SM, Koonce JF, Ausmus BS. (1979) Consumer regulation of nutrient cycling. *BioScience* **29**: 28–34. [18.8]

Klotz RL. (1985) Factors controlling phosphorus limitation in stream sediments. *Limnology and Oceanography* **30**: 543–53. [18.3, 18.6]

Klotz RL. (1988) Sediment control of soluble reactive phosphorus in Hoxie Gorge Creek, New York. *Canadian Journal of Fisheries and Aquatic Sciences* **45**: 2026–34. [18.6]

Knorr DF, Fairchild GW. (1987) Periphyton, benthic invertebrates and fishes as biological indicators of water quality in the East Branch Brandywine Creek. *Proceedings of the Pennsylvania Academy of Science* **61**: 61–6. [18.3]

Krogstad T, Løvstad Ø. (1989) Erosion, phosphorus and phytoplankton response in rivers of South-Eastern Norway. *Hydrobiologia* **183**: 33–41. [18.3, 18.6]

Kunishi HM, Taylor AW, Heald WR, Gburek WJ, Weaver RN. (1972) Phosphate movement from an agricultural watershed during two rainfall periods. *Journal of Agricultural and Food Chemistry* **20**: 900–5. [18.6]

Kuwabara JS, Helliker P. (1988) Trace contaminants in streams. In: Cheremisinoff PN, Cheremisinoff NP, Cheng SL (eds) *Civil Engineering Practice*, Vol. 5, pp 739–65. Technomic Publishing, Lancaster, Pennsylvania. [18.4]

Ladle M, Bass JAB, Jenkins WR. (1972) Studies on the production and food consumption by the larval Simuliidae (Diptera) of a small chalk stream. *Hydrobiologia* **39**: 429–48. [18.8]

Lamberti GA, Resh VH. (1983) Stream periphyton and insect herbivores: an experimental study of grazing by a caddisfly population. *Ecology* **64**: 1124–35. [18.8]

Lamberti GA, Ashkenas LR, Gregory SV, Steinman AD. (1987) Effects of three herbivores on periphyton communities in laboratory streams. *Journal of the North American Benthological Society* **6**: 92–104. [18.8]

Lamberti GA, Gregory SV, Ashkenas LR, Steinman AD, McIntire CD. (1989) Productive capacity of periphyton as a determinant of plant herbivore interactions in streams. *Ecology* **70**: 1840–56. [18.8]

Lane P, Levins R. (1977) The dynamics of aquatic systems. 2. The effects of nutrient enrichment on model plankton communities. *Limnology and Oceanography* **22**: 454–471. [18.8]

Lean DRS. (1973) Movements of phosphorus between its biologically important forms in lake water. *Journal of the Fisheries Research Board of Canada* **30**: 1525–36. [18.6]

Lean DRS, Nalewajko C. (1976) Phosphate exchange and organic phosphorus excretion by freshwater algae. *Journal of the Fisheries Research Board of Canada* **33**: 1312–23. [18.6]

Leonard RL, Kaplan LA, Elder JF, Coats RN, Goldman CR. (1979) Nutrient transport in surface runoff from a subalpine watershed, Lake Tahoe Basin, California. *Ecological Monographs* **49**: 281–310. [18.6]

Li WC, Armstrong DE, Williams JDH, Harris RF, Syers JK. (1972) Rate and extent of inorganic phosphate exchange in lake sediments. *Soil Science Society of America, Proceedings* **36**: 279–85. [18.6]

Lock MA, Ford TE, Hullar MAJ, Kaufman M, Vestal JR, Volk GS, Ventullo RM. (1990) Phosphorus limitation in an arctic river biofilm—a whole ecosystem experiment. *Water Research* **24**: 1545–9. [18.3]

McAuliffe JR. (1984) Resource depression by a stream herbivore—effects on distributions and abundances of other grazers. *Oikos* **42**: 327–33. [18.8]

McCallister DL, Logan TJ. (1978) Phosphate adsorption–desorption characteristics of soils and bottom sediments in the Maumee River Basin of Ohio. *Journal of Environmental Quality* **7**: 87–92. [18.6]

McColl RHS. (1974) Self-purification of small freshwater streams: phosphate, nitrate, and ammonia removal. *New Zealand Journal of Marine Freshwater Research* **8**: 375–88. [18.6, 18.7]

McCullough DA, Minshall GW, Cushing CE. (1979) Bioenergetics of a stream 'collector' organism, *Tricorythodes minutus* (Insecta: Ephemeroptera). *Limnology and Oceanography* **24**: 45–58. [18.8]

McCutcheon S. (1987) Laboratory and instream nitrification rates for selected streams. *Journal of Environmental Engineering* **113**: 628–46. [18.7]

McCutcheon SC. (1989) *Water Quality Modeling. Vol. 1. Transport and Surface Exchange in Rivers.* CRC Press, Boca Raton, Florida. [18.4]

Maciolek JA, Tunzi MG. (1968) Microseston dynamics in a simple Sierra Nevada lake-stream system. *Ecology* **49**: 60–75. [18.8]

Marker AFH, Gunn RJM. (1977) The benthic algae of some streams in southern England. III. Seasonal variations in chlorophyll *a* in the seston. *Journal of Ecology* **65**: 223–34. [18.5]

Mason JW, Wegner GD, Quinn GI, Lange EL. (1990) Nutrient loss via groundwater discharge from small watersheds in southwestern and south central Wisconsin. *Journal of Soil and Water Conservation* **45**: 327–31. [18.6]

Matulewich VA, Finstein MS. (1978) Distribution of autotrophic nitrifying bacteria in a polluted river (the Passaic). *Applied and Environmental Microbiology* **35**: 67–71. [18.7]

Mayer LM, Gloss SP. (1980) Buffering of silica and phosphate in a turbid river. *Limnology and Oceanography* **25**: 12–22. [18.6]

Megard RO. (1981) Effects of planktonic algae on water quality in impoundments of the Mississippi River in Minnesota. In: Stefan HG (ed.) *Symposium on Surface Water Impoundments*, 2–5 June 1980, Minneapolis, pp 1575–84. American Society of Civil Engineers, New York. [18.3]

Meybeck M. (1982) Carbon, nitrogen, and phosphorus transport by world rivers. *American Journal of Science* **282**: 401–50. [18.6, 18.7]

Meyer JL. (1979) The role of sediments and bryophytes in phosphorus dynamics in a headwater stream ecosystem. *Limnology and Oceanography* **24**: 365–75. [18.6]

Meyer JL. (1980) Dynamics of phosphorus and organic matter during leaf decomposition in a forest stream. *Oikos* **34**: 44–53. [18.6]

Meyer JL, Edwards RT. (1990) Ecosystem metabolism and turnover of organic carbon along a blackwater river continuum. *Ecology* **71**: 668–77. [18.5]

Meyer JL, Johnson C. (1983) The influence of elevated nitrate concentration on rate of leaf decomposition in a stream. *Freshwater Biology* **13**: 177–83. [18.3]

Meyer JL, Likens GE. (1979) Transport and transformation of phosphorus in a forest stream ecosystem. *Ecology* **60**: 1255–69. [18.6]

Meyer JL, McDowell WH, Bott TL, Elwood JW, Ishizaki

C, Melack JM *et al.* (1988) Elemental dynamics in streams. *Journal of the North American Benthological Society* 7: 410–32. [18.1]

Minear RA. (1972) Characterization of naturally occuring dissolved organophosphorus compounds. *Environmental Science & Technology* 6: 431–7. [18.6]

Minshall GW, Petersen RC, Cummins KW, Bott TL, Sedell JR, Cushing CE, Vannote RL. (1983) Interbiome comparison of stream ecosystem dynamics. *Ecological Monographs* 53: 1–25. [18.5]

Minshall GW, Petersen RC, Bott TL, Cushing CE, Cummins KW, Vannote RL, Sedell JR. (1992) Stream ecosystem dynamics of the Salmon River Idaho: an 8th order drainage system. *Journal of the North American Benthological Society* 11: 111–37. [18.5]

Mulholland PJ. (1981) Organic carbon flow in a swamp–stream ecosystem. *Ecological Monographs* 51: 307–22. [18.5]

Mulholland PJ, Watts JA. (1982) Transport of organic carbon to the oceans by rivers of North America: a synthesis of existing data. *Tellus* 34: 176–86. [18.5]

Mulholland PJ, Newbold JD, Elwood JW, Hom CL. (1983) The effect of grazing intensity on phosphorus spiralling in autotrophic streams. *Oecologia* 58: 358–66. [18.8]

Mulholland PJ, Elwood JW, Newbold JD, Webster JR, Ferren LA, Perkins RE. (1984) Phosphorus uptake by decomposing leaf detritus: effect of microbial biomass and activity. *Verhandlungen Internationale Vereinigung für Theoretische und Angewandt Limnologie* 22: 1899–905. [18.6]

Mulholland PJ, Elwood JW, Newbold JD, Ferren LA. (1985a) Effect of a leaf-shredding invertebrate on organic matter dynamics and phosphorus spiralling in heterotrophic laboratory streams. *Oecologia* 66: 199–206. [18.8]

Mulholland PJ, Newbold JD, Elwood JW, Ferren LA, Webster JR. (1985b) Phosphorus spiralling in a woodland stream: seasonal variations. *Ecology* 66: 1012–23. [18.6]

Mulholland PJ, Minear RA, Elwood JW. (1988) Production of soluble, high molecular weight phosphorus and its subsequent uptake by stream detritus. *Verhandlungen Internationale Vereinigung für Theoretische und Angewandt Limnologie* 23: 1190–7. [18.6]

Mulholland PJ, Steinman AD, Elwood JW. (1990) Measurement of phosphorus uptake length in streams: comparison of radiotracer and stable PO_4 releases. *Canadian Journal of Fisheries and Aquatic Sciences* 47: 2351–7. [18.4, 18.6]

Munn MD, Osborne LL, Wiley MJ. (1989) Factors influencing periphyton growth in agricultural streams of central Illinois. *Hydrobiologia* 174: 89–97. [18.3]

Munn NL, Meyer JL. (1990) Habitat-specific solute re-

tention in two small streams: an intersite comparison. *Ecology* 71: 2069–82. [18.6, 18.7]

Munn N, Prepas E. (1986) Seasonal dynamics of phosphorus partitioning and export in two streams in Alberta, Canada. *Canadian Journal of Fisheries and Aquatic Sciences* 43: 2464–71. [18.6]

Naiman RJ, Melillo JM, Lock MA, Ford TE, Reice SR. (1987) Longitudinal patterns of ecosystem processes and community structure in a subarctic river continuum. *Ecology* 68: 1139–56. [18.5]

Nakajima T. (1979) Denitrification by the sessile microbial community of a polluted river. *Hydrobiologia* 66: 57–64. [18.7]

Newbold JD. (1987) Phosphorus spiralling in rivers and river-reservoir systems: implications of a model. In: Craig JF, Kemper JB (eds) *Regulated Streams: Advances in Ecology*, pp 303–27. Plenum Press, New York. [18.3]

Newbold JD, Elwood JW, O'Neill RV, Van Winkle W. (1981) Measuring nutrient spiralling in streams. *Canadian Journal of Fisheries and Aquatic Sciences* 38: 860–3. [18.2]

Newbold JD, Mulholland PJ, Elwood JW, O'Neill RV. (1982a) Organic carbon spiralling in stream ecosystems. *Oikos* 38: 266–72. [18.5]

Newbold JD, O'Neill RV, Elwood JW, Van Winkle W. (1982b) Nutrient spiralling in streams: implications for nutrient limitation and invertebrate activity. *American Naturalist* 120: 628–52. [18.3, 18.8]

Newbold JD, Elwood JW, O'Neill RV, Sheldon AL. (1983a) Phosphorus dynamics in a woodland stream ecosystem: a study of nutrient spiralling. *Ecology* 64: 1249–65. [18.2, 18.6, 18.8]

Newbold JD, Elwood JW, Schulze MS, Stark RW, Barmeier JC. (1983b) Continuous ammonium enrichment of a woodland stream: uptake kinetics, leaf decomposition, and nitrification. *Freshwater Biology* 13: 193–204. [18.3, 18.7]

O'Connor DJ. (1988a) Models of sorptive toxic substances in freshwater systems. I. Basic equations. *Journal of Environmental Engineering* 114: 507–32. [18.4]

O'Connor DJ. (1988b) Models of sorptive toxic substances in freshwater systems. II. Lakes and reservoirs. *Journal of Environmental Engineering* 114: 533–51. [18.4]

O'Connor DJ. (1988c) Models of sorptive toxic substances in freshwater systems. III. Streams and rivers. *Journal of Environmental Engineering* 114: 552–74. [18.4]

O'Neill RV. (1979) A review of linear compartmental analysis in ecosystem science. In: Matis JH, Patten BC, White GC (eds) *Compartmental Analysis of Ecosystem Models*, pp 3–28. International Co-operative Publishing House, Fairland, Maryland. [18.4]

Omernik JM. (1977) *Nonpoint source-stream nutrient*

level relationships: a nationwide survey. EPA-600/3-77-105, Ecological Research Series, United States Environmental Protection Agency, Washington, DC. [18.6, 18.7]

Orlob GT (ed.). (1983) *Mathematical Modeling of Water Quality: Streams, Lakes, and Reservoirs.* John Wiley and Sons, New York. [18.4]

Paerl HW, Downes MT. (1978) Biological availability of low versus high molecular weight reactive phosphorus. *Journal of the Fisheries Research Board of Canada* **35**: 1639–43. [18.6]

Paul BJ, Duthie HC. (1989) Nutrient cycling in the epilithon of running waters. *Canadian Journal of Botany* **67**: 2302–9. [18.6]

Paul BJ, Corning KE, Duthie HC. (1989) An evaluation of the metabolism of sestonic and epilithic communities in running waters using an improved chamber technique. *Freshwater Biology* **21**: 207–17. [18.6]

Peters RH. (1978) Concentrations and kinetics of phosphorus fractions in water from streams entering Lake Memphremagog. *Journal of the Fisheries Research Board of Canada* **35**: 315–28. [18.6]

Peters RH. (1981) Phosphorus availability in Lake Memphremagog and its tributaries. *Limnology and Oceanography* **26**: 1150–61. [18.6]

Petersen RC, Cummins KW. (1974) Leaf processing in a woodland stream. *Freshwater Biology* **4**: 343–68. [18.8]

Petersen RC, Cummins KW, Ward GM. (1989) Microbial and animal processing of detritus in a woodland stream. *Ecological Monographs* **59**: 21–39. [18.8]

Peterson BJ, Hobbie JE, Hershey AE, Lock MA, Ford TE, Vestal JR, McKinley VL *et al.* (1985) Transformation of a tundra river from heterotrophy to autotrophy by addition of phosphorus. *Science* **229**: 1383–6. [18.3, 18.6]

Pomeroy LR. (1970) The strategy of mineral cycling. *Annual Review of Ecology and Systematics* **1**: 171–90. [18.2]

Pomeroy LR, Smith EE, Grant CM. (1965) The exchange of phosphate between estuarine water and sediments. *Limnology and Oceanography* **10**: 167–72. [18.6]

Power ME. (1990) Effects of fish in river food webs. *Science* **250**: 811–14. [18.8]

Power ME, Matthews WJ, Stewart AJ. (1985) Grazing minnows, piscivorous bass, and stream algae: dynamics of a strong interaction. *Ecology* **66**: 1448–56. [18.8]

Power ME, Stewart AJ, Matthews WJ. (1988) Grazer control of algae in an Ozark Mountain stream: effects of short-term exclusion. *Ecology* **69**: 1894–98. [18.8]

Prairie YT, Kalff J. (1988) Particulate phosphorus dynamics in headwater streams. *Canadian Journal of Fisheries and Aquatic Sciences* **45**: 210–15. [18.6]

Pringle CM. (1987) Effects of water and substratum nutrient supplies on lotic periphyton growth: an integrated bioassay. *Canadian Journal of Fisheries and Aquatic Sciences* **44**: 619–29. [18.3]

Pringle CM, Bowers JA. (1984) An *in situ* substratum fertilization technique: diatom colonization on nutrient-enriched, sand substrata. *Canadian Journal of Fisheries and Aquatic Sciences* **41**: 1247–51. [18.3]

Pringle CM, Paaby-Hansen P, Vaux PD, Goldman CR. (1986) *In situ* nutrient assays of periphyton growth in a lowland Costa Rican stream. *Hydrobiologia* **134**: 207–13. [18.3]

Redfield AC, Ketchum BH, Richards FA. (1963) The influence of organisms on the composition of sea water. In: Hill MN (ed.) *The Sea*, Vol. 2, pp 26–77. Interscience, New York. [18.7]

Rekolainen S. (1989) Effect of snow and soil frost melting on the concentrations of suspended solids and phosphorus in two rural watersheds in western Finland. *Aquatic Sciences* **51**: 211–23. [18.6]

Rhee G-Y. (1974) Phosphate uptake under nitrate limitation by *Scenedesmus* sp. and its ecological implications. *Journal of Phycology* **10**: 470–5. [18.6]

Richey JE, Hedges JI, Devol AH, Quay PD, Victoria R, Martinelli L, Forsberg BR. (1990) Biogeochemistry of carbon in the Amazon River. *Limnology and Oceanography* **35**: 352–71. [18.5]

Richey JS, McDowell WH, Likens GE. (1985) Nitrogen transformations in a small mountain stream. *Hydrobiologia* **124**: 129–39. [18.7]

Rigler FH. (1966) Radiobiological analysis of inorganic phosphorus in lakewater. *Verhandlungen Internationale Vereinigung für Theoretische und Angewandt Limnologie* **16**: 465–70. [18.5, 18.6]

Ruttner F. (1963) *Fundamentals of Limnology* 3rd edn. University of Toronto Press, Toronto. [18.3]

Saunders JF III, Lewis WM Jr. (1988) Transport of phosphorus, nitrogen, and carbon by the Apure River, Venezuela. *Biogeochemistry* **5**: 323–42. [18.6]

Sayler GS, Puziss M, Silver M. (1979) Alkaline phosphatase assay for freshwater sediments: application to perturbed sediment systems. *Applied and Environmental Microbiology* **38**: 922–7. [18.6]

Schlessinger WH, Melack JM. (1981) Transport of organic carbon in the world's rivers. *Tellus* **33**: 172–87. [18.5]

Sebetich MJ, Kennedy VC, Zand SM, Avanzino RJ, Zelweger GW. (1984) Dynamics of added nitrate and phosphate compared in a northern California woodland stream. *Water Resources Bulletin* **20**: 93–101. [18.7]

Seitzinger SP. (1988) Denitrification in freshwater and coastal marine ecosystems: ecological and geochemical significance. *Limnology and Oceanography* **33**: 702–24. [18.7]

Simmons BL, Cheng DMH. (1985) Rate and pathways of phosphorus assimilation in the Nepean River at

Camden, New South Wales. *Water Research* **19**: 1089–96. [18.6]

Smart MM, Jones JR, Sebaugh JL. (1985) Stream-watershed relations in the Missouri Ozark Plateau Province. *Journal of Environmental Quality* **14**: 77–82. [18.6]

Smith REH, Harrison WG, Harris L. (1985) Phosphorus exchange in marine microplankton communities near Hawaii. *Marine Biology* **86**: 75–84. [18.6]

Smock LA, Roeding CE. (1986) The trophic basis of production of the macroinvertebrate community of a southeastern USA blackwater stream. *Holarctic Ecology* **9**: 165–74. [18.8]

Søballe DM, Kimmel BL. (1987) A large-scale comparison of factors influencing phytoplankton abundance in rivers, lakes, and impounds. *Ecology* **68**: 1943–54. [18.3]

Sprent JI. (1987) *The Ecology of the Nitrogen Cycle.* Cambridge University Press, Cambridge. [18.7]

Stabel H-H, Geiger M. (1985) Phosphorus adsorption to riverine suspended matter. Implications for the P-budget of lake Constance. *Water Research* **19**: 1347–52. [18.6]

Stainton MP. (1980) Errors in molybdenum blue methods for determining orthophosphate in freshwater. *Canadian Journal of Fisheries and Aquatic Sciences* **37**: 472–8. [18.6]

Stanley DW, Hobbie JE. (1981) Nitrogen recycling in a North Carolina coastal river. *Limnology and Oceanography* **26**: 30–42. [18.7]

Stewart AJ. (1987) Responses of stream algae to grazing minnows and nutrients: a field test for interactions. *Oecologia* **72**: 1–7. [18.8]

Stockner JG, Shortreed KRS. (1978) Enhancement of autotrophic production by nutrient addition in a coastal rainforest stream on Vancouver Island. *Journal of the Fisheries Research Board of Canada* **35**: 28–34. [18.3]

Stream Solute Workshop. (1990) Concepts and methods for assessing solute dynamics in stream ecosystems. *Journal of the North American Benthological Society* **9**: 95–119. [18.1, 18.6]

Streeter HW, Phelps EB. (1925) A study of the pollution and natural purification of the Ohio River. III. Factors concerned in the phenomena of oxidation and reaeration. Bulletin 146, US Public Health Service, Washington DC. pp 1–75. [18.5]

Strickland JDH, Parsons TR. (1972) *A Practical Handbook of Seawater Analysis.* Bulletin 167, 2nd edn. Fisheries Research Board of Canada. Ottawa. [18.6]

Sumner WT, McIntire CD. (1982) Grazer–periphyton interactions in laboratory streams. *Archives in Hydrobiology* **93**: 135–57. [18.8]

Swank WT, Caskey WH. (1982) Nitrate depletion in a second-order mountain stream. *Journal of Environmental Quality* **11**: 581–4. [18.7]

Tarapchak SJ. (1983) Soluble reactive phosphorus measurements in lake water: evidence for molybdate-enhanced hydrolysis. *Journal of Environmental Quality* **12**: 105–8. [18.6]

Tarapchak SJ, Herche LR. (1988) Orthophosphate concentrations in lake water: analysis of Rigler's radiobioassay method. *Canadian Journal of Fisheries and Aquatic Sciences* **45**: 2230–7. [18.6]

Taylor AW, Kunishi HM. (1971) Phosphate equilibria on stream sediment and soil in a watershed draining an agricultural region. *Journal of Agricultural and Food Chemistry* **19**: 827–31. [18.6]

Thomann RV. (1984) Physico-chemical and ecological modeling the fate of toxic substances in natural water systems. *Ecological Modelling* **22**: 145–70. [18.4]

Thomann RV, Mueller JA. (1987) *Principles of Surface Water Quality Modeling and Control.* Harper and Row Publishers, New York. [18.4]

Triska FJ, Kennedy VC, Avanzino RJ, Zellweger GW, Bencala KE. (1989a) Retention and transport of nutrients in a third-order stream: channel processes. *Ecology* **70**: 1877–92. [18.3, 18.7]

Triska FJ, Kennedy VC, Avanzino RJ, Zellweger GW, Bencala KE. (1989b) Retention and transport of nutrients in a third-order stream in northwestern California: hyporheic processes. *Ecology* **70**: 1893–905. [18.3]

Triska FJ, Duff JH, Avanzino RJ. (1990) Influence of exchange flow between the channel and hyporheic zone on nitrate production in a small mountain stream. *Canadian Journal of Fisheries and Aquatic Sciences* **47**: 2099–111. [18.7]

van Kessel JF. (1977) Factors affecting the denitrification rate in two water-sediment systems. *Water Research* **11**: 259–67. [18.7]

Vannote RL, Minshall GW, Cummins KW, Sedell JR, Cushing CE. (1980) The river continuum concept. *Canadian Journal of Fisheries and Aquatic Sciences* **37**: 130–7. [18.5]

Verhoff FH, Melfi DA. (1978) Total phosphorus transport during storm events. *Journal of the Environmental Engineering Division, Proceedings of the American Society of Civil Engineers.* **104**: 1021–6. [18.6]

Verhoff FH, Melfi DA, Yaksich SM. (1979) Storm travel distance calculations for total phosphorus and suspended materials in rivers. *Water Resources Research* **15**: 1354–60. [18.6]

Viner AB. (1988) Phosphorus on suspensoids from the Tongariro River (North Island, New Zealand) and its potential availability for algal growth. *Archiv für Hydrobiologie* **111**: 481–89. [18.6]

Vollenweider RA. (1976) Advances in defining critical loading levels for phosphorus in lake eutrophication. *Memorie dell'Istituto Italiano di Idrobiologia* **33**: 53–83. [18.3]

Wallace JB, Webster JR, Woodall WR. (1977) The role

of filter feeders in flowing waters. *Archives in Hydrobiologie* **79**: 506–32. [18.1]

Wallace JB, Webster JR, Cuffney TF. (1982) Stream detritus dynamics: regulation by invertebrate consumers. *Oecologia* **53**: 197–200. [18.5, 18.8]

Waters TF. (1961) Standing crop and drift of stream bottom organisms. *Ecology* **42**: 532–7. [18.5]

Waters TF. (1965) Interpretation of invertebrate drift in streams. *Ecology* **46**: 327–34. [18.5]

Webster JR. (1975) *Analysis of potassium and calcium dynamics in stream ecosystems on three southern Appalachian watersheds of contrasting vegetation.* PhD thesis, University of Georgia, Athens. [18.1]

Webster JR. (1983) The role of benthic macroinvertebrates in detritus dynamics of streams: a computer simulation. *Ecological Monographs* **53**: 383–404. [18.8]

Webster JR, Benfield EF. (1986) Vascular plant breakdown in freshwater ecosystems. *Annual Review of Ecology and Systematics* **17**: 567–94. [18.5]

Webster JR, Patten BC. (1979) Effects of watershed perturbation on stream potassium and calcium dynamics. *Ecological Monographs* **49**: 51–72. [18.1, 18.2]

Webster JR, D'Angelo DJ, Peters GT. (1991) Nitrate and phosphate uptake in streams at Coweeta Hydrologic Laboratory. *Verhandlungen Internationale Vereinigung für Theoretische und Angewandt Limnologie* **24**: 1681–6. [18.7]

White E, Payne G, Pickmere S, Pick FR. (1981) Ortho-phosphate and its flux in lake waters. *Canadian Journal of Fisheries and Aquatic Sciences* **38**: 1215–19. [18.6]

White RE, Beckett PHT. (1964) Studies on the phosphate potentials of soils. Part I. The measurement of phosphate potential. *Plant and Soil* **20**: 1–16. [18.6]

Whitford LA, Schumacher GJ. (1961) Effect of current on mineral uptake and respiration by a fresh-water alga. *Limnology and Oceanography* **6**: 423–5. [18.3]

Whitford LA, Schumacher GJ. (1964) Effect of a current on respiration and mineral uptake in *Spirogyra* and *Oedogonium*. *Ecology* **45**: 168–70. [18.3]

Wiley MJ, Osborne LL, Larimore RW. (1990) Longitudinal structure of an agricultural prairie river system and its relationship to current stream ecosystem theory. *Canadian Journal of Fisheries and Aquatic Sciences* **47**: 373–84. [18.3]

Wong SL, Clark B. (1976) Field determination of the critical nutrient concentrations for *Cladophora* in streams. *Journal of the Fisheries Research Board of Canada* **33**: 85–92. [18.3]

Wuhrmann K, Eichenberger E. (1975) Experiments on the effects of inorganic enrichment of rivers on periphyton primary production. *Verhandlungen Internationale Vereinigung für Theoretische und Angewandt Limnologie* **19**: 2028–34. [18.3]

Yoshinari T, Hynes R, Knowles R. (1976) Acetylene inhibition of nitrous oxide reduction and measurement of denitrification and nitrogen fixation in soil. *Soil Biology and Biochemistry* **9**: 177–83. [18.7]

Part 4: Examples

Here, we move from the general principles of river ecosystem structure and function enunciated in Parts 1–3 to particular examples; in fact, case studies are presented on several rivers. In doing this we cover all major geographical regions—subarctic to tropics—and cover examples from both mountain and lowland regions. Large rivers dominate, primarily because small rivers and streams have been given emphasis in other books and reviews. As will become clear, there is still much to be done in understanding the ecology of these large systems.

19: La Grande Rivière:
A Subarctic River and a
Hydroelectric Megaproject[*]

P. P. HARPER

19.1 INTRODUCTION

Canada and Russia have the world's greatest reserves of fresh water, both being largely set in the mid-latitude high pressure zones of heavy precipitation; Québec alone possesses 16% of the world reserves (Hydro-Québec 1990a). Both countries have great rivers flowing north into the Arctic Ocean and both have undertaken or are considering large hydroelectric developments and/or diversions of water (Neu 1982; Rogel 1985). The ecology of these northern rivers is still poorly known and the effects of large projects cannot always be predicted (Hecky et al 1984).

Yet, the engineering projects are often the only opportunity to study these complex ecosystems, under the best conditions with observations before, during and after the engineering work. It is perhaps through the consequences of such operations that the natural processes of the rivers are best illustrated and the megaprojects can in a way be regarded as continent-size experiments.

The present review deals with the large-scale hydroelectric development undertaken in North and Central Québec since 1972 on the drainage of La Grande Rivière, and generally referred to as phase I of the James Bay Hydroelectric Project. A number of ecological studies were conducted on the three rivers involved and these provide some understanding of the economy of large northern rivers and their relations with adjoining eco-

systems. Except where specifically mentioned, the data for this discussion are taken from the three major reviews already available on the systems (Magnin 1977; Anonymous 1978; Roy et al 1986).

19.2 THE HISTORICAL BACKGROUND

Inhabited by the Cree since prehistoric times, this territory was first visited by Europeans in the early 17th century. In 1670, Charles II founded the Hudson Bay Company to administer all the lands draining into James and Hudson Bays. The Treaty of Ryswick in 1697 ceded the territory to France, but British presence continued nevertheless.

From the 1763 Treaty of Paris, after the conquest of Canada by Great Britain, the territory, now Rupert Land, was administered again by the Hudson Bay Company until 1870. The Rupert Land Act of 1868 gave the territory to Canada. The southern portion draining into James Bay was annexed to the Province of Québec in 1898, and the northern portion draining into Hudson Bay was added in 1912.

The main economic activities of the local populations include fishing and hunting for both food and sport, and in the south, forest exploitation and mining activity, especially of gold, copper and zinc (Anonymous 1978).

The development of the hydroelectric potential of the area was put forward by the Québec government in 1971 and work began in 1972. The Société d'Energie de la Baie James thus became the administrator of a territory of some 350 000 square kilometres between the 49th and 55th parallels.

[*] Dedicated to the memory of the late Professor Etienne Magnin (1922–90), sometime Director of the Laboratoire d'Ecologie de la Société d'Energie de la Baie James (Université de Montréal), who pioneered many of the freshwater studies in the James Bay area.

The project was contested in court by the native populations (then some 6500 strong) and an injunction obtained, in effect interrupting work until territorial rights had been settled (Malouf & Gagnon 1973). The injunction was rejected by higher courts within a week, but an agreement was eventually reached between the various government agencies, the Cree Grand Council and the Association of Inuit in 1975 (Anonymous 1976).

The first phase of the project is nearly completed and produces some 10 282 MW, and LG2A power station is due to begin operations in 1991-2 for an extra 1900 MW. The second phase is scheduled to be completed in 1994-5 and it includes three additional stations, LG1, Laforge I and Brisay, with respective peak powers of 1310, 820 and 380 MW. Future projects include two additional stations, Laforge 2 (for 1996) and Eastmain 1 (for 2004), with potentials of, respectively, 270 and 470 MW. A further mega-project on La Grande Baleine River to the north with a combined power of its three reservoirs and stations of 3060 MW is now being considered for 1998-2000, and impact assessment hearings are presently being scheduled. Also contemplated is an extensive development on the more southerly Nottaway−Broadback−Rupert systems (Complexe NBR) with a projected power of 8400 MW. The Québec utility has also considered plans to dam the great rivers flowing into Ungava Bay, particularly the Rivers George, la Baleine (Whale River) and aux Feuilles (Leaf River) (Gauguelin 1978) for an additional potential of some 9000-11 000 MW, although these schemes do not appear in the most recent development plan (Hydro-Québec 1990b).

19.3 THE GEOGRAPHICAL SETTING

La Grande Rivière is an important tributary of James Bay on the western coast of the Québec−Labrador Peninsula. Its catchment basin is of 97 640 km² and its average discharge before diversions was 1700 m³ s⁻¹ (range 320−6710). This compares well with the Rivière des Outaouais (Ottawa River) which has a similar discharge of 1970 m³ s⁻¹ but a larger drainage area of 146 335 km². There are a few large lakes in the basin, in particular Lake Sakami (530 km², maximum depth 110 m), numerous smaller lakes, and innumerable bogs of various sizes and origins.

The adjoining river systems of the Eastmain to the south (46 360 km³, 933 m³ s⁻¹) and of the Caniapiscau to the northeast (89 615 km², 1716 m³ s⁻¹) will also be considered because much of their water has been diverted into the La Grande Complex. The affected zone therefore covers an immense territory from 49 to 58°N and from 68 to 78°W (Fig. 19.1).

The underlying rock is mostly Precambrian granite and gneiss from the Canadian Shield with surface glacial deposits over much of the upper drainages. Sedimentary deposits from glacial freshwater lakes (Lake Ojibway-Barlow) and marine invasions (Tyrell Sea) occur in the lower drainages of both La Grande and the Eastmain. The Caniapiscau flows north off the Canadian Shield along the Mistassini-Otish Highlands into Ungava Bay through the Koksoak River. Soils everywhere are very shallow, generally less than 20 cm, except in bog areas.

Most of the regions fall within the taiga forest zone, although tundra occurs in the highlands along the Caniapiscau. Climatic zones range from the subarctic in the south to the arctic in the north, mean annual temperatures ranging from −1°C in the south to −7°C in the north; mean temperatures in July vary from 16°C in the south to about 10°C in the north.

Water temperatures reach 20−25°C by mid-July for a period of 3−4 weeks, but the upper basin waters do not go beyond 16°C. The thermal stratification in lakes during summer is never very clear. The ice-free period lasts between 150 days downstream to about 100 days upstream.

Water quality is generally characterized by a low pH (6.0−6.9) and low conductivity (8−38 μS cm⁻¹), both parameters increasing gradually downstream. The waters are soft: 0.5−2.0 mg Ca^{2+} l⁻¹ and 0.1−0.6 mg Mg^{2+} l⁻¹; hardness usually increases slightly downstream and is maximum in late winter. Nutrient levels are very low, generally with less than 0.1 mg l⁻¹ nitrates, 0.02 mg l⁻¹ phosphates, 0.5−1 mg l⁻¹ inorganic carbon, but with 3.5−7 mg l⁻¹ organic carbon. The high levels of silicates are related to the composition of the underlying rock. Oxygen

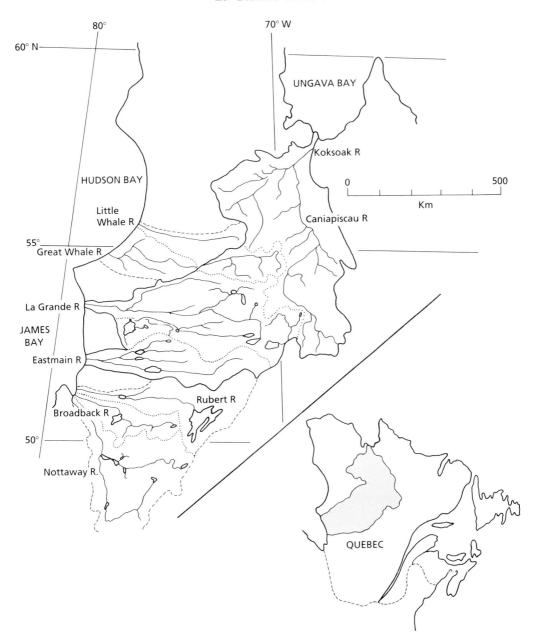

Fig. 19.1 Map of La Grande Rivière drainage and adjacent drainages involved in the James Bay Hydroelectric Project in Québec, Canada. Inset map shows extent of presently affected drainages in Québec.

is always plentiful even in the hypolimnion of lakes, and concentrations rarely dip below 6 mg l^{-1}. Table 19.1 summarizes some of the general characteristics. Lakes tend to be somewhat more mineralized than the running waters, but all parameters remain on the low end of the scale compared with mean values of freshwaters (Livingstone 1963).

Bogs of various sizes and configurations are numerous, and their water quality (Table 19.1) differs from other habitats by a lower pH, low nutrients, but higher levels of organic carbon (19–41 mg l^{-1} versus 4–19 in lakes and 3.5–7 in rivers).

Phytoplankton is dominated by diatoms, green algae and blue-green algae. Contant *et al* (1974) recognized three main types of lake on the basis on their algal communities: diatom lakes dominated by *Melosira*, *Tabellaria* and *Asterionella*; chrysophyte lakes characterized by *Dinobryon bavaricum*; and cyanophyte lakes dominated by various species, such as *Anabaena flos-aquae*, *Aphanizomenon flos-aquae*, *Aphanothece* sp., and *Chroococcus limneticus*. Phytoplankton in rivers is extremely varied, but generally low in densities: diatoms (*Eunotia*) often predominate, but desmids, chrysophytes (*Synura*, *Uroglena*), and even cyanophytes (*Anabaena*) were very abundant in some sites. Productivity is of the order observed of other freshwater habitats on the Shield (126–183 mg C m^{-2} day^{-1}) (Ostrofsky 1974).

Zooplankton communities are naturally poor and the number of species reduced; some 60 species were collected, 20 Copepoda, 27 Cladocera, and 13 Rotifera. Seven species clearly dominate in both lakes and rivers: *Leptodiaptomus minutus*, *Epischura lacustris*, *Diacyclops bicuspidatus* in the Copepoda, *Holopedium gibberum*, *Bosmina longirostris*, *Daphnia longiremis* in the Cladocera, and *Kellicottia longispina* in the Rotifera.

The macrobenthic fauna often have low densities, generally fewer than 500 individuals m^{-2} in lakes and slightly larger numbers in littoral weedy areas. The dominant groups are Chironomidae (Diptera), Amphipoda (*Hyallela azteca*), Sphaeriidae (Mollusca) and Oligochaeta in lakes, Simuliidae (Diptera) in streams, and Baetidae (Ephemeroptera) and Hydropsychidae (Trichoptera) in rivers. Chaoboridae appear to be either rare or absent. Magnin (1977) even considered the possibility of introducing species such as the crustaceans *Mysis relicta* and *Pontoporeia affinis* which inhabit the hypolimnion of deep lakes and are an important part of the diet of fish in the western parts of the Canadian Shield. The fauna, particularly of insects, is none the less quite varied, with southern and northern elements abundant, and some western elements present (Harper 1989, 1990).

Fish are represented by 31 species in the La Grande drainage and only 16 in the Caniapiscau; by contrast, on the western slope of James Bay, an additional 13 species have been collected. Growth is generally slow, but a long life allows most species to reach a respectable size; there are, none the less, differences between species, and some (lake whitefish (*Coregonus clupeaformis*), longnose (*Catostomus catostomus*) and black

Table 19.1 Summary of physicochemical characteristics: pH, conductivity, hardness, sulphates and silicates and phosphates

Habitat	pH	Conductivity μs cm^{-1}	Hardness mg CaCo$_3$ l^{-1}	Sulphates (mg l^{-1})	Silicates (mg l^{-1})	Phosphates (μg l^{-1})
Bogs ($n = 14$)	3.7–4.7	18–37	1.6–11.9	2.4–5.5	0.3–2.4	2–6
Lakes ($n = 59$)	6.4–6.8	22–38	4.6–13.1	2.8–3.4	2.1–4.5	7
Rivers ($n = 97$)	6.2–6.9	11–17	3.5–4.8	1.6–2.2	1.5–3.0	10
Reservoir LG2 (flooded)	6.3–6.4	16–22	5.9–7.6	0.8–2.1	0.5–2.5	15–23
Reduced flow river	6.7–6.9	21–39	10.9–12.9	2.1–5.6	2.9–4.7	22–49

After Magnin (1977), Anonymous (1978) and Roy *et al* (1986).

suckers (*C. commersoni*)) have near normal rates, while others (sturgeon (*Acipenser fulvescens*), goldeye (*Hiodon alosoides*), sauger (*Stizostedion canadense*) and walleye (*S. vitreum*)) have very reduced growth. There is often great individual variations in rates, and populations are often dominated by fish of similar size but of widely differing ages. Fecundity is low, sexual maturation retarded, and reproduction episodes in an individual may occur only at every second to sixth year. This leads to low population densities, often of the order of a few adult fish per hectare.

19.4 THE DEVELOPMENT SCHEME

It was only 2 years after the launching of the project, in 1973, that the final choice was made as to site. The La Grande drainage was preferred to the more southerly Nottaway–Broadback–Rupert system, but it was expanded by the creation of diversions. It encompassed the construction of three dams (LG2, LG3 and LG4) and adjoining power plants (a fourth is planned) on La Grande ('The Great River') and the diversion of water from two adjacent drainages, one from the south, the River Eastmain (90% of discharge), and the other from the north, the River Caniapiscau (30% of discharge) (Figs 19.1 & 19.2). The first two flow west into James Bay and the third flows north into Ungava Bay. The predicted power was 10 216 MW and an annual production of 68 billion kilowatt-hours (Gauquelin 1978), and the actual values are 10 282 MW.

La Grande Rivière has little slope, falling only 376 m over 860 km. Up to six dams were envisaged, flooding up to 7% of the drainage basin

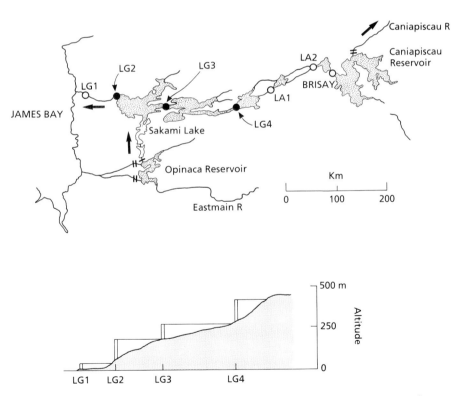

Fig. 19.2 Hydroelectric scheme involving the La Grande, Eastmain and Caniapiscau Rivers, with position of power stations: La Grande 1–4 (LG1, LG2, LG3, LG4), Laforge 1–2 (LA1, LA2) and Brisay. Black dots represent functional plants, open dots those presently in construction or planned. Lower figure shows slope of main river, and profiles of reservoirs LG1–4.

(Gauquelin 1979). Most of these are now completed, or will be by 1993–4 (Hydro-Québec 1990b).

Reservoir LG2, the first reservoir to be built, began filling up in November 1978 and became operational the following October. In 1980, the waters diverted from the south (Eastmain) began flowing into LG2, and since 1984 waters diverted from the north (Caniapiscau) entered the system.

19.5 MONITORING AND IMPACT ASSESSMENT

The environmental impact of the project was predicted to be rather limited, given the restricted areas directly affected by the constructions. No great climatic change was forecasted, except perhaps locally; anyway, nothing comparable to the effects of deforestation on the Saint Lawrence Valley after colonization by Europeans was contemplated (Michel Jurdant, cited by Gauquelin 1978). The main concern was for the native peoples who might lose their traditional way of life.

From 1973 to 1976, numerous baseline studies were conducted on lakes and rivers, and they covered all components of the ecosystem. These have been summarized by Magnin (1977).

A monitoring scheme was tried out on a small lateral reservoir (Desaulniers) which was flooded in 1977, and from this a scheme was extended to the whole system (Bachand & Fournier 1977; Lemire 1978).

Monitoring of flooded and control sites mainly on LG2, with occasional samples on LG3 and LG4, was conducted by the Société d'Energie de la Baie James (SEBJ) from 1978 until 1985, when responsibility for the environmental follow-up was transferred to Hydro-Québec. The objectives of the monitoring scheme were threefold: (1) to evaluate physical, chemical and biological changes in the reservoirs; (2) to rationalize management of the installations and to minimize consequences; and (3) to acquire tools for predicting changes in future reservoirs (Roy *et al* 1986).

The original scheme called for monitoring to start 2 years before operation and to continue for 5 years afterwards, then with checks at every third year. As many of the projects were started earlier than originally planned, the 2-year period was not always observed. A panel of experts, mainly from universities, was entrusted with quality control.

Studies were eventually concentrated on the areas of LG2, Opinaca and Caniapiscau, and the rivers affected, with a control station in each region. Some 27 stations were thus regularly maintained. Occasional checks were none the less performed on other sites.

Monitoring at each station included extensive chemical analyses and collections of zooplankton, benthic fauna (partially abandoned in 1982) and fish.

It was then suggested that future monitoring be restricted to physicochemical characteristics and fish communities (Roy *et al* 1986). Continuation of this environmental follow-up is planned in the latest development plan of the utility (Hydro-Québec 1990a) and an evaluation report is apparently due within a year or two.

The James Bay agreement (Anonymous 1976) with the native peoples provides for special monitoring of fish stocks and fish exploitation, as well as of the effects of diversions on the flora and the fauna.

19.6 THE FLOODED AREAS

Water quality in surface waters in the reservoirs did not change appreciably after flooding during the summer months, except in littoral areas. The most notable differences were a decrease in pH (from 6.9 to 6.4) and in conductivity (from 29 to 16 µs/cm) consecutive to a decrease of most ions. In winter under the ice, however, the changes were more dramatic: oxygen levels declined as low as 17% saturation, the pH dropping to 5.6; conductivity was maintained (20–30 µs/cm); and total phosphorus concentration increased from 7 to up to 49 µg l^{-1}. In littoral stations, far from the main water body, phosporus levels generally doubled or tripled in the first years of operation, from 5–9 to 8–33 µg l^{-1}.

The flooding of the generally acidic soils (pH 4.5–5.5) resulted at the beginning in a rapidly decreased water acidity and an input of organic matter from soils and flooded vegetation. As well, erosion of the newly inundated soils by wave action contributed to an additional liberation of organic matter. By 1985, it was felt that most of the erosion was under control and that it was not to be a continuing problem as in other northern reservoirs, such as Southern Indian Lake in Manitoba which is located in a sedimentary area (Hecky *et al* 1984).

The decomposition of organic matter in LG2 probably peaked in 1982 when lowest oxygen saturation (66%) and highest phosphorus concentrations (23 µg l^{-1}) were observed in the open water (upper 10 m). This resulted in an increase of chlorophyll concentrations 2 years after flooding; blooms of phytoplankton developed early in the season, but aborted later, apparently due to lack of silicates (Roy *et al* 1986).

Deep waters were strongly deoxygenated during summer stratification and especially during the 6 months of ice cover; this resulted in liberation into the water column of phosphates and other ions and reduced molecules. Such deoxygenation peaked in the first 2 years after flooding, except in Caniapiscau Reservoir where it was maintained for a third year, because of the low turnover rate of the water (Table 19.2).

Two factors controlled water quality, namely the decomposition of organic matter and the water turnover rate. Given the generally short turnover period, deoxygenation was not as important as it would have been normally in lakes of similar magnitude.

Phytoplankton biomass was assessed on the basis of chlorophyll concentration, but no attempt was made to evaluate periphyton, especially given the vastly increased colonization surface resulting from the flooded forests. Production increased during the first 3 years after flooding and remained high in the subsequent years, often at a level of three to five times over controls, particularly in LG2 reservoir (21–63 versus 7–20 g C m^{-2} year^{-1}); however, chlorophyll concentrations decreased during the first year (1.1–1.2 µg l^{-1}), but increased thereafter (3.2–4.2 µg l^{-1}). Production and biomass increased less in Caniapiscau, the reservoir set on the poorest soils.

Zooplankton biomass and species composition was followed over the years, as well as gross seston. Net seston decreased from 36–80 to 10–21 mg m^{-2} due to the greater sedimentation of mineral particles, and the fraction of organic matter therefore increased from 18% to 65% (Pinel-Alloul & Méthot 1984). Populations of zooplankton peaked in LG2 on the fourth year after flooding, mainly owing to the proliferation of rotifers, probably because of their detritus feeding habits (Pinel-Alloul *et al* 1989). In the other two reservoirs, populations peaked during the first 2 years and then returned to normal levels. The greatest increases were observed in sheltered bays where water residence time was highest, and the lowest were in the main channel of the reservoir. During the early years, communities were dominated by pioneer species with high growth rates. These were replaced by normal slow-growing lake species, but they had not in 1985 reached the densities found in adjacent unperturbed lakes. Less than 50% of the zooplankton flowed through the turbines and these were mostly small organisms; 90% of larger

Table 19.2 Characteristics of the reservoirs

Reservoir	Area (km^2)	Flooded area (km^2)	Mean depth (m)	Discharge (m^3 s^{-1})	Power (MW)	Turnover (months)
LG2	2835	2630	21.8	3400	5328	6.3
LG3	2420	2176	24.8	2050	2304	11.3
LG4	765	699	25.4	1505	2650	5.6
Opinaca	1040	738	8.2	810	—	4.1
Caniapiscau	4285	3432	12.6	776	—	24.6
Desaulniers	5.2	5.1	3.3	1.7	—	—
Grande Baleine	3143	865	—	—	3060	—

After Roy *et al* (1986)

animals were not carried down. No concentration of zooplankton was observed near inflows, but the large number of ciscos (see below) carried through the turbines may be a consequence of such a concentration of zooplankton.

Benthic biomass was studied using artificial substrates which proved to be unrepresentative; large accumulations of debris on the bottom hindered attempts to make quantitative estimations by standard methods. Effects of flooding varied and were less important in sites that were previously aquatic. Mollusca (both bivalves and gastropods), Amphipoda, Ephemeroptera and Trichoptera were negatively affected. By contrast, Oligochaeta and Chironomidae (Diptera) increased considerably. Although numbers were often greater in flooded zones, the generally smaller size of the organisms accounted for a lower biomass. Densities were highest near shore, but the communities were dominated by small vagile species. Densities were lowest in areas of low oxygen concentration. Logs and dead trees replaced macrophytes as substrates for aquatic invertebrates; long-term high densities are expected, but no study of the phenomenon has been undertaken.

Fish were sampled with a series of experimental gill nets; attempts to use a seine met with little success and much difficulty. Seventeen species of fish were collected in LG2, but seven species made up 95% of the total; these were lake cisco (*Coregonus artedii*), lake whitefish, northern pike (*Esox lucius*), longnose sucker, white sucker, burbot (*Lota lota*) and walleye. One year after the dam was filled, yields in experimental fishing declined to one-third of their previous value, probably due to a phenomenon of dilution. In a few years, yields returned to their previous value and have remained constant since (16.5–17.5 fish per net per day), a value comparable to those in large natural lakes of the region; biomasses were, however, higher (17.5 kg per net per day). The species which seem to have best adapted to the reservoir are northern pike and lake whitefish, and yields are commonly 5–6 and 3–4 times, respectively, those of surrounding natural lakes. As well, their growth rate has increased significantly. Lake cisco and burbot, although still representing only 5% and 7% of total catches,

increased their yield by a factor of 2 and 4. Walleye and white sucker have fared rather poorly, and yields were only half those normally encountered in the region. Most species had an increased condition factor after the flooding of the reservoir, and in 1985 these were still 10–20% superior to preimpoundment values.

The same species dominated the Opinaca Reservoir, with similar success rates. In Caniapiscau Reservoir, walleye was absent and salmonids were present (lake charr, *Salvelinus namaycush*; landlocked salmon or ouananiche, *Salmo salar*; and brook charr, *Salvelinus fontinalis*). Yields declined during the filling up of the reservoir, but only by 10–20%.

In general, fish biomass in the reservoirs is comparable or higher than that of large lakes in the region, and natural recruitment of the populations is ensured without any further human intervention (Lalumière & Brouard 1984).

Models were developed to predict physical parameters (temperature, ice patterns, decomposition of organic matter) and have proven reliable; of the others dealing with biological parameters, only that on phytoplankton has had any success.

One of the major problems was the destruction of the littoral zone consequent on the flooding and particularly the fluctuating levels which impede all regeneration. Attempts to replant some of the denuded littoral areas met with mixed success. It has been estimated that some 95% of the riparian zones have been destroyed (Julien *et al* 1985). This affected waterfowl and it was thought that some 7000 pairs of birds had to seek nesting sites elsewhere (Hydro-Québec 1990a). The La Grande project generated some 83 300 km of shoreline, of which about 20% provides good faunal habitats (Hydro-Québec 1990a). This is perhaps of lesser consequence than could have been expected, because waterfowl prefer to use coastal marine habitat and make little use of inland freshwater shore habitats, even under ideal conditions (Julien & Laperle 1986).

It was also feared that the loss of littoral weed beds would imperil the reproduction of northern pike, walleye and the suckers. To counter this, a project was undertaken to seal five small bays from water level fluctuations and to provide them with a fish ladder; only one was eventually set up. The experiment gave mixed results and the plan was abandoned. The littoral vegetation did in fact regenerate, but the fish made little use of it (Lalumière & Brouard 1984). Pike reproduction was, however, good in the reservoir itself (Roy *et al* 1986).

There was concern that the bog areas would be uplifted and would float around, and a study was made of this problem. Trees were generally not cut, the wood being of little commercial value and far from markets. Some clearing was carried out in areas close to roads (for scenic purposes) or in zones of high boat traffic. An evaluation of floating and accumulating debris on the beaches was also made. The main clearing activities (1464 ha) were at the mouths of tributaries to allow spawning migrations of the fish, particularly of walleye (Lalumière & Brouard 1984). Despite this effort, walleye reproduction has met with little success during the first 5–6 years after flooding (Roy *et al* 1986).

It was felt that many of the trees would be caught in the ice and pulled out by the rising waters in the spring. Leaves decomposed in about 200 days (Richard 1981) and small branches in 1–4 years depending on exposure (Crawford & Rosenberg 1984). Decreasing influences of organic matter from flooded terrestrial areas were noted in the third year and mathematical models predict little influence after 7–8 years (10 years at Opinaca) (Grimard & Jones 1982). The organic load depended on the thickness and quality of the flooded soils: therefore, LG2 and Caniapiscau Reservoirs received more per unit than Opinaca where soils were shallow and the forest sparse. In LG2, however, the influence of other large and more recent reservoirs upstream was also apparent, so effects are expected to last longer.

19.7 REDUCTIONS IN FLOW

Reduction in flow occurred in three river systems: on a temporary basis on La Grande during the filling of LG2, and on a permanent basis, due to diversions, on the southern Eastmain River and the northern Caniapiscau.

During the completion of the LG2 dam and its subsequent filling up, discharge in La Grande Rivière was reduced by 97%. Massive saltwater invasions were expected as a consequence and a weir was planned at 16.5 km, but this was found not to be necessary because discharge was maintained at a sufficient level ($8.5 \, \mathrm{m}^3 \, \mathrm{s}^{-1}$ at 20 km) to prevent the intrusion of marine waters. Special efforts were made to protect cisco and anadromous whitefish populations, which are both heavily fished by native populations. Fish ladders were provided, allowing the fish to migrate upstream into the reservoir.

The Eastmain River and two of its tributaries, the Opinaca River and the Little Opinaca River, were diverted north to LG2 through Ruisseau Boyd and Rivière Sakami which saw their flows increased by a factor of up to 6. At the estuary of the Eastmain, only about 10% of the average natural flow still remains. Four weirs were constructed on the river to regulate flow and aid navigation. Impacts of the reduction were studied by SEBJ-SOTRAC (1985). Predicted effects included reduced nutrients and primary production as well as saltwater invasion into the river. Erosion of shores has been important, and plantations of shrubs and other plants not entirely successful to prevent it.

The Caniapiscau River was interrupted and lost 40% of its discharge at its junction with the Koksoak, which itself lost 28% of its own discharge as a consequence. All the water was diverted into La Grande, but there is provision for flood control. Impacts were studied by Shooner (1985). Saltwater reaches 50–60 km upstream on the Koksoak and tidal influence is felt for some 87 km. Fish yields on the river did not change ($10\,000 \pm 3400$ per year), but the percentage of salmonids (Arctic charr, *Salvelinus alpinus*) went down dramatically, from 5500 ± 200 in 1977–81 to some 3700 ± 400 in 1982–5 after the diversion.

Water quality evolved in a similar manner in all three instances. The rivers received relatively more water from the sedimentary coastal areas through their remaining tributaries, and this was no longer diluted by the nutrient-poor upstream water (now diverted) from the Canadian Shield and its overlying glacial deposits. The phenomenon was less marked in the Caniapiscau system where the river must still cross some 300 km of Shield country below the dam before reaching sedimentary zones; the water chemistry therefore changed relatively little. Erosion of the new banks increased the silt loads significantly, although the total sediment loads of the rivers actually decreased. Finally, the residence time of water at any one point increased because of lower discharge. Oxygen concentrations remained high.

Primary production increased markedly despite the higher turbidity. This was due to higher nutrient concentrations and longer residence times, especially in Eastmain where the construction of weirs raised residence time from less than 1 day to more than 43 days. This also explains the larger zooplankton populations.

Benthic fauna were reduced near the shore because of high erosion and siltation. Lotic species gave way to lentic assemblages, but densities were comparable with those of adjacent natural lakes (Roy *et al* 1986). The fauna was dominated by Chironomidae (Diptera) (43−58%), Oligochaeta (12−36%) and Amphipoda (17−20%), although the last group represented between 33% and 75% of biomass.

Fish yields increased (by an average factor of 5) with the reduction of the habitats, at least at first, but soon declined. In 1985, yields were, however, still superior to those before flow reduction. Sturgeon and lake whitefish were the species most severely affected.

Reduced flow in the rivers allowed an invasion of marine water high up into the estuaries (Messier 1985), increasing the salinity but lowering the temperature. Freshwater species were either eliminated or pushed upstream, and their habitats taken over by brackish water species. In the Eastmain estuary, flow inversions occur throughout on a distance of 27 km. The influence of wind on the river has increased because of the shallower water. Saltwater invasion in the Koksoak probably boosted the productivity of the estuary, because the waters from Ungava are relatively rich (Messier 1985). The invasion of seawater also brought in marine macroalgae which the local populations complain clog the nets.

Overall consequences of the temporary reduction in discharge in La Grande during construction of the LG2 dam appear to be negligible (Roy 1982).

19.8 INCREASES IN FLOW

Since the completion of the dams and diversions, the discharge of La Grande Rivière has doubled from 1800 to 3400 m^3 s^{-1} (900−4300 m^3 s^{-1} yearly variation). But the greatest change has been the increase of winter flow from a mean of 500 to 4000 m^3 s^{-1} and the loss of the spring flood (from 5000 to 1500 m^3 s^{-1}), thereby reversing the natural flow cycle of the river. In summer, the discharge remains more or less unchanged.

As water from the reservoir passes through the turbines, carbon dioxide is expelled and the pH rises a little. Water remains cool (the temperature of the upper 40 m of the reservoir) and the annual heating cycle in early summer is postponed by 2−3 weeks. Ice develops 2−3 weeks later and breaks up some 3−4 weeks earlier than it did under natural conditions. Summer water temperatures are typically 3−5°C lower in summer and 1°C higher in winter. A new shoreline is formed because of the higher water level and erosion is important, which increases turbidity. In winter, ice forms 2−3 m higher up on the shore than it did previously.

The riparian vegetation has been severely affected, especially the shrub zones (Ouzilleau & Brunelle 1988). This is due to the fluctuating levels which follow an unnatural cycle, determined by exploitation needs of the power stations, as well as the increased erosion and the naturally nutrient-poor substrates (Julien *et al* 1985).

The increase in discharge and the high flow rates impede phytoplankton development, and the high chlorophyll concentrations reflect high populations in the upstream reservoir. The same holds true for zooplankton, although many of the larger species, mainly Cladocera and Copepoda, seem to be able to resist being dragged downstream. Erosional and depositional zones with their associated faunas were displaced along the river according to the new dynamics of the system.

Fisheries seem to have improved in the river following the doubling of the discharge, probably due to a large number of fish being sucked down through the outflows of the reservoir. The colder water is, however, detrimental to some species, in particular walleye which takes refuge in the tributaries (Roy *et al* 1986) and northern pike (Messier 1985). For most species, however, there has been an increase in the condition factor.

The terrestrial fauna make little use of the shore habitats, but there has been some negative impact on local populations of beaver, muskrat, hare and ptarmigan (Messier 1985).

Increased discharge also occurs on a permanent basis in the diversion canals and streams which link the Eastmain to the LG2 reservoir. The consequences are less clear, but somewhat parallel to what was observed in the lower course of La Grande.

19.9 THE ADJACENT OCEAN

The mean discharge of La Grande Rivière now represents 34% of the total discharge into James Bay (10 000 m^3 year^{-1} (Kranck & Ruffman 1982)). James Bay is shallow (generally <40 m) and relatively unproductive due to cold water and low nitrogen (Messier 1985) and phosphorus (Legendre & Simard 1978) levels. Half of the discharge now comes between December and April, and the plume in the bay is increased in winter by a factor of three. There is little evidence that the productivity has changed, nor do the fisheries seem affected. The increased organic matter in the water does not cause much oxygen depletion because of the fast turnover of water.

The dynamics of the phytoplankton do not appear to have been affected greatly, although local and temporary effects due to changes in hydrodynamic patterns have been observed (Ingram *et al* 1985).

However, the dilution of coastal waters affected the littoral eel-grass (*Zostera*) marshes. The geese have also been affected because the latter is an important diet item, and because they use the coastal regions to feed and breed (Julien & Laperle 1986). As well, the reduction of the ice in spring has decreased the hunting success of native populations (Messier 1985).

19.10 THE SURROUNDING FOREST AND TUNDRA

An increase of approximately 1°C in mean

temperatures was predicted by Perrier (1975) for the first 15 km surrounding the reservoirs.

Zones affected by operations, some 7000 ha, were subjected to regeneration with local and introduced plants, a difficult enterprise on shallow soils (Gauquelin 1978).

The effects on the terrestrial mammal fauna are not easy to evaluate. The much publicized drowning of some 10 000 caribou in September 1984 has been variously interpreted, either as an unfortunate natural phenomenon (SEBJ 1987) or as a direct consequence of the project (MLCP 1985). In recent years the caribou herds have greatly increased (Huot & Paré 1986; Marsan *et al* 1987) and they have been making extensive use of the areas adjacent to the reservoirs during winter (Hydro-Québec 1990a,b), although many excellent calving sites have been lost.

High-quality moose habitat has been lost, particularly due the destruction of littoral zones, but good-quality second-growth forest habitats, and therefore good feeding grounds, have increased (Julien & Nault 1986).

A general loss of habitat for hares and ptarmigan has occurred along river margins and reservoir shorelines, particularly near large rivers and hydroelectric stations, with the consequence that native populations must go a longer distance to have access to hunt small game (Julien 1986).

19.11 OTHER CONSEQUENCES

One of the most serious environmental consequences, the gravity of which was not totally foreseen, was the release of mercury. Other pollutants and toxic substances were deemed to be of no significance (Roy *et al* 1986). The introduction of mercury into the food chain is associated with methylation of mercury under the action of bacteria associated with the decomposition of organic matter in the reservoirs (Hydro-Québec 1989).

Four years after the completion of reservoir LG2, mercury levels in fish tissue had increased considerably, by a factor of five in lake whitefish (to 0.51 mg kg^{-1}), of three in longnose sucker (to 0.40 mg kg^{-1}) and walleye (to 2.06 mg kg^{-1}), and of two in northern pike (to 1.24 mg kg^{-1}), thus often exceeding the Canadian standard of 0.5 mg kg^{-1}

for edible fish. The increase was greater in young fish whose growth occurred after the completion of the reservoir. Concentrations generally increased from 1982 to 1984 except in whitefish. Similar increases were observed in both Opinaca and Caniapiscau, but at a reduced rate in the latter instance. The effects of mercury contamination are expected to peak after 5–7 years, but to be felt for at least 15 years after flooding in non-piscivorous fish and for some 20 years in fish-eating species (Roy *et al* 1986). A more extended flooding period would probably have reduced peak concentrations, but would also have prolonged the recovery period (Messier 1985).

Mercury contamination did not occur in rivers with reduced discharge following diversions, such as the Eastmain (Roy *et al* 1986). In LG2, however, following the doubling of discharge, mercury concentration in fish flesh soared, reaching up to twice the amounts in the upstream reservoirs, particularly in the case of lake whitefish and longnose sucker. Roy and colleagues (1986) interpreted this as indicating that both species have become more piscivorous in the river than they normally are in lakes and reservoirs. Semi-anadromous fish in the estuaries, which feed mainly in the sea, are not affected because mercury enters the fish through the freshwater food chain.

Numerous other environmental problems have occurred on a smaller scale. Road construction required filling of bogs. Insecticides were sprayed to control mosquitoes, but this met with little success. Particulate insecticides were introduced at breeding sites for the control of blackflies (Gauquelin 1978). There has been used oil disposal, the use of oil on roads for dust abatement, and the elimination and recycling of human waste and sewage for up to 16 000 workers (Gauquelin 1978).

19.12 THE HUMAN POPULATIONS

The Cree population is presently of some 9593 (8281 residents) (Bisson 1990) and they have exclusive fishing and hunting rights on some 5546 km².

A joint Cree-SEBJ group called SOTRAC (Société des Travaux de Correction du Complexe La Grande) was created in the mid-1970s to correct some of the negative consequences of the development, particularly as they affected the traditional hunting and fishing activities of the Indians (Gauquelin 1978): relocation of beaver, redefinition of trapping territories, waterfowl feeding grounds, navigation on the rivers, clearing of spawning grounds . . .

There is a general perception in local populations that fish stocks have dwindled considerably. This may in part be due to the fact that the fish are no longer in the traditional river sites because of alterations in discharge. There are also shifts in species composition, which alters fishing patterns. On the other hand, the development of the road system has given access to new possibilities, namely to fishing in lakes. Reduced productivity of wetlands is commonly reported, but this has not been studied systematically (Berkes 1989).

The problem of mercury contamination of fish is a serious one and it affects the population to a great degree, given the high proportion of fish in the diet. In 1986 an agreement between the natives and the Québec government was signed to study the problem and its implications on health, society, culture and the environment, and to find solutions. By 1977, 30% of the population at Chisasibi, on La Grande estuary, had high levels of mercury in their blood, increasing to 64% by 1984. Mercury levels in humans have been declining since 1986. No solution has been found, except the recommendations to reduce fish in the diet, to avoid piscivorous species, and to fish preferentially in natural lakes rather than in impoundments. Commercial uses for non-edible fish are also envisaged, such as fertilizer and animal feed (Hydro-Québec 1990a).

Native populations grow by 2–4% per year and there is a risk of overexploitation of fish and game (Gauquelin 1978). Government programmes of Income Security have also been encouraging natives to maintain or to return to their traditional way of life, which puts extra pressure on the resources (Marsan *et al* 1987).

Some integration of natives in the new economic life is reflected in their founding of companies dealing with construction and transportation. However, they were largely barred from construction sites, because of language, union and race problems, and did not profit from the high salaries and technology available on the site (Gauquelin 1978).

Clearly, the effects of the project have been serious for the local population and the utility is presently preparing a review of all socioeconomic consequences of its projects over the past 20 years (Hydro-Québec 1990a).

19.13 CONCLUSION

Except in the case of mercury, impacts have been less than predicted (Roy *et al* 1986), and in many cases recuperation was faster than expected. Extrapolation of the observations made on the Complexe La Grande to other systems can be made only with care. Comparisons with Southern Indian Lake Reservoir in Manitoba show that the main problem there was erosion of fine materials which increased turbidity and hindered primary production (Hecky *et al* 1984), a problem that was much reduced at La Grande.

The prospect of further developments on the La Grande Baleine drainage just north of the La Grande Rivière complex raises anew all the issues of environmental impact. Extensive flooding will again occur with the consequent introduction of large amounts of methylated mercury into the food chains. The population of Cree and Innu at Kuujjuarapik (Poste-de-la-Baleine) will be affected directly, since freshwater fish is an important part of their diet. Additional problems, not present in the earlier project, are also predicted. Will the populations of marine mammals in the Manitounouk Straits be affected? What will happen to the freshwater seal populations at Lac aux Loups marins? To what extent will migratory bird habitats be destroyed (Hydro-Québec 1990a)? The case of beluga whales should also be considered (Watts & Draper 1988) as well as that of polar bears (Goddard 1990).

The same situation also occurs on the west shore of James and Hudson Bays where Ontario Hydro has a 25-year plan for the erection of six dams on the Moose River and the expansion of six existing structures in order to generate some 2000 MW (Goddard 1990).

The future of hydroelectric development in the north depends on the interactions between the native populations who strive to keep their traditional way of life and cultural heritage and the southern populations whose demand for electricity is continuously growing. The production of electricity for export to the USA is also particularly appealing to some governments, particularly that of Québec. By some politicians and managers, Québec is seen as a huge hydroelectric producing plant, and all water flowing freely to the sea is regarded as wasted potential energy (Noël 1990). An overall assessment of environmental effects is requested by the natives, rather than a piecemeal evaluation, because cumulative effects of several marginally acceptable individual projects can result in major deleterious consequences (Goddard 1990).

Arguments are being heard again on what importance such projects ultimately have and assessments range from 'minimal' to 'considerable' (Berkes 1989). This will be the basis on which further projects will be analysed and decided upon. The decision will ultimately be political and the bottom line becomes 'How does one compare the value of a kilowatt-hour and that of a ouananiche or a moose?' (Noël 1990). How does one evaluate losses of a way of life and of a culture?

Environmental awareness seems to have increased in most societies, and it can only be hoped that, whatever happens, the stakes will be assessed carefully and that some of our great northern rivers, and their intricate ecosystems, including the human component, will be preserved for future generations. The danger lies in that experiments of the magnitude that is envisaged can always have unpredicted large scale effects, given our poor knowledge of northern river systems. And the danger of this happening increases exponentially as new 'developments' are continually added to these fragile ecosystems.

19.14 FINAL REMARKS

The ecological studies completed during this megaproject have provided much information on the biota of this little-known area. Most of the work, however, has been descriptive, and this was no doubt necessary as a preliminary measure.

A number of interesting experiments were also performed on particular subsystems, for instance on the Desaulniers Reservoir.

It was also important to follow up the effects of the project on the various ecosystems, and this should be continued over the next decades.

Despite all these efforts, little insight has been gained on the ecological processes themselves in the reservoirs and the rivers. This is a major drawback and the opportunity should be taken to go beyond the present state of knowledge into an understanding of the ecosystems. This is particularly important, given the magnitude of the projects under study for these northern regions. The existence of a series of large reservoirs of different ages and situations offers a unique combination of experimental conditions that should be taken up by the scientific community with the help of the government agencies involved.

ACKNOWLEDGEMENTS

I wish to thank the Centre de Documentation, Direction Environnement, Hydro-Québec in Montréal for help in gathering information.

REFERENCES

Anonymous (1978) *Connaissance du milieu des territoires de la Baie James et du Nouveau-Québec.* Société d'Energie de la Baie James, Montréal. [19.1, 19.2, 19.3]

Anonymous (1976) *Convention de la Baie James et du Nord québécois.* Editeur officiel du Québec, Québec. [19.2, 19.5]

Bachand CA, Fournier JJ. (1977) *Réseau de surveillance écologique du Complexe La Grande.* Société d'Energie de la Baie James, Service Environment, Division Ecologie, Montréal. [19.5]

Berkes F. (1989) Impacts of James Bay development. *Journal of Great Lakes Research* 15: 375. [19.12, 19.13]

Bisson B. (1990) Le Québec ne sera pas 'redéfini' sans les Indiens. *La Presse* (Montréal), 11 August 1990, B1, B3. [19.12]

Contant H, Alaert-Smeesters E, Plinski M, Venne L. (1974) *Phytoplancton récolté en 1973 dans les lacs et les rivières du Territoire de la Baie James.* Laboratoire d'Ecologie de la SEBJ, Rapport de Recherche, 19. [19.3]

Crawford PJ, Rosenberg DM. (1984) Breakdown of conifer needle debris in a new northern reservoir, Southern Indian Lake, Manitoba. *Canadian Journal of Fisheries and Aquatic Sciences* 41: 649–58. [19.6]

Gauquelin M. (1978) La Baie James pour le meilleur ou pour le pire. *Québec Science* 17: 15–26 & 33–9. [19.2, 19.4, 19.5, 19.10, 19.11, 19.12]

Goddard J. (1990) Damned if they do – proposed hydro projects imperil northern rivers and a way of life. *Harrowsmith* 93: 40–51. [19.13]

Grimard Y, Jones HG. (1982) Trophic upsurge in new reservoirs: a model for total phosphorus concentrations. *Canadian Journal of Fisheries and Aquatic Sciences* 39: 1473–83. [19.6]

Harper PP. (1989) Zoogeographical relationships of aquatic insects (Ephemeroptera, Plecoptera, and Trichoptera) from the eastern James Bay drainage. *Canadian Field-Naturalist* 103: 535–46. [19.3]

Harper PP. (1990) Associations of aquatic insects (Ephemeroptera, Plecoptera, and Trichoptera) in a network of subarctic lakes and streams in Québec. *Hydrobiologia* 199: 43–64. [19.3]

Hecky RE, Newbury RW, Bodaly RA, Patalas K, Rosenberg DM. (1984) Environmental impact prediction and assessment: the Southern Indian Lake Experience. *Canadian Journal of Fisheries and Aquatic Sciences* 41: 720–32. [19.1, 19.6, 19.13]

Huot J, Paré M. (1986) *Surveillance écologique du Complexe La Grande – synthèse des études sur le caribou de la région de Caniapiscau.* Société d'Energie de la Baie James, Direction Ingéniérie et Environnement, Montréal. [19.10]

Hydro-Québec (1989) *Hydro-Québec et l'environnement. Plan de développement d'Hydro-Québec 1989–1991 – Horizon 1998.* Hydro-Québec, Vice-présidence Environnement, Montréal. [19.11]

Hydro-Québec (1990a) *Hydro-Québec et l'environnement. Proposition de plan de développement d'Hydro-Québec 1990–1992 – Horizon 1999.* Hydro-Québec, Vice-présidence Environnement, Montréal. [19.1, 19.5, 19.6, 19.12, 19.13]

Hydro-Québec (1990b) *Proposition de plan de développement d'Hydro-Québec 1990–1992 – Horizon 1999.* Hydro-Québec, Montréal. [19.2, 19.4]

Ingram RG, Legendre L, Simard Y, Lepage S. (1985) Phytoplankton response to freshwater runoff: the diversion of the Eastmain River, James Bay. *Canadian Journal of Fisheries and Aquatic Sciences* 42: 1216–21. [19.9]

Julien M. (1986) *Surveillance écologique du Complexe La Grande – Synthèse des études sur le lièvre et les lagopèdes.* Société d'Energie de la Baie James, Direction Ingéniérie et Environnement, Montréal. [19.10]

Julien M, Laperle M. (1986) *Surveillance écologique du Complexe La Grande – Synthèse des études sur la sauvagine.* Société d'Energie de la Baie James, Direction Ingéniérie et Environnement, Montréal. [19.6, 19.9]

Julien M, Nault R. (1985) *Surveillance écologique du Complexe La Grande—Synthèse des études sur l'original*. Société d'Energie de la Baie James, Direction Ingéniérie et Environnement, Montréal. [19.10]

Julien M, Lemieux M, Ouzilleau J. (1985) *Surveillance écologique du Complexe La Grande—Synthèse des études sur les zones riveraines*. Société d'Energie de la Baie James, Direction Ingéniérie et Environnement, Montréal. [19.6, 19.8]

Kranck K, Ruffman A. (1982) Sedimentation in James Bay. *Naturaliste Canadien* **109**: 353–61. [19.9]

Lalumière R, Brouard D. (1984) *Synthèse des aménagements piscicoles: réservoirs du Complexe La Grande (Phase I)*. Gilles Shooner Inc., prepared for Société d'Energie de la Baie James, Direction Ingéniérie et Environnement, Montréal. [19.6]

Lemire R. (1978) *Rapport d'étape sur les études écologiques réalisées au bassin de la rivière Desaulniers de mai á octobre 1976*. Société d'Energie de la Baie James, Service Environnement, Division Ecologie, Montréal. [19.5]

Livingstone DA. (1963) *Chemical Composition of Rivers and Lakes: Data of Geochemistry* 6th edn. Geological Survey Professional Paper 440. US Geological Surveys, Washington [19.3]

Magnin E. (1977) *Ecologie des eaux douces du territoire de la Baie James*. Société d'Energie de la Baie James, Montréal. [19.1, 19.3, 19.5]

Malouf A, Gagnon A. (1973) *La Baie James indienne. Texte intégral du jugement du juge Albert Malouf*. Editions du Jour, Montréal. [19.2]

Marsan A et Associés (1987) *Faune terrestre du territoire de la Baie James—Document synthèse des connaissances et orientations de gestion*. Report prepared for Société d'Energie de la Baie James, Direction Ingéniérie et Environnement, Montréal. [19.10, 19.12]

Messier D. (1985) *Modifications océanographiques physiques et biologiques à la suite de l'aménagement du Complexe La Grande*. Société d'Energie de la Baie James, Direction Ingéniérie et Environnement, Montréal. [19.7, 19.8, 19.9, 19.11]

MLCP (1985) *Considérations relatives à la noyade de caribous du fleuve George sur la rivière Caniapiscau (Septembre 1984)*. Ministère du Loisir, de la Chasse et de la Pêche, Direction générale de la Faune, Québec. [19.10]

Neu HJA. (1982) Man-made storage of water resources—a liability to the ocean environment? Part II. *Marine Pollution Bulletin* **13**: 44–7. [19.1]

Noël A. (1990) Les plans d'Hydro: faire du Québec une immense centrale. *La Presse* (Montréal), 17 March 1990, p. B5. [19.13]

Ostrofsky ML. (1974) *Primary productivity of four shield lakes in the James Bay drainage of Quebec*. Sheppard T. Powell Consultants (report). [19.3]

Ouzilleau J, Brunelle J. (1988) *Suivi des habitats riverains de La Grande Rivière, 1981–1988*. Formamec Inc., prepared for Société d'Energie de la Baie James, Montréal. [19.8]

Perrier MG. (1975) *Influences de modifications de surface sur le climat (Région de la Baie James)*. Publication in Meteorology, No. 116, McGill University, Montréal. [19.10]

Pinel-Alloul B, Méthot G. (1984) Etude préliminaire des effets de la mise en eau du Réservoir de LG-2 (Territoire de la Baie James, Québec) sur le seston grossier et le zooplancton des rivières et lacs inondés. *Internationale Revue der Gesamten Hydrobiologie* **69**: 57–78. [19.6]

Pinel-Alloul B, Méthot G, Florescu M. (1989) Zooplankton species dynamics during impoundment stabilization in a subarctic reservoir. *Archiv für Hydrobiologie, Ergebnisse Limnologie* **33**: 521–37. [19.6]

Richard Y. (1981) *Colonisation des feuilles de différentes espèces d'arbres (peuplier, aulne, myrique) par le macrobenthos et les micro-organismes dans les eaux oligotrophes du Bouclier canadien*. Unpublished MSc thesis, Université Laval, Québec. [19.6]

Rogel JP. (1985) Histoires d'eau: les grandes manoeuvres de M. Bourassa. *Québec Science* (October), 17–23. [19.1]

Roy D. (1982) Répercussions de la coupure de la Grande Rivière à l'aval de LG2. *Naturaliste Canadien* **109**: 883–91. [19.7]

Roy D, Boudreault J, Boucher R, Schetagne R, Therrien N. (1986) *Réseau de surveillance écologique du Complexe La Grande 1978–1984. Synthèse des observations*. Société d'Energie de la Baie James, Direction de l'Ingéniérie et de l'Environnement, Montréal. [19.1, 19.3, 19.5, 19.6, 19.7, 19.8, 19.11, 19.13]

SEBJ (1987) *Causes probables de la noyade des caribous à la Chute du Calcaire en Septembre 1984*. Société d'Energie de la Baie James, Montréal. [19.10]

SEBJ-SOTRAC (1985) *Etude des effets du détournement des rivières Eastmain et Opinaca en aval des ouvrages de dérivation—synthèse des résultats du suivi environnemental de 1980 à 1984*. Société de travaux de correction du Complexe La Grande (SOTRAC) and Société d'Energie de la Baie James, Montréal. [19.7]

Shooner G, Inc. (1985) *Répercussions d'une réduction de débit sur les rivières Caniapiscau et Koksoak*. Rapport au Groupe d'étude conjoint Caniapiscau–Koksoak. Société d'Energie de la Baie James, Montréal. [19.7]

Watts PD, Draper BA. (1988) Beluga whale ecology: a concern of freshwater ecologists? *Freshwater Biology* **20**: 211–13. [19.13]

20: The Rhône River:
A Large Alluvial Temperate River

J.P.BRAVARD, A.L.ROUX, C.AMOROS
AND J.L.REYGROBELLET

20.1 INTRODUCTION

The Rhône is a 812-km long nine-order river. Arising in Switzerland, from the Furka Glacier (2431 m), it is mainly an Alpine river, i.e. 66% of the watershed (95 500 km²) is situated above 500 m and 33% is above 1500 m (Fig. 20.1). The eastern tributaries (Arve, Isère) maintain the influence of glaciers and snowmelt on the seasonal regime of flow (Fig. 20.2) and on water temperature (less than 20°C during summer months downstream of the Isère confluence) well downstream. Other features of the regime are the increasing influence of oceanic winter rains as far as Lyon (overland flow in the mountains, pure oceanic regime of the Saône) and the influence of mediterranean rains on the downstream tributaries such as the Ardèche, the Gard and the Durance. Before the industrial era, the Rhône had one of the most complex flow regimes in the world (Pardé 1933). Floods occurred in every season and low flows shifted from winter in the upper course to autumn in the lower course (Fig. 20.2 (d)). With such characters, the Rhône exemplifies most temperate rivers: seasonal and yearly fluctuations of flow and physical parameters that are moderate and rather predictable, compared with tropical or arctic rivers. Nevertheless, long-term changes of climate and geomorphic processes at the watershed scale during the quaternary era have been responsible for the complexity of external controls on river behaviour, such as the presence of glacial lakes, the steep slope of the valley floor and the coarse size of bedload. They explain longitudinal, lateral and vertical discontinuities alongside the alluvial plains of the Rhône.

Moreover, like most of European temperate rivers, the Rhône experienced the early developments of an industrial era seeking multiple uses of running waters. Flood embankments at the end of the 18th century, navigation training works as early as the 1840s and power plants as early as the 1870s progressively changed the Rhône into a tamed river. Complete harnessing of this stream and the long history of human impacts profoundly influenced the functioning of the ecosystem and the quality of water.

20.2 THE PRESENT SPATIAL HETEROGENEITY AS A CONSEQUENCE OF WATERSHED AND RIVER DEVELOPMENT

At the end of the preindustrial era, i.e. during the early 19th century, the Rhône was braided from the headwaters to the upper delta in Camargue, except for some gorges and narrows. This pattern, typical of a mountain and foreland river, is very representative of the geological history of the Alpine range. Moreover, relict geomorphic features provide evidence of Holocenic river metamorphosis in response to climate and changes induced by humans at the watershed scale.

Geological setting

In Switzerland the Upper Rhône flows between the crystalline Pennine Range and the northern Prealps made up of folded sedimentary rocks.

Downstream from Geneva the river course is probably antecedent to the folding of the Jura Mountains, the paroxysmic phase of which is dated end of the Tertiary era.

From the Jura Mountains to the Mediterranean

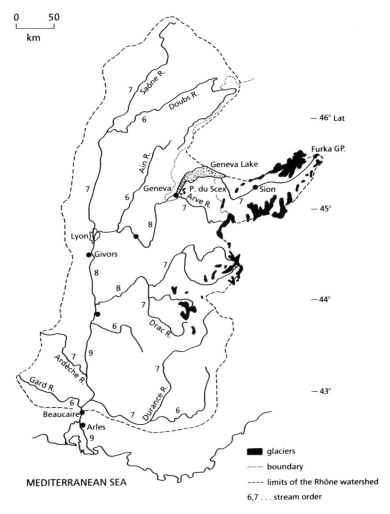

Fig. 20.1 The Rhône watershed.

Sea the Rhône valley corridor has experienced a complex geological history at the boundary of the hercynian Massif Central and the tertiary Alpine range. During the late Oligocene and the early Miocene, the Alpine surrection was coupled with a perialpine subsidence. The deposition of marine sand and then Alpine gravels built the huge alluvial fan of the Isère river. During the late Miocene the Alpine foothill was surrected up to 350–400 m and the river network settled (Mandier 1988); the incision of the Rhône cut narrow defiles into the crystalline outcrops of the eastern border of the

Massif Central, and shaped plains in the tertiary sedimentary deposits of the Alpine foothills.

Most of the present characteristics of the river originate in the glaciations which concerned the Alpine range at least four times during the quaternary era. The outermost limit of these glaciers was the Ain confluence area during the Würm glaciation until 20 000–15 000 BP. The main consequences were the following:
1 The glaciers overdeepened large valleys in the Alps, namely in the synclines filled in with weak mollassic sandstone; the present course of the

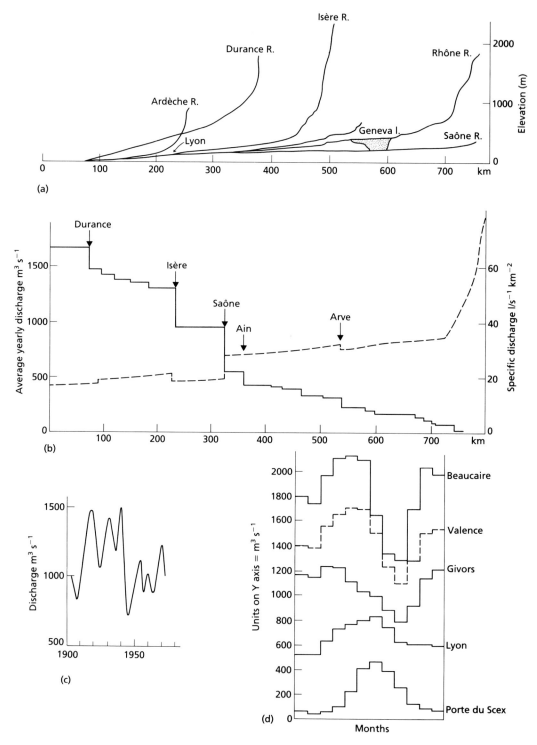

Fig. 20.2 Hydrological characteristics of the Rhône River. (a) Longitudinal profile of the Rhône and its main tributaries. (b) Average yearly and specific discharge. (c) Average yearly discharge during the 20th century at Givors. (d) Flow regimes from upstream to downstream.

Rhône downstream from Geneva to Lyon follows one of the glacial troughs carved into the detritic deposits of the Alpine foothill and through the folds of the Jura Mountains. Lake Geneva and Lake Bourget are remnants of a former extended pattern of glacial lakes. These wide upstream valleys are of prime interest in the reduction of peak floods which threaten cities such as Lyon; for instance, the Chautagne and Lavours marshes, Lake Bourget and floodplains provide a 150 km^2 floodable area that reduces the Q_{100} peak flood from 4100 at the Fier River confluence to 2450 m^3 s^{-1} at Sault-Brénaz (Table 20.1).

2 The youth of the river course implies that the equilibrium profile was far from being reached before the training of the river. Lake Geneva and the Sault-Brénaz rocky rapids were base levels for the Rhône as well as the exhumed outcrops of limestone of the downstream course.

3 The Alpine glaciers produced large amounts of outwash gravel deposits extending to the Mediterranean Sea (Bourdier 1961; Mandier 1988). The oldest terraces were surrected by tectonics downstream from Lyon; downstream from Orange these coarse deposits were fossilized by Holocenic sediments because the former long profiles were adjusted to low sea levels (−100 m during the Würm glaciation). As a consequence of intense Holocene erosion of the Alpine watershed and the reworking of glacial and fluvioglacial deposits by the rivers, the Rhône carries coarse bedload in Switzerland, downstream from Geneva to the Basses Terres floodplain, and from the Ain confluence to the head of its delta. At Donzère, the average flux approximated 360−400 × 10^3 m^3 in the early 1950s (Savey & Deleglise 1967). The large quantity of coarse alluvial deposits provides

excellent opportunities for water supply but the phreatic waters are sensitive to surficial pollutions.

The limestone of the Jura and Prealps Mountains bordering the Rhône encloses rather well-defined karstic reservoirs, the springs of which contribute to the regularization of the river discharge. So the Bugey Massif is drained by the Rhône in its northern part, while the eastern border receives some water losses from the river (Bureau de Recherches Géologiques et Minières 1970). Downstream from Lyon, the river is accompanied by a linear aquifer consisting of alluvial parts included in areas deprived of large water tables (crystalline border of the Massif Central, clays and mollassic sandstones of the Alpine piedmont).

The importance of the quaternary fluvioglacial deposits is not uniform all along the valley. These sandy–gravelly deposits were left first by the glaciers, and then by the river and its tributaries. The water-tables are included in this quaternary substratum, and their outflows depend on the geometry of the floodplain.

The embedding of ancient alluvial terraces and recent lower terraces partitions the water-table (Gemaehling 1969). So the aquifer included in the ancient terraces is more influenced by precipitations than that in the lower terraces whose level fluctuates with the level of the river. In that case, a large vertical discontinuity can be observed in the Rhône alluvium because gravelly layers with a high porosity alternate with sandy–muddy layers with a low porosity. Consequently the horizontal porosity of the aquifer is higher than the vertical one (Henry 1963; Gemaehling 1969).

Briefly, the rhodanian aquifers may allow an

Table 20.1 Hydrological characteristics of the Rhône River at four stations

	Annual discharge (m^3 s^{-1})	Q_1 (m^3 s^{-1})	Q_5 (m^3 s^{-1})	Q_{10} (m^3 s^{-1})	Q_{100} (m^3 s^{-1})	Q_{1000} (m^3 s^{-1})
Porte du Scex	175					
Sault-Brénaz	456	740	1620	1800	2450	3150
Lyon	603	1490	2900	3240	4420	5550
Valence	1409	2500	5000	5600	7600	9600
Beaucaire	1695	3600	7300	8200	11 000	13 800

Q_1, 1-year flood; Q_5, 5-year flood etc.

overall continuity but the global heterogeneity induces preferential fluxes which clearly determine discontinuous water-bearing formations (Engalenc 1982).

The Holocene history of the river

It is well established that geomorphic features of rivers and associated floodplains are responses to long-term changes in sediment inputs (Q_S) and water discharge (Q_L) into the fluvial system (Schumm 1977; Starkel 1983). Climate changes, combined with anthropogenic impacts over several millenia, through direct (increased or decreased precipitations) or indirect (nature, density and elevation of forest cover) influences have changed the balance of Q_S and Q_L several times, resulting in diversified patterns of alluvial plain construction.

The upstream overdeepened basins registered a net gain of alluviation through dominant periods of aggradation and probable short periods of incision. From Neolithic times at least, archaeological remains such as cities, quarries, artefacts and subfossil trees are buried below coarse alluvion; the aggradation of the active tract of the Rhône is up to 8 m in Chautagne and Lavours backswamps and accounts for the development of hundreds of hectares of peat bogs along the margin of the valley. The steep (slope 0.7×10^{-3} m) wedge of coarse sediments reached the southern end of the Jura Mountains c. 8000 BP, and incompletely filled the Basses Terres alluvial plain; the distal part of the reach upstream from Sault-Brénaz has fine-grained sediments and low gradient (slope $< 1 \times 10^{-4}$ m).

Downstream from the Ain River confluence, i.e. the ice-shaped area, the Rhône was sensitive to alternate periods of bed degradation and aggradation. Several low terraces, partly floodable for the Q_{100} discharge, were built up at different periods of the Holocene and incised during periods of relatively low discharge of coarse sediments.

The river responses in these floodplains were also changes in fluvial patterns as demonstrated, for instance, by the alternation of meandering and braiding. In the Basses Terres floodplain, meander belts have been dated (c. 3800−3400 BP and c. 1800−1500 BP).

Braided patterns have been documented c. 2700−2500 BP and c. 500−100 BP; the latest coincides perfectly with the so-called Little Ice Age, a period of climate deterioration and intense slope erosion, all the more so as the Alpine slopes had been deforested by humans during the climatic optimum of the 10−12th centuries (Le Roy Ladurie 1983; Grove 1987; Bravard 1990).

The geomorphic heritage of river metamorphosis, i.e. off-channels of different shapes and ages and variable surface extent of floods over the sedimentary patches, has important ecological consequences.

The fluxes of suspended sediments are also influenced by the history of the watershed. During the early 20th century, the Rhône downstream from Lyon carried only 3×10^6 t per year due to the trapping of fine sediments in Geneva Lake and the low inputs of the Ain (karstic watershed) and the Saône (the gradient of this river, 1×10^{-4} due to tectonics, promotes heavy siltation in the floodplain during floods). The Isère, an eighth-order tributary, delivered up to $30-35 \times 10^6$ t per year and the Durance up to 13×10^6 t per year (Pardé 1925). The sedimentary inputs of these tributaries may have considerable influence on the ecology of the Rhône along its downstream course.

For more than a decade, archaeological investigations have been documenting strong evidence of human settlement in the Rhône floodplains; increased fluvial activity obliged the first settlers to give up their settlements in rural areas (Salvador *et al* 1989) but humans were protected from the floods in the cities. Quarters of early Roman cities were built along the river at the end of the first century BC after a period of high discharge and floods which lasted at least one century and obliged the inhabitants to raise the bottom level by 2−3 m in Vienne (Bravard *et al* 1990).

Thus, the changeable extent and wetness of floodplains over time provided changing ecological conditions, landscapes and resources. The last occurrence of degrading conditions of the milieu took place during the Little Ice Age; humans had to face river bed aggradation, increased floods, land erosion, crop siltation and marsh extension. In the Alpine valleys the authorities forbade the clearance of the remnants of the alluvial forests, and protected meadows against tillage in order to prevent a weakening of the river banks; rural communities and land-owners built stone or wooden spurs early, but unfloodable dykes were not erected before the second half of the 18th century. The braided pattern, dated Little Ice Age, was inconvenient for navigation (shallow riffles less than 0.5 m deep) and agriculture (extensive grazing by cattle such as horses in the Swiss Valais, cows and horses in the French Upper Rhône; wood cutting; wheat crops were flooded by summer high waters). Then fluvial dynamics and humans interfered to increase the activity of the Rhône over its floodplain; maps of the time depict large cobble bars, up to 10–15 fast-flowing channels, islands and banks covered with willows and alders at early stages due to the truncation of terrestrial successions by fluvial dynamics and humans; the deposition of sand by fast-flowing overbank waters was intense and widespread (Bravard 1987).

20.3 A RIVER INFLUENCED BY HUMANS

Until the early 19th century, humans had to adapt to fluctuating natural constraints when settling along the Rhône. Since that time, the improvement of technology has progressively improved conditions of navigation and for energy supply, while pollution has increased.

Improvement of the conditions for navigation along the Rhône downstream from Geneva

Up until the early 19th century, the Rhône was used mainly for rafting logs from the Alps to the cities or for transporting stone on board flat boats, 20–30 m long. At that time the journey from Lyon to Arles lasted for 3 days, but for 20 days when boating upstream (310 km) owing to the speed of the current. Then the traffic from Lyon to Arles was 270 000 t a year out of 300 000 for the total; most goods and people were carried by more than 12 000 carriages, 14–15 000 stage-coaches and 48 000 horses, to be compared with 4000 small boats (Rivet 1962).

However, the Rhône downstream from Lyon carried 25% of the goods transported on French waterways. There was then fierce competition between the railway and river boats in the years 1825–55. The administration of Ponts et Chaussées, in charge of the Rhône, interfered by systematically taming the river until the beginning of the 20th century, in order to keep pace with the improvement of rail transportation. The first experimental steamboat was launched on the Saône at Lyon in 1783, and later on the fast-flowing Rhône. The first private companies founded in 1825 by the engineer Seguin, and in 1829 by the American Edward Church, were not successful before 1837–8. They had a serious impact on transportation but were severely affected by the construction of the Lyon–Arles railway (1852–6) which drew most of the traffic at the end of the 19th century (Lentheric 1892).

The Rhône River downstream from Lyon sums up the progressive improvement of training technology during the 19th century, as successive generations of engineering works contributed to the complete harnessing of the river (Fig. 20.3). The first dykes were too remote to be efficient (1840), and sets of low embankments increased the depth and the speed of the water, and worsened the pool-riffle morphology. Then the engineers adopted the German techniques of 'drowned spurs' before promoting the so-called Girardon system which completely stabilized the river bed. This was successful because the number of riffles shallower than 1.40 m decreased from 126 in the 1880s to zero in 1944 (Béthemont 1972).

But all this was detrimental to the ecological balance: the Rhône was changed into a single-channel river and most of the cutoff side arms were quickly silted up and the lateral wandering of the former braided channels completely disappeared.

The situation was completely different along the section between Seyssel and Lyon, for the following reasons.

First, until the French Revolution and between 1815 and 1860, the Rhône was a boundary between France and Savoy. The King of Piedmont, Duke of Savoy, promoted the embankment of the left bank of the river as early as 1780 to protect rural communities. This policy failed because peasants intended to allow the Rhône to continue flooding the marshes in which they cut hay for cattle and for green manure. Degradation, as a

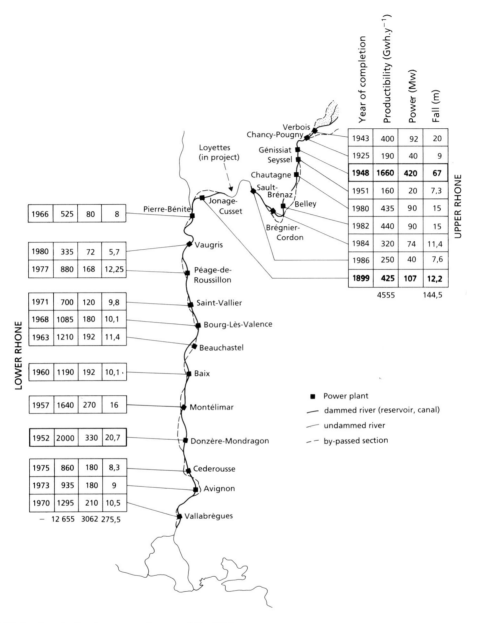

Fig. 20.3 Hydroelectric development schemes of the Rhône.

response to embankments, is more than 200 years old along this reach and was up to 4 m deep.

Second, in 1858, a French law was enacted that forbade any new embankment against flooding upstream from Lyon, one of the French cities to have been damaged by the Q_{100} flood of 1856. Until recently this law has been applied to the benefit of the river and of the spatial extent of the floods over the alluvial plains.

Third, engineering works to improve shipping

upstream from Lyon were late and relatively light owing to the scarcity of boats and a lack of public funding. Except for a 20-km reach upstream of Lyon, which experienced a dramatic long profile tilting (5 m), the embankments consisted of low stone dykes protecting the concave banks and blocking the upstream end of side arms. These works, completed in the years 1880–90, were not responsible for the siltation of all side channels; many kept flowing during floods and high waters. Moreover, the cessation of shipping in the early 20th century and the dereliction of agriculture up to the 1960s led to the development of a wide riparian forest along the channel.

The Rhône upstream from Lake Geneva: the late conquest of a floodplain

The Rhône flowed free and braided until the beginning of the 1860s. The 1860 flood motivated engineering work (between 1864 and 1877) between Brig and the mouth of the Rhône into Geneva Lake (140 km). Two dykes limited a flood channel 70-m wide upstream and 120-m wide downstream; the single low-flow channel was constricted inside orthogonal spurs and was 30-m wide upstream, 60-m wide downstream. As a consequence bedload aggraded the river bed and, despite the surelevation of dykes and dredgings, floods demolished the dykes while the raising of the water-table changed the protected alluvial plain into upstream marshes, especially from the confluence of tributaries. These impacts have ceased, largely as a result of a shortage of coarse bedload (as a result of mountain reforestation), the training of torrents during the late 19th century, and the trapping of sediments in upland reservoirs. The completion of embanking by longitudinal low dykes and the systemic drainage of the plain ensure excellent conditions for fruit production. However, this is responsible for the suppression of wetlands and alluvial forests (3% of the total area, i.e. a narrow corridor along the dykes) and has detrimental effects on river ecology (fast-flowing water, lack of shelters, intensity of gravel reworking; the productivity is 30 kg fish ha^{-1} year^{-1}).

Power production and economic development
(Figs 20.3 & 20.4)

In the Swiss Valais, the Rhône does not produce energy (except in the small power plant of Lavey downstream from Martigny); this is due to the fact that since 1902 the water authorities have equipped the best sites along the alpine tributaries, as exemplified by the Grande Dixence reservoir (1955: 150×10^6 m^3; 1970: 1400×10^6 m^3). During the early 1980s the so-called 'HYDRORHONE project' (1984) planned the construction of ten dams along an 80-km reach of the Rhône, to produce 700 Gwh.y^{-1}. Objections from farmers and ecologists has recently delayed these works.

Downstream of Lake Geneva, the Rhône was one of the first rivers to be used for generation of modern hydraulic energy (Tournier 1952; Faucher 1968; Lafoucrière *et al* 1987). The first reach to be developed was the outlet of Lake Geneva complemented by the power plant of Coulouvrenière which supplied electricity to Geneva from 1886 on; the gorges were damned in 1871 at Bellegarde, along a reach which has been drowned by the Génissiat reservoir; the turbines of a power plant provided mechanical energy to an early industrial estate. Another early development scheme was the 18-km artificial canal of Jonage and the power plant of Cusset (1892–9) which was built in the close vicinity of Lyon by engineers of the Suez canal for supplying electricity to Lyon; it is the oldest low-fall power plant in France.

The French section of the Rhône has been dammed except for a 30-km reach at the confluence of the Ain River. After the completion of the Cusset dam, every new development scheme was delayed by the administration of Ponts et Chaussées in Lyon which intended to preserve the Rhône for large-scale investments instead of piecemeal development. This policy was confirmed in 1921 when the parliament voted the Law of the Rhône which kept the river for future multipurpose development (i.e. energy, navigation, irrigation and regional development). It was not before 1933 that the river was consigned to the Compagnie Nationale du Rhône (CNR) that performed a programme similar to the Tennessee Valley project. The first site chosen

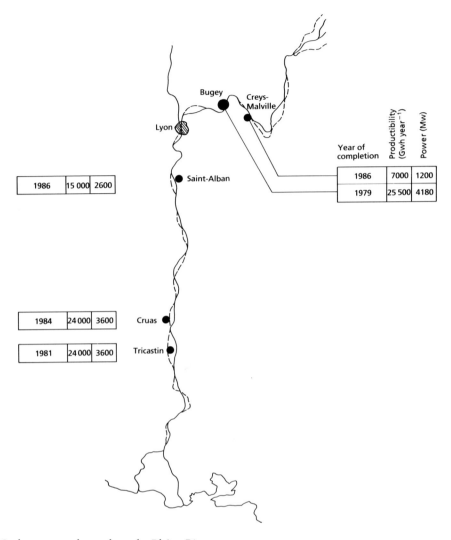

Fig. 20.4 Nuclear power plants along the Rhône River.

was Génissiat in the upper gorges and the scheme of Seyssel, built for compensating daily high-flow discharges. Then attention was turned to the lower Rhône was undertaken (1952–80): while Génissiat was built for the production of energy, this 310-km section was developed to fulfil three main objectives:

1 The production of peak energy averages 12 655 Gwh.y^{-1} out of the 16 000 Gwh.y^{-1} year^{-1} of the French section of the Rhône, i.e. 6% of the French production of energy or 27% of hydroelectricity.

2 The improvement of navigation. However, the waterway is not used at full capacity (3–4 mt year^{-1}, i.e. 2–3% of the total traffic of the Rhône valley) because the connection with the Seine basin and the Rhine basin is not yet completed.

3 The improvement of agriculture through the protection of land against floods (13 470 ha totally, 27 750 partially out of the 42 350 ha of floodplains downstream of Lyon), through the irrigation of 35 000 ha which compensate for the loss of 7500 ha of agricultural land. In the future, 350 000 ha may be irrigated by pumping 175 m^3 s^{-1}. Irrigation was developed at first along the tail races of the development schemes which are responsible for the lowering of the water-table.

The upper section of the French Rhône was developed between 1980 and 1985 after the energy crisis of 1973 which raised the price of fuel. Production of electricity, coupled with some tourism, was the main objective in this depressed rural area.

The development of hydroelectricity in the watershed of the Rhône had several consequences. First, the successive improvements of the retention dam of Geneva, which controls the level of the lake, influenced the Rhône downstream: water from Valais filled the lake in spring and early summer, and was then kept until winter. This policy resulted in decreased values of discharge at low flow (60–80 m^3 s^{-1} in August and September), while discharge increased during winter and spring for the production of valuable energy; then the Rhône upstream of Lyon lost part of its Alpine character. Second, since 1902 water retention in the high-altitude reservoirs had considerable impact on the flow regime of the Rhône in Switzerland. Regulation was markedly increased because water storage from March to September endangers the filling of Lake Geneva. These anthropogenic changes, coupled with the natural reduction of flow from alpine glaciers since the beginning of the 20th century, affected the flow regime from Geneva (Vivian 1989).

Major physicochemical characteristics

The physicochemical characteristics of the water indicate that the French Rhône can be split into two major parts: upstream of Lyon (Upper Rhône) and downstream of Lyon (Lower Rhône). The French Upper Rhône is in a relatively good state: dissolved oxygen values show saturation or nearly saturation; BOD$_5$ is about 2 mg l^{-1} or lower, and the concentration of nutrients is moderate, 0.1–0.2 mg l^{-1} nitrogen in the form of ammonia, 1 mg l^{-1} in the form of nitrate, lower than 0.2 mg l^{-1} phosphorus (orthophosphates). Nevertheless, the quality of the water is severely

affected every 3 years by the release of the reservoirs of Verbois and Genissiat (Roux 1984). Downstream of Lyon, due to the inputs of its main tributary (the Saône River) and the impacts of the city and the chemical factories in the south of Lyon, the quality of the water is poor and remains so until the next confluence of the Isère River where the improvement comes from dilution. Nevertheless, organic pollution is moderate, dissolved oxygen values are over 80% of saturation, and BOD$_5$ is about 3 mg l^{-1}. Inorganic inputs are more important: 0.3 mg l^{-1} nitrogen in the form of ammonia, 1.5 mg l^{-1} in the form of nitrate, 0.5–0.6 mg l^{-1} phosphorus (orthophosphates).

20.4 THE PRESENT RIVER ECOLOGY AS A CONSEQUENCE OF NATURAL CHANGES AND THOSE INFLUENCED BY HUMANS

Rivers flowing through broad alluvial floodplains have to be considered not only along the upstream–downstream direction of running water, but with a broader spatiotemporal perspective according to the extensive interactions between lotic and surrounding patches. The concept of fluvial hydrosystem (see Part 1) presumes rivers to be four-dimensional complex systems of interacting patches (Amoros *et al* 1987; Ward 1989): the longitudinal dimension (upstream–downstream gradient), the transverse dimension (interactions between the active river, backwaters, abandoned channels, marshes, carrs, riparian ecosystems), the vertical dimension (surface and groundwater) and the temporal dimension (natural changes and anthropogenic impacts).

Longitudinal zonation

Longitudinal zonation is important in the structure and functioning of hydrosystems. It can be either continuous (according to the river continuum concept; see Chapter 11) or discontinuous owing to natural modifications such as confluences, special geomorphic situations ... or anthropic perturbations such as fluvial development schemes, dams, reservoirs etc. The latter situation is the one that pertains for the Rhône. The French Rhône River (downstream from Lake Geneva) is now completely harnessed with a set of three reservoir-dams and 18 diversion dams. This development scheme diverts up to 700 m³ s⁻¹ into a canal comprising a headrace, a power plant and a tail race; the former river bed, called the bypassed section, receives only a compensation flow outside the periods of spates, but all the flow during spates when the discharge exceeds 700 m³ s⁻¹.

So, in the former channel, during most of the year, the water level and the discharge are reduced far below the previous lowest waters. In contrast, in the artificial, newly created canals and reservoirs the water level is almost the same throughout the year, with annually regulated discharge but with daily fluctuations of level due to the production of peak energy (Fig. 20.5).

The impact on fauna will be illustrated by examples taken from one group of invertebrates

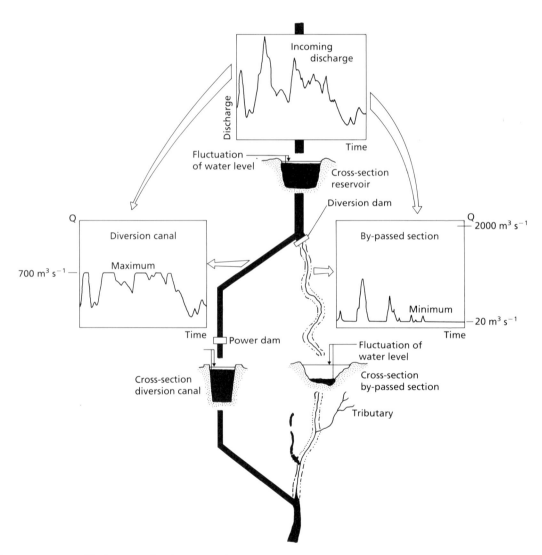

Fig. 20.5 Pattern of hydropower development scheme characteristic of the Rhône (discharge exemplifies the Upper Rhône).

(the Trichoptera) and one group of vertebrates (the fishes).

The longitudinal distribution of the *Hydropsyche* species (Trichoptera) along the river continuum is well documented, and well established on the Rhône for a set of five congeneric species (Bournaud *et al* 1982; Roux *et al* 1992; Tachet *et al* unpublished data): *H. pellucidula* and *H. siltalai* are rare in the Upper Rhône (generally they inhabit streams and small rivers). *H. exocellata* is present in the Upper Rhône where it coexists with *H. contubernalis* and *H. modesta*, but only *H. modesta* persists in the Lower Rhône (Fig. 20.6). It is difficult to specify the impact of the management of the river on this group of Trichoptera and more generally on the invertebrate fauna.

The situation is different concerning the fishes. Results on fish populations, between Lake Geneva and the Mediterranean Sea, sampled by different research centres over the last 10 years, have been arranged into a single database (more than 90 000 fishes caught in about 1000 fish samples were included in this) in order to test whether a longi-

tudinal zonation of the communities exists or not. The water-course has been subdivided into sectors corresponding to the successive management schemes developed all along the river (five on the Upper Rhône, 13 on the Lower Rhône; Fig. 20.7). Four main sectors can be distinguished along the upstream–downstream gradients. The major break appears between the communities of the Upper Rhône and those of the Lower Rhône. Three running water species, the dace (*Leuciscus leuciscus*), the trout (*Salmo trutta*) and the stream bleak (*Alburnoides bipunctatus*), are limited to the Upper Rhône. The other rheophilic species, the nase (*Chondrostoma nasus*), the barbel (*Barbus barbus*), the gudgeon (*Gobio gobio*) and the perch (*Perca fluviatilis*), may be present downstream all along the water-course. The partitioning of the Lower Rhône seems to be due mainly to the confluence of major tributaries, the Saône River and the Isère River, and to the abundance of the eel (*Anguilla anguilla*). The first of the three sectors of the Lower Rhône is greatly influenced by the species of the Saône River: bream species (*Abramis brama* and *Blicca bjoerkna*), the catfish

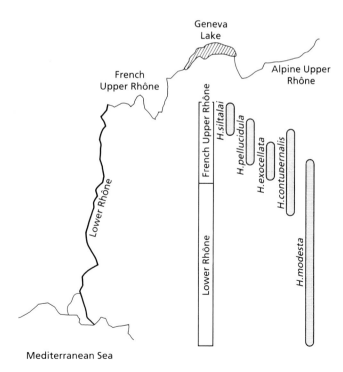

Fig. 20.6 Distribution of five species of *Hydropsyche* (Trichoptera) along the Rhône River.

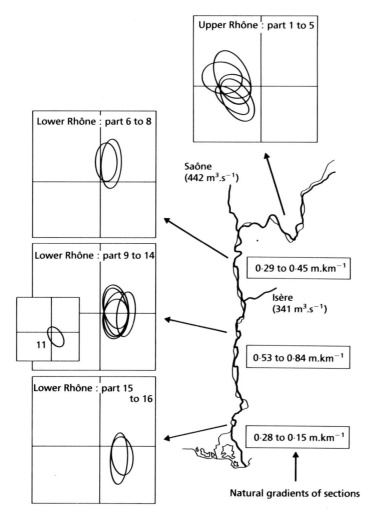

Fig. 20.7 Longitudinal zonation of the fishes. The distribution of the samples in each sector is depicted by an ellipse, the centre of which is pointed out by the mean co-ordinates of the samples on the two principal axes of a principal components analysis. The size is a function of the variance of the co-ordinates of the samples and the slope a function of the covariance.

(*Ictalurus melas*), the tench (*Tinca tinca*). The second sector is influenced by the Isère River characterized by the roach (*Rutilus rutilus*), bleak (*Alburnus alburnus*) and the eel. In the third sector the frequencies of eel and bream increase while those of bleak and roach decrease (Pont & Persat 1990).

So, the longitudinal zonation of fish communities seems to be due largely to the influence of both general morphological features and major tributaries, but it is also influenced by the partitioning of the course by hydroelectric developments of the river (dams). For instance, anadromic migrators such as the shad (*Alosa alosa*) and the

lamprey (*Lampetra fluviatilis*) have disappeared or are confined to the lowest part between the last dam and the sea, or are naturally reduced to the lowest sector for the eel (eels present in the upstream sectors are due to restocking by fishermen).

Alluvial floodplain

The habitat diversity of the Rhône floodplain, like other medio-european large rivers (Danube, Rhine), is due to a combination of Holocene heritage (fluvial dynamics, river metamorphosis) as well as human impact. The habitat diversity of the plain rules the diversity of communities, and thus the nature and intensity of patch interactions within the fluvial hydrosystem.

Because the Rhône River is more harnessed downstream from Lyon, the diversity of the alluvial floodplain is greater in the upstream reach between Geneva and Lyon. However, the diversity is not homogeneous all along the French Upper Rhône. This reach may be split into several sectors determined by geomorphic features and additionally by human impact. Within each, the habitat and community diversity, as well as the patch interactions, depend on: (1) fluvial dynamics; (2) the development of ecological successions; and (3) the connectivity with running water of the main stream and with underground water of the alluvial aquifer.

Fluvial dynamics and habitat diversity

Within gorge sectors (e.g. from Geneva to Seyssel; see Fig. 20.1), no alluvial plain occurs. In other single-channel sectors (e.g. from Sault-Brénaz to the River Ain confluence), the alluvial plain is reduced to terraces used as field for crops.

In braided sectors (e.g. La Chautagne; Figs. 20.1 & 20.8), the patch mosaic is composed of coarse sediment (mainly sand and gravel) bars, islands and levees, more or less narrow side-channels and former arms. Because of their soil texture and the high occurrence of fluvial erosion and deposition, the terrestrial habitats were used only as extensive pastures until the beginning of the 20th century. Now, agriculture has made derelict most of these lands except on the margins of the plain where silt deposition occurred. The narrow (5–15 m) former braided channels are shaded by the riparian canopy, which supplies the aquatic habitat with organic matter while reducing solar radiation inputs and thus aquatic primary productivity. So, these former channels function as heterotrophic ecosystems such as headwaters and low-order streams. These characteristics are reinforced by the high occurrence of groundwater supply and, due to the high porosity of this kind of alluvial aquifer (pebbles, gravels, coarse sand), limnocrene springs. In one of these sectors, the Bregnier-Cordon area (Fig. 20.8), the aggradation of the plain causes the development of highly sinuous

anastomosed channels on the depressed margins of the plain (Bravard *et al* 1986). These anastomosed channels are deeper and wider and less reworked by fluvial dynamics than the braided ones. Other features, such as slower flow velocity, finer bed sediment and biotic components, make them similar to some extent to meanders.

In meander sectors (Fig. 20.8), the fine-textured soils of the flats are generally cultivated while the former channels remain as oxbow lakes or more or less forested wetlands according to the development of ecological successions. These former meanders are wide (about 100 m) and deep (3–5 m); thus, the riparian canopy plays a minor role and the oxbow lake ecosystems function mainly as autotrophic standing water.

Ecological successions

The co-occurrence of patches that are at different stages of development increases the habitat and community diversity. The development stage is ruled by the cut-off date of former channels and the rate of the successional processes involved.

In the braided sectors, up until the embankments of the 19th century, the active fluvial dynamics instigated frequent shifts in bars, islands and channels. This topographical instability impeded the long-term development of terrestrial as well as aquatic ecological successions. Only the pioneer stages could occur on sand bars and in cutoff side arms. The embankment stopped the lateral erosion of the river and then allowed the succession developments. The successions occurring within braided areas, driven mainly by allogenic processes (silt deposition), contrast with those of meander areas.

As in the former braided channels of the Rhine (Carbiener *et al* 1990), a few dead-arms of the Rhône River, supplied by oligotrophic groundwater, are colonized by *Potamogeton coloratus*, succeeded by mesotrophic communities with *Berula erecta* and *Mentha aquatica*. As eutrophication and siltation proceed, the following stages are marked by *Callitriche* communities, then *Phalaris arundinacea*. The terrestrial stage, corresponding to full siltation of braided arms, is a wooded community with *Ulmus minor*.

The coarse sediment bars and levees are first

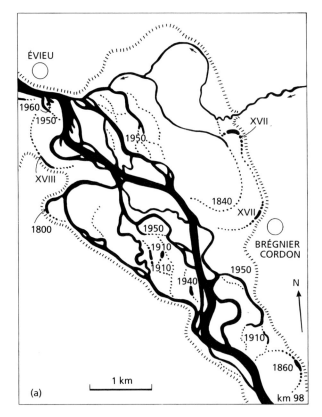

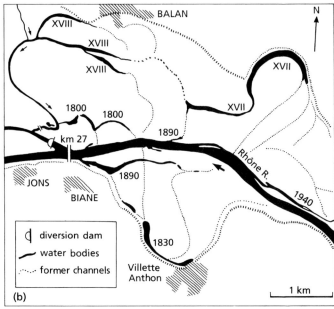

Fig. 20.8 The network of the side arms and former channels, with the dates of their upstream disconnection in two sectors of the French Upper Rhône: (a) the Brègnier–Cordon area, a braided anastomosed sector (river km: 98–91), (b) the Jons–Jonage area, a former meandering sector (river km: 27–32) a part of the Jonage–Cusset development scheme.

colonized by *Melilotus alba* and *Epilobium dodonaei*, followed by shrubs of *Salix purpurea* and *S. eleagnos*, then trees of *Salix alba* that are replaced by *Alnus incana* communities, then by *Quercus robur* and *Populus alba* (Pautou & Décamps 1985; Bravard *et al* 1986).

In the former meanders, as well as in the former anastomosed channels, the first stage of the succession corresponds to the open-water area of the oxbow lakes. While eutrophication proceeds, the open-water area is progressively reduced by the extension of submerged and floating leaved plants (*Myriophyllo–Nupharetum* community). The next stage is a reed swamp (*Phragmites australis* and *Carex elata*) that is succeeded by a fen stage (*Salix cinerea*) then by a carr stage (*Alnus glutinosa*). Such a succession in former meanders is driven mainly by autogenic processes, namely *in situ* production and accumulation of organic matter (Pautou & Décamps 1985; Rostan *et al* 1987).

Connectivity

The still or standing-water bodies of the floodplain, receiving nutrients and organic matter from both running water and surrounding land ecosystems, are much more productive than the main river. A large part of this production can be washed away and drifts into the main river when those water bodies are connected to the main river and scoured by running water. In a reciprocal way, through these connections, lotic fishes can use the floodplain water bodies as spawning and growing areas or as shelters during high floods or accidental pollution of the main river. This connectivity changes seasonally, as well as historically. Some water bodies remain connected throughout the year while others are connected during high water level or only during floods. The connections depend on the alluvial plugs constructed by the river at the ends of the side arms and the cutoff meanders. Then these alluvial plugs are colonized by helophytes (*Phalaris arundinacea*) that slow down the flow and thus increase the alluvial deposition. The aggradation of the plug allows the development of shrubs (*Salix purpurea*) that increase the positive feedback (alluvial deposition and aggradation) that reduces the frequency of connections. So, the ecological successions reduce the connectivity on a historical timescale (decades to a century).

Connectivity between groundwater and surficial water plays an important role in so far as the constant and low temperature of the groundwater, and sometimes its oligotrophic character, reduces the productivity of floodplain water bodies. In addition, the groundwater discharge can wash a part of the surficial production away from the former channels through outlets that remain connected to the main river. However, the underground–surficial connections are reduced by the organic matter deposition on the bottom of the former channels during the development of ecological successions.

Alluvial aquifer

All the rhodanian aquifers converge toward the river, which unites them in a so-called aquifer/river global system, where the direction of groundwater runoff depends on the piezometric contact position and this in turn depends on the water level of the main channel. So, two functioning patterns can be defined, acting alternatively or permanently: a 'draining pattern' when the channel drains the aquifer because its water level is lower, and an 'infiltrating pattern' when the surface water feeds the depressed water-table.

The major consequences of these two functioning types are silting or not-silting of the river banks and bottom. The silting may occur in infiltrating conditions but has never been recorded in a draining one (Henry 1963). When an alternative process occurs, the silting decreases or may disappear during the drainage period, and is restored, partly or totally, as soon as the first flood occurs, carrying the major part of suspended load (Castany 1985).

In the original natural state, the braided pattern induced a very slow siltation by a high level of aquifer/river exchanges, and maintained a maximum vertical and horizontal connectivity between all the patches of the floodplain. After the construction of embankments, lateral wandering of the channel was no longer possible and the patches of the floodplain were fixed and rapidly silted (section 3), except for such sectors as Brégnier-Cordon and Jons (Fig. 20.8), where horizontal connectivity is still preserved under the newly built soil. There, the former channels, colonized by alder (see above), are of great influence on the remaining groundwater flow. These 'functional relicts' effectively improve the groundwater quality, acting as a very efficient sink for nitrates of agricultural origin and contaminating all the lateral aquifers of the floodplain, as demonstrated by Pinay and Décamps (1988) on the Garonne River.

The major problem at present is heavy silting of the head and tailraces of the 18 diversion dams built on the channel. The long-term impact of such schemes is obviously the deterioration of the quality of the water resources of the floodplain aquifer situated in the influence zone of these reservoirs.

From a biological point of view, many studies have demonstrated that the interface zone between the surface of the floodplain (water and soil) and the groundwater is of essential functional importance. For example, the specialized (stygobiont) fauna inhabiting the rhodanian aquifer (such as the Archiannelid *Troglochaetus beranecki*; the Crustacea *Parastenocaris* sp., *Bathynella* sp., *Microcharon* sp., *Proasellus walteri*, *Salentinella* sp., *Niphargopsis casparyi*, *Niphargus rhenorhodanensis*; the Coleoptera *Siettitia avenionensis*) can mix in this zone with superficial species that colonize easily the upper part of the interstitial biotope when the sediment size allows it, permanently (Cladocera, Copepoda, *Gammarus fossarum*, *Asellus aquaticus* ... Mollusca, Oligochaeta, etc.) or temporarily (amphibiotic insects such as early stages of Trichoptera, Ephemeroptera, Plecoptera larvae, and adults and larvae of Coleoptera *Esolus parallelepipedus*, *Bidessus cf. minutissimus* ...), in the channel or all the parts of the floodplain where the piezometric level outcrops. The structures and the vertical distribution of these particular communities have been well established since the work, for example, of Danielopol (1976, 1989) and Bretschko and Klemens (1986) on the Danube system, Gibert *et al* (1977, 1981), Reygrobellet and Dole (1982), Dole (1983, 1985) on the Rhône, or Pennak and Ward (1986), Stanford and Ward (1988) and Rouch (1988) on smaller rivers. The flood dependence of this vertical structure has been well known on the Rhône system since the work of Marmonier and Dole (1986) and Dole and Chessel (1986).

One of the long-term impacts of the regulation of the Rhône is to have modified the vertical structure of the interstitial communities, and so their functional part in this human-influenced hydrosystem. This is particularly obvious in bypassed sections. For example, one of the oldest of these is the Miribel Canal, in the Jonage-Cusset sector (Fig. 20.8) where the dam began operating in 1938. This is one of the arms of the previous braided pattern. It receives at present a reserved discharge of 30 m^3 s^{-1}, acting as a spillway above the possible working maximum of the power station (Cusset) which is 650 m^3 s^{-1}. Consequently, the artificially low waters are maintained throughout most of the year, except when highly eroding flash floods wash the channel. So, the longitudinal profile is totally modified, as this development increases the effects of previous embankments (Winghart & Chabert 1965; Poinsart *et al* 1989), i.e. the upper part is very degraded (more than 4 m in one century) while aggradation has raised the lower part by 5.5 m (see Figs 20.10 (a) & (b)). The equilibrium point, at present around 16.5 km upstream from Lyon, is still moving. This new profile exacerbates the river/aquifer relationships in two ways: as was demonstrated by a multivariate analysis applied to the physicochemistry of the interstitial water (Creuze des Chatelliers & Reygrobellet 1990), the first half of the channel course acts as an almost permanent draining system, whereas the downstream part maintains a permanent supply to the aquifer, acting as an infiltrating system.

The ecological consequences of this kind of impact have been studied in 16 sampling sites. These were confined to the riffles of the canal, since these are the only geomorphological units which can be found from one end to the other, both on right and left banks. Each site was subdivided in two substations, the upstream-riffle and downstream-riffle zones, in order to integrate both the geographical and the riffle scales. Benthic and hyporheic biota were sampled, and the results were subjected to multivariate analysis.

On the geographical scale, rheophilic benthos dominates in the superficial layer of the degraded

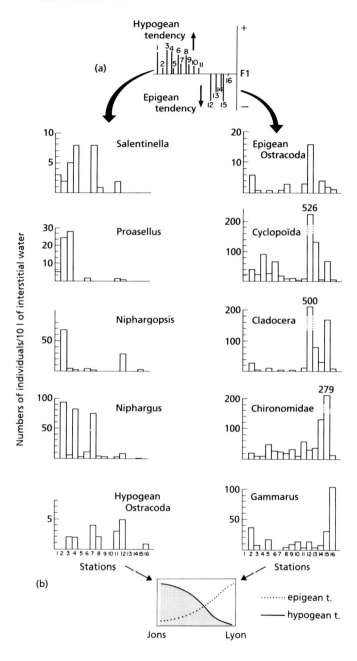

Fig. 20.9 Miribel canal: communities distribution in the hyporheic layer. Factorial correspondence analysis after addition of the samples from upstream-riffle and downstream-riffle zones (the riffle is taken as a single morphological unit). (a) The F1 axis clearly discriminates the stations located in the upper part of the canal (positive co-ordinates) from those located in the lower part (negative co-ordinates). (b) Abundance histograms of some representative taxa explain this separation of the stations: high rate of stygobiont taxa in the hyporheic communities of the upper part; high rate of superficial taxa in the lower part (from Creuzé des Châtelliers & Reygrobellet 1990).

zone (great abundance of *Psychomyia pusilla, Polycentropus flavomaculatus, Hydropsyche exocellata* aged and juveniles, *Heptagenia sulphurea*). In the aggraded part, the species are more lentic and eutrophic, mainly because of the great abundance of microcrustacea (*Biapertura*

affinis, Illiocryptus sordidus, Alona quadrangularis, Illiocypris sp., etc.), Oligochaeta, Mollusca and Chironomidae.

The hyporheic layer communities are also separated by the equilibrium point: upstream, where the canal drains the lateral aquifers, a diversified

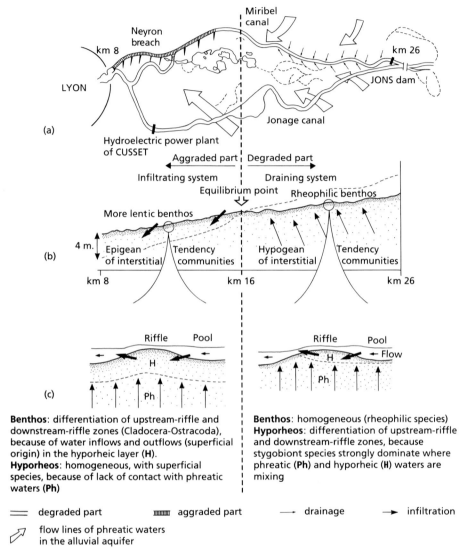

Fig. 20.10 Fluxes acting in the Miribel canal (in terms of physicochemical and faunal characteristics). (a) & (b) On the scale of the geographical sector. (c) On the scale of the geomorphological unit (riffle) (after Creuzé des Châtelliers & Reygrobellet 1990). =, degraded part; ▦, aggraded part; →, drainage; →, infiltration; ⇦, flow lines of phreatic waters in the alluvial aquifer.

and abundant stygobiont fauna is well established (*Salentinella*, *Proasellus*, *Niphargopsis*, *Niphargus* and hypogean Ostracoda). This community tends to be 'hypogean'. In contrast downstream, the strong domination of the superficial faunal groups defines the community as 'epigean'. These results are summarized in Figs 20.9 and 20.10.

Thus, the fauna reflects in the two layers the human-induced processes of degradation/aggradation of the canal, as well as the chemistry of the water fluxes. After recent sediment extractions, the profile is quickly changing by erosion of the lower part of the canal (Poinsart *et al* 1989; Creuzé des Châtelliers 1991), and this threatens to transform all the channel to an 'open aquifer' as in the upper part.

20.5 THE HYDROSYSTEM CONCEPT APPLIED TO MANAGEMENT OF TEMPERATE RIVERS AND THEIR FLOODPLAINS

If we postulate that environmental management must preserve the diversity of habitats and communities at the scale of the fluvial hydrosystem, we should focus on two points: the connectivity between the hydrosystem patches and the ecological successions of the patches towards a terrestrial state.

Connectivity preservation and restoration

In most cases, human impacts on rivers reduce the connectivity between the main channel and the water bodies of the floodplain. For example, embankments or channelization prevent natural rejuvenation of the connections by natural fluvial dynamics. Consequently, an ecological management of fluvial hydrosystems must include the conservation or the rejuvenation of fluvial dynamics processes, or the reconnection of former channels.

In this context, several strategies have been developed on the Rhône River. During the recent completion of the hydroelectric plant at Brégnier-Cordon (see Fig. 20.8), some of the submersible embankments built in the 19th century to improve navigation (no longer in service) were opened to allow the connections between the upstream end of side arms and the main river again. Weirs were constructed to maintain the mean water level as well as the water-table. In this way, connectivity between former channels and the main river, between groundwater and surficial waters, were preserved. Fish ladders were also constructed for the preservation of the longitudinal connectivity.

Ecosystem development and reversible processes

Because the river is regulated, fluvial dynamics can no longer compensate (by lateral shifting and the cutting-off of channels) for the terrestrial-ization of former channels. Consequently, in the long term (one or two centuries), most of the floodplain water bodies and wetlands would be changed into terrestrial ecosystems without any new production of this kind of habitat by fluvial dynamics. For the preservation of floodplain habitat diversity, two cases have to be considered in relation to the degree of reversibility of the developmental processes involved in the ecological successions (Amoros *et al* 1987).

If the processes are reversible (i.e. most of the allogenic processes such as alluvial deposition within the former channels), an operation to rejuvenate the former channel should be considered. For example, in the French Upper Rhône, alluvial plugs deposited at the ends of former braided channels have been easily removed, so that the running water washed away most of the organic matter stored in the previously isolated former channel.

In contrast, most of the autogenic processes seem irreversible, at least without any input of energy that would be considered as unrealistic in our socioeconomic system. In the case of the former channels, whose development is driven mainly by such irreversible autogenic processes, the management principle should be the preservation of the ecosystem. Generally, such autogenic-driven successions develop more slowly (over several centuries) than the allogenic-driven successions (over several decades).

REFERENCES

Amoros C, Rostan JC, Pautou G, Bravard JP. (1987) The reversible process concept applied to the environmental management of large river systems. *Environmental Management* **11**: 607–17. [20.4, 20.5]
Béthemont J. (1972) *Le thème de l'eau dans la vallée du Rhône*. Thèse lettres. Saint-Etienne: Le Feuillet Blanc. [20.3]
Bourdier F. (1961) *Le bassin du Rhône au Quaternaire – Géologie et Préhistoire*, Vol. 2. CNRS, Paris. [20.2]
Bournaud M, Tachet H, Perrin JF. (1982) Structure et fonctionnement des écosystèmes du Haut-Rhône français. 32 – Les Hydropsychidae (Trichoptera) du Haut-Rhône entre Genève et Lyon. *Annales de Limnologie* **18**: 61–80. [20.4]
Bravard JP. (1987) *Le Rhône, du Léman à Lyon*. La Manufacture, Lyon. [20.2]
Bravard JP. (1990) La métamorphose des rivières des

Alpes françaises à la fin du Moyen-Age et à l'époque moderne. *Bulletin de la Société Geographique de Liège* **25**: 145–57. [20.2]

Bravard JP, Amoros C, Pautou G. (1986) Impacts of civil engineering works on the succession of communities in a fluvial system: a methodological and predictive approach applied to a section of the Upper Rhône River. *Oikos* **47**: 92–111. [20.4]

Bravard JP, Lebot Helly A, Helly B, Savay-Guerraz H. (1990) Le site de Vienne (38), Saint-Romain (69), Sainte-Colombe (69). L'évolution de la plaine alluviale du Rhône, de l'Age du Fer à la fin de l'Antiquité: proposition d'interprétation. In: *Archéologie et Espaces*, Xè Rencontres Internationales d'Archéologie et Histoire d'Antibes. [20.2]

Bretschko G, Klemens W. (1986) Quantitative methods and aspects in the study of the interstitial fauna of running waters. *Stygologia* **2**: 297–305. [20.4]

Bureau de Recherches Géologiques et Minières. (1970) *Atlas des Eaux Souterraines de la France*. Rhône-Alpes, BRGM. [20.2]

Carbiener R, Trémolières M, Mercier JL, Ortscheit A. (1990) Aquatic macrophyte communities as bioindicators of eutrophication in calcareous oligosaprobe stream waters (Upper Rhine plain, Alsace). *Vegetatio* **86**: 71–88. [20.4]

Castany G. (1985) Relations hydrauliques entre les aquifères et les cours d'eau: système global aquifère/rivière. *Stygologia* **1**: 1–27. [20.4]

Creuzé des Châtelliers M. (1991). *Dynamique de répartition des biocénoses interstitielles du Rhône en relation avec des caractéristiques géomorphologiques*. Thèse, University of Lyon. [20.4]

Creuzé des Châtelliers M, Reygrobellet JL. (1990) Interactions between geomorphological processes in benthic and hyporheic communities: first results on a by-passed canal of the French Upper Rhône River. *Regulated Rivers: Research and Management* **5**: 139–58. [20.4]

Danielopol D. (1976) The distribution of the fauna in the interstitial habitats of riverine sediments of the Danube and the Piesting (Austria). *International Journal of Speleology* **8**: 23–51. [20.4]

Danielopol D. (1989) Groundwater fauna associated with riverine aquifers. *Journal of the North American Benthological Society* **8**: 18–35. [20.4]

Dole MJ. (1983) Le domaine aquatique souterrain de la plaine alluviale du Rhône à l'Est de Lyon. 1—Diversité hydrologique et biocénotique de trois stations représentatives de la dynamique fluviale. *Vie et Milieu* **33**: 219–29. [20.4]

Dole MJ. (1985) Le domaine aquatique souterrain de la plaine alluviale du Rhône à l'Est de Lyon. 2—Structure verticale des peuplements des niveaux supérieurs de la nappe. *Stygologia* **1**: 270–91. [20.4]

Dole MJ, Chessel D. (1986) Stabilité physique et bio-

logique des milieux interstitiels. Cas de deux stations du Haut-Rhône. *Annales de Limnologie* **22**: 69–81. [20.4]

Engalenc M. (1982) Hétérogénéité des formations alluviales. Colloque sur les milieux discontinus en Hydrogéologie, Orléans, November 1982: Jubilé Castany, 283–290, BRGM. [20.2]

Faucher D. (1968) *L'Homme et le Rhône*. NRF, Gallimard, Paris. [20.3]

Gemaehling MC. (1969) Drainage of the alluvial plains of the lower Rhône protected by embankments. International Commission on Irrigation and Drainage, 7th Congress, Mexico 1969. Question No. 25, 23–40. [20.2]

Gibert J, Ginet R, Mathieu J, Reygrobellet JL, Seyed-Reihani A. (1977) Structure et fonctionnement des écosystèmes du Haut-Rhône français, IV—Le peuplement des eaux phréatiques, premiers résultats. *Annales de Limnologie* **13**: 83–97. [20.4]

Gibert J, Ginet R, Mathieu J, Reygrobellet JL. (1981) Structure et fonctionnement des écosystèmes du Haut-Rhône français, IV—Analyse des peuplements de deux stations phréatiques alimentant des bras morts. *International Journal of Speleology* **11**: 141–58. [20.4]

Grove J. (1987) *The Little Ice Age*. Methuen, London. [20.2]

Henry M. (1963) Contribution à l'étude des nappes phréatiques. *Annales des Ponts et Chaussées* No. 5. Imprimerie Nationale, Paris. [20.2, 20.4]

Lafoucrière J, Varaschin A, Varaschin D. (1987) Le Haut-Rhône, force et lumière. *La Luiraz*, Hauterville. [20.3]

Lentheric P. (1892) *Le Rhône, Histoire d'un Fleuve*. Plon, Paris. [20.3]

Le Roy Ladurie E. (1983) *Histoire du Climat depuis l'An Mil*. Flammarion, Paris. [20.2]

Mandier P. (1988) *Le relief de la moyenne vallée du Rhône au Tertiaire et au Quaternaire. Essai de synthèse paléogéographique*. Documents du BRGM No. 151. 3 vol., BRGM Editions, Orleans. [20.2]

Marmonier P, Dole MJ. (1986) Les Amphipodes des sediments d'un bras court-circuité du Rhône: logique de répartition et réaction aux crues. *Sciences de l'Eau* **5**: 461–86. [20.4]

Pardé M. (1925) *Le régime du Rhône, Etude hydrologique*. Etudes et Travaux de l'Institut des Etudes Rhodaniennes, Lyon. [20.2]

Pardé M. (1933) *Fleuves et Rivières*. A. Colin, Paris. [20.1]

Pautou G, Décamps H. (1985) Ecological interactions between the alluvial forests and hydrology of the Upper Rhône. *Archiv für Hydrobiologie* **104**: 13–37. [20.4]

Pennak KW, Ward JV. (1986) Interstitial fauna communities of the hyporheic and adjacent groundwater biotopes of a Colorado mountain stream. *Archiv Für*

Hydrobiologie **74**(Suppl.): 356−96. [20.4]

Pinay G, Décamps H. (1988) The role of riparian woods in regulating nitrogen fluxes between the alluvial aquifer and surface water: a conceptual model. *Regulated Rivers* **5**: 507−16. [20.4]

Poinsart D, Bravard JP, Caclin-Brüser MC. (1989) Profil en long et granulométrie du lit des cours d'eau aménagés: l'exemple du Canal de Miribel (Haut-Rhône). *Revue Géographie Lyon* **64**: 240−51. [20.4]

Pont D, Persat H. (1990) Spatial variability of fish community in major central European regulated river. Symposium on Floodplain Rivers, 9−11 April 1990, Baton Rouge, Louisiana. [20.4]

Reygrobellet JL, Dole MJ. (1982) Structure et fonctionnement des écosystèmes du Haut-Rhône français, XVII−Le milieu interstitiel de la lône du Grand Gravier; premiers résultats hydrologiques et faunistiques. *Polski Archiwum Hydrobiologii* **29**: 485−500. [20.4]

Rivet F. (1962) *La navigation à vapeur sur la Saône et le Rhône (1783−1863)*. Presses Universitaires de France, Paris. [20.3]

Rostan JC, Amoros C, Juget J. (1987) The organic content of the surficial sediment: a method for the study of ecosystems development in abandoned river channels. *Hydrobiologia* **148**: 45−62. [20.4]

Rouch R. (1988) Sur la répartition spatiale des Crustacés dans le sous-écoulement d'un ruisseau des Pyrénées. *Annales de Limnologie* **24**: 213−34. [20.4]

Roux AL. (1984) Impacts of emptying and cleaning reservoirs on the downstream Rhône physicochemical and biological water quality. In: Lillehammer A, Saltvern SJ (eds) *Regulated Rivers*, pp 61−70. Oslo. 61−70. [20.3]

Roux C, Tachet H, Bournaud M. (1992) Stream continuum and metabolic rate of five species of *Hydropsyche* larvae (Trichoptera). *Holarctic Ecology*. **15**: 70−76. [20.4]

Salvador PG, Bravard JP, Vital J, Voruz JL. (1989) Archaeological evidence for Holocene floodplain development in the Rhône Valley, France. 2nd Congress of Geomorphology, Francfort, 1989. *Zeitschrift für Geomorphologie* (in press). [20.2]

Savey P, Deleglise R. (1967) Les incidences de l'aménagement du Tiers central du Bas-Rhône sur les transports solides par charriage. *Assosiation International des Sciences Hydrologiques* **75**: 449−61. [20.2]

Schumm SA. (1977) *The Fluvial System*. John Wiley & Sons, New York. [20.2]

Stanford JA, Ward JV. (1988) The hyporheic habitat of river ecosystems. *Nature* **335**: 213−34. [20.4]

Starkel L. (1983) The reflection of hydrologic changes in the fluvial environment of the temperate zone during the last 15 000 years. In: Gregory KJ (ed.) *Background to Palaeohydrology, a Perspective*, pp 213−35. John Wiley & Sons, Chichester, New York. [20.2]

Tachet H, Pierrot JP, Roux C, Bournaud M. (submitted to *Oikos*) Current velocity, longitudinal distribution and net-building behaviour of six species of *Hydropsyche* (Trichoptera). [20.4]

Tournier G. (1952) *Rhône, dieu conquis*. Plon, Paris. [20.3]

Vivian H. (1989) Hydrological changes of the Rhône river. In: Petts G, Möller H, Roux AL (eds) *Historical Change of Large Alluvial Rivers, Western Europe*, pp 57−77. John Wiley & Sons, New York. [20.3]

Ward JV. (1989) The four-dimensional nature of lotic ecosystems. *Journal of the North American Benthological Society* **8**: 2−8. [20.4]

Winghart J, Chabert J. (1965) Haut-Rhône à l'amont de Lyon: étude hydraulique de l'île de Miribel Jonage. *La Houille Blanche* **7**: 1−21. [20.4]

21: The Orinoco:
Physical, Biological and Cultural Diversity
of a Major Tropical Alluvial River

E. VASQUEZ AND W. WILBERT

12.1 INTRODUCTION

Since the 1970s, when the systematic research of the limnology and fisheries of the Orinoco River first began, a variety of analytical works have appeared that critically review the advances made over the past two decades, and suggest new areas of investigations based on their observations (Novoa 1982; Vásquez 1989a; Colonnello 1990). To facilitate the exchange of scientific data concerning the Orinoco, a symposium was dedicated to the river in 1986 (The Orinoco Ecosystem: Present Knowledge and the Need for Future Studies), the proceedings of which were published in 1990 (Weibezahn et al 1990). In 1990 a second major symposium was held in Venezuela (International Symposium on Major Latin American rivers) in which the Orinoco again figured as a major topic of discussion (Colonello 1990; Hamilton & Lewis 1990; Lewis et al 1990b; Machado-Allison 1990; Novoa & Ramos 1990; Paolini 1990).

The investigation of the social and ecological aspects of the Orinoco system designed to generate baseline data and environmental impact studies related to the exploitation of oil reserves has received considerable national support from the Venezuelan oil industry and the Ministry of the Environment and Renewable Resources (Ministerio del Ambiente y de los Recursos Naturales Renovables, MARNR). As a matter of policy the latter government agency has placed a special emphasis on the developing of sound land use and management strategies for the Basin as a whole (MARNR 1982). In spite of the enormous dimensions of the Orinoco Basin, research in this region has shown some advances, principally due

to the merging of national and international research teams which are collaborating to address the multifaceted complexity of the Basin.

This chapter examines the available ecological data concerning the Orinoco River within a broad and holistic framework that includes basic scientific knowledge, historical and current human interactions with the Basin environment, as well as the current management plans for the area. We shall analyse the data in terms of present-day limnological concepts such as the flood pulse (Junk et al 1989). However, our main goal is to provide a review of the Basin that will be of use to scientists, managers, politicians and students dedicated to the functioning and conservation of this major tropical river.

12.2 THE ORINOCO BASIN

The Amazon, the Orinoco and the Magdalena are the largest tropical floodplain river systems in South America. The upper sections of the La Plata Basin are located in a tropical climatic region where as the lower sections are temperate. In general, most South American systems demonstrate a tropical unimodal flood regime which is also typical of most African rivers. The Amazon is the largest hydrographic basin in the world. It is comprised of some 7000000 km², of which approximately 5000000 km² are covered by tropical rain forests. Second in ranking in South America is the La Plata Basin (3000000 km²) of which the Paraná and Paraguay river systems represent 84% (Roberts 1973; Quirós 1988; Welcomme 1990).

The Orinoco Basin extends over some 1100000 km². It is shared by Venezuela (70%)

and Colombia (30%) (Fig. 21.1). The main stem of the Orinoco is 2060 km long with an average water discharge in excess of 36 000 m^3 s^{-1} (1.1 × 10^{12} m^3 year^{-1}) which ranks the river third largest on a global scale after the first-ranked Amazon (6.3 × 10^{12} m^3 year^{-1}) and the second-ranked Zaire (1.3 × 10^{12} m^3 year^{-1}) (Milliman & Meade 1983). The Orinoco River receives its waters from a complex hydrographic system composed of a mosaic of landscapes, geoformations and biotopes under the influence of such factors as the geological and geomorphological make-up, soil types, climates and vegetation cover of the Basin (Weibezahn 1990). This spatial heterogeneity, together with the fact that the Orinoco is one of the few remaining rivers of the world draining a minimally human-altered basin, makes it an area of enormous interest for the investigation of its biological and physical processes.

Geology and physiography

The major physiographic landscapes drained by the Basin can be classified into three well-defined units: (1) the Guayana Shield to the south; (2) the Venezuelan and Colombian Andes to the west and north-west; and (3) the alluvial plains (Llanos) located to the north and west of the river (Figs 21.2 & 21.3). The Guayana Shield is one of the world's oldest geological formations (principally composed of early or middle Proterozoic origin (Gibbs & Barron 1983)) which covers almost 50% of the Venezuelan territory. Four geological provinces (the Imataca, Pastora, Cuchivero and the Roraima) are situated on the shield. According to González de Juana and colleagues (1980), Gibbs and Barron (1983), and Yanes and Ramírez (1988), the Imataca Province to the north is formed by rock of a sedimentary origin with metamorphic degrees corresponding to the granulite and amphibolite facies. The ages of these formations are 3.5–3.6 Ga (1 Ga = 10^9 years). The lower sections of the Caroní, Aro and Caura rivers are situated in the Imataca Province. The Pastora Province is located south of the Imataca complex and situated principally in the Cuyuni River Basin. It is integrated by volcanic rock metamorphosed to the degree of the amphibolite and green

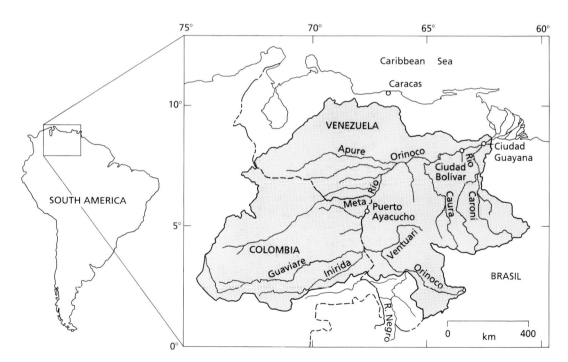

Fig. 21.1 The Orinoco River Basin.

450 *Examples*

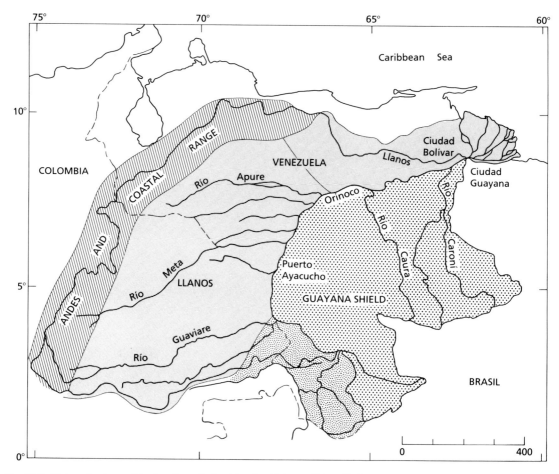

Fig. 21.2 Major physiographical regions drained by the Orinoco River Basin.

schists. The estimated age of the Pastora Province is c. 2.1 Ga. The Cuchivero Province is located on the western side of the shield. It includes vast extensions of plutonic rocks, a series of volcanic rocks of Proterozoic age (2.1–1.4 Ga) and a lesser proportion of sedimentary rocks, all of which are metamorphosed to the facies of green schists. Most of this province is found in the Paragua, Caura, Cuchivero and Aro river basins. The Roraima Province (Proterozoic age, 1.6 Ga) is located south of the Pastora Province and includes a large sedimentary layer of continental origin which covers most of the Shield. The Roraima Province is mainly found in sections of the Caroní and Paragua rivers. The Andes and the

Llanos belong to more recent ages of the Tertiary and Quaternary. They are comprised of schists, gneisses and intrusive granites covered by sandstones, lutites, slates and limestones (González de Juana *et al* 1980). The Llanos and the deltaic plain occupy some 300 000 km², a third of the Venezuelan territory (Carpio 1981). They separate the Andes Mountain Range and the Coastal Mountain Range from the Guayana Shield. The relief of the Llanos is neither uniform nor horizontal. From 200 m asl in the Andes and Coastal ranges, the relief slowly descends to the level of the Orinoco with an approximate slope of one per one thousand. The 100-m above sea level (asl) elevation is agreed upon as distinguishing

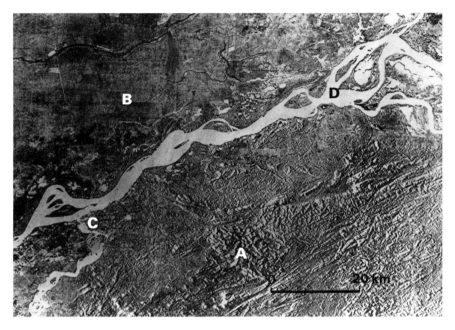

Fig. 21.3 Section of the Lower Orinoco and the Delta recorded by scanning radar showing: (A) relief of the Guayana Shield; (B) relief of the Llanos; (C) lower course of the Caroní River; (D) apex of the Orinoco Delta (scale 1:250 000, reduced). (Courtesy of Petróleos de Venezuela, S.A.; photograph taken in December 1977.)

the Upper Llanos from the Lower Llanos. In practice, however, the Lower Llanos are referred to as those that are seasonally inundated (Sarmiento 1983). To the south of the river are also found plain regions of erosive origin such as the peniplains of the Ventuari and Casiquiare rivers.

Climate and vegetation

Precipitation constitutes the major climatic factor within the tropics (Boadas 1983). In Venezuela, the meridional migration of the Intertropical Convergence Zone leads to a dry (February–May) and a wet season (August–October) north of about 5°N. In the Orinoco Basin, precipitation tends to increase towards the south and east (Fig. 21.4). Approximately 40% of the Venezuelan territory receives a mean annual precipitation >2000 mm. In areas south of the Orinoco, precipitation may surpass 3600 mm year^{-1}, while in its delta it may exceed 2000 mm year^{-1} (Instituto Venezolano del Petróleo 1979). According to Koeppen's climatic

classification, most of the basin experiences three climate types: (1) forest; (2) tropical monzonic; and (3) savannas (MARNR 1979). The forest climate is characterized by high temperatures throughout the year, intense precipitation and a lack of a well-defined dry season. Generally, this climate is encountered below the sixth parallel, south of the basin. The tropical monzonic type climate is found in the Orinoco Delta and northeast of the Guayana Shield. This climate is characterized by high precipitation and a short dry season. The savanna climate, generally registered north of the sixth parallel (Hernández 1987), exhibits high temperatures throughout the year with a dry season extending from December to March.

Ewel *et al* (1976) indicated that the Llanos region north of the Orinoco is dominated by a life zone of tropical dry forests. South of the Orinoco, however, the situation is more complex. This region exhibits humid tropical forests, humid premontane forests, very humid premontane forests and pluvial premontane forests.

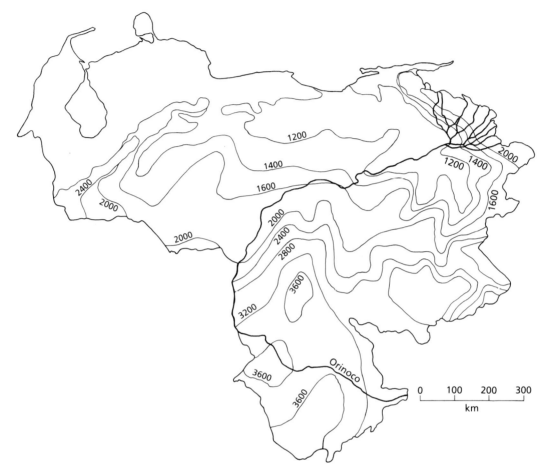

Fig. 21.4 Precipitation isobars (units = mm) in the Orinoco River Basin in Venezuela (after MARNR 1979).

Owing to the seasonal variation in rainfall, there is a pronounced annual water level fluctuation of the Orinoco River (Fig. 21.5), a contrasted or unimodal hydrological regime which is typical for most Venezuelan rivers. High waters are generally registered between June and November. Peak levels have been observed in August, with average oscillations (in Ciudad Bolívar) between 29 and 30 m asl. The lowest water levels are most often recorded in March with average values between 16 and 17 m, suggesting a mean amplitude approximating 13 m (Zinck 1977).

21.3 THE ORINOCO AND ITS FLOODPLAINS

The course of the Orinoco is usually divided into four sections: the Upper, Middle, Lower and the Delta. With the exception of the Delta, however, researchers differ as to the limits of these divisions. The source of the river is located at 1047 m asl on the western flank of the Sierra de Parima (63° 21' 42" W, 2° 19' 05" N). Vila (1960) states that the Upper Orinoco extends from its source to the hydrological knot of the Atabapo–Guaviare rivers. An analysis by Weibezahn (1990) agrees with the latter. On the other hand, the

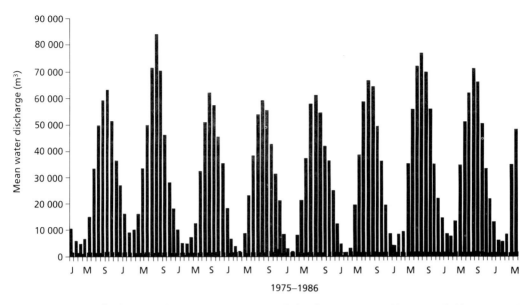

Fig. 21.5 Mean water discharge of the Orinoco River at Ciudad Bolívar, 1975–86 (data provided by MARNR).

Comisión del Plan Nacional de Aprovechamiento de los Recursos Hidráulicos (1969, 1972) has proposed the confluence of the Casiquiare River as the inferior limit of the Upper Orinoco, while Hernández (1987) has placed this limit at the Guaharibos rapids. Given these contradictory delimitations, we propose to redefine the Upper, Middle and Lower Orinoco based upon the four major physiconatural landscape units identified by Boadas (1983) for the Venezuelan Amazon and other pertinent physiographic and hydrographic criteria (Fig. 21.6).

The Upper Orinoco extends from its source to a region just prior to the Mavaca River. Here the river drains a landscape of precambrian highlands characterized by a dense forest (mainly evergreen submontane rain forest) (Boadas 1983; Huber & Alarcón 1988). In the same stretch the river flows through a well-defined valley with pronounced topographic unevenness (Zinck 1977).

The Middle Orinoco extends from the proximity of the Mavaca River down to the mouth of the Atabapo River. This stretch is characterized by a landscape of alteration and erosion plains (i.e. peniplains) with a predominance of medium and dense forest (evergreen rain forest partially subject to floodings and a complex of transitional forests

between the rain forest and the Amazonian caatinga) (Boadas 1983; Huber & Alarcón 1988). Physiographically, the Middle Orinoco region shows a marked relief dissymmetry between its right and left margins (Zinck 1980). On its right, the river borders are high land sections while on the left side there is a predominance of peniplains which extend down to the Atabapo confluence. From this point, the river passes through the alluvial plain landscape of the Llanos and receives the waters of the Guaviare River, the first tributary draining the Andes and the Llanos.

From its source, down to the Atabapo–Guaviare hydrological knot, the Upper and Middle Orinoco receive, from the north, the waters of tributaries draining the Guayana Shield (Ocamo, Padamo, Kunukunuma and Ventuari) and, from the south, the waters of rivers draining the alteration and erosion plains originating from the granite rock of the Shield. At the inferior limit of the Middle Orinoco, its waters show a high transparency and low sediment load (Weibezahn 1985).

This situation changes radically with the inflow of the Llanos rivers in the Lower Orinoco. Meade *et al* (1990) indicated that out of the 150×10^6 t year^{-1} of suspended sediments transported by the Orinoco River to the Delta, some

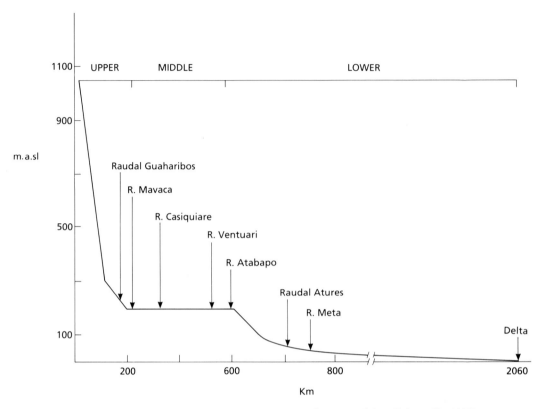

Fig. 21.6 Gradient of the Orinoco River from its source at Sierra de Parima (after Colonnello 1990).

90% is introduced by the Andean and Llanos rivers (the Meta, 50%; the Guaviare, 20%; and the Apure, 20%). The remaining 10% is introduced via the remainder of the tributaries. Only *c.* 5% of suspended sediment in the Orinoco is derived from the Guayana Shield (Meade *et al* 1990). As a consequence of the mentioned geological characteristics and the nature of the soils, the waters draining the Guayana Shield are poor in electrolytes (conductivity <20 µS cm^{-1}), exhibit a low suspended load (<50 mg l^{-1}) and high concentrations of dissolved organic carbon which usually confers a dark coloration to the waters (blackwaters). However, waters draining the Andes and the Llanos have a higher electrolyte content (conductivity >70 µS cm^{-1}) and a higher suspended load (>100 mg l^{-1}) (Weibezahn 1987). In consequence, they are referred to as whitewater rivers. The Lower Orinoco extends from the Atabapo–Guaviare knot to the apex of the delta.

Along this stretch the Orinoco first receives the waters of three important tributaries (Guaviare, Vichada y Meta) originating from the Andes, the western Llanos and the Amazon plains of Colombia, respectively. Further downriver, additional significant tributaries (Capanaparo, Arauca y Apure) join the Orinoco after draining sections of the Colombia and Venezuelan Llanos. In general, the Llanos rivers have poorly defined river beds which tend to change position during the inundations. In the Lower Orinoco, the slope is weak, falling to *c.* 4–5 m per 100 km on the 490-km stretch between Caicara and Ciudad Bolívar (MARNR 1979); water velocity is also slow (1–2 m s^{-1}), but the flow is high. Along much of its length alluvial fans of the Andean and Llanos tributaries maintain the course of the Lower Orinoco along the Guayana Shield. On this stretch the river flows in alluvial channels, and bed controls are normally encountered. At low water, sandwaves and bars may be observed along large areas of the exposed river bed (Nordine & Pérez-Hernández 1989).

The Delta is a fan of alluvial deposits situated between the northern coastal range of Venezuela and the Guayana Shield margin to the south, with an expanse of some $22\,500$ km^2 (Wilbert 1979). The deltaic plain is formed by very low lands (slope <1% and elevation asl <10 m). It is drained by nine major distributaries and many interconnecting smaller canals, 'caños', resulting in numerous islands and temporary as well as permanent lakes (van Andel 1967; MARNR 1979). The region is characterized by arbustive formations with a predominance of middle and low forests, palm trees and mangroves (Canales 1985). The Río Grande is the principal distributary, discharging 84% of the Orinoco River flow. The Delta is usually divided into Upper, Middle and Lower Delta (Canales 1985). Flooding of the Upper Delta is closely linked to the seasonal inundations of the Orinoco. At the apex, the tidal fluctuation in water level is only 60 cm, while the seasonal river fluctuation reaches *c.* 9 m (van Andel 1967). Fluviomarine and marine processes predominate in the Middle and Lower Delta (Canales 1985). In the latter two regions, flooding is either prolonged or permanent as a result of the low slope (0–1%), higher precipitation and tidal dynamics which act twice upon most of the deltaic floodplain. In general, the tidal range in the delta is relatively small in comparison with the differences between high and low waters of the river itself (van Andel 1967). At high water the entire delta almost floods.

21.4 CHEMICAL ASPECTS OF THE ORINOCO RIVER WATERS AND ITS MAJOR TRIBUTARIES

Recent hydrochemical and sediment research of the Orinoco River system include, among others, those of Lewis and Weibezahn (1981), Nemeth *et al* (1982), Paolini *et al* (1983), Weibezahn (1985, 1990), Sánchez and Vásquez (1986, 1989), Yanes and Ramírez (1988), Lewis and Saunders (1989, 1990) and Meade *et al* (1990). Physicochemical data from several blackwater rivers from the Orinoco and the Amazon Basins have been made available by Vegas-Vilarrúbia *et al* (1988), while

Sánchez (1990) has contributed the, thus far, only seasonal and longitudinal water quality study of the Lower Orinoco.

Previously available estimates of suspended sediments transported to the Delta by the Orinoco have proved to be inconsistent (Monente 1989). Recently, however, a revised estimate of the mean suspended-sediment discharge conducted by Meade *et al* (1990) demonstrated a value of 150×10^6 t year^{-1} with an error estimate of approximately 30–50%. According to this new value, the Orinoco ranks tenth on a global scale. Out of the total of 150×10^6 ($\pm 50 \times 10^6$) t year^{-1} transported by the river, 85–90% is contributed by three large tributaries (Meta, Guaviare and Apure) draining the Andes and the Llanos, while <5% is derived from tributaries draining the Guayana Shield. In contrast, water sources contributing to the mean annual discharge of 1.1×10^{12} m^3 ($36\,000$ m^3 s^{-1}) of the Orinoco system are evenly divided between the Andes, the Llanos and the Guayana Shield tributaries (Meade *et al* 1983, 1990). Approximately 73% of the Orinoco waters reaching its mouth originate from above the Apure River. The latter contributes 7% to the Orinoco while the Caura and Caroní Rivers contribute 9% and 11%, respectively (Lewis 1988).

In a study of 16 basin rivers including the Orinoco main stem, the lowest mean conductivity values were 7.0–25 µS cm^{-1} for rivers draining the Guayana Shield, while highest values were generally observed in rivers draining the Andes and the Llanos (55.3–200 µS cm^{-1}) (Yanes & Ramírez 1988). The Orinoco main stem demonstrates the influence of the various morphotectonic regions along its course with mean conductivity values of 13.3 µS cm^{-1} in its upper course, to 21.6 and 60 µS cm^{-1} on its lower course (Puerto Ayacucho and Ciudad Bolívar, respectively). There was also a strong negative correlation (>99.9%) between conductivity and water discharge along the main stem (Weibezahn 1985). Orinoco waters are slightly acid to slightly alkaline, ranging between a pH value of 5.6 and 7.5 (Sánchez & Vásquez 1986). Weibezahn (1990) found a significant relationship between pH and water discharge in the Orinoco main stem. A similar observation had previously been reported by Gessner (1965) and Sánchez and Vásquez (1986). In general, the tributaries from Guayana Shield show the lowest mean pH values in comparison with those of the Andes and the Llanos.

Sodium is the dominant cation in all the rivers draining the Guayana Shield and the Orinoco itself before merging with the Andean and Llanos tributaries (Yanes & Ramírez 1988). The most common orders of cations by concentration in the Shield rivers (Caroní and Caura Rivers) are $Na > Ca > K > Mg$ and $Na > Ca > Mg > K$, respectively (Sánchez & Vásquez 1986; Lewis & Saunders 1990). Calcium dominates the series in the left margin tributaries and in the Orinoco main stem, following the influence of the Andean and Llanos waters. In the Lower Orinoco, the most common order in cation concentration is $Ca > Na > Mg > K$ (Sánchez & Vásquez 1986; Lewis & Saunders 1990), in the Apure River the concentration is $Ca > Mg > Na > K$ (Lewis & Saunders 1990), and in the Guaviare and Meta Rivers the order is $Ca > Na > K > Mg$ and $Ca > Na > Mg > K$, respectively. The highest concentrations of calcium in these rivers reflect the influence of the lithology on the chemical composition of these waters. The high Ca/Na relationships indicates the presence of sedimentary rock in their catchment areas (Yanes & Ramírez 1988). In the Orinoco, bicarbonate was found to be the principal component of anions (Sánchez & Vásquez 1986). The total organic carbon yield for the river has been estimated at 6.8×10^6 metric tons per year, which represents 1.6% of the global carbon flow to the oceans. While nitrogen yield is high (0.54×10^6 metric tons per year), phosphorus concentration is low (0.072×10^6 metric tons per year) (Lewis & Saunders 1990).

In sum, the whitewater rivers draining the Andes as well as the alluvial plains of the Llanos located between the Andes and the Guayana Shield contribute the greatest quantity of sediments transported by the Orinoco, followed by the blackwater tributaries draining the lithologically complex Precambrian rock of the Shield to the south. In comparison to whitewaters, blackwaters are characterized by a lower content of dissolved and suspended matter, nutrients and electrolytes. Their dark colour is caused by a relatively higher concentration of dissolved organic matter.

21.5 SUSPENDED ORGANISMS OF THE ORINOCO RIVER SYSTEM

Before receiving the waters from the Andes and the Llanos, the Chlorophyta generally dominate the phytoplankton species composition in rivers draining the Guayana Shield and in the Orinoco proper. In the Lower Orinoco, and particularly in the Apure, there exists a clear dominance of diatoms over other algae groups (Vásquez & Sánchez 1984; Carvajal-Chitty 1989; Lewis et al 1990a). The phytoplankton density and species composition of the Orinoco main stem and its major tributaries are influenced by the annual hydrological cycle, with the highest number of organisms per unit volume of water present during the period of low water discharge. In the Lower Orinoco, for example, Vásquez and Sánchez (1984) reported a peak phytoplankton abundance of $53\,485$ organisms l^{-1} at low water, which decreases to 429 organisms l^{-1} at high water. Similarly, in the Caroní River, the highest phytoplankton density ($12\,020$ organisms l^{-1}) was recorded at low water, and the lowest at high water (1022 organisms l^{-1}) (Sánchez & Vásquez 1989). Even though algal abundance per unit volume of water is higher at low water, transport of phytoplankton is higher at high water. Generally the biomass of suspended algae in the Orinoco is low. Levels of chlorophyll *a* along the lower main stem have been reported to average from 0.11 to 0.19 μg l^{-1}, corresponding to c. 10 μg l^{-1} algal carbon. In general, chlorophyll concentrations are much lower for rivers draining the Guayana Shield (0.01–0.04 μg l^{-1} chlorophyll *a* for the Caura and Caroní Rivers) than for the Apure River (mean 0.36 μg l^{-1} chlorophyll *a*) (Lewis et al 1990a). The lowest algal abundance and biomass values for Shield rivers are usually explained by their low nutrient concentration, limited water transparency and current speed (Sánchez & Vásquez 1989). In whitewaters such as those of the Apure River and in the Lower Orinoco, phytoplankton abundance is not limited by scarcity of nutrients but rather by low water transparency and current speed (Vásquez & Sánchez 1984; Lewis *et al* 1990a). Gross primary production in the lower Orinoco system shows a similar marked seasonality as that observed for algal biomass and abundance. Mean annual gross primary production has been found to be lower for the Shield rivers (Caura, 13 mg C m^{-2} day^{-1}; Caroní, 4 mg C m^{-2} day^{-1}) than for the Apure River (26 mg C m^{-2} day^{-1}) and along the Orinoco main stem where productivity varied by station, from 19 to 43 mg C m^{-2} day^{-1} (Lewis 1988). Suppression of potential production by dissolved and suspended matter in the Orinoco main stem ranges from 60% to 80% (Lewis *et al* 1990a).

In terms of numbers of species, the zooplankton of the Orinoco main stem is dominated by rotifers. In terms of biomass, however, cladocerans are more important (Vásquez & Sánchez 1984; Lewis *et al* 1990). A similar observation was reported for the Paraná River (Paggi & José de Paggi 1990). In a longitudinal study along the Lower Orinoco and the Delta, 100 rotifer taxa and 48 cladoceran taxa were identified (Vásquez & Rey 1989). Among the rotifers, the highest number of species was represented by *Brachionus*, followed by *Lecane* and *Trichocerca*. Cladocerans were mainly represented by species of Chydoridae, followed by Sididae, Daphnidae, Bosminidae, Macrothricidae and Moinidae. In spite of the relatively high number of zooplankton species in the main stem of the Orinoco, only a small number of species has been found to be frequent and numerically important (primarily *Keratella americana*, *Lecane proiecta*, *Ploesoma lenticulare* and *Polyarthra vulgaris*, among the rotifers, and *Bosmina tubicen*, *Bosminopsis deitersi*, *Diaphanosoma birgei* and *Moina minuta*, among clado cerans). In the river, naupliar stages are always more abundant than copepodites and adults. The dominance of rotifers is also characteristic of the Shield rivers and of rivers draining the Andes and the Llanos (Vásquez 1984; Vásquez & Sánchez 1984; Saunders & Lewis 1988a, 1988b). The major difference between these two groups of rivers is the abundance of organisms, being higher in rivers draining the Andes, the Llanos and in the Lower Orinoco main stem than in those draining the Shield. Attempts have been made to differentiate both groups of waters on the basis of their cladoceran and rotiferan species associations (Vásquez 1984; Rey & Vásquez 1986; Vásquez & Rey 1989). For example, the main stem of the Lower Orinoco is characterized by the dominance of the cladoceran association of the *Moina minuta*, *Ceriodaphnia cornuta*, *Bosmina tubicen* and *Bosminopsis deitersi*, while the Lower Caroní is dominated by species of the Bosminidae.

Similar to the phytoplankton data, the abundance of zooplankton in the Lower Orinoco system is closely related to hydrological events, showing an inverse relationship to water discharge. For example, in the Orinoco main stem and its delta, the mean low-water zooplankton density has been reported to be eight times higher than at high water (Vásquez & Rey 1989). Contrary to phytoplankton, highest transport of zooplankton in the river takes place at low water. In the Lower Orinoco main stem, the total annual transport of zooplankton biomass has been estimated to be 0.32×10^6 kg C year^{-1} (Lewis *et al* 1990), while that of phytoplankton has been estimated at 2.4×10^6 kg C year^{-1} (Lewis 1988). Considering that both density and transport of plankton are sustained or reach highest values at low water when the river is isolated from the floodplain, it appears that stagnant incubation areas in the river channel or areas where current speed is slow are very important as sources of organisms (Lewis *et al* 1990a). However, Vásquez and Rey (1989) found a high proportion of egg-carrying cladocerans and rotifers at low water, suggesting an ability of the species for growth and reproduction in the river itself.

The mean annual respiration rate in the Orinoco main stem attributable to suspended organisms has been estimated at 80 μg C l^{-1} day^{-1}. Some 75% of this amount is accounted for by bacteria and 25% by phytoplankton, while that of zooplankton is negligible. The Shield rivers (Caura and Caroní) have lower respiration rates in comparison to the Orinoco main stem and the Apure River. The low respiration rates do not account for significant changes in the total organic carbon transported by the Orinoco River (Lewis 1988; Lewis *et al* 1990a).

21.6 ECOLOGICAL ASPECTS OF THE FLOODPLAIN OF THE ORINOCO RIVER

As is the case for Africa, South American river systems are characterized by the presence of extensive floodplains. For equations establishing the relationships between basin-area and main channel length of South American and African rivers, see Welcomme (1990). On both continents, tributary development and other geomorphological aspects show a high degree of similarity. Junk and colleagues (1989) proposed in the flood pulse concept that floodplains share many characteristics of productivity and function regardless of their position in the river. This proposal contrasts with the river continuum concept which supports the theory of a continuous transition in the functioning and productivity along the main river channel (Vannote *et al* 1980). Welcomme (1990), however, presents evidence from both African and South American rivers that supports the former concept.

In the Orinoco system there are three types of floodplains present (fringing, internal delta and deltaic floodplains) (Welcomme 1979). The first two types are found in the Lower Orinoco, while the delta constitutes the third. This section of the river is geologically and ecologically distinct from the other floodplain types of the same river (Hamilton & Lewis 1990). The internal deltaic floodplain (70 000 km^2) is located in the vicinity of the Apure and Arauca Rivers (Welcomme 1979). Hamilton and Lewis (1990) reported that, during peak floods, the area of the fringing floodplain covers *c.* 7000 km^2 and extends approximately 600 km (from the Meta River to the apex of the Orinoco Delta). This floodplain is larger to the north of the river, given that sediment transport is greater due to the loading from the turbid rivers draining the Andes and the Llanos. The extension of inundated floodplain varies seasonally by a factor of 10–15. At low water, some 2300 permanent floodplain lakes cover 5–12% of the Orinoco fringing floodplain. Their mean area varies from 0.17 to 0.26 km^2 (Hamilton & Lewis 1990). At high water, floodplain lakes lose their identities as separate water bodies and often merge into a continuous sheet of water that covers the entire floodplain (Vásquez 1989b). In a sample of 349 of these lakes, Vásquez (1988) estimated that at low water 77% of the lakes had a perimeter ranging between 100 and 200 m and the majority showed maximum lengths ranging from 100 to 600 m. In 63% of the lakes the maximum breadth ranged between 10 and 200 m. The shore-line development factor was low (1.0–1.6 in 60% of the lakes). Comparing the morphological features of six floodplain lakes at low and high water, Vásquez (1989b) reported that water depth showed the highest percentage of variability, ranging between 0.1 and 5.8 m with a mean of 1.6 m. An analysis of the relationship between the estimated surfaces at high and low water and the maximum and minimum water depth values revealed that for relatively small increments in water depths large extensions of land and forest adjacent to the lakes are inundated.

The continuous modelling of the floodplain is under the dynamic influence of erosive and sedimentary processes leading to the formation of a complex floodplain composed of islands, bars, levees, floodplain lakes and swamps, which are in turn permanently modified by continuous erosion and sedimentation (Figs 21.7 & 21.8). According to their origins, Drago (1976, 1989) classified floodplain lakes of the Paraná River as obstruction lakes, levee lakes, lateral expansion lakes, interbar lakes, overflow lakes, annexation lakes and swamps. In the fringing floodplain of the Orinoco River, all of these types are present even though the lateral expansion and overflow floodplain lakes are particularly numerous (Vásquez 1988). In addition to the main river channel, there exist smaller channels (caños), some of which act as either permanent or temporary connections between the lakes and the river. Connections between the floodplain lakes and the main river channel may be either direct or indirect (Drago 1989). In the former situation communication is established by means of a mouth or a short channel, while in the latter, communication is established through longer channels. In the latter, intermediate lakes or swamps may also be encountered. At high water, timing and intensity of mixing of the river with the lake waters are strongly influenced by these types of connections, given that river water displaces differentially

Fig. 21.7 Relief of the fringing floodplain landscape showing the floodplain forest, lakes and side branches of the Lower Orinoco (Courtesy of L. Pérez).

across the floodplain (Drago 1989; Vásquez 1989b).

Hamilton and Lewis (1987) distinguished four hydrological phases in the floodplain lakes of the Orinoco: (1) the filling phase — the time at which river waters first begin to enter and fill the lakes; (2) the through-flow phase — a simultaneous flow of river water into the lakes and lake water into the river. At this phase the mixing of river and lake waters reaches its maximum; (3) the drainage phase — when the inflow of river water ceases but the lakes continue to flow into the river; and (4) the isolation phase — when communication between the river and the lakes ceases completely.

These phases coincide well with those described by Drago (1989) for the Paraná floodplain system (bankfull rising-water, over bank, bankfull falling-water and isolation phases). The timing of these phases in the different floodplain lakes will vary according to their origin and connection degrees. Figure 21.9 shows water depths for three flood-plain lakes on the Lower Orinoco. In general, the waters of the river flow through most of the floodplain between July and October. The lakes generally reach their maximum depths in August and September. The waters of the lakes drain into the river during November and December, following which the lakes become isolated and lake volume decreases by evaporation (Hamilton & Lewis 1987; Vásquez 1989b).

The vegetation cover of the fringing floodplain of the Orinoco is dominated by trees and herbaceous plants, which may be classified as a seasonal várzea forest according to Prance's (1979) classification for the Amazon (Colonnello 1990). The plant cover of the different regions of the floodplain depends on factors such as age, soil texture and the periodicity of flooding (Junk 1980, 1984). For example, in areas of the Orinoco which experience long periods of inundation, the diversity of trees (some 40 species) is lower than in non-flooded forests. Those areas with a long period of inundation are dominated by species such as *Piranhea trifoliata*, *Homalium racemosum*, *Pterocarpus* sp., and *Symmeria paniculata*. In areas where flooding is less severe there are species such as *Ceiba pentandra* and *Spondias mombin*, characteristic of non-flooded forests (Colonnello 1990). All plant species from the floodplain are adapted to the seasonal inundation of the floodplain. The adaptations may be morphological and metabolic. For example, trees may extend lenticels and adventitious roots whose abundance is closely linked to the levels of the inundation. The trees may also develop metabolic mechanisms to counteract the action of toxic substances produced as a result of the anoxic conditions in the roots. During inundation most of the species were observed to continue producing leaves, flowers and fruits (Colonello

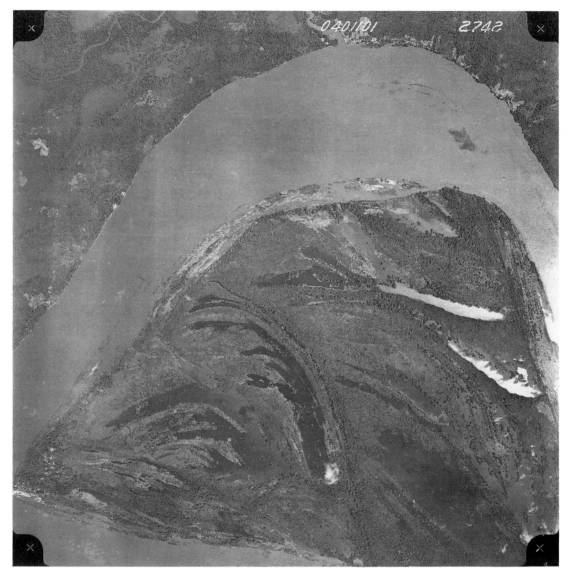

Fig. 21.8 Aerial photograph of a channel island in the Lower Orinoco showing floodplain lakes formed by the downstream evolution of the island (photograph courtesy of MARNR, Dirección de Cartografía Nacional).

1990). Similar adaptations have been reported for flood-plain areas in the Paraná River (Neiff 1990). In the water bodies of the Orinoco floodplain, aquatic macrophytes are important primary producers and they usually compete with phytoplankton for nutrients and light (Junk 1984). Large amounts of aquatic macrophytes are produced in the Orinoco floodplain. *Eichhornia crassipes*,

Paspalum repens and *Oxycaryum cubense* are the most frequent and dominant species. Aquatic macrophytes are mainly annual plants which reproduce during the period of inundation while at low water growth and plant cover is greatly reduced (Sánchez & Vásquez 1986). Macrophyte mats act in the fringing floodplain of the Orinoco have been found to be particularly important in

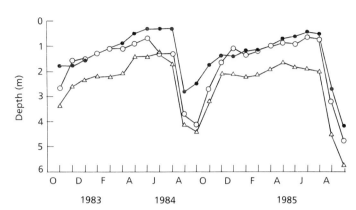

Fig. 21.9 Seasonal water depth oscillations in three floodplain lakes of the Lower Orinoco. ●, Playa Blanca; ○, Jobera; △, Orsinera.

the processing efficiency of the floodplain. They retain plankton passing between lakes (Hamilton *et al* 1990) and support large densities of periphytic organisms (Blanco 1989).

The hydrochemistry of floodplain areas is under the influence of the annual flood pulse of entering river water and internal biotic and abiotic processes which modify river water quality (Junk 1984). In the Orinoco, for example, floodplain lakes are annually flushed by riverine water leading to a convergence of their chemical conditions. After the inundation, the chemistry of the lakes diverges under the influence of factors such as basin morphology, sedimentation and decomposition processes, water−sediment interactions, and uptake and release of substances through the biota (Junk 1984; Hamilton & Lewis

1987). In Lake Mamo, the largest floodplain lake of the Lower Orinoco (18 km²), the lowest values of dissolved oxygen, pH and turbidity are measured during the earliest stages of the inundation. Similarly, phytoplankton and zooplankton densities are lowest during this phase. At low water, this situation reverses and dissolved oxygen, pH and turbidity values increase. Dissolved oxygen shows marked seasonal oscillations among the different lakes (Vásquez & Sánchez 1984; Hamilton & Lewis 1987) (Fig. 21.10). The magnitude of these variations seems to be influenced by the basin morphology, degree of connection between the lake and the river and the presence of other inflowing water sources. Generally, lowest dissolved oxygen concentration is found at the onset of the inundation. This

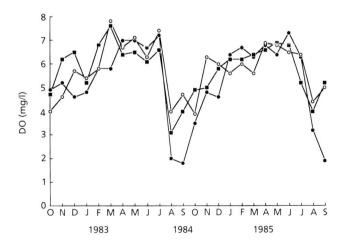

Fig. 21.10 Seasonal dissolved oxygen (DO) oscillations (surface) in three floodplain lakes of the Lower Orinoco. ●, Playa Blanca; ■, Orsinera; ○, Jobera.

decrease is associated with the large amounts of organic matter and detritus that become available for decomposition at the high-water phase. Throughout most of the year, surface waters of the lakes are generally below saturation levels. With increasing water levels, aquatic vegetation flourishes and the dissolved oxygen below these mats is severely depleted (Vásquez 1989a). Decomposition of large amounts of organic material in the Amazon floodplain during inundation leads to a high oxygen consumption. Anoxic, and most often hypoxic, conditions are frequently found together with H_2S and CH_4 production (Junk 1984; Junk *et al* 1989).

While diatoms dominate the phytoplankton in the river throughout the annual cycle, in the lakes blue-green algae dominate at low water when highest phytoplankton densities are recorded (Vásquez & Sánchez 1984; Hamilton & Lewis 1987). In Lake Tineo, Hamilton and Lewis (1987) found aquatic macrophytes and phytoplankton to cause a depletion of inorganic N and P during the inundation phase. At low water, high densities of blue-green algae were found to fix N and increase its concentration in the lake. In this case the conversion of inorganic N and P to organic forms was found to be the major alteration of river water in the lake. While floodplains have been considered as sources of carbon for the main river channel, other evidence suggests that floodplains may in fact act as closed systems in which carbon is internally recycled (Junk 1984; Junk *et al* 1989). In the Orinoco, organic carbon and fixed nitrogen seem to be efficiently metabolized within the floodplain in a closed system of synthesis and degradation (Lewis *et al* 1990b). Retention mechanisms such as the stranding of aquatic macrophytes at low water, and the stranding of organisms by macrophyte mats, contribute to reduce carbon export to the main channel. In two interconnected lakes of the Orinoco floodplain, Hamilton *et al* (1990) found a severe reduction of plankton density due to the passage of water through macrophyte beds. The stranding of organisms contributed to increased planktonic production in the epiphytic and benthic habitats, thus reducing export of organisms to the main river.

Fish communities from both South American and African river systems are highly complex. Welcomme (1990) provides information on the number of fish species as related to basin area in a number of rivers from both continents. Fish yields have also been shown to be related to the strength of the flood pulse in previous years to the capture of fish (Junk *et al* 1989). While some 2000 fish species have been reported for the Amazon Basin, some 450 species and subspecies have been reported for the Orinoco River (Novoa 1990; Welcomme 1990). In the Orinoco main stem, reports reveal the presence of a very diverse and productive fauna with dominance of electric fishes (Gymnotiformes) and catfishes (Siluriformes) (López-Rojas *et al* 1984). In floodplain lakes of the Lower Orinoco and Lower Caura Rivers, up to 170 fish species have been found dominated by the Characiformes and Siluriformes, followed by Perciformes and Gymnotiformes (Lasso 1988; Rodríguez & Lewis 1990). Species richness was found to be substantially lower in the blackwater lakes of the Caura River in comparison with the whitewater lakes of the Orinoco (Rondríguez & Lewis 1990). In general, fish species of the floodplain have one annual reproductive phase synchronized with the onset of the inundations. This reproductive phase coincides with the presence of abundant food supply in terms of plankton, insect larvae, and seeds and fruits produced by trees from the flooded forest. Fish species show a marked tendency to be euryphagous; some, however, such as the prochilodontids, show certain feeding specificity, their diet being based mainly on clay and associated detritus (Novoa 1990). Mats of rooted and floating aquatic vegetation form the most important nursery grounds for fish species in the floodplain, providing the larvae and juveniles with protection and an abundant food supply (Machado-Allison 1987, 1990). Novoa (1989) classified fish species from the Orinoco into three major groups. (1) Species which complete their entire biological cycles in the floodplain (*Cichla ocellaris* and *Hoplias malabaricus*). Species from this group generally show parental care. (2) Those that only partially accomplish their biological cycles in floodplain areas (*Prochilodus mariae, Semaprochilodus laticeps, Pseudoplatystoma fasciatum, P. tigrinum* and *Piaractus brachypomum*). These species generally spawn in the main river channel and their eggs and larvae are transported to flood-plain areas where they develop further. Most species from this group do not show parental care and the group includes most of the commercially valuable species. (3) Species restricted to the main channel (*Brachyplatystoma* sp., *B. vaillantii* and *Goslinea platynema*).

Species from all groups may show morphological, physiological and nutritional adaptations to the constantly changing conditions in flood-plain areas. For example, some fish species enlarge their lower lip under hypoxic conditions, thereby improving their uptake of oxygen from the surface layer of the water. As in other floodplain systems, extensive migratory movements have been reported for fish species in the Orinoco. Generally, the best areas for spawning are not so for feeding and growth. For this reason, migratory movements are made between these two areas. Normally, migration is established laterally between the floodplain and the main stem (lateral migrations) or between upstream and downstream areas (or vice versa) in the main stem (longitudinal migrations). In general, the proximity of the inundation phase leads to longitudinal migrations of some fish species which spawn before or during the inundation phase. When floodplain areas are connected to the main river, lateral migrations begin from the floodplain to the river leading to a wide dispersion of fish larvae, juveniles and adults (Machado-Allison 1990; Novoa 1990).

In general, floodplain areas provide habitats for a variety of wildlife, including currently endangered or threatened species. Colonnello (1990) has documented that a high percentage of the Venezuelan vertebrate terrestrial fauna is found in the Orinoco Basin. Some 1200 bird species, 300 mammals, and 300 reptiles and amphibians have been reported for this particular region. Most of the terrestrial fauna is present in the floodplain at low water. At high water, however, only some arboreal species and most of the birds remain in the inundated floodplain, and some species such as the freshwater dolphin (*Inia geofrensis*) and the manatees (*Trichechus manatus*) migrate into the floodplain in search of food, protection and reproduction grounds.

21.7 HUMAN PRESENCE IN THE ORINOCO BASIN: PAST AND PRESENT

Currently, 26 tribes inhabit the political boundaries of Venezuela, 16 of which (78 695 individuals) are situated within the Orinoco Basin. Six of the latter, i.e. the Warao, Kariña, Piaroa, Guajibo, Yanomami and the Pemon, are of a combined population size of 61 889 individuals, or nearly 79% of the total indigenous basin inhabitants (Oficina Central de Estadística e Informática (OCEI 1985). These population figures are low in relation to the expanse and population-sustaining potential of the basin, and are in fact misleading owing to a tremendous depopulation process that occurred as a consequence of the Contact and Post-contact eras during which an estimated 99% of the Amerindian populations were decimated through warfare, slavery and, above all, disease (Crosby 1973; Morey 1979; Murdock 1980). Although accurate figures are difficult to obtain, it is estimated that at the time of Contact, the Llanos population alone comprised 513 500 individuals (Denevan 1976). The current distribution within the confines of the basin demonstrates that of its 16 tribal societies, 12 live south of the Orinoco, two south of the Apure river, one in the Delta, and one in the state of Monagas (Kariña) north of the Lower Orinoco (OCEI 1985) (Fig. 21.11). The modern distribution pattern also demonstrates clearly that only those territories that have been inaccessible and/or economically 'undesirable' to western civilization continue to provide sanctuary to the Indians of contemporary tribal societies.

As to the origin of humans in the Orinoco Basin, there are good archaeological, cultural and linguistic data to suggest an early occupation of the region by preagricultural societies. Warao tradition, for example, has it that their ancestors occupied the Orinoco Delta at a time when the mainland was still connected with, or at least in close proximity to, the island of Trinidad, between 5000 and 7000 years ago (van Andel & Postma 1954; Wilbert 1979). And indeed, the Warao seem to be survivors of a very early immigration wave into the Orinoco Basin. Linguistically, they are not related to the predominant language phyla of

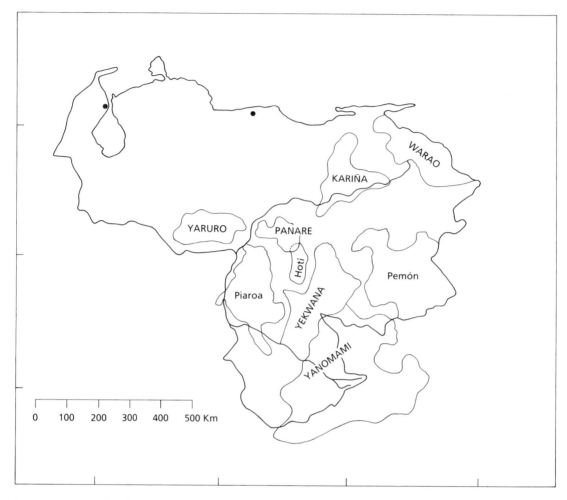

Fig. 21.11 Current distribution of major indigenous tribes of the Orinoco Basin (after OCEI 1985).

Greater Guyana, and their distinct mode of sub-sistence based on fishing, foraging and palm sago recovery, as well as their general sea-orientedness, including seaworthy canoes and star navi-gation, suggest that they represent an ancient littoral tradition of South America (Greenberg 1959; Loukotka 1968; Wilbert 1972, 1979; Heinen 1987).

However, it was the combination of the dugout canoe and the practice of agriculture which pro-duced a linguistic and demographic dispersal of indigenous populations throughout the Orinoco Basin and beyond, turning the river system into a network of wide-flung communication.

For example, theory has it that one such com-munication route led along the Magdalena, the Meta and Apure rivers to the middle and the lower Orinoco. Five thousand-year-old archaeo-logical evidence from the lower Magdalena con-nects ancient populations from northwestern South America with apparently 3000-year-old Arawakan cultures along the lower Orinoco in the east (Reichel-Dolmatoff 1965; Olsen 1974; Rouse 1985).

Another route of dispersal was apparently followed (about 3500 years ago) by Arawakan food-plain horticulturalists from the Amazon valley via the Casiquiare River, down the

Orinoco, reaching the Lesser Antilles several centuries before the time of Discovery (Steward & Faron 1959; Olsen 1974; Rouse 1985).

Finally, there is a postulated route of immigration by Cariban tribes from the north-east Amazon, across the Guyana Shield into the Orinoco Basin. Between AD 400 and 500, these horticultural traders are believed to have come along the Guaniamo, Cuchivero, Caura, Aro and Caroní Rivers. Once in the Orinoco Basin they appear to have been travelling in a fulminating fashion southward towards the Ventuari, westward up the Apure and its tributaries (i.e. Guarico, Pao, Portuguesa, Guanare), and northward to the apex of the Orinoco Delta (Zucchi 1985). Among those river nomads appear to have been the ancestors of such modern Indian populations as the Kariña, Panare, Yabarana, Yekuana and the Pemon.

Thus, interpreting the Orinoco Basin as a mosaic of diverse and interacting ecosystems one must include the human factor as an integral component of ecological cause−effect relationships. Generally, literature tends to isolate the tribal basin communities in terms of linguistic affiliation (Independent, Carib, Arawak, etc.) and subsistence activities (hunter−gatherers, fishermen and horticulturalists), distinctions which individual tribal members often make themselves. Nevertheless, on a macro-level it becomes evident that, to varying degrees, the hunter−gatherers also fish and practice swidden agriculture, the fishermen forage and hunt, and the horticulturalists also hunt and fish. In addition the existence of an extensive trade network between local groups and certain regions of the basin has been amply demonstrated (e.g. Wilbert 1979; Henley 1985; Rouse 1985; Tarble 1985; Zucchi 1985; Mansutti 1986). Based on an exchange network of products originating from pluvial forests, floodplains, savannas, gallery forests and deltaic environments, individual tribes were able to adopt regional specializations which would otherwise have been impossible. At the time of Contact, for example, the Orinoco (especially the Middle Orinoco) served as a commerce centre for up to 60 tribal units (Mansutti 1991), bringing into contact even groups which under normal conditions considered each other as enemies. Peak trading periods were established during the dry seasons which facilitated travelling, thereby allowing for the convergence of diverse tribes or their representatives in specially designated commercial centres on the Orinoco. The principal items of exchange were dyes, yuca grinders, turtle eggs (a resource estimated at 33 million units per year), cultigens, game, furs and ceramics (Morey & Morey 1980; Mansutti 1992). It has been further postulated that trade opportunities were even a major component in stimulating the expansion of the 'waring' Caribs, some of the principal 'merchants' of the basin. Apparently trade was such a basic and vital enterprise on the Orinoco proper that, to facilitate trading relationships, Salivan (the language of the predominant population of the middle Orinoco at the time) was adopted as an official *lingua franca* and trade language. For a more detailed description of the production of the Orinoco Basin and exchange networks, see Morey and Morey (1980).

In summary, the Orinoco River has provided a principal means of communication and migration. Along its main course and its net of tributaries it has served the Indians as a dynamic lifeline since prehistoric times, allowing for the exchange of food items, material culture and, undoubtedly, ideological beliefs. The successful trade network established by the indigenous sector of the basin also demonstrates the economic 'unifying potential' which the basin offers western society.

21.8 RESOURCES AND MANAGEMENT PLANS FOR THE ORINOCO BASIN

Until recently, western society perceived the Orinoco River and its basin as a barrier to development rather than an integral part of it. This ignorance becomes evident when examining the current demographic pattern of the country (MARNR 1982; Vásquez 1989a). The territory encompassing the left banks of the Orinoco to the coastal zones, known as the central−north−coastal axis, represents 55% of the country. This region is inhabited by 97% of the total population and it generates 94% of the gross national product.

In drains only 16% of the country's surface waters. On the eastern side of the Orinoco lives 3% of the population in the remaining 45% of the country which drains 84% of Venezuela's surface-drained waters. Given this occupational pattern, environmental consequences in the central–north–coastal axis have resulted in an ever-decreasing quality of life in the urban centres, an excessive demand for water, and a loss of agricultural land potential to urban development. With the intention to diversify the Venezuelans' oil-based economy, the government has turned to the development of the enormous natural resources found in the central and southern territories. In fact, the intention is to develop an economic corridor employing the Apure and Orinoco Rivers as linking elements between the western Andean region and Puerto Ordaz to the east (the largest industrial complex in the country). This major project has been supported by the Ministry of the Environment and Natural Renewable Resources and the national oil industry. Resources available for exploitation include oil, minerals, hydroelectricity, agriculture and cattle, and fishing. Since the early 1950s, the exploitation of minerals has led to the establishment of extensive industrial installations in Puerto Ordaz. Concerning hydroelectricity, 76% of the gross national hydroelectrical potential is concentrated on the eastern side of the Orinoco. Currently, major hydroelectrical developments are taking place in the lower section of the Caroní River, along some 215 km. On this river, the Guri Dam recently started operation at full capacity, forming a reservoir of 4250 km^2 at normal maximum level. Another dam is currently being enlarged and two more are being built.

The navigational potential of the Orinoco River has created a great interest in the instrumentation of an Orinoco–Apure corridor. At present, ocean-going vessels may navigate up the Orinoco as far as Puerto Ordaz (some 300 km). Beyond this city, navigation, so far, is restricted to smaller vessels such as the trains of barges which transport bauxite. Dredging is in progress in the Apure and in the Orinoco Rivers as a first step toward the implementation of the mentioned corridor. (For a more detailed discussion of government and private plans for the basin, see Vásquez (1989a)).

Traditional uses of the Orinoco River have given rise to a fluvially adapted western society which practices a mixed exploitation of available resources. These people may simultaneously be fishermen, farmers and shepherds. The Orinoco fisheries are subsistence oriented and characterized by their multispecific nature, i.e. the simultaneous exploitation of various species by means of distinct fishing gear and corresponding techniques (Novoa 1986). Commercial catches include about 60 species. This fishery has been in an expansion phase since the early 1970s. Actual landings are about 12 000 t year^{-1}, which is still much lower than the estimated potential of 45 000/year^{-1}. Increased fishing intensity in certain sections of the Lower Orinoco has led to changes in the species composition of the catch. Thus, some species of high economic demand, such as the large catfishes as well as cachamas (*Colossoma macropomum*) and morocotos (*Piaractus brachipomum*), are being replaced by high-yield, fast-growing species such as *Prochilodus mariae* and *Semaprochilodus laticeps* (Novoa 1989). Similar changes in the composition of fishery species have also been reported in African rivers and in South American rivers such as the Magdalena and the Amazon (Bayley & Petrere 1989; Valderrama & Zarate 1989; Welcomme 1990).

21.9 SUMMARY

The Orinoco Basin is a complex and heterogeneous array of differing relief, landscapes and climates. The river itself may be considered as part of a complex ecosystem (Lewis *et al* 1990b), formed by functionally coupled segments of landscapes. The river and its floodplains are the major receptors of this heterogeneity of watersheds draining the Guayana Shield, the alluvial plains of the Llanos and the Andes. The floodplains of the Orinoco (fringing floodplain, internal delta and deltaic floodplain) may be considered as a cluster of ecosystems (Lewis *et al* 1990b) formed by floodplain lakes, rivers, channels, backwaters, islands and levees. In floodplain systems, the flood pulse (Junk *et al* 1989) is considered the major driving force responsible for the existence, productivity and interactions of this cluster of ecosystems. The rejuvenation (reset process) of communities in floodplains has been considered as a consequence of the periodic floods. This concept and its limitations are discussed extensively by Neiff (1990) in the context of the Paraná River system. Owing to the large water-level fluctuations there is a continuous interaction between terrestrial and aquatic environments, leading to the formation of particular and dynamic habitat patches with particular biota and processes. These habitats may be difficult to define in terms of classical limnology (Junk 1980; Paggi & José de Paggi 1990). Organisms have developed specific adaptations to cope with the changing conditions of the floodplain. The study of this cluster of ecosystems demands a perspective of analysis beyond the classical limnological paradigms which led to an understanding of river floodplain systems as formed by two important components: the river channel (lotic component) and the floodplain (lakes and flooded forests as the lentic component). As research in large unmodified floodplain systems advances in the tropics, concepts such as the river continuum concept and the flood pulse concept may be complemented and enriched. The study of the Orinoco River system offers the opportunity to contribute to develop the new paradigms, which Neiff (1990) refers to as fluviology.

Pringle *et al* (1988) recognized three processes in the modification of floodplain dynamics (i.e. colonization, utilization and destruction). The Orinoco River system as a whole may be considered to be at the early utilization phase, characterized by extensive rural use involving agriculture, cattle raising and fisheries activities. Urban centres and industrial activities are found in restricted areas adjacent to the river. Current development plans for the basin may imply a future shift from this phase to the destruction phase where the fragile river system would be so altered that very few of the original features of the system would remain. The current trend of management plans in the Orinoco is striving to understand the various and distinct elements of the ecosystem complex of the river. The conservation of the Orinoco as one of the few remaining pristine river floodplains will depend on a continuous scientific activity to provide information on the uses of the system and to mitigate the changes. At the political and economic level there must be a continuous interaction among politicians, managers and scientists to evaluate development programmes in terms of their economic, social and environmental benefits. This is of paramount importance for both national and international development agencies which historically have tended to maximize short-term economic returns, more often than not to the detriment of environmental and social well-being.

REFERENCES

Bayley PB, Petrere M Jr. (1989) Amazon fisheries: assessment methods, current status and management options. In: Dodge DP (ed.) *Proceedings of the International Large Rivers Symposium*, pp 385–98. Canadian Special Publication of Fisheries and Aquatic Sciences, 106. [21.8]

Blanco L. (1989) Estudio de las Comunidades de invertebrados asociados a las macrofitas acuáticas de tres lagunas de inundación de la sección baja del río Orinoco, Venezuela. *Memoria de la Sociedad de Ciencias Naturales La Salle 49* 71–107. [21.6]

Boadas AR. (1983) *Geografía del Amazonas Venezolano*. Editorial Ariel-Seix Barral Venezolana, Caracas. [21.2, 21.3]

Canales H. (1985) *La Cobertura Vegetal y el Potencial Forestal del Territorio Federal Delta Amacuro*. Ministerio del Ambiente y de los Recursos Naturales Renovables, Serie de Informes Técnicos, Zona 12 (IT) 270 Maturín. [21.3]

Carpio R. (1981) *Geopolítica de Venezuela*. Editorial Ariel-Seix Barral Venezolana, Caracas. [21.2]

Carvajal-Chitty H. (1989) *Estudio sistemático del fitoplancton del río Orinoco en su cuenca alta y media y su variación cualitativa estacional*. Tesis de Licenciatura, Universidad Simón Bolívar, Caracas. [21.5]

Colonnello G. (1990) Elementos fisiográficos y ecológicos de la cuenca del rio Orinoco y sus rebalses. *Interciencia* **15**: 476–85. [21.1, 21.3, 21.6]

Comision del Plan Nacional de Aprovechamiento de los Recursos Hidraulicos (1969) *Inventario Nacional de Aguas Superficiales*, Vol. 1. COPLANARH, Caracas. [21.3]

Comision del Plan Nacional de Aprovechamiento de los Recursos Hidraulicos (1972) *Plan Nacional de Aprovechamiento de los Recursos Hidráulicos, Tomo 1. El Plan*. COPLANARH, Caracas. [21.3]

Crosby A. (1973) *The Colombian Exchange: the Biological and Cultural Consequence of 1492*. Contribution in American Studies 2. Greenwood Press,

Westport. [21.7]

Denevan WM. (1976) The aboriginal population of Amazonia. In: Denevan WM (ed.) *The Native Population of the Americas in 1492*, pp 205–34. University of Manitoba Press, Madison. [21.7]

Drago E. (1976) Origen y clasificación de ambientes leníticos en 11 anuras aluviales. *Revista de la Asociación de Ciencias Naturales del Litoral* 7: 123–37. [21.6]

Drago E. (1989) Morphological and hydrological characteristics of the floodplain ponds of the Middle Paraná River (Argentina). *Revue d'Hydrobiologie Tropicale* 22: 183–90. [21.6]

Ewel JJ, Madriz A, Tosi JA Jr. (1976) *Zonas de Vida de Venezuela*. Ediciones del Fondo Nacional de Investigaciones Agropecuarias, Ministerio de Agricultura y Cría, Caracas. [21.2]

Gessner F. (1965) Zur limnologie des unteren Orinoco. *Internationale Revue der gesamten Hydrobiologie* 50: 305–33. [21.4]

Gibbs AK, Barron CN. (1983) The Guiana Shield reviewed. *Episodes* 1983: 7–14. [21.2]

González de Juana C, Iturralde JM, Picard X. (1980) *Geología de Venezuela y de sus Campos Petrolíferos*. Ediciones Foninves, Caracas. [21.2]

Greenberg JH. (1959) Tentative linguistic classification of Central and South America. In: Steward JH, Faron LC (eds) *Native Peoples of South America*, pp 22–3. McGraw-Hill, New York. [21.7]

Hamilton SK, Lewis WM Jr. (1987) Causes of seasonality in the chemistry of a lake on the Orinoco River floodplain, Venezuela. *Limnology and Oceanography* 32: 1277–90. [21.6]

Hamilton SK, Lewis WM Jr. (1990) Physical characteristics of the fringing floodplain of the Orinoco River, Venezuela. *Interciencia* 15: 491–500. [21.1, 21.6]

Hamilton SK, Sippel SJ, Lewis WM Jr, Saunders JF III. (1990) Zooplankton abundance and evidence for its reduction by macrophyte mats in two Orinoco floodplain lakes. *Journal of Plankton Research* 12: 345–63. [21.6]

Heinen HD. (1987) Los Warao. In: Coppens W, Escalante B (eds) *Los Aborígenes de Venezuela*, pp 585–689. Fundación La Salle/Monte Avila Editores, Caracas. [21.7]

Henley P. (1985) Reconstructing Chaima and Cumangoto kinship categories: an excercise in 'tracking down' ethnohistorical connections. *Antropológica* 64–65: 151–96. [21.7]

Hernández R. (1987) *Geografía del Estado Bolívar*. Academia Nacional de la Historia, Caracas. [21.2, 21.3]

Huber O, Alarcón C. (1988) *Mapa de la Vegetación de Venezuela*. Ministerio del Ambiente y de los Recursos Naturales Renovables, Dirección de Suelos Vegetación y Fauna, División de Vegetación Caracas. [21.3]

Instituto Tecnológico Venezolano del Petróleo (1979) *Caracterización Climatológica del Area de Interés para el Plan de Desarrollo del Sur de Monagas*. Informe Técnico elaborado para Lagoven, Caracas. [21.2]

Junk WJ. (1980) Areas inundáveis—um desafio para limnologia. *Acta Amazonica* 110: 775–95. [21.6, 21.9]

Junk WJ. (1984) Ecology of the várzea, floodplain of Amazonian whitewater rivers. In: Sioli H (ed.) *The Amazon: Limnology and Landscape Ecology of a Mighty Tropical River and its Basin*, pp 215–43. Dr W. Junk Publishers, Dordrecht. [21.6]

Junk WJ, Bayley PB, Sparks RE. (1989) The flood pulse concept in river—floodplain systems. In: Dodge DP (ed.) *Proceedings of the International Large River Symposium*, pp 110–27. Canadian Special Publication of Fisheries and Aquatic Sciences, 106. [21.1, 21.6, 21.9]

Lasso C. (1988) Inventario de la ictiofauna de nueve lagunas de inundación del Bajo Orinoco, Venezuela. Parte I: Batoidei—Clupeomorpha—Ostariophysi (Characiformes). *Memoria de la Sociedad de Ciencias Naturales La Salle* 48: 121–41. [21.6]

Lewis WM Jr. (1988) Primary production in the flowing waters of the Orinoco in relation to its tributaries and floodplain. *Ecology* 69: 679–92. [21.4, 21.5]

Lewis WM Jr, Saunders JF III. (1989) Concentration and transport of dissolved and suspended substances in the Orinoco River. *Biogeochemistry* 7: 203–40. [21.4]

Lewis WM Jr, Saunders JF III. (1990) Chemistry and element transport by the Orinoco main stem and lower tributaries. In: Weibezahn FH, Alvarez H, Lewis WM Jr. (eds) *El Río Orinoco como Ecosistema/The Orinoco River as an Ecosystem*, pp 211–39. Editorial Galac, Impresos Rubel, Caracas. [21.4]

Lewis WM Jr, Weibezahn FH. (1981) The chemistry and phytoplankton of the Orinoco and Caroní rivers, Venezuela. *Archiv für Hydrobiologie* 91: 521–8. [21.4]

Lewis WM Jr, Saunders JF III, Dufford R. (1990a) Suspended organisms and biological carbon flux along the lower Orinoco River. In: Weibezahn FH, Alvarez H, Lewis WM Jr. (eds) *El Río Orinoco como Ecosistema/The Orinoco River as an Ecosystem*, pp 269–300. Editorial Galac, Impresos Rubel, Caracas. [21.5]

Lewis WM Jr, Weibezahn FH, Saunders JF III, Hamilton SK. (1990b) The Orinoco River as an ecological system. *Interciencia* 15: 346–57. [21.1, 21.6, 21.9]

López-Rojas H, Lundberg JG, Marsh E. (1984) Design and operation of a small trawling apparatus for use with dugout canoes. *North American Journal of Fisheries Management* 4: 331–4. [21.6]

Loukotka C. (1968) *Classification of South American Indian Languages*. Reference Series, University of California, Latin American Centre, Vol. 7, 452. [21.7]

Machado-Allison A. (1987) *Los Peces de los Llanos de Venezuela: Un Ensayo sobre su Historia Natural*. Universidad Central de Venezuela, Caracas. [21.6]

Machado-Allison A. (1990) Ecología de los peces de las áreas inundables de los Llanos de Venezuela. *Interciencia* **15**: 411–23. [21.1, 21.6]

Mansutti A. (1986) Hierro, barro cocido, curare y cerbatanas: el comercio intra e interétnico entre los Uwotjuja. *Antropológica* **65**: 3–76. [21.7]

Mansutti A. (1992) Enfermedades exógenas, mortalidad y panorama poblacional en la cuenca del Medio Orinoco durante los siglos XVII y XVIII. *Antropológica* Supplement 4: (in press). [21.7]

Meade RH, Nordin CF, Pérez-Hernández D, Mejia A, Pérez-Godoy JM. (1983) Sediment and water discharge in río Orinoco, Venezuela and Colombia. In: *Proceedings of the Second International Symposium on River Sedimentation*, pp 1134–44. Water Resources and Electric Power Press, Beijing. [21.4]

Meade RH, Weibezahn FH, Lewis WM Jr, Pérez-Hernández D. (1990) Suspended sediment budget for the Orinoco River. In: Weibezahn FH, Alvarez H, Lewis WM Jr, (eds) *El Río Orinoco como Ecosistema/The Orinoco River as an Ecosystem*, pp 55–79. Editorial Galac, Impresos Rubel, Caracas. [21.3, 21.4]

Milliman JD, Meade RH. (1983) World-wide delivery of river sediment to the oceans. *Journal of Geology* **91**: 1–21. [21.2]

Ministerio del Ambiente y de los Recursos Naturales Renovables. (1979) *Atlas de Venezuela*. Dirección de Cartografía Nacional, Caracas. [21.2, 21.3]

Ministerio del Ambiente y de los Recursos Naturales Renovables (1982) *Proyecto Orinoco-Apure, Bases Conceptuales y Operativas*. Serie de Informes Técnicos, Caracas. [21.1, 21.8]

Monente JA. (1989) Materia en suspensión transportada por el río Orinoco. *Memoria de la Sociedad de Ciencias Naturales La Salle* **49**: 5–13. [21.4]

Morey R. (1979) A joyful harvest of souls: disease and the destruction of the Llanos Indians. *Antropológica* **52**: 77–108. [21.7]

Morey NC, Morey RV. (1980) Los Saliva. In: Coppens W, Escalante B (eds) *Los Aborigenes de Venezuela*, Monograph 26, pp 241–306. Fundación La Salle, Instituto Caribe de Antropología y Sociología, Caracas. [21.7]

Murdock GP. (1980) *Theories of Illness: A World Survey*. University of Pittsburgh Press, Pittsburg. [21.7]

Neiff JJ. (1990) Ideas para la interpretación ecológica del Paraná. *Interciencia* **15**: 424–41. [21.6, 21.9]

Nemeth A, Paolini J, Herrera R. (1982) Carbon transport in the Orinoco River: preliminary results. In: Degens ET (ed.) *Transport of Carbon and Minerals in Major World Rivers*, Part 1, pp 357–64. Universitat Hamburg Geologisch–Palaöntologischen Institute der Universitat, SCOPE/UNEP Sond 52. [21.4]

Nordine CF, Pérez-Hernández D. (1989) *Sand Waves, Bars, and Wind-Blown Sands of the Rio Orinoco, Venezuela and Colombia*. United States Geological Survey Water-Supply, Paper 2326-A, Denver. [21.3]

Novoa D (ed.) (1982) *Los Recursos Pesqueros del Río Orinoco y su Explotación*. Editorial Arte, Caracas. [21.1]

Novoa D. (1986) Una revisión de la situación actual de las pesquerías multiespecíficas del río Orinoco y una propuesta del ordenamiento pesquero. *Memoria de la Sociedad de Ciencias Naturales La Salle* **46**: 167–91. [21.8]

Novoa D. (1989) The multispecies fisheries of the Orinoco River: development, present status, and management strategies. In: Dodge DP (ed.) *Proceedings of the International Large River Symposium*, pp 422–8. Canadian Special Publication of Fisheries and Aquatic Sciences, 106. [21.6, 21.8]

Novoa D. (1990) El río Orinoco y sus Pesquerías; estado actual, perspectivas futuras y las investigaciones necesarias. In: Weibezahn FH, Alvarez H, Lewis WM Jr. (eds) *El Río Orinoco como Ecosistema/The Orinoco River as an Ecosystem*, pp 387–406. Editorial Galac, Impresos Rubel, Caracas. [21.6]

Novoa D, Ramos F. (1990) Las pesquerías comerciales del río Orinoco: su ordenamiento vigente. *Interciencia* **15**: 486–90. [21.1]

Oficina Central de Estadística e Informática (1985) *Censo Indígena de Venezuela: Nomenclador de Comunidades y Colectividades*. Taller Gráfico de la Oficina Central de Estadística e Informática, OCEI, Caracas. [21.7]

Olsen F. (1974) *On the Trail of the Arawaks*. University of Oklahoma Press, Norman. [21.7]

Paggi JC, José de Paggi S. (1990) Zooplankton of the lotic and lentic environments of the middle Paraná River. *Acta Limnologica Brasileira* **3**: 685–719. [21.5, 21.9]

Prance GT. (1979) Notes on the vegetation of Amazonia. 3. The terminology of Amazonian forest types subject to inundation. *Brittonia* **31**: 26–38. [21.6]

Paolini J, Herrera R, Nemeth A. (1983) Hydrochemistry of the Orinoco and Caroní rivers. In: Degens ET, Kenpe S, Soliman H (eds) *Transport of Carbon and Minerals in Major World Rivers*, Part 2, pp 223–36. Universitat Hamburg Geologisch–Palaeöntologischen–Institute der Universitat, SCOPE/UNEP Sond 55. [21.4]

Pringle CM, Naiman RJ, Bretschko G, Karr JR, Oswood MW, Webster JR, Welcomme RL, Winterbourn MJ. (1988) Patch dynamics in lotic systems: the stream as a mosaic. *Journal of the North American Bentho-*

logical Society **7**: 503–24. [21.9]

Quirós R. (1988) Resultados del Simposio Internacional sobre Grandes Ríos y su aplicabilidad a los grandes ríos de América Latina. *COPESCAL Documentos Ocasionales* **5**: 70 pp. [21.2]

Reichel-Dolmatoff G. (1965) *Colombia: Ancient Peoples and Places*, Vol. 44. Frederick A. Praeger, New York. [21.7]

Rey J, Vásquez E. (1986) Cladocères de quelques corps d'eaux du bassin moyen de l'Orénoque (Vénézuela). *Annales de Limnologie* **22**: 137–68. [21.5]

Roberts TR. (1973) Ecology of fishes in the Amazon and Congo basins. In: Meggers BJ, Ayers ES, Duckworth WD (eds) *Tropical Forest Ecosystems in Africa and South America: A Comparative Review*, pp 239–54. Smithsonian Institution Press, Washington. [21.2]

Rodríguez MA, Lewis WM Jr. (1990) Diversity and species composition of fish communities of Orinoco floodplain lakes. *National Geographic Research* **6**: 319–28. [21.6]

Rouse I. (1985) Arawakan phylogeny, Caribbean chronology, and their implications for the study of population movement. *Antropológica* **63–64**: 9–20. [21.7]

Sánchez JC. (1990) La calidad de las aguas del río Orinoco. In: Weibezahn FH, Alvarez H, Lewis WM Jr. (eds) *El Río Orinoco como Ecosistema/The Orinoco River as an Ecosystem*, pp 241–68. Editorial Galac, Impresos Rubel, Caracas. [21.4]

Sánchez L, Vásquez E. (1986) Notas sobre las macrofitas acuáticas de la sección baja del río Orinoco, Venezuela. *Memoria de la Sociedad de Ciencias Naturales La Salle* **46**: 107–25. [21.4, 21.6]

Sánchez L, Vásquez E. (1989) Hydrochemistry and phytoplankton of a major blackwater river (Caroní) and a hydroelectric reservoir (Macagua), Venezuela. *Archiv für Hydrobiologie Beiheft Engebnisse der Limnologie* **33**: 303–13. [21.4, 21.5]

Sarmiento G. (1983) The savannas of tropical America. In: Bourliere F (ed.) *Tropical Savannas. Ecosystems of the World 13*, pp 245–88. Elsevier, Amsterdam. [21.2]

Saunders JF III, Lewis WM Jr. (1988a) Zooplankton abundance in the Caura River, Venezuela. *Biotropica* **20**: 206–14. [21.5]

Saunders JF III, Lewis WM Jr. (1988b) Zooplankton abundance and transport in a tropical white-water river. *Hydrobiologia* **162**: 147–55. [21.5]

Steward JH, Faron LC. (1959) *Native Peoples of South America*. McGraw-Hill, New York. [21.7]

Tarble K. (1985) Un nuevo modelo de expansión Caribe para la época prehispánica. *Antropológica* **64–65**: 45–82. [21.7]

Valderrama M, Zarate M. (1989) Some ecological aspects and present state of the fishery of the Magdalena river basin, Colombia, South America. In: Dodge DP (ed.) *Proceedings of the International Large Rivers Symposium*, pp 409–21. Canadian Special Publication of

Fisheries and Aquatic Sciences, 106. [21.8]

van Andel TjH. (1967) The Orinoco Delta. *Journal of Sedimentary Petrology* **37**: 297–310. [21.3]

van Andel TjH, Postma H. (1954) Recent sediments of the Gulf of Paria: Reports of the Orinoco Shelf Expedition. *Verhandelingen der Koninklijke Nederlandse Akademie van Wetenschappen, Afdeeling Naturkunde* **20**: 1–245. [21.7]

Vannote RL, Minshall GW, Cummins KW, Sedell JR, Cushing CE. (1980) The river continuum concept. *Canadian Journal of Fisheries and Aquatic Sciences* **37**: 130–7. [21.6]

Vásquez E. (1984) Estudio de las comunidades de rotíferos del Orinoco Medio, Bajo Caroní y algunas lagunas de inundación (Venezuela). *Memoria de la Sociedad de Ciencias Naturales La Salle* **44**: 95–108. [21.5]

Vásquez E. (1988) Morfometría de un conjunto de lagunas de inundación del Bajo Orinoco, Venezuela. *Pantepuy* **4**: 31–7. [21.6]

Vásquez E. (1989a) The Orinoco River: a review of hydrobiological research. *Regulated Rivers* **3**: 381–92. [21.1, 21.6, 21.8]

Vásquez E. (1989b) Características morfométricas de seis lagunas de la planicie aluvial del Bajo Río Orinoco, Venezuela. *Memoria de la Sociedad de Ciencias Naturales La Salle* 309–27. [21.6]

Vásquez E, Rey J. (1989) A longitudinal study of zooplankton along the lower Orinoco River and its Delta (Venezuela). *Annales de Limnologie* **25**: 107–20. [21.5]

Vásquez E, Sánchez L. (1984) Variación estacional del plancton en dos sectores del río Orinoco y una laguna de inundación adyacente. *Memoria de la Sociedad de Ciencias Naturales La Salle* **44**: 11–31. [21.5, 21.6]

Vegas-Vilarrúbia T, Paolini J, Herrera R. (1988) A physico-chemical survey of blackwater rivers from the Orinoco and the Amazon basins in Venezuela. *Archiv für Hydrobiologie* **111**: 491–506. [21.4]

Vila P. (1960) *Geografía de Venezuela*. Ministerio de Educación, Dirección de Cultura y Bellas Artes, Caracas. [21.3]

Weibezahn FH. (1985) *Concentraciones de Especies Químicas Disueltas y Transporte de Sólidos Suspendidos en el Alto y Medio Orinoco y sus Variaciones Estacionales* (Febrero 1984–Febrero 1985). Ministerio del Ambiente y de los Recursos Naturales Renovables y Petróleos de Venezuela, Caracas. [21.3, 21.4]

Weibezahn FH. (1987) *Sólidos Suspendidos y Disueltos en el Alto y Medio Orinoco (Abril 1986–Marzo 1987): Concentraciones, Transporte y Drenaje Específico*. Universidad Simón Bolívar, Instituto de Recursos Naturales, Caracas. [21.3]

Weibezahn FH. (1990) Hidroquímica y sólidos suspendidos en el Alto y Medio Orinoco. In: Weibezahn FH, Alvares H, Lewis WM Jr. (eds) *El Río Orinoco como*

Ecosistema/The Orinoco River as an Ecosystem, pp 150–210. Editorial Galac, Impresos Rubel, Caracas. [21.2, 21.3, 21.4]

Weibezahn FH, Alvarez H, Lewis WM Jr. (ed.) (1990) *El Río Orinoco como Ecosistema/The Orinoco River as an Ecosystem*. Editorial Galac, Impresos Rubel, Caracas. [21.1]

Welcomme RL. (1979) *Fisheries Ecology of Floodplain Rivers*. Longman, New York. [21.6]

Welcomme RL. (1990) Status of fisheries in South American rivers. *Interciencia* **15**: 337–45. [21.2, 21.6, 21.8]

Wilbert J. (1972) *Survivors of Eldorado*. Praeger Publishers, New York. [21.7]

Wilbert J. (1979) Geography and telluric lore of the Orinoco Delta. *Journal of Latin American Lore* **5**: 129–50. [21.3, 21.7]

Yanes CE, Ramírez AJ. (1988) Estudio geoquímico de grandes ríos venezolanos. *Memoria de la Sociedad de Ciencias Naturales La Salle* **48**: 41–58. [21.2, 21.4]

Zinck A. (1977) *Ríos de Venezuela*. Cuadernos Lagoven, Cromotip, Caracas. [21.2, 21.3]

Zinck A. (1980) *Valles de Venezuela*. Cuadernos Lagoven, Cromotip, Caracas. [21.3]

Zucchi A. (1985) Evidencias arqueológicas sobre grupos de posible lengua Caribe. *Antropológica* **63–64**: 23–44. [21.7]

22: The River Murray, Australia: A Semiarid Lowland River

K.F.WALKER

22.1 INTRODUCTION

Arid and semiarid regions occupy about one-third of the world's land area and are home to more than half the human population (Thomas 1989). These figures are approximate, because the boundaries of semiarid regions change from year to year, but they highlight our reliance on areas where water resources are meagre. The main sources of supply are river systems such as the Nile and Orange–Vaal in Africa, the Columbia and Colorado in North America, the Ganges, Indus and Tigris–Euphrates in Asia, and the Murray–Darling system in Australia.

The term 'semiarid river' is convenient, but perhaps a misnomer because the discharge is likely to be much greater than could be sustained by regional rainfall: semiarid rivers generally are fed from nearby, more humid, areas. Nevertheless, they are subject to wide fluctuations in discharge associated with floods and droughts. Flow variability is typical, and pervades every feature of the physical, chemical and biological environment.

The coupling of variable discharge and a persistently high demand for water is a natural precondition for flow regulation. Hence water resource developments, particularly for agriculture, are a common feature of semiarid rivers (Graf 1987). Indeed, the regulated regime is so different from the natural regime that the environmental impact of regulation in semiarid regions may be greater than for rivers elsewhere.

The nature of this conflict is well shown by the Murray–Darling river system of south-east Australia. This chapter reviews the Murray–Darling environment, and the River Murray in particular, with regard to the effects of flow regulation. It updates earlier reviews (Walker 1979, 1985, 1986a) and omits some topics for which there are no substantial new data (e.g. environmental impact of dams, ecology of vertebrates other than fish).

22.2 PHYSICAL ENVIRONMENT

Geography

The Murray–Darling Basin extends over 1.073 million km^2 and occupies most of southeast Australia (Fig. 22.1). It is contained by the Eastern Highlands to the south and east, and by vast expanses of arid land to the north and west. The basin is home to 1.6 million people, mostly in rural areas, and is a vital resource for eastern Australia (Murray–Darling Basin Ministerial Council (MDBMC) 1987a). It supports 25% of the nation's cattle and dairy farms, 50% of sheep and cropland, and 75% of irrigated land, and returns more than A$10 billion in annual production. Irrigation accounts for about 90% of annual water consumption and so largely determines the pattern of regulation. The river system also supplies 16 cities and provides more than 40% of the water needs of South Australia. The system is among the longest in the world (Murray 2560 km, Darling 2740 km) and includes 20 major rivers. The annual discharge is comparatively small, however, and remains variable even under intensive flow regulation (e.g. at Blanchetown, 1950–80: mean 318.2, range 19.6–1562.4 m^3 s^{-1}; Walker 1986a).

The Murray lies near the southernmost perimeter of the basin (Fig. 22.1). It rises at

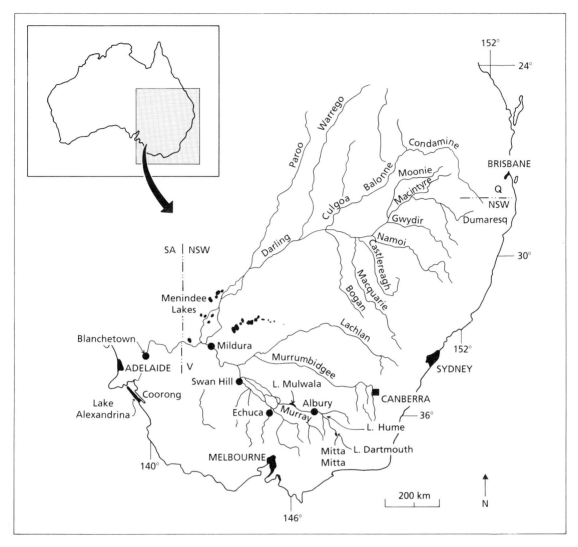

Fig. 22.1 Some geographical features of the Murray–Darling Basin.

1430 m in the Snowy Mountains and flows north-west as the border between New South Wales and Victoria. In its middle reaches the river is met by its main tributaries, the Ovens, Goulburn, Murrumbidgee and Darling rivers; it then continues westerly into South Australia until near Morgan, where it turns south towards Lake Alexandrina and the sea.

Landform evolution

The genesis of the river system was more than 50 million years ago, when the Eastern Highlands began to rise and the basin subsided, forming extensive swamplands. The climate was subtropical and much wetter than at present, and the basin was covered by rainforest (Martin 1989). In the late Tertiary the sea flooded the basin several times, burying the swamplands under marine sediments. At its maximum extent, about 20 million years ago, the sea extended to the present Murray–Murrumbidgee confluence (Fig. 22.1). Around the margins of the 'Murravian Gulf' there was a complex drainage system fanning out from the mountains eastward, including a precursor of the Murray (Stephenson & Brown 1989).

In the Pleistocene, beginning 2–3 million years ago, the highlands were further uplifted and the marine sediments buried under newly eroded silts and gravels. Glaciation and vulcanism further modified the landscape. The sea returned to flood the lower valley, leaving deep layers of limestone. Tectonic movements near the South Australian border impounded the drainage system as a large inland lake (Stephenson 1986), and a series of movements along the Cadell Fault thwarted drainage lines along the middle reaches of the Murray (Fig. 22.1; Bowler 1978).

Between 30 000 and 15 000 years ago the climate was glacial, with temperatures up to 10°C lower than now (Bowler 1986, 1990). The land surface was unstable and sparsely vegetated (Martin 1989), and groundwater tables were elevated. The sediments associated with the 'prior streams' were deeply incised by newly formed 'ancestral rivers', the direct antecedents of the modern rivers. With recession of the glaciers, stream discharges moderated and sediment loads increased. Stream banks and land surfaces were stabilized, soil mantles developed and groundwater tables fell as the climate became drier, approaching present conditions. The sea receded, causing the river to incise a shallow gorge through the limestones in its lower reaches (Twidale *et al* 1978). Ten thousand years ago the sea rose to its present level and the lower valley was infilled with coarse sands and later clays and silts.

The ancient history of the Murray is reflected in its generally low gradient. The *Headwaters Tract* is remarkably short (350 km) and the riverbed gradient declines sharply to about 16 cm km^{-1}. Near Albury (Fig. 22.1) the river enters 'Lake' Hume, a reservoir that impounds flows from the upper Murray and Mitta Mitta rivers. Beyond, in its *Riverine Plains Tract*, the Murray spreads out across a broad floodplain (up to

25 km) with many anabranches and billabongs (oxbow lakes), and the gradient declines to about 8 cm km^{-1} as the river approaches the Murrumbidgee junction. In the *Mallee Plains Tract*, extending to near the Darling junction, the river resumes a single winding channel and the gradient is about 5 cm km^{-1}. In the *South Australian Tract* the river enters a limestone gorge (1–9 km) before entering the sea via Lake Alexandrina. The gradient of the Murray below the Darling junction is less than 5 cm km^{-1}, and the river environment is dominated by a series of weirs (section 22.3).

Climate

The Murray–Darling Basin extends from latitude 24–37°S and longitude 138–151°E, and regional climates vary accordingly (MDBMC 1987a). The northern region is subtropical, the east is cool and humid, the south is temperate, and the intervening area, inland from the ranges, is dry and hot. Mean annual temperatures range from 13.7°C in the highlands to 20.2°C in the Darling region. Seasonal water temperatures in the Murray vary from 6–28°C above Lake Hume to 10–30°C in South Australia (Mackay *et al* 1988). Annual rainfall varies from more than 1400 mm in the highlands to less than 300 mm in the north-west. Rainfall variability increases inland, and is reflected in variable stream flows (see below). The eastern area has a reasonably uniform seasonal pattern, but peak rainfall is in winter–spring in the south and summer in the north. Annual evaporation generally is far in excess of rainfall, except in highland areas.

The consequences of the greenhouse effect are difficult to forecast, partly because past climatic trends have been erratic: rainfall across the basin is subject to secular changes in the general pattern of atmospheric circulation. Since the 1940s summer rainfall in central New South Wales has risen by up to 30%, and stream flows have increased accordingly (Cornish 1977; Riley 1988). As global warming becomes pronounced, flows in the Murray are likely to increase; this may ease salinity problems and possibly improve the security of supplies for water consumers (MDBMC 1987a; Close 1988; *cf.* Walker 1989).

Hydrology

Most of the Murray–Darling catchment is arid or semiarid land that contributes no significant run-off. The mean annual runoff from the basin (14 mm) is very low, and only about 3% of the average annual rainfall (MDBMC 1987a).

Typically, about half the discharge of the system originates within 500 km of the source of the Murray, when there is rain and snowfall in winter and spring. Flows in the middle Murray are supplemented by the Ovens, Goulburn and Murrumbidgee rivers, also fed by winter–spring precipitation. Flows in the Darling, on the other hand, are determined by unreliable summer monsoons in south-east Queensland. Although the Darling drains about half of the total Murray–Darling catchment, it contributes only about 12% of the long-term average discharge.

The variability of streamflow increases westwards across the basin, reflecting changes in the temporal and geographical distribution of rainfall (Riley 1988). Thus, the greatest recorded variations are for the Darling and its tributaries (Fig. 22.2). The Darling's influence ensures that flows in the lower Murray are more variable than those in the middle Murray (Walker *et al* 1991). The average annual flow of the lower Murray at Blanchetown is 10 035 GL, within a range of 0.617–49 271 GL (data for 1950–80: Walker 1986a).

Australia generally has more variable annual river flows than other continents (Finlayson & McMahon 1988). The variability increases with catchment area, and so is greatest in rivers allied to the Murray–Darling system. This cannot be explained wholly in terms of rainfall, because other areas of the world experience rainfall that is no less variable. Rather, it may arise from the low relief of the continent and variability in the transfer of effective rainfall to runoff, in turn related to the comparatively high evaporative power of the atmosphere.

Flows in the Murray–Darling system may take a long time to travel from headwaters to the sea. For example, water in the Murray takes 6 weeks to travel from Albury to the Murray–Darling junction, and another 2 weeks to reach the sea. Floods may take still longer (Jacobs 1989). This is partly a consequence of gentle gradients, and partly because the channel capacities of some rivers decrease as they traverse the basin (MDBMC 1987a). This causes overbank flows during floods and encourages the development of

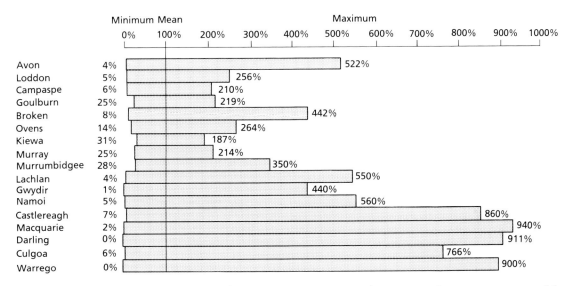

Fig. 22.2 Flow variability in the Murray–Darling river system. Minima and maxima are shown as percentages of the mean (100%) (from MDBMC 1987a).

wetlands. The floods are vital for maintenance of floodplain communities, although their frequency and magnitude have been diminished by flow regulation (section 22.3).

22.3 HUMAN ENVIRONMENT

Clash of cultures

Two hundred years ago, when Europeans first arrived in Australia, the Murray–Darling Basin was populated by people whose history extended back for more than 40 000 years. The native tribes were concentrated along the rivers where there were reliable food supplies, and they maintained a nomadic, hunter–gatherer lifestyle that was well adapted to the vagaries of the Australian environment (Blainey 1976). They maintained extensive trade networks, so that river tribes could barter resources with people from inland areas. The Murray–Darling system was no less important for the aboriginals than it is for modern society.

The 19th-century European invasion vanquished the Murray–Darling tribes in little more than 100 years: people were dispossessed of their land and many were lost to epidemics of disease. Little remains of contemporary aboriginal culture and traditions apart from fragments in the art and literature of the colonists. Other relics include scarred trees, stone quarries, shell middens and burial grounds (MDBMC 1987a). Archaeology has flourished, however, since the discovery of ancient occupational residues along the lower Darling River (e.g. Flood 1983).

History of flow regulation

After about 1830, following the paths of explorers such as Charles Sturt (who named the Murray), the colonists opened up inland stock routes and developed riparian lands as pastures for sheep and cattle. After 1850 the rivers became avenues for commerce, supporting a flourishing paddle-steamer industry that was to decline slowly from 1880, as railways expanded. Large areas of eucalypt scrub were cleared for wheat, and swamps along the lower Murray were drained and reclaimed for pasture. Large-scale irrigation of

citrus, vines and stone-fruit crops began in the late 1880s (Eastburn 1990).

The federation of Australia (1901) was followed by an Interstate Royal Commission (1902) concerned with sharing the Murray's resources among the states. This culminated 13 years later in the signing of the *River Murray Waters Agreement* (1915) and establishment of a management authority, the River Murray Commission (1917). Part of the Commission's charter was to oversee the construction of dams and weirs, beginning a new era of flow regulation.

The first of 26 planned weirs (with adjacent locks) was completed at Blanchetown in 1922 (Fig. 22.3). Construction continued at other sites until 1937, but after 1930 the emphasis in river management had turned from navigation to irrigation, and plans for some weirs were dropped. Those completed were locks 1–10, all below the Murray–Darling junction, and locks 11 (Mildura), 15 (Euston) and 26 (Torrumbarry).

In 1919 work began on Hume Dam near Albury (see Fig. 22.1), still the only major storage on the Murray main stem. The first stage was completed in 1936, and later works (1950–61) increased its capacity to 3038 GL. Other dams, all on tributaries to the Murray, include Eildon (Goulburn River, 3390 GL: 1951–5), Dartmouth (Mitta Mitta River, 4000 GL: 1972–9) and many smaller storages (to 1700 GL). Other structures include Lake Victoria (1928) and the Menindee Lakes (1963–8), which regulate flows to the lower Murray, a large irrigation-diversion weir at Yarrawonga (1939), and barrages across Lake Alexandrina (1940), preventing sea-water inflows to the lower river.

Pipelines were constructed to transport Murray water to remote areas of South Australia (1944–70). The Snowy Mountains Scheme (1949–74) was engineered to divert water from the south-east coastal drainage to augment flows in the upper Murray and Murrumbidgee Rivers and to generate hydroelectric power. Irrigation received a boost after the Second World War, and although South Australia has since declared a moratorium on further development (1968), new schemes are still being established on the Darling River in New South Wales.

In 1982 the *River Murray Waters Agreement* was amended to give the River Murray Com-

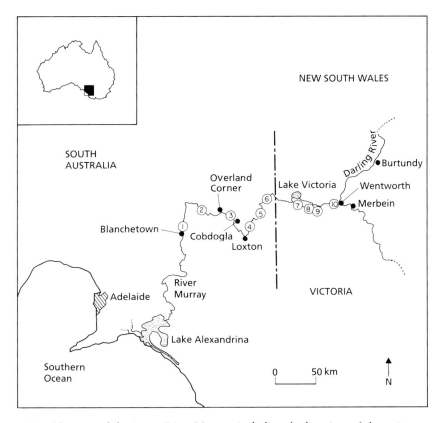

Fig. 22.3 Geographical features of the Lower River Murray, including the locations of the weirs.

mission jurisdiction for the first time over water quality and other environmental issues. These changes were consolidated in 1985 with establishment of an overriding authority, the Murray–Darling Basin Ministerial Council, and again in 1988 when the River Murray Commission was renamed. The Murray–Darling Basin Commission now administers a revised *Murray–Darling Basin Agreement* (1987) on behalf of the Commonwealth and three state governments.

Flow management

Natural flows in the Murray were variable, but showed a distinct long-term seasonal pattern (Baker & Wright 1978). Peak flows in the upper Murray were in winter and spring, reaching South Australia in spring and early summer. The seasonal pattern was superimposed on erratic sequences of floods and droughts.

The main objective in developing the water resources of the system has been to secure reliable supplies for irrigated agriculture. Winter and spring flows from the high-country catchments are retained in reservoirs such as Lake Hume, and released in summer and autumn as natural flows decline and the demand for irrigation water increases. The Murray–Darling Basin Commission uses a complex set of operating rules to determine flow allocations under different conditions (Jacobs 1989, 1990). A proviso is that South Australia should receive a minimum annual 'entitlement flow' of 1850 GL.

From an operational viewpoint, regulation has decreased the frequency of overbank flows and increased the frequency of flows at or near channel capacity (Bren 1988). It has also changed the frequency, extent and duration of floods (Baker & Wright 1978; Jacobs 1989; Close 1990a), although the pattern is somewhat obscured by the effects of long-term climatic variability. In the Lower Murray, small floods (e.g. return period of less than 7 years) have been almost eliminated, but there is insufficient storage to check large floods. At Barmah Forest (near Echuca; see Fig. 22.1) the duration of seasonal floods has decreased by 1–2 months since construction of Hume Dam (Bren 1987); this has affected regeneration in the commercial red gum forests (Chesterfield 1986; Dexter *et al* 1986). These changes are evident in the recession limb of the flood stage hydrograph. At Overland Corner (South Australia) in 1885 and 1986 (before and after regulation) the times for recession of two floods of similar magnitude were 59 and 4 days respectively (Thoms & Walker 1989). A similar degree of attenuation is evident in the Murray at Corowa, near Albury (Jacobs 1989).

The seasonality of regulated flows varies along the river as water is diverted for irrigation (e.g. Margules *et al* 1990). At Albury there has been a large decrease in average flows from June to November and an increase from December to May. This increase is much less at Yarrawonga, below the diversion weir that impounds Lake Mulwala (see Fig. 22.1). At Euston, downstream of other major diversions, there are decreased flows throughout the year, and particularly from July to December, but the natural pattern has been retained to the extent that winter–spring flows are higher than those in summer and autumn. The impact of regulation on the flow regime probably has been greatest over the last 30–40 years, reflecting the expansion of irrigation areas in New South Wales after about 1950 (Mackay *et al* 1988).

In the lower Murray the pattern remains similar to that at Euston, reflecting smaller irrigation diversions in South Australia (average 500 GL, *cf.* 3400 GL in New South Wales and Victoria). The average annual regulated flow to South Australia (6650 GL) is about 42% less than under natural conditions (Jacobs 1989). Regulated flows to South Australia contain disproportionate volumes of highly turbid water from the Darling (Mackay *et al* 1988; Walker *et al* 1991).

Flows in the Lower Murray are further controlled by a series of 3-m weirs (Fig. 22.3). The weir pools are nearly overlapping, so that the entire lower river is continuous pool environments (indeed, weir pools occupy 52% of the river below Lake Hume; Pressey 1986). In floodplain areas below normal pool levels many once-temporary wetlands have become permanently flooded. The weirs are operated so that the impounded water remains near a specified 'pool level' except during floods. These operations induce daily fluctuations (typically ±20 cm) in the river downstream, although they are progressively diminished in the approach to the next downstream weir. The changeable water levels and high turbidities exert a strong influence on littoral communities (Walker *et al* 1991).

For most rivers in the system the regulated flow is about half the annual discharge or less. Annual water use is not much less than the regulated flow, and the volumes of storages on the upper Murray, Goulburn, Gwydir and Murrumbidgee rivers are significantly greater than the mean annual flows from their catchments. Given also that the total storage volume in the system is 136% of the summed mean annual flows of individual basins (30 million ML), there appears to be little scope for more development. From a purely economic viewpoint, new storages are unlikely to be feasible and increased benefits may have to be won largely by improved efficiency and by allocating water to higher value products (MDBMC 1987a).

Channel changes

The gentle gradient of the lower Murray indicates a depositional fluvial regime that has been stable over a long period. Stability is further indicated by low stream energy, low sinuosity and cohesive bank sediments (Thoms & Walker 1992a). The morphology of the channel does change, however, in response to short-term changes in the flow regime. Indeed, semiarid rivers typically have complex cross-sections, with benches representing adjustments to extreme discharges (Graf 1987).

Flow regulation may also cause channel changes. The annual sediment trap efficiencies of lakes Hume and Mulwala are 35% and 52% respectively (16-year means; Thoms & Walker

1992b), indicating a large reduction in the downstream supply of sediment. Given the changes induced by regulation in both the sediment and water regimes, it is likely that the Murray's channel has changed significantly throughout its middle and lower reaches (*cf.* Rutherfurd 1990).

Channel changes are clearly evident in response to weir construction along the Lower Murray (Thoms & Walker 1989, 1992a). Although the configuration of the channel has remained constant, a complex 'stepped' gradient has been imposed upon the river planform, corresponding to stepped changes (3 m) in water elevation. Typically there is an eroding zone below each weir, grading to a depositional zone behind the next weir. Bank erosion is a conspicuous problem: in 1988–9 an estimated 1.8 million tonnes of bank material was lost over a 153-km section of the lower river. Erosion is promoted by rapid recessions of high flows, which render the bank material vulnerable to collapse, and may be encouraged also by changes in water level associated with weir operations.

Channel adjustments in response to the weirs have involved redistribution of sediment rather than changes in the sediment budget (Thoms & Walker 1992a). Over 1906–88 less than 1% of the sediment input (3.174 million tonnes) to the lower Murray was retained. At individual sites, however, changes in bed elevation ranged from −11.53 to 10.09 m. It is remarkable that, nearly 70 years after the first weir was commissioned, the process of adjustment is not yet complete.

Salinization

Irrigation causes a massive disturbance in the local hydrological environment, causing the groundwater table to rise (Grieve 1987). Because groundwater in the Murray Valley is rendered saline by residual marine sediments (Evans & Kellet 1989), the effect of irrigation has been to cause widespread salinization of land and water resources (e.g. Williams 1987; Allison *et al* 1990; Macumber 1990). The problem is exacerbated by the hydraulic effects of weirs, rapid recessions following floods, excessive clearing of vegetation, waterlogging in poorly drained soils, inefficient irrigation techniques and generally inadequate land and water management (MDBMC 1987b).

In the river, typical salinities range from <30 mg l^{-1} in the headwaters to more than 450 mg l^{-1} near Lake Alexandrina (Mackay *et al* 1988; Close 1990b). At Echuca (see Fig. 22.1) the salinity is about 60 mg l^{-1}, but downstream salinities approach 250 mg l^{-1} following large inflows from irrigation areas near Kerang. The Murrumbidgee and Darling inflows are comparatively fresh and cause slight reductions in salinity, but there are further salt accessions in South Australia. At Morgan (Fig. 22.3); salinities in 1978–86 varied from 146 to 723 mg l^{-1}, depending on discharge. In 1962–87 the average salinity at Morgan increased annually by 2.6 mg l^{-1} (Mackay *et al* 1988), although the increment varied widely from year to year (*cf.* Morton & Cunningham 1985).

Increased salinities may not have had a great impact on the native plants and animals because many species (at least the adults of many species) are able to tolerate levels above those prevailing in the river (e.g. fish: Langdon 1987; Williams & Williams 1991; crayfish: Mills & Geddes 1980; freshwater mussels: Walker 1981; see further Hart *et al* 1990). There are obvious exceptions, however, in evaporation basins on the floodplain, particularly along the lower river (e.g. Evans 1989). These are wetlands used to impound and concentrate irrigation return water, pending release to the river during high flows. Depending on their salinity regime, the basins support sparse communities of salt-tolerant plants and animals. The surrounding trees are often killed by the raised groundwater table, and the understorey is replaced by halophilic plants (e.g. chenopods: *Halosarcia* spp.). Other exceptions are in floodplain areas where rising saline groundwater has caused the decline of riparian vegetation, especially red gum and black box (Margules *et al.* 1990; O'Malley & Sheldon 1990).

The effect of salinization on the irrigation industry is a major concern in river management. Salinized land may be useless for agriculture, and water more saline than 300–500 mg l^{-1} is unusable for irrigation (depending on the crop) because it causes crop losses through diminished productivity or defoliation. In 1987 the area of irrigated land affected by salinization was 96000 ha, and more than 500000 ha may be affected in future (MDBMC 1987b). The total economic cost of salinization throughout the Murray Valley in 1987 was more than A$100 million, measured in lost agricultural production and other costs, including water-quality problems experienced by domestic and industrial consumers in South Australia.

The Murray–Darling Basin Ministerial Council has initiated an elaborate strategy to ameliorate these problems (MDBMC 1987b). This includes schemes to intercept and divert irrigation return water from the Murray, to increase dilution flows, to change irrigation practices and to make other improvements in land and water management. As the strategy has evolved, the emphasis has turned from a specific salinity objective (that by 1995 salinities in the Murray at Morgan should remain below 480 mg l^{-1} for 95% of the time) to one where action is to be determined by economic costs and benefits (MDBMC 1987b). Future implementation of the strategy will need to be monitored to ensure that adverse environmental effects are minimized (*cf.* O'Malley & Sheldon 1990).

Water quality

Before 1978, few water-quality data (other than salinity) were gathered routinely for the Murray, although there are some fragmentary records in the archives of state authorities and other agencies (e.g. Walker & Hillman 1982). Since 1978 the Murray–Darling Basin Commission has maintained a wide-ranging monitoring programme; summaries are published in the *Annual Report* of the Commission and data for 1978–86 have been reviewed in detail (Mackay *et al* 1988). Other data are contained in studies related to Murray–Darling reservoirs (e.g. Boon 1989; Banens 1990).

The ionic composition of Murray water is strongly influenced by inflows from tributaries and irrigation areas downstream of Echuca (see Fig. 22.1). In round figures, the proportion of sodium among total cations increases from 35% in the headwaters region to 60% downstream of Echuca (Mackay *et al* 1988: Fig. 4.3); potassium also increases (generally less than 10%) and calcium and magnesium both decline (from 35% to 15% and from 30% to 20%, recpectively). Chloride rises (from 10 to 65%), bicarbonate falls

(from 80 to 20%), and sulphate remains constant (7–10%). Typical pH values are from about 7.4 in the headwaters to 8.0 in the Lower Murray.

Turbidity is a significant water-quality problem in the Murray (Mackay *et al* 1988), although it may provide some compensatory relief from problems of algal growth. Turbidities increase from less than 3 NTU in the headwaters to about 70 NTU is the lower reaches. Each of the major tributaries contributes a significant increase, but there is generally a fall of 10–20 NTU where the Murray enters the weir pool of Lock 15 (Euston; Fig. 22.3), before the Darling junction. This may be caused by a change in the nature of the suspended sediment: finely divided clays predominate in the Riverine Plains, whereas coarse marine sands occur in the Mallee Plains.

The Darling carries a suspended load of very fine clay particles that do not settle under normal flow conditions (Woodyer 1978; Thoms & Walker 1992a). During the irrigation season (November to May), the average daily flow to South Australia is 7000 ML, made up of 2000 ML from the Murray and 5000 ML from Lake Victoria (Jacobs 1989), which impounds water transferred from the Menindee Lakes on the lower Darling (Fig. 22.3). Thus, summer flows in the Lower Murray are mostly from the Darling, and generally are highly turbid (Walker *et al* 1991).

In the Headwaters and Riverine Plains tracts of the Murray phosphorus is potentially a limiting nutrient for algal growth (Walker & Hillman 1982), but levels increase substantially downstream, probably from agricultural fertilizers. Total phosphorus concentrations increase from about 0.02 mg l^{-1} in the headwaters to 0.15 mg l^{-1} in the lowest reaches (Mackay *et al* 1988). Nitrogen also is generally in abundant supply; for example, Kjeldahl-nitrogen levels increase from about 0.25 mg l^{-1} in the headwaters to 0.9 mg l^{-1} in South Australia. Seasonal shortages of nitrate may occur in the lower river, however, and it is possible that nitrate may then limit algal growth. Silica concentrations also vary (0–10 mg l^{-1}) with the density of diatom populations.

Pesticide concentrations in the river water are negligible (Mackay *et al* 1988), but significant levels of contamination may occur in the sediments of the Lower Murray (M.C. Thoms & K.F.

Walker, unpublished data). Pesticides may be responsible for malformed mouthparts in larval midges (Bennison *et al* 1989; Pettigrove 1989). There are no indications of severe pollution by heavy metals (e.g. Mackay *et al* 1988), although high levels of iron, manganese and zinc are accumulated by hyriid mussels (e.g. Walker 1986b).

Coliform counts in the Murray often exceed potable water standards (Mackay *et al* 1988), although this implies pollution from agricultural as well as domestic sources. Most towns use land-based disposal systems for treated sewage effuent (*cf.* Chapman & Simmons 1990; Hopmans *et al* 1990), and there are strict controls for waste disposal from boats (e.g. MDBMC 1987a).

22.4 BIOLOGICAL ENVIRONMENT

Riparian vegetation

The riparian vegetation of the Murray is dominated by two eucalypts: the river red gum *Eucalyptus camaldulensis* and the black box *E. largiflorens*. Red gums occur on riverbanks and in low-lying areas, and black box are in less-often flooded areas (e.g. higher ground or around the floodplain margins). Although the low diversity might suggest uniformity, the floodplain is a complex mosaic of communities superimposed on gradients determined by flooding frequency, and the distributions of the various species are strongly influenced by local differences in soils, landforms, climate and groundwater hydrology (Margules *et al* 1990).

Red gums form dense forests near Echuca, but more commonly occur in open woodlands, where they attain heights of 15–25 m (Margules *et al* 1990). Older trees are prone to drop branches, providing a habitat for many small animals. The understorey is mainly herbaceous and includes a mixture of annuals, postflooding ephemerals and perennials. Black box also occur as woodlands, generally with a lower (10–14 m), more open tree layer, and relatively high proportions of shrubs and dryland species in the understorey. One of the most prominent understorey species among stands of both trees is lignum (Polygonaceae: *Muehlenbeckia florulenta*), a woody shrub that

is widespread on the lower Murray flood-plain (Craig *et al* 1991). Exotic willows (*Salix babylonica*) are also conspicuous; they were planted extensively to stabilize the riverbanks after weir and levee construction, and have largely displaced red gums along the weir pools in South Australia.

Intensive land use in the Murray Valley has degraded the riparian vegetation, particularly the red gum woodlands of the lower Murray (Margules *et al* 1990; O'Malley & Sheldon 1990). Some of the factors responsible are salinization, drowning by impoundments, changes in groundwater tables, overgrazing by stock and feral animals (especially rabbits), fire, clearing and logging, and damage associated with recreational use. In addition, erosion has undermined many trees along the riverbank in South Australia (Lloyd *et al* 1991b; Walker *et al* 1991). The extent of disturbance is indicated by the general dominance of weeds in understorey communities.

The decline of the vegetation has been accelerated by river regulation, as the regeneration of both red gums and black box is largely dependent on flooding (e.g. Dexter *et al* 1986; Mackay & Eastburn 1990). The impact probably is least for black box, as they are associated with drier areas.

Macrophytes

The variable flows, high turbidities and unstable sediments of the Murray's main channel preclude development of extensive macrophyte beds. Macrophytes do occur, however, in sheltered littoral habitats and floodplain wetlands (e.g. Sainty & Jacobs 1990). The ecology, distribution and conservation status of many species are poorly known, and a general survey (*cf.* Margules *et al* 1990) would be invaluable. Several native and exotic species cause localized problems by obstructing irrigation channels, but the river generally remains free of major 'weed' problems.

The distributions of emergent and submerged species along the river in South Australia are broadly correlated with gradients of water-level fluctuations between weirs, and influenced also by high turbidities and local variations in bank slope and sediment stability (Walker *et al* 1991). There are useful ecological data for two of the most conspicuous emergent species, the com-

mon reed, *Phragmites australis*, and cumbungi
(bulrush), *Typha* spp. (e.g. Roberts & Ganf 1986;
Hocking 1989).

Plankton

Phytoplankton

The Murray supports a diverse phytoplankton
flora of more than 150 taxa, although only about
14 are abundant (Sullivan *et al* 1988). Throughout
much of the river, the dominant alga is the filamentous centric diatom *Melosira granulata*.
Blooms of this species develop in winter in the
river below Lake Hume, when the reservoir is
drawn down and destratified at the end of the
irrigation season. In the middle reaches the
blooms persist through spring (when *M. distans*
may also occur), late summer and autumn, and in
downstream areas they are restricted to early
winter and spring peaks associated with rising
floods from the upper Murray.

The centric diatoms *Cyclotella* and *Stephanodiscus* are abundant throughout the middle
and lower reaches, and dominate in the Murray
below the Darling junction. These species are
associated with areas of low turbulence (as in
weir pools), whereas high turbulence apparently
favours filamentous species (Sullivan *et al* 1988).
Species of *Asterionella*, *Atteya*, *Nitzschia*,
Rhizosolenia and *Synedra* are widespread in
the river. Other common groups include Cryp-
tomonadales (*Cryptomonas*, *Rhodomonas*),
Euglenales (*Euglena*, *Phacus*, *Trachelomonas*)
and Ochromonadales (*Mallomonas*, *Synura*);
these are mainly motile organisms able to persist
under conditions of high turbidity.

The commonest green algae, although in relatively low numbers, are chlorococcaleans (*Actin-
astrum*, *Ankistrodesmus*, *Scenedesmus*). The
filamentous species *Planctonema lauterbornei*
(Ulothricales) is seasonally abundant in the
Mallee Plains and South Australian tracts of the
river, particularly when turbidity is high. In early
1981, when there was a marked decrease in the
turbidity of Lake Alexandrina, the usual summer
peak of *P. lauterbornei* there was supplanted by a
cyanobacterial bloom (Geddes 1988; Sullivan *et
al* 1988).

The common Cyanobacteria (blue-green algae)
are species of *Anabaena*, *Anabaenopsis*,
Aphanizomenon and *Microcystis*. The high
turbidity of the lower Murray mitigates against
blooms, but they are prone to occur during low
flows in summer, when turbidity decreases via
settling of particulate matter and flocculation
caused by high salinity. Under these conditions,
given abundant nutrients, blooms have periodically clogged filters and caused shutdowns of
irrigation pumps and interruptions to town water
supplies.

In general, the composition and productivity of
the phytoplankton vary with season and location
(Sullivan *et al* 1988). Diversity and abundance are
highest in winter and spring, when diatoms are
dominant. Winter–spring assemblages are transported along the river, causing peaks at successive
downstream stations, but populations in summer
and autumn are less widely distributed. Flow is
clearly a dominant influence, but it is not easy to
distinguish the effects of nutrients, turbidity and
other correlated factors. Increased flows bring
higher levels of nutrients, turbidity and turbulence, and dilution of algal cells, and are less
favourable for growth than low to moderate flows.
The impact of high turbidities may not be simply
an optical effect (*cf.* Oliver 1990), but may be
related to the nature of the suspended particles
(Sullivan *et al* 1988).

Zooplankton

The zooplankton in the middle reaches of the
Murray is a typically limnetic assemblage, dominated by Cladocera and Copepoda, whereas the
Darling has a potamoplankton, dominated by
Rotifera (e.g. Shiel & Walker 1985). The differences reflect the prevalence of wetlands and
impoundments along the Murray, compared with
the generally dry floodplain and long, unregulated
reaches of the Darling. The zooplankton of the
Lower Murray includes species from both the
Darling and the Middle Murray. Tributaries and
reservoirs throughout the system tend to have
individually distinctive assemblages, although
there is wide temporal as well as geographical
variability (Shiel 1986, 1990).

Benthos and littoral communities

The sediments of the open river are barren in comparison to the littoral zone and floodplain wetlands (see below), presumably due to strong currents, unstable sediments, negligible underwater light and other manifestations of the harsh main-channel environment. Dredge samples from mid-channel generally yield only small numbers of chironomids and oligochaetes. The species richness of samples increases greatly, however, towards the river margins.

It appears that most biological diversity is concentrated in the littoral zone, especially among stands of water plants and in the vicinity of snags. This is most evident in the Lower Murray, where the littoral zone is marked by a narrow band of emergent plants (e.g. *Phragmites australis*, *Typha domingensis*) that provides a refuge for many terrestrial and aquatic animals. It is influenced by high turbidities in summer and autumn (owing to the way in which water from the Darling and Murray is mixed to provide irrigation flows), and by water-level fluctuations associated with weir operations. The distributions of littoral plants are broadly correlated with gradients in water-level changes between weirs, and there is evidence of secondary zonation in the distributions of some invertebrates.

Snags are another important littoral microhabitat (Lloyd *et al* 1991b). Along the river banks fallen branches and trunks, particularly of red gums, provide shelter from the current and encourage the development of littoral plant and animal communities. They are also a refuge, spawning site and feeding ground for native fish. In water deeper than about 0.5 m (where there is little light), the snags support dense populations of fungi and nematodes (*cf.* Nicholas *et al* 1991). The littoral algae and fungi are grazed by large numbers of prawns and shrimps (*Macrobrachium australiense*, *Paratya australiensis* and *Caridina mccullochi*). In general, the diversity of macroinvertebrates at any site varies with a number of factors, including the location, configuration and orientation of the snag and the complexity of its surface (Lloyd *et al* 1991b; O'Connor 1991). Extensive desnagging carried out in the past to assist navigation and improve channel hydraulic efficiency has undoubtedly had adverse effects, but these works are now more restrained, for economic and environmental reasons.

A survey in 1980–5 of macroinvertebrates along the length of the Murray recorded 439 taxa (Bennison *et al* 1989). Most were insect larvae, but crustaceans, molluscs and oligochaetes were also well represented. Coleoptera, Ephemeroptera, Plecoptera and Trichoptera were numerically dominant in the upper reaches, whereas Diptera (Chironomidae) dominated at most other sites. Odonates were generally common. Analyses of 'functional feeding groups' revealed a gradual change in proportional abundances along the river (an increase in Crustacea at the expense of Insecta), consistent with the River Continuum Concept. An apparent reduction in diversity at stations in South Australia was attributed to the effects of high turbidity and predominantly lacustrine conditions.

A striking feature of the Lower Murray fauna is the scarcity of freshwater snails (Walker *et al* 1991). Although about 15 species were present in the period before weir construction, there have been sporadic records of only six species over the past decade and only two, the ancylid *Ferrissia petterdi* and the planorbid *Glyptophysa gibbosa*, are consistently present (but in small, patchily distributed populations). The decline of the snails may have various causes, including the isolation of wetlands and the direct and indirect effects of high turbidity and changeable water levels. There is no reason to suppose that there has been a change in the availability of calcium over the period of the decline. The sporadic resurgences of snail populations over the last 10 years may represent short-lived recoveries linked to floods.

Freshwater mussels (Hyriidae) are common throughout the Murray (e.g. Walker 1981, 1986a). The range of the obligate riverine species *Alathyria jacksoni* has contracted since construction of weirs, and *Velesunio ambiguus*, a species typical of floodplain wetlands, has invaded the weir pools and the sheltered margins of the main channel. *A. jacksoni* may be excluded from floodplain environments mainly because it is unable to tolerate low oxygen levels, and *V. ambiguus* may be excluded from the faster flowing sections of the river by its relatively weak burrowing ability (Sheldon & Walker 1989).

A similar redistribution has occurred among the crayfish (Parastacidae) (Walker 1986a). The yabbie (*Cherax destructor*) is typical of still-water floodplain habitats, but is common also in weir pools and along the river margins of the lower river. The Murray crayfish (*Euastacus armatus*) is a riverine species and virtually extinct in the lower river, although still common in New South Wales. The decline of the Murray crayfish in the lower river is circumstantially related to the construction of weirs, and the Murray−Darling Basin Commission is presently sponsoring research to determine the feasibility of restocking populations in South Australia (M.C. Geddes, University of Adelaide, personal communication).

Fish and fisheries

The Murray−Darling system supports only about 24 native freshwater fish (or 40−50 including marine and estuarine species). This is a tiny complement by world standards, but consistent with a small continental fauna (about 150 freshwater species) that is a legacy of changeable, often harsh climates and long geological isolation. The Murray−Darling fish are all recent evolutionary derivatives of marine families; this is reflected, for example, in broad salinity tolerances (e.g. Langdon 1987; Williams & Williams 1991). Another ten species have been introduced to the system (Morison 1989), increasing the freshwater fauna by about 40%. These include members of the families Cobitidae, Cyprinidae, Percidae, Poeciliidae and Salmonidae, all absent from the original fauna.

The life histories of the native fish are broadly of four kinds, depending on the response to floods (Lloyd *et al* 1991a):

1 Species such as callop (or golden perch; Percichthyidae: *Macquaria ambigua*) and silver perch (Teraponidae: *Bidyanus bidyanus*) spawn only when a flood occurs, provided that the temperature is sufficiently high. These are highly mobile species, able to exploit localized flooding and then disperse widely. The eggs are buoyant and may be spawned in the river to develop as they float downstream.

2 Small, short-lived species such as the crimson-spotted rainbowfish (Melanotaeniidae: *Melanot-*

aenia splendida fluviatilis) and western carp gudgeon (Eleotridae: *Hypseleotris klunzingeri*) spawn seasonally, and more intensely (perhaps repeatedly) when there is a flood.

3 Murray cod (Percichthyidae: *Maccullochella peeli*) and freshwater catfish (Plotosidae: *Tandanus tandanus*) also spawn seasonally, independently of floods. Breeding appears to depend on temperature and day length, but the young depend upon food from flooded wetlands and recruitment is likely to fail in the absence of a flood. Cod are territorial and often occupy sheltered holes in lowland reaches; their eggs are attached to fallen wood ('snags') or other solid surfaces. Catfish spawn in shallow gravel nests.

4 Some species (e.g. galaxiids, lampreys) rely on estuarine or marine environments for spawning and recruitment.

Nearly all the native fish have declined in range and abundance over the past century. The decline is variously attributed to environmental changes, interactions with exotic species and overfishing (Walker 1983; Cadwallader 1986a, 1986b; Lawrence 1989; Rowland 1989; Cadwallader & Lawrence 1990).

Habitat changes include the effects of flow regulation, notably changes in the flooding regime (Geddes & Puckridge 1989; Rowland 1989; Thoms & Walker 1989; Gehrke 1990), the downstream effects of dams on temperatures, oxygen and water levels (Walker 1985), barriers to migration (Mallen-Cooper 1989), desnagging (Lloyd *et al* 1991b), impoundments and siltation.

The exotic species include common carp (*Cyprinus carpio*), gambusia (*Gambusia holbrooki*), European perch (*Perca fluviatilis*) and trout (*Salmo gairdneri*, *S. trutta*). These are variously implicated in the decline of native species, through predation, competition and habitat destruction (Fletcher 1986; Lloyd *et al* 1986). Deliberate and accidental introductions have occurred throughout the past century (Morison 1989), and the establishment of tilapia (*Oreochromis mossambicus*) in at least part of the Murray−Darling system now seems inevitable, following its escape from captivity in Queensland (Arthington 1986; Bluhdorn & Arthington 1990).

The commercial and recreational fisheries associated with the Murray are administered by three state governments and have a complex history (Walker 1983; O'Connor 1989; Rohan 1989; Rowland 1989) that is obscured by changes in management and a general lack of statistical data. Murray cod was the principal catch until the 1950s, when populations declined sharply. Callop became most important through the 1960s and common carp in the 1970s. The main commercial species now are bony bream, callop and common carp. Native fish such as the catfish and silver perch presently have little or no commercial significance, although they have periodically been important. The case for over-fishing is speculative, but is based on the observation that previously unfished populations of long-lived fish including many age classes are susceptible to catastrophic declines. This may account for the decline in catches of Murray cod (Rowland 1989). The practice of concentrated fishing during periods of drought may also be implicated (Walker 1983). The Murray–Darling fisheries have never been of major commercial value, and with the decline of the most favoured species the state governments are allowing the professional fishery to be overtaken gradually by the recreational fishery. There are restrictions on fishing methods, and the closed season for Murray cod in South Australia (Rohan 1989) has been extended to a year-round moratorium.

Five Murray–Darling species are threatened by extinction in the foreseeable future and a further nine may soon become endangered (Lloyd *et al* 1991a). Perhaps the only native species to have increased in numbers is the bony bream, a gizzard shad that thrives in the impounded waters of the Lower Murray (Puckridge & Walker 1990). The situation is most precarious in South Australia, where there have already been several regional extinctions (Lloyd & Walker 1986). Official recognition may be too conservative: it is ominous that there have been no records of the western Chanda perch (*Ambassis castelnaui*) for nearly 20 years. Nevertheless, various management initiatives have been proposed (e.g. Lawrence 1989).

Wetlands

The extensive wetlands associated with the Murray are a vital part of the river-floodplain ecosystem because the wetland communities depend upon the river for replenishment, and because the riverine biota depend on the wetlands for food, breeding sites and refuges (Walker 1986a). They have been degraded by flow regulation, and several government agencies are actively concerned with wetland restoration and conservation, particularly the prospects for manipulating water levels to enhance breeding of fish and waterfowl.

The floodplains of the Murray and its major anabranches (8900 km^2) contain 6433 billabongs, backwaters, lakes, swamps and other wetlands with a total area of 1040 km^2 (Pressey 1986). Seventy-one per cent of the floodplain wetlands have areas of 5 ha or less, and 91% are 20 ha or less. Other major wetland areas (1187 km^2) are associated with lakes Alexandrina and Albert and the Coorong lagoon, near the river mouth (Pressey 1986; Thompson 1986; Geddes 1987).

Flow regulation has extended the area covered by permanent wetlands, but decreased the area subject to occasional flooding (Pressey 1986). Before regulation, South Australian wetlands that are now connected to the river at minimum flows (pool level) were subject to larger, more frequent, water-level changes, and would have dried periodically . Wetlands above regulated river levels also have been affected, particularly those exposed to minor floods and droughts under the pre-regulation regime (*cf.* section 22.2). Thus, impermanent wetlands are more representative of the natural (unregulated) regime than those sustained by regulated flows. There are useful ecological data for temporary wetlands on the Lachlan River (see Fig. 22.1) (Briggs & Maher 1985; Briggs *et al* 1985, Crome 1986, Crome & Carpenter 1988) and for the floodplain of the Murray at Chowilla, South Australia (O'Malley & Sheldon 1990; Boulton & Lloyd 1991; Roberts & Ludwig 1991), where once-temporary wetlands have been made permanent by flow regulation. Other data are included in studies of emergent plants in wetlands associated with irrigation areas near the Murrumbidgee River (Roberts & Ganf 1986; Hocking 1989).

Concern about the long-term implications of hydrological changes for wetlands has prompted a number of surveys. All the Murray wetlands have been mapped and assigned to geomorphological and hydrological categories, and some have been recommended for independent hydrological management to restore and maintain more nearly natural conditions (Pressey 1986, 1989; Thompson 1986). More work is needed to confirm that these categories are ecologically significant, and so appropriate for management of the biological resources associated with wetlands.

The physical diversity of wetlands, including various lentic channel forms, floodplain channels, scroll swales and slackwater areas, is not adequately conveyed by colloquial terms such as 'billabong'. This often refers to an oxbow lake associated with a former river channel (e.g. Boon *et al* 1990), but may be loosely applied to other lentic riverine wetlands. Many of the so-called billabongs along the Murray gorge in South Australia are 'channel-margin swales' (Pressey 1986); they are constrained by artificial levees, and are remnants of once-extensive riverside swamps that are now used for agriculture.

There are many typical oxbow billabongs along the Murray between Lake Hume and the Murrumbidgee junction (see Fig. 22.1), but even within this apparently cohesive group there is a great diversity of sizes, shapes and flooding regimes (Boon *et al* 1990). Billabong environments are highly variable, spatially and temporally, and even closely adjacent water-bodies may behave quite differently in physical, chemical and biological terms. Many of the plant and animal inhabitants are opportunistic, and the composition of the communities may change completely within only a few days. The patchiness of billabong resources may partly explain why species richness and biomass in these communities are much greater than in the river (e.g. bacteria: Boon 1990; aquatic plants: Boon *et al* 1990; micro- and macroinvertebrates: Boon & Shiel 1990; Hillman & Shiel 1991; Boulton & Lloyd 1991). Much of the biodiversity associated with the Riverine Plains Tract of the Murray occurs on the floodplain, and particularly in wetlands.

Billabongs share many plant and animal species with the river, but generally have a distinctive biota. Species of the angiosperms *Myriophyllum*, *Potamogeton* and *Vallisneria* occur in both environments near Albury, but in the river they apparently reproduce vegetatively rather than by flowering (Boon *et al* 1990). Many species of caddisflies (Trichoptera) and midges (Chironomidae) occur in both environments, and other insects (e.g. Ephemeroptera, Megaloptera, Plecoptera) are restricted mainly to flowing water. The prawn *Macrobrachium australiense* also is principally a lotic species. The zooplankton (Shiel 1990), crayfish and freshwater mussels include specialized lentic and lotic species (see above), and there may be a similar separation among the atyid shrimps (F. Sheldon, University of Adelaide, personal communication).

22.5 RESEARCH AND MANAGEMENT

In the past 5 years environmental research on the Murray has been greatly expanded by contributions from government agencies (*cf.* Walker 1986c). Indeed, the volume of literature may have more than doubled in this time. Part is so-called 'grey literature', but this is often valuable and cannot be ignored. It is cited here freely provided that the article is a formal publication and readily available through Australian (if not international) libraries.

Some of the impetus for the expansion has come from the Murray–Darling Basin Ministerial Council and its agency the Murray–Darling Basin Commission, and from a grants scheme administered by Australian Water Research Advisory Council (now the Land and Water Resources Research and Development Corporation). Several newly founded research centres are involved at least partly in work on the Murray; the Murray–Darling Freshwater Research Centre at Albury is particularly significant. State government departments, the Commonwealth Scientific and Industrial Organization (CSIRO) and the University of Adelaide also have long-term, continuing interests.

The new commitment to research reflects the importance of the Murray, and the Murray–Darling Basin in general, as a national resource. The commitment is long overdue, given so many indications of degradation in the environment, and it may be precarious because, unfortunately, research does not guarantee immediate solutions to the problems.

If scientists regard environmental research as a

long-term, *proactive* occupation that anticipates the future needs of management, this perception is not always shared by management (Cullen 1990). Rather, it is often seen as a short-term, *reactive* process, seeking solutions to problems at hand. Research and management therefore may still be 'estranged partners' (Walker 1985), because the potential relationship between the two is not fully realized. Long-term perspectives do not easily find favour in the prevailing climate of economic rationalism, although long-term sustainable use is part of the management strategy for the Murray–Darling Basin (MDBMC 1989).

Good information is an essential part of good management. The water resources of the system are regulated using sophisticated models developed from extensive historical records of river flow and levels. The basis for ecological management, however, is very slender: there are few biological data before the mid-1970s, and it is therefore difficult to determine trends. This situation is likely to persist until there is a commitment to comprehensive monitoring of the biological, physical and chemical environment. This needs to be a long-term commitment, as short-term data clearly are inadequate for an environment that is highly variable.

Despite encouraging advances, there remains a tendency to resolve the Murray's complex environmental problems into single 'management issues' such as native fish, red gums and waterbirds, and to develop proposals for each. Yet waterbirds clearly are not the only element of wetland communities, and management specifically in the interests of waterbird populations is likely to prejudice the other important ecological functions that wetlands perform. The same argument might apply to management directed specifically at red gums, or native fish, or even humans. These initiatives need to be implemented in the context of a broader, holistic view of the dynamics of the river-floodplain ecosystem. This kind of understanding is merely rudimentary at present.

Although the Murray has been profoundly changed by flow regulation, with numerous indications of decline in the native environment, rivers generally have a remarkable capacity for recovery from disturbance (Sparks *et al* 1990). Over the past 20 years major floods in the lower

Murray (e.g. 1974–5, 1983, 1990) have brought significant resurgences in the floodplain vegetation and populations of waterbirds, fish and invertebrates. There are limits to the resilience of the system, however, and the challenge for management and research is to determine levels of short- and long-term exploitation that are commensurate with the capacity of the river to absorb the effects.

22.6 CONCLUSION

This chapter began with the implicit assumption that the Murray could be regarded as a typical semiarid river. In fact, such a claim might be difficult to defend because the Murray is also a floodplain river. The floodplain is well watered, compared with the surrounding semiarid land, and there are more-or-less frequent exchanges between the river and its forests and wetlands. The Darling does not have comparable floodplain communities and would have engendered quite a different picture of a semiarid river. There is very little ecological information for the Darling, however, and even basic exploratory work remains to be done. Whereas neither the Murray nor Darling alone could be regarded as representative of semiarid rivers elsewhere in the world, together they indicate some of the diversity that exists among rivers of this genre. The Murray–Darling system may be most closely compared, perhaps, with the Orange–Vaal system of southern Africa (Cambray *et al* 1986; Allanson *et al* 1990).

The salient feature of all semiarid rivers is their highly variable discharge. This governs the physical, chemical and biological environment, and also determines the way that the water resources are utilized for industry. From an ecological viewpoint, floods are of paramount importance. Many of the native plants and animals depend upon periodic (not necessarily seasonal) floods to cue reproduction and sustain recruitment. Inevitably, regulation interferes with the flooding regime and may cause a decline among these species. The hydrological character of floods, and the nature of biological responses to them, are extremely complex and the discipline of river ecology is only just beginning to develop an adequate basis for management of flood-dominated systems. The Food Pulse Concept (Junk *et al* 1989) recognizes the biological significance of floods, and presents an exciting opportunity for further conceptual development.

ACKNOWLEDGEMENTS

This review draws upon work supported by the Australian Water Research Advisory Council, Commonwealth Department of Primary Industries and Energy (projects 85/45, 86/40).

REFERENCES

Allanson BR, Hart RC, O'Keeffe JH, Robarts RD. (1990) *Inland Waters of Southern Africa: An Ecological Perspective.* Kluwer Academic Publishers, Dordrecht. [22.6]

Allison GB, Cook PG, Barnett SR, Walker GR, Jolly JD, Hughes MW. (1990) Land clearance and river salinisation in the western Murray Basin, Australia. *Journal of Hydrology* **119**: 1–20. [22.3]

Arthington AH. (1986) Introduced cichlid fish in Australian inland waters. In: De Deckker P, Williams WD (eds) *Limnology in Australia,* pp 239–48. CSIRO, Melbourne and Dr W. Junk, Dordrecht. [22.4]

Baker BW, Wright GL. (1978) The Murray Valley: its hydrological regime and the effects of water development on the river. *Proceedings of the Royal Society of Victoria* **90**: 103–10. [22.3]

Banens RJ. (1990) The limnology of Dumaresq Reservoir: a small monomictic upland lake in northern New South Wales, Australia. *Archiv für Hydrobiologie* **117**: 319–55. [22.3]

Bennison GL, Hillman TJ, Suter PJ. (1989) Macroinvertebrates of the River Murray, South Australia (survey and monitoring, 1980–85). *Water Quality Report* 3. Murray–Darling Basin Commission, Canberra. [22.3, 22.4]

Blainey G. (1976) *Triumph of the Nomads. A History of Ancient Australia.* Sun Books, Melbourne. [22.3]

Bluhdorn DR, Arthington AH. (1990) Somatic characteristics of an Australian population of *Oreochromis mossambicus* (Pisces, Cichlidae). *Environmental Biology of Fishes* **29**: 277–91. [22.4]

Boon PI. (1989) Relationships between actinomycete populations and organic matter degradation in Lake Mulwala, southeastern Australia. *Regulated Rivers: Research and Management* **4**: 409–18. [22.3]

Boon PI. (1990) Organic matter degradation and nutrient regeneration in Australian freshwaters, II. Spatial and temporal variation, and relation with environmental conditions. *Archiv für Hydrobiologie* **117**: 405–36. [22.4]

Boon PI, Shiel RJ. (1990) Grazing on bacteria by zooplankton in Australian billabongs. *Australian Journal of Marine and Freshwater Research* **41**: 247–57. [22.4]

Boon PI, Frankenberg J, Hillman TJ, Oliver RL, Shiel RJ. (1990) Billabongs. In: Mackay NJ, Eastburn D. (eds) *Understanding the Murray,* pp 183–200. Murray–Darling Basin Commission, Canberra. [22.4]

Boulton AJ, Lloyd LN. (1991) Macroinvertebrate assemblages in floodplain habitats of the lower River Murray, South Australia. *Regulated Rivers: Research and Management* **6**: 183–201 [22.4]

Bowler JM. (1978) Quaternary climates and tectonics in the evolution of the Riverine Plain, southeastern Australia. In: Davies JL, Williams MAJ (eds) *Landform Evolution in Australasia,* pp 70–112. Australian National University Press, Canberra. [22.2]

Bowler JM. (1986) Quaternary landform evolution. In: Jeans DN (ed.) *Australia—A Geography,* Vol. 1, pp 117–47. Routledge & Kegan Paul, London. [22.2]

Bowler JM. (1990) The last 500 000 years. In: Mackay JN, Eastburn D (eds) *Understanding the Murray,* pp 95–110. Murray–Darling Basin Commission, Canberra. [22.2]

Bren LJ. (1987) The duration of inundation in a river red gum forest. *Australian Forest Research* **17**: 191–202. [22.3]

Bren LJ. (1988) Effects of river regulation on flooding of a riparian red gum forest on the River Murray, Australia. *Regulated Rivers: Research and Management* **2**: 65–77. [22.3]

Briggs SV, Maher MT. (1985) Limnological studies of waterfowl habitat in south-western New South Wales, II. Aquatic macrophyte productivity. *Australian Journal of Marine and Freshwater Research* **36**: 707–15. [22.4]

Briggs SV, Maher MT, Carpenter SM. (1985) Limnological studies of waterfowl habitat in south-western New South Wales, I. Water chemistry. *Australian Journal of Marine and Freshwater Research* **36**: 59–67. [22.4]

Cadwallader PL. (1986a) Fish of the Murray–Darling system. In: Davies BR, Walker KF (eds) *The Ecology of River Systems,* pp 679–94. Dr W. Junk, Dordrecht. [22.4]

Cadwallader PL. (1986b) Flow regulation in the Murray River system and its effect on the native fish fauna. In: Campbell IC (ed.) *Stream Protection. The Management of Rivers for Instream Uses,* pp 115–33. Water Studies Centre, Chisholm Institute of Technology, Victoria. [22.4]

Cadwallader PL, Lawrence B. (1990) Fish. In: Mackay NJ, Eastburn D (eds) *Understanding the Murray,* pp 317–36. Murray–Darling Basin Commission, Canberra. [22.4]

Cambray JA, Davies BR, Ashton PJ. (1986) The Orange–Vaal River system. In: Davies BR, Walker KF (eds)

Ecology of River Systems, pp 89–122. Dr W. Junk, Dordrecht. [22.6]

Chapman JC, Simmons BL. (1990) The effects of sewage on alpine streams in Kosciusko National Park, NSW. *Environmental Monitoring and Assessment* **14**: 275–95. [22.3]

Chesterfield EA. (1986) Changes in the vegetation of the river red gum forest at Barmah, Victoria. *Australian Forestry* **49**: 4–15. [22.3]

Close AF. (1988) Potential impact of greenhouse effect on the water resources of the River Murray. In: Pearman GI (ed.) *Greenhouse – Planning for Climatic Change*, pp 312–23. CSIRO & E.J.Brill, Melbourne. [22.2]

Close AF. (1990a) The impact of man on the flow regime. In: Mackay NJ, Eastburn D (eds) *Understanding the Murray*, pp 61–76. Murray–Darling Basin Commission, Canberra. [22.3]

Close AF. (1990b) River salinity. In: Mackay NJ, Eastburn D (eds) *Understanding the Murray*, pp 127–46. Murray–Darling Basin Commission, Canberra. [22.3]

Cornish PM. (1977) Changes in seasonal and annual rainfall in New South Wales. *Search* **8**: 38–40. [22.2]

Craig AE, Walker KF, Boulton AJ. (1991) Effects of edaphic factors and flood frequency on the abundance of lignum (*Muehlenbeckia florulenta* Meissner) (Polygonaceae) on the River Murray floodplain, South Australia. *Australian Journal of Botany* **39**: 431–43. [22.4]

Crome FHJ. (1986) Australian waterfowl do not necessarily breed on a rising water level. *Australian Wildlife Research* **13**: 461–80. [22.4]

Crome FHJ, Carpenter SM. (1988) Plankton community cycling and recovery after drought–dynamics in a basin on a flood plain. *Hydrobiologia* **164**: 193–211. [22.4]

Cullen P. (1990) The turbulent boundary between water science and water management. *Freshwater Biology* **24**: 201–9. [22.5]

Dexter BD, Rose HJ, Davies N. (1986) River regulation and associated forest management problems in the River Murray red gum forests. *Australian Forestry* **49**: 16–27. [22.3, 22.4]

Eastburn D. (1990) *The River Murray: History at a Glance.* Murray–Darling Basin Commission, Canberra. [22.3]

Evans RS. (1989) Saline water disposal options. *BMR Journal of Australian Geology and Geophysics* **11**: 167–85. [22.3]

Evans WR, Kellett JR. (1989) The hydrogeology of the Murray Basin, southeastern Australia. *BMR Journal of Australian Geology and Geophysics* **11**: 147–66. [22.3]

Finlayson BL, McMahon TA. (1988) Australia *v* the world: a comparative analysis of streamflow charac-teristics. In: Warner RF (ed.) *Fluvial Geomorphology of Australia*, pp 17–39. Academic Press, Sydney. [22.2]

Fletcher AR. (1986) Effects of introduced fish in Australia. In: De Deckker P, Williams WD (eds) *Limnology in Australia*, pp 231–8. CSIRO, Melbourne and Dr W. Junk, Dordrecht. [22.4]

Flood J. (1983) *Archaeology of the Dreamtime.* Collins, Sydney. [22.3]

Geddes MC. (1987) Changes in salinity and in the distribution of macrophytes, macrobenthos and fish in the Coorong lagoons, South Australia, following a period of River Murray flow. *Transactions of the Royal Society of South Australia* **111**: 173–81. [22.4]

Geddes MC. (1988) The role of turbidity in the limnology of Lake Alexandrina, River Murray, South Australia: comparisons between clear and turbid phases. *Australian Journal of Marine and Freshwater Research* **39**: 201–9. [22.4]

Geddes MC, Puckridge JT. (1989) Survival and growth of larval and juvenile native fish: the importance of the floodplain. In: Lawrence B (ed.) *Proceedings of the Native Fish Management Workshop*, pp 101–14. Murray–Darling Basin Commission, Canberra. [22.4]

Gehrke PC. (1990) Spatial and temporal dispersion patterns of golden perch, *Macquaria ambigua*, larvae in an artificial floodplain environment. *Journal of Fish Biology* **37**: 225–36. [22.4]

Graf WL. (1987) *Fluvial Processes in Dryland Rivers.* Springer-Verlag, Berlin. [22.1, 22.3]

Grieve AM. (1987) Salinity and waterlogging in the Murray–Darling Basin. *Search* **18**: 72–4. [22.3]

Hart BT, Bailey P, Edwards R, Hortle K, James K, McMahon A *et al.* (1990) Effects of salinity on river, stream and wetland ecosystems in Victoria, Australia. *Water Research* **24**: 1103–17. [22.3]

Hillman TJ, Shiel RJ. (1991) Micro- and macro-invertebrates in Australian billabongs. *Verhand-lungen Internationale Vereinigung für Theoretische und Angewandte Limnologie* **24**: (in press). [22.4]

Hocking PJ. (1989) Seasonal dynamics of production, and nutrient accumulation and cycling by *Phragmites australis* (Cav.) Trin. *ex* Stuedel in a nutrient-enriched swamp in inland Australia, I. Whole plants. *Australian Journal of Marine and Freshwater Research* **40**: 421–44. [22.4]

Hopmans P, Stewart HTL, Flinn DW, Hillman TJ. (1990) Growth, biomass production and nutrient accumu-lation by seven tree species irrigated with municipal effluent at Wodonga, Australia. *Forest Ecology and Management* **30**: 203–11. [22.3]

Jacobs T. (1989) Regulation of the Murray–Darling river system. In: Lawrence B (ed.) *Proceedings of the Native Fish Management Workshop*, pp 55–96. Murray–Darling Basin Commission, Canberra. [22.2, 22.3]

Jacobs T. (1990) River regulation. In: Mackay NJ,

Eastburn D (eds) *Understanding the Murray*, pp 39–60. Murray–Darling Basin Commission, Canberra. [22.3]

Junk WJ, Bayley PB, Sparks RE. (1989) The flood pulse concept in river-floodplain systems. In: Dodge DP (ed.) *Proceedings of the International Large River Symposium*, pp 110–27. Canadian Special Publications in Fisheries and Aquatic Sciences 106. [22.6]

Langdon JS. (1987) Active osmoregulation in Australian bass, *Macquaria novemaculeata* (Steindachner), and golden perch, *Macquaria ambigua* (Richardson) (Percichthyidae). *Australian Journal of Marine and Freshwater Research* **38**: 771–6. [22.3, 22.4]

Lawrence B. (ed.) (1989) *Proceedings of the Workshop on Native Fish Management*. Murray–Darling Basin Commission, Canberra. [22.4]

Lloyd LN, Walker KF. (1986) Distribution and conservation status of small freshwater fish in the River Murray, South Australia. *Transactions of the Royal Society of South Australia* **106**: 49–57. [22.4]

Lloyd LN, Arthington AH, Milton DA. (1986) The mosquitofish—a valuable mosquito-control agent or a pest? In: Kitching RL (ed.) *The Ecology of Exotic Animals and Plants. Some Australian Case Histories*, pp 7–25. Wiley, Brisbane. [22.4]

Lloyd LN, Puckridge JT, Walker KF. (1991a) Fish populations of the Murray–Darling river system and their requirements for survival. In: Dendy T, Coombe M (eds) *Conservation and Management of the River Murray: Making Conservation Count*. Department of Environment and Planning, South Australia pp 86–99. [22.4]

Lloyd LN, Walker KF, Hillman TJ. (1991b) *Environmental significance of snags in the River Murray*. Project Completion Report 85/45, Australian Water Research Advisory Council. Department of Primary Industries and Energy, Canberra pp 1–33. [22.4]

Mackay NJ, Eastburn D. (eds) (1990) *Understanding the Murray*. Murray–Darling Basin Commission, Canberra. [22.4]

Mackay NJ, Hillman TJ, Rolls J. (1988) Water quality of the River Murray: review of monitoring, 1978 to 1986. *Water Quality Report 1*. Murray–Darling Basin Commission, Canberra. [22.2, 22.3]

Macumber P. (1990) The salinity problem. In: Mackay NJ, Eastburn D (eds) *Understanding the Murray*, pp 111–26. Murray–Darling Basin Commission, Canberra. [22.3]

Mallen-Cooper M. (1989) Fish passage in the Murray–Darling Basin. In: Lawrence B (ed.) *Proceedings of the Native Fish Management Workshop*, pp 123–35. Murray–Darling Basin Commission, Canberra. [22.4]

Margules & Partners P, Smith J, Department of Conservation, Forests and Lands, Victoria (1990) *River Murray Riparian Vegetation Study*. Murray–Darling Basin Commission, Canberra. [22.3, 22.4]

Martin HA. (1989) Vegetation and climate of the late Cainozoic in the Murray Basin and their bearing on the salinity problem. *BMR Journal of Australian Geology and Geophysics* **11**: 291–9. [22.2]

Murray–Darling Basin Ministerial Council (MDBMC) (1987a) *Murray–Darling Basin Environmental Resources Study*. State Pollution Control Commission, Sydney, for MDBMC, Canberra. [22.2, 22.3]

MDBMC (1987b) *Salinity and Drainage Strategy*. Background Paper 87/1. Murray–Darling Basin Ministerial Council, Canberra. [22.3]

MDBMC (1989) *Murray–Darling Basin Natural Resources Management Strategy*. Background papers. Murray–Darling Basin Ministerial Council, Canberra. [22.5]

Mills BJ, Geddes MC. (1980) Salinity tolerance and osmoregulation of the Australian freshwater crayfish *Cherax destructor* Clark (Decapoda, Parastacidae). *Australian Journal of Marine and Freshwater Research* **31**: 667–76. [22.3]

Morison AK. (1989) Management of introduced species in the Murray–Darling Basin—a discussion paper. In: Lawrence B (ed.) *Proceedings of the Native Fish Management Workshop*, pp 149–61. Murray–Darling Basin Commission, Canberra. [22.4]

Morton R, Cunningham RB. (1985) Longitudinal profile of trends in salinity in the River Murray. *Australian Journal of Soil Research* **23**: 1–13. [22.3]

Nicholas WL, Bird AF, Beech TA, Stewart AC. (1991) The nematode fauna of the Murray River estuary: the effect of the barrages across its mouth. *Hydrobiologia* (in press) [22.4]

O'Connor NA. (1991) The effects of habitat complexity on the macroinvertebrates colonising wood substrates in a lowland stream. *Oecologia (Berlin)* **85**: 504–12. [22.4]

O'Connor PF. (1989) Fisheries management in inland New South Wales. In: Lawrence B (ed.) *Proceedings of the Native Fish Management Workshop*, pp 19–23. Murray–Darling Basin Commission, Canberra. [22.4]

Oliver RL. (1990) Optical properties of waters in the Murray–Darling Basin, south-eastern Australia. *Australian Journal of Marine and Freshwater Research* **41**: 581–601. [22.4]

O'Malley C, Sheldon F. (eds) (1990) *Chowilla Floodplain Biological Study*. Hyde Park Press (Richmond, South Australia) for Nature Conservation Society of South Australia, Adelaide pp 1–224. [22.3, 22.4]

Pettigrove V. (1989) Larval mouthpart deformities in *Procladius paludicola* Skuse (Diptera: Chironomidae) from the Murray and Darling rivers, Australia. *Hydrobiologia* **179**: 111–17. [22.3]

Pressey R. (1986) *Wetlands of the River Murray below Lake Hume*. River Murray Commission, Canberra. [22.3, 22.4]

Pressey R. (1989) Wetland evaluation: a review of

approaches and their relevance to water management. In: Teoh CH (ed.) *Proceedings of the Specialist Workshop on Instream Needs and Water Uses.* Australian Water Resources Council Conference Series 18, 13 pp. [22.4]

Puckridge JTW, Walker KF. (1990) Reproductive biology of the bony bream, *Nematalosa erebi* (Günther), in the River Murray, South Australia. *Australian Journal of Marine and Freshwater Research* **41**: 695–712. [22.4]

Riley SJ. (1988) Secular change in the annual flows of streams in the NSW section of the Murray–Darling Basin. In: Warner RF (ed.) *Fluvial Geomorphology of Australia*, pp 245–66. Academic Press, Sydney. [22.2]

Roberts J, Ganf GG. (1986) Annual production of *Typha orientalis* Presl. in inland Australia. *Australian Journal of Marine and Freshwater Research* **37**: 659–68. [22.4]

Roberts J, Ludwig JA. (1991) Riparian vegetation along current exposure gradients in floodplain wetlands of the River Murray, Australia. *Journal of Ecology* **79**: 117–27. [22.4]

Rohan G. (1989) River fishery (South Australia) — review of management arrangements. In: Lawrence B (ed.) *Proceedings of the Native Fish Management Workshop*, pp 37–53. Murray–Darling Basin Commission, Canberra. [22.4]

Rowland SJ. (1989) Aspects of the history and fishery of the Murray cod, *Maccullochella peeli* (Mitchell) (Percichthyidae). *Proceedings of the Linnean Society of New South Wales* **111**: 201–13. [22.4]

Rutherfurd I. (1990) Ancient river, young nation. In: Mackay NJ, Eastburn D (eds) *Understanding the Murray*, pp 17–38. Murray–Darling Basin Commission, Canberra. [22.3]

Sainty GR, Jacobs SWL. (1990) Waterplants. In: Mackay NJ, Eastburn D (eds) *Understanding the Murray*, pp 265–74. Murray–Darling Basin Commission, Canberra. [22.4]

Sheldon F, Walker KF. (1989) Effects of hypoxia on oxygen consumption by two species of freshwater mussel (Unionacea: Hyriidae) from the River Murray. *Australian Journal of Marine and Freshwater Research* **40**: 491–9. [22.4]

Shiel RJ. (1986) Zooplankton of the Murray–Darling system. In: Davies BR, Walker KF (eds) *The Ecology of River Systems*, pp 661–77. Dr W. Junk, Dordrecht. [22.4]

Shiel RJ. (1990) Zooplankton. In: Mackay NJ, Eastburn D (eds) *Understanding the Murray*, pp 275–86. Murray–Darling Basin Commission, Canberra. [22.4]

Shiel RJ, Walker KF. (1985) Zooplankton of regulated and unregulated streams: the Murray–Darling river system, Australia. In: Lillehammer A, Saltveit S (eds) *Regulated Rivers*, pp 263–70. Universitetsforlaget, Oslo. [22.4]

Sparks RE, Bayley PB, Kohler SL, Osborne LL. (1990) Disturbance and recovery of large floodplain rivers. *Environmental Management* **14**: 699–709. [22.5]

Stephenson AE. (1986) Lake Bungunnia: a Plio-Pleistocene megalake in southern Australia. *Palaeogeography, Palaeoclimatology, Palaeoecology* **57**: 137–56. [22.2]

Stephenson AE, Brown CM. (1989) The ancient Murray River system. *BMR Journal of Australian Geology and Geophysics* **11**: 387–95. [22.2]

Sulivan C, Saunders J, Welsh D. (1988) Phytoplankton of the River Murray, 1980–1985. *Water Quality Report 2.* Murray–Darling Basin Commission, Canberra. [22.4]

Thomas DSG. (1989) The nature of arid environments. In: Thomas DSG (ed.) *Arid Zone Geomorphology*, pp 1–10. Belhaven Press, London & Halsted Press, New York. [22.1]

Thompson MB. (1986) *River Murray Wetlands: Their Characteristics, Significance and Management.* Department of Environment & Planning, Adelaide. [22.4]

Thoms MC, Walker KF. (1989) Preliminary observations on the environmental effects of flow regulation on the River Murray, South Australia. *South Australian Geographical Journal* **89**: 1–14. [22.3, 22.4]

Thoms MC, Walker KF. (1992a) Channel changes related to low-level weirs on the River Murray, South Australia. In: Carling PA, Petts GE (eds) *Lowland Floodplain Rivers. Geomorphological Perspectives* pp 235–49. J. Wiley and Sons, Chichester. [22.3]

Thoms MC, Walker KF. (1992b) Sediment transport in a regulated semi-arid river: the River Murray, Australia. *Canadian Special Publications in Fisheries and Aquatic Sciences* (in press). [22.3]

Twidale CR, Lindsay JM, Bourne JA. (1978) Age and origin of the Murray River and gorge in South Australia. *Proceedings of the Royal Society of Victoria* **90**: 27–42 [22.2]

Walker KF. (1979) Regulated streams in Australia: the Murray–Darling river system. In: Ward JV, Stanford JA (eds) *The Ecology of Regulated Streams*, pp 143–63. Plenum, New York. [22.1]

Walker KF. (1981) *Ecology of freshwater mussels in the River Murray.* Australian Water Resources Council Technical Paper 63. [22.3, 22.4]

Walker KF. (1983) Impact of Murray–Darling Basin development on fish and fisheries. *FAO Fisheries Reports* **288**: 139–49. [22.4]

Walker KF. (1985) A review of the ecological effects of river regulation in Australia. *Hydrobiologia* **125**: 111–29. [22.1, 22.4, 22.5]

Walker KF. (1986a) The Murray–Darling river system. In: Davies BR, Walker KF (eds) *The Ecology of River Systems*, pp 631–59. Dr W. Junk, Dordrecht. [22.1, 22.4]

Walker KF. (1986b) The freshwater mussel *Velesunio ambiguus* as a biomonitor of heavy metals associated with particulate matter. In: Hart BT (ed.) *Water Quality Management: The Role of Particulate Matter in the Transport and Fate of Pollutants*, pp 175–85. Water Studies Centre, Chisholm Institute of Technology, Melbourne. [22.3]

Walker KF. (1986c) The state of ecological research on the River Murray. In: De Deckker P, Williams WD (eds) *Limnology in Australia*, pp 637–48. CSIRO, Melbourne and Dr W. Junk, Dordrecht. [22.5]

Walker KF. (1989) Implications of the greenhouse effect for wetlands in South Australia: the River Murray. In: Dendy T (ed.) *Greenhouse '88: Planning for Climatic Change*, pp 132–4. Department of Environment and Planning, Adelaide. [22.2]

Walker KF, Hillman TJ. (1982) Phosphorus and nitrogen loads in waters associated with the River Murray near Albury-Wodonga, and their effects on phytoplankton populations. *Australian Journal of Marine and Freshwater Research* **33**: 223–43. [22.3]

Walker KF, Thoms MC, Sheldon F. (1991) The effect of weirs on the littoral environment of the River Murray, South Australia. In: Boon PJ, Petts GE, Calow P (eds) *River Conservation and Management*. pp 271–92. John Wiley and Sons, Chichester. [22.2, 22.3, 22.4]

Williams MD, Williams WD. (1991) Salinity tolerances of four species of fish from the Murray–Darling system. In: Williams WD (ed.) *Salt Lakes and Salinity*. Dr W. Junk, Dordrecht. [22.3, 22.4]

Williams WD. (1987) Salinization of rivers and streams: an important environmental hazard. *Ambio* **16**: 180–5. [22.3]

Woodyer KD. (1978) Sediment regime of the Darling River. *Proceedings of the Royal Society of Victoria* **90**: 139–47. [22.3]

23: A Mountain River

J.V.WARD

23.1 INTRODUCTION

Mountain rivers are one of our most valued resources. The term, mountain river, conjures up visions of crystal-clear waters in a pristine setting. Indeed, most undisturbed running water segments occur in mountainous regions at high elevations. Mountain rivers share common features worldwide and have served as the primary site for the evolution of many lotic organisms. It is in such settings that the aquatic biota most clearly exemplify the remarkable adaptations to running water conditions. For the purposes of this chapter, the term 'mountain river' refers to running waters of high altitudes, the upper reaches of river systems. Although mountain rivers may extend downward to sea level in areas of recent coastal subsidence or in geologically youthful regions, such low elevation mountain rivers will be considered only briefly. The emphasis will be on high-gradient streams and rivers at high elevations in the temperate zone.

The ensuing material begins with the idealized features of mountain rivers, including typical environmental conditions, energy resources, biotic communities, and adaptations of the biota. This is followed by a general discussion of the types of running waters occurring at high elevations and their longitudinal patterns. The remainder of the chapter is devoted to the St. Vrain, an archetypical mountain river of the Colorado Cordillera.

23.2 GENERAL FEATURES OF MOUNTAIN RIVERS

Environmental conditions

The high gradient (slope) of the typical mountain river, coupled with a highly irregular bed structure, results in rapid current and high turbulence. Consequently oxygen levels are near saturation and the stream water tends to be chemically and thermally homogeneous.

Only recently have lotic ecologists begun to characterize precisely the microcurrent regime near the surface of the streambed where many organisms reside (Wetmore et al 1990; Davis and Barmuta 1989; Nowell and Jumars 1984; Statzner et al 1988; see Chapter 5). Near-bed conditions may be hydraulically smooth (e.g. over flat sheets of bedrock) or hydraulically rough (over irregular stream bottoms). Under hydraulically smooth conditions there is a 'laminar sublayer' consisting of a thin zone of non-turbulent water above the bottom surface (Davis & Barmuta 1989; see Chapter 5). Under the hydraulically rough conditions that typify mountain streams and rivers, the near-bed current microenvironment is normally turbulent, although stable eddies may develop between closely spaced substrate particles as the current skims across the tops of those particles. Fluid dynamics depend not only on the spacing of substrate elements, but also on the height of individual particles above the streambed and 'relative roughness' (height of particles/water depth). In high-gradient riffles at low water, near-bed current patterns are extremely complex and high velocities occur close to the bed. None the less, the microcurrent regime of such riffles is highly heterogeneous and includes low-velocity microdepositional habitats. In addition, water velocities rapidly approach zero with increasing depth within the interstitial spaces of the hyporheic zone (Williams & Hynes 1974).

Mountain streams are 'erosional' habitats in

the sense that fine mineral particles tend to be eroded and transported downstream to 'depositional' habitats (e.g. lakes, lowland rivers). At smaller spatial scales, however, both erosional (e.g. riffles) and depositional (e.g. pools) habitats co-occur. None the less, coarse materials such as cobble (64–256 mm), pebble (32–64 mm), gravel (2–32 mm) and wood characterize the substrata of mountain streams and rivers (see Chapter 3). The composition and distribution of substrata within the channel are largely structured by physical processes related to the unidirectional flow of water interacting with basin geology and allochthonous organic debris (Keller & Swanson 1979; Bilby & Likens 1980). In old growth forests, wood debris may exert major controls on channel morphology (Harmon *et al* 1986) and the retention of organic matter (Bilby & Likens 1980; Speaker *et al* 1984). In contrast, wood debris plays a very minor role in streams above treeline where the source of wood is restricted to small riparian shrubs (e.g. *Salix* spp.). Tundra streams are often lined by willows and dense herbaceous vegetation, however, so relatively large standing crops of organic detritus may accumulate in the streambed during the growing season (Ward 1986).

High water clarity is a distinguishing feature of most mountain streams and rivers. Despite this, however, light for *in situ* primary production may be limiting. In densely forested headwater streams, the light available for autotrophic production is limited by canopy development. In exposed streams above treeline, autotrophic production is limited by low summer water temperatures and nutrient levels, and by the short growing season. Small rivers of middle elevations typically provide optimal conditions for instream primary production because of an open canopy, adequate levels of plant nutrients, and shallow, clear water.

Elevation exerts a primary influence on the temperatures of mountain waters through its influence on air temperatures, as reflected by the lapse rate (an average decrease in air temperature of about 6.5°C for every 1000-m increase in elevation). Stream temperatures tend to track air temperatures except when ice covered or during periods of snowmelt or spate (Ward 1985). Because groundwater temperatures approximate mean annual air temperatures, even mountain streams fed by groundwater are under the influence of the lapse rate. Factors related to insolation (aspect and canopy) can modify the influence of air temperature, as exemplified by the higher temperatures of some exposed streams above treeline compared with their canopied reaches at lower elevations (Ulfstrand 1968).

Typical river water is a dilute solution of calcium bicarbonate. The total dissolved solids (TDS) content of the world's rivers averages 120 mg l^{-1} (Livingston 1963; see Chapter 4). The ionic content of the river generally increases downstream as the water ages and is contributed from an increasingly larger catchment (Hynes 1970). TDS values <50 mg l^{-1} usually indicate drainage from igneous rock (Wetzel 1983). If the high elevation portion of the catchment consists of relatively insoluble crystalline bedrock, whereas soluble sedimentary strata occur at lower elevations, the downstream trend of increasing ionic strength is accelerated. Water hardness also tends to increase downstream in river systems. Therefore, mountain streams, unless originating as calcareous springs, are characterized by soft water of low ionic strength.

Because of the turbulent nature of high-gradient mountain streams, dissolved gases are well mixed and in equilibrium with the atmosphere. The decrease in gas solubility with increasing elevation (decreasing atmospheric pressure) is countered by the increase in solubility associated with the low temperature of the water.

Energy resources

Both *in situ* primary production (autochthonous production) and allochthonous production derived from the surrounding landscape contribute organic matter to running waters (see Chapter 17). In high-gradient mountain streams bryophytes (mosses and lichens) and attached algae constitute the main photosynthetic organisms. The high current velocity and paucity of soft bottom sediment severely restricts or excludes floating or rooted angiosperms. Mountain rivers do not provide suitable conditions for the development of phytoplankton.

Much of the organic matter in mountain rivers is detritus of allochthonous origin and this is especially marked in forested headwaters. In a canopied deciduous forest stream, leaf litter and other allochthonous detritus accounted for >99% of the annual energy budget (Fisher & Likens 1973). In coniferous forest streams examined by Naiman and Sedell (1979), more than 90% of the benthic detritus was wood. Mountain streams lacking fully developed canopies, such as alpine streams, depend to a greater extent on autochthonous production (Minshall 1978; Winterbourn *et al* 1981). For example, the fish fauna of high-gradient Andean headwaters above the forest consists mainly of algal-grazing catfishes (Lowe-McConnell 1987). The limited fauna inhabiting the sources of glacial-fed streams feed on wind-blown organic particles entrapped on the surface of the glacier and later released to the stream (Steffan 1971). Kohshima (1984) documented the occurrence of blue-green algae and bacteria in the meltwater drainage channels of a Himalayan glacier and in the guts of chironomid larvae.

Biotic communities

The majority of organisms in high-gradient mountain rivers are benthic in the sense of being closely associated with the substratum. In contrast, organisms residing in the open water (plankton, nekton) or associated with the air–water interface (pleuston) are poorly represented, as are higher plants that require soft bottom sediments for rooting.

Benthic community types

The benthic communities of running waters include biofilm assemblages, attached macrophytes, zoobenthos, hyporheos and bottom fishes. Biofilm, the organic layer that coats all solid surfaces, is a heterogeneous assemblage of attached algae, bacteria, fungi, protozoans and micrometazoans in a polysaccharide matrix (Lock *et al* 1984; see Chapter 15). In mountain streams, the primary sites of biofilm development are the surfaces of stones (epilithon) and submerged wood (epixylon). In addition to photosynthetic production, the biofilm is a major uptake site for dissolved organic matter, thereby adding to the food available to

grazing fishes and invertebrates (Rounick & Winterbourn 1983).

Bryophytes are the primary attached macrophytes in many mountain rivers. Although not generally consumed as food, mosses and liverworts provide important physical habitat structure (Glime & Clemons 1972). The largely tropical attached angiosperms (Podostemaceae and Hydrostychaceae) are adapted to torrential mountain streams (Gessner 1955).

The distinctive and highly adapted zoobenthic community of high-gradient mountain streams is remarkably similar worldwide and is dominated by insects except in calcareous streams where crustaceans and molluscs are also abundant (Hynes 1970). For example, insects constitute more than 95% of total zoobenthos numbers in streams in the Rocky Mountains of North America (Ward 1975) and the Tien Shan Mountains of Asia (Brodsky 1980).

The hyporheic community consists of invertebrates that inhabit the interstitial spaces between sediment grains within the streambed and laterally under the banks (Williams & Hynes 1974; Pennak & Ward 1986; Stanford & Ward 1988; Danielopol 1989; see Chapter 11). The hyporheic community consists of selected components from the zoobenthic community that spend a portion of their lives in the hyporheic zone (e.g. early instars, diapausing nymphs) as well as permanent residents that are highly adapted for a hypogean existence. The few data available suggest that altitudinal distribution patterns of the hyporheos are at least partially decoupled from altitude, depending more on site-specific geomorphic features of the fluvial landscape (Ward & Voelz 1990).

Both resident and migratory fishes occur in mountain rivers. Migrants that spawn in the headwaters may come from lower riverine reaches (potamodromy), lakes (adfluvial), or the sea (anadromy). Catadromous eels (*Anguilla*) ascend rivers to complete growth and sexual maturity. Whereas most migratory species are nektonic, many resident fishes of mountain streams are truly benthic. Only the strongest swimmers, such as salmonids, can maintain a nektonic existence in high-gradient mountain streams.

Adaptations of the biota

Running waters are physiologically richer than still waters (Whitford & Schumacher 1961; Hynes 1970) and lotic organisms have evolved various adaptations enabling them to inhabit high-gradient streams and take advantage of the benefits conferred by current. Morphological, behavioural and physiological adaptations often function in concert but are discussed under separate headings.

Morphological adaptations

Sustained swimming is not commonly employed by the organisms of high-gradient streams. Only nektonic fishes, fusiform-shaped and elliptical in cross-section, are able to swim against the current for extended periods. Most fishes and other aquatic organisms use other adaptations to maintain position in fast water or to avoid the full force of the current.

Sessile forms use various means of attachment to anchor themselves to solid surfaces. Examples include algal holfasts, the rhizoids of mosses, the modified roots of attached angiosperms, and the salivary secretions used to affix the shelters of certain caddisflies and midges. Motile aquatic insects of mountain streams employ modified gills, hair fringes and hydraulic suckers to increase frictional resistance with the substratum. Dorsoventral flattening of the body, well exemplified by heptageniid mayflies and water pennies (Psephenidae) may serve several functions. It enables organisms to feed on algae attached to the tops of rocks in rapid water, to move into cervices, and to be less vulnerable to predation. Many lotic invertebrates use hooks and claws to obtain purchase in rapid water. Riffle beetles (Elmidae), for example, crawl along the bottom using well-developed tarsal claws, quite in contrast to the nektonic beetles of lowland rivers that use hair fringes for swimming. Unlike most beetles, which are positively buoyant and must periodically renew air stores, the plastron of elmids frees them from surface visits, a dangerous and energetically expensive proposition for inhabitants of torrential streams (Ward 1992).

Benthic fishes of high-gradient mountain streams also exhibit a remarkable array of morphological adaptations (Allen 1969). They are dorsoventrally flattened or have an arched profile with a flat ventrum, in contrast to the laterally flattened fishes of lakes and lowland rivers. The paired fins tend to be positioned more laterally on the body and are more muscular than those of related slow water species. The pectoral fins may function as reverse hydrofoils; when the leading edge faces into the current, a negative pressure under the fin apresses the fish to the bottom. Bottom dwellers typically have reduced swim bladders and decreased buoyancy. Eyes are more dorsal, gill openings are more lateral, and the mouth of bottom dwellers is more ventral than in nektonic fishes. Benthic fishes of torrential streams exhibit various adaptations that serve to increase frictional resistance with the substratum. In Sculpins (Cottidae), the pelvic fins are far forward on the body and with the pectoral fins form a ventral friction device. A few other North American and European fishes exhibit friction devices similar to sculpins, but in African, Asian and South American torrential waters much more elaborate organs of attachment have been described (Hora 1930; Marlier 1953; Kleerekoper 1955). Fishes from several families have independently evolved mouth suckers or hydraulic discs near the mouth enabling them to attach very firmly to rocks.

Behavioural adaptations

Because of the highly heterogeneous current regimes of torrential reaches, mobile organisms can exert choice over the current to which they are exposed. At the microspatial scale, organisms exhibit cryptic behaviour by moving into crevices or seeking refuge in moss tufts; others position themselves on rock faces. Even strong-swimming nektonic forms such as salmonids spend much time in areas of reduced current. Invertebrates poorly adapted to resist current are concentrated in the quieter margins, whereas other species are more abundant near the stream's centre (Needham & Usinger 1956). Differences in the morphological adaptations of congeneric fishes are consistent with their exposure to and ability to resist current (Mathews 1985). The more rheophilic dace, *Rhinichthys cataractae*, not only occurs in faster current than its sympatric congener (*R. falcatus*), it also has a smaller swim bladder (Gee & Northcote 1963).

Behavioural drift, the downstream transport of benthic organisms by current, is a functional attribute of running waters, although whether drift entry is passive or active is contentious (Allan *et al* 1986; Brittain & Eikeland 1988). The colonization-cycle hypothesis has been proposed as a drift compensation mechanism for aquatic insects with a winged adult stage (Müller 1982); upstream flight by ovigerous females prevents faunal depletion of the headwaters and provides a means of exploiting different habitats during different seasons or lifecycle stages. Upstream migration within the water has been documented for a variety of stream invertebrates (Söderström 1987).

Another behavioural adaptation to torrential conditions is employed by black fly larvae (Simuliidae), which use silk threads as 'life-lines', enabling them to regain their original position if displaced accidently or to avoid a predator (Wotton 1986). Certain caddisflies add mineral particles to their cases to increase ballast, especially before the quiescent pupal stage.

Respiratory physiology of aquatic insects

Aquatic insects perhaps best exemplify, through their respiratory physiology, the special conditions of high-gradient mountain streams. Insects tend to be the most abundant and diverse metazoans in such habitats; entire orders of insects evolved in cool headwater streams; and a great deal is known of their respiratory physiology, which has common features with other faunal groups characteristic of high-gradient mountain streams (e.g. salmonid fishes; Allen

1969). The fauna of high-gradient mountain streams evolved in an environment — cool, turbulent, oxygen-saturated waters — that does not necessarily require special physiological adaptations. Adaptive radiation occurred later as lowland rivers and lentic habitats were colonized (Ward & Stanford 1982). It is, in fact, the lack of adaptations that is most striking when the respiratory physiology of insects from cool lotic waters is considered and contrasted with related forms from more stagnant waters (Table 23.1).

Insects from running waters generally, and high-gradient headwaters in particular, are unable to regulate oxygen consumption, which varies directly with the oxygen concentration of the medium (dependent-type respiration). In addition, insects from running waters tend to have high metabolic rates, are intolerant of low oxygen levels, and generally lack the ability to use ventilatory movements to enhance oxygen uptake. Insects and other animals with an evolutionary history tied to mountain streams have had little or no selective pressures to develop adaptations to low oxygen conditions. Obligatory rheophiles, while able to survive temporary oxygen deficits in moving water, succumb in still water even at high concentrations of oxygen (Philipson 1954; Hynes 1970).

Temperature is also intimately related to the respiratory physiology of aquatic insects. For example, onset of death occurred at an oxygen concentration 2.4 times higher when stonefly nymphs were held at 15.6°C than at 10°C (Knight & Gaufin 1964). The cold stenothermous nature of mountain stream animals probably relates as much to high oxygen requirements as it does to low temperature (Hynes 1970).

Table 23.1 Contrasting features in some general respiratory characteristics of insects in running and standing waters

Characteristic	Standing waters	Running waters
Type of respiration	Independent[a]	Dependent[a]
Oxygen consumption	Low	High
Tolerance of low oxygen concentration	High	Low
Ventilatory movements	Well developed	Poorly developed[b]

[a] See text for definition; [b] well developed in case- and tube-dwelling forms.

Longitudinal patterns

Spatial changes in abiotic and biotic variables along the course of river systems have long attracted the attention of river ecologists (e.g. Shelford 1911; Thienemann 1912). Illies (1961) proposed a universal zonation scheme for running waters based on the rather sudden transition in zoobenthic community composition that corresponded to fish zonation patterns (Müller 1951) and to shifts in factors such as temperature and current. The most marked faunal changes occurred at the lower end of the salmonid zone; areas upstream from this point were designated 'rhithral', those below were designated 'potamal'. Rhithral was defined as the upper reaches where the annual range of monthly mean water temperatures does not exceed 20°C, dissolved oxygen levels are continuously high, the substratum is predominantly coarse, the current is turbulent, and the water is clear (Table 23.2). The annual range of monthly mean water temperatures in the potamal commonly exceed 20°C, oxygen levels are low at times, finer substratum materials are abundant, the current is less turbulent, and water clarity is lower. Illies and Botosaneanu (1963) expanded the zonation scheme to include an upper krenal zone for streams originating as springs (summer-cool, winter-warm). Steffan (1971) added yet another upper zone, the kryal, to encompass streams arising from the meltwater of glaciers and permanent snowfields ($T_{max} < 4°C$). The distinctiveness of the biota of the major zones has focused on differences in temperature regimes. Statzner and Higler (1986) proposed that stream hydraulics also play an important role in structuring zonation patterns (see Chapters 3 and 11).

The river continuum concept (RCC), developed in North America (Vannote et al 1980), perceives the downstream changes along rivers from a clinal rather than a zonal perspective (see Chapter 17). According to the RCC, river systems are longitudinal resource gradients along which stream organisms are predictably structured. This conceptual model was initially formulated for undisturbed deciduous forest streams with headwaters that are fully canopied by riparian vegetation and dominated by groundwater. The upper reaches are, therefore, light-limited heterotrophic systems (allochthonous leaf litter providing the major energy source) with temperature, discharge, and chemical regimes that are relatively stable and predictable (Table 23.3). As the canopy opens in the middle reaches, where the water is shallow and remains clear, primary production is maximized and autotrophy prevails ($P/R > 1$). The stream is no longer buffered by a dense canopy or groundwater dominance, and environmental heterogeneity reaches a maximum (e.g. diel amplitudes of temperature). The large volume of

Table 23.2 Idealized features of the four major zones of river systems as interpreted from the zonation perspective

	Major zones			
	Kryal	Krenal	Rhithral	Potamal
Temperature	<4°C	Summer-cool, winter-warm	$T_{max} > 4°C \leq 20°C$	$T_{max} > 20°C$
Transparency	Clear or turbid	Clear	Clear	Turbid
Substratum	Coarse	Fine to coarse	Coarse	Fine
Oxygen	Saturated	Variable	Saturated	Periodic deficits
Gradient	High	Variable	High	Low
Plankton	Absent	Absent	Absent	Present
Algae	Absent	Variable	Epilithon, Epixylon	Epipelon, Epiphyton
Bryophytes	Absent	Variable	Abundant	Sparse/absent
Rooted plants	Absent	Abundant	Sparse/absent	Variable
Zoobenthos	*Diamesa, Prosimulium*	Non-insects/ insects	Insects predominate	Insects/non-insects
Fish fauna	Absent	Variable	Rheophilic	Limnophilic

Table 23.3 Idealized features of the three reaches of river systems as interpreted from the perspective of the river continuum concept (as initially formulated)

	River reaches		
	Upper	Middle	Lower
Temperature	Cool, low amplitude	High amplitude	Moderate amplitude
P/R	<1.0	>1.0	<1.0
Energy source	Terrestrial detritus	*In situ* PP*	Transport detritus
Bottom light	Low	High	Low
Nutrient availability	Low	High	Low
Attached algae	Sparse	Abundant	Sparse
Submerged angiosperms	Absent	Abundant	Sparse
Plankton	Absent	Absent	Present
Leaf litter	Abundant	Sparse	Negligible
Invertebrates			
Shredders	Co-dominant	Rare	Absent
Collectors	Co-dominant	Co-dominant	Dominant
Grazers	Sparse	Co-dominant	Absent
Predators	Low	Low	Low
Fish fauna	Cool-water invertivores	Piscivores and invertivores	Planktivores and bottom feeders
Environmental heterogeneity	Low	High	Low
Biotic diversity	Low	High	Low

* PP = primary production

water in the lower reaches reduces environmental heterogeneity. The lower reach river is again heterotrophic ($P/R < 1$), being light limited by reduced water transparency and greater depth. The RCC stresses the interdependence of organic resources and invertebrate functional feeding groups, and how these change along the continuum. Criticisms of the RCC (e.g. Winterbourn *et al* 1981; Barmuta & Lake 1982) stem only partly from the fact that not all rivers are expected to exhibit the structural and functional features of undisturbed deciduous forest streams exemplified by Vannote *et al* (1980). The river continuum should be perceived as a 'sliding scale', whereby adjustments are made to accommodate different climatic/vegetational/hydrological settings (Minshall *et al* 1983) or anthropogenic disturbances (Ward & Stanford 1983). A more basic difficulty relates to the essentially deterministic nature of the RCC model, since many lotic ecologists view stream communities as stochastic phenomena (e.g. Grossman *et al* 1982; Reice

1985). In addition, interactions between the channel and adjacent floodplain and aquifer systems should be fully integrated into any holistic perspective of river ecosystems (Ward 1989).

23.3 THE ST VRAIN: AN ARCHETYPICAL MOUNTAIN RIVER

The St Vrain River, a major tributary of the South Platte River system, rises on the eastern slope of the Continental Divide in the Southern Rocky Mountain Physiographic Province (Fig. 23.1). The confluence of the north and south branches, fourth-order segments, occurs at the village of Lyons at the base of the foothills. The catchment above this point (the mountain catchment) encompasses 522 km². The headwaters of the North St Vrain lie within Rocky Mountain National Park; the headwaters of the middle and south branches are in the Indian Peaks Wilderness area. Much of the remaining catchment is in National

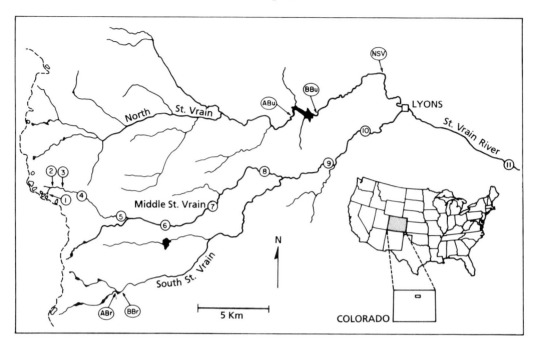

Fig. 23.1 The St Vrain River Basin showing the 11 sampling sites along the elevation gradient and other sampling locations referred to in the text. The dashed line along the western margin is the Continental Divide, along the eastern side of which are shown extant cirque glaciers. The village of Lyons is located at 40°13′N latitude and 105°16′W longitude.

Forest land. Research has been conducted at the sites shown in Fig. 23.1 (Ward 1975, 1981, 1982, 1984, 1986, 1987; Ward & Berner 1980; Ward & García de Jalón 1991). Studies conducted at the site on the North St Vrain (NSV), in the foothills above Lyons at 1677 m a.s.l., dealt with macro-invertebrate distribution patterns on various natural substrata. The influence of a deep-release storage impoundment on downstream conditions was examined by studying stream sites above (ABu, 1951 m) and below (ABu, 1890 m) Button-rock Reservoir. The influence of a surface-release storage impoundment, Brainard Lake, located in the headwater of the South St Vrain, was examined from studies conducted at sites ABr (3153 m) and BBr (3146 m).

The most intensive and extensive research, the results of which provide the focus for the re-mainder of this chapter, was conducted at the 11 sites (arabic numerals in Fig. 23.1) located along the longitudinal river profile (see p. 504). From site 1, a first-order alpine tundra stream at 3414 m

a.s.l., the pristine Middle St Vrain drops precipi-tously to site 11 on the Great Plains (1544 m) in less than 55 km. The plains river segment from Lyons to site 11 has a relatively high gradient (15.9 m km^{-1}), a rocky substratum, is well oxygenated, and has good water quality, thereby providing an interesting contrast with the mountain river (sites 1–10).

Bedrock geology and glacial history

At the sources of the St Vrain River system there are 16 mountain peaks exceeding 4000 m and the largest extant glaciers in Colorado (Madole 1969). Above the foothills the exposed bedrock is largely igneous and metamorphic rock of Precambrian age, consisting primarily of the gneisses and schists of the Idaho Springs formation and Longs Peak–St Vrain granite. The lower foothills are composed of sedimentary rocks of the late Paleozoic and Mesozoic.

Four major glaciations occurred in this region:

Pre-Bull Lake, Bull Lake, Pinedale and the Neo-glaciation (Madole 1969; White 1982). Pre-Bull Lake deposits are sparse and their ages are uncertain. Bull Lake glaciers, also of uncertain age, extended 13 km down the South St Vrain Valley to 2700 m, 15 km down the Middle St Vrain Valley to 2580 m, and 14 km down the North St Vrain Valley to 2400 m (White 1982). Pinedale glaciers advanced to nearly the same elevations as Bull Lake glaciers and Pinedale till now covers 90% of the glaciated areas. Pinedale glaciation began more than 30 000 years BP; the glaciers began retreating from their outermost positions about 14 000 years ago, receding to the cirques about 10 000 years BP. There is no evidence of glaciers or perennial snowfields in the Front Range during the Altithermal maximum (*c.* 7500–6000 BP). During the last 6000 years there have been several minor advances of cirque glaciers, collectively termed the Neoglaciation, the most recent of which occurred from 300 to 100 years BP. Extant glaciers are tiny and are confined to north- and east-facing cirques along the Continental Divide (Fig. 23.1).

Mountain valleys, formed by stream incision in the early Tertiary, are broad and U-shaped at higher elevations, but are narrow and deep below the maximum extent of Pleistocene glaciation. The Neoglacial ice remnants are not active, so that even streams originating directly from these cirque glaciers contain clear water.

The climate/vegetation gradient

Climatic conditions and terrestrial vegetation communities change dramatically along the steep elevation gradient (Marr 1961; Peet 1981). The continental climate is subject to extreme fluctuations. At higher elevations, appreciable snowfall can occur during all months of the year and high winds are common. Mean values of climatological data from five locations along the elevation gradient are shown in Table 23.4.

The forest vegetation of the mountain landscape can be illustrated as a community mosaic, with gradients of moisture-topography superimposed on the elevation gradient (Fig. 5 in Peet 1981). Dry foothill locations are typified by open ponderosa pine (*Pinus ponderosa*) woodlands. More mesic or higher elevation foothill sites exhibit higher tree densities and the addition of Douglas fir (*Pseudotsuga menziesii*). Dense stands of lodgepole pine (*Pinus contorta*), a post-fire successional species, occur at many middle elevation sites. Lodgepole pine is succeeded by Douglas fir at lower mid-elevations and by Engelmann spruce (*Picea engelmannii*) and sub-alpine fir (*Abies lasiocarpa*) at higher mid-elevation sites. In the subalpine zone, Engelmann spruce and subalpine fir are successional and climax forest co-dominants, except on the driest sites where limber pine (*Pinus flexilis*) occurs. Alpine tundra begins at approximately 3500 m elevation.

Site 1 on Middle St Vrain creek (Fig. 23.1) is located in alpine tundra on a north-facing slope where treeline extends only to 3325 m elevation. A small cirque lake, fed by the St Vrain Glaciers, is the source for the stream. During the summer, lush tundra vegetation, consisting of low willow shrubs (*Salix*) and herbs, forms a nearly complete canopy over portions of the first-order stream.

Table 23.4 Long-term climatological data means from five locations along the eastern slope of the Front Range, Colorado

	Elevation (m a.s.l.)				
	1603	2195	2591	3048	3750
Annual air temperature (°C)	8.8	8.3	5.6	1.7	−3.3
Daily minimum temperature (°C)	—	1.4	−0.8	−4.4	−7.3
Daily maximum temperature (°C)	—	14.6	12.1	6.9	−0.2
Annual precipitation (cm)	40	58	58	77	102
Days per year >0°C (days)	135	125	104	59	47

From Barry (1973) and Peet (1981).

Willows (*Salix* spp.) occur in the riparian zone along the entire elevation gradient (sites 1−11). Rocky Mountain alder (*Alnus tenuifolia*) is an important riparian species at middle elevations, where it co-occurs with willows and Rocky Mountain birch (*Betula occidentalis*). Scattered cottonwood trees (*Populus sargentii* and *P. angustifolia*) occur in the stream corridor in the foothills and lower montane and extend on to the plains.

Physicochemical gradients

The St Vrain is a snowmelt river (*sensu* Poff & Ward 1989). Although the magnitude of snowmelt runoff varies between years, the temporal pattern is predictable (Fig. 23.2). Dissolved oxygen is near saturation at all mountain stream sites. Oxygen supersaturation is observed at the plains site, where dense macrophyte beds and filamentous algae occur. Summer values of nitrate-nitrogen ranged from 0.10 to 0.15 mg l^{-1} with no discernible altitudinal pattern. Orthophosphate levels are below detection limits (5 µg l^{-1}).

Other physicochemical variables exhibit longitudinal patterns (Fig. 23.3). Water temperature maxima (T_{max}) exhibit a progressive downstream increase. Because the water temperature drops to 0°C at all sites, T_{max} is also the annual temperature range. At the highest elevations the stream is

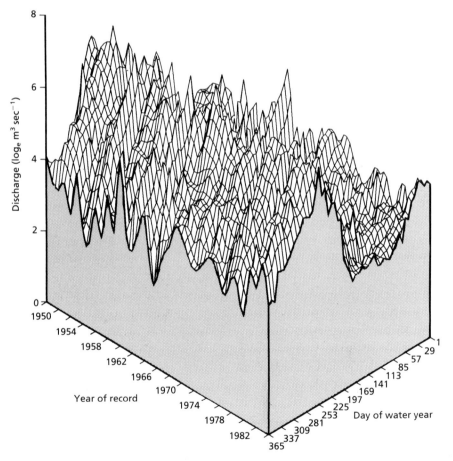

Fig. 23.2 Long-term hydrograph based on daily mean discharge records from a location 0.6 km downstream from the confluence of the north and south branches of the St Vrain River. Year of record is the water year, 1 October−30 September.

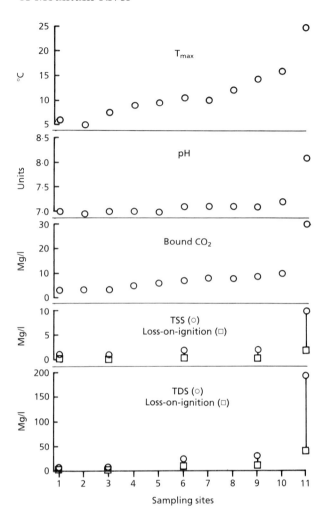

Fig. 23.3 Longitudinal patterns of maximum temperature, pH, bound carbon dioxide (methyl orange alkalinity) and total suspended and dissolved solids (showing loss-on-ignition fractions) along the course of the St Vrain River.

snow covered for more than half the year, whereas site 11 drops to 0°C for only a short period in January when ice forms only along the edges. Annual thermal summation estimates range from 450 degree-days (>0°C) at site 1 to 4220 degree-days at site 11. Values of the remaining variables in Fig. 23.3 increase only slightly from the tundra to the lower foothills, then abruptly increase to the plains site.

All sampling sites were located on rubble riffles to minimize between-site differences in bed materials. In the smaller mineral fraction (<32 mm), gravel predominates (78–87%) in the mountain river. Sand (62.5 μm to 2 mm) and gravel are equivalent in the plains river. Silts (3.9–62.5 μm)

and clays (<3.9 μm) constitute <3% of the smaller size fraction at mountain and plains sites. During summer, fine particles (0.05–1 mm) constitute the majority of the sedimentary detritus (66–96%) on riffles at mountain and plains locations. Coarse detritus reflects the terrestrial vegetation, with grass and sedge fragments common in the tundra stream, conifer needles at middle elevations, and sparse deciduous plant detritus in the plains river.

Biotic gradients

The distribution and abundance patterns of aquatic flora, fishes and zoobenthos are included in this section. Emphasis will be placed on the zoobenthos, which have been most intensively studied. General trends and idealized patterns are emphasized.

Aquatic flora

The biomass of the attached flora attains greatest values in the headwaters, where bryophytes (mosses and liverworts) and lichens are well developed, and in the plains river, where the filamentous chlorophyte *Cladophora glomerata* proliferates (Fig. 23.4(a)). Aquatic lichens were collected only from the two headwater sites, whereas bryophytes occur at all mountain stream locations. *Cladophora* is restricted to the plains river. At mid-elevation sites, where biomass levels are low, a variety of epilithic microalgae predominate. Distribution patterns of individual taxa are best demonstrated with diatoms (Bacillariophyta), the most intensively studied group (63 species and varieties). Some species of diatoms occur at all sites along the elevation gradient (e.g. *Achnanthes lanceolata*); others are found only in upper (e.g. *Diatoma anceps*) or lower reaches (e.g. *Nitzschia palea*). No common species are restricted to middle elevations. The algal flora of the St Vrain River, however, does not exhibit the distinct zonation patterns demonstrated by Kawecka (1974) for streams of the High Tatra and Rila mountains. Aquatic angiosperms are restricted to the plains river, where dense beds of *Elodea canadensis* and *Ranunculus aquatilus* occur from July through October.

Fish fauna

Salmonid fishes occupy the entire elevation gradient and extend a short distance into the plains river (Ellis 1914; Propst 1982). Cut-throat trout (*Oncorynchus clarki*), the only indigenous salmonid of this region, is now confined to high elevation tributaries, having been largely displaced by introduced species. Brook char (*Salvelinus fontinalis*) now occupies the same headwater habitat as cut-throat trout. Rainbow trout (*Oncorynchus mykiss*) is planted in most accessible reaches of Colorado streams. Brown trout (*Salmo trutta*) is more tolerant of warmer

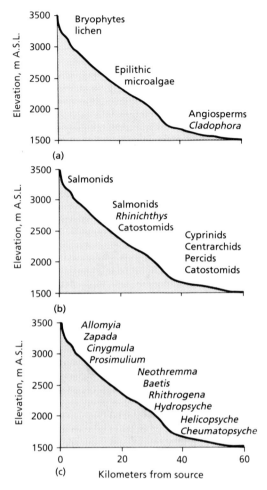

Fig. 23.4 The long profile of the St Vrain River showing dominant floral and faunal components of the headwater, middle elevation, and plains river segments. (a) Aquatic flora. (b) Fish fauna. (c) Zoobenthic fauna.

waters than most salmonids and extends from mid-elevations downward on to the plains. A depauperate fish fauna characterizes Front Range river systems (Fig. 23.4(b)). Sculpins (Cottidae), typically co-dominant with salmonids in high-elevation segments, are historically absent from the eastern slope of Colorado. Two catostomids (*C. catostomus* and *C. commersoni*) and the long-nose dace (*Rhinichthys cataractae*) are the only other fishes commonly encountered in mountain streams of this region. A somewhat richer fauna occurs in the plains river where Propst (1982) recorded 16 species of fishes, mainly cyprinids (ten species, including *R. cataractae*), but also including the two catostomids, two centrarchids, a percid, and (rarely) brown trout.

Zoobenthic fauna

Insects dominate the zoobenthos, constituting 89–99% of the densities, 84–99% of the biomass, and 79–93% of the taxa at the 11 sites. The numerically dominant species at each site are all insects (Fig. 23.4(c)) from only four orders: Ephemeroptera (mayflies), Plecoptera (stoneflies), Trichoptera (caddisflies) and Diptera (true flies). Additional orders (Odonata, Lepidoptera, Collembola) were collected from the plains river, and aquatic beetles (Coleoptera) occur at low densities at sites 2–11. Some non-insect groups are widely distributed along the elevation gradient (nematodes, oligochaetes, mites, planarians); others are confined to site 11 on the plains (amphipods, isopods, snails, leeches).

Zoobenthic abundance is low in the head-waters, increases from the montane to the upper foothills, levels off in the high-gradient foothills segment, and markedly increases in the plains river (Fig. 23.5). In the tundra stream (site 1), zoobenthic density averages 1856 organisms m^{-2} and biomass averages 652 mg dry weight m^{-2}. In the plains river, mean values are 6355 organisms and 3646 mg m^{-2}.

Zoobenthic diversity exhibits a sigmoid pattern along the elevation gradient (Fig. 23.5). From a total of 210 taxa, 34 occur at site 1, and 106 at site 11. The major faunal components exhibit one of the seven species richness curves illustrated in Fig. 23.6. Pattern number 7 is for those groups that are absent from the mountain river. The other six patterns all exhibit low species richness in the headwaters. Pattern 5 organisms (plan-arians, mites) maintain low levels of diversity along the entire gradient. Mayfly diversity (pat-tern 4) progressively increases downstream over most of the river's course. Dipterans (pattern 1) and beetles (pattern 6) show progressively greater lag phases before the start of diversity increases. Caddisfly species richness (pattern 3) reaches maximum values in the upper montane zone, then remains at similar levels over the remaining sites. Only stoneflies (pattern 2) exhibit declining diversity downstream.

Zoobenthos occurring in the tundra stream in-clude a few special headwater elements, but most abundant species are euryzonal forms. The two

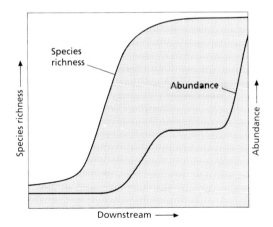

Fig. 23.5 Idealized altitudinal patterns of species richness and abundance for total zoobenthos along the course of the St Vrain River (from Ward 1986; ©1986 by E. Schweizerbart'sche Verlagsbuchhandlung).

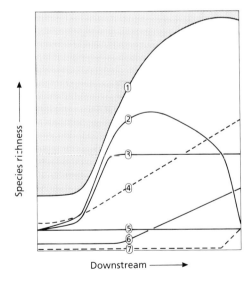

Fig. 23.6 Idealized types of species richness curves exhibited by the various faunal components along the course of the St Vrain River. See text (from Ward 1986; ©1986 by E. Schweizerbart'sche Verlagsbuchhandlung).

most abundant stoneflies at site 1, for example, extend downstream to site 10 (*Megarcys signata*) or site 9 (*Zapada oregonensis*). The addition of species downstream, without loss of those present at higher elevations, generally characterizes the chironomid midge fauna, mayflies, beetles and oligochaetes. Even in these groups, however, many species do not extend their range into the plains river. Conversely, many species that occur at site 11 are not found at any mountain site. Euryzonality is less well developed in the caddisfly fauna, which contains distinct headwater (e.g. *Allomyia tripunctata*) and plains elements (e.g. *Heliopsyche borealis*) as well as species restricted to middle elevations (e.g. *Neothremma alicia*). The two planarians of the St Vrain River exhibit remarkably distinct distributions, with *Polycelis coronata* occurring at all mountain sites, but not in the plains river to which *Dugesia dorotocephala* is restricted.

23.4 SUMMARY AND CONCLUSIONS

A rhithral character prevails along the elevation gradient of the St Vrain River from sites 1–10,

followed by a rather abrupt transition to potamal conditions at site 11. This is best exemplified by the sharp faunal discontinuity between sites 10 and 11. Site 10 in the lower foothills marks the downstream distribution limit of many zoobenthic species. Conversely, numerous species of the plains river segment, including some of extreme abundance, do not penetrate the mountain river. Faunal overlap between adjacent mountain stream sites (along the diagonal in Fig. 23.7) is high. Even the tundra stream (site 1) exhibits a relatively high overlap (39.2%) with site 2 located below treeline. Site 10 exhibits a 58.4% faunal similarity with site 9, but only a 17.8% overlap with site 11, the plains river.

The St Vrain River, although glacial fed, lacks the distinct kryon biocoenosis present at slightly higher elevations in an adjacent drainage (Elgmork & Saether 1970). Rhithral conditions extend to the headwater of the Middle St Vrain. Some rhithral features are maintained even at the plains site (coarse substrata, relatively high gradient and turbulence, well oxygenated water of high transparency). What factors then are responsible for the spatial patterns of zoobenthos observed along the course of the St Vrain River?

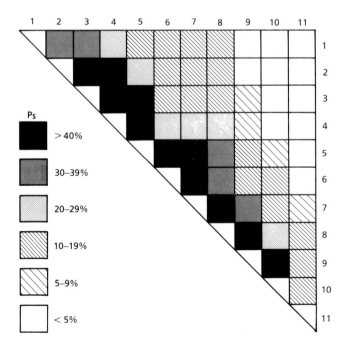

Fig. 23.7 Matrix of percentage similarity (Whittaker 1975) between sampling sites based on total zoobenthic species (from Ward 1986; ©1986 by E. Schweizerbart'sche Verlagsbuchhandlung).

Many researchers have espoused changes in temperature as a major, if not primary, factor responsible for the altitudinal changes in zoobenthic communities (e.g. Dodds & Hisaw 1925; Kownacka & Kownacki 1972; Brodsky 1980). Both the universal zonation scheme proposed by Illies and Botosaneanu (1963) and the river continuum concept of Vannote *et al* (1980) ascribe an important role to the thermal regime. The low faunal diversity at the highest elevation of the St Vrain is largely attributable to temperature (including 7+ months of ice and snow cover, $T_{max} \leq 6°C$, <500 annual degree-days), as only the most cold-adapted forms can tolerate such extreme thermal conditions. Additional species appear downstream as the thermal amplitude expands and degree-days increase. Even at site 10 in the lower foothills, after descending more than 1700 m in elevation, the annual temperature range is only 0–16°C. At site 11, however, the plains river inhabitants are exposed to an annual range of 0–25°C. Because the zoobenthos of mountain streams tend toward cold stenothermy, it is not surprising that site 10 marks the lower elevation limit of many species and that there is a sharp transition from rhithron to potamon elements between the lower foothills and the plains. A diverse and distinct warm-adapted fauna occurs in the plains river.

Factors other than temperature also contribute to the distinctive nature of the plains fauna. Beds of aquatic angiosperms, absent from the mountain river, increase habitat heterogeneity, provide current refugia and protection from predation, supply case-building materials, and serve as attachment sites for epiphytic algae. Site 11 exhibits similarities with the autotrophic middle reaches of the river continuum (see Table 23.3), given the high amplitude temperature regime, the open canopy, the relatively shallow and clear water, the abundant algae and angiosperms, and the high levels of environmental heterogeneity and biotic diversity. Additional factors that may have structured the plains zoobenthic community include the presence of a more diverse, and largely invertivorous, fish fauna (Propst 1982), and invertebrate predator–prey and competitive interactions that differ from those occurring in the mountain river.

ACKNOWLEDGEMENTS

E. Bergey, S. Canton, L. Cline, J. Harvey and R. Martinson provided laboratory assistance. Dr N. LeRoy Poff kindly supplied the computer plot of the St Vrain River hydrograph. The following specialists graciously provided taxonomic expertise: Drs R.W. Baumann, G.W. Byers, W.P. Coffmann, D.G. Denning, R.G. Dufford, G.F. Edmunds Jr, D.C. Ferguson, J.D. Haddock, C.L. Hogue, J.R. Holsinger, S.G. Jewett, B. Smith, S.D. Smith, J.A. Stanford, K.W. Stewart, R.F. Surdick, S.W. Szczytko, W.C. Welbourn, D.S. White and G.B. Wiggins. Supported in part by grants from the Colorado Experiment Station and the US Forest Service. Mrs Nadine Kuehl typed the manuscript.

REFERENCES

Allan JD, Flecker AS, McClintock NL. (1986) Diel epibenthic activity of mayfly nymphs, and its noncordance with behavioral drift. *Limnology and Oceanography* **31**: 1057–65. [23.2]

Allen KR. (1969) Distinctive aspects of the ecology of stream fishes: a review. *Journal of the Fisheries Research Board of Canada* **26**: 1429–38. [23.2]

Barmuta LA, Lake PS. (1982) On the value of the river continuum concept. *New Zealand Journal of Marine and Freshwater Research* **16**: 227–9. [23.2]

Barry RG. (1973) A climatological transect on the east slope of the Front Range, Colorado. *Arctic and Alpine Research* **5**: 89–110. [23.3]

Bilby RE, Likens GE. (1980) Importance of organic debris dams in the structure and function of stream ecosystems. *Ecology* **61**: 1107–13. [23.2]

Brittain JE, Eikeland TJ. (1988) Invertebrate drift—a review. *Hydrobiologia* **166**: 77–93. [23.2]

Brodsky KA. (1980) *Mountain Torrent of the Tien Shan.* Dr W. Junk, The Hague. [23.2, 23.4]

Danielopol DL. (1989) Groundwater fauna associated with riverine aquifers. *Journal of the North American Benthological Society* **8**: 18–35. [23.2]

Davis JA, Barmuta LA. (1989) An ecologically useful classification of mean and near-bed flows in streams and rivers. *Freshwater Biology* **21**: 271–82. [23.2]

Dodds GS, Hisaw FL. (1925) Ecological studies of aquatic insects. IV. Altitudinal range and zonation of mayflies, stoneflies, and caddisflies in the Colorado Rockies. *Ecology* **6**: 380–90. [23.4]

Elgmork K, Saether OR. (1970) Distribution of invertebrates in a high mountain brook in the Colorado Rocky Mountains. *University of Colorado Studies,*

Series in Biology **31**: 1–55. [23.4]

Ellis MM. (1914) Fishes of Colorado. *University of Colorado Studies* **11**: 1–136. [23.3]

Fisher SG, Likens GE. (1973) Energy flow in Bear Brook, New Hampshire: an integrative approach to stream ecosystem metabolism. *Ecological Monographs* **43**: 421–39. [23.2]

Gee JH, Northcote TG. (1963) Comparative ecology of two sympatric species of dace (*Rhinichthys*) in the Fraser River system, British Columbia. *Journal of the Fisheries Research Board of Canada* **20**: 105–18. [23.2]

Gessner F. (1955) Die limnologischen Verhältnisse in den Seen und Flüssen von Venezuela. *Internationale Vereinigung für Theoretische und Angewandte Limnologie Verhandlungen* **12**: 284–95. [23.2]

Glime JM, Clemons RM. (1972) Species diversity of stream insects on *Fontinalis* spp. compared to diversity on artificial substrates. *Ecology* **53**: 458–64. [23.2]

Grossman GD, Moyle PB, Whitaker JO. (1982) Stochasticity in structural and functional characteristics of an Indiana stream fish assemblage: a test of community theory. *American Naturalist* **120**: 423–54. [23.2]

Harmon ME, Franklin UF, Swanson FJ, Sollins P, Gregory SV, Lattin JD *et al.* (1986) Ecology of coarse woody debris in temperate ecosystems. *Advances in Ecological Research* **15**: 133–302. [23.2]

Hora SL. (1930) Ecology, bionomics and evolution of the torrential fauna, with special reference to the organs of attachment. *Philosophical Transactions of the Royal Society of London, Section B* **218**: 171–282. [23.2]

Hynes HBN. (1970) *The Ecology of Running Waters.* University of Toronto Press, Toronto. [23.2]

Illies J. (1961) Versuch einer allgemeinen biozönotischen Gliederung der Fliessgewässer. *Internationale Revue Gesampten Hydrobiologie* **46**: 205–13. [23.2]

Illies J, Botosaneanu L. (1963) Problèmes et méthodes de la classification et de la zonation écologique des eaux courantes, considerees surtout du point de vue faunistique. *Internationale Vereinigung für Theoretische und Angewandte Limnologie Mitteilungen* **12**: 1–57. [23.2, 23.4]

Kawecka B. (1974) Vertical distribution of algae communities in Maljovica Stream (Rila-Bulgaria). *Polskie Archiwum Hydrobiologii* **21**: 211–28. [23.3]

Keller EA, Swanson FJ. (1979) Effects of large organic material on channel form and fluvial processes. *Earth Surface Processes* **4**: 361–80. [23.2]

Kleerekoper H. (1955) Limnological observations in northeastern Rio Grande do Sul, Brazil. *Archiv für Hydrobiologie* **50**: 553–67. [23.2]

Knight AW, Gaufin AR. (1964) Relative importance of varying oxygen concentration, temperature, and water flow on the mechanical activity and survival of the Plecoptera nymph, *Pteronarcys californica* Newport. *Proceedings of the Utah Academy of Science, Arts and Letters* **41**: 14–28. [23.2]

Kohshima S. (1984) A novel cold-tolerant insect found in a Himalayan glacier. *Nature* **310**: 225–7. [23.2]

Kownacka M, Kownacki A. (1972) Vertical distribution of zoocoenoses in the streams of the Tatra, Caucasus and Balkan Mts. *Internationale Vereinigung für Theoretische und Angewandte Limnologie Verhandlungen* **18**: 742–50. [23.4]

Livingston DA. (1963) *Chemical composition of rivers and lakes.* United States Geological Survey Professional Paper 440–G. US Government Printing Office, Washington, DC. [23.2]

Lock MA, Wallace RR, Costerton JW, Ventullo RM, Charton SE. (1984) River epilithon: toward a structural–functional model. *Oikos* **42**: 10–22. [23.2]

Lowe-McConnell RH. (1987) *Ecological Studies in Tropical Fish Communities.* Cambridge University Press, Cambridge. [23.2]

Madole RF. (1969) Pinedale and Bull Lake glaciation in upper St Vrain drainage basin, Boulder County, Colorado. *Arctic and Alpine Research* **1**: 279–87. [23.3]

Marlier G. (1953) Etude biogéographique du bassin de la Ruzizi, basée sur la distribution des poissons. *Annales de Société Royale de Zoologie Belgiques* **84**: 175–224. [23.2]

Marr JW. (1961) Ecosystems of the east slope of the Front Range in Colorado. *University of Colorado Studies, Series in Biology* **8**: 1–134. [23.3]

Mathews WJ. (1985) Critical current speeds and microhabitats of the benthic fishes *Percina roanoka* and *Etheostoma flabellare.* *Environmental Biology of Fishes* **12**: 303–8. [23.2]

Minshall GW. (1978) Autotrophy in stream ecosystems. *BioScience* **28**: 767–71. [23.2]

Minshall GW, Petersen RC, Cummins KW, Bott TL, Sedell JR, Cushing CE, Vannote RL. (1983) Interbiome comparison of stream ecosystem dynamics. *Ecological Monographs* **53**: 1–25. [23.2]

Müller K. (1951) Fische und Fischregionen der Fulda. *Berlin Limnologie Flussstation Freudenthal* **2**: 18–23. [23.2]

Müller K. (1982) The colonization cycle of fresh-water insects. *Oecologia* **52**: 202–7. [23.2]

Naiman RJ, Sedell JR. (1979) Benthic organic matter as a function of stream order in Oregon. *Archiv für Hydrobiologie* **87**: 404–22. [23.2]

Needham PR, Usinger RL. (1956) Variability in the macrofauna of a single riffle in Prosser Creek, California, as indicated by the Surber sampler. *Hilgardia* **24**: 383–409. [23.2]

Nowell ARM, Jumars PA. (1984) Flow environments of aquatic benthos. *Annual Review of Ecology and Systematics* **15**: 303–28. [23.2]

Peet RK. (1981) Forest vegetation of the Colorado Front Range — composition and dynamics. *Vegetatio* **45**: 3–75. [23.3]

Pennak RW, Ward JV. (1986) Interstitial faunal communities of the hyporheic and adjacent groundwater biotopes of a Colorado mountain stream. *Archiv für Hydrobiologie Monographische Beiträge, Supplementbänd* **74**: 356–96. [23.2]

Philipson GN. (1954) The effect of water flow and oxygen concentration on six species of caddis fly (Trichoptera). *Proceedings of the Zoological Society of London* **124**: 547–64. [23.2]

Poff NL, Ward JV. (1989) Implications of streamflow variability and predictability for lotic community structure: a regional analysis of streamflow patterns. *Canadian Journal of Fisheries and Aquatic Sciences* **46**: 1805–18. [23.3]

Propst DL. (1982) *Warmwater Fishes of the Platte River Basin, Colorado.* Doctoral Dissertation, Colorado State University, Fort Collins, Colorado. [23.3, 23.4]

Reice SR. (1985) Experimental disturbance and the maintenance of species diversity in a stream community. *Oecologia* **67**: 90–7. [23.2]

Rounick JS, Winterbourn MJ. (1983) The formation, structure and utilization of stone surface organic layers in two New Zealand streams. *Freshwater Biology* **13**: 57–72. [23.2]

Shelford VE. (1911) Ecological succession I. Stream fishes and the method of physiographic analysis. *Biological Bulletin* **21**: 9–35. [23.2]

Söderström O. (1987) Upstream movements of invertebrates in running waters — a review. *Archiv für Hydrobiologie* **111**: 197–208. [23.2]

Speaker RW, Moore KM, Gregory SV. (1984) Analysis of the process of retention of organic matter in stream ecosystems. *Internationale Vereinigung für Theoretische und Angewandte Limnologie Verhandlungen* **22**: 1835–41. [23.2]

Stanford JA, Ward JV. (1988) The hyporheic habitat of river ecosystems. *Nature* **335**: 64–6. [23.2]

Statzner B, Higler B. (1986) Stream hydraulics as a major determinant of benthic invertebrate zonation patterns. *Freshwater Biology* **16**: 127–39. [23.2]

Statzner B, Gore JA, Resh VH. (1988) Hydraulic stream ecology: observed patterns and potential applications. *Journal of the North American Benthological Society* **7**: 307–60. [23.2]

Steffan AW. (1971) Chironomid (Diptera) biocoenoses in Scandinavian glacier brooks. *Canadian Entomologist* **103**: 477–86. [23.2]

Thienemann A. (1912) Der Bergbach des Sauerlandes. *Internationale Revue Gesampten Hydrobiologie, und Hydrographie Supplementband* **4**: 1–125. [23.2]

Ulfstrand, S. (1968) Life cycles of benthic insects in Lapland streams (Ephemeroptera, Plecoptera, Trichoptera, Diptera Simuliidae). *Oikos* **19**: 167–90.

[23.2]

Vannote RL, Minshall GW, Cummins KW, Sedell JR, Cushing CE. (1980) The river continuum concept. *Canadian Journal of Fisheries and Aquatic Sciences* **37**: 130–7. [23.2, 23.4]

Ward JV. (1975) Bottom fauna–substrate relationships in a northern Colorado trout stream: 1945 and 1974. *Ecology* **56**: 1429–34. [23.2, 23.3]

Ward JV. (1981) Altitudinal distribution and abundance of Trichoptera in a Rocky Mountain stream. In: Moretti GP (ed.) *Proceedings of the Third International Symposium on Trichoptera*, pp 375–81. Dr W. Junk Publishers, The Hague. [23.3]

Ward JV. (1982) Altitudinal zonation of Plecoptera in a Rocky Mountain stream. *Aquatic Insects* **4**: 105–10. [23.3]

Ward JV. (1984) Diversity patterns exhibited by the Plecoptera of a Colorado Mountain stream. *Annales de Limnologie* **20**: 123–8. [23.3]

Ward JV. (1985) Thermal characteristics of running waters. *Hydrobiologia* **125**: 31–46. [23.2]

Ward JV. (1986) Altitudinal zonation in a Rocky Mountain stream. *Archiv für Hydrobiologie Monographische Beiträge, Supplementbänd* **74**: 133–99. [23.2, 23.3, 23.4]

Ward JV. (1987) Trichoptera of regulated Rocky Mountain streams. In: Bournaud M, Tachet H (eds) *Proceedings of the Fifth International Symposium on Trichoptera*, pp 375–80. Dr W. Junk Publishers, Dordrecht, The Netherlands. [23.3]

Ward JV. (1989) The four-dimensional nature of lotic ecosystems. *Journal of the North American Benthological Society* **8**: 2–8. [23.2]

Ward JV. (1992) *Aquatic Insect Ecology. 1. Biology and Habitat.* John Wiley and Sons, New York. [23.2]

Ward JV, Berner L. (1980) Abundance and altitudinal distribution of Ephemeroptera in a Rocky Mountain stream. In: Flannagan JF, Marshall KE (eds) *Advances in Ephemeroptera Biology*, pp 169–78. Plenum Press, New York. [23.3]

Ward JV, García de Jalón D. (1991) Ephemeroptera of regulated mountain streams in Spain and Colorado. In: Alba-Tercedor J, Sanchez-Ortega A (eds) *Overview and Strategies of Ephemeroptera and Plecoptera*, pp 567–78. Sandhill Crane Press, Florida. [23.3]

Ward JV, Stanford JA. (1982) Thermal responses in the evolutionary ecology of aquatic insects. *Annual Review of Entomology* **27**: 97–117. [23.2]

Ward JV, Stanford JA. (1983) The serial discontinuity concept of lotic ecosystems. In: Fontaine TD, Bartell SM (eds) *Dynamics of Lotic Ecosystems*, pp 29–42. Ann Arbor Science Publishers, Ann Arbor, Michigan. [23.2]

Ward JV, Voelz NJ. (1990) Gradient analysis of interstitial meiofauna along a longitudinal stream profile. *Stygologia* **5**: 93–9. [23.2]

Wetmore SH, MacKay RJ, Newbury RW. (1990) Characterization of the hydraulic habitat of *Brachycentrus occidentalis*, a filter-feeding caddisfly. *Journal of the North American Benthological Society* **9**: 157–69. [23.2]

Wetzel RG. (1983) *Limnology*. Saunders, Philadelphia. [23.2]

White SE. (1982) Physical and geological nature of the Indian Peaks, Colorado Front Range. *Institute of Arctic and Alpine Research Occasional Paper* **37**: 1–12. [23.3]

Whitford LA, Schumacher GJ. (1961) Effect of current on mineral uptake and respiration by a fresh-water alga. *Limnology and Oceanography* **6**: 423–5. [23.2]

Whittaker RH. (1975) *Communities and Ecosystems*. Macmillan, New York. [23.4]

Williams DD, Hynes HBN. (1974) The occurrence of benthos deep in the substratum of a stream. *Freshwater Biology* **4**: 233–56. [23.2]

Winterbourn MJ, Rounick JS, Cowie B. (1981) Are New Zealand streams really different? *New Zealand Journal of Marine and Freshwater Research* **15**: 321–8. [23.2]

Wotton RS. (1986) The use of silk life-lines by larvae of *Simulium noelleri* (Diptera). *Aquatic Insects* **8**: 255–61. [23.2]

Index